Sand Dunes of the Northern Hemisphere

Distribution, Formation, Migration and Management

Volume 1

Formation and Evolution of Dune Landscapes

Edited by

Lu Qi

Institute for Desertification Studies, Beijing China

Mahesh K. Gaur

ICAR-Central Arid Zone
Research Institute, Jodhpur, India

and

Victor R. Squires

Institute for Desertification Studies, Beijing China

CRC Press
Taylor & Francis Group
Boca Raton London New York

CRC Press is an imprint of the
Taylor & Francis Group, an **informa** business
A SCIENCE PUBLISHERS BOOK

First edition published 2023
by CRC Press
6000 Broken Sound Parkway NW, Suite 300, Boca Raton, FL 33487-2742

and by CRC Press
4 Park Square, Milton Park, Abingdon, Oxon, OX14 4RN

CRC Press is an imprint of Taylor & Francis Group, LLC

Library of Congress Cataloging-in-Publication Data (applied for)

ISBN: 978-0-367-64605-9 (hbk)
ISBN: 978-0-367-64607-3 (pbk)
ISBN: 978-1-003-12542-6 (ebk)

DOI: 10.1201/9781003125426

Typeset in Times New Roman
by Radiant Productions

Foreword

Our deserts have been with us for a very long time. The most recent phase of aridity began about 24 million years ago and is evident in the deposits of wind-blown desert dust in central Asia and in similar wind-blown sediments recovered off the Atlantic coast of North Africa. Aridity proceeded in fits and starts, with major expansion of Northern Hemisphere deserts between 8 and 7 million years ago, when global cooling reduced evaporation from the inter-tropical oceans and greatly reduced precipitation on land. The outcome was a major change in global ecosystems at about the time that our remote ancestors first began to appear. The accumulation of huge ice caps over North America and parts of Eurasia at intervals from about 2.6 million years ago onwards (which is also when early humans first began to make stone tools) steepened the temperature gradient between equator and pole and greatly strengthened the wind systems responsible for blowing sand from desert river channels and fashioning them into a variety of desert dunes. The deserts have waxed and waned ever since, with dunes becoming covered in plants and stabilized during wetter climatic phases and active once more as aridity set in and the plants died.

Having worked in desert regions for over fifty years, it gives me great pleasure to introduce this book. The three editors, who are also co-authors of certain chapters in this volume, have a long and distinguished history of contributing to our knowledge of desert ecosystems, rangeland and dryland land management, and the origin and evolution of the great deserts of China, India, the Middle East and parts of Saharan Africa. The authors of the individual chapters have adopted a range of approaches, all of which help us to appreciate just how dynamic are the inland and coastal dune systems of these regions, how they have exerted a sustained and often subtle influence upon human activities and how they have in turn experienced the impact of human activities, sometimes to the detriment of both humans and dune systems. The case studies presented in this book are not simply of local interest but contain lessons that are applicable far more widely. Only by learning to understand in greater depth the long history of interactions between humans and desert and coastal dunes can we be confident of managing our uncertain future. Humans have long adapted to living in desert regions and will need to show the same skills in the future, aided by far more effective management tools.

Dr Professor Martin (M.A.J.) Williams
Department of Earth Sciences
University of Adelaide, Adelaide, Australia

Preface

Scientific interest in sand dunes is not new. The Romans as early as 2000 years ago undertook numerous expeditions into the Sahara. Unfortunately, the accounts of the Roman geographers were lost or ignored. Arab scientists of a thousand years ago had a good deal of exact knowledge, but because of their language and script their writings were relatively unknown in Europe. A period of systematic research by European scientists, especially German and French, began in the 1850s. Latter day researchers such as the British geographer R.A. Bagnold, studied the physics of blown sand and desert dune formation and the mechanisms involved. While W.S. Chepil and his colleagues in USA studied wind erosion. Subsequently, there has been an increasing stream of literature—references to which, appear throughout this book.

With advances in technology has come the application of remote sensing and Geographic Information Systems (GIS) to the detailed study of sand dunes. Several chapters in this book explain the application of satellite imagery and LIDAR (a detection system which works on the principle of radar, but uses light from a laser) has made it possible to study the evolution and development of sand dunes from even the remotest part of the landscape. For the more detailed study of sand grains such techniques as x-ray crystallography, and other such high-tech equipment has been used to good effect. Studies of the origin of dune systems and the impact of climate variability over millennia can be traced with the aid of such equipment as X-ray diffraction (XRD), X-ray fluorescence, photon-induced X-ray emission spectroscopy (PIXE), photon-induced gamma emission spectroscopy (PIGE), infrared (IR) spectroscopy, ultraviolet-visible (UV-vis) spectroscopy, and scanning electron microscopes.

The journey of sand particles through different ecological and geomorphological environments is reflected (like finger prints) on the physical and chemical properties of the sand dune particles. The morphological properties of the sand dunes give a good indication about the wind regime in the area. For example, the presence of barchans with tight horns indicates very high wind speed which makes the area appropriate for initiating windfarm-based industry. This is just one example of how better knowledge about the dunes can be of value to the broader community, at a time when alternative energy (solar and wind-power) is high on the agenda of many countries.

Scope and purpose of the book

The aim of this book is to provide an authoritative source of information on the genesis, evolution and development of sand dunes in the mid latitudes of the northern hemisphere.[1] The book has contributions in a variety of disciples and from many different countries. It aims at assembling and synthesizing wide-ranging contributions from case studies, reviews, reports on technological developments, outputs of research/studies and examples of successful research projects, as well conceptual approaches, which record current knowledge. We use an extensive bibliography (some in languages other than English) to provide readers with a comprehensive menu of further readings.

[1] North American dunes are not included in this book because of the excellent book "Inland Dunes of North America" edited by N. Lancaster and P. Hesp. 2020 (Springer, ISBN: 978-3-030-40498-7).

Most sand dunes occur in sandy deserts that are a widespread type of landscape each having a completely different genesis. Generally, they are developed over thick, loose sandy sediments, mostly of river or marine origin. By origin of sediments they may be subdivided to several subtypes:

(i) sands of ancient alluvial plains

(ii) sands of marine lowlands that may incorporate sands formed as result of deflation of ancient or recent marine deposits

(iii) sands on piedmont plans often seen as separate, large, sand massifs. In their geological structure the proluvial[2] deposits (sands, loamy sands, loams) dominate

(iv) sands of ancient structural plains composed of Paleogene, Neogene and Cretaceous sandstones. Closer to the surface these plains are composed of mostly Neogene sandstones striking horizontally. More seldom, outcrops of Paleogene and Upper Cretaceous sandstones and limestones are observed on large areas. Sands of this subtype are widespread.

Sandy deserts, irrespective of their genesis, are subdivided into three groups based on sand mobility and thickness of vegetation cover:

Bare sands are represented by single and group barchans, complexes of barchans, barchan chains, and barchan-ridge sands. They cover extensive areas in many deserts. Barchan sands also surround settlements, or develop along point and linear objects. Sand dune formation is connected with a number of physiographical and anthropogenic factors. Under the influence of strong prevailing, winds they can move and are the cause of disruption to the operation of some economic units including oil fields, irrigation canals, and transport networks.

Semifixed dunes occur in several varieties: ridge, barchan mounds, ridge mound, cellular, ridge-basin, and ridge-honeycomb sands. In general, they do not move. Barchan-like mobile forms are found only on tops of ridges in areas with very sparse vegetation cover. Some mobile forms may occur even among fixed dunes as a result of anthropogenic impact.

Fixed dunes This group incorporates stable, often immoveable forms of a sand relief mound, vegetated sand dunes, honey-combed and slightly rolling sands generally well fixed by vegetation. If vegetation is destroyed these sands will turn within a short time period into drifting barchan forms.

Sandy-gravel deserts

This type of desert is confined to ancient accumulative or denudational plains, the surface of which was formed over sandstones, conglomerates and pebbles of the Neogene and more ancient period. In some places they are developed on meriodinally, extending ridges—kyrs, as well as in deep inter-kyr hollows of different sizes. Sometimes they can be 20–30 km long. In between large kyrs, one can see small sand relief forms (e.g., honey-comb and cone-like hollows). They occur at different levels which creates an impression of terraced relief. Their width is different varying from 1–200 m up to 1–5 km. On the surface they have a sandy, often mound-shaped form with barchans in some places. Many residual uplands are broken by arch-like valleys and deep erosion gullies. Valleys are characterized by the presence of wide, flat bottoms covered by crushed stone-sandy material. Residual uplands may be separated by sandy massifs. Aeolian sands are sometimes underlain by a gypsum bed from 20–30 cm to 1–4 m thick. The top of the sandstones, conglomerates and pebbles has, in some places, sand admixtures or even interbeds with significant massifs of aeolian sands. Aeolian forms of mounds, vegetated sand dunes, honeycomb and slightly rolling sands are largely the products of weathering and deflation of the underlying Pliocene and Lower Quaternary deposits.

---

[2] A complex, friable, deltaic sediment accumulated at the foot of a slope as a result of an occasional torrential washing of fragmental material.

We give emphasis to case studies and practical initiatives to deal with migration of dunes and encroachment onto cropland, onto infrastructure such as roads, railways, pipelines and canals—all of which may help to address a problem with a global scope, but the impacts of which are mostly local. There is a vital need for reliable information and data to support the efforts pursued by local governments, NGOs and other organizations to address the problems associated with the management of sand dunes, including sand storms. Thanks to its interdisciplinary focus, the book aims to concretely contribute to a better understanding of the state-of-the-art of sand dune fixation, and of the tools with which it can be implemented on the ground. Attention is given to ecological restoration as means of fixing mobile dunes and protecting the livelihoods of the dryland users.

We are particularly pleased with the level of international cooperation. The three co-editors are from widely different backgrounds and geographic locations. The authorship of chapters spans several continents and we have benefited from the inclusion of the work of our Russian colleagues who have shared their research and experience from both Russia itself but also work done in countries in Central Asia that were once part of the Soviet Union. To quote Dr. Ali Al Dousari from the Kuwait Institute of Scientific Research (KISR).

> "Regional scale co-operation such as that in the chapter that combines the work of researchers from Kuwait, Saudi Arabia and Iraq takes a lot of effort to achieve and involves unprecedented regional scale co-operation. It is hoped that the coming together for this book project will ensure on-going cooperation and serve as a model for the neighbors who share common problems".

The book consists of 39 peer-reviewed chapters written by over 120 contributors from 17 different countries that are arranged in two volumes.

Volume 1 of 23 chapters is sub-titled **Formation and evolution of dune landscapes** and focuses on Methodologies for the study and utilization of sand dunes; Dunes and dune systems in China; Dunes and dune systems in India's Thar desert; and Coastal dune systems of the Indian subcontinent.

Part 1 of 7 chapters focuses on the application of technology to enhance our understanding of the structure and function of sand dunes. Satellite and near-earth air borne imagery combined with geographic information systems (GIS) have enhanced the ability to analyze changes in spatial and temporal relationships among and between dunes at the mineralogical level. Sophisticated equipment and procedures such as x-ray flourescence have allowed scientists to derive information about the age and origin of sand and their resultant dune systems. Studies in this Part seek to elaborate on the methodological aspects with studies from Saudi Arabia, Egypt, and Algeria but also consider economic aspects of dune sand utilization in Egypt.

Part 2 of 7 chapters focuses on the Dunes and dunes systems in China. It presents a general description of the environmental status of Chinese deserts, describing their extent, location, uniqueness, vulnerability, and status of their natural resources. It begins with a commentary on China's extensive and unique dunes systems—their genesis, evolution and forcing mechanisms. Detailed analysis of dunes in the 6 major deserts (including coastal) follows.

Part 3 of 9 chapters turns the spotlight on India's Thar desert, examining *inter alia*, late quaternary paleoenvironments, chronology, geospatial diversity and geological significance in terms of landform evolution as well as a review of the role played by the Thar in migration corridors for early humans as they dispersed eastward. But four of the chapters in this Part focus exclusively on coastal dune systems with studies from western India, South India, Sri Lanka and Bangladesh. Both ecology and morphology receive attention here.

Volume 2 is sub-titled **Characteristics, dynamics and provenance of sand dunes**

It brings together a vast body of information and insight into sand dune and desert systems from North Asia, Central Asia, North Africa and the Middle East. Chapters from the Russian Federation include studies on dune systems within the permafrost zone and a case study from temperate zone dune system in coastal Japan.

Volume 2 of 15 chapters in three Parts focuses on Saharan Africa, Egypt, and Middle East and gives attention to sand mobility and encroachment with case studies from a number of countries where these matters are of concern. We also include chapters on the remote dunes in the permafrost zone and in the hyper-arid deserts of Iran. Case studies are used to highlight the characteristics if dunes and their interaction with humans in several widely divergent settings. Volume 2 concludes by some musing on the value of study of the past as key to the future and speculates on what the future might hold in the light of a warmer and drier Earth and a rise in sea level that threaten large tracts of low-lying land with marine incursions and destruction from storm surge.

PART 4 Sand Mobility and Encroachment

The seven chapters in this Part examine the real-world impact of sand encroachment and dune migration on people and their economic activities and the health, welfare and financial implications related to destruction of infrastructure, including human habitations. It also covers sand encroachment and its potential threats to the human life as well as the ecosystem, including the socio-economic system of the region. Anthropogenic activities combined together with the climatic phenomenon accentuate the mobility of the sand dunes, thus, causing environmental hazards. Case studies from Egypt, Israel, Morocco, Sudan, and Nigeria have been presented covering these issues.

PART 5 Sand Dune Landscapes Distribution, Formation and Management

Seven Case studies from several geographic regions in Africa, south-west Asia, north-east Asia are presented here to demonstrate the underlying mechanisms in dune formation and the diverse approaches to their management. Human impacts such sand mining, tourism development, combined with natural forces such as climate variability to challenge the realization of an optimum management strategy. Each chapter describes a unique and little-known dune system that because of remoteness or the unusualness are rarely reported on in English language publications. We are fortunate to be able to bring together experts who can give an authorative account of dunes in remote regions within the Artic circle or in Iran's desert interior. The studies presented on the dunes that formed on the temperate coastal regions in Japan provides an opportunity to explore the anthropogenic impact on the geomorphology, and ecology of the dunes.

PART 6 Concluding Thoughts: Coping an Uncertain Future from a Little-Known Past

The two chapters in Part have special roles. We are privileged to publish new research findings that are summarized here in Chapter 38 from extensive and detailed work conducted in the Kyzyl Kum, and Karakum sand seas of Central Asia. There is a strong belief that further study of the little-known past such as revealed in this study could unlock clues as to what a future Earth might look like.

This leads on to speculation in Chapter 39 about the implications of what we already know about global change (not only climate change) and the impact of the Anthropocene on the dune systems, both inland coastal. Sea Level rise, marine incursions and an increase in extreme weather events will affect dune systems and sand seas on the Northern Hemisphere (and beyond).

Acknowledgements

Some of the research reported here and this entire publication received support from the Central Public-Interest Scientific Institution Basal Research Fund of China (No. CAFYBB2021ZD003, CAFYBB2019ZD002 and CAFYBB2018ZE002). We Editors are also grateful for the inputs from Dr. Aayush Srivastava, Dr. Dmitry Konyushkov and from Dr. Cui Mengchen who helped take some of the load at critical times. The input of Dr. Sudesh K. Wadhawan and Dr. Neelratan Singh was appreciated. The 71 authors/co-authors and the institutions who provided support and encouragement are to be commended. Thanks are due to Dr. Bing Qi Zhu for the photograph used on the cover of this volume.

Contents

Volume 2

Characteristics, Dynamics and Provenance of Sand Dunes in the Northern Hemisphere [*IN PRESS*]

List of Contributors

Abdelhak, Maazouzi
Chemistry and Environmental Sciences Laboratory (CESL) Bechar University 08000, Algeria.
Email: maazdz@yahoo.fr

Acharjee, Shukla
Centre for Studies in Geography, Dibrugarh University, Dibrugarh, India.

Achyuthan, Hema
Professor, Institute for Ocean Management, Anna University, Chennai 600025, Tamil Nadu.
Email: hachyuthan@gmail.com

Ahmed, Kettab
Associate Professor University of Bouira-Algeria and Professor/Research Director of the National Polytechnic School-Algiers. Algeria.

Aina, Yusuf A.
Department of Geomatics Engineering Technology, Yanbu Industrial College, Yanbu, Saudi Arabia.
Email: yusufaina@gmail.com

Alappat, Linto
Department of Geology and Environmental Science, Christ College Autonomous, Irinjalakuda, Thrissur, Kerala, 680 125 India. Email: alappatlinto@christcollegeijk.edu.in

Alqurashi, Abdullah F.
Geography Department, Umm Al-Qura University, Makkah Al-Mukarramah, Kingdom of Saudi Arabia.

Balogun, Abdul-lateef
Geospatial Analysis and Modelling (GAM) Research Group, Department of Civil and Environmental Engineering, Universiti Teknologi PETRONAS (UTP), 32610 Seri Iskandar, Perak, Malaysia.

Bo, Wu
Institute of Desertification Studies, Chinese Academy of Forestry, Beijing 100091, China.

Brahim, Zahraoui
Chemistry and Environmental Sciences Laboratory (CESL)Bechar University 08000, Algeria.

Chakraborty, Elora
Centres for Study of Regional Development, Jawaharlal Nehru University, New Delhi, India.

Chang, Xi
Frontiers Science Center for Critical Earth Material Cycling, School of Geography and Ocean Science, Nanjing University, Nanjing 210023, China.

Chopra, Sundeep
Geochronology Group, Inter University Accelerator Centre, New Delhi, India.

Cui, Mengchun
Frontiers Science Center for Critical Earth Material Cycling, School of Geography and Ocean Science, Nanjing University, Nanjing 210023, China.

Dandabathula, Giribabu
Regional Remote Sensing Centre – West, NRSC/ISRO, Department of Space, Jodhpur, India.
Email. giribabu.d@nrsc.gov.in

Deshmukh, B.
Geology Division, School of Sciences, IGNOU New Delhi, India.

Dhir, R.P.
498, Defense Colony, Jodhpur-342008. India..

Dong, Zhibao
School of Geography and Tourism, Shaanxi Normal University, Xi 'an, Shaanxi, China.

Dong, Yuxiang
School of Geography and Planning, Guangdong, Sun Yat-Sen University, Guangzhou 510275, China. Email: eesdyx@mail.sysu.edu.cn

Driss, Lahcene
Chemistry and Environmental Sciences Laboratory (CESL) Bechar University 08000, Algeria.

EL-Bady, Mohamed S.
Geological Sciences Department, National Research Center, Dokki, Cairo, Egypt.
Email: elbady_nrc@yahoo.com

Feng, Han
Frontiers Science Center for Critical Earth Material Cycling, School of Geography and Ocean Science, Nanjing University, Nanjing 210023, China.

Fu, Shuyi
School of Geography and Planning, Guangdong, Sun Yat-Sen University, Guangzhou 510275, China.

Gaber, Ahmed
Geology Department, Port Said University, Egypt.

Gaur, Mahesh K.
ICAR-Central Arid Zone Research Institute, Jodhpur-342003, India.
Email: maheshjeegaur@gmail.com

Guipeng, Cui
Institute of Desertification Studies, Chinese Academy of Forestry, Beijing 100091, China.

Hasi, Eerdun
School of Natural Resources, Faculty of Geographical Science, Beijing Normal University, Beijing 100082, China.

Hermas, ElSayed
The National Authority for Remote Sensing and Space Sciences, Cairo, Egypt and Geography Department, Umm Al-Qura University, Makkah Al-Mukarramah, Kingdom of Saudi Arabia.
Email: eaibrahim@uqu.edu.sa

Hua, Limin
College of Grassland Science, Gansu Agricultural University, Lanzhou, China.
Email: hualm@gsau.edu.cn

Islam, M. Shahidul
Coastal Research Unit (CRU), Department of Geography and Environment, University of Dhaka, Bangladesh. Email: shahidul.geoenv@du.ac.bd

Jayawardena, U.de S.
Department of Civil Engineering, Faculty of Engineering, University of Peradeniya, Peradeniya, Sri Lanka. Email: udesja@gmail.com, udsj@pdn.ac.lk

Jianmin, Chu
Research Institute of Forestry, Chinese Academy of Forestry, Beijing 100091, China.

Kumar, Pankaj
Geochronology Group, Inter University Accelerator Centre, New Delhi, India.

Kumar, Rajesh
Centres for Study of Regional Development, Jawaharlal Nehru University, New Delhi, India and Department of Geography - Sikkim University, Gangtok, India.

Lu, Huayu
Frontiers Science Center for Critical Earth Material Cycling, School of Geography and Ocean Science, Nanjing University, Nanjing 210023, China. Email: huayulu@nju.edu.cn

Lv, Nana
Frontiers Science Center for Critical Earth Material Cycling, School of Geography; and Ocean Science, Nanjing University, Nanjing 210023, China.

Majumdar, Dipanjan Das
Post-Doctoral Fellow, DACEE, Faculty of Interdisciplinary Studies, Law and Management, Jadavpur University, West Bengal India; and Assistant Professor, Department of Geography, Netaji Satabarshiki Mahavidyalaya, Ashoknagar, North 24 Parganas-743223, West Bengal, India.

Mengchun, Cui
Institute of Desertification Studies, Chinese Academy of Forestry, Beijing 100091, China.

Niu, Yujie
College of Grassland Science, Gansu Agricultural University, Lanzhou, China and Department of Disturbance Ecology, BayCEER, University of Bayreuth, Bayreuth, Germany.
Email: yujie.niu@uni-bayreuth.

Pan, Gao
Institute of Desertification Studies, Chinese Academy of Forestry, Beijing 100091, China.

Paul, Anurupa
Senior Research Fellow (SRF), Department of Remote Sensing & GIS, Vidyasagar University, Midnapore-721102, West Bengal, India.

Paul, Ashis Kumar
Professor, Department of Geography, Vidyasagar University, Midnapore 721102, West Bengal, India. Email: akpaul_geo2007@yahoo.co.in

Qi, Lu
Institute of Desertification Studies, Chinese Academy of Forestry, Beijing 100091, China and Experimental Center of Desert Forestry, Chinese Academy of Forestry, Dengkou 015200, Inner Mogolia, China. Email: luqi@caf.ac.cn

Rajendra, K.
Amity School of Earth & Environmental Science, Amity University, Haryana, India.

Ren, Xiaozong
School of Geography Science, TaiYuan Normal University, Jinzhong 030619, China.

Rioual, Patrick
Key Laboratory of Cenozoic Geology and Environment, Institute of Geology and Geophysics, Chinese Academy of Sciences, Beijing 100029, China.

Sharma, Milap C.
Centres for Study of Regional Development, Jawaharlal Nehru University, New Delhi, India

Shekhar, Shubhendu
Department of Geology, Center for Advanced Studies, University of Delhi, Delhi, India.

Singh, Atul Kumar
Geochronology Group, Inter University Accelerator Centre, New Delhi, India.

Singh, Neelratan
Geochronology Group, Inter University Accelerator Centre, New Delhi, India. And CSIR - Central Institute of Mining & Fuel Research, Nagpur Research Centre (Mining Technology), (Ministry of Science & Technology, Govt. of India), Nagpur - 440001, Maharashtra, India.

Singh, Yamuna
Centre for Earth Sciences, University of Hyderabad, Hyderabad, India.

Singhvi, Ashok Kumar
AMOPH Division, Physical Research Laboratory, Ahmedabad, 380 009, India.

Squires, Victor R.
Institute of Desertification Studies, Beijing, China and College of Grassland Science, Gansu Agricultural University, Lanzhou, China. Email: dryland1812@internode.on.net

Srinivasalu, S.
Department of Geology, Anna University, Chennai, Tamil Nadu, 600 025 India.

Srivastava, Aayush
Environmental Sciences, University of St Andrews, UK. and School of Geography and the Environment, University of Oxford, UK.

Sun, Chen
Key Laboratory of Water Cycle and Related Land Surface Processes, Institute of Geographic Sciences and Natural Resources Research, Chinese Academy of Sciences, Beijing 100101, China.

Wadhawan, Sudesh K.
Adjunct Professor, Amrita Vishwa Vidyapeetham University, Kerala-690525, and Former Director General, Geological Survey of India. Email: wadhawansk.leo@gmail.com

Wang, Hanlin
Frontiers Science Center for Critical Earth Material Cycling, School of Geography and Ocean Science, Nanjing University, Nanjing 210023, China.

Wang, Xianyan
Frontiers Science Center for Critical Earth Material Cycling, School of Geography and Ocean Science, Nanjing University, Nanjing 210023, China.

Wang, Xiaoyong
Frontiers Science Center for Critical Earth Material Cycling, School of Geography and Ocean Science, Nanjing University, Nanjing 210023, China.

Wang, Xunming
Key Laboratory of Water Cycle and Related Land Surface Processes, Institute of Geographic Sciences and Natural Resources Research, Chinese Academy of Sciences, Beijing 100101, China.

Wang, Yao
Frontiers Science Center for Critical Earth Material Cycling, School of Geography and Ocean Science, Nanjing University, Nanjing 210023, China.

Wu, Bo
Institute of Desertification Studies, Chinese Academy of Forestry, Beijing 100091, China.

Xu, Zhiwei
Frontiers Science Center for Critical Earth Material Cycling, School of Geography and Ocean Science, Nanjing University, Nanjing 210023, China.

Xue, Dong
Experimental Center of Desert Forestry, Chinese Academy of Forestry, Dengkou, Inner Mongolia, China. Email: dongxue98765@126.com

Yamuna, Singh
Centre for Earth Sciences, University of Hyderabad, Hyderabad, India.

Yang, Yanyan
Institute of Ecological Conservation and Restoration, Chinese Academy of Forestry.

Yekeen, Shamsudeen Temitope
Geospatial Analysis and Modelling (GAM) Research Group, Department of Civil and Environmental Engineering, Universiti Teknologi PETRONAS (UTP), 32610 Seri Iskandar, Perak, Malaysia and Department of Geography, Environment and Geomatics, University of Guelph, Guelph, ON, Canada.

Yi, Shuangwen
Frontiers Science Center for Critical Earth Material Cycling, School of Geography and Ocean Science, Nanjing University, Nanjing 210023, China.

Zhang, Jiaxin
Key Laboratory of Water Cycle and Related Land Surface Processes, Institute of Geographic Sciences and Natural Resources Research, Chinese Academy of Sciences, Beijing 100101, China.

Zhang, Shaoyun
School of Geography and Planning, Guangdong, Sun Yat-Sen University, Guangzhou 510275, China.

Zhang, Xiaojian
Frontiers Science Center for Critical Earth Material Cycling, School of Geography and Ocean Science, Nanjing University, Nanjing 210023, China.

Zhang, Hanzhi
Frontiers Science Center for Critical Earth Material Cycling, School of Geography and Ocean Science, Nanjing University, Nanjing 210023, China.

Zhang, Wenchao
Frontiers Science Center for Critical Earth Material Cycling, School of Geography and Ocean Science, Nanjing University, Nanjing 210023, China.

Zhou, Yali
School of Geography and Tourism, Shaanxi Normal University, Xi 'an 710062, China.

Zhu, Bing-Qi
Key Laboratory of Water Cycle and Related Land Surface Processes, Institute of Geographic Sciences and Natural Resources Research, Chinese Academy of Sciences, Beijing 100101, China. Email: zhubingqi@1gsnrr.ac.cn

Part 1

Methodologies for the Study and Utilization of Sand Dunes

The seven chapters in this Part focus on ways and means of examining sand and sand dunes. The application of remote sensing (including acquisition of near-Earth imagery) laboratory techniques to determine physico-chemical properties and the application of this knowledge are at the heart of these contributions that demonstrate, through the various case studies, the wide variety of techniques and the sophisticated instrumentation associated with the contemporary study of sand and sand dunes.

Chapter 1: *Lu Qi, V. R. Squires* and *M. K. Gaur.* This short overview precedes the specialist chapters and touches on the key issues around deserts and sand dunes that are dealt with in detail in the chapters that follow. Since pioneering studies in the mid-part of the twentieth century there has been much progress in understanding the true extent of deserts and how dunes form and develop. The application of remote sensing and near-earth scanning including the use of drones, has revealed much. And, invention of an array of new analytical tools at the researchers' disposal have allowed many of the sand's secrets to be revealed. The results of these research efforts in some of the world's deserts are elaborated upon in this book.

Chapter 2: *ElSayed Hermas* and his co-authors provide an overview of advances in the study of sand dune migration that is both comprehensive and detailed. Each technique has a role to determine dune movement and direction from a technical and scientific perspective, some are better than others. Although each technique has its own shortcomings and limitations, however, each technique represented a remarkable step in determining, understanding, and modeling the magnitudes and directions of dune movements either at a single-dune or at dune field scales. Presenting and discussing all the criteria of the previous techniques may help outline propositions for future development of these techniques. This chapter is a great example of cross-border cooperation. Information and data generated from five countries in the Arabian Peninsula and Middle East was shared.

Chapter 3: *Shamsudeen Temitope Yekeen* and his international team review the application of remote sensing in its many forms that range from large scale synoptic overviews such as generated by NOAA's and NASA's various to more recent high resolution various satellites images and other images collected by unmanned aircraft (drones of various sizes and sophistication). Remote sensing provides access to hitherto easily unavailable data and when paired with geographic information systems (GIS) can 'unlock' much information that can assist academicians, scientists, policy makers, field researchers and others.

Chapter 4: *Giribabu Dandabathula* and *Mahesh K. Gaur* dwell upon the applications of ICESat-2 Laser Altimeter Data for Sand Dunes Morphology and Dynamics Studies. This is a new study on the sand dunes of Thar Desert of India. Geophysical parameters can be generated by applying Laser Altimetry data for the study of the shape, structure and evolution of the dunes. Revelations and insights are useful and provide a platform for future applications and understanding the morphological parameters.

Chapter 5: *Mohamed S. El-Bady* studied the properties of sand on the beach and in the dunes in the Nile delta region of Egypt and showed the benefits of applying new techniques to measure the various attributes of the sand. From these measurements and observations new insights were developed into the chronology and development of dunes.

Chapter 6: *Maazouzi Abdelhak* and his co-authors focus on the characterization of sand dunes in the western Erg of the Algerian Sahara. A representative sand sample of Algerian Sahara sand dunes (western Erg) was collected and analyzed by several physicochemical methods; which involved the setting up and utilization of a multi technical methodology. Granular analysis permitted determination of some parameters such as the uniformity coefficient (CU) and equivalent diameter. Scanning Electron Microscopy (SEM) observations and X-Ray analysis are presented.

Chapter 7: *Mohamed S. El-Bady* applied modern techniques and instrumentation to assess the economic heavy metals in the coastal area between Burg El-Burullus and Baltim, north of the Nile Delta. Deposition and migration patterns of the various economically-valuable mineral sands such as illite were elucidated.

Chapter 1

A Synoptic View of Northern Hemisphere Deserts and Sand Dunes

Perceptions, Misconceptions and Distinctions

Victor R. Squires,[1,*] *Mahesh K. Gaur*[2] and *Lu Qi*[1]

1. Development of research into deserts and sand dunes

Although deserts and dunes in North Africa were investigated by Roman and Greek scholars from the earliest times and in the past 100 years or so, by many geographers and 'specialists' (geomorphologists, geologists, stratigraphers, sedimentologists, archaeologists, anthropologists, physicists, ecologists and those who study aeolian landforms), there is still much to be discovered. Notable early work was conducted by German geographers (Hermann Barth in 1850 and Gustav Nachtigal in 1863) and the French researcher (Theodore Monod). A French mining engineer (Rolland, 1881) published papers on the sources of sand, the distribution of ergs, the relationship of dunes to wind direction and the nature of dune mobility. Chudeau (1909) gave a more general account of Saharan dunes in the Sudan and in 1920 described the nature of sand, its possible sources, the localization of sand in great fluvial basins, and the main dune forms (those parallel to the wind), and those that are transverse to the wind, barchans, and small obstacle dunes of various types, including *nebkas* and *zemlat*. Collectively, their findings on sediment transport, dune genesis and development reflect the situation in North Africa and the Sahel. Russian scientists A.S. Babiev and N.S. Orlovosky extended the geographical range by studying deserts and dune systems in the former Soviet Republics of Turkmenistan, Kazakhstan and Uzbekistan.

The knowledge gained (and published) specifically on deserts and sand dunes has come from relatively few studies in a limited range of settings. North American dunes have received much more attention and are now the subject of a new book (Lancaster and Hesp, 2020). Exceptions however can be found and well-known studies on deserts in Africa, Asia and elsewhere, include the works of (Goudie, 2002). Williams, 2014, Lancaster (1994), Tsoar (2005), Middleton (2019), Thomas (2019), Zhu Zhenda (1984), Babiev (1986), Orlovosky (1986), Dhir (2018), Singh (1982), Singhvi (2004), Kar (2002) and so on. Some specific sites have received more attention and publicity than others.

[1] Institute of Desertification Studies, Beijing, China.
[2] ICAR-Central Arid Zone Research Institute, Jodhpur, India.
* Corresponding author: dryland1812@internode.on.net

Modern desert research is concerned with investigating the desert's function on Earth's history and its effect on the evolution of life forms. The difficult and complex questions being raised can be answered only by interdisciplinary endeavors, and geomorphologists, geologists, paleontologists, climatologists, ecologists and others, working together in desert research. A science that could comprehend and describe the phenomenon of deserts, would possess the key to the past and future of our planet.

2. Key characteristics of selected Northern Hemisphere deserts

The location and extent of arid zones, especially those classified as deserts, have shifted through time owing to changes in continental configuration, position and relief. On the basis of climatic characteristics one-third of the Earth's land surface can be classified as desert of one type or another (not all are sandy deserts). Table 1.1 is a list of major Northern Hemisphere inland deserts, with some details about location and tectonic setting (see below for an elaboration of the significance of tectonics). Those inland deserts that are the subject of this book are highlighted in bold typeface in Table 1.1.

It is not surprising that there should be a great diversity of desert landscapes. Desert landscapes and desert processes vary a lot. There are distinctive characteristics of each desert—they are recognizably different from each other. The diversity of desert landscapes is affected by their climatic history. All deserts have been subjected to substantial degrees of climatic change. In particular, the history of climate in the deserts of China and the north-western Indian subcontinent has been dominated by the effects of the uplift history of the Himalayas and the Tibetan Plateau on monsoonal circulation. Sometimes even the impact of very early climatic changes can be discerned in desert landscapes today—the Sahara, for example, shows the imprint of Ordovician glaciation (Goudie, 2002). Desert landscapes have the history of climate change buried in their stratigraphy. Another aspect of landscape diversity between different deserts is the extent to which drainage is to the interior (i.e., endoreic) or directly to the sea. For example, the western Sahara in Morocco has several rivers that flow into the Atlantic Ocean (Draa and Sebu) while in NW China inland rivers (e.g., Tarim, Hei He, etc.,) are endoreic and end in terminal lakes (like Lop Nor) or just discharge to the desert.

Climate, vegetation and fauna have all been used to delimit desert boundaries. Deserts may be divided into categories based on their temperature (hot, temperate, coastal) or moisture (hyper arid, arid, semiarid) characteristics. In many instances, the drier areas of the Earth are simply divided into two groups: arid and semiarid (or synonymously, desert and semidesert, or desert and steppe).

Within the broader arid zones are many individual deserts with distinctive characteristics that are the subject of this book. Their nature is determined by physiography, ecozone, geology, edaphic characteristics aridity, and these factors affect their mapped extent.

2.1 The causes of aridity

The arid regions of the world, other than those in the high Arctic, owe their origin to climatic, topographic and oceanographic factors that prevent the incursion of moisture-bearing weather systems. Most deserts are hot. Within the latitudes of the Tropics of Cancer and Capricorn, thirty degrees on either side of the equator, the sun appears directly overhead at noon on at least one day of the year. They are located in the *Horse Latitudes*, where constant high-pressure systems drive away the rain clouds, and swirl above the Earth to the tune of global temperature variations and the *Coriolis Effect* produced by the Earth's rotation in space. It should be noted that most deserts are arid because of a combination of factors, like (i) atmospheric high-pressure zones (Hadley Cells), (ii) continentality or (distance from oceans), (iii) coastal cooling: deserts may occur if air is cooled, and then rewarmed, before reaching the region, and (iv) rain shadow effects. Sometimes these factors work singly, or more often in combination, as primary forces producing arid lands. Continental interiors are arid owing to the distance from the sea and sources of moisture. In NW China and central Asia, for example, a great mountain arc prevents the deep incursion of moisture-laden air

Table 1.1. The major deserts of Northern Hemisphere.

Desert	Area (Km²)	Tectonic Setting	Location	Remark
Arabian	1,855,470	Cratons	Arabian Peninsular	Occupies most of the Arabian Peninsula. The 5th largest desert.
Badain Jaran	49,000	Older orogenic belt	China	Endoreic river basin
Tengger	36,700	Older orogenic belt	China	Endoreic river basin
Gobi	1,295,000	Older orogenic belt	China and Mongolia	Cenozoic orogenic belts
Taklamakan	560,000	Older orogenic belt	China	Endoreic river basin
Kumtag	22,800	Older orogenic belt	China	Endoreic river basin. It is a section of the Taklamakan Desert.
Hexi Corridor	215,000	Older orogenic belt	China	Endoreic river basins
Tibetan Plateau	2,500,000	Cenozoic orogenic belts	China	Alpine desert
Ladakh	~ 97,000	Active continental margins and Cenozoic orogenic belts	India	Alpine desert
Thar	238,254	Cenozoic orogenic belts	India	Adjoins Thal desert of Pakistan
Cholistan Thal (Rohi)	25,800	Active continental margins and Cenozoic orogenic belts	Pakistan	Extension of Thar desert
Registan Desert (Rigestan Desert)	146,000	Cenozoic orogenic belts	Afghanistan	A sandy desert consisting of small, red sand hills about 50–100 feet (15–30 m) high
Dasht-e Margo (also *Dasht-e Mārgow* or *Dasht-e Margoh*)	150,000	Cenozoic orogenic belts	Afghanistan	The desert's name means "Desert of Death" in Dari, with dasht meaning "plain" and margo meaning "death". The world's 20th largest desert.
Sahara	9,200,000	Cenozoic orogenic belts	Algeria, Chad, Egypt, Libya, Mali, Mauritania, Morocco, Niger, Western Sahara, Sudan and Tunisia	Nile valley. It is mainly rocky *hamada* (stone plateaus); *ergs* (sand seas – large areas covered with sand dunes) form only a minor part, but many of the sand dunes are over 180 meters high.
Dasht-e-Lut	51,800	Older orogenic belt	Iran	Endoreic River basin
Lake Urmia	5,200	Cenozoic orogenic belts	Iran	Drying lake
Dasht-e Kavir	77,600	Older orogenic belt	Iran	
Sinai-Negev	73,000	Active continental margin	Palestine/Israel Egypt	
Egypt Western Desert (Nile Delta)	700,000	Active continental margin	Egypt	
Western Sahara	266,000	Active continental margin	Morocco	
Algerian Sahara	1,905,393	Older orogenic belt	Algeria	
Libyan	1,100,000	Older orogenic belt	Libya/Egypt	
Eastern Egypt	221,940	Older orogenic belt	Egypt	Consists of a rolling sandy highland
Aralkum	17,000	Older orogenic belt	Uzbekistan Kazakhstan	Man-made desert from Aral Sea, world's youngest desert
Kara Kum	350,000	Older orogenic belt	Turkmenistan	Endoreic River basin
Kyzyl Kum	298,000	Older orogenic belt	Uzbekistan	Endoreic River basin. Its name means Red Sand in Turkic languages.

of the summer monsoon. In winter, a vast high-pressure cell develops, with very dry subsiding air. Interior deserts have a much greater range in annual temperature than those closer to a coast. Despite the high latitudes of Asian deserts, the summers are extremely hot, with maximum temperatures in excess of 38°C at all lowland stations. Winters are very cold, with minima ranging from –30 to −50°C. Some deserts are underlain by permafrost (Chapters 13 and 34).

According to Laity (2008), the tectonic setting of deserts plays a role in determining the relative and absolute relief of the region, affecting sediment sources and sinks, influencing timescales of relative stability or instability and, over the long term, the degree of aridity as mountain systems rise and block incoming moisture. Arid zones provide an important setting for tectonic study. These tectonic settings play a major role in determining the character of deserts.

Five types of tectonic setting are identified: cratons (shield and platform areas), which have shown relative stability since the late Tertiary; active continental margins, associated with Cenozoic orogenic belts, in a compressional setting, with thrust and transcurrent faulting; older, Phanerozoic orogenic belts, in which there is some reactivation of existing fault zones; inter-orogenic basin and range and inter-cratonic rift zones, with "pull-apart" basins in an extensional tectonic setting; and passive continental margins (Rendell, 1997). Examples of deserts in these settings are shown in Table 1.1. The Saharan Desert is of such physical extent that it is included in more than one category.

The location and extent of arid zones have shifted through time, owing to changes in continental configuration, position and relief. The onset of modern aridity can be traced back millions of years, with the development of drier conditions as long ago as 80 Ma. At a global level, it is thought that aridity was in place by 3 Ma (Williams, 1994), related to declining temperatures in the late Cenozoic associated with the movement of continents towards their modern latitudes.

2.2 Deserts and sand dunes

According to (Tsoar, 2005), the accumulation of windblown sand creates dunes which are one of nature's most dynamic and intriguing phenomena. Sand dunes are found in most climates of the world as coastal dunes and in some arid regions. Grains of sand between 0.062 and 2.0 mm in diameter are not cohesive and therefore are easily carried by the wind. This property of sand is reflected in the wind threshold speed curve for sand transport (Fig. 1.1) explaining why dune sand, in most cases, is composed of fine particles between 0.125–0.250 mm.

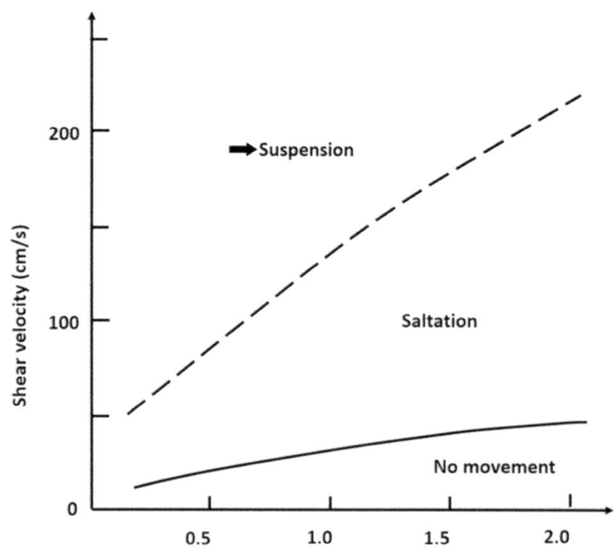

Figure 1.1. A typical wind threshold speed curve for sand transport.

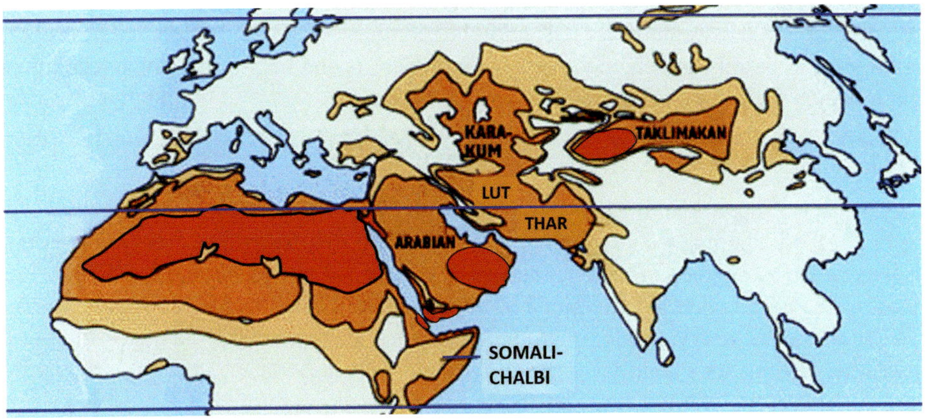

Figure 1.2. Major deserts of the Northern Hemisphere that are the subject of this book.

Lancaster (1994) said "Quartz sand dominates aeolian deposits in arid regions, as dust-sized particles are carried out of the desert entirely" or are trapped by vegetation or rough surfaces on desert margins (Tsoar and Pye, 1987). Most (> 95%) sand occurs in accumulations known as sand seas or ergs (Wilson, 1973) that comprise areas of dunes of varying morphological types and sizes as well as areas of sand sheets (Fig. 1.2). Smaller dune areas are known informally as dune fields.

Based on both natural agents including wind speed and direction, precipitation and evapotranspiration rates, sand supply, the kind and density of vegetation cover, topographic barriers and reliefs and anthropogenic effects, including livestock over-grazing and groundwater over-exploitation, the sand seas and sand fields are categorized as active or inactive (Tsoar, 2001, 2005; Lancaster, 1988, 1994). Vegetation cover is also a crucial controller of sand movement and dune development. Except along coasts, active dunes tend to occur in areas where mean annual precipitation is less than 100–250 mm, and the specific forms of dunes may be related to vegetation cover. Parabolic forms, for example, seem to be characteristic of more vegetated areas.

3. Desert dunes: Diverse, distinctive or unique

While many more individual studies should be undertaken in a wide variety of settings before a truly comprehensive understanding of desert dunes is achieved, it is possible to recognize some regional similarities as well as contrasts. There is a range of landforms in desert areas. However none is unique. A continuum of aeolian environments exist, it is important to note that climate is just one criterion influencing dune systems.

The science of desert dunes consists of study of geomorphology, geology, sedimentology, aeolian influences, physics and mineralogy and some or all of these are used in paleoenvironmental reconstructions in this volume (and elsewhere) (see Chapters 8, 16, 18 and 38). These sources, from the present volume, provide information and greater detail than that presented here.

The complexity of trying to derive clear and unequivocal explanation as to the genesis and development because of the recognition that certain landscapes in a given climatic region may be relict or complex, partly inherited from earlier climates and there can be certain equifinality[1] in landforms evolving under different climates (Chorley et al., 1984; Summerfield, 1991). Dunes however, adjust more rapidly and completely to climate change than do whole landscapes. It is possible, therefore, that if climate change does influence deserts and their associated landforms and sedimentary successions could adjust rapidly to reflect essentially contemporary climatic conditions,

[1] Equifinality is the principle that in open systems a given end state can be reached by many potential means.

although the possibility of equifinality remains. Any attempt to categorize deserts and sand dunes as a distinctive group, thus remains an intriguing but complex problem. To do so requires both an appreciation of the controls and characteristics of both deserts and their dunes and a recognition of the extent to which these controls and characteristics are unique.

3.1 Climate and its role

In any assessment of deserts and dunes as a distinct group, an appreciation of the role of climate is paramount, as it is the main determinant of precipitation and runoff. Deserts are characterized by high (but variable) degrees of aridity, with aridity being promoted by various factors such as wind. Most of the deserts that are examined in this book are located in the subtropics where they are under the influence of dry, stable, high-pressure airmasses that inhibit the penetration of rain-bearing weather systems for a significant part of the year. Some deserts, e.g., the Taklamakan Desert or Tibetan plateau deserts of NW China, are located in rain shadows formed by topographic barriers, or are in the center of large continental land masses (e.g., central Eurasian deserts). The genesis of the Thar's aeolian landforms is the product of a delicately balanced relationship between wind intensity and length, supply of sediments, rainfall, cover of vegetation and conditions of the land surface. Not only do these variables decide the formation of the landform, but they are also vital to its resilience over time. With the arrival of monsoon rains, the wind and the sand dynamics cease (along the eastern margin of the desert at end of June, and in the western part around mid-July).

Clearly, average climatic parameters are often not a particularly useful guide for defining environments that are characterized as deserts. Climate can directly and indirectly, influence weathering processes and rates. When these climatically influenced variables are combined with those that are independent of contemporary climate, yet can also influence landscape form and process, such as physiography, geological structure and lithology, tectonism and paleoclimate, it is not difficult to see why the classification of desert and dune types based on climate alone becomes a complex issue. Any characterization of deserts and their dune systems as a group must reflect both climate and the range of other variables with which climate may or may not interact.

Notably, there is a wide range of hydrological conditions in most deserts. Such diversity is reflected in terms of the almost continuous spectrum of flow occupancy of rivers and streams, varying from rivers in hyper arid deserts that can remain dry for decades, through to allogenic rivers that sustain perennial flow or yet again endoreic inland rivers with source bordering dunes (see Zhu et al., this volume) As with climate, this diversity makes it difficult to classify deserts and dune systems as a distinctive group on the basis of hydrology alone.

3.2 Sediment transport

Desert regions typically support sparse, unevenly distributed or temporally available vegetation cover, which is usually thought to affect sediment supply. Sediment yield influences the supply of sediment to valley floors and streams. Studies show that many desert streams transport large quantities of sediment during flood events, both as suspended and bed load. These high sediment loads occur partly because of the ready availability of sediment but also because of infrequent flows greatly concentrate the period during which this available sediment is transported (Cooke et al., 1993). For example the Thar sand dunes have mean grain sizes of 2.30 to 2.74 φ. Quartz is dominant and felspar is also in abundance (Singh, 1982; Kar, 1996, 2002; Wadhawan and Kumar, 1996) and other major sources of sand are the alluvium and colluvium, and the pre-existing landforms.

4. Waterways and aeolian dune interactions

A distinctive characteristic of some streams in deserts is that they interact with adjacent aeolian dune fields. Very often the irregular flood flows twist and turn through the dune fields and a continuous channel disappears, becoming instead a series of flood basins and playas. Luminescence dating shows that during the late Quaternary, source bordering transverse dunes in the Tarim basin of Xinjiang, China were deflated from seasonally dry sections of the river. The result is a situation where sand is deflated from the stream bed to form large dunes, only to be eroded and re-worked by the river during wetter phases. A similar situation has been described by Makaske (1998) for the upper Inland Niger Delta in central Mali. Here Pleistocene and Holocene climate changes resulted in substantial interaction between the river and the adjacent aeolian dunes, with blocking dune cordons later being breached by the rivers. In the middle Holocene as climate became wetter, stream powers increased but sand supply from the vegetated dunes declined, causing the rivers to incise. In the late Holocene, a drying climate and increased human activity again enhanced sand supply, causing the rivers to aggrade.

5. Sensitivity of deserts and dune systems to changing environments

If deserts and their dunes systems are strongly reflective of their particular environment (e.g., climate, hydrology, sediment supply, vegetation) then it might be expected that they will respond to changing environmental conditions (Gaur and Squires, 2016). There is some disagreement, however, as to just how sensitive they are to change. Some are of the view that they are comparatively insensitive to changes in climate, unless these changes are substantial (see discussion on global warming in Chapter 39 at the end of Volume 2).

6. Conclusions

Research on deserts and their dune systems in the Northern Hemisphere is still in its infancy. However, there have been numerous studies in the American south-west,[2] part of North Africa and a small area of the Middle East (see chapters in this book for further reading). With the threat of accelerated climate change, the burgeoning human and livestock populations there is some urgency in gaining a better understanding of system function and process within dune-dominated deserts.

Clements et al. (1957) did an analysis of physiography of Northern Hemisphere deserts and concluded that "eastern hemisphere deserts had more sand dunes, than North American deserts" (Table 1.2). The greatest contrasts between the deserts of the United States and those of the Eastern Hemisphere (Saharan, Libyan and Arabian) are found in the amounts and distribution of dune areas, fans and bajadas and bedrock fields. In North Africa and Arabia, the dune and bedrock categories constitute much more of the desert surfaces than in the United States. On the other hand, fans and bajadas comprise a much larger percentage of American deserts than they do of foreign deserts.

Since these studies in the mid twentieth century there has been much progress in understanding the true extent of deserts and how dunes form and develop. The application of remote sensing and near-earth scanning using drones and tethered balloons, etc., has revealed a great deal, Hemas et al., this volume. And the invention of an array of new analytical tools at the disposal of researchers have allowed many of the sand's secrets to be revealed. The results of these research efforts in some of the world's deserts are revealed in this book.

Table 1.2. Proportions of selected landform types in several Northern Hemisphere deserts.

Landform Type	South-western USA	Sahara Desert	Libyan Desert	Arabian Desert
Dunes	0.6	28	22	26
Desert mountains	38.1	43	39	47
Desert flats	20.5	10	18	16

[2] Nick Lancaster and Patrick Hesp, 2020. Dunes of inland North America, Springer.

References

Adams, W.M., Goudie, A.S. and Orme, A.R. (eds.). 1996. The Physical Geography of Africa (Oxford University Press, Oxford) 429p.

Babiev, A.G. 1986. Principles and Methods of Shifting Sands Fixation. United Nations Environment Programme; USSR Commission for UNEP.

Burkinshaw, J.R., Illenberger, W.K. and Rust, I.C. 1993. The dynamics and environmental context of aeolian sedimentary systems. *In*: Pye, K. (ed.). Geological Society, Bath 72: 25–36.

Chorley, R.j., Schumm, S.A. and Sugden, D.E. 1984. Geomorphology. Methuen. London and New York.

Chudeau, R. 1909. Ammonites Du Damergou (Sahara meridional). Bulletin de la Société Géologique de France 4(9): 67e71. pls. 1e3.

Clements, T., Merriam, R.H., Stone, R.O., Reade, E.L. and Eymann, J.L. 1957. A study of desert surface conditions. Headquarters Quartermaster Research and Development Command. Environmental Protection Research Division. Technical Report EP-53. Environmental Protection Research Division, Natick, MA.

Cooke, R., Warren, A. and Goudie, A. 1993, Desert Geomorphology: London, Press Limited with University of California, 526 p.

Dhir, R.P., Joshi, D.C. and Kathju, S. 2018. Thar Desert in Retrospect and Prospect. Scientific, Jodhpur, India.

Gaur, Mahesh K. and Squires, Victor R. 2016. Geographic extent and characteristics of the world's arid zones and their peoples. pp. 3–20. *In*: Mahesh K. Gaur and Victor R. Squires (eds.). Climate Variability Impacts on Land Use and Livelihoods in Drylands. Springer International Publishing.

Glantz, M.H. and Orlovsky, N.S. 1986. Desertification: Anatomy of a complex environmental process. pp. 213–29. *In*: Dahlberg, K.A. and Bennet, J.W. (eds.). Natural Resources and People: Conceptual Issues in Interdisciplinary Research. Boulder, CO: Westview Press.

Goudie, A.S. 1996. Climate: Past and present Chapter 3. *In*: Adams, W.M., Goudie, A.S. and Orme, A.R. (eds.). The Physical Geography of Africa. Oxford University Press. Oxford.

Goudie, 2002. Great Warm Deserts of the World. Oxford University Press.

Goudie, A.S. 2009. Global deserts and their geomorphological diversity. *In*: Parsons, A.J. and Abrahams, A.D. (eds.). Geomorphology of Desert Environments. Springer, Dordrecht. https://doi.org/10.1007/978-1-4020-5719-9_2.

Haynes, C.V. Jr. 2001. Geochronology and climate change of the Pleistocene-Holocene transition in the Darb el Arba'in Desert, eastern Sahara. Geoarchaeology: An International Journal 16: 119–41.

Kar, A. 1996. Morphology and evolution of sand dunes in the Thar desert as key to sand control measures. Indian Journal of Geomorphology 1(2): 177–206.

Kar, A. 2002. Granulometry of aeolian bedforms in the extremely arid Jaisalmer region of Thar Desert. pp. 261–286. *In*: Tandon, S.K. and Thakur, B. (eds.). Recent Advances in Geomorphology, Quaternary Geology and Environmental Geosciences: Indian Case Studies (New Delhi: Manisha Publications).

Laity, J.J. 2008. Deserts and Desert Environments. Wiley.

Laity, J.J. 2009. Deserts and Desert Environments (Vol. 3). John Wiley & Sons.

Lancaster, N. 1994. Dune morphology and dynamics. *In*: Abrahams, A.D. and Parsons, A.J. (eds.). Geomorphology of Desert Environments. Springer, Dordrecht. https://doi.org/10.1007/978-94-015-8254-4_18.

Lancaster, N. 1988. Development of linear dunes in the Southwestern Kalahari, Southern Africa. Journal of Arid Environments 14: 233–244.

Lancaster, N. and Hesp, P. 2020. Inland Dunes of North America. Springer.

Makaske, B. 1998. Anastomosing Rivers: Forms, Processes and Sediments. Faculteit Ruimtelijke, Wetenschappen, University of Utrecht, Utrecht,

McTainsh, G.H. and Walker, P.H. 1982. Nature and distribution of Harmattan Dust. Zeitschrift Für Geomorphologie 26: 417–35.

Middleton, N.J. 2019. Desert dust: its perils and its importance. Global Asia 14(4): 20–26.

Raisa, Erwin. 1952. Landform Maps of Arabia, and of the Near East. Prepared for the Environmental Protection Section, Office of The Quartermaster General.

Ramale, F. 1952. World Deserts: Limits and Environmental Characteristics. Special Report No. 57, Environmental Protection Section, Office of The Quartermaster General, 1952.

Rendell, H. 1997. Tectonic frameworks. pp. 13–22. *In*: Thomas, D.S.G. (ed.). Arid Zone Geomorphology: Process, Form and Change in Drylands, 2nd edn. John Wiley & Sons, Chichester.

Rolland, G. 1881. Sur Les Grandes dunes de sable du Sahara. Bulletin de la Societe Geologigue de France series 3; 10: 30–47.

Sarnthein, M. 1978. Sand deserts during glacial maximum and climatic optimum. Nature 272: 43–6.

Singh, S. 1982. Types and formation of sand dunes in Rajasthan desert. pp. 165–183. *In*: Sharma, H.S. (ed.). Perspectives in Geomorphology. Concept. New Delhi.

Singh, S. 1992. Geomorphology and Remote Sensing in Environmental Management. Scientific, Jodhpur, India.
Singhvi, A.K. 2004. Quaternary history and palaeoenvironmental record of the Thar Desert in India-Preface. Proceedings of the Indian Academy of Sciences-Earth and Planetary Sciences 113(3): 367–368.
Stokes, S., Thomas, D.S.G. and Washington, R. 1997. Multiple episodes of aridity in southern Africa since the last interglacial period. Nature 388: 154–8.
Summerfield, M.A. 1991. Global Geomorphology. Longman, London.
Thomas, D.S.G. 2019. Quaternary climate variation in southern Africa. Oxford Research Encyclopedia of Climate Science. https://doi.org/10.1093/acrefore/9780190228620.013.527.
Tsoar, H. and Pye, K. 1987. Dust transport and the question of desert loess formation. Sedimentology 34(1): 139–153.
Tsoar, H. 2005. Sand dunes mobility and stability in relation to climate. Physica A: Statistical Mechanics and its Applications 357(1): 50–56.
Wadhawan, S.K. and Kumar, V. 1996. Subsurface quaternary aeolian stratigraphy in the Ghaggar basin of Thar Desert, India. Jour. Arid Environment. 32(1): 37–51.
Williams, M.A.J. 2014. Climate Change in Desert: Past, Present and Future. Cambridge University Press, Cambridge.
Wilson, I.G. 1973. Aeolian bedforms, their development and origins. Sedimentology 19: 173–210.
Zhenda, Z. 1984. Aeolian landforms in the Taklimakan Desert. *In*: El-Baz, F. (ed.). Deserts and Arid Lands. Remote Sensing of Earth Resources and Environment, vol 1. Springer, Dordrecht. https://doi.org/10.1007/978-94-009-6080-0_7.

Chapter 2

Advances in Measurement Techniques for Estimating Sand Dune Movements

ElSayed Hermas,[1,2,*] *Ahmed Gaber*[3] and *Abdullah F. Alqurashi*[2]

1. Introduction

Sand accumulations occupy ~ 20% of arid zones (Pye and Tsoar, 2008) along with significant spatial extents in many areas of humid regions such as North America and Europe (Tsoar, 2004). Dune movements are considered one of the fundamental aspects of sand accumulations. Since the early work on the aeolian processes by Bagnold (1941), the determination of the rates of dune movement has received considerable attention since it is directly related to the dynamics and development of dunes. This interest has been represented by the development of many empirical models to estimate the rates of dune movement (e.g., Bagnold, 1941; Finkel, 1959; Liu, 1960; Wilson, 1972; Greeley and Iversen, 1987). However, the results of these empirical models are always constrained by various boundary conditions (Hunter et al., 1983; Livingstone and Warren, 1996). To avoid errors in the predictions of dune movements using empirical models, single-dune observation for the determination of the rates of dune movements in the field was one of the important trends in aeolian geomorphology between 1970 and 2000.

The ever-increasing interest of geomorphic research in developing field tools for the determination of dune movements during that period was a two-fold issue. First, dune movement was considered an essential indicator of the activities of aeolian sand transport (Howard et al., 1978; Fryberger et al., 1979; Lancaster, 1989; Wiggs et al., 1996). Second, the prediction of hazards associated with dune movements against various anthropogenic activities represented a challenge in many desert areas all around the world that had experienced considerable urban expansion since the 1970s. The problems of dune movements in these areas have been reviewed for the purpose of establishing mitigation measures by many pioneers in the field of arid geomorphology (Watson, 1985; Livingstone, 1989; Cooke and Doornkamp, 1990; Stipho, 1992). Evidently, developing techniques for determining and measuring the rates and directions of dune movement and directions is a great concern of geomorphologists and environmental scientists.

The early studies of measuring sand dune movements were conducted on a single dune during the 1970s, 1980s and 1990s using various tools. These included linen tapes, fluorescent dyes, steel

[1] The National Authority for Remote Sensing and Space Sciences, Cairo, Egypt.
[2] Geography Department, Umm Al-Qura University, Makkah Al-Mukarramah, Kingdom of Saudi Arabia.
[3] Geology Department, Port Said University, Egypt.
* Corresponding author: eaibrahim@uqu.edu.sa

and wooden stakes and pegs (e.g., Calkin and Rutford, 1974; Tsoar, 1974). These techniques became more sophisticated with the increasing advancement of geodetic tools such as theodolites, total stations, data loggers and Global Positioning Systems (GPS) (e.g., Dong et al., 2000) for the purpose of topographic mapping rather than 2D horizontal measurement of dune movements. Although different field techniques enabled the scientific knowledge of dune dynamics and development to be enhanced during the time of single dune studies, they oversimplified the aeolian processes of dune fields or sand seas (Livingstone et al., 2007). For instance, wind regimes vary significantly through dune fields from unimodal at the peripheries of the dune field to wide unimodal to bimodal inside the dune field (Lancaster, 1983). These changes in local wind regimes result in changes in dune forms, morphology and dune patterns, which, in turn, significantly influence the rates of dune movements. Accordingly, field techniques lack the ability to constraint the spatial and temporal variability in dune movements on the field scale.

Remote sensing images can be used to develop various techniques to potentially document this variability. Early acquisition of aerial photographs in the 1950s, 1960s and 1970s over many geographic regions worldwide allowed the scientific community to map and categorize aeolian sand dunes for the first time. Successive acquisitions of the aerial photographic images over these regions were originally used to study many morphological and dynamic aspects of sand dunes (Finkel, 1959). However, because of the high cost of acquiring aerial photographs, these images were used in a limited number of geographical locations. However, in the last six decades, different satellite remote sensing systems representing a potential source of satellite images are available at low cost, and now are even free of charge. In addition, they provided an intensive stream of data over the globe with short and regular temporal intervals (Wulder et al., 2016). Accordingly, the satellite images enabled potential documentation and enhanced understanding of various aspects and evolution of aeolian sand dunes. Conceptually, the synoptic perspective and the multi-temporal advantages of remote sensing imagery have recently stimulated a shift from the single dune studies dominant in the 1970s, 1980s and 1990s (Livingstone et al., 2007) toward dune-field scale studies.

In the past, studies using satellite remote sensing in the field of aeolian sand dunes focused on dune mapping and dune taxonomy (McKee, 1979), and on the driving forces that control the formation of these dunes, such as source areas of sands, pathways, depositional environments, vegetation cover and wind analysis (Fryberger et al., 1979; Wasson and Hyde, 1983). These studies were extended to incorporate spatial analysis of boundary conditions, dune activity, dune pattern and hierarchies and dune–dune relationships (e.g., Tsoar and Blumberg, 2002; Hugenholtz and Wolfe, 2005; Kocurek and Ewing, 2005; Mitasova et al., 2005; Ewing et al., 2006; Bishop, 2007; Wilkins and Ford, 2007; Ewing and Kocurek, 2010a; Ewing and Kocurek, 2010b; Hugenholtz and Barchyn, 2010; Kocurek et al., 2010). However, rapid advances in computer hardware and software in the 1990s enabled scientists to study the reflectance characteristics of dune surfaces (Paisley et al., 1991; Lancaster et al., 1992; Tsoar and Karnieli, 1996; Walden and White, 1997; White et al., 1997; Blumberg, 1998; Pease et al., 1999). Furthermore, during this period, innovative approaches for quantifying historical changes in dune activity were implemented (e.g., Anthonsen et al., 1996; Brown and Arbogast, 1999). In the last two decades, additional progress has been made in the quantitative analysis of dune morphodynamics (Levin et al., 2004; Vermeesch and Drake, 2008; Bourke et al., 2009; Necsoiu et al., 2009; Hermas et al., 2012) and dune field pattern analysis (Kocurek and Ewing, 2005; Ewing et al., 2006; Bishop, 2007; Wilkins and Ford, 2007; Mason et al., 2008; Bishop, 2010; Ewing and Kocurek, 2010a; Ewing and Kocurek, 2010b; Hugenholtz and Barchyn, 2010; Kocurek et al., 2010).

In this chapter, various techniques and tools for detecting and monitoring sand dune movements are reviewed in detail. The chapter covers two main aspects: field techniques and remote sensing data and techniques (Fig. 2.1). For field techniques, both simple tools, such as *in situ* field markers and advanced tools, such as geodetic techniques, are described and evaluated. The use of remote sensing is discussed comprehensively, both traditional techniques that depend on the user manual interaction and the more advanced techniques that automatically detect the changes in sand dune

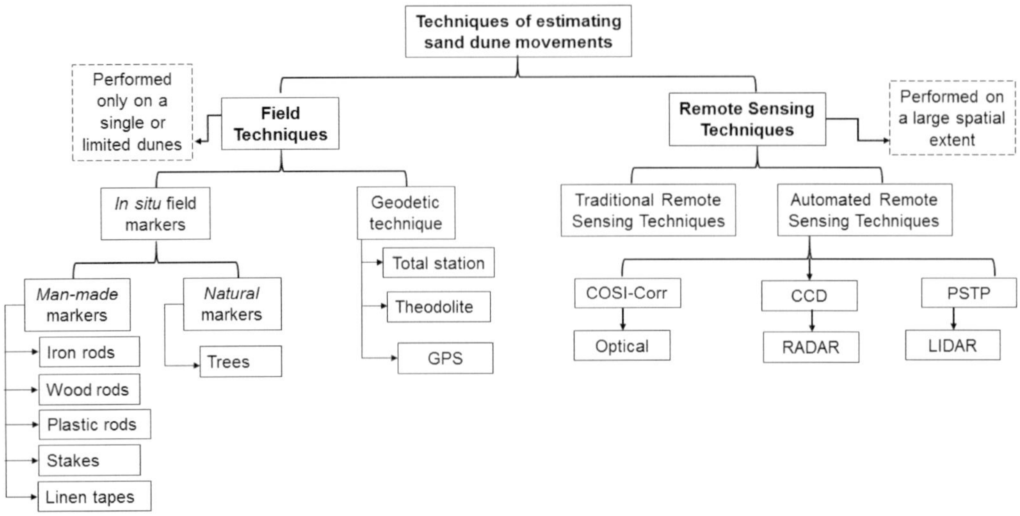

Figure 2.1. A flow chart for various tools, data and techniques for measuring dune movements.

movements. The use of optical, radar and recently light detection and ranging (LiDAR) remote sensing data is also described and evaluted. This chapter allows the reader to assess the data, tools and challenges in sand dune movement detection during the last six decades.

2. Field techniques

Field techniques use various markers or geodetic tools to monitor and measure different dune movements in different geographic areas. According to the tools used, field techniques can tentatively be categorized into *in situ* field marker techniques and geodetic field techniques. Both have been widely used in single dune studies, especially before the widespread use of remote sensing in the field of aeolian research in the last four decades. Field techniques are usually conducted on a single dune or a limited number of dunes, preferably at an accessible site of sand seas, such as coastal areas or close to desert tracks or major paved roads. To the best of our knowledge, field techniques have not been carried out inside large dune fields or sand seas. However, these techniques have been carried out on dunes in different climates.

2.1 In-situ field marker techniques

These techniques rely on using man-made and natural field markers. Man-made field markers include iron, wood and plastic rods or stakes along with linen tapes to be established on different locations over the dune surface or at a distance from the targeted dune (Fig. 2.2). Regular revisits are organized on a monthly, seasonally or yearly basis to monitor changes in position between the markers and the moving parts of dunes rather than the whole body of the dune. The natural field markers use natural features as an index of dune movements. Trees of natural vegetation are a good example of this category (Fig. 2.3), where there is a strong relationship between the vegetation and climatic conditions that control dune dynamics. The response of the growth of the vegetation trees and tree-rings to dune dynamics can be documented in a variety of ways. This concept has been used to monitor dune dynamics in areas of high latitude, where forests can often be found.

Numerous studies have been conducted in the Sahara Desert using man-made field markers. For example, Tsoar (1974) used field techniques to measure the migration rates of three crescentic dunes of different heights in El-Arish region along the Mediterranean coast. The migration rates of these barchan dunes were 6.16 m/yr, 6.42 m/yr, and 13.08 m/yr, with an average of 8.6 m/yr. Misak and El Shazly (1982) used linen tape in the El-Arish area and estimated the migration rate

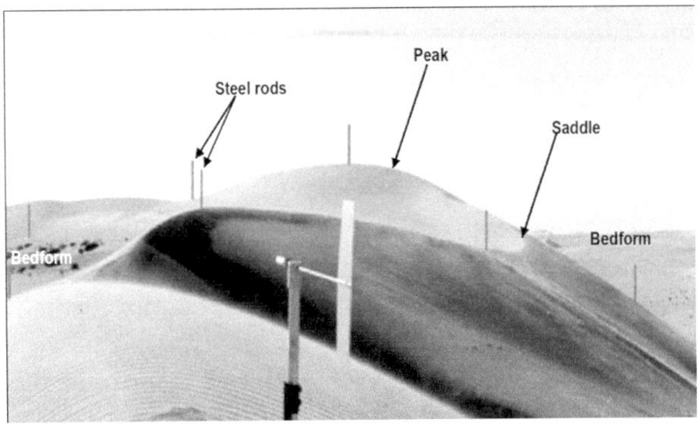

Figure 2.2. Iron rods installed over the various parts of a linear dune on the North Sinai Peninsula (modified from Khidr, 2006).

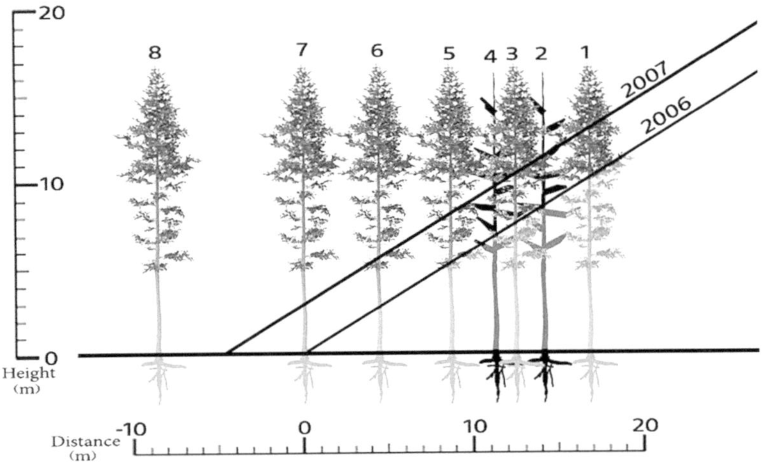

Figure 2.3. Tree–dune interaction that has been used as a field technique to measure dune migration in the dune field of Slowinski National Park, Poland (modified from Koprowski et al., 2010).

of barchan dunes over nine months to be 13.3 m/yr. El-Kayali (1984) used steel and wooden rods to measure the migration rates at Bir Al-Musama (North Sinai) over nine months for two sets of barchan dunes: a set of small dunes and a set of relatively large dunes. The small set of barchan dunes migrated at a rate of 16.0 m/yr, whereas the migration rate of the large barchan dunes was 10.0 m/yr. Assal (1999) studied two crescentic dunes in the ElSadat area (North Sinai) using a grid of 20 steel posts and estimated a migration rate ranging from 6.0 to 7.0 m/yr for the small dune and 1.5 m/yr for the larger one, with an average of 4.0 m/yr. Aql (2002) fixed wood posts on five barchan dunes east of Bitter Lakes, North Sinai Peninsula, and measured the movement of dunes every three months between June 2001 and June 2002. The rate of the recorded movements ranged from 9.6 to 15.3 m/yr, with an average of 13.4 m/yr. A similar approach has been adopted on linear dunes in desert environments.

In desert environments, linear dunes are dominant (Tsoar, 1978; Breed et al., 1979). However, few studies have measured the rates of movement of the linear dunes using field techniques (El-Banna, 2004; Embabi, 2004; Khidr, 2006). This is mainly attributed to the complex set of movements and the long spatial extent of the linear dunes that may reach several hundreds or thousands of meters or kilometers. Tsoar (1978) used fluorescent dye to determine the rates of the movement of a linear dune in the North Sinai Peninsula. The average of the lateral movements of peaks and saddles was in the order of 6.0 m/yr, the average of the elongation of the studied linear dune was in the order of 14.5 m/yr, and the average of the longitudinal displacements of peaks and saddles along the studied linear dune was in the order of 8.4 m/yr. Tsoar (1983) used field techniques to measure the rates of elongation and the longitudinal displacements of peaks and saddles of a linear dune in North Sinai. The average rate of elongation was 14.7 m/yr, whereas the rate of longitudinal displacements was in the order of 8.2 m/yr. Assal (1999) fixed a grid of 20 steel posts on two linear dunes in different locations in North Sinai to measure the lateral and elongation movements of these dunes. The lateral movements of the peaks and saddles were in the order of 6.0 m/yr for one linear dune and 4.5 m/yr for the other one, whereas the elongation rates were in the order of 13.0 m/yr for the former and 8.5 m/yr for the latter. Khidr (2006) established steel posts on the surface of a linear dune in North Sinai (Fig. 2.2) to measure both the lateral movements of peaks and saddles and the elongation movements of a linear dune. He estimated an average of lateral movements in the order of 10.6 m/yr and an average of elongation movements in the order of 27.0 m/yr. Generally, the lateral movements along the crest lines of the linear dunes ranged from 0.7 to 10.6 m/yr, with an average of 5.2 m/yr. The elongation movements of all the studied linear dunes in North Sinai ranged from 2.3 to 27.0 m/yr, with an average of 12.9 m/yr, whereas the longitudinal displacement of peaks and saddles of these dunes ranged from 7.5 to 10.9 m/yr, with an average of 8.8 m/yr. Man-made field markers have been used in various desert areas across the world, even in the Arctic and sub-Arctic regions at the North and South Poles. These primitive man-made field techniques became more sophisticated with the advancement in the technologies of geodetic field tools.

2.2 Geodetic field technique

The most common tools in the geodetic technique are total station, theodilite and GPS. All these tools are able to collect data that can be processed to produce topographic profiles rather than topographic mapping by interpolating the elevations of the collected points. The comparison of the multi-temporal topographic profiles or maps could identify changes in the dune surfaces that imply dune displacement through the time intervals of data acquisitions.

Compared to the *in situ* field marker technique, a limited number of earlier studies were carried out using the geodetic field technique. This may be attributed to the high cost of geodetic equipments and their required data processing software. Neverthless, a number of studies successfully used these tools to measure sand dune movement in different areas. Among them, Dong et al. (2000) conducted a topographic survey using a theodolite to monitor sand dune advances in late 1991, 1992 and 1993 in the Taklimakan Desert, China. The mean advance rates of the dunes were 7.29 m/yr and 5.56 m/yr in 1992 and 1993, respectively. El-Banna (2004) used the total station tool to measure the lateral and elongation movements of two linear dunes in different locations in North Sinai over a one-year interval, 2001–2002. The lateral movements ranged from 0.7 m/yr in one location (Bir El-Abd) to 2.0 m/yr in another location (Wadi El Gady), whereas the elongation movements ranged from 2.3 m/yr to 13.0 m/yr in the two locations, respectively. Barchan dunes recorded a rate of movements in the order of 3.5 m/yr using the same total station technique (El-Banna, 2004). Muñoz-Perez et al. (2009) used topographic data collected from the total station tool (June 1995, May 1999, November 1999, June 2000, November 2003) and from the Differential Global Positioning System (DGPS) (May 2006). These long-term collections of topographic data were carried out on the same five profiles across the Valdevaqueros transgressive active dune in southern Spain. The study recorded an average migration rate of 17.5 m/yr, with a maximum of 38.0 m/yr in the most active profile. The DGPS can provide topographic accuracy to the centimeter level and is currently one of the best geodetic techniques for collecting topographic data for the purpose of dune monitoring, especially when the morphology of the dunes is complex (Navarro et al., 2011). Navarro et al. (2011) used the DGPS technique to measure the displacement rate of a highly active dune field at Valdevaqueros (Tarifa) in southern Spain at very short (hourly, 14–15 August 2009), short (daily, 5–22 July 2006), medium (bimonthly, September 2008 to August 2009) and monthly (May 2006 to December 2006) intervals. Based on the topographic data that were collected from the DGPS and reconstructed using Empirical Orthogonal Functions, the brink of the studied dune experienced a fast advance migration rate, reaching 1.75 m/day during a severe easterly sandstorm.

Several important points that can be highlighted from the above review of field techniques studies are outlined below.

- *Dune scale and spatial extent*

  Most field studies were conducted on barchan dunes because of their small size. Although linear dunes are the dominant dunes in all deserts (Tsoar, 1978; Breed et al., 1979), few studies have measured the rates of movement of linear dunes using various field techniques (El-Banna, 2004; Embabi, 2004; Khidr, 2006). Moreover, some studies have been conducted on reversing and transverse dunes. The limited spatial extent of the measurement of dune migration rates represents a fundamental gap in field techniques as well. The majority of field techniques were performed on sand dunes that are located on the periphery of dune fields or sand seas. Accordingly, the migration rates of dunes inside these fields or seas cannot be monitored using different field techniques.

- *The temporal intervals*

  Most of the earlier field studies were short-term (months) or medium-term studies (one to ten years) while a few of them were long-term studies (e.g., Livingstone, 2003; Navarro et al., 2011). However, these long-term studies did not involve a regular annual measurement over a decade or more in addition to using different field techniques during each survey (Livingstone, 2003).

It is possible to conclude that although field techniques show a high level of accuracy, they are mainly used in a small-scale targeted dune(s) in short-term to medium-term studies with limited spatial extents. Therefore, these methods inherently lack the ability to monitor dune movements with wide spatial coverage and long temporal intervals (Embabi and Ashour, 1985). In addition, they are costly and time-consuming.

3. Remote sensing techniques

The above-mentioned limitations of field techniques can be overcome to a large extent with remote sensing in which the sand dune migration over wide spatial coverage and long temporal intervals can be measured. The use of remote sensing techniques in the detection and measurement of dune movement has evolved from the traditional remote sensing approaches to more advanced automated approaches. While the traditional approaches involve user interaction through the process of dune definition to final determination of dune migration, the advanced approaches rely on automated correlation for the purpose of detecting surface signals related to dune movements. The advanced automated approaches include the Co-registration of Optical Satellite Imagery and Correlation (COSI-Corr) technique, the Coherence Change Detection (CCD) technique, and the Pairs of Source and Target Points (PSTP) technique. In the following sections, the various remote sensing approaches used in the detection of dune movements will be presented and assessed.

3.1 Traditional remote sensing techniques

Traditional remote sensing techniques for measuring sand dune movements depend on comparing two or more photographic or satellite images acquired at different times of a given geographic region. The process involves geo-referencing the images followed by tracing dune boundaries or axes in the successive images for eventual determination and measurement of the displacement of dune boundaries or axes in meters (Fig. 2.4). The amount of displacement is equivalent to the movements of the same dunes relative to each other. This process has been used to measure the movement rates of sand dunes in many geographic regions across the world.

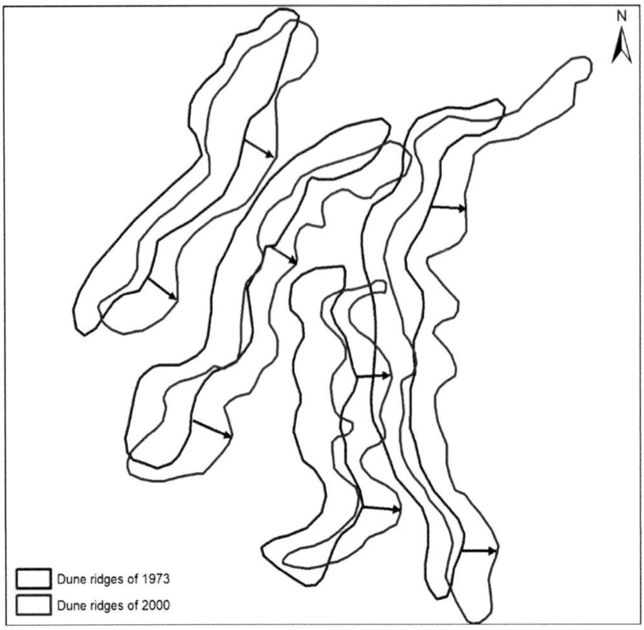

Figure 2.4. Estimating migration rates through the comparison of the positions of dune ridges extracted from the multispectral scanner (MSS) and Landsat ETM images acquired in 1973 and 2000, respectively (modified from Yao et al., 2007).

The early use of remote sensing techniques for measuring sand dune movements was attributed to Finkel (1959), who compared two aerial photographs acquired in 1955 and 1958 to measure the migration rates of 75 barchan dunes on the desert coast of southern Peru. A net displacement of 15.4 m was estimated for all barchan dunes. Moreover, Long and Sharp (1964) monitored the movements of 47 barchan dunes on the western side of Salton, California, between 1956 and 1963 using aerial photographs. They estimated migration rates of these barchans in the order of 99.0–282.0 m over the seven years of monitoring, with an average of 25.0 m/yr. During a period of 15 years from 1941 to 1956, the migration rates of 34 barchan dunes in the same area were in the order of 106.7–365.8 m, with an average of 15.2 m/yr. Yao et al. (2007) used Landsat images acquired in 1973, 1991 and 2000 to measure the migration rates in the Alxa Plateau of Inner Mongolia, China (Fig. 2.4). The migration rates ranged from 4.0 to 7.4 m/yr, with an average of 5.3 m/yr over the entire period. Between 1973 and 1991, the migration rates ranged between 3.4 and 5.9 m/yr, but between 1991 and 2000, they ranged between 3.7 and 12.9 m/yr. Many other studies used this approach in the Sahara Desert of the Middle East.

Maxwell and Haynes (2001) used multi-temporal repeat orbital images and field surveys to measure the migration rates of one of the largest hyper arid environments in the world, which is Selima Sand Sheet. The authors reported a migration rate as much as 500–1000 m/yr. In Egypt, many studies have been conducted to determine the migration rates of sand dunes using the classical remote sensing techniques. For instance, Tsoar (1978) used fine-resolution aerial photographs to measure the rates of longitudinal movements of peaks and saddles of a linear dune in North Sinai Peninsula, which were in the order of 7.5 m/yr. Aql (2002) assessed the rates of sand dune movements between 1984 and 2001 using two Landsat TM images. The rates of movement varied from ~ 50 m to ~ 186 m during the measurement period, with an average of 6.18 m/yr. El-Banna (2004) compared an aerial photograph acquired in 1958 with a Landsat TM image acquired in 1998 to measure the displacement of a number of barchan dunes located northwest of the study area. He estimated migration rates of barchan dunes in the order of 1.1 to 6.8 m/yr. The author used the same images to estimate the rate of longitudinal movements of peaks and saddles of linear dunes east of Ismailia, Egypt. The average rate of longitudinal movement of these linear dunes was as much as 10.9 m/yr. Tsoar (2004) estimated the lateral movements of the peaks and saddles for a linear dune in North Sinai using three aerial photographs from 1973, 1982, and 1999. The average rate of the lateral movements ranged from 3.8 to 5.8 m/yr during the period 1973 to 1982 and from 6.1 to 7.5 m/yr during the period 1982 to 1999.

Many advantages of measuring dune movements using traditional remote sensing techniques can be achieved. First, the wide spatial coverage of successive aerial and satellite images enables the measurement of a high number of dunes, for example, as many as 50 dunes or even more distributed throughout the dune field. Second, the availability of successive images over a long temporal interval makes the investigation of temporal changes of dune movements possible. Both advantages can help in understanding the driving forces controlling the spatial and temporal variations of dune movements, such as the regional and local wind regimes, dune morphometry and patterns, and the various environmental conditions that may influence the rates of dune movements (e.g., vegetation and human activities). Nevertheless, many disadvantages can also be highlighted. Aerial photographs are expensive and consume time through anlyzing them. In addition, the geospatial accuracy of co-registration of the satellite images is greatly affected by satellite geometry, sun azimuth and zenith angles and topographic conditions (Liu et al., 1997). Furthermore, the classical definition of dune boundaries and axes from the multiple satellite images can be largely affected by personal bias through the interpretation and digitization of dune boundaries. The above-mentioned considerations make determination of sand dune movements from satellite images inaccurate (Liu et al., 1997). Consequently, developing an automated vectorization approach using pixel-based change detection techniques that may help in avoiding the gaps of the earlier mentioned techniques is needed.

3.2 Advanced remote sensing techniques

3.2.1 The co-registration of optically sensed images and correlation (COSI-Corr) technique

This automated vectorization process aimed at the correlation of pixels of two or more optical satellite images acquired at different times for the same locations. The correlation produces spatial pixel offsets that are represented as vectors with an orientation and length. The orientations of the vectors refer to the direction of terrain displacement, while the lengths represent the magnitude of displacement in meters. The detection of terrain changes through the determination of the spatial pixel offsets (i.e., fault displacements, sand dune movements) has followed three approaches in the last four decades: the imageodesy technique (Crippen and Blom, 1991), Wavelet transform (Strang and Nguyen, 1997), and Fourier Transform (Van Puymbroeck et al., 2000).

Using the imageodesy technique, which was first introduced by Crippen and Blom (1991), the subtle terrain changes can be detected and measured at sub-pixel resolution accuracy. The technique relies on statistical comparison of thousands of pixels in the "before" image to thousands of pixels in the "after" image to detect pixel offset vectors. This procedure is repeated many times to resolve spatial offsets in the order of 1/128 of the image spatial resolution, which is equivalent to ~ 8 cm in the case of the input images of 10 m spatial resolution. The length of each offset vector represents the amount of the spatial change between the corresponding nodes of the compared images. Crippen and Blom (1991) applied the technique on two SPOT panchromatic images of 10 m spatial resolution to measure the sub-resolution spatial changes of a barchan dune located in the area west of Superstition Hills, Imperial County, California (Fig. 2.5). The longest offset vector refers to sub-pixel resolution spatial changes of 8 m dune movement over the time span between the two acquired SPOT-4 panchromatic images. Although the location of the offset vectors remarkably characterizes the targeted dune and points in the downwind direction, and their lengths can be compared with the amount of movement of the nearby dunes, the accuracy of the measurements could be highly influenced by different sources of errors (Crippen and Blom, 1991). The impact of these errors could influence the spatial pixel offsets that belong to the real environmental changes of the studied features, such as sand dune movements (Crippen and Blom, 1991).

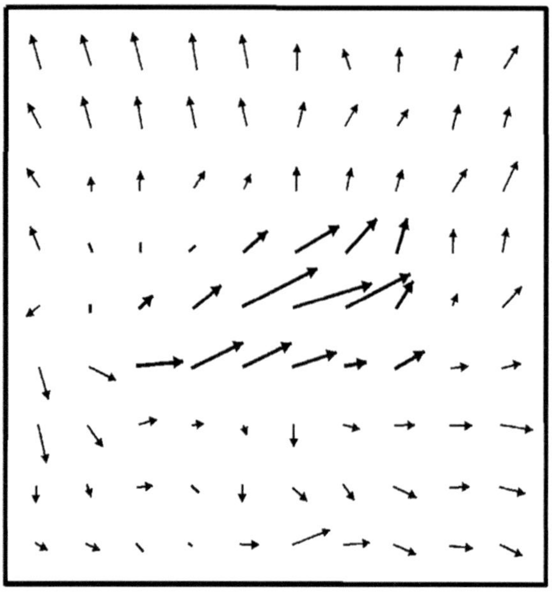

Figure 2.5. Displacement vectors produced from the imageodesy correlation of two SPOT-4 panchromatic images. Thick and long vectors are equivalent to the barchan movements (the longest vector is equivalent to 8 m dune movement), and thin and short vectors represent the background areas (modified from Crippen and Blom, 1991).

Van Puymbroeck et al. (2000) identified the sources of the various spatial pixel offsets produced when two images of different orbits are correlated. These sources include the distance between the orbits of the two images, their different viewing angles, the topography of the imaged area, the attitude of the platform during the acquisition of each image (i.e., roll, yaw, pitch) and the geometry of the satellite detector along with the terrain environmental changes (i.e., fault displacement, sand dune migration, coastal sediment transport). In addition, Van Puymbroeck et al. (2000) proposed models for compensating the influence of these sources so that the residual spatial pixel offsets represent only the terrain environmental changes. After the compensation of error-induced spatial offsets, Van Puymbroeck et al. (2000) used Fourier Transform to detect and measure the displacement of the ground associated with the 7.3 M_W Landers earthquake that occurred in June 1992 in California. The Fast Fourier Transform analysis conducted by Van Puymbroeck et al. (2000) measured fault surface displacement associated with the Landers earthquake in the order of ~ 80 m, which was fairly consistent with the measurements of the elastic models designed for this seismic event from the large amount of geodetic data collected in the area. Using the same compensation models of removing false pixel offsets (Van Puymbroeck et al. 2000) and Fast Fourier Transform, Schiek and Hurtado (2006) measured surface displacement along the Kunlun faults (China) associated with the Kokoxili earthquake of M_S 8.1 that occurred on November 14, 2001. The imageodesy and Fast Fourier Transform automatic change detection techniques represented the seminal works by Leprince et al. (2007) to introduce the Co-registration of Optically Sensed Imagery and Correlation (COSI-Corr) technique for the detection and measurement of subtle terrain changes.

The COSI-Corr technique was developed to detect and measure subtle crustal deformation in the tectonically active regions using multi-temporal aerial photographs and optical satellite imagery (Avouac et al., 2006; Leprince et al., 2007; Ayoub et al., 2009a; Avouac et al., 2014; Milliner et al., 2015). The COSI-Corr technique had been developed as plug-in ENVI© digital image processing software. Along with its ability to use SPOT images for detecting subtle surface changes, the software is able to process a variety of optical remote sensing imagery, such as Landsat, ASTER, WorldView-1, FORMOSAT-2 and QuickBird images. Two or more satellite images can be used to automate the COSI-Corr technique. The processing protocol involves multiple steps of pre-

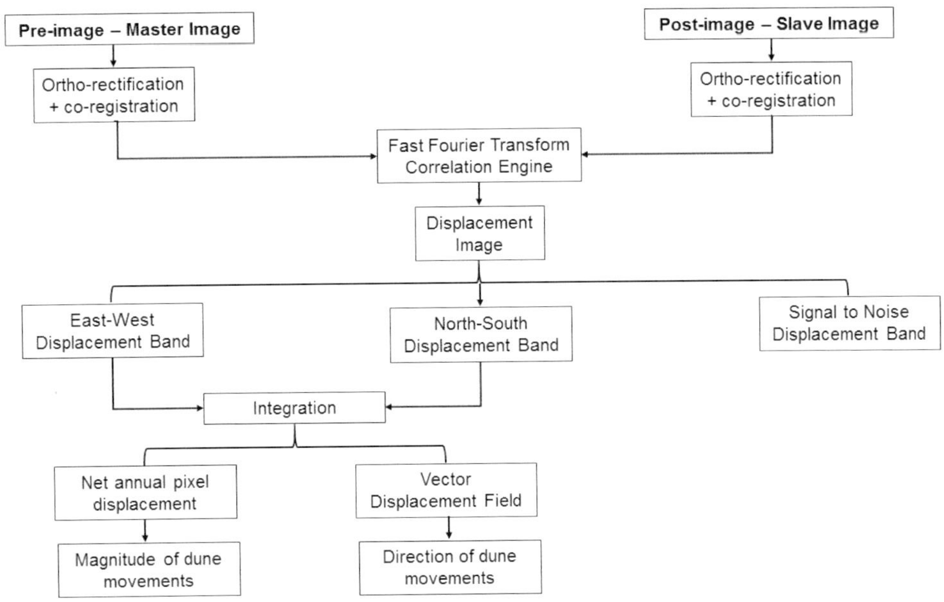

Figure 2.6. A flow chart for the main COSI-Corr pre-processing and processing steps for the estimation of the magnitude and direction of dune movement.

processing and processing of the optical images for eventual retrieval of change signals on the Earth's surface during the selected time interval (Fig. 2.6).

The pre-processing step involves ortho-rectification and co-registration processes of the selected optical sensing images (Ayoub et al., 2009b). These processes include removal of various radiometric and geometric errors using the ancillary file of each input image and a sink-free Digital Elevation Model (DEM) (e.g., Shuttle Radar Topographic Mapping DEM). After the ortho-rectification and the co-registration processes have been accomplished, the processing step can be conducted to estimate the terrain environmental changes (e.g., fault displacement, glacial movements, sand dune migration) that occurred during the time interval between the two ortho-rectified images (Fig. 2.6).

The processing step comprises four processes: correlation of the ortho-rectified images to produce the displacement bands and Signal to Noise Ratio (SNR) band, filtering of the displacement bands, calculation of the net displacement image from the displacement bands and production of the vector displacement field (Ayoub et al., 2009b). The COSI-Corr Fast Fourier Transform correlation engine is automated to correlate the "before" ortho-rectified image with the "after" ortho-rectified image. The correlation process produces a horizontal displacement image composed of three components: the East/West (EW) and North/South (NS) components of displacement and the SNR (Leprince et al., 2007) (Fig. 2.6). The EW displacement band comprises pixels that contain values and signs. The values represent the magnitude of surface displacement, and the signs denote the direction of the displacement. The positive values refer to displacements that take place in the east direction, and the negative values indicate displacements that occur in the west direction. Similarly, the NS displacement band comprises pixels that contain values and signs. The values denote the magnitude of displacement, and the signs indicate the direction of the displacement. Positive values reflect the northward direction of the displacement, whereas negative values reflect the southward direction of the displacement. The SNR band assesses the quality of the measures. The EW and NS displacement bands are integrated to calculate the net displacement magnitude during the period of study and the vector displacement field. To avoid the overestimation of the displacement magnitude using the Euclidian norm of the EW and NS components, Necsoiu et al. (2009) proposed the projection of each component on a robust local migration direction, which is independent of the noise. The main objective of this projection is maintaining the original noise characteristics, namely zero mean and white Gaussian noise.

The robust direction unitary vectors are then deduced and are eventually integrated with the original EW and NS components to calculate the Displacement Magnitude (DM) as follows (Necsoiu et al., 2009):

$$DM = d_{EW} \ x \cos\theta + d_{NS} \, x \sin\theta \tag{1}$$

where d_{EW} and d_{NS} are the original east–west and north–south displacement measurements, and $\cos\theta$ and $\sin\theta$ are the robust local migration direction.

In addition, the EW and NS displacement bands could be integrated to automatically produce a vector displacement field, which is a set of arrows. The length of each arrow refers to the magnitude of the displacement (dune movements) and the direction of each arrow denotes the direction of the displacement (dune movement direction). Since its development, the COSI-Corr technology has been used to estimate dune movements in a variety of geographic areas across the world.

Vermeesch and Leprince (2012) applied the COSI-Corr technology to measure dune movements in Bodele Depression, Chad. The authors used one Landsat 4 image (1984), two SPOT-3 images (1990, 1996) and four ASTER images (2000, 2003, 2007, 2010) of the depression. The COSI-Corr technology was conducted on the orthro-rectified and co-registered images of these scenes to produce migration rates ranging from 0 to 50 m/yr over a 26-year period from 1984 to 2010. The manual determination of dune movements in the same area was carried out by comparing the 1984 Landsat image with the 1965 and 1970 images acquired by the declassified CORONA spy satellite. The extracted rates of dune movements using the manual measurement were similar to the estimated rates from the COSI-Corr measurements. After studying the temporal variability of the rates of dune movements over a 45-year period, no significant trends were noticed in these rates, implying a steady magnitude of winds over this period.

Hermas et al. (2012) used two 10-m SPOT-4 panchromatic images collected 229 days apart between May 2007 and January 2008. The two images were ortho-rectified and co-registered using a Landsat 7 band 8 of 15 m spatial resolution and then passed through the Fast Fourier Transform of COSI-Corr technology to estimate the displacement bands and SNR band. The EW and NS displacement bands were then projected onto a robust local migration direction using the approach of Necsoiu et al. (2009) to calculate the resultant pixel-wise displacements (in meters) (Fig. 2.7). The measurements were then extrapolated to one year. The approach indicated that the areas of barchan dunes were characterized by a migration rate of 7.7 m/yr. The migration rate of a sample of selected barchan dunes was 11.9 m/yr. The lateral movements of crest lines in the areas occupied by linear dunes recorded 6.8 m/yr. For a selected sample of linear dunes, the rate of the lateral movement of peaks and saddles was 12.4 m/yr. The integration of the original EW and NS bands using COSI-Corr technology was also able to calculate the vector displacement field over the study area (Fig. 2.8). The arrows of the vector displacement field indicate that the direction of dune movements in the study area of the North Sinai Peninsula was from east to southeast. The study enabled monitoring the high spatial variability in dune movements across wide spatial coverage in North Sinai. The measured rates of dune movements from the automated pixel-wise COSI-Corr technology were close to the earlier field-measured rates of both the linear and the barchan dunes in North Sinai. In addition, the directions of dune movements extracted from the integration of EW and NS displacements using COSI-Corr technology in the study area were aligned with the sand dune drifting potential calculated by the previous measurements in North Sinai.

Al-Ghamdi and Hermas (2015), Al-Mutiry et al. (2016) and Hermas et al. (2019) followed the same approach as Hermas et al. (2012) in the area northwest of Al-Lith city, Irq Al-Rethmah, Riyadh, and Wadi Fatmah, Saudi Arabia, respectively, with an assessment of dune hazards against roads, recreational camps and agricultural farms. Most of these earlier studies applying the COSI-Corr technique in the field of the aeolian processes used data from two satellite platforms: ASTER and SPOT. New datasets can also be used as a potential input for the COSI-Corr processing.

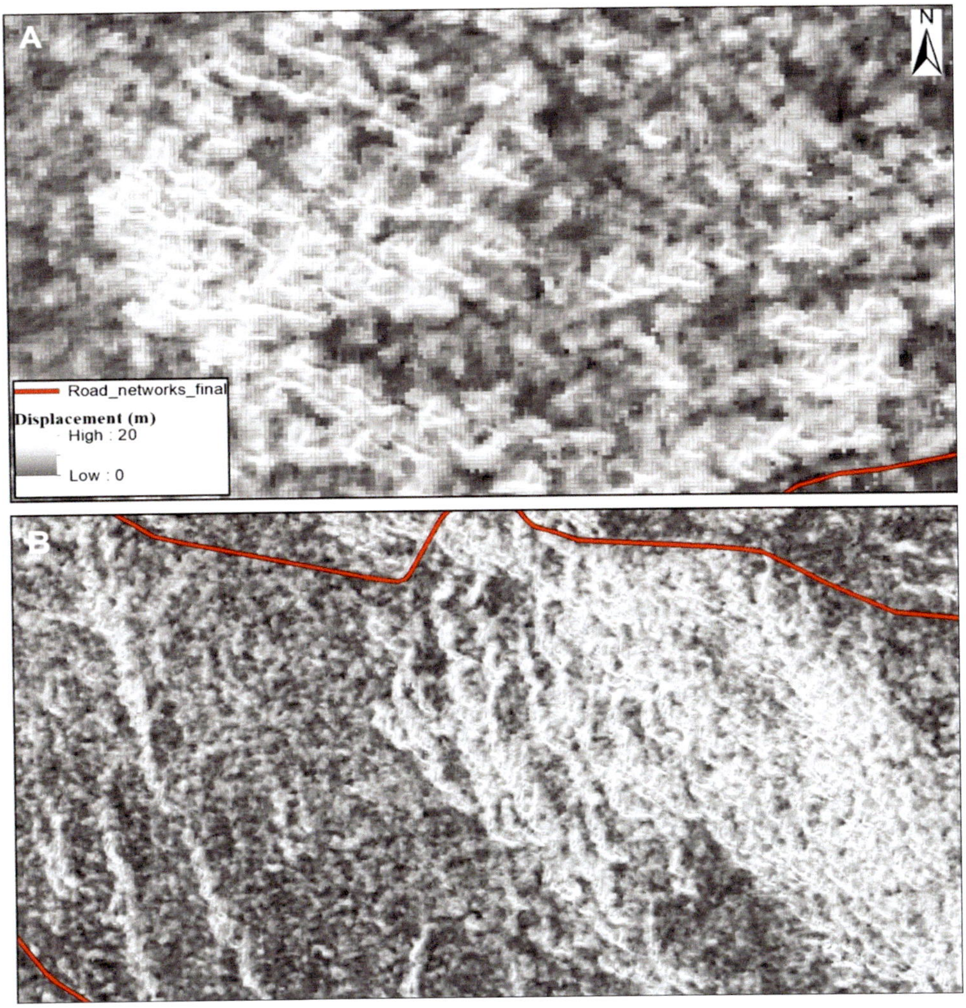

Figure 2.7. The net lateral pixel displacement of a barchan dune field (A) and a linear dune field (B) in North Sinai Peninsula, Egypt.

Recently, new datasets have been used as an input for the COSI-Corr processing. For example, Sam et al. (2015) used COSI-Corr technology and data from the Indian satellite Cartosat-1 platform to determine both the migration rates and the sand flux in the Barmer district of the Thar Desert, India, between March 2010 and March 2011. They used COSI-Corr technology to determine dune movements and to isolate the areas of these movements from the interdune areas. Along with the DEM, the areas of dune were integrated with its corresponding areas of the DEM to calculate the sand flux. The measured mean dune shift was in the order of 1.2 m/yr, and the mean sand flux was in the order of 5.67 m^3/m/yr. Analysis of the dune shift rates found that these rates were within the cross-correlation errors, which means that the dune did not move significantly during the period of study because of stabilization processes associated with habitation and arid vegetation in the study area. The earlier studies showed the ability of COSI-Corr technology to process different datasets from various platforms for the purpose of automatically producing a raster map of sand dune movements. However, the temporal and spatial resolutions of the images produced from these platforms along with the costs may limit the development of automatic monitoring of dune movements at long and regular temporal intervals and over the spatial scale of dune fields or sand seas. Reliable and cost-effective data (e.g., Landsat data) can overcome these gaps for detection of long-term change in sand dune movements.

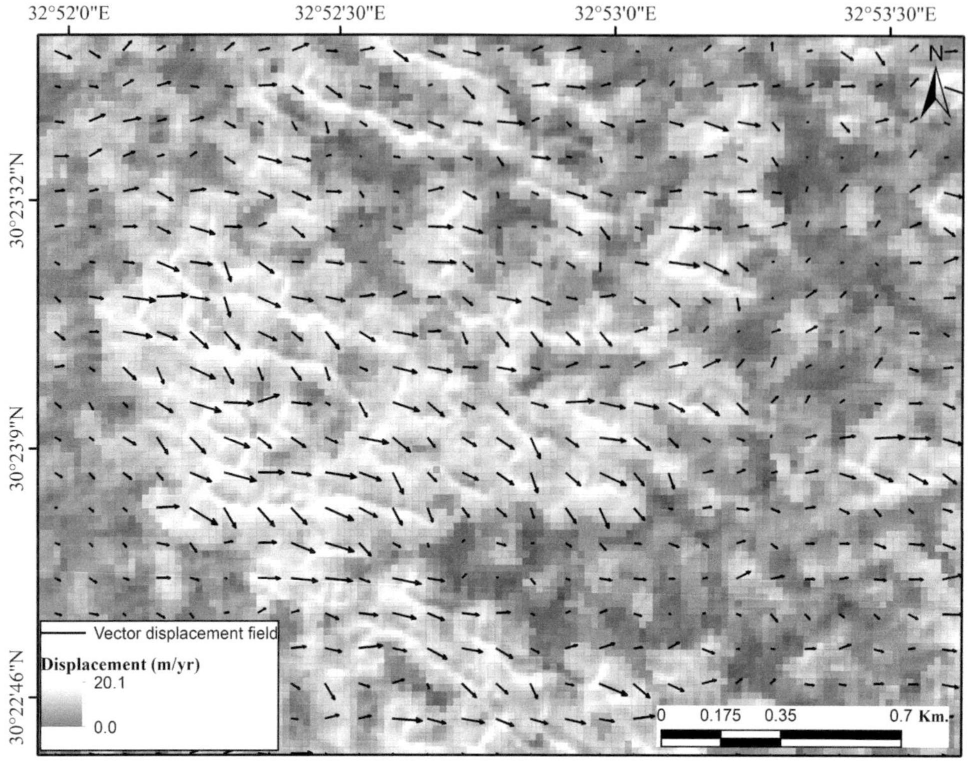

Figure 2.8. The vector displacement field displayed over the calculated net pixel displacement image of a barchan dune field in North Sinai Peninsula, Egypt.

Landsat satellite systems have the longest archived data back to 1972 (Wulder et al., 2016). The Landsat images cover the globe with regular monthly acquisition. In addition, images from different imaging systems of the Landsat satellites (i.e., Landsat MSS, TM, ETM+, OLI) are characterized by high radiometric and geometric quality, which means that they can be used in automatic change detection studies (Crippen and Blom, 1991; Van Puymbroeck et al., 2000; Wulder et al., 2016; Baird et al., 2019). With these advantages of the Landsat data and the continuous progress and improvements of the COSI-Corr technology, it has become possible to monitor dune movements at unprecedented temporal and spatial extents (Baird et al., 2019). Baird et al. (2019) sought to design a robust COSI-Corr workflow for Landsat 8 L1TP scenes and combine it with an innovative tool for automatic definition of dunes. The designed COSI-Corr workflow was successfully applied on a "big dataset" from the archived Landsat extending from 1987 to 2009 for the purpose of analyzing the spatiotemporal characteristics of dune movements across the Bodele Depression, Chad. The COSI-Corr workflow was repeated on a significant area across the Grand Erg Oriental, Algeria, using a Landsat dataset from 1987 to 2005 in two contiguous frames. The COSI-Corr workflow on the dataset in the Bodele Depression measured average dune movement in the order of 15.8 m/yr. In addition to its contribution to determining the spatial variability of dune movements along wide spatial coverage, the COSI-Corr workflow was able to characterize the temporal changes of dune movements from 1987 to 2009 (Baird et al., 2019). For instance, the aeolian activity increased by an average of 2.56 ± 12.48 m/yr over the study period. In the area of highest aeolian activities, dune migration rates increased by 53.4 m during the period 1987–1998 to 81.7 m/yr during the period 1998–2009. Although the measurement of dune migrations was determined in the Bodele Depression using the designed COSI-Corr workflow, it was not possible to determine dune migrations in Grand Erg Oriental, Algeria, because of the errors produced from geo-registration of the Landsat L1TP.

Evidently, continuous work on improving the technical capabilities of the proposed COSI-Corr workflow will enhance the automatic correlation of the archived Landsat images for the purpose of conducting spatiotemporal analysis of sand dune movements in the Sahara on an unprecedented global scale (Baird et al., 2019).

3.2.2 The coherence change detection (CCD) InSAR technique

Unlike the passive optical satellite systems, the Synthetic Aperture Radar (SAR) sensors transmit their own electromagnetic waves in the microwave portions onto the Earth's surface. The received signals from the features on the Earth's surface are recorded in SAR images as complex pixel numbers in terms of intensity and phase. The intensity depends on the collective physical characteristics of the scatterers within each corresponding pixel of SAR images while the phase represents the geometry and position of these scatterers (Liu et al., 1997). Target parameters (e.g., orientation, roughness and moisture content) and the radar system parameters (e.g., look angle, wavelength, polarization, mode and polarization) determine the intensity and the phase of the backscatter radar signals stored in SAR image pixels. Based on a priori knowledge of both parameters, SAR images helped distinguish between different features on the Earth's surface better than the optical images, especially in the case of the targets that might have similar spectral reflectance characteristics and small sizes, such as sand dune features. The sensitivity of radar signals to small geometric changes in the various targets on the Earth's surface enabled SAR images to define and map different dune forms and their spatial characteristics, such as distributions, patterns and extents (Blom and Elachi, 1981; Blom and Elachi, 1987; Lancaster et al., 1992; Blumberg and Greeley, 1993; Blumberg, 1998).

Since dune surfaces always experience changes in the geometry and position in response to aeolian erosional and depositional processes, the phase of the returned radar signals also changes in SAR images. The InSAR technology is an advanced technique that exploits the phase differences in the corresponding pixels of multi-temporal SAR images to constrain subtle changes in topography and other terrain deformations associated with seismicity, land subsidence and volcanic and glacial flow with unprecedented accuracy in the order of centimeter level or less (Zebker and Goldstein, 1986; Gabriel et al., 1989; Goldstein et al., 1993; Zebker et al., 1994a). To create a significant interferogram for these measurements with high accuracy, two SAR images acquired at different times should have relatively high coherence (Liu et al., 2001).

The coherence represents the similarity of the intensity and phase between the two SAR images (Corr et al., 1996). Accordingly, the produced InSAR coherence image represents the cross-correlation between the two co-registered complex-valued SAR images (Zebker et al., 1994b). From this perspective, if there is no change on the surface, the phase difference between the two SAR images will remain constant, resulting in full coherence (Liu et al., 2001). Conversely, if the surface features experienced any small changes on the scale of radar wavelength (e.g., 5 cm in the case of C-band) during the period between the two acquired SAR images, phase differences will take place, resulting in coherence reduction (loss of coherence or decorrelation). Losing coherence is referred to as decorrelation (Zebker and Villasenor, 1992). Specifically, the InSAR coherence images can be considered a measure of terrain stability over the time intervals separating the acquisition of two SAR images, which means that they can be used in change detection studies. These InSAR coherence images have been tested in a wide range of change detection studies, such as the study of surface erosion in desert areas. The technique has been referred to as the Coherence Change Detection (CCD) technique.

As the CCD technique efficiently detects and monitors subtle changes, it has been applied in the field of aeolian processes. For instance, Liu et al. (2001) used three ERS-1 SAR images acquired on September 8, 1992 (image 1), October 13, 1992 (image 2) and September 28, 1993 (image 3) that cover desert areas in Algeria. The study indicated that the coherence images were able to delineate dune boundary and mobile sand sheets, especially in the coherence image produced from interfering SAR image 1 and SAR image 2, 35 days apart. The low coherence values were limited to areas where sand occurs, whereas the higher coherence values were limited to the areas of stable terrain. The low coherence values in sandy areas are attributed to random changes in the micro-geometry of the scatterers on the sand-covered surfaces, resulting in loss of coherence over a very short period of time (Liu et al., 2001). Lee and Liu (2001) used the same ERS-1 SAR images of the desert of Algeria to investigate the influence of both topographic and temporal decorrelation on the coherence images to introduce a new method to analyze the coherence images referred to as ratio coherence imagery. The authors concluded that the low coherence values in sand dune areas are attributed to the rapid temporal decorrelation and volume scattering (Lee and Liu, 2001). The rapid temporal decorrelation was produced by the continuous micro-scale movement of the sand particles by the wind, which continuously changed the position of scatterers on the surface of dunes. Volume scattering can also contribute significantly to the temporal decorrelation.

Bodart and Ozer (2007) used eight tandem pairs of ERS-1 and ERS-2 SAR images at 24-hour intervals to produce an InSAR coherence image from each pair. To facilitate the interpretation, an RGB composite image was produced from the combination of the InSAR coherence image in Red color, an intensity SAR image of the pair in Green color, and another intensity image in Blue color. The dark colors characterized the areas of sand dunes, where a great loss of coherence was recorded, implying that there were random fast changes on the surface of sand areas on the scale of a one-day time interval.

This approach was extended to include interdune depressions in the desert areas of Niger, Mauritania and Morocco by Bodart et al. (2009). They used ERS-1 and ERS-2 SAR tandem pairs to study desertification and sand encroachment in a selected area across these countries. These tandem pairs included one pair on the study area of Niger acquired on April 25–26, 1996, two tandem pairs on the study area of Mauritania acquired on 2–3 November, 1995 and 7–8 November, 1995, and one tandem pair on the study area in Morocco acquired on 18–19 May, 1996. The authors used the same approach described by Bodart and Ozer (2007). Using this approach, they were able to detect and locate changes in the areas of sand dunes in the three studied locations. The result of applying the CCD technique was validated using fine-resolution images, such as SPOT-5, IKONOS and QuickBird, along with field work verifications.

Gaber et al. (2018) used the CCD technique at selected locations across the eastern desert and western desert of Egypt to investigate the potentiality of groundwater. Two pairs of Sentinel-1 SAR images were selected. The SAR images of the first pair were acquired on September 16, 2016 and December 28, 2017 for Wadi Al-Muttalah in the eastern desert, and the SAR images of the second pair were acquired on September 28, 2016 and December 28, 2017 for El-Gallaba plain in the western desert. Conducting the InSAR image processing on the two pairs produced coherence images for Wadi Al-Mutalla and another one for the El-Gallaba plain. The low coherence values in Wadi El-Mutalla were found to be restricted to surface changes produced by anthropogenic activities (e.g., urban infrastructure, agricultural activities and quarry processes) and geomorphic processes (e.g., sand dune encroachment and active fluvial processes), as well as changes in the natural vegetation cover along the Wadi courses. In the western desert, the low coherence was limited to the area occupied by the active linear dunes, reflecting strong surface changes in response to aeolian processes. The study results indicated the strong potential of the CCD technique to detect and map the surface changes of sand-enriched areas over the period separating the acquired SLC SAR images. The previous outputs of this study were validated using fine-resolution satellite images obtained from Google Earth and field work.

Havivi et al. (2018) used the InSAR coherence to study the dynamics of the biggest sand accumulations on the Israeli coastal plain. These accumulations comprised parabolic dunes that differed in their vegetation cover and thereby varied in their stability in response to the predominant westerly wind regime. Unlike the earlier studies, the authors used SAR data from a different radar system platform, TerraSAR-X, which provided SAR images with a temporal revisit of 11 days and pixel size of 1.8 m on the vertical axis and 0.9 m on the horizontal axis. The authors acquired 14 images starting on February 9, 2012 and ending on July 1, 2012. All the images were precisely co-registered to the master SAR image acquired on February 9, 2012 with a sub-pixel level of accuracy. Thirteen coherence images were produced from the InSAR processing of the master image, and each provided the ability to explore temporal and spatial variations of sand accumulation in the study area. The coherence values ranged from 0, indicating an unstable area (areas of high surface change) to 1, indicating a fully stable area (areas of low or null surface change) during the time separating the acquisitions. The authors integrated the coherence maps with the average computed SAR amplitude image (from the 13 SAR images) and a visible fine-resolution ortho-photo image to define both the windward side and dune crests of 51 sand dunes in the study area for eventual on-screen digitizing of them to produce 102 polygons. Zonal statistics was then used to extract the median of coherence values of each coherence image (13 coherence images) using the polygons of the windward slopes and crest line to study the temporal and spatial distribution of the dunes over the six-month period of the study. Generally, the coherence values of the crests were always higher than the coherence values of the windward slopes because of the smooth surface of the crests that produced a bright signal (Blumberg, 1998). However, the coherence values of dune crests showed a temporal loss of coherence over time from winter to summer, indicating that significant changes occurred in the crests as a result of the changes in sand transportation and vegetation cover over time. The relative high coherence values of crests in winter were attributed to the high moisture content when rainfall is high. The low coherence values of the crests were associated with spring and summer because of the shortage of rainfall and accordingly, the dominance of strong wind. The coherence values of the windward slopes implied less stability over time than the coherence values of the crests in the studied dunes. Although there were a few exceptions that were justified by the authors, a high match between the drifting potential values and the temporal changes of the coherence values was recorded by them, indicating significant short-term changes during the studied period from winter to summer. Furthermore, the CCD technique applied by the authors was able to analyze the spatial variability of the stability of both the dune crests and the windward slopes of sand dunes by analyzing the transect that extended from the Mediterranean Coast to the inland areas of the studied sand accumulation. It was evident from the analysis that the low coherence values were found beside the coastal areas and gradually increased inland as the distance from the coastal area increased. The low coherence values on the coastal areas implied higher dune activities. Conversely, the relative high coherence values increased inland away from the coast, where the dune activities on the dune crests and windward slopes increased. This study is considered a pioneer by introducing a framework for analyzing spatial and temporal changes of sand dune activities based on the CCD technique (Havivi et al., 2018).

Recently, Song et al. (2020) added another advanced step in the application of the CCD technique in the field of aeolian processes. The perspective of the authors relied on the proposition of Zebker and Villasenor (1992), who demonstrated that the loss of coherence, referred to as decorrelation, can be the result of thermal, spatial and temporal decorrelations. Only the temporal decorrelation between two SLC SAR images is related to the environmental terrain changes, such as changes on sandy surfaces. Both thermal and spatial decorrelations are related to radar system parameters or spatial parameters of the target, such as topography (Zebker and Villasenor, 1992; Martone et al., 2016). Both decorrelations are rarely considered in the environmental studies, such as those on sand dune changes (Gaber et al., 2018; Havivi et al., 2018). Song et al. (2020) isolated the temporal decorrelation from both the thermal and spatial decorrelations in the InSAR coherence images produced from 28 Sentinel-1A SLC SAR images acquired at 12 days of orbit revisit. The acquired images represent the period from May 22, 2017 to May 29, 2018. The approach comprised the calculation of the overall decorrelation, thermal decorrelation, spatial decorrelation and temporal decorrelation of the test area over a significant period. By eliminating the thermal and spatial decorrelations from the overall decorrelation, the temporal decorrelation was significantly improved (Song et al., 2020). The authors then validated the modeled temporal decorrelation using sand drift potential and precipitation in the study area. The validated model was applied to map temporal and spatial changes in stability of crescentic dunes on both sides of a road in the Taklemakan Desert from May 22, 2017 to May 29, 2018. The temporal investigation implied that the dune stability changes on a seasonal scale, where spring is characterized by the lowest stability in response to the high dune activity on both sides of the road. The highest stability of the dunes was recorded in winter when the dune activity was at its lowest level. The outputs of applying the temporal decorrelation model highly matched with the calculated synthetic sand drift potential in the study area (Song et al., 2020). The temporal model proposed by Song et al. (2020) was able to capture the spatial variability of dune stability along the road in the study area. The spatial variability was attributed to the different sizes of the crescentic dunes along the road. The unstable and relatively stable areas were easily distinguished from the temporal decorrelation images. The concordance of the high- and low-stability areas with the spatial distribution of the large and small dunes was validated using the fine-resolution optical image of the area. Although the earlier studies indicated that the CCD technique can potentially detect subtle micro-scale changes occurring on the surface of sand dunes, the technique cannot measure dune movement in the 2D dimension (Liu et al., 1997; Liu et al., 2001; Gaber et al., 2018; Havivi et al., 2018; Song et al., 2020).

In summary, the principal criteria of the COSI-Corr and CCD techniques and their contributions to the scientific understanding of dynamic aspects of sand accumulations are outlined below.

- Satellite data platforms

  While the COSI-Corr technology relies on the optical remote sensing imagery, the CCD technique works with the radar images. This variation creates differential potentiality for both of them to document the aeolian processes on a global scale. For instance, the Landsat sensor is considered one of the oldest Earth observation satellite systems that has acquired images on wide global coverage since 1972 (Wulder et al., 2016). This satellite system and other optical systems mean that the COSI-Corr technique is a powerful monitoring system at monthly temporal acquisitions, or recently less than a month, for many geographic regions over a long period of time. In addition, the COSI-Corr technique works well with aerial photographs, which qualifies it to extend the temporal scale back further than 1972 for the areas that were imaged using aerial photograph technologies (Ayoub et al., 2009a). Unlike optical remote sensing, one of the radar systems' obstacles was the lack of spatial coverage of the Earth's surface. However, the launch of Sentinel satellite sensors in 2013 may have helped enhance the temporal and spatial characteristics of the radar CCD technique.

- Temporal scale

  As the COSI-Corr technology relies on the optical remote sensing images that could be acquired monthly for decades or more, the technique can document dune movements and directions in both the medium term and the long term (Vermeesch and Leprince, 2012; Baird et al., 2019) by using a regular and consistent series of acquisitions on given geographic areas. In addition to the ability of the CCD technique to document dune movement in both the medium term and the long term (Lee and Liu, 2001; Liu et al., 2001; Gaber et al., 2018; Havivi et al., 2018; Song et al., 2020), it can provide measurements in both the short term and the very short term (Bodart and Ozer, 2007; Bodart et al., 2009). The ability to measure dune movements and surface changes of dunes is critical for studying the temporal changes in the aeolian processes, which may be linked to climatic (Vermeesch and Leprince, 2012; Baird et al., 2019) or environmental changes (Sam et al., 2015) in given geographic areas.

- Spatial scale

  One of the main contributions of the COSI-Corr and CCD techniques is providing measurement over a wide spatial area and at a sub-pixel level of accuracy. The spatial variability of measurements by both techniques across dune fields or sand seas can lead to an insight into the impact of regional and local wind regimes, topography, dune pattern, dune morphometry, dune form and environmental conditions on the rates of the aeolian processes (Hermas et al., 2012; Al-Ghamdi and Hermas, 2015; Al-Mutiry et al., 2016; Baird et al., 2019).

- Dune measurements

  The COSI-Corr and CCD techniques measure different representative characteristics associated with the aeolian processes. The COSI-Corr technique measures the horizontal pixel displacements that represent dune movement that occurred between two satellite acquisitions. In addition, it produces a vector displacement in which the length of each arrow represents the magnitude of dune movements while the direction of each arrow represents the direction of the movements. The CCD technique does not measure dune movement or its direction; rather, it measures surface changes over the dune in response to the continuous movements of sand grains associated with the aeolian erosional and depositional processes. However, the full capacity of the CCD technique has not been comprehensively explored in the field of aeolian geomorphology (Havivi et al., 2018).

- Elevation data of sand dunes

  Neither the COSI-Corr technique nor the CCD technique provides measurements on the elevation of sand dunes. They only provide measurements of the dune location or surface changes in the 2D domain. This can be considered a gap in the measurements provided by both techniques. While the geodetic field technique dominates in this aspect in the field of aeolian geomorphology, it provides only topographic measurements along the transects of a targeted dune(s). Therefore, the area between the transects lacks topographic observations rather than the entire dune field or sand sea. These gaps in the COSI-Corr, CCD and geodetic field techniques can potentially be obviated with the recent LiDAR technology.

3.2.3 The pair of source and target point (PSTP) LiDAR technique

Along with the optical and microwave radar remote sensing, the high-resolution airborne Light Detection And Ranging (LiDAR) remote sensing data with its high spatial resolution of a few tenths of centimeters is considered a potential advanced source of remote sensing imagery for characterizing dune morphologies and processes, such as erosion and deposition (Hugenholtz et al., 2012). The fundamental output of the LiDAR high spatial resolution imagery is a DEM with high vertical accuracy. Usually, digital topographic data extracted from hardcopy aerial photographs are used to study 3D change associated with dune activity (e.g., Brown and Arbogast, 1999; Ojeda et al., 2005). Brown and Arbogast (1999) applied digital photogrammetric methods to scan stereographic aerial photographs to quantify erosion and deposition at a dune complex along the eastern shore of Lake Michigan. Despite relatively poor accuracy in the digital elevation models produced by the photogrammetric models, the data provided a reasonable estimate of course-scale topographic changes during the time span between aerial photographs (1965–1987). Generally, highly accurate photogrammetrically generated digital elevation models of sand dunes from hardcopy aerial photographs are rare (Hugenholtz et al., 2012). However, direct collection of topographic data from the field is apparently leading to an understanding of dune morphology, volumetric changes and accurate measurements of dune migration. The new LiDAR technology is a potential alternative for collection of topographic data with high spatiotemporal resolution not only for studying dune migration, but also for performing volumetric analysis of dune fields. Nevertheless, few studies have been conducted using LiDAR technology because of the high cost and complex requirements for planning new acquisitions.

The earlier studies that have been carried out based on LiDAR data only or by integrating them with other data for detecting and measuring dune movements relied on the manual definition and extraction of dune crests from the different acquisitions for eventual determination of dune movement rates. In addition, these measurements have been carried out on a limited number of dunes since manual tracing is time-consuming. Dong (2015) introduced an advanced methodology for automatic detection and measurement of sand dune movements and directions. The methodology is referred to as the PSTP technique. The methodology uses multi-temporal LiDAR data to automatically detect and measure dune migration rates and direction over wide spatial coverage with high accuracy. The method is implemented in a Geographic Information System (GIS) environment. The conceptual framework and the designed procedures are discussed next.

As sand saltates up on the windward side and accumulates over the crest, a short slip face is formed on the leeward side. This slip face maintains a 34-degree angle of repose under dry desert conditions. Dunes move in the inclination direction of the slip face in response to avalanching and slumping along the slip face (Bagnold, 1941). The PSTP technique relies on the determination of the centerlines of the slip faces in two acquisitions and measures the vector distance between them (Dong, 2015), as outlined below.

From the topographic profiles produced from the LiDAR extracted DEM, it is possible to develop these equations (Dong, 2015):

$$d_1 = d_0 + \frac{w_1}{2} - \frac{w_2}{2} \tag{2}$$

$$d_2 = d_0 + \frac{w_1}{2} - \frac{w_2}{2} \tag{3}$$

From equations 2 and 3, d_0 can be expressed as

$$d_0 = \frac{d_1 + d_2}{2} \tag{4}$$

where, d_1 and d_2 are the horizontal distances between slip faces of the same dune in two acquisitions and d_0 is the average of d_1 and d_2. The equation implies that d_0 represents sand dune migration distance and can be measured by determining the distance between the centerlines of the slip faces in two LiDAR acquisitions (Dong, 2015). Subsequently, d_0 can be converted to the migration rate r_0 (m/yr) using the following equation (Dong, 2015):

$$r_0 = (\frac{356}{t}) \times d_0 \tag{5}$$

where t is the time interval between two dates for sand dune elevation data acquisition.

The automated PSTP technique was proposed to measure dune migration rates on a field scale (Dong, 2015). The concept suggested that the centerlines of the old slip faces can be referred to as source lines, whereas the centerlines of the new slip faces represent the target lines. For any target point t (target point) on a target line, there might be a nearest point s (source point) on a source line within a defined search radius (Dong, 2015). The line connecting the two points is called the vector. The length of the vector is the dune migration distance, and the direction from the source point s relative to the target point t is called the source direction. Therefore, many random pairs of target and source points can be identified for the determination of migration distances and directions of numerous dunes across the dune field (Dong, 2015).

The implementation of the PSTP method comprises three main steps: the generation of the centerlines from the "before" and "after" LiDAR extracted DEMs, processing of sources and target lines (centerlines of the slip faces), and identification of pairs of source and target points (Dong, 2015). This procedure eventually produces a raster dataset for sand dune migration rates of a given sand dune field. The methodology was performed in detail on two LiDAR datasets acquired on January 24, 2009 and June 6, 2010. Each data point cloud was interpolated using the inverse distance weighting to produce two DEMs with a 1 m pixel size. Slope rasters were then generated from both DEMs. Slip faces were extracted from the slope rasters as a binary raster by assigning slopes ranging from 30 to 35 to 1. Other cell values of slopes were assigned to 0. Noise was eliminated using a 5×5 kernel (Dong, 2015). The vectorization operation available in ArcGIS was then used to convert the raster slip faces into polylines to extract the centerlines of these slip faces. To ensure that the source and target points on the centerlines were selected properly, buffering with 20 m distance from each centerline was conducted for eventual clipping of dangle errors from the unmatched source and target points. The trimmed source and target lines were then ready to define the matched and the unmatched points on both lines using a search circle with a radius of 20 m (Dong, 2015). This step was followed by the identification of pairs of source and target points valid for the determination of the dune migration rates and directions.

For each target point on the target lines, a search circle with a radius of 20 m was used to locate the nearest source point on the source line. Therefore, the attribute table of each matched target point on the target centerline included the x and y coordinates of the corresponding source point, the distance from the target point to the source point, and the angle between the x-axis and the line connecting the target point and the source point. Defining these three parameters as attributes for the target points enabled the calculation of migration distance of dunes and the direction of dune migration, and provided a raster dataset for the whole field to determine the spatial variability of dune migration across wide spatial coverage.

Dune migration rates can be measured automatically as well using the PSTP technique for each target point according to equation 3. The measured magnitude of the migration rates varies spatially on the target points along the target lines, implying the spatial variation of the aeolian processes over the same dune. Across the study area, the migration rate of dunes ranged from 4 to 7 m/yr, with an average of 5.56 m/yr. The spatial variability may be attributed to dune forms, changing wind intensity and direction across the study area or dune pattern interaction (Dong, 2015; Ewing et al., 2015). The raster coverage of the study area can be produced by interpolating the values of the migration rates of the target points. The PSTP technique proved its potential to measure the migration directions and rates of sand dunes using fine spatial resolution DEM extracted from LiDAR datasets. The approach was expanded by Xia and Dong (2016) into a generic ArcGIS add-in toolbar using the Python programming language. The plug-in software will help scientists gain an understanding of the driving processes controlling the erosional and depositional processes that result in dune migration on both a single dune and the field scale. However, many limitations can be highlighted (Dong, 2015; Xia and Dong, 2016), including dune form and dune interactions, as well as the suitable temporal intervals between acquisitions.

The PSTP technique works well with crescentic dunes, such as barchan, barchanoid, transverse and parabolic dunes, as these dunes have a single slip face. However, the automatic performance on linear and star dunes is more difficult since these dunes have more than one slip face (Dong, 2015). It might require more user interactions through performing the PSTP routines. The continuous changes in the spatial scales of dunes and the locations and patterns of the various bedforms in the dune field (Ewing and Kocurek, 2010a; Ewing et al., 2015) may represent other sources of challenges to the automatic PSTP routines and even in the crescentic dune areas (Dong, 2015). However, the technique contributes significantly to the progress of remote sensing techniques that rely on multiple acquisitions of the elevation data extracted from fine-resolution sources such as LiDAR data and aerial photographs.

4. Conclusions and suggestions

One of the fundamental dynamic aspects of dune systems is the continuous movement of sand dunes, particularly in bare deserts. These movements change significantly both temporally and spatially in response to variations in wind regimes, dune morphology and the surrounding conditions (Livingstone and Warren, 1996), as well as the changes in sand supply, vegetation cover and human activities (Lancaster, 2009; Yizhaq et al., 2009). As the magnitude and direction of dune movement is significant for understanding the processes of aeolian sand transport and for avoiding or minimizing the potential hazards of moving dunes, numerous techniques for the determination of the magnitude of sand movements have been developed since the end of the 1950s. These techniques vary significantly in terms of tools, methods of data collection, data processing, concepts and accuracy of measurement. The most common aspect of these techniques is their increasing sophistication due to technological advancement of tools, data and algorithms. For instance, traditional field techniques that rely on primitive man-made markers or markers from the local environment have been transformed into more sophisticated geodetic field techniques. Similarly, traditional optical remote sensing techniques that involve intensive labor have been developed into automated techniques using sophisticated algorithms and a wide variety of satellite imagery from different platforms. Unlike the optical remote sensing technique, the radar technique relies on other data, concepts and algorithms. Whereas these remote sensing techniques rely on optical and radar data in the determination of the magnitude of dune movement, the LiDAR technology relies on elevation data.

In summary, each technique has the ability to determine dune movement and direction from different technical and scientific perspectives than the other. Although each technique has its own shortcomings and limitations, there is no doubt that each technique represents a remarkable step in determining, understanding and modeling the magnitude and directions of dune movements either on a single-dune scale or on a dune-field scale. Presenting and describing all the criteria of the earlier techniques may help in developing propositions for future development of these techniques.

Among the propositions, we suggest the integration of the various techniques and the calibration of the remote sensing measurements by field work. The integration of the different techniques may help constrain dune movement in a more reliable way. For instance, the traditional and COSI-Corr optical remote sensing techniques detect and measure the horizontal displacement of dunes, but does not provide an insight into the aeolian processes responsible for these horizontal displacements. In contrast, the CCD technique has the potential to detect the micro-scale changes on the surface of dunes in the very short term, and to some extent in the medium term. These surface changes are strongly associated with the aeolian processes responsible for dune mobility (Gaber et al., 2018; Havivi et al., 2018; Song et al., 2020). Nevertheless, the CCD technique has not yet been developed to measure the horizontal displacement of sand dunes. Therefore, the integration of measurements from the COSI-Corr and CCD techniques on the same dune area may help not only in detecting the magnitude of lateral movements, but also in enhancing our understanding of the aeolian processes responsible for these movements.

Increasingly, the calibration of the measurement of dune movements achieved from remote sensing techniques by the measurements of field techniques may represent a significant contribution in improving the measurements of the various remote sensing techniques. Although many of the measurements of dune movement by remote sensing techniques in earlier studies have been compared with the measurements of dune movements acquired from previous field works, this comparison was just an indicative approach and lacks many parameters of a well-designed deductive calibration approach. We assume that designing a calibration study of remote sensing measurements of dune movements by measurements of a field technique for the same geographic area over at least one year may help refine remote sensing measurements. In this perspective, many calibration criteria should be established first, such as the spatial extent of the dune field, locations where field tools will be established, the temporal interval of field revisits for measurements, which satellite platforms will provide monthly or seasonal images, the spatial and temporal resolutions of remote sensing images and the proposed algorithm for detecting and measuring sand dune movements.

References

Ahlbrandt, T.S. and Brookfield, M.E. 1983. Eolian Sediments and Processes. Elsevier.

Al-Ghamdi, K. and Hermas, E. 2015. Assessment of dune migration hazards against landuse northwest al-lith city, saudi arabia, using multi-temporal satellite imagery. Arabian Journal of Geosciences 8(12): 11007–11018. Available from: https://doi.org/10.1007/s12517-015-1947-8.

Al-Mutiry, M., Hermas, E.A., Al-Ghamdi, K.A. and Al-Awaji, H. 2016. Estimation of dune migration rates north riyadh city, ksa, using spot 4 panchromatic images. Journal of African Earth Sciences 124: 258–269. Available from: http://www.sciencedirect.com/science/article/pii/S1464343X16303284.

Andrews, B.D., Gares, P.A. and Colby, J.D. 2002. Techniques for gis modeling of coastal dunes. Geomorphology 48(1): 289–308. Available from: http://www.sciencedirect.com/science/article/pii/S0169555X02001861.

Anthonsen, K.L., Clemmensen, L.B. and Jensen, J.H. 1996. Evolution of a dune from crescentic to parabolic form in response to short-term climatic changes: Råbjerg mile, skagen odde, Denmark. Geomorphology 17(1): 63–77. Available from: http://www.sciencedirect.com/science/article/pii/0169555X9500091I.

Aql, M.T. 2002. Movement of sands east suez canal and its impact on human activities: A study in applied geomorphology (in arabic). Al-Ensaniat 19: 1–75.

Assal, E.M. 1999. Sedimentological Studies on the Quaternary Sand Dunes and Sabkhas, Northern Sinai, Egypt. Unpublished M. Sc. (Geology), University of Mansura, Damiatta Faculty of Science, 268.

Avouac, J., Ayoub, F., Leprince, S., Konca, O. and Helmberger, D.V. 2006. The 2005, mw 7.6 Kashmir earthquake: Sub-pixel correlation of aster images and seismic waveforms analysis. Earth and Planetary Science Letters 249(3): 514–528. Available from: http://www.sciencedirect.com/science/article/pii/S0012821X06004572.

Avouac, J., Ayoub, F., Wei, S., Ampuero, J., Meng, L., Leprince, S., Jolivet, R., Duputel, Z. and Helmberger, D.V. 2014. The 2013, mw 7.7 Balochistan earthquake, energetic strike-slip reactivation of a thrust fault. Earth and Planetary Science Letters 391: 128–134. Available from: http://www.sciencedirect.com/science/article/pii/S0012821X14000521.

Ayoub, F., Leprince, S. and Avouac, J. 2009a. Co-registration and correlation of aerial photographs for ground deformation measurements. ISPRS Journal of Photogrammetry and Remote Sensing 64(6): 551–560. Available from: http://www.sciencedirect.com/science/article/pii/S0924271609000483.

Ayoub, F., Leprince, S. and Keene, L. 2009b. Cosi-corr user's guide. Co-registration of optically sensed images and correlation. Calif. Inst. of Technol., Pasadena, Calif.

Bagnold, R.A. 1941. The Physics of Blown Sand and Desert Dunes: New york. William Morrow & Company.

Baird, T., Bristow, C.S. and Vermeesch, P. 2019. Measuring sand dune migration rates with cosi-corr and Landsat: Opportunities and challenges. Remote Sensing 11(20): 2423. Available from: https://www.mdpi.com/2072-4292/11/20/2423.

Bishop, M.A. 2007. Point pattern analysis of north polar crescentic dunes, Mars: A geography of dune self-organization. Icarus 191(1): 151–157. Available from: http://www.sciencedirect.com/science/article/pii/S0019103507001984.

Bishop, M.A. 2010. Nearest neighbor analysis of mega-barchanoid dunes, Ar rub' al Khali, sand sea: The application of geographical indices to the understanding of dune field self-organization, maturity and environmental change. Geomorphology 120(3): 186–194. Available from: http://www.sciencedirect.com/science/article/pii/S0169555X10001480.

Blom, R. and Elachi, C. 1981. Spaceborne and airborne imaging radar observations of sand dunes. Journal of Geophysical Research: Solid Earth 86(B4): 3061–3073. Available from: https://agupubs.onlinelibrary.wiley.com/doi/abs/10.1029/JB086iB04p03061.

Blom, R. and Elachi, C. 1987. Multifrequency and multipolarization radar scatterometry of sand dunes and comparison with spaceborne and airborne radar images. Journal of Geophysical Research: Solid Earth 92(B8): 7877–7889. Available from: https://agupubs.onlinelibrary.wiley.com/doi/abs/10.1029/JB092iB08p07877.

Blumberg, D.G. and Greeley, R. 1993. Field studies of aerodynamic roughness length. Journal of Arid Environments 25(1): 39–48. Available from: http://www.sciencedirect.com/science/article/pii/S0140196383710414.

Blumberg, D.G. 1998. Remote sensing of desert dune forms by polarimetric synthetic aperture radar (SAR). Remote Sensing of Environment 65(2): 204–216. Available from: http://www.sciencedirect.com/science/article/pii/S0034425798000285.

Bodart, C., Gassani, J., Salmon, M. and Ozer, A. 2009. Contribution of sar interferometry (from ers1/2) in the study of aeolian transport processes: The cases of Niger, Mauritania And Morocco. pp. 129–136. *In*: Marini, A. and Talbi, M. (eds.). Desertification and Risk Analysis Using High and Medium Resolution Satellite Data. NATO Science for Peace and Security Series C: Environmental Security. Springer, Dordrecht. https://doi.org/10.1007/978-1-4020-8937-4_11.

Bodart, C. and Ozer, A. 2007. The use of sar interferometric coherence images to study sandy desertification in southeast niger: Preliminary results. Proc. Envisat Symposium 2007, Montreux, Switzerland (ESA SP-636), 23–27.

Bourke, M.C., Ewing, R.C., Finnegan, D. and Mcgowan, H.A. 2009. Sand dune movement in the victoria valley, Antarctica. Geomorphology 109(3): 148–160. Available from: http://www.sciencedirect.com/science/article/pii/S0169555X09000889.

Breed, C.S., Fryberger, S.G., Andrews, S., Mccauley, C., Lennartz, F., Gebel, D. and Horstman, K. 1979. Regional studies of sand seas using Landsat (ERTS) imagery. A study of global sand seas. US Geological Survey, Professional Paper, 305–397.

Brown, D.G. and Arbogast, A.F. 1999. Digital photogrammetric change analysis as applied to active coastal dunes in Michigan. Photogrammetric Engineering and Remote Sensing 65: 467–474.

Calkin, P.E. and Rutford, R.H. 1974. The sand dunes of Victoria Valley, Antarctica. Geographical Review 64: 189–216.

Cooke, R.V. and Doornkamp, J.C. 1990. Geomorphology in Environmental Management: A New Introduction. Oxford University Press (OUP).

Corr, Douglas G., Simon W. Whitehouse, David H. Mott and Jim F. Baldwin. 1996. Automatic change detection in space borne SAR imagery. In Algorithms for Synthetic Aperture Radar Imagery III 2757: 58-69. SPIE.

Crippen, R.E. and Blom, R.G. 1991. Measurement of subresolution terrain displacements using spot panchromatic imagery. Conference Paper 19920052569 Jet Propulsion Lab. Pasadena.

Dong, P. 2015. Automated measurement of sand dune migration using multi-temporal Lidar data and GIS. International Journal of Remote Sensing 36(21): 5426–5447. Available from: https://doi.org/10.1080/01431161.2015.1093192.

Dong, Z., Wang, X. and Chen, G. 2000. Monitoring sand dune advance in the Taklimakan desert. Geomorphology 35(3): 219–231. Available from: http://www.sciencedirect.com/science/article/pii/S0169555X00000398.

El-Banna, M.S. 2004. Geological Studies Emphasizing the Morphology and Dynamics of Sand Dunes and Their Environmental Impacts on the Reclamation and Developmental Areas in Northwest Sinai, Egypt. Sci., Cairo University.

El-Kayali, M. 1984. Geomorphologic Studies of the Coastal Plain, North Sinai Peninsula. PhD Thesis, Ain Shams University.

Embabi, N.S. and Ashour, M.M. 1985. Sand Dunes in Qatar Peninsula. University of Qatar.

Embabi, N.S. 2004. Geomorphology of Egypt: Landforms and Evolution, vol I. The Egyptian Geographical Society, Cairo, The Nile Valley and the Western Desert.

Ewing, R.C., Kocurek, G.A. and Lake, L.W. 2006. Pattern analysis of dune-field parameters. Earth Surface Processes and Landforms 31(9): 1176–1191.

Ewing, R.C., Peyret, A.B., Kocurek, G. and Bourke, M.C. 2010. Dune field pattern formation and recent transporting winds in the Olympia Undae Dune Field, north polar region of Mars. Journal of Geophysical Research: Planets 115(E8).

Ewing, R.C. and Kocurek, G.A. 2010a. Aeolian dune-field pattern boundary conditions. Geomorphology 114(3): 175–187. Available from: http://www.sciencedirect.com/science/article/pii/S0169555X09002621.

Ewing, R.C. and Kocurek, G.A. 2010b. Aeolian dune interactions and dune-field pattern formation: White sands dune field, new Mexico. Sedimentology 57(5): 1199–1219. Available from: https://onlinelibrary.wiley.com/doi/abs/10.1111/j.1365-3091.2009.01143.x.

Ewing, R.C., Mcdonald, G.D. and Hayes, A.G. 2015. Multi-spatial analysis of aeolian dune-field patterns. Geomorphology 240: 44–53. Available from: http://www.sciencedirect.com/science/article/pii/S0169555X14005935.

Finkel, H.J. 1959. The barchans of southern Peru. The Journal of Geology 67(6): 614–647.

Frank, A.J. and Kocurek, G.A. 1996a. Airflow up the stoss slope of sand dunes: Limitations of current understanding. Geomorphology 17(1): 47–54. Available from: http://www.sciencedirect.com/science/article/pii/0169555X9500094L.

Frank, A.J. and Kocurek, G.A. 1996b. Toward a model for airflow on the lee side of aeolian dunes. Sedimentology 43(3): 451–458. Available from: https://onlinelibrary.wiley.com/doi/abs/10.1046/j.1365-3091.1996.d01-20.x.

Fryberger, S.G., Dean, G. and Mckee, E.D. 1979. Dune forms and wind regime. A Global Study of Sand Seas 1052: 137–170.

Gaber, A., Abdelkareem, M., Abdelsadek, I.S., Koch, M. and El-Baz, F. 2018. Using INSAR coherence for investigating the interplay of fluvial and aeolian features in arid lands: Implications for groundwater potential in Egypt. Remote Sensing 10(6): 832. Available from: https://www.mdpi.com/2072-4292/10/6/832.

Gabriel, A.K., Goldstein, R.M. and Zebker, H.A. 1989. Mapping small elevation changes over large areas: radar interferometry. Journal of Geophysical Research: Solid Earth 94(B7): 9183–9191. Available from: https://agupubs Differential.onlinelibrary.wiley.com/doi/abs/10.1029/JB094iB07p09183.

Goldstein, R.M., Engelhaamrdt, H., Kb, B. and Frolich, R.M. 1993. Satellite radar interferometry for monitoring ice sheet motion: Application to an Antarctic ice stream. Science 262(5139): 1525. Available from: http://science.sciencemag.org/content/262/5139/1525.abstract.

Greeley, R. and Iversen, J.D. 1987. Wind as a Geological Process: On Earth, Mars, Venus and Titan: CUP Archive.

Havholm, K.G. and Kocurek, G.A. 1988. A preliminary study of the dynamics of a modern draa, algodones, southeastern California, USA. Sedimentology 35(4): 649–669. Available from: https://onlinelibrary.wiley.com/doi/abs/10.1111/j.1365-3091.1988.tb01242.x.

Havivi, S., Amir, D., Schvartzman, I., August, Y., Maman, S., Rotman, S.R. and Blumberg, D.G. 2018. Mapping dune dynamics by INSAR coherence. Earth Surface Processes and Landforms 43(6): 1229–1240. Available from: https://onlinelibrary.wiley.com/doi/abs/10.1002/esp.4309.

Hermas, E., Leprince, S. and El-Magd, I. 2012. Retrieving sand dune movements using sub-pixel correlation of multi-temporal optical remote sensing imagery, northwest Sinai peninsula, Egypt. Remote Sensing of Environment 121: 51–60. Available from: http://www.sciencedirect.com/science/article/pii/S0034425712000144.

Hermas, E., Alharbi, O., Alqurashi, A., Niang, A., Al-Ghamdi, K., Al-Mutiry, M. and Farghaly, A. 2019. Characterisation of sand accumulations in wadi fatmah and wadi ash Shumaysi, KSA, using multi-source remote sensing imagery. Remote Sensing 11(23): 2824. Available from: https://www.mdpi.com/2072-4292/11/23/2824.

Howard, A.D., Morton, J.B., Gad-El-Hak, M. and Pierce, D.B. 1978. Sand transport model of barchan dune equilibrium. Sedimentology 25(3): 307–338.

Hugenholtz, C.H. and Wolfe, S.A. 2005. Biogeomorphic model of dunefield activation and stabilization on the northern great plains. Geomorphology 70(1): 53–70. Available from: http://www.sciencedirect.com/science/article/pii/S0169555X05001224.

Hugenholtz, C.H. and Barchyn, T.E. 2010. Spatial analysis of sand dunes with a new global topographic dataset: New approaches and opportunities. Earth Surface Processes and Landforms 35(8): 986–992. Available from: https://onlinelibrary.wiley.com/doi/abs/10.1002/esp.2013.

Hugenholtz, C.H., Levin, N., Barchyn, T.E. and Baddock, M.C. 2012. Remote sensing and spatial analysis of aeolian sand dunes: A review and outlook. Earth-Science Reviews 111(3): 319–334. Available from: http://www.sciencedirect.com/science/article/pii/S001282521100170X.

Hunter, R., Richmond, B. and Rho Alpha, T. 1983. Storm-controlled oblique dunes of the Oregon coast. GSA Bulletin 94(12): 1450–1465. Available from: https://doi.org/10.1130/0016-7606(1983)94<1450:SODOTO>2.0.CO;2 [Accessed 8/7/2020].

Khidr, M. 2006. Aeolian Forms and their Hazards in the West of Wadi El-Arish: A Geomorphological Study. PhD Dis-sertation, Geography Department, Ain Shams University, Egypt.

Kocurek, G.A. 1996. Desert aeolian systems. pp. 125–153. *In*: Reading, H.G. (ed.). Sedimentary Environments: Processes, Facies and Stratigraphy. Blackwell Oxford, UK.

Kocurek, G.A. and Ewing, R.C. 2005. Aeolian dune field self-organization – implications for the formation of simple versus complex dune-field patterns. Geomorphology 72(1): 94–105. Available from: http://www.sciencedirect.com/science/article/pii/S0169555X05001674.

Kocurek, G.A., Ewing, R.C. and Mohrig, D. 2010. How do bedform patterns arise? New views on the role of bedform interactions within a set of boundary conditions. Earth Surface Processes and Landforms 35(1): 51–63. Available from: https://onlinelibrary.wiley.com/doi/abs/10.1002/esp.1913.

Koprowski, M., Winchester, V. and Zielski, A. 2010. Tree reactions and dune movements: Slowinski National Park, Poland. Catena 81: 55–65.

Lancaster, N. 1983. Controls of dune morphology in the Namib sand sea. Developments in Sedimentology. Elsevier, 261–289.

Lancaster, N. 1989. The dynamics of star dunes: An example from the gran desierto, Mexico. Sedimentology 36(2): 273–289. Available from: https://onlinelibrary.wiley.com/doi/abs/10.1111/j.1365-3091.1989.tb00607.x.

Lancaster, N., Gaddis, L. and Greeley, R. 1992. New airborne imaging radar observations of sand dunes: Kelso dunes, California. Remote Sensing of Environment 39(3): 233–238. Available from: http://www.sciencedirect.com/science/article/pii/0034425792900882.

Lancaster, N. 1995. Geomorphology of Desert Dunes: Psychology Press.

Lancaster, N., Nickling, W.G., Neuman, C.K.M. and Wyatt, V.E. 1996. Sediment flux and airflow on the stoss slope of a barchan dune. Geomorphology 17(1): 55–62. Available from: http://www.sciencedirect.com/science/article/pii/0169555X9500095M.

Lancaster, N. 2009. Dune morphology and dynmaics. *In*: Parsons, A.J. and abrahams, A.D. (eds.). Geomorphology of Desert Environments, 2nd ed. Springer Science + Business Media B.V.

Lee, H. and Liu, J.G. 2001. Analysis of topographic decorrelation in SAR interferometry using ratio coherence imagery. IEEE Transactions on Geoscience and Remote Sensing 39(2): 223–232.

Leprince, S., Barbot, S., Ayoub, F. and Avouac, J. 2007. Automatic and precise orthorectification, coregistration, and subpixel correlation of satellite images, application to ground deformation measurements. IEEE Transactions on Geoscience and Remote Sensing 45(6): 1529–1558.

Levin, N., Ben-Dor, E. and Karnieli, A. 2004. Topographic information of sand dunes as extracted from shading effects using Landsat images. Remote Sensing of Environment 90(2): 190–209. Available from: http://www.sciencedirect.com/science/article/pii/S0034425703003791.

Liu, J.G., Capes, R., Haynes, M. and Moore, J.M. 1997. Ers sar multi-temporal coherence image as a tool for sand desert study (dune movement, sand encroachment and erosion). Proceedings of the 12th International Conference and Workshop on Applied Geologic Remote Sensing, Denver, Colorado, 17–19 November 1997.

Liu, J.G., Black, A., Lee, H., Hanaizumi, H. and Moore, J.M. 2001. Land surface change detection in a desert area in Algeria using multi-temporal Ers SAR coherence images. International Journal of Remote Sensing 22(13): 2463–2477. Available from: https://doi.org/10.1080/01431160119991.

Liu, Z. 1960. Movement of dunes under the action of wind. Acta Meteorologica Sinica 31: 84–91.

Livingstone, I. 1989. Monitoring surface change on a namib linear dune. Earth Surface Processes and Landforms 14(4): 317–332. Available from: https://onlinelibrary.wiley.com/doi/abs/10.1002/esp.3290140407.

Livingstone, I. and Warren, A., 1996. Aeolian Geomorphology: An Introduction: Longman.

Livingstone, I. 2003. A twenty-one-year record of surface change on a Namib linear dune. Earth Surface Processes and Landforms 28(9): 1025–1031. Available from: https://onlinelibrary.wiley.com/doi/abs/10.1002/esp.1000.

Livingstone, I., Wiggs, G.F.S. and Weaver, C.M. 2007. Geomorphology of desert sand dunes: A review of recent progress. Earth-Science Reviews 80(3): 239–257. Available from: http://www.sciencedirect.com/science/article/pii/S0012825206001449.

Long, J.T. and Sharp, R.P. 1964. Barchan-dune movement in imperial valley, California. GSA Bulletin 75(2): 149–156. Available from: https://doi.org/10.1130/0016-7606(1964)75[149:BMIIVC]2.0.CO;2 [Accessed 7/28/2020].

Martone, M., Bräutigam, B., Rizzoli, P., Yague-Martinez, N. and Krieger, G. 2016. Enhancing interferometric SAR performance over sandy areas: Experience from the Tandem-X mission. IEEE Journal of Selected Topics in Applied Earth Observations and Remote Sensing 9(3): 1036–1046.

Mason, J.A., Swinehart, J.B., Lu, H., Miao, X., Cha, P. and Zhou, Y. 2008. Limited change in dune mobility in response to a large decrease in wind power in semi-arid northern China since the 1970s. Geomorphology 102(3): 351–363. Available from: http://www.sciencedirect.com/science/article/pii/S0169555X08001335.

Maxwell, T.A. and Haynes, C.V. 2001. Sand sheet dynamics and quaternary landscape evolution of the selima sand sheet, southern Egypt. Quaternary Science Reviews 20(15): 1623–1647. Available from: http://www.sciencedirect.com/science/article/pii/S0277379101000099.

Mckee, E.D. 1979. Introduction to a study of global sand seas: A study of global sand seas. Professional Paper, 1–19.

Michel, R. and Avouac, J. 2002. Deformation due to the 17 august 1999 izmit, Turkey, earthquake measured from spot images. Journal of Geophysical Research: Solid Earth 107(B4): ETG 2-1–ETG 2-6. Available from: https://agupubs.onlinelibrary.wiley.com/doi/abs/10.1029/2000JB000102.

Milliner, C.W.D., Dolan, J.F., Hollingsworth, J., Leprince, S., Ayoub, F. and Sammis, C.G. 2015. Quantifying near-field and off-fault deformation patterns of the 1992 mw 7.3 landers earthquake. Geochemistry, Geophysics, Geosystems 16(5): 1577–1598. Available from: https://agupubs.onlinelibrary.wiley.com/doi/abs/10.1002/2014GC005693.

Misak, R.F. and El Shazly, M. 1982. Studies on blown sand at some localities in Sinai and northern desert, Egypt. Journal of Geology, Special Issue 1: 115–131.

Mitasova, H., Drake, T.G., Bernstein, D. and Harmon, R.S. 2004. Quantifying rapid changes in coastal topography using modern mapping techniques and geographic information system. Environmental and Engineering Geoscience 10(1): 1–11. Available from: https://doi.org/10.2113/10.1.1 [Accessed 7/28/2020].

Mitasova, H., Overton, M. and Harmon, R.S. 2005. Geospatial analysis of a coastal sand dune field evolution: Jockey's ridge, North Carolina. Geomorphology 72(1): 204–221. Available from: http://www.sciencedirect.com/science/article/pii/S0169555X0500173X.

Muñoz-Perez, J.J., Navarro, M., Roman-S., J., Tejedor, B., Rodriguez, I. and Gomez-Pina, G. 2009. Long-term evolution of a transgressive migrating dune using reconstruction of the EOF method. Geomorphology 112(1): 167–177. Available from: http://www.sciencedirect.com/science/article/pii/S0169555X09002311.

Navarro, M., Muñoz-Pérez, J.J., Román-Sierra, J., Tsoar, H., Rodríguez, I. and Gómez-Pina, G. 2011. Assessment of highly active dune mobility in the medium, short and very short term. Geomorphology 129(1): 14–28. Available from: http://www.sciencedirect.com/science/article/pii/S0169555X11000262.

Necsoiu, M., Leprince, S., Hooper, D.M., Dinwiddie, C.L., McGinnis, R.N. and Walter, G.R. 2009. Moniktoring migration rates of an active subarctic dune field using optical imagery. Remote Sensing of Environment 113: 2441–2447.

Neuman, C.M., Lanaster, N. and Nickling, W.G. 1997. Relations between dune morphology, air flow, and sediment flux on reversing dunes, silver peak, Nevada. Sedimentology 44(6): 1103–1111. Available from: https://onlinelibrary.wiley.com/doi/abs/10.1046/j.1365-3091.1997.d01-61.x.

Neuman, C.M., Lancaster, N. and Nickling, W.G. 2000. The effect of unsteady winds on sediment transport on the stoss slope of a transverse dune, silver peak, NV, USA. Sedimentology 47(1): 211–226. Available from: https://onlinelibrary.wiley.com/doi/abs/10.1046/j.1365-3091.2000.00289.x.

Nickling, W.G. 1986. Aeolian Geomorphology: Proceedings of the 17th Annual Binghamton Geomorphology Symposium, September 1986: Allen & Unwin.

Ojeda, J., Vallejo, I. and Malvarez, G.C. 2005. Morphometric evolution of the active dunes system of the Doñana National Park, Southern Spain (1977–1999). Journal of Coastal Research 49: 40–45.

Paisley, E.C.I., Lancaster, N., Gaddis, L.R. and Greeley, R. 1991. Discrimination of active and inactive sand from remote sensing: Kelso dunes, Mojave Desert, California. Remote Sensing of Environment 37(3): 153–166. Available from: http://www.sciencedirect.com/science/article/pii/003442579190078K.

Pease, P.P., Bierly, G.D., Tchakerian, V.P. and Tindale, N.W. 1999. Mineralogical characterization and transport pathways of dune sand using Landsat TM data, Wahiba sand sea, Sultanate of Oman. Geomorphology 29(3): 235–249. Available from: http://www.sciencedirect.com/science/article/pii/S0169555X9900029X.

Pye, K. and Tsoar, H. 2008. Aeolian Sand and Sand Dunes. Springer Science & Business Media.

Pye, K. and Tsoar, H. 2009. Aeolian bed forms. Aeolian Sand and Sand Dunes. Berlin, Heidelberg: Springer Berlin Heidelberg, 175–253.

Reitz, M.D., Jerolmack, D.J., Ewing, R.C. and Martin, R.L. 2010. Barchan-parabolic dune pattern transition from vegetation stability threshold. Geophysical Research Letters 37(19). Available from: https://agupubs.onlinelibrary.wiley.com/doi/abs/10.1029/2010GL044957.

Sam, L., Gahlot, N. and Prusty, B.G. 2015. Estimation of dune celerity and sand flux in part of west Rajasthan, Gadra area of the Thar desert using temporal remote sensing data. Arabian Journal of Geosciences 8(1): 4. Available from: https://doi.org/10.1007/s12517-013-1219-4.

Scheidt, S.P. and Lancaster, N. 2013. The application of COSI-CORR to determine dune system dynamics in the southern Namib desert using ASTER data. Earth Surface Processes and Landforms 38(9): 1004–1019. Available from: https://onlinelibrary.wiley.com/doi/abs/10.1002/esp.3383.

Schiek, C.G. 2004. Terrain Change Detection Using Aster Optical Satellite Imagery Along The Kunlun Fault, Tibet. The University of Texas at El Paso.

Schiek, C.G. and Hurtado, J.M. 2006. Slip analysis of the Kokoxili earthquake using terrain-change detection and regional earthquake data. Geosphere 2(3): 187–194. Available from: https://doi.org/10.1130/GES00033.1 [Accessed 7/28/2020].

Song, Y., Chen, C., Xu, W., Zheng, H., Bao, A., Lei, J., Luo, G., Chen, X., Zhang, R. and Tan, Z. 2020. Mapping the temporal and spatial changes in crescent dunes using an interferometric synthetic aperture radar temporal decorrelation model. Aeolian Research 46: 100616. Available from: http://www.sciencedirect.com/science/article/pii/S1875963720300677.

Stipho, A.S. 1992. Aeolian sand hazards and engineering design for desert regions. Quarterly Journal of Engineering Geology and Hydrogeology 25(2): 83–92.

Strang, G. and Nguyen, T. 1997. Handbook of Wavelets and Filter Banks. Wellesley-Cambridge Press, 2nd Ed.

Tsoar, H. 1974. Desert dunes morphology and dynamics, El Arish, Northern Sinai. Zeitschrift fur Geomorphologie 20: 41–61.

Tsoar, H. 1978. The Dynamics of Longitudinal Dunes: Final Technical Report: European Research Office, United States Army.

Tsoar, H. 1982. Internal structure and surface geometry of longitudinal (seif) dunes. Journal of Sedimentary Research 52(3): 823–831. Available from: https://doi.org/10.1306/212F8062-2B24-11D7-8648000102C1865D [Accessed 7/28/2020].

Tsoar, H. 1983. Dynamic processes acting on a longitudinal (seif) sand dune. Sedimentology 30(4): 567–578. Available from: https://onlinelibrary.wiley.com/doi/abs/10.1111/j.1365-3091.1983.tb00694.x.

Tsoar, H. and Karnieli, A. 1996. What determines the spectral reflectance of the Negev-Sinai sand dunes. International Journal of Remote Sensing 17(3): 513–525. Available from: https://doi.org/10.1080/01431169608949024.

Tsoar, H. and Blumberg, D.G. 2002. Formation of parabolic dunes from barchan and transverse dunes along Israel's Mediterranean coast. Earth Surface Processes and Landforms 27(11): 1147–1161. Available from: https://onlinelibrary.wiley.com/doi/abs/10.1002/esp.417.

Tsoar, H. 2004. Sand dunes. pp. 462–471. *In*: Hillel, D. (ed.). Encyclopedia of Soils in the Environment. Elsevier, Oxford.

Van Puymbroeck, N., Michel, R., Binet, R., Avouac, J. and Taboury, J. 2000. Measuring earthquakes from optical satellite images. Applied Optics 39(20): 3486–3494. Available from: http://ao.osa.org/abstract.cfm?URI=ao-39-20-3486.

Vermeesch, P. and Drake, N. 2008. Remotely sensed dune celerity and sand flux measurements of the world's fastest barchans (bodélé, chad). Geophysical Research Letters 35(24). Available from: https://agupubs.onlinelibrary.wiley.com/doi/abs/10.1029/2008GL035921.

Vermeesch, P. and Leprince, S. 2012. A 45-year time series of dune mobility indicating constant windiness over the central Sahara. Geophysical Research Letters 39(14). Available from: https://agupubs.onlinelibrary.wiley.com/doi/abs/10.1029/2012GL052592.

Walden, J. and White, K. 1997. Investigation of the controls on dune colour in the Namib sand sea using mineral magnetic analyses. Earth and Planetary Science Letters 152(1): 187–201. Available from: http://www.sciencedirect.com/science/article/pii/S0012821X97001544.

Wang, X., Dong, Z., Qu, J., Zhang, J. and Zhao, A. 2003. Dynamic processes of a simple linear dune—a study in the Taklimakan sand sea, China. Geomorphology 52(3): 233–241. Available from: http://www.sciencedirect.com/science/article/pii/S0169555X02002581.

Wang, X., Dong, Z., Zhang, J. and Qu, J. 2004. Formation of the complex linear dunes in the central Taklimakan sand sea, China. Earth Surface Processes and Landforms 29(6): 677–686. Available from: https://onlinelibrary.wiley.com/doi/abs/10.1002/esp.1063.

Warren, A. and Knott, P. 1983. Desert dunes: A short review of meeds in desert dune research and a recent study of micrometeorological dune-initiation mechanisms. pp. 343–352. *In*: Brookfield, M.E. and Ahlbrandt, T.S. (eds.). Developments in Sedimentology. Elsevier.

Warren, A. and Kay, S. 1987. Dune networks. Geological Society, London, Special Publications 35(1): 205–212. Available from: https://sp.lyellcollection.org/content/specpubgsl/35/1/205.full.pdf.

Warren, A. 1988. The dunes of the Wahiba sands. Journal of Oman Studies 3: 1985–1987.

Wasson, R.J. and Hyde, R. 1983. Factors determining desert dune type. Nature 304(5924): 337–339.

Watson, A. 1985. The control of wind blown sand and moving dunes: A review of the methods of sand control in deserts, with observations from Saudi Arabia. Quarterly Journal of Engineering Geology and Hydrogeology 18(3): 237. Available from: http://qjegh.lyellcollection.org/content/18/3/237.abstract.

Weydahl, D.J. 2001. Analysis of Ers SAR coherence images acquired over vegetated areas and urban features. International Journal of Remote Sensing 22(14): 2811–2830. Available from: https://doi.org/10.1080/01431160010006412.

White, K., Walden, J., Drake, N., Eckardt, F. and Settlell, J. 1997. Mapping the iron oxide content of dune sands, Namib sand sea, Namibia, using Landsat thematic mapper data. Remote Sensing of Environment 62(1): 30–39. Available from: http://www.sciencedirect.com/science/article/pii/S0034425797000680.

Wiggs, G.F.S. 1993. Desert dune dynamics and the evaluation of shear velocity: An integrated approach. Geological Society, London, Special Publications 72(1): 37–46. Available from: https://sp.lyellcollection.org/content/specpubgsl/72/1/37.full.pdf.

Wiggs, G.F.S., Livingstone, I. and Warren, A. 1996. The role of streamline curvature in sand dune dynamics: Evidence from field and wind tunnel measurements. Geomorphology 17(1): 29–46. Available from: http://www.sciencedirect.com/science/article/pii/0169555X9500093K.

Wilkins, D.E. and Ford, R.L. 2007. Nearest neighbor methods applied to dune field organization: The coral pink sand dunes, Kane county, Utah, USA. Geomorphology 83(1): 48–57. Available from: http://www.sciencedirect.com/science/article/pii/S0169555X06002406.

Wilson, I.G. 1972. Aeolian bedforms—their development and origins. Sedimentology 19(3-4): 173–210. Available from: https://onlinelibrary.wiley.com/doi/abs/10.1111/j.1365-3091.1972.tb00020.x.

Wulder, M.A., White, J.C., Loveland, T.R., Woodcock, C.E., Belward, A.S., Cohen, W.B., Fosnight, E.A., Shaw, J., Masek, J.G. and Roy, D.P. 2016. The global landsat archive: Status, consolidation, and direction. Remote Sensing of Environment 185: 271–283. Available from: http://www.sciencedirect.com/science/article/pii/S0034425715302194.

Xia, J. and Dong, P. 2016. A GIS add-in for automated measurement of sand dune migration using Lidar-derived multitemporal and high-resolution digital elevation models. Geosphere 12(4): 1316–1322. Available from: https://doi.org/10.1130/GES01329.1 [Accessed 8/1/2020].

Yao, Z.Y., Wang, T., Han, Z.W., Zhang, W.M. and Zhao, A.G. 2007. Migration of sand dunes on the northern Alxa plateau, inner Mongolia, China. Journal of Arid Environments 70(1): 80–93. Available from: http://www.sciencedirect.com/science/article/pii/S0140196307000092.

Yizhaq, H., Ashkenazy, Y. and Tsoar, H. 2009. Sand dune dynmaic and climate change: A modeling approach. Journal of Geophysical Research 114: 1–11.

Zebker, H.A. and Goldstein, R.M. 1986. Topographic mapping from interferometric synthetic aperture radar observations. Journal of Geophysical Research: Solid Earth 91(B5): 4993–4999. Available from: https://agupubs.onlinelibrary.wiley.com/doi/abs/10.1029/JB091iB05p04993.

Zebker, H.A. and Villasenor, J. 1992. Decorrelation in interferometric radar echoes. IEEE Transactions on Geoscience and Remote Sensing 30(5): 950–959.

Zebker, H.A., Rosen, P.A., Goldstein, R.M., Gabriel, A.K. and Werner, C.L. 1994a. On the derivation of coseismic displacement fields using differential radar interferometry: The landers earthquake. Journal of Geophysical Research: Solid Earth 99(B10): 19617–19634. Available from: https://agupubs.onlinelibrary.wiley.com/doi/abs/10.1029/94JB01179.

Zebker, H.A., Werner, C.L., Rosen, P.A. and Hensley, S. 1994b. Accuracy of topographic maps derived from Ers-1 interferometric radar. IEEE Transactions on Geoscience and Remote Sensing 32(4): 823–836.

Chapter 3

A Review of the Development in the Remote Sensing of Sand Dunes

Shamsudeen Temitope Yekeen,[1,2] *Abdul-lateef Balogun*[1] and *Yusuf A. Aina*[3,*]

1. Introduction

Globally, a sand dune is one of the most prominent geomorphological arid features. Despite its dynamic nature, the study of a sand dune can provide information on the climatic and geological landscape of an area (Michel et al., 2018). Sand dunes serve as a modern and ancient global environmental landmark (Lancaster, 1989). The formation of the dune is primarily based on the accumulation of transported sediments by wind force (Tsoar, 2001). Dunes could be bare and mobile or vegetated and stable (fixed), while the former is mostly a result of suppressed deposition and erosion activities the latter are commonly populated by different vegetation types (Dong et al., 2013). The absence of vegetation, landform, topography, hydrological and tectonic level, as well as local and regional climate, contribute greatly to the formation of a sand dune (Rahimzadeh et al., 2019). Despite the negative effects, the presence of sand dunes serves as an important habitat for psammophilic fauna and flora as well as natural groundwater storage. With the improvement in detection-technology, areas with sand dunes can also be important sources of petroleum (oil and gas). Exploration of desert areas, for petroleum and mineral deposits, has accelerated from the 1970s.

Understanding the morphodynamics of these dune features requires sophisticated tools. In the past, comparison between observed geometry and changes in dune bedform in addition to sand flux predicted with General Circulation Models (GCM) were used to identify sand dune dynamics (Bazzichetto et al., 2016; Newman et al., 2017; Duffy et al., 2018) with recorded inaccuracy due to a number of factors like uncleared wind speed threshold (Michel et al., 2018). Due to this limitation, remote sensing technology which has the ability to capture data without directly coming into contact with the object has been identified as a promising tool for sand dune study (see also Hermas et al., and Dandabathula and Gaur, this volume). Earlier studies in the use of remote sensing were mainly for dune taxonomy and mapping (McKee, 1979). The advancement in computer software and hardware in the 1990s also led to the use of spectral attribute for dune surface monitoring (Pease et al., 1999; Blumberg, 1998), surface change (Brown and Arbogast, 1999) quantitative morpho-dynamic analysis (Bourke et al., 2009; Necsoiu et al., 2009) and dune pattern analysis (Kocurek et al., 2010; Hugenholtz and Barchyn, 2010). In this chapter, we review the state-of-the-art remote

[1] Geospatial Analysis and Modelling (GAM) Research Group, Department of Civil and Environmental Engineering, Universiti Teknologi PETRONAS (UTP), 32610 Seri Iskandar, Perak, Malaysia.
[2] Department of Geography, Environment and Geomatics, University of Guelph, Guelph, ON, Canada.
[3] Department of Geomatics Engineering Technology, Yanbu Industrial College, Yanbu, Saudi Arabia.
* Corresponding author: yusufaina@gmail.com

sensing technology for sand dune studies that include identification, movement monitoring and topographic change analysis.

2. Environmental impact of sand dunes

Globally, approximately 54 million km^2 is classified as dryland, with a high percentage of 58.5% located in Asia and Africa. Sand dunes are seen as an important habitat for specialized flora and fauna as well as a source of different mineral resources such as lithium and boron, among others. The absence of high surface runoff contributes to the preservation of these resources in the sand dune region (Varma et al., 2014). However, dune sands can have low soil nutrient level and low water retaining capability (Todd, 2006; Tarhouni et al., 2010). The low fertility of sands, coupled with the presence of wind soil erosion, may lead to land degradation and be a threat to food security if sand encroaches onto cropland (Montgomery, 2012). Equally, the presence of mobile sand dunes affects existing infrastructures like railway, roads and human settlement (Amin and Seif, 2019).

3. Remote sensing of sand dunes

A sand dune is one of the most amazing landforms globally and remote sensing has been identified as a promising tool for understanding the formation, movement and changes (Dong, 2015; Bourke et al., 2010). This is due to its ability to cover a large land area and is time-saving and cost-efficient (Lam et al., 2011). Remote sensing technology for sand dune monitoring could be optical and microwave sensors which could be airborne or space-borne (van der Meer et al., 2012; Ben-Dor et al., 2008b; Li et al., 2013). The enhancement in the development of these remote sensing tools has led to various standardized geospatial databases for free data download that aids the study of a sand dune as illustrated in Table 3.1. The visible region of sand dune varies from each sensor,

Table 3.1. Different sand dune remote sensing sensors.

S. No.	Sensor	Sensor period	Resolution	Bands	Operator
1	Corona	1960–1972	2–7.5 m	Panchromatic	Central Intelligence Agency Directorate of Science and Technology
2	Landsat 1, 2, 3, 4, 5, 6, 7, 8, and proposed 9	1972-present (Landsat 9 is proposed to start operation by early 2021)	15–120 m	Up to 12 bands	National Aeronautics and Space Administration (NASA)
3	Moderate Resolution Imaging Spectroradiometer (MODIS)	1999-present	250–1000 m	36 bands	National Aeronautics and Space Administration (NASA)
4	Advance Spaceborne Thermal Emission and Reflection Radiometer (ASTER)	1999-present	15–90 m	14 bands	National Aeronautics and Space Administration (NASA) and Japan's Ministry of Economy, Trade, and Industry (METI)
5	Shuttle Radar Topography Mission (SRTM)	2000-present	30–90 m	C and X-band	National Aeronautics and Space Administration (NASA)
6	ASTER Global Digital Elevation Model (ASTER GDEM)	2009–present	30 m	C-band	National Aeronautics and Space Administration (NASA) and Japan's Ministry of Economy, Trade and Industry (METI)
7	Sentinel 1 and 2	2013 till date	10–60 m	Up to 12 bands	European Space Agency

for example, multi-spectral Landsat thematic mapper imagery at less than 62 μm, SPOT $_{PAN}$ 0.48–0.71 μm, ASTER VNIR 0.52–0.86 μm (Necsoiu et al., 2009). Radebaugh et al. (2010) also identified that sand dunes are domiciled within the ± 30° equation on a synthetic apparatus radar spacecraft, visual and infrared mapping spectrometer at 1–5 μm and imaging science system at 0.94 μm.

3.1 Remote sensing of sand dune activities and characteristics

The monitoring of sand dune activities and characteristics involves the observation and quantification of surface changes due to the different transportation agents. Sand dune change either referred to as *migration* or *activities* which are mostly used interchangeably (McKee, 1979; Hugenholtz and Barchyn, 2010). Studies have stated that change does not always mean migration but behavior dynamics which could mean the extension for a linear dune and/or accumulation for a star dune (Hugenholtz and Barchyn, 2010). This indicates that the redistribution of sand dune mainly leads to formation changes but not necessarily dune migration. To understand sand dune activities, remote sensing technology has been indicated as a productive tool for the quantification and monitoring of changes in the dune giving information on the trajectory in relation to different condition factors. For monitoring to be achieved, identification of the appropriate remote sensing technologies used for examining the presence of a dune, morphological movement and topographical changes needs to the undertaken.

3.1.1 Sand dune detection

The development of remote sensing technology has made it possible for the study of sand dunes present over a large area at lower cost and time. A sand dune possesses a unique attribute in the visible (VIS) 0.4–0.7 μm, near-infrared (NIR) 0.7–11 μm and short-wave infrared (SWIR) 1.1–2.5 μm spectral region of remote sensing (Karnieli et al., 2001; Karnieli et al., 2002). The application of these sensors has been demonstrated in different studies (Stevens et al., 2010; Ben-Dor et al., 2008a; Gomez et al., 2008; Stevens et al., 2008; Ben-Dor et al., 2009), however, a major shortfall is the dependence on the sun as the energy source affecting its applicability at all times under any weather condition. In recent year, thermal infrared (TIR) wavelength regions between 8 and 14 μm (longwave infrared) LWIR are being used equally for the identification of sand dune mapping due to its high sensitivity to different soil properties, in addition to the inherent self-emission making it possible to capture images under any weather condition and at any period (Eisele et al., 2012; Eisele et al., 2015). The difference in the land surface temperature of the sand dune region from other cover types also contributes to its applicability. The application of LWIR has been well documented in literature (Yitagesu et al., 2011; Notesco et al., 2014; Hecker et al., 2012; Hecker et al., 2010; Eisele et al., 2015; Eisele et al., 2012).

The instability of a sand dune, influenced by its interaction with crust and vegetation, causes reduction in dune detection accuracy from remote sensing images. Image classification and segmentation algorithms such as, maximum likelihood, minimum distance, parallelepiped, object-based image analysis, edge-based segmentation and point-based segmentation have been applied to solve the problem of dune detection and identification (Flor-Blanco et al., 2013; Azzaoui et al., 2019). In recent times, artificial intelligence tools especially machine and deep learning algorithms have been introduced in sand dune studies to improve the detection and mapping of sand dune dynamics (Aydda et al., 2020; Azzaoui et al., 2019; Horrall et al., 2019; Kochanski et al., 2019; Roushangar et al., 2017). Machine learning models like Support Vector Machine (SVM), Artificial Neural Network (ANN), logistic regression, deep learning Convolutional Neural Network (CNN) are mostly used for the automation of the sand dune detection study due to their ability to provide high detection accuracy with low data (Bandeira et al., 2011). However, the spatial and spectral property of satellite images used equally contributes to the higher accuracies attained by the different machine learning models. While the results of the adoption of machine and deep learning models have been successful, the field is still evolving and might take some time to be fully integrated into sand dune studies.

3.1.2 Sand dune morphological movement

The dynamic of a landscape is attributed to different Earth surface processes and human activities. A major dynamic of the Earth's surface is the movement of a sand dune by wind erosion, which indicates active sand transportation. Sand dunes move at different rates and directions due to the aeolian transportation (typically the main process that results in the change in sand dune topography). Equally, the large difference between evaporation and precipitation rate as well as the scarcity in vegetation contributes to the morphological movement of a sand dune. It is therefore important to determine the movement rate and the spatial variability, to aid the conservation of both the natural and artificial resources present in a region. Earlier the use of ground-based techniques was predominant which involves the measurement of the distance between two lines that present the successive positions of the dune (Parker Gay, 1999), which is also known as nose to nose by (Bailey and Bristow, 2004), through the use of linen tape, steel, iron, plastic rope and GPS integrated total station (Hermas et al., 2012). Although a high level of accuracy was recorded in the use of this method, a major constraint is the high cost and time consumed in addition to a limited area of coverage. This makes it difficult to evaluate the temporal and spatial dynamics of sand dune movement. Thus, the use of remote sensing technology due to its wide spatial extent, is time saving and less expensive.

Remote sensing for dune migration is based on the comparison of different multi-temporal satellite images of a particle's place. Both optical and microwave sensors are being adopted due to the crucial insight on the climatic and geological conditions that resulted in its migration and formation. Examples of optical remote sensing images that have been adopted for sand dune movement monitoring include the study of Michel et al. (2018) who used SPOT images of two different years at an interval of 10 years to identify the movement of a sand dune in Qatar and Ghadiry et al. (2012) used SPOT images of 1995 and 2007 to identify sand dune movement in Dakhla Oases, western desert of Egypt. Xia and Dong (2016) used the Digital Elevation Model (DEM) derived from a light detection and ranging (LiDAR) data for evaluating annual sand dune estimation migration rate at the white sand dune field in Mexico. Dong (2015) also used LiDAR generated terrain model to monitor the movement direction, migration distance and rate of a sand dune. The study of Sam et al. (2013) used 2-year Catosat-1 scenes (Stereopairs) in the measurement of sand dune migration rate in the Thar (Gaur, this volume). Levin and Ben-Dor (2004) used gray aerial photographs of 12 years between 1944 and 1999 to examine the sand dune movement in Ashdod-Nizanim, Israel (Hezi et al., this volume). Other studies like Baird et al. (2019) used Landsat image indicating that the standardized acquisition scene enables image mosaicking for the evaluation of sand dune movement over a larger area. Despite the recorded success rate, a major concern is the difficulty in identifying small changes in term of the physical geometry of sand dunes due to the lower sensitivity of an optical sensor as sand encroachment also has similar spectral reflectance (Song et al., 2020).

In contrast to optical remote sensing, the microwave sensor, most especially radar, has a high level of sensitivity, making it easy to identify small changes. This is because the radar backscattering is highly sensitive to the geometry of an object which enables it to identify small changes in objects. This has relevance to a sand dune. The microwave sensor also has an all-weather and all-time capability. The development of Synthetic Aperture radar (SAR) has contributed to the mapping of sand dune movement. The application of SAR imagery for sand dune movement has been widely recorded in literature (Gómez et al., 2018; Abdelkareem and El-Baz, 2017; Ma et al., 2020; Schmugge et al., 1992; Necsoiu et al., 2013). The interferometric synthetic aperture radar (InSAR) is widely used today due to its powerful capability to describe the movement of sand dunes temporally and spatially (Havivi et al., 2018; Gaber et al., 2018; Maghsoudia et al., 2013; Song et al., 2020).

3.1.3 Sand dune topographical changes

Aeolian transportation is usually the main process that results in the change of sand dune topography. LiDAR airborne remote sensing has been predominantly used because of less human effort that is required for the production of dune topographic change unlike conventional analytical (digital photogrammetry). LiDAR enables the detail and accurate acquisition of sand dune topographic data over a broad spatial extent (Brock et al., 1999; Stockdonf et al., 2002; Gutierrez et al., 2001; Sallenger et al., 2003). Its application has been widely accepted over time which is evident by the study of Mitasova et al. (2009) that detects, analyzes and quantifies the change in the topography in a rapidly evolving landscape using multitemporal sets of LiDAR data. Revell et al. (2011), Loftis et al. (2016), Allen et al. (2012) also used airborne LiDAR for change detection in the topography of dune sites. Unlike other studies, where in order to provide a more accurate and robust monitoring tool, Brownett and Mills (2017) integrated both LiDAR image and Compact Airborne Spectrographic Imager (CASI) to map and understand the changes in sand dune sites around the UK. Despite the relative significant advantage of the airborne LiDAR, its high cost of operation remains a major limitation.

Advancement in technology led to the use of an Umanned Aerial Vehicle (UAV) for sand dune topography monitoring due to its high level of efficiency, cost-saving, deployment flexibility and large spatial-temporal extent. Studies on the use of UAVs have mainly centered on the use of the Structure from Motion (SfM) approach for topographic modeling which is based on the computer vision process of image-to-image registration method (Sofonia et al., 2019; Mancini et al., 2013). Sofonia et al. (2019), Mancini et al. (2013), Fonstad et al. (2013), Verhoeven et al. (2009), D'Oleire-Oltmanns et al. (2012), Mathews and Jensen (2013), Wallace et al. (2012) documented that the use of its generated digital surface models (DEM) with a sub-meter resolution for the effective creation, detection and measurement of changes in a dune. The high spatial resolution of the UAV enables the penetration through dune vegetation to create an accurate digital terrain model that facilitates more direct measurement of dune geomorphic change.

4. Lessons learnt

The monitoring of sand dune activities deals with the understanding of the present, and any changes that could occur. Although a sand dune has both positive and negative effects, its adverse effect has been more widely reported due to its large implication on human existence. The paradigm shift in the understanding of sand dune activities has been centered on the advancement of remote sensing technology. Limitation in accuracy, efficiency, cost, time and spatial-extent and temporal has contributed to the use of remote sensing technology (Bello and Aina, 2014). Remote sensing technology can either be a microwave sensor or an optical sensor. While the former is not affected by operation time, weather conditions and cloud cover, the latter is largely limited in application due to its reliance on the sun for illumination to acquire images. The relative short-temporal capability of tradition field survey resulted in the use of long archive aerial photography for the identification of sand dune presence and for assessing change. Although while VIS, NIR and SWIR spectral are able to identify the existence of a sand dune, limited use-period led to the adoption of the TIR (LWIR) sensor. However, a major constraint still remains in the poor parameterization of areas where vegetation interferes with sand dune resulting in mixing pixel problems (Aina et al., 2019). Knowing that remote sensing technology and its application is advancing, seasonal changes in the appearance of a sand dune is still limited in existing studies which give more information on the temporal identification.

Active sand dunes are identified due to their obvious movement pattern. While the use of ground-based techniques is invaluable for determining the movement pattern and change detection, its limited area coverage, high cost and time consumption affect its use. Remote sensing enables the understanding of long temporal migration patterns in terms of distance and rate. Successful applications of SPOT, Landsat and LiDAR sensor have been well documented but the low sensitivity to small changes resulted in the use of radar SAR images with a high level of sensitivity. Image segmentation and artificial intelligence models (machine learning and deep learning) have shown a large capacity in providing accurate detection of a sand dune. Although artificial intelligence models have been identified to be promising techniques for sand dune detection, however, its application is still very limited compared to other areas of remote sensing. The substantial advancement in the area of remote sensing has enhanced the ability to capture the change in sand dune topography that allows one to understand the dynamics of sand and migration of sand dunes. LiDAR airborne remote sensing was mostly used in the past replacing the conventional digital photogrammetry. However, its high cost of operation and limited spatial resolution led to the use of unmanned aerial vehicle (UAV). The UAV's used have been centered on the structure from motion (SfM) approach which is based on the computer vision process of image-to-image registration method.

5. Future outlook and conclusion

This chapter has confirmed the significant development in the use of remote sensing technology for sand dune study that includes the identification, movement and the topographical change. Over time, substantial progress has been made to enhance the study of a sand dune in a more effective, cost and time saving and wide spatial extent ways. However, each of the remote sensing technology used today is limited in its own way. As a future direction, there is a need for the development of remote sensing technology that will aid the parametrization of the differences between sand dune and vegetation. The need for a more high-resolution sensor for the capturing of relatively short interval migration of sand dune is also there. Equally, the accessibility and availability of high-quality data will further enhance research on sand dunes. The application of machine and deep learning models in sand dune studies is a promising area that should be further explored in reducing the current knowledge gap as well as the use of cloud-based systems like the evolving Google Earth Engine.

References

Abdelkareem, M. and El-Baz, F. 2017. Remote sensing of paleodrainage systems west of the Nile River, Egypt. Geocarto International 32: 541–555.

Aina, Y.A., Adam, E., Ahmed, F., Wafer, A. and Alshuwaikhat, H.M. 2019. Using multisource data and the VIS model in assessing the urban expansion of Riyadh city, Saudi Arabia. European Journal of Remote Sensing 52(1): 557–571.

Allen, T.R., Oertel, G.F. and Gares, P.A. 2012. Mapping coastal morphodynamics with geospatial techniques, Cape Henry, Virginia, USA. Geomorphology 137: 138–149.

Amin, A. and Seif, E.-S.S.A. 2019. Environmental hazards of sand dunes, South Jeddah, Saudi Arabia: An assessment and mitigation geotechnical study. Earth Systems and Environment 3: 173–188.

Aydda, A., Althuwaynee, O. and Pokharel, B. 2020. An easy method for barchan dunes automatic extraction from multispectral satellite data. IOP Conference Series: Earth and Environmental Science, IOP Publishing, 012015.

Azzaoui, M.A., Masmoudi, L., El Belrhiti, H. and Chaouki, I.E. 2019. Segmentation of crescent sand dunes in high resolution satellite images using a support vector machine for allometry. International Journal of Advanced Computer Science and Applications, 191–198.

Bailey, S.D. and Bristow, C.S. 2004. Migration of parabolic dunes at Aberffraw, Anglesey, north Wales. Geomorphology 59: 165–174.

Baird, T., Bristow, C.S. and Vermeesch, P. 2019. Measuring sand dune migration rates with COSI-Corr and Landsat: Opportunities and challenges. Remote Sensing 11(20): 2423.

Bandeira, L., Marques, J.S., Saraiva, J. and Pina, P. 2011. Automated detection of martian dune fields. IEEE Geoscience and Remote Sensing Letters 8: 626–630.

Bazzichetto, M., Malavasi, M., Acosta, A. and Carranza, M.L.J.E.I. 2016. How does dune morphology shape coastal EC habitats occurrence? A remote sensing approach using airborne LiDAR on the Mediterranean coast. Ecological Indicators 71: 618–626.

Bello, O.M. and Aina, Y.A. 2014. Satellite remote sensing as a tool in disaster management and sustainable development: towards a synergistic approach. Procedia-Social and Behavioral Sciences 120: 365–373.

Ben-Dor, E., Taylor, R., Hill, J., Demattê, J., Whiting, M., Chabrillat, S. and Sommer, S. 2008a. Imaging spectrometry for soil applications. Advances in Agronomy 97: 321–392.

Ben-Dor, E., Taylor, R.G., Hill, J., Demattê, J.A.M., Whiting, M.L., Chabrillat, S. and Sommer, S. 2008b. Imaging Spectrometry for Soil Applications. Advances in Agronomy. Academic Press.

Ben-Dor, E., Chabrillat, S., Demattê, J., Taylor, G., Hill, J., Whiting, M. and Sommer, S. 2009. Using imaging spectroscopy to study soil properties. Remote Sensing of Environment 113: S38–S55.

Blumberg, D.G. 1998. Remote sensing of desert dune forms by polarimetric synthetic aperture radar (SAR). Remote Sensing of Environment 65: 204–216.

Bourke, M.C., Ewing, R.C., Finnegan, D. and Mcgowan, H.A. 2009. Sand dune movement in the Victoria Valley, Antarctica. Geomorphology 109: 148–160.

Bourke, M.C., Lancaster, N., Fenton, L.K., Parteli, E.J., Zimbelman, J.R. and Radebaugh, J. 2010. Extraterrestrial dunes: An introduction to the special issue on planetary dune systems. Geomorphology 121: 1–14.

Brock, J., Sallenger, A., Krabill, W., Swift, R., Manizade, S., Meredith, A., Jansen, M. and Eslinger, D. 1999. Aircraft laser altimetry for coastal process studies. Coastal Sediments. ASCE, 2414–2428.

Brown, D.G. and Arbogast, A.F. 1999. Digital photogrammetric change analysis as applied to active coastal dunes in Michigan. Photogrammetric Engineering Remote Sensing 65: 467–474.

Brownett, J.M. and Mills, R.S. 2017. The development and application of remote sensing to monitor sand dune habitats. Journal of Coastal Conservation 21: 643–656.

D'oleire-Oltmanns, S., Marzolff, I., Peter, K.D. and Ries, J.B. 2012. Unmanned aerial vehicle (UAV) for monitoring soil erosion in Morocco. Remote Sensing 4: 3390–3416.

Dong, P. 2015. Automated measurement of sand dune migration using multi-temporal lidar data and GIS. International Journal of Remote Sensing 36: 5426–5447.

Dong, Z., Qian, G., Lv, P. and Hu, G. 2013. Investigation of the sand sea with the tallest dunes on Earth: China's Badain Jaran Sand Sea. Earth-Science Reviews 120: 20–39.

Duffy, J.P., Shutler, J.D., Witt, M.J., Debell, L. and Anderson, K.J.R.S. 2018. Tracking fine-scale structural changes in coastal dune morphology using kite aerial photography and uncertainty-assessed structure-from-motion photogrammetry. Remote Sensing 10: 1494.

Eisele, A., Lau, I., Hewson, R., Carter, D., Wheaton, B., Ong, C., Cudahy, T.J., Chabrillat, S. and Kaufmann, H. 2012. Applicability of the thermal infrared spectral region for the prediction of soil properties across semi-arid agricultural landscapes. Remote Sensing 4(11): 3265–3286.

Eisele, A., Chabrillat, S., Hecker, C., Hewson, R., Lau, I.C., Rogass, C., Segl, K., Cudahy, T.J., Udelhoven, T., Hostert, P. and Kaufmann, H. 2015. Advantages using the thermal infrared (TIR) to detect and quantify semi-arid soil properties. Remote Sensing of Environment 163: 296–311.

Flor-Blanco, G., Flor, G. and Pando, L.J.G.-M.L. 2013. Evolution of the Salinas-El Espartal and Xagó beach/dune systems in north-western Spain over recent decades: evidence for responses to natural processes and anthropogenic interventions. Geo-Marine Letters 33: 143–157.

Fonstad, M.A., Dietrich, J.T., Courville, B.C., Jensen, J.L. and Carbonneau, P.E. 2013. Topographic structure from motion: a new development in photogrammetric measurement. 38: 421–430.

Gaber, A., Abdelkareem, M., Abdelsadek, I.S., Koch, M. and El-Baz, F. 2018. Using InSAR coherence for investigating the interplay of fluvial and aeolian features in arid lands: implications for groundwater potential in Egypt. Remote Sensing 10: 832.

Ghadiry, M., Shalaby, A. and Koch, B. 2012. A new GIS-based model for automated extraction of Sand Dune encroachment case study: Dakhla Oases, western desert of Egypt. The Egyptian Journal of Remote Sensing and Space Science 15: 53–65.

Gomez, C., Viscarra Rossel, R.A. and Mcbratney, A.B. 2008. Soil organic carbon prediction by hyperspectral remote sensing and field vis-NIR spectroscopy: An Australian case study. Geoderma 146: 403–411.

Gómez, D., Salvador, P., Sanz, J., Casanova, C. and Casanova, J.L. 2018. Detecting areas vulnerable to sand encroachment using remote sensing and GIS techniques in Nouakchott, Mauritania. 10: 1541.

Gutierrez, R., Gibeaut, J., Smyth, R., Hepner, T., Andrews, J., Weed, C., Gutelius, W. and Mastin, M. 2001. Precise airborne lidar surveying for coastal research and geo-hazards applications. International Archives of Photogrammetry Remote Sensing Spatial Information Sciences 34: 185–194.

Havivi, S., Amir, D., Schvartzman, I., August, Y., Maman, S., Rotman, S.R. and Blumberg, D.G. 2018. Mapping dune dynamics by InSAR coherence. Earth Surface Processes Landforms 43: 1229–1240.

Hecker, C., Van Der Meijde, M. and Van Der Meer, F.D. 2010. Thermal infrared spectroscopy on feldspars—Successes, limitations and their implications for remote sensing. Earth-Science Reviews 103: 60–70.

Hecker, C., Dilles, J.H., Van Der Meijde, M. and Van Der Meer, F.D. 2012. Thermal infrared spectroscopy and partial least squares regression to determine mineral modes of granitoid rocks. Geochemistry, Geophysics, Geosystems 13.

Hermas, E., Leprince, S. and El-Magd, I.A. 2012. Retrieving sand dune movements using sub-pixel correlation of multi-temporal optical remote sensing imagery, northwest Sinai Peninsula, Egypt. Remote Sensing of Environment 121: 51–60.

Hezi, Y., Dubinin, V. and Yosef Ashkenazy, Y. (in press). Human and climate impacts on the mobility of Nizzana sand dunes in Israel (volume 2 of this book).

Horrall, J., Mohan, D., Kochanski, K. and Rountree, B.J.A. 2019. Machine Learning Application to Sand Dune Model Prediction Using Generative Adversarial Networks (GANs), 2019, EP54C-14.

Hugenholtz, C.H. and Barchyn, T.E. 2010. Spatial analysis of sand dunes with a new global topographic dataset: new approaches and opportunities. Earth Surface Processes Landforms 35: 986–992.

Karnieli, A., Gabai, A., Ichoku, C., Zaady, E. and Shachak, M. 2002. Temporal dynamics of soil and vegetation spectral responses in a semi-arid environment. International Journal of Remote Sensing 23: 4073–4087.

Karnieli, A., Kokaly, R., West, N. and Clark, R. 2001. Remote Sensing of Biological Soil Crusts. Biological Soil Crusts: Structure, Function, and Management. Springer.

Kochanski, K., Mohan, D., Horrall, J., Rountree, B. and Abdulla, G.J.A.P.A. 2019. Deep learning predictions of sand dune migration. arXiv:191210798 [cs LG].

Kocurek, G., Ewing, R.C. and Mohrig, D. 2010. How do bedform patterns arise? New views on the role of bedform interactions within a set of boundary conditions. Earth Surface Processes Landforms 35: 51–63.

Lam, D.K., Remmel, T.K. and Drezner, T.D. 2011. Tracking desertification in California using remote sensing: A sand dune encroachment approach. Remote Sensing 3: 1–13.

Lancaster, N. 1989. The Namib Sand Sea: Dune Forms, Processes and Sediments. Balkema.

Levin, N. and Ben-Dor, E. 2004. Monitoring sand dune stabilization along the coastal dunes of Ashdod-Nizanim, Israel, 1945–1999. Journal of Arid Environments 58: 335–355.

Li, Z.-L., Tang, B.-H., Wu, H., Ren, H., Yan, G., Wan, Z., Trigo, I.F. and Sobrino, J.A. 2013. Satellite-derived land surface temperature: Current status and perspectives. Remote Sensing of Environment 131: 14–37.

Loftis, J.D., Wang, H.V., Deyoung, R.J. and Ball, W.B. 2016. Using lidar elevation data to develop a topobathymetric digital elevation model for sub-grid inundation modeling at langley research center. Journal of Coastal Research 134–148.

Ma, D., Liu, Y., Li, Z., Bao, H. and Jin, Z. 2020. Recognition of the coastal dune migration micro-deformation in Changli Gold Coast of China based on GB-InSAR. Marine Georesources & Geotechnology 1–9.

Maghsoudia, M., Hajizadehb, A., Nezammahallehb, M., Seyedrezaic, H., Jalalid, A. and Mahzoune, M. 2013. Interferometric synthetic aperture radar (INSAR) technology and geomorphology interpretation. International Archives of the Photogrammetry, Remote Sensing Spatial Information Sciences 1: W3.

Mancini, F., Dubbini, M., Gattelli, M., Stecchi, F., Fabbri, S. and Gabbianelli, G. 2013. Using unmanned aerial vehicles (UAV) for high-resolution reconstruction of topography: the structure from motion approach on coastal environments. Remote Sensing 5: 6880–6898.

Mathews, A.J. and Jensen, J.L. 2013. Visualizing and quantifying vineyard canopy LAI using an unmanned aerial vehicle (UAV) collected high density structure from motion point cloud. Remote Sensing 5: 2164–2183.

Mckee, E.D. 1979. A study of global sand seas. Professional Paper 1052 USGS, Washington.

Michel, S., Avouac, J.-P., Ayoub, F., Ewing, R.C., Vriend, N. and Heggy, E. 2018. Comparing dune migration measured from remote sensing with sand flux prediction based on weather data and model, a test case in Qatar. Earth and Planetary Science Letters 497: 12–21.

Mitasova, H., Overton, M.F., Recalde, J.J., Bernstein, D.J. and Freeman, C.W. 2009. Raster-based analysis of coastal terrain dynamics from multitemporal lidar data. Journal of Coastal Research 25: 507–514.

Montgomery, D.R. 2012. Dirt: The Erosion of Civilizations. Univ of California Press.

Necsoiu, M., Leprince, S., Hooper, D.M., Dinwiddie, C.L., Mcginnis, R.N. and Walter, G.R. 2009. Monitoring migration rates of an active subarctic dune field using optical imagery. Remote Sensing of Environment 113: 2441–2447.

Necsoiu, M., Longépé, N. and Hooper, D.M. 2013. A new methodology to monitor soil moisture over a complex arctic environment, Kobuk River valley, Alaska. Remote Sensing Letters 4: 251–260.

Newman, C.E., Gómez-Elvira, J., Marin, M., Navarro, S., Torres, J., Richardson, M.I., Battalio, J.M., Guzewich, S.D., Sullivan, R. and De La Torre, M. 2017. Winds measured by the Rover Environmental Monitoring Station (REMS) during the Mars Science Laboratory (MSL) rover's Bagnold Dunes Campaign and comparison with numerical modeling using MarsWRF. Icarus 291: 203–231.

Notesco, G., Kopačková, V., Rojík, P., Schwartz, G., Livne, I. and Dor, E.B. 2014. Mineral classification of land surface using multispectral LWIR and hyperspectral SWIR remote-sensing data: a case study over the Sokolov lignite open-pit mines, the Czech Republic. Remote Sensing 6: 7005–7025.

Parker Gay, S. 1999. Observations regarding the movement of barchan sand dunes in the Nazca to Tanaca area of southern Peru. Geomorphology 27: 279–293.

Pease, P.P., Bierly, G.D., Tchakerian, V.P. and Tindale, N.W. 1999. Mineralogical characterization and transport pathways of dune sand using Landsat TM data, Wahiba Sand Sea, Sultanate of Oman. Geomorphology 29: 235–249.

Radebaugh, J., Lorenz, R., Farr, T., Paillou, P., Savage, C. and Spencer, C. 2010. Linear dunes on Titan and earth: Initial remote sensing comparisons. Geomorphology 121: 122–132.

Rahimzadeh, N., Khormali, F., Gribenski, N., Tsukamoto, S., Kehl, M., Pint, A., Kiani, F. and Frechen, M. 2019. Timing and development of sand dunes in the Golestan Province, northern Iran—Implications for the Late-Pleistocene history of the Caspian Sea. Aeolian Research 41: 100538.

Revell, D.L., Dugan, J.E. and Hubbard, D.M. 2011. Physical and ecological responses of sandy beaches to the 1997–98 El Niño. Journal of Coastal Research 27: 718–730.

Roushangar, K., Saghebian, S.M. and Mouaze, D.J.I.J.O.S.R. 2017. Predicting characteristics of dune bedforms using PSO-LSSVM. 32: 515–526.

Sallenger, A.H. Jr., Krabill, W.B., Swift, R.N., Brock, J., List, J., Hansen, M., Holman, R.A., Manizade, S., Sontag, J., Meredith, A., Morgan, K., Yunkel, J.K., Frederick, E.B. and Stockdon, H. 2003. Evaluation of airborne topographic lidar for quantifying beach changes. Journal of Coastal Research 19: 125–133.

Sam, L., Gahlot, N. and Prusty, B.G. 2013. Estimation of dune celerity and sand flux in part of West Rajasthan, Gadra area of the Thar Desert using temporal remote sensing data. Arabian Journal of Geosciences 8: 4.

Schmugge, T., Jackson, T.J., Kustas, W.P. and Wang, J.R. 1992. Passive microwave remote sensing of soil moisture: results from HAPEX, FIFE and MONSOON 90. ISPRS Journal of Photogrammetry and Remote Sensing 47: 127–143.

Sofonia, J., Phinn, S., Roelfsema, C. and Kendoul, F. 2019. Observing geomorphological change on an evolving coastal sand dune using SLAM-based UAV LiDAR. Remote Sensing in Earth Systems Sciences 2: 273–291.

Song, Y., Chen, C., Xu, W., Zheng, H., Bao, A., Lei, J., Luo, G., Chen, X., Zhang, R. and Tan, Z. 2020. Mapping the temporal and spatial changes in crescent dunes using an interferometric synthetic aperture radar temporal decorrelation model. Aeolian Research 46: 100616.

Stevens, A., Udelhoven, T., Denis, A., Tychon, B., Lioy, R., Hoffmann, L. and Van Wesemael, B. 2010. Measuring soil organic carbon in croplands at regional scale using airborne imaging spectroscopy. Geoderma 158: 32–45.

Stevens, A., Van Wesemael, B., Bartholomeus, H., Rosillon, D., Tychon, B. and Ben-Dor, E. 2008. Laboratory, field and airborne spectroscopy for monitoring organic carbon content in agricultural soils. Geoderma 144: 395–404.

Stockdonf, H.F., Sallenger Jr, A.H., List, J.H. and Holman, R.A. 2002. Estimation of shoreline position and change using airborne topographic lidar data. Journal of Coastal Research 502–513.

Tarhouni, M., Salem, F.B., Belgacem, A.O. and Neffati, M. 2010. Acceptability of plant species along grazing gradients around watering points in Tunisian arid zone. Flora-Morphology, Distribution, Functional Ecology of Plants 205: 454–461.

Todd, S.W. 2006. Gradients in vegetation cover, structure and species richness of Nama-Karoo shrublands in relation to distance from livestock watering points. Journal of Applied Ecology 43: 293–304.

Tsoar, H. 2001. Types of aeolian sand dunes and their formation. *In*: Balmforth, N.J. and Provenzale, A. (eds.). Geomorphological Fluid Mechanics. Berlin, Heidelberg: Springer Berlin Heidelberg.

Van Der Meer, F.D., Van Der Werff, H.M.A., Van Ruitenbeek, F.J.A., Hecker, C.A., Bakker, W.H., Noomen, M.F., Van Der Meijde, M., Carranza, E.J.M., Smeth, J.B.D. and Woldai, T. 2012. Multi- and hyperspectral geologic remote sensing: A review. International Journal of Applied Earth Observation and Geoinformation 14: 112–128.

Varma, S., Shah, V., Banerjee, B. and Buddhiraju, K.M. 2014. Change detection of desert sand dunes: a remote sensing approach. Advances in Remote Sensing 3(01): 10.

Verhoeven, G.J., Loenders, J., Vermeulen, F. and Docter, R. 2009. Helikite aerial photography—a versatile means of unmanned, radio controlled, low-altitude aerial archaeology. Archaeological Prospection 16: 125–138.

Wallace, L., Lucieer, A., Watson, C. and Turner, D. 2012. Development of a UAV-LiDAR system with application to forest inventory. Remote Sensing 4: 1519–1543.

Xia, J. and Dong, P. 2016. A GIS add-in for automated measurement of sand dune migration using LiDAR-derived multitemporal and high-resolution digital elevation models. Geosphere 12: 1316–1322.

Yitagesu, F.A., Van Der Meer, F., Van Der Werff, H. and Hecker, C. 2011. Spectral characteristics of clay minerals in the 2.5–14 μm wavelength region. Applied Clay Science 53: 581–591.

Chapter 4

Applications of ICESat-2 Laser Altimeter Data for Sand Dunes

Morphology and Dynamics Studies

Giribabu Dandabathula[1,*] and *Mahesh K. Gaur*[2]

1. Introduction

The consequences of nature's course results in changes of landforms. The flow of water in the river carries silt and accumulates the suspended sediment fluxes in river catchments. Volcanoes allow molten rock or magma to create rock formations, and in the same way, stimulated by the forces of wind on the dry sands of deserted environments and coastal areas that will induce the sand accumulation and transportation activity (Bagnold, 1941). Sand dunes are developed due to the exposure of sand to an agitating medium such as wind that lifts the sand grains from the ground (as a part of an aeolian process) and entrains them into a surface flow (Kroy et al., 2002; Pye and Tsoar, 2008). Sand dunes are the results of accumulation of mounds or ridges formed by deposition of loose sand and can range from 1 m to several kilometers. In general, sand dunes are classified based on their structure that are influenced by the strength and direction of the interacting wind and the type of sand; as either transverse to the wind, longitudinal or lengthwise in the direction of wind or solitary dunes related to varying wind directions (Edgell, 2006). The shape of active dunes can be mainly in any one of the forms such as a crescent, parabolic, linear and star (Bagnold, 1941; Wasson and Hyde, 1983). Analyzing the cross-section profiles of sand dunes help in assessing the impact of the wind force on its morphological parameters, stability parameters and mechanisms of the dune process (Tsoar, 1985; Zimbelman et al., 2012; Courrech du Pont, 2015).

1.1 Background

Remote sensing technology has enabled one to understand the evolution, morphometry, dynamics and organization of aeolian sand dunes (Hugenholtz et al., 2012). Research related to sand dunes and their related landforms has been promoted due to the use of satellite data (Gutiérrez and Benito, 2005; Hermas et al., this volume). Surface changes act as a proxy to indicate parameters of aeolian sand transport process and remote sensing data provides an excellent platform to map the surface changes. Optical remote sensing data have determined the synoptic view of sand dune diversity not

[1] Regional Remote Sensing Centre – West, NRSC/ISRO, Department of Space, Jodhpur, India.
[2] ICAR-Central Arid Zone Research Institute, Jodhpur, India.
* Corresponding author: giribabu.d@nrsc.gov.in

only on Earth but also on other planets (Hayward et al., 2007). Researchers used microwave remote sensing data to retrieve surface geometrical characteristics (Stephen and Long, 2005). Substantial advances in remote sensing techniques such as Photogrammetry and LiDAR can give quantitative information about the parameters of aeolian sand transport mechanism (Hugenholtz et al., 2012).

Satellite laser altimeter transmits light in the form of energy pulses towards a planetary surface. The instrument measures the round-trip time of the light pulse between the emitted and the returned surface reflection to determine the distance of the spacecraft with respect to the surface and yields the true height of the topographic features on the ground (Falkner and Schulz, 2015). Space-based laser altimeters are effective in providing topographic measurements that are critically important to the understanding of formation and early evolution of planetary bodies. Using laser altimetry data, geophysical parameters can be generated which provide significant insight into the shape, structure and evolution of the subject body (Cole, 1997). In terms of sand dunes, the surface measurements retrieved from consecutive orbits can be used to understand the morphological parameters.

2. Short introduction about NASA's ICESat-2 laser altimeter

The Ice, Cloud and land Elevation Satellite (ICESat) mission is a laser altimetry-based system to determine the changes in elevations of glaciers and ice sheets, as well as sea ice thickness distribution. Measurements from the ICESat mission, which operated from 2003 to 2009 have provided vital information on the response of the Earth's frozen surface (Cryosphere) with respect to the changes of atmosphere and ocean parameters (Markus et al., 2017). The primary instrument of ICESat satellite is the Geoscience Laser Altimeter (GLAS) that carries 532 nm (visible) and 1064 nm (near infrared) laser light to propitiate both atmospheric and topographic investigations. A full-waveform detection strategy is used to measure the two-way travel time between the sensor platform and the Earth's surface (Martino et al., 2019). The full-waveform GLAS LiDAR transmitted pulses at 40 times per second, yielding a 170 m along-track spacing of ~ 65 m diameter footprint on Earth (Schutz et al., 2005). The ICESat mission has provided large altimetry-based measurements not only to the cryosphere scientific community, but also to the ocean, terrestrial and atmospheric scientific communities. Earlier, researchers had used ICESat laser altimetry data to study sand dune migration vectors (Dabboor et al., 2013). Despite ICESat's success, the science community identified some limitations that prohibited the full exploitation of the dataset for scientific applications in terms of its spatial resolution and also, the concept of single beam laser in ICESat failed to consider slope effects of true elevation changes on an orbit-by-orbit basis (Abdalati et al., 2010; Markus et al., 2017).

A new space-based laser altimetry mission called the Ice, Cloud and land Elevation Satellite-2 (ICESat-2) was launched by the National Aeronautics and Space Administration (NASA) in September 2018 as continuity to the ICESat mission. As with its predecessor, ICESat-2 carries a single laser altimeter, called the Advanced Topographic Laser Altimeter System (ATLAS). ATLAS is a photon-counting laser altimeter operating on a visible wavelength. The ATLAS instrument and ICESat-2 observatory utilize a photon-counting LiDAR and other ancillary systems (like Global Position System (GPS) and star cameras) to measure the time a photon takes to travel from ATLAS to Earth and back again and to determine the photon's geodetic latitude and longitude. Approximately 10^{14} photons begin the journey from ATLAS sensor, traveling through the atmosphere to reflect off the Earth's surface, return through the atmosphere and back into the ATLAS telescope. Of these, only few of the photons complete the journey and are recorded by instrument electronics (Markus et al., 2017).

The ATLAS laser will produce a visible (532 nm) laser pulse at a repetition rate of 10 kHz. Each laser pulse will be split into six separate beams by a diffractive optical element, resulting in six individual footprints, each ~ 14 m in diameter arranged in three pairs of beams. The 10 kHz laser repetition rate coupled with the ~ 500 km orbit produces an along-track sampling interval of nearly 70 cm (0.7 m). The three beam pairs, each consisting of one high-energy beam and one low-energy beam (with an energy ratio between them of approximately 1:4), will be separated in the cross-track direction by nearly 3 km on the Earth's surface, and the beam tracks within each pair will be separated by 90 m (Leigh et al., 2014; Markus et al., 2017). Each ground track is numbered according to the laser spot number that generates it, with ground track 1L (gt1l) on the far left and ground track 3R (gt3r) on the far right.

As per the mission objectives, ICESat-2 will disseminate numerous levels of data products to the scientific fraternity and one being a level 2A data product namely, the ATL03 product that was designed to be a single source for all the geo-location photon events and related ancillary information needed by other higher-level ATLAS/ICESat-2 products. However, it also includes spacecraft and instrument parameters and ancillary data not explicitly required for ATL03 (Neumann et al., 2019). Primarily ATL03 contains geolocated ellipsoidal heights for each time tagged photon event downlinked from the ATLAS sensor. These heights are corrected for several geophysical phenomena (e.g., atmospheric refraction tides) and are classified as a confident signal photon event or background (noise) photon events.

The other higher levels of products that are derived from ATL03 are ATL06 for land-ice height, ATL07 for sea-ice elevation and ATL08 for land-canopy height. The photon-counting technology of the ATLAS sensor in ICESat-2 has enabled to map and measure the changes in various thematic areas like glaciers, terrain, water-levels (like sea and inland waters), vegetation (canopy structure and tree height) (Brown et al., 2016). Data sets of ICESat-2 are being released and disseminated in the HDF format at a web portal of NASA's Distributed Active Archive Centre (DAAC) and maintained by National Snow & Ice Data Centre (NSIDC) which is available at 'https://nsidc.org/data/icesat-2/data-sets'. Another web-portal namely, 'OpenAltimetry.com' is a cyber-infrastructure platform for discovery, access and visualization of data from NASA's ICESat and ICESat-2 missions; through this web portal one can visualize and download various thematic products of ICESat mission. The applications from ICESat-2 photon data can be extended to study the characteristic of sand dunes and measure the migrations of sand dune variations by retrieving the surface elevation.

3. Cross-section profiles of barchan dunes from ICESat-2 data

Barchan dunes are crescent shaped dunes that form in areas of unidirectional winds with limited amount of sand and usually display an asymmetric shape with one limb extended downwind (Fryberger and Dean, 1979; Parteli et al., 2014). Figure 4.1(A) shows the satellite imagery of Maghreb region of the Western Sahara Desert area where various forms of sand dune exist, and the image highlights two barchan sand dunes that are selected to retrieve the elevation profile form ICESat-2 photon-based ground track. Figure 4.1(B) shows the cross-section profile retrieved from the ICESat-2 ground track passing through the twin barchan sand dunes. The height of the slip face and the length of main body of the barchan sand dune is clearly represented in the elevation profile. Parameters like height of the slip face, length of the main body and the total cross-section profile helps in computing the mobility of dunes (OuldAhmedou et al., 2007). Persistent unidirectional winds with limited sand supply may preserve the shape of barchan dunes, but the migration process will continue and its speed varies approximately as the inverse of the height; thus, height of the barchan sand dune is extremely useful during the computation of migration process dynamics as a part of dune celerity (Haff and Presti, 1995; OuldAhmedou et al., 2007).

3.1 Cross-section profiles of barchanoids

Coalesced barchans referred to as barchanoid ridges are essentially straight crested transverse dunes. Barchanoid ridges occur when the sand supply is greater than in the conditions that create a barchan dune. Figure 4.2 shows the elevation profile for a heterogeneous area containing barchanoid ridges and stable longitudinal sand sheets with sparse vegetation in the western part of Jaisalmer city, Rajasthan, India. Earlier research conducted using satellite stereo photogrammetric techniques in this area had revealed that these barchanoids ranges between 30 to 45 meters high and are closely

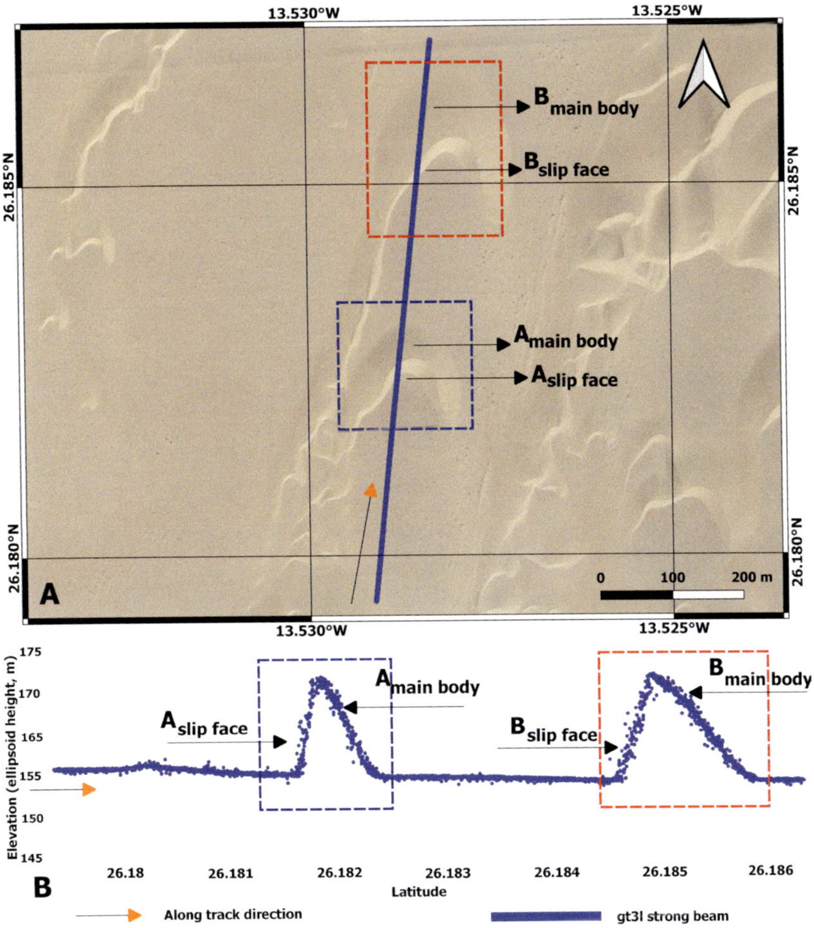

Figure 4.1. Elevation profile generated from ICESat-2 ground track over two barchan sand dunes. (A) Satellite data (Maxar's imagery base-map with spatial resolution of 50 cm in Google Earth) showing the area containing barchan sand dunes in the part of Maghreb region of Western Sahara. Red and blue dashed boxes are the area of interests where elevation profiles have been retrieved. The image is overlaid with the ICESat-2 ground track with id 1030 (gt3l strong beam) dated 5th March, 2019. (B) Elevation profile of the two barchan sand dunes with slip face and main body of the sand dunes captured by the profile in the along-track direction.

spaced (Kar, 1990). The elevation profile retrieved from ICESat-2 ground track photon data for this region leads to similar information and is shown in Fig. 4.2(B). The elevation profile retrieved from ICESat-2 photon data helps in distinguishing the stable longitudinal sand sheets (with vegetation) and sand dunes. Dune heights and lengths retrieved from these profiles are very useful in computation of sediment transport over a time period (Tjerry and Fredsøe, 2005).

Figure 4.3 shows the elevation profile generated from ICESat-2 ATL03 product over a series of barchan sand dunes in the southwestern parts of Jaisalmer city, Rajasthan, India. The profiles, generated independently from two data sets acquired on April 1, 2019 and May 29, 2020, detected changes in the peak heights in the order ranging from ~ 0.5 to 1 m (within a time span of nearly one year duration). These kinds of sand dunes are highly influenced by dust storms that occur during the months of March till June (Kumari et al., 2020). Dust storms occur when strong or turbulent winds combine with exposed loose sands. Sand storms interact closely with the ground surface, but finer dust particles may be lifted kilometers high into the atmosphere, where strong winds aid in transporting them for long distance travel, even across continents and can influences the structure of sand dunes in the process (Shepard et al., 2016; Middleton and Kang, 2017). A series of high

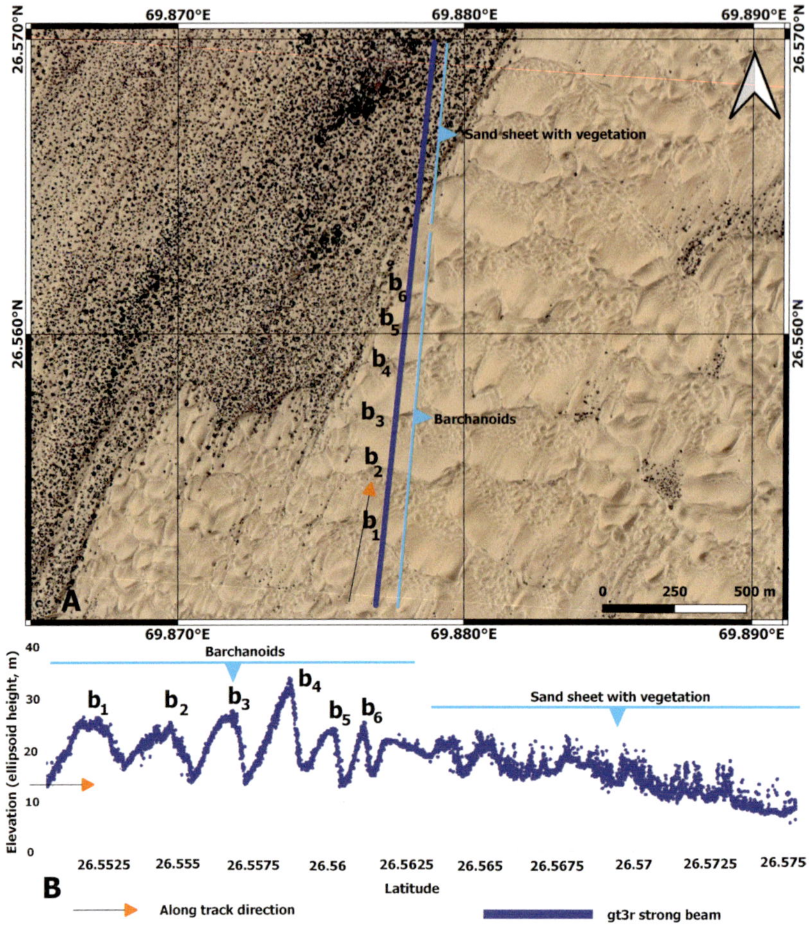

Figure 4.2. Elevation profile generated over a heterogeneous landscape containing a sequence of barchanoids and a sable longitudinal sand sheet with sparse vegetation. (A) Satellite data (Maxar's imagery base-map with spatial resolution of 50 cm in Google Earth) showing a sequence of barchanoids (b1, b2, b3, b4, b5 and b6) and a sand sheet with sparse vegetation in the western parts of Jaisalmer, Rajasthan state, India. The image is overlaid with the subset of ICESat-2 ground track with id 51 (gt3r strong beam) dated 29th March, 2020. (B) Elevation profile over group of barchanoids and sand sheets comprising sparse vegetation. The profile clearly distinguishes the elevation profile over barchanoids and the sand sheet containing sparse vegetation.

intensity sand storms had occurred during the month of April and May in the year 2019 in this area, and resulting in the changes of sand dune heights that are notable from elevation profiles.

3.2 Cross-section profiles of megabarchanoids

Megabarchanoids, which are compound forms of sand dunes, consist of superimposed barchans of different sizes; where the windward slope is covered with certain number of tiers of nested barchans roughly forming from the lower mid-slope to the crustal segment (Kar, 1990; Elbelrhiti and Hargitai, 2014). The structure of these compound megabarchans can be studied using the profiles generated from the ICESat-2 ground track data. Figure 4.4 shows the perspective view with high resolution satellite data acquired over a sequence of megabarchanoids in the southwestern parts of Jaisalmer city and respective profile retrieved from the ICESat-2 ground track. As shown in Fig. 4.4(B) the fish-scale pattern in the low-mid slope to the crustal segment for each megabarchanoid (marked as b_1 and b_2) are evident in the profile with height ranging from 2 to 6 m.

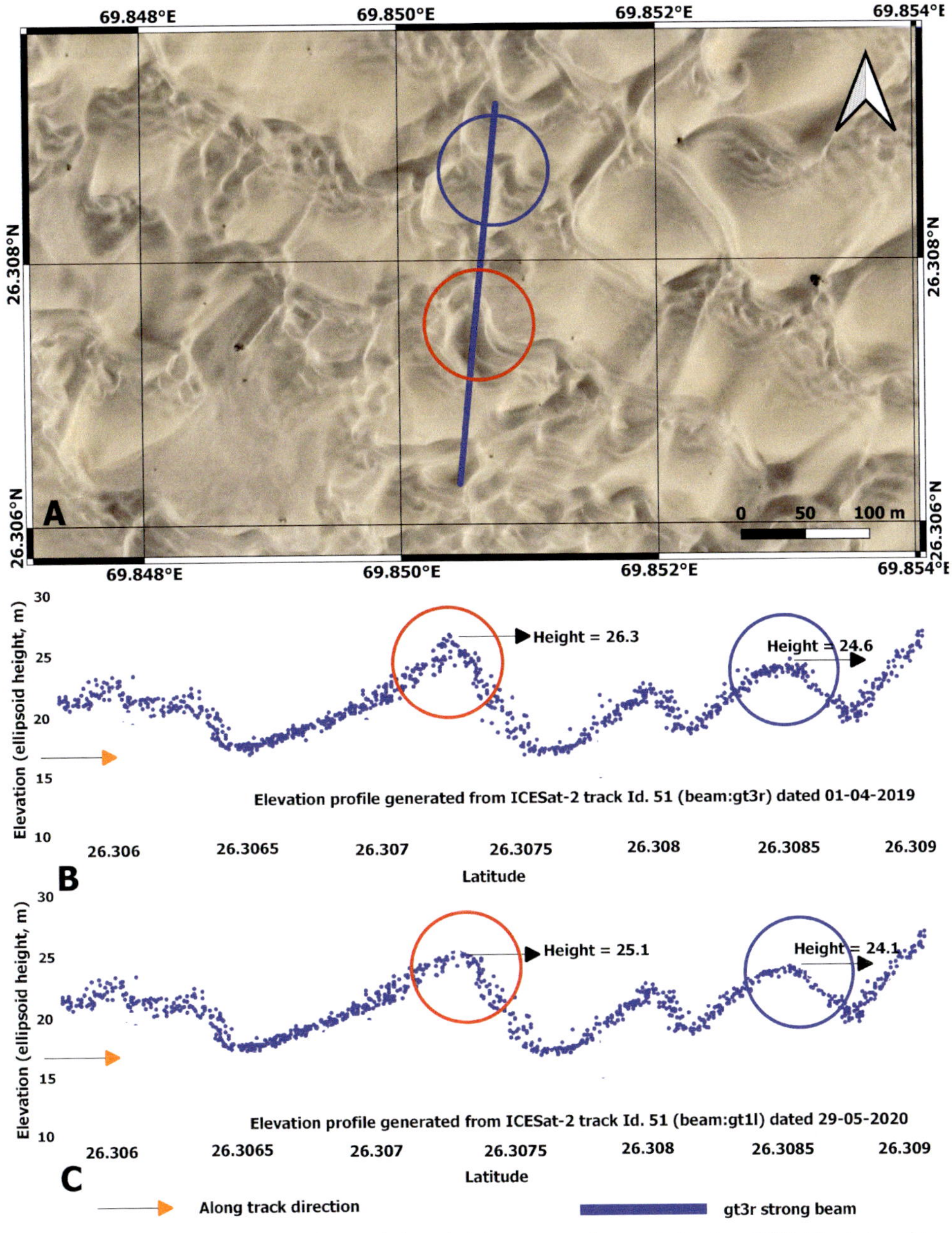

Figure 4.3. Detection of changes in the sand dune heights over a time period using ICESat-2 photon data. It is evident that differences in the heights at the crest ranging from 0.5 m to 1 m were noted in a span of one year duration. (A) Satellite data (Maxar's imagery base-map with spatial resolution of 50 cm in Google Earth) and area showing barchanoids in the West-southern parts of Jaisalmer, Rajasthan state, India. The image is overlaid with the subset of ICESat-2 ground track with id 51 (gt3r strong beam) dated 29th March, 2020. (B) Elevation profile over group of barchanoids generated from ICESat-2 ground track (beam gt1l) dated 1st April, 2019. (C) Elevation profile over group of barchanoids generated from ICESat-2 ground track (beam gt3r) dated 29th May, 2020.

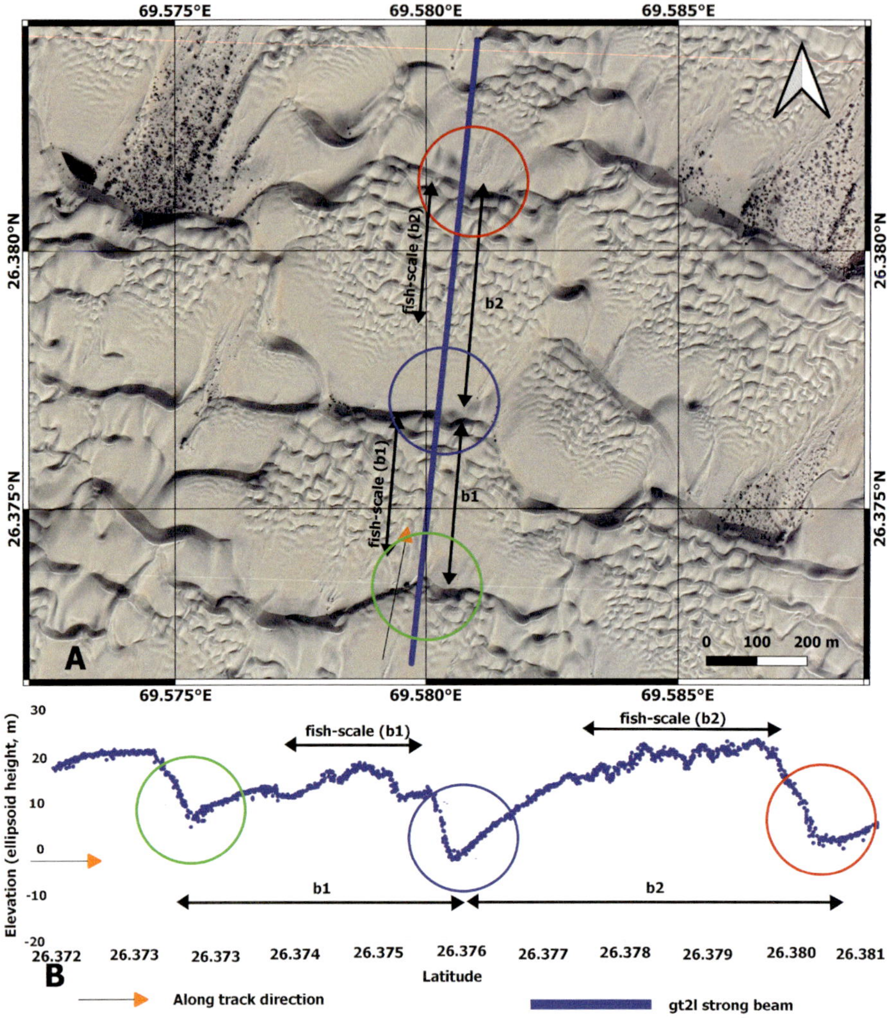

Figure 4.4. Analysing the structure of megabarchanoids formed by superimposed barchans of different sizes using the profiles generated from ICESat-2 ground track. (A) Perspective view with high resolution satellite data (Maxar's imagery base-map with spatial resolution of 50 cm in Google Earth) showing megabarchanoids located in the Southwestern parts of Jaisalmer, Rajasthan state, India. The image is overlaid with the subset of ICESat-2 ground track with id 996 (gt2lr strong beam) dated 03rd March, 2019. (B) Elevation profile over two megabarchanoids (marked as b1 and b2). Note the smaller barchanoids of varying size and in fish-scale pattern forming at the lower mid-slope to the crustal segment on each megabarchanoid.

4. Cross-section profiles of parabolic dunes from ICESat-2 data

Parabolic dunes are scoop shaped hollow/blowout of sand with points tapering to the windward side and the ground plan approximates the parabola shape; these forms of sand dunes are also known as U dunes, upsiloidal dunes or hairpin dunes (Hesp, 2011; May, 2014). Parabolic dunes originate due to strong unidirectional winds and indicates the presence of a stabilizing agent like vegetation cover (May, 2014). These kinds of dunes can get degraded by continuous deflation in the absence of sustained sediment supply (Pye, 1982). Morphological modifications can be monitored using LiDAR based data. Figure 4.5 shows high resolution satellite data acquired over a hairpin type of parabolic dune located at the southern part of Jaisalmer city. Figure 4.5(B) shows the profile retrieved

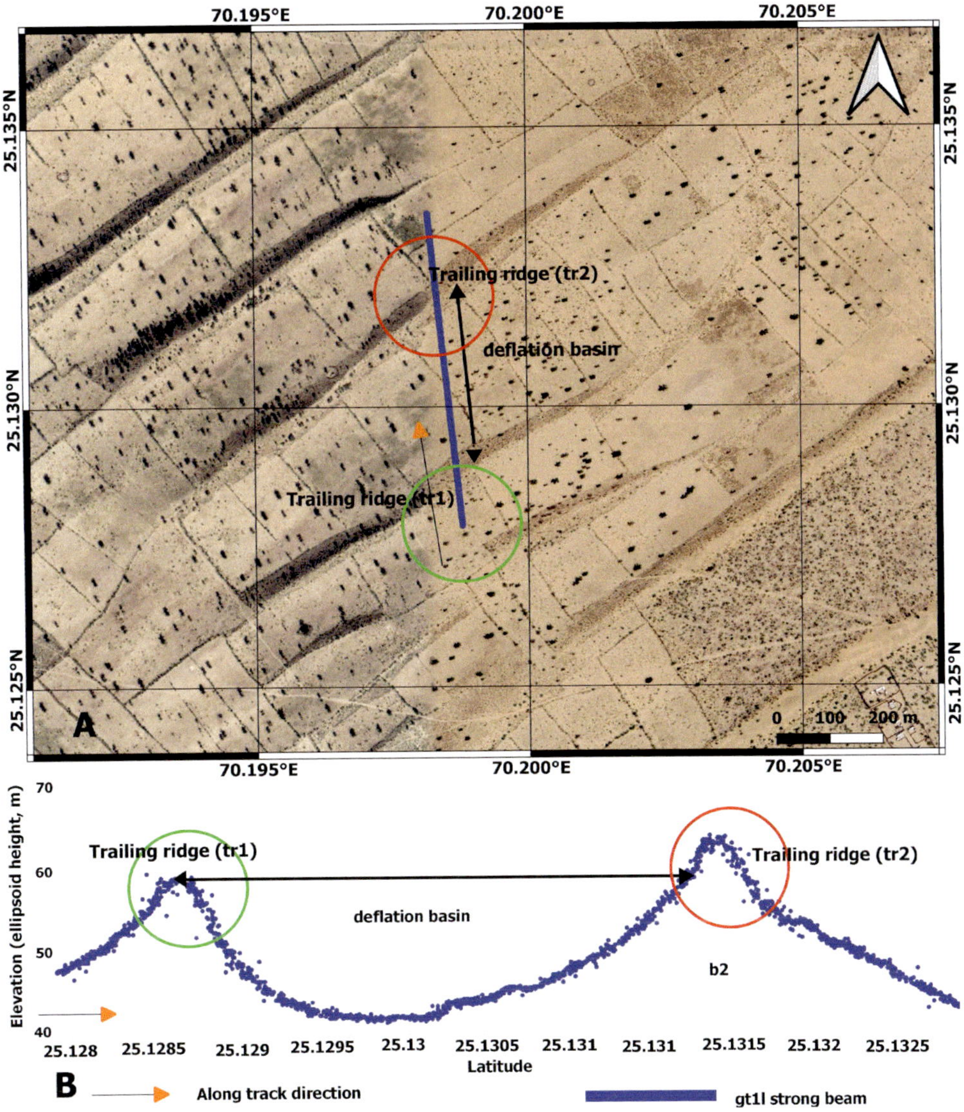

Figure 4.5. Retrieval of the cross-section profile for a hairpin type of parabolic sand dune. (A) High resolution satellite data (Maxar's imagery base-map with spatial resolution of 50 cm in Google Earth) showing hairpin type of parabolic sand dune located in the Southern parts of Jaisalmer, Rajasthan state, India. The image is overlaid with the subset of ICESat-2 ground track with id 760 (gt1l strong beam) dated 18th May, 2019. The sparse vegetation seen in this high resolution data indicates the stability of the sand dune. (B) Cross-section profile over the hairpin type of parabolic sand dune. Note the length of deflation basin and height of the trailing ridges at the cross-sections.

from a subset of the ICESat-2 ground track and represents the deflation basin of ~ 370 m length with trailing ridges of approximately 59 m height. Digital Elevation Models (DEM) can be generated for stabilized sand dunes using high confidence photon heights derived from multiple ground tracks of ICESat-2 Lidar data using advanced photogrammetric and interpolation techniques. Figure 4.6 shows a reconstructed DEM generated from high confidence photon heights from 10 beams of 5 ICESat-2 ground tracks and draped with image high resolution satellite data (WMS service of Maxar's imagery base map with 50 cm spatial resolution).

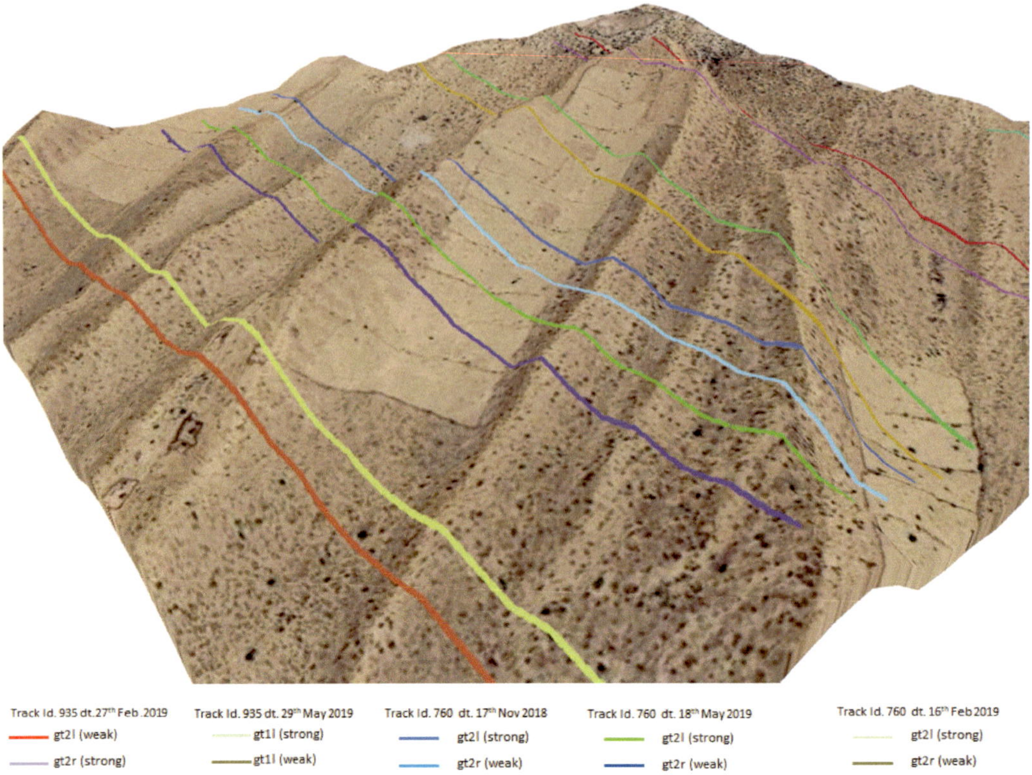

Figure 4.6. A Digital Elevation Model (DEM) representing the depletion basin that was reconstructed using photon data from multiple ICESat-2 Lidar ground tracks and advanced photogrammetric/interpolation methods. In this image, a high resolution satellite data (through Web Map Service (WMS) service of Maxar's imagery base map of 50 cm spatial resolution) was draped on the DEM generated from 10 beams of ICESat-2 data.

5. Cross-section profiles of linear dunes from ICESat-2 data

In general, the shape of dunes depends on the history of wind flow and sand availability. Most often dunes in the semi-arid regions are found to be linear ridges either perpendicular or parallel to the mean wind direction, are known as longitudinal and transverse dunes (Reffet et al., 2010). Knowledge about their dimensions is especially useful to investigate their evolution and dynamics. The crest lines of longitudinal dunes are broadly straight unless they are of the seif dune variety which have meandering crests (Radebaugh et al., 2014). Longitudinal dunes exhibit symmetrical patterns (Pye and Tsoar, 1990). Figure 4.7(A) shows high resolution satellite data of typical longitudinal dune located south of Jaisalmer, Rajasthan, India where the crest lines aligned along northeast and southwest direction can be seen. Figure 4.7(B) shows the profile retrieved from the ICESat-2 photons based LiDAR ground track. Near symmetrical heights are evident for each pair of crests in juxtaposition (R_1 & R_2), (R_3 & R_4) and (R_5 & R_6) and can be derived from this profile analysis.

6. Summary

Stimulated by the force of winds on the dry sand of the desert environments as a part of aeolian processes will induce sand accumulation and transportation activity; during which, landforms like sand dunes get evolved and developed. Sand dunes are classified based on their shape forms. Three dimensional (3D) cross-sectional profile analysis of sand dunes helps in retrieving the morphological parameters of the landform and corresponding height information about the sand

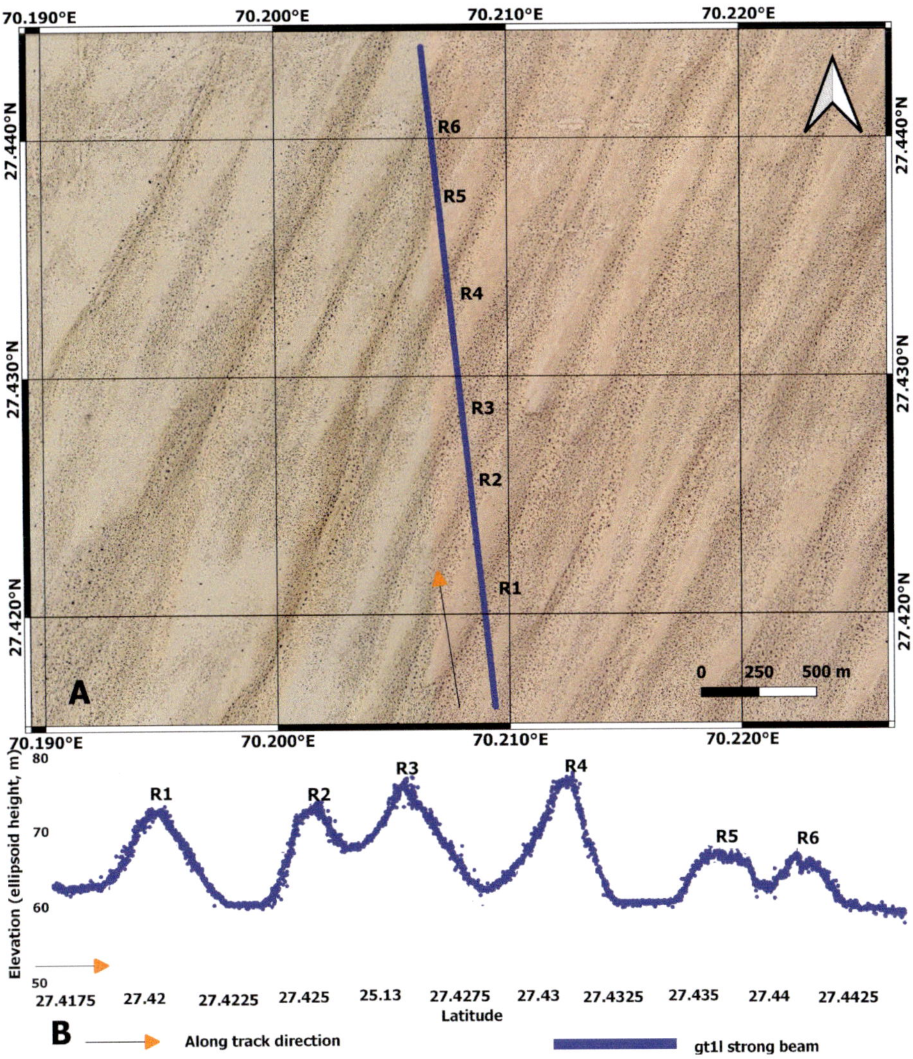

Figure 4.7. Analysis of crest pair pattern in a linear sand dunes system using ICESat-2 Lidar ground track data. (A) High resolution satellite data (Maxar's imagery base-map with spatial resolution of 50 cm in Google Earth) showing linear sand dunes located in the Western part of Jaisalmer, Rajasthan state, India. The image is overlaid with a subset of ICESat-2 ground track data. (B) Elevation profile over a series of linear sand dunes. Crest points at R1, R2, R3, R4, R5, and R6 can be noted in the satellite data and the corresponding elevation profile. Even though crests occur as a single linear sand dune, the crest pair in juxtapositions (R1 & R2), (R3 & R4) and (R5 & R6) can be noted in the profile analysis.

dune plays a pivotal role in deriving its dynamics in relation with time. Substantial advances in the laser-based altimetry techniques provide quantitative information about the parameters of sand dunes. National Aeronautics and Space Administration's (NASA) latest laser-based earth observation platform, namely Ice, Cloud and land Elevation Satellite-2 (ICESat-2) is a new genre of LiDAR based altimeter that transmits light in the form of energy pulses towards the Earth's surface. A photon-counting laser altimeter titled Advanced Topographic Laser Altimeter System (ATLAS) in ICESat-2 measures the round-trip time of the light pulse between emitted and the returned surface reflection to determine the distance of the spacecraft with respect to the surface and yields the true height of the topographic features on the ground. The ATLAS laser will provide data with a ground tracks containing six beams arranged in three pairs of strong and weak energies. The 10 kHz laser repetition rate coupled with the ~ 500 km orbit produces an along-track sampling interval of nearly 70 cm (0.7 m). This chapter provided insights about the use of geolocated photons height information from ICESat-2 laser altimeter to analyze cross-section profiles of barchan, barchanoids, megabarchanoids, parabolic and linear type of sand dunes.

References

Abdalati, W., Zwally, H.J., Bindschadler, R., Csatho, B., Farrell, S.L., Fricker, H.A., Harding, D., Kwok, R., Lefsky, M., Markus, T., Marshak, A., Neumann, T., Palm, S., Schutz, B., Smith, B., Spinhirne, J. and Webb, C. 2010. The ICESat-2 laser altimetry mission. Proceedings of the IEEE 98: 735–751.

Bagnold, R.A. 1941. The Physics of Blown Sands and Desert Dunes. Methuen, London, England.

Brown, M.E., Delgodo Arias, S., Neumann, T., Jasinski, M.F., Posey, P., Babonis, G., Glenn, N.F., Birkett, C.M., Escobar, V.M. and Markus, T. 2016. Applications for ICESat-2 data: From NASA's early adopter program. IEEE Geosci. Remote Sens. Mag. 4: 24–37.

Cole, T.D. 1997. Spaceborne laser altimetry. *In*: Advancement of Photonics for Space: A Critical Review. 10288. International Society for Optics and Photonics.

Courrech du Pont, S. 2015. Dune morphodynamics. ComptesRendus Physique 16(1): 118–138.

Dabboor, M.D., Braun, A. and Kneen, M.A. 2013. Tracking sand dune migration in the Rub Al-Khali with ICESat laser altimetry. Intl. J. Rem. Sens. 34(11): 3832–3847.

Edgell, H.S. 2006. Arabian Deserts. Nature, Origin and Evolution. Springer, Dordrecht, Netherlands.

Elbelrhiti, H. and Hargitai, H. 2014. Barchan. pp. 1–11. *In*: Hargitai, H. (ed.). Encyclopedia of Planetary Landforms. Springer, New York.

Falkner, P. and Schulz, R. 2015. Instrumentation for planetary exploration missions. *In*: Schubert, G. (ed.). Treatise on Geophysics. Elsevier.

Fryberger, S.G. and Dean, G. 1979. Dune forms and wind regimes. *In*: McKee, M.D. (ed.). A Study of Global Sand Seas, United States Geological Survey Professional Paper 1052.

Gutiérrez, M. and Benito, G. 2005. Climatic Geomorphology. Amsterdam: Elsevier.

Haff, P.K. and Presti, D.E. 1995. Barchan dunes of the Salton Sea Region, California. *In*: Tchakerian, V.P. (ed.). Desert Aeolian Processes. Springer, Dordrecht.

Hayward, R.K., Mullins, K.F., Fenton, L.K., Hare, T.M., Titus, T.N., Bourke, M.C., Colaprete, A. and Christensen, P.R. 2007. Mars global digital dune database and initial science results. J. Geophys. Res. 112(E11).

Hesp, P. 2011. Dune coasts. *In*: Wolanski, E. and McLusky, D.S. (eds.). Treatise on Estuarine and Coastal Science. 3: 193–221. Waltham, Academic Press.

Hugenholtz, C.H., Levin, N., Barchyn, T.E. and Baddock, M.C. 2012. Remote sensing and spatial analysis of aeolian sand dunes: A review and outlook. Earth-Sci. Rev. 111(3-4): 319–334.

Kar, A. 1990. Megabarchanoids of the Thar: Their environment, morphology and relationship with longitudinal dunes. Geogr. J. 156(1): 51–61.

Kroy, K., Sauermann, G. and Herrmann, H.J. 2002. Minimal model for aeolian sand dunes. Phys. Rev. E. 66(3).

Kumari, S., Lakhani, A. and Kumari, K.M. 2020. Transport of aerosols and trace gases during dust and crop-residue burning events in Indo-Gangetic Plain: Influence on surface ozone levels over downwind region. J. Atmosenv. 241: 117829.

Leigh, H.W., Magruder, L.A., Carabajal, C.C., Saba, J.L. and McGarry, J.F. 2014. Development of onboard digital elevation and relief databases for ICESat-2. IEEE Trans. Geosci. Remote Sensing 53: 2011–2020.

Markus, T., Neumann, T., Martino, A., Abdalati, W., Brunt, K., Csatho, B., Farrell, S., Fricker, H., Gardner, A., Harding, D., Jasinski, M., Kwok, R., Magruder, L., Lubin, D., Luthcke, S., Morison, J., Nelson, R., Neuenschwander, A., Palm, S., Popescu, S., Shum, C., Schutz, B.E., Smith, B., Yang, Y. and Zwally, J. 2017. The Ice, Cloud, and land Elevation Satellite-2 (ICESat-2): Science requirements, concept, and implementation. Remote Sensing of Environment 190: 260–273.

Martino, A.J. Neumann, T.A., Kurtz, N.T. and McLennan, D. 2019. ICESat-2 mission overview and early performance. Sensors, Systems, and Next-Generation Satellites XXIII 11151: 111510C.

May, J.H. 2014. Parabolic Dune. Encyclopedia of Planetary Landforms. 1–5. Springer, New York.

Middleton, N. and Kang, U. 2017. Sand and dust storms: impact mitigation. Sustainability 9(6): 1053.

Neumann, T., Brenner, A., Hancock, D., Robbins, J., Saba, J., Harbeck, K. and Gibbons, A. 2019. Ice, cloud, and land elevation satellite–2 (ICESat-2) Project: algorithm theoretical basis document (ATBD) for global geolocated photons (ATL03). National Aeronautics and Space Administration, Goddard Space Flight Center.

OuldAhmedou, D., OuldMahfoudh, A., Dupont, P., Ould El Moctar, A., Valance, A. and Rasmussen, K.R. 2007: Barchan dune mobility in Mauritania related to dune and interdune sand fluxes. J. Geophys. Res. 112(F2).

Parteli, E.J.R., Durán, O., Bourke, M.C., Tsoar, H., Pöschel, T. and Herrmann, H. 2014. Origins of barchan dune asymmetry: Insights from numerical simulations. Aeolian Res. 12: 121–133.

Pye, K. 1982. Morphological development of coastal dunes in a humid tropical environment, Cape Bedford and Cape Flattery, North Queensland. Geogr. Ann. 64(213–227).

Pye, K. and Tsoar, H. 2008. Aeolian Sand and Sand Dunes. Springer Science & Business Media.

Radebaugh, J., Sharma, P., Korteniemi, J. and Fitzsimmons, K.E. 2014. Longitudinal dunes (or linear dunes). pp. 1–11. *In*: Encyclopedia of Planetary Landforms. Springer New York.

Reffet, E., Courrech du Pont, S., Hersen, P. and Douady, S. 2010. Formation and stability of transverse and longitudinal sand dunes. Geol. 38(6): 491–494.

Schutz, B.E., Zwally, H.J., Shuman, C.A., Hancock, D. and DiMarzio, J.P. 2005. Overview of the ICESat mission. Geophys. Res. Lett. 32(21).

Shepherd, G., Terradellas, E., Baklanov, A., Kang, U. et al. 2016. Global assessment of sand and dust storms. UNEP, WMO, UNCCD; United Nations Environment Programme. ISBN: 978-92-807-3551-2.

Stephen, H. and Long, D.G. 2005. Modeling microwave emissions of erg surfaces in the Sahara desert. IEEE Trans. Geosci. Remote Sensing 43(12): 2822–2830.

Tjerry, S. and Fredsøe, J. 2005. Calculation of dune morphology. J. Geophys. Res. 110(F4).

Tsoar, H. 1985. Profiles analysis of sand dunes and their steady state signification. Geogr. Ann. A - Physical Geography 67: 47–59.

Wasson, R.J. and Hyde, R. 1983. Factors determining desert dune type. Nature 304: 337–339.

Zimbelman, J.R., Williams, S.H. and Johnston, A.K. 2012. Cross-sectional profiles of sand ripples, megaripples, and dunes: a method for discriminating between formational mechanisms. Earth Surf. Process. Landforms 37(10): 1120–1125.

Chapter 5

New Perspectives on the Occurrence and Characteristics of the Coastal Sand Dunes, North of Nile Delta, Egypt

Mohamed S. EL-Bady

1. Introduction

The occurrence and the characteristics of coastal sand dunes in the Nile Delta and along the Mediterranean shoreline depends on the impact of natural and human activities. To determine occurrences and characteristics of these sand dunes many parameters should be studied. These definitions are grain sizes, mineralogy, waves, tides, wind, the available materials, mud, sand, cobbles, boulders, shells, etc. Moreover, these parameters of coastal sand dunes are also dependent on the sedimentology and mineralogy of the Nile Delta itself.

Before the construction of Aswan High Dam in 1964 the Mediterranean Sea received 134 million tons of sediments and about 55.5 billion m^3 of water annually (Abu-Zeid and El-Shibini, 1997; Sharaf El Din, 1977). After the construction of the Aswan High dam, the quantity of sediments discharged to the Mediterranean Sea decreased dramatically, with negative effects on agricultural lands and on most of the coastal geomorphologic units.

Many authors studied the sedimentology and mineralogy of the Nile Delta coast in relation with accretion and erosion of beaches (Anwar et al., 1979; El-Bouseily and Frihy, 1984; Frihy and Komar, 1993; EL-Asmar, 2000; Ahmad, 2002 and others). They suggested that near the river promontories the sand is finer with coarse sand being transported by longshore currents to accreted areas. They noted that the heavy minerals were deposited with the finer sediments in the eroded areas near the river mouth.

The northern coastal zone of the Nile Delta (between Port Said and Alexandria) includes a strip of coastal sand dunes and many coastal formations. This strip consists of five classes such as coastal dunes and barren soil, water body, urban areas, agricultural land and fish farms (El-Banna and Frihy, 2009). The areas classified as coastal sand dunes and barren soil have decreased by about 45% from 1990 to 2014 (from 1897 km^2 to 846 km^2) (Ali and Abu El-Magd, 2016).

Ali and Abou EL-Magd (2016) suggested that the area of erosion was closely related to the area of accretion in the shoreline of the Nile Delta. They evaluated the eroded and accreted areas

Geological Sciences Department, National Research Center, Dokki, Cairo, Egypt.
Email: elbady_nrc@yahoo.com

Table 5.1. Quantitative evaluation of land use changes during the period from 1990 to 2014 (Ali and Abu EL-Magd, 2016).

Class	Area 1990 (Km^2)	Area 2014 (Km^2)	Area increase/decrease (Km^2)	% of changes	Yearly rate
Water	1099.8	617.2	−482.6	− 44%	−1.8%
Fish-farm	81.3	936.8	+855.6	+1052%	+44%
Urban	249.5	1685.5	+1436	+576%	+60%
Agriculture	3232.9	2267.6	−965.2	−30%	−1.25%
Coastal Sand and Bare Soil	1897.4	1050.9	−846.5	−45%	−1.9%
Total	6560.8	6558.1	0.0	0.0	0.0
Deficit		2.7	Overall shoreline retreat		

along the shoreline of the Nile Delta, and determined that the eroded area was about 19.64 km^2 and the accreted area was 16.94 km^2 during the period from 1990 to 2014. At both the promontories Rosetta and Damietta the erosion rates were high, with the maximum retreat of shoreline at Rosetta (823.2 m) and at Damietta (1330 m) during the period from 1990 to 2014. The area between the two promontories was subjected to a smoothing process, where the eroded areas were 4.44 km^2 and the accreted areas were 6.53 km^2, thus the deposition rate was greater than the erosion rate creating an area of newly added land (2.09 km^2) from 1990 to 2104 (Ali and Abou EL-Magd, 2016).

El-Fishawi and El-Askary (1981) suggested that west of Gamasa (31.44° N, 31.53° E), the southern parts of longitudinal dunes tend to become a series of isolated barchan dunes. Bagnold (1971) believed that barchan dune types can only occur when the wind is almost unidirectional, and the longitudinal (linear) dunes are formed when strong bidirectional wind blows. He also proposed that barchan dunes can be formed in the longitudinal troughs between multiple seif chains where the effect of cross winds is excluded. Glennie (1970) indicated that barchan dunes develop at lower wind velocities and longitudinal dunes are formed when the wind velocities are higher. Such modification occurs with the direction of prevailing winds. Borsy (1976) suggested that the longitudinal dune created by vortex currents within a unidirectional wind regime. It is known that the waves, currents and tides transfer the sands from the bottom of the sea to the beach, where the NW winds are very important for this process. On the beach the sands transferred by winds and deposited due to decreasing the wind velocity or/and due to the obstacles in the ground such as vegetation, broken rocks and changes, especially local relief in the landscape.

In the earlier century, the northern coastal strip had large numbers of the coastal sand dunes especially from Gamasa to Edku. Many authors in the past classified these dunes into older and younger or into lower and upper dunes (El Fishawi and El Askary, 1981; Stanley et al., 1992; El-Bady, 2001). The areas between Baltim and EL-Burg contains many dunes with longitudinal or linear type that are aligned sub-parallel to the dominant NW–SE winds. In the area without sand dunes there are small sand mounds, especially around bushes or vegetation on the beach parallel to the shoreline (EL-Banna, 2004). According to O'Keefe (1978), berms increased in width and height and formed coastal dunes due to the obstacles on berms. Where, the foredunes were built due to growing the berms and the presence of vegetation were seen as obstacles. Hesp (1999) defined the foredunes as vegetated sand dunes formed on the backshore by aeolian sand deposition within vegetation. In the eroded coast, foredunes occur either with other types or singly (Hesp, 2002). They are distinct from beach ridges that form during storm wave and elevated water level events (Psuty, 1965).

2. Occurrence and origin of coastal sand dunes

To determine the occurrences and characteristics of the coastal sand dunes in the northern Nile Delta along the Mediterranean Coast, the sedimentology of the beaches and sand dunes must be carefully studied. The main sites of the coastal sand dunes occur in the area between Baltim and Burg

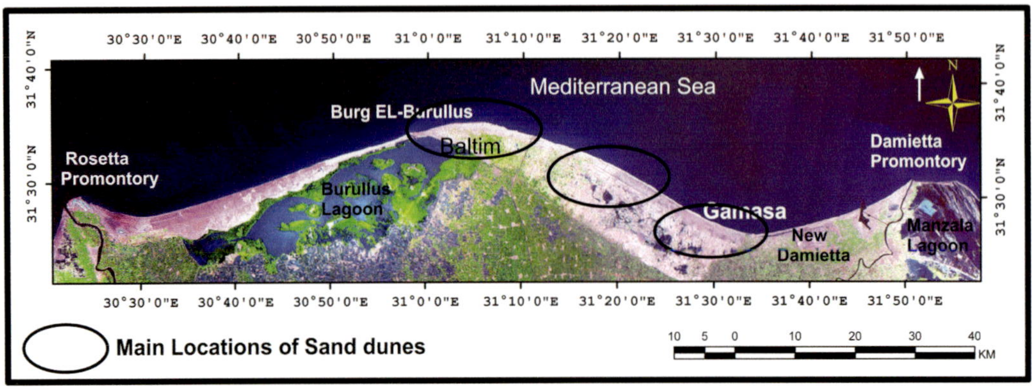

Figure 5.1. Location map of the study sites along the Mediterranean Coast.

ELBurullus, as well as some in the area between Gamasa and Baltim (Fig. 5.1). Longitudinal dunes are mainly found, of east Burg EL-Burullus towards Baltim, also small barchans and longitudinal dunes occur in the area west Gamasa towards Baltim.

There are three main sources of sediments in the Nile Delta called Rosetta, Damietta and Sebennitic branches where the Rosetta and Damietta branches are still present as well as the Sebennitic branch which was one of the principal Nile River distributaries during 7000 to 3000 B.C. (Stanley et al., 1992). Sediment was transported over centuries from west to the east by longshore currents and by aeolian processes.

The prevailing wind in this area comes from WNW, NW and N and the strong wind occurs during the winter (from December to March), especially during December and January with a minor trend from E and NE (El-Fishawi and El_Askary, 1981). The annual frequency shows that wind comes from the W to N sector with predominately NW direction, while strong storms advance the coast from W to NE (El-Asmar, 1994; Tetra Tech, 1984). Sesteni (1992) stated that the months of winter are the windiest with dominant winds from NW and WNW, less frequently from N and NE over the eastern Mediterranean Coast and in summer, as a result average wind speeds during summer are greater, in July the dominant winds are from WNW and W over central and eastern Mediterranean Coast of the Nile Delta.

The coastal sand dunes in the study area were formed from sand transferred from the bottom of the sea to the beach by waves and currents (longshore current) as well as by the wind on the beach. The sands that move to the beach are deposited as mounds sand and berms. Berms grow due to the advancing of the shoreline to form many dunes parallel to the shoreline. Where the small dunes (embryo dunes) formed on the backshore then large dunes (coastal sand dunes) formed later.

Steps of coastal sand dunes formation (Fig. 5.2a) as follows:

1. Sand that is from offshore and the eroded beaches is deposited on the beach (foreshore) due to the action of currents and waves
2. Due to the action of currents, waves and winds, the sand is transported from the foreshore to backshore to form dunes, berms and sand mounds parallel to the shoreline (Fig. 5.2c).
3. Obstacles, including vegetation can stop the migration of sand on the backshore and retard or prevent formation of such formations as dunes, berms and mounds during calm weather
4. Berms are the initial stage for dune formation parallel to the shoreline
5. Berms grow, where its widths and heights increase

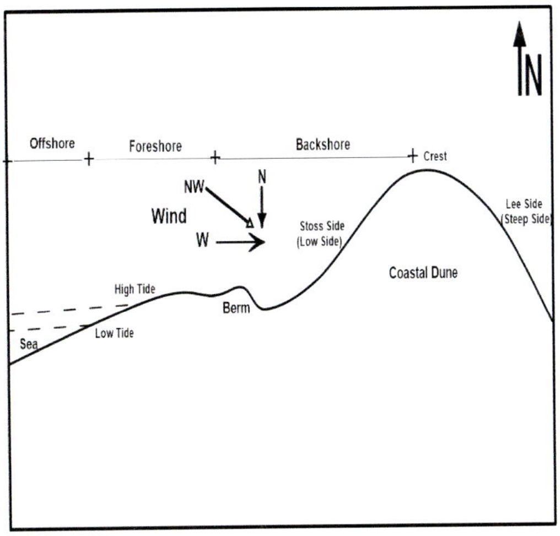

Fig. (a): formation of coastal sand dunes in he study area

Fig. (b): effects of waves and winds on the coastal sand dunes and beach

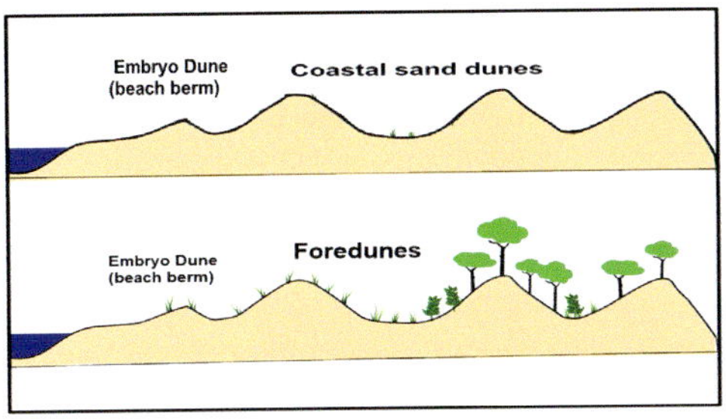

Fig. (c): profiles of coastal sand dunes and foredunes

Fig. (d): Barchans dunes between Baltim and Burg el-Burullus

Fig. (e): Longitudinal Gamsa dunes before 2000

Fig. (f): Small Gamasa barchan dunes after 2000

Fig. (g): Small Longitudinal or Linear Gamasa dunes after 2000

Fig. (h): belts of longitudinal dunes east of Burg EL-Burullus

Figure 5.2. (a–h) Illustrating mechanisms of dune formation and the resultant dune types.

6. By growing berms, coastal sand dunes or foredunes can be formed. Foredunes appear as the vegetated sand dunes and are parallel to the shoreline. They are formed on the backshore zone by deposition of sand within vegetation, where they are more or less completely stabilized by vegetation. Not all sand dunes on the backshore or foreshore zones are foredunes, but there are many types of coastal landforms such as coastal sand dunes types and ridges.
7. The coastal sand dunes that occurred in the study area are formed due to the accumulation of sand around different obstacles including sparse vegetation, thus they are not mainly stabilized by vegetation as in foredunes.
8. Due to coalescence of berms and their expansion, many coastal land forms are created (Coastal sand dunes and foredunes) (Fig. 5.2b).

In the area between New Damietta and Baltim, the coastal sand dunes and sand mounds were common before 2000 but now, due to urbanization, industrial and agricultural activities, the coastal sand dunes have disappeared. The sand mound and small longitudinal and barchan dunes formed but were removed within a short time due to the acceleration of human activities. Before 2000, longitudinal sand dunes and few barchan dunes were found in the area between New Damietta and Baltim, where high sand supply due to the strand plain is very wide with variable winds (bidirectional wind). These days, there are small mounds of sand and small isolated barchans on the wide strand plain. The coastal sand dunes (longitudinal) have occurred in the narrow strand plain, where, there is a very little sand supply (strand plain is very narrow) and almost unidirectional winds in the area between Baltim and Burg EL-Burullus. Longitudinal or linear sand dunes formed due to the growing of berms parallel to the shoreline. Longitudinal or linear dunes occur sub-parallel to the dominant NW winds. In the area between Baltim and Burg EL-Burullus, the longitudinal belts of coastal dunes occur on the beach. These longitudinal coastal sand dunes have been found adjacent to the shoreline for a long time ago. The strand plain of the area between New Damietta and Baltim is wide and without sand dunes (rarely small dunes do occur). This strand/beach supports many human activities, while the strand plain of the area between Baltim and Burg EL-Burullus is very narrow with a continous belt of longitudinal coastal sand dunes on the shoreline.

The movement of El-Burg dunes is at the rate of about 3 m/yr and the average movement of Gamasa–Baltem dunes is about 8 m/yr.

According to Lancaster (1995) there are five factors that control the type, alignment and size of dunes:

1. The characteristics of wind regime, especially its directional variability. This is the main factor determining the dune type.
2. The amount of sand supply is also a major reason to determine the types of dunes
3. The grain size of sand dunes
4. Vegetation cover
5. The role of time

2.1 The old situation

The results extracted from earlier studies (Glennie (1970); Bagnold (1971); Goldsmith et al. (1977); Rubin and Hunter (1982); Wasson and Hyde (1983)), and from many other authors such as El Fishawi and El Askary, 1981; Stanley et al., 1992; El-Bady, 2001; EL-Banna, 2004 show that there are mainly two reasons that determine the dune type, these are sand supply and wind velocities. Before 2000, barchan dunes were located mainly adjacent to the shoreline in the area between Burg El-Burullus and near the Kitchener drain (Fig. 5.2c) and present in small numbers near Gamasa where a limited amount of sand is available and strong winds occur. The longitudinal dunes on the other hand are recorded mainly west of Gamasa far from the shoreline (Fig. 5.2d), owing to large amounts of sand supply and low velocities of winds, they are also recorded in small numbers in the area between El-Burullus and the Kitchener drain.

The sand supply grows with the increase of the width of the strand plain from the Kitchener drain to the west of Gamasa city. The longitudinal dunes found near Gamasa, in this area the sand supply is increased and the wind velocities is lower than that at Baltim and El-Burge (El-Asmar, 2000). While, the sand supply is low in the area between Kitchener and Burg EL-Burullus due to the narrow strand plain, and the wind velocity is higher than that at Gamasa. Thus barchan dunes could have formed. Field observations show that barchan dunes occur near Gamasa, close the shoreline where the sand supply is low and the wind velocity is relatively high. Thus, the distance between dunes and shoreline is an important role to determine the type of dune.

The longitudinal dunes that have occurred in this area are probably due to the modification (by wind) of barchan dunes to longitudinal dunes where the sand supply is relatively high and wind velocity is relatively low. The dune becomes longitudinal in shape when the distance between the dunes and shoreline increases or due to the modification of barchan dunes.

2.2 *The current situation*

The study area between New Damietta City and Burg EL-Burullus is considered to be a very sensitive area, because of dune mobility brought on from many natural and human activities, especially on the narrow coastal strip where sand dunes are prevalent. With respect to coastal sand dunes, many of its formation factors have changed especially due to sand supply, vegetation and wind direction. The formation factors of the coastal sand dunes have alternated because of the speed of human activities. Most of these formation factors are the same in many of the studied areas but may have changed.

At present, the area between New Damietta City and Baltim does not allow dune formation as the dune formation factors have changed—the strand plain is now very wide. The impact of human activities has increased significantly from about 15 years ago. The area between Baltim and Burg EL-Burullus still contain the longitudinal belt dunes adjacent to the shoreline. The dune formation factors changed a little due to the narrow strand plain, where, the coastal sand dunes still retain their types, shapes and sizes. In the area between New Damietta City and Baltim there are many sand mounds and small barchans dunes near Gamasa (Fig. 5.2e) and, rarely, some small longitudinal dunes (Fig. 5.2h). The strong dunes belts occur in the area east of Burg El-Burullus, where they are connected with each other as longitudinal belts (Figs. 5.2f and 5.2g), here, these belts are considered the first defense against the sea.

Approximately 80% of coastal sand dunes in the study area are devoid of vegetation or vegetation is very sparse. Dunes are now suffering from destruction by human activities and many may disappear in the near future (El-Banna, 2004).

2.3 *Dune and beach relationship*

2.3.1 Materials and methods

About 40 samples were collected from coastal sand dunes in Gamasa, Baltim and Burg EL-Burullus and from the beaches in the same areas (27 samples of beaches and 14 samples of dunes). Beach samples collected from the shoreline and the dunes samples were collected from the bottom stoss side, top stoss side, bottom lee side, top lee side and the crest of coastal sand dunes.

About 100 grams of each sample were first soaked in water to facilitate their de-segradation. The carbonate content was removed using HCl (10%), each sample was treated with H_2O_2 (Fuchtbauer and Muller, 1970) to remove organic matter. The sand was separated by sieving using 0.063 mm sieve. The sand fraction was weighed and subjected to dry sieving using Ro-Tap shaker and a set of standard sieves of one Phi intervals. The values of mean sizes (Mz) and sorting (ʒ) were obtained from the formulas of Folk and Ward (1957) and were classified according to Folk (1968).

2.3.2 Sedimentological properties

At present, there are many belts of coastal sand dunes in the area between Baltim and Burg EL-Burullus. These dunes with longitudinal shape have formed adjacent to the beach. There are also some sand mounds and small longitudinal and barchans dunes in the area between new Damietta City and Baltim. In this study the longitudinal or linear dunes belts and the beach between Baltim and Burg EL-Burullus were carefully studied. In addition, we acquired some data details for dunes and beach between New Damietta City and Balim. From the sedimentological point of view, the collected surface sedimentary samples (beach and dunes) were subjected to carbonate – sand – mud analyses. The sand fraction reached to 99.99% in samples from the beach and 100% from samples collected in the dunes.

1. Average Grain Size of sand

Folk and Ward (1957) proposed the graphic mean size (Mz) which is much higher to the median as it takes the mean of diameter at three points on the cumulative curve as follow:

$$M_z = 1/3\ (\phi_{16} + \phi_{50} + \phi_{84})$$

El-Asmar (2000) suggested that there is a relation between threshold velocity and mean diameter of sand grains, where he stated that the lowest calculated wind velocity was detected at the Gamasa dunes and the highest values at the Baltim dunes. He also suggested the wind velocity at both Gamasa and Baltim in the range of 4 to 16 KT (205.6 to 822.2 cm/sec) indicating that these wind velocities are roughly able to transport equal amounts of sand with grain diameter in the range of 177 μm to 250 μm to the height between 5 to 10 m.

El-Burg beaches (3 samples) have Mz values that range from 2.13 ϕ to 2.53 ϕ (fine sand). The Baltim beach between two detached breakwater samples have Mz values from 1.55 ϕ to 2.4 ϕ. The samples were collected from east to west between two detached breakwaters (between two Tombola) (Table 5.2). The coarser and medium sand occurred adjacent to the detached break water and near to the center between two tombola, while the fine sand occurred near the body of a detached breakwater away from the center at each side.

The Gamasa beach is represented by five samples, they are fine sand where Mz values range from 2.06 ϕ to 2.65 ϕ, the samples near Kitchener (east of Kitchener drain) are medium sand where the Mz is 1.83 ϕ (Table 5.2 and Fig. 5.2h), this location is considered the beginning of erosion towards Baltim. El-Burg coastal sand dune samples showed that Mz of top lee side is 2.67 ϕ, the bottom lee side is 2.15, Mz at top of stoss side is 2.52 ϕ, and the bottom of stoss side is 2.15 ϕ (Table 5.3 and Fig. 5.2i). In the Baltim dune, the Mz of bottom of lee side is 2.25 ϕ, Mz of top lee side is 2.32 ϕ, the Mz of bottom stoss side is 1.56 ϕ, and the Mz of top of stoss side is 2.23 ϕ. Thus, finer sand occurs in the top and in the lee side of the dunes more than that at the bottom and in the stoss side of the dunes (Table 5.3 and Fig. 5.2d).

The Gamasa dunes are represented by six samples. Four samples for the linear dune and two samples for the small barchan dune. The linear dune of Gamasa show that the mean size (Mz) at the bottom of the lee side is about 2.37 ϕ and the top of lee side is about 2.62 ϕ, the bottom stoss side 2.25 ϕ and the top of stoss side is 2.53 ϕ (Table 5.3 and Fig. 5.2d).

In small barchan dunes, the Mz of stoss side is 2.45 ϕ and the lee side is 2.56 ϕ. Thus, finer sand occurs at the top and in the lee side of dunes more than that in the bottom and in stoss side of dunes. The Mz value of the beach is not an indicator for accretion or erosion, in general, the beach samples in the study area range from medium to very fine sand. The finer sand in the dunes occurs in the lee side and at the top of dunes more than in the stoss side and bottom of dunes.

The relation between Mz of the beach and Mz of the dune in each area showing that the Mz of beach is coarser than the corresponding dunes as in the Gamasa due to the large distance between the Gamasa beach and Gamasa dunes. On the other hand, the beaches and dunes of Baltim and El-Burg have nearly the same Mz due to the very small distance between the shoreline and dunes.

Table 5.2. The grain size parameters of the beach samples.

Samples	Location of beach	Md	Mz		δ	
1	EL-Burg Beach	2.1	2.13	F.S	0.54	M.W.S.
2		2.5	2.53	F.S	0.37	W.S.
3		2.45	2.48	F.S	0.57	M.W.S.
4	Baltim beach	2.35	2.4	F.S	0.51	M.W.S.
5		1.65	1.66	M.S	0.58	M.W.S.
6		1.8	1.85	M.S	0.39	W.S.
7		1.6	1.6	M.S	0.48	W.S.
8		1.3	1.33	M.S	0.5	M.W.S.
9		2.05	2.15	F.S	0.59	M.W.S.
10		1.85	1.91	M.S	0.45	W.S.
11		1.65	1.65	M.S	0.32	V.W.S.
12		1.55	1.55	M.S	0.37	W.S.
13		1.75	1.81	M.S	0.46	W.S.
14		1.8	1.88	M.S	0.42	W.S.
15		1.9	2	F.S	0.46	W.S.
16		1.65	1.65	M.S	0.26	V.W.S.
17		2	2.06	F.S	0.41	W.S.
18	West Baltim Beach	2	1.95	M.S	0.33	V.W.S.
19		2	2	F.S	0.30	V.W.S.
20	East Baltim Beach	2.05	1.99	M.S	0.65	M.W.S.
21		2.05	1.89	M.S	0.53	M.W.S.
22		2.2	2.21	F.S	0.31	V.W.S.
23	Gamasa Beach	1.8	1.83	M.S	0.57	M.W.S.
24		2.55	2.65	F.S	0.69	M.W.S.
25		2.1	2.13	F.S	0.81	M.S.
26		2.55	2.51	F.S	0.65	M.W.S.
27		2.45	2.45	F.S	0.34	V.W.S.

F.S. = fine sand, M.S. = medium sand, V.W.S. = very well sorted, W.S. = well sorted, M.W.S. = moderately well sorted

Thus, from the beach passing through the bottom and up to the top of dunes, the percentages of fine sand increases.

2. Sorting (measure of uniformity)

The values of sorting were obtained from the cumulative curves and calculated according to formulas of Folk and Ward (1957):

$$Sorting\ (\delta_I) = \phi_{84} - \phi_{16}/4 + \phi_{95} - \phi_5/6.6$$

The sorting of the collected samples classified according to Folk (1968) is as follows: In the El-Burg beach, the sorting ranges between moderately well sorted to well sorted (0.61 ϕ to 0.37 ϕ) (Table 5.2 and Fig. 5.2d). In the El-Burg dunes, the sorting of samples range between moderately well sorted to very well sorted (0.45 ϕ to 0.34 ϕ) (Table 5.3 and Fig. 5.2d). In the west Baltim beach, the sorting of samples ranges from 0.3 ϕ to 33 ϕ (very well sorted) (Table 5.3). In the Baltim beach, the sorting ranges between moderately sorted to very well sorted to (0.65 ϕ to 0.3 ϕ). In the east Baltim beach, the sorting ranges between 0.65 ϕ and 0.31 ϕ (from moderately well sorted to very well sorting) (Table 5.2 and Fig. 5.2i). In Baltim dunes, the sorting of samples ranges between

Table 5.3. The grain size parameters of the dune samples.

Samples		Location of Dunes		Md	Mz		δ	
1	Burg el-Burullus Dunes	Linear	Stoss side bottom	2.2	2.15	F.S.	0.44	W.S.
2			Stoss side top	2.62	2.52	F.S.	0.35	V.W.S.
3			Lee side bottom	2.23	2.15	F.S.	0.45	W.S.
4			Lee side top	2.9	2.67	F.S.	0.38	W.S.
5	Baltim Dunes	Linear	Stoss side bottom	1.53	1.56	M.S.	0.37	W.S.
6			Stoss side top	2.11	2.23	F.S.	0.47	W.S.
7			Lee side top	2.22	2.32	F.S.	0.45	W.S.
8			Lee side bottom	2.45	2.25	F.S.	0.56	M.W.S.
9	Gamasa Dunes	Dune 1	Lee side	2.6	2.57	F.S.	0.39	W.S.
10		Small barchan	Stoss side	2.65	2.45	F.S.	0.38	W.S.
11		Dune 2 Linear	Stoss side top	2.52	2.53	F.S.	0.34	V.W.S.
12			Stoss side bottom	2.22	2.25	F.S.	0.54	M.W.S.
13			Lee side top	2.62	2.62	F.S.	0.32	V.W.S.
14			Lee side bottom	2.44	2.37	F.S.	0.44	W.S.

moderately well sorted to well sorted (0.56 ϕ to 0.37 ϕ) (Table 5.3 and Fig. 5.2d). In the Gamasa beach the sorting of samples ranges from moderately sorted to very well sorted (0.81 ϕ to 0.34 ϕ) (Table 5.2 and Fig. 5.2c). In the Gamasa dunes, the sorting of samples ranges between moderately well sorted to very well sorted to (0.54 ϕ to 0.32 ϕ) (Table 5.3), where the sorting improved at the top and in the lee side of dunes (Table 5.3 and Fig. 5.2d)

Anwar et al. (1979), Frihy and Lotfy (1994) and EL-Askary and Lotfy (1995) recorded that the accreted sands are coarser and less well sorted than the eroded sands. Each beach has the eroded and accreted areas adjacent to each other, and thus the sorting changes many times in the same beach from one locality to another. Thus, there is not a stable site in the study area and it is difficult to detect the change in the sorting from Rosetta to Ras EL-Bar, but each one can be studied as an individual beach.

The level of sorting improved from the beach to the dune as in, Gamasa (improved from Gamasa beach to Gamasa dunes), Baltim (improved from Baltim beach to Baltim dunes) and El-Burg (improved from EL-Burg beach to EL-Burg dunes). The sorting may be developing on the top and lee side of dunes or may be the same without any changes on each side of and on the top and bottom of the dune. The sorting advances from the beach to the dune because of the wind action on the fine grains where, the wind moves them from the beach to the dune and from the dune bottom to the top (Fig. 5.3, Fig. 5.4).

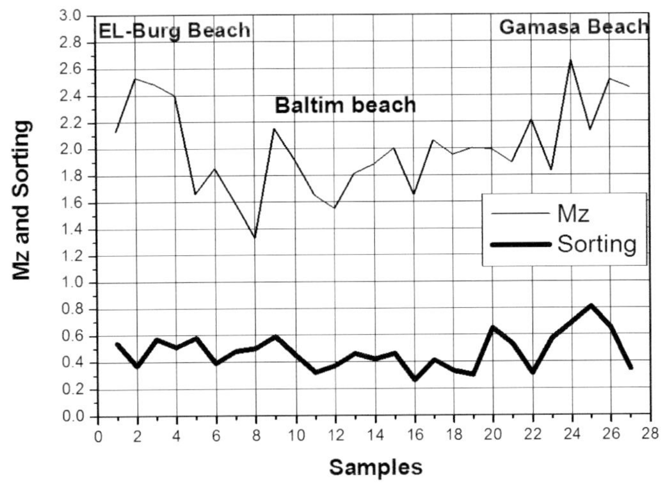

Figure 5.3. Mz and Sorting of the beaches of the study area.

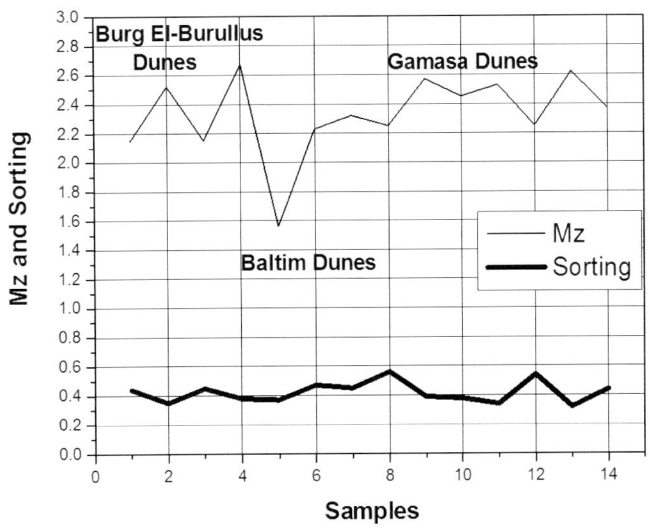

Figure 5.4. Mz and sorting of the coastal sand dunes of the study area.

3. Conclusions

The sand transferred from the bottom of the sea to the beach is drifted by waves, winds and currents. These sands move to the beach to be deposited as sand mounds, sand sheets and berms, these berms grow due to the advancing of the shoreline to form many dunes parallel to the shoreline.

In the study area, there are different beaches and dunes in many sub areas. Each sub area has its beach and dune containing a special beach/dune system. In the sub area between Gamasa and Kitchener, the beach is almost flat and wide. The width of the beach and strand plan as well as the fast drying of the sand of both the beach and strand plan by winds and the sun led to the formation and movement of longitudinal dunes. In the sub area between the Kitchener and Burg El-Burullus, the conditions are different where the beach is very narrow and steeper. Thus, the sand in this area dries faster by winds and the sun, the wind velocity in this area is also higher than that at Gamasa. Accordingly, the sand moves up and the barchan dune in this area are higher than the longitudinal dune at Gamasa. Now there are longitudinal dunes in the area between Baltim and Burg EL-Burullus, while small dunes and sand mounds occur in the area between Gamasa and Baltim. The Mz of beach is coarser than the corresponding dunes as in Gamasa. On the other hand, the beaches and dunes of Baltim and El-Burg have nearly the same Mz due to the very small distance between the shoreline and dunes. Sorting has improved from beaches to the sand dunes on the backshore.

Recently, most of coastal sand dunes have disappeared due to the agricultural, industrial and urbanization activities. The remaining coastal sand dunes are now concentrated in the coastal strip between Baltim and EL-Burg with the coastal area without any dunes such as in New Damietta and Gamasa as well as the coastal area between Gamasa and Baltim. All coastal areas are under threat from impacts of climate change (sea level rise and storm surge) and from human activities (Ali and El-Magd, 2016).

References

Abu-Zeid, M.A. and El-Shibini, F.Z. 1997. Egypt's High Aswan Dam. Int. J. Water Resour. Dev. 13(2): 219–217.

Ahmad, M.H. 2002. Multi-Temporal Conflict of the Nile Delta Costal Changes, Egypt National Authority for Remote Sensing and Space Science (NARSS). Littoral 2002, The Changing Coast, EuroCoast/EUCC, Porto-Portugal, ISBN 972-8558-09-0.

Ali, E.M. and El-Magd, I.A. 2016. Impact of human interventions and coastal processes along the Nile Delta coast, Egypt during the past twenty-five years. Egyptian Journal of Aquatic Research (2016), http://dx.doi.org/10.1016/j.ejar.2016.01.00.

Anwar, Y.M., Gindy, A.R., El-Askary, M.A. and El-Fishawi, N.M. 1979. Beach accretion and erosion, Brullus-Gamasa coast, Egypt. Marin Geol. 30: M1–M7.

Bagnold, R.A. 1971. The Physics of Blown Sand and Desert Dunes. 5th Impression, Chapman and Hall. London. P. 265.

Borsy, Z. 1976. Relief forms of windblown sand. pp. 134–137. *In*: Geomorphology and Palaeogeography, Section 1, I. p. Gerasimov, ed., Moscow: 23rd International Geographical Congress. Distributed by Pergamon Press, Ltd., Oxford.

El-Askary, M.A. and Lotfy, M.F. 1995. The use of texture and heavy mineral properties of sand as indicators of the advancing and receding of the Nile Delta beaches, Egypt. N.jb. Geol. Palaont. Mh. 257–270 p.

El-Asmar, H.M. 1994. Severe coastal Damage along Ras El Bar shoreline, North of Nile Delta: An effect of the construction of the detached breakwater system. Journ. Geol. 38(2): 793–812.

El-Asmar, H.A. 2000. Geoenvironmental studies on the coastal area between Gamasa and Baltim, North of the Nile Delta. Z. Geomorph. N.F. 44(1): 59–73.

EL-Bady, M.S.M. 2001. Geoenvironmental Study on the Coastal Area between Gamasa and Baltim, Northeast of the Nile Delta, Egypt. M.Sc. Thesis, Damietta Fac. of Sci., Mansoura Uni. Egypt.

El Banna, M.M. 2004. Nature and human impact on Nile Delta coastal sand dunes, Egypt. Environmental Geology 45: 690–695.

El-Banna, M.M. and Frihy, O.E. 2009. Human-induced changes in the geomorphology of the northeastern coast of the Nile Delta, Egypt, Geomorphology 107: 72–78.

El-Bouseily, A. and Frihy, O. 1984. Textural and Nile branch on the Mediterranean coast, Egypt. Journal of African Earth Sciences 2: 103–107.

El Fishawi, N.M. and El Askary, M.A. 1981. Characteristic features of coastal sand dunes along the Burullus-Gamasa stretch, Egypt.

Folk, R.L. and Ward, W.C. 1957. Brazos river bar: A study in the significance of grain size parameters. J. Sed. Petrol. 27: 3–26.

Folk, R.L. 1968. A review of grain size parameters. Sedimentology 6: 73–93.

Frihy, O. and Komar, P.D. 1993. Long-term shoreline changes and the concentration of heavy minerals in beach sands of the Nile Delta, Egypt. Mar. Geol. 115: 253–261.

Frihy, O. and Lotfy, M. 1994. Mineralogy and textures of beach sands in relation to erosion and accretion along the Rosetta promonotory of the Nile Delta, Fgypt. J. Coast. Res. 10: 588–599.

Fuchtbauer, H. and Muller, G. 1970. Sediments and Sedimentgesteine. Springerver Lag. Stuttgart, 726 pp.

Glennie, K.W. 1970. Desert Sedimentary Environments, Developments in Sedimentology. Elsevier 14, p. 222.

Goldsmith, V., Hennigar, H.F. and Gutman, A.L. 1977. The VAMP coastal dune classification. pp. 26-1–26-20. *In*: Goldsmith, V. (ed.), Coastal Processes and Resulting Forms of Sediment Accumulation, Currituck Spit, Virginia/North Carolina. SRAMSOE No. 143, Virginia Institute of Marine Science, Gloucester Point, VA.

Hesp, P.A. 1999. The beach backshore and beyond. pp. 145–170. *In*: Short, A.D. (ed.). Handbook of Beach and Shoreface Morphodynamics. Chichester. J Wiley & Sons, NY.

Hesp, P.A. 2002. Foredunes and Blowouts: initiation, geomorphology and dynamics. Geomorphology 48: 245–268.

Lancaster, N. 1995. Geomorphology of Desert Dunes. Routledge, London. 290 pp.
OKeefe, P.D. 1978. Sediment budgeting, beach conservation. Beach Protection Authority, Queensland, Australia.
Psuty, N.P. 1965. Beach-ridge development in Tabasco, México. Annals Assoc. Amer. Geog. 55: 112–124.
Rubin, D.M. and Hunter, R.E. 1982. Bedform climbing in theory and nature. Sedimentology 29: 121–138.
Sestini, G. 1992. Implication of climatic changes for the Nile Delta. pp. 555–601. *In*: Jefftic, L.J., Milliman, J.D. and Sestini, G. (eds.). Climatic Changes and the Mediterranean. Edward Arnold, New York.
Sharaf El Din, S. 1977. Effects of Aswan high Dam on the Nile flood and on the estuarine and coastal circulation pattern along the Mediterranean Egyptian coast. Limnol. Oceanogr. 22(2): 194–207.
Stanley, D.J., Warne, A.G., Hugh, D.R., Bernasconi, M.P. and Chen, Z. 1992. The late Quaternary north-central Nile Delta from Manzala to Burullus lagoons, Egypt. National Geographic Research and Exploration 8(1): 22–51.
Tetra Tech. 1984. Shoreline master plan for the Nile Delta coast. Progress Report No. 1.
Wasson, R.J. and Hyde, R. 1983. Factors determining desert dune type. Nature 304: 337–339.

Chapter 6

Algerian Sahara Sand Dune (Western Erg) Characterization

An Experimental Study

Maazouzi Abdelhak,[1,*] *Zahraoui Brahim,*[1] *Lahcene Driss*[1] and *Kettab Ahmed*[2,3]

1. Introduction

The western erg in the Algerian Sahara spreads on the right strands (east) of the Zousfana and the Saoura Rivers, southbound, to the west it is attached to the erg El Atchane, that links up more at the west to the erg El Raoui in the region of the Daoura River on its turn and until the Algerian-Moroccan border, close to the anti-Atlas. More to the south, these ergs are less accessible to the different national roads and spread until the Algerian Mauritanian and Malian borders, these are: the *Chech* erg and the *Iguidi* erg (Fig. 6.1). Geologically these ergs dated from the Plioquaternaries and spread over a large area. Although aspects of the dune sand have been very extensively studied (Ahammed et al., 1996; Mantz and Eng, 2004; Hamoda et al., 2004; Maazouzi et al., 2007; Maazouzi et al., 2008; Zri et al., 2009) one objective of the work reported here is to promote the need for a

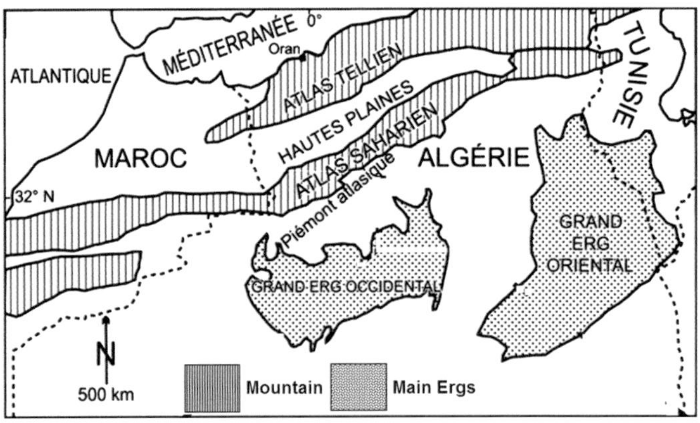

Figure 6.1. Localization of western erg (Yann, 1991).

[1] Chemistry and Environmental Sciences Laboratory (CESL) Bechar University 08000, Algeria.
[2] Associate Professor University of Bouira-Algeria.
[3] Professor/Research Director of the National Polytechnic School-Algiers. Algeria.
* Corresponding author: maazdz@yahoo.fr

more comprehensive scientific survey of this vast area. This should be undertaken soon to determine the nature and extent of economic and various other potentials that the dune systems can offer.

2. Materials and methods

The suitability of the sand from the western erg for wastewater pretreatment was assessed. The sand used in this work, is from sites in Beni Abbes, Taghit and Timimoun (Fig. 6.1) that comes directly from the western erg (sand dunes) in the Algerian Sahara.

The sand was physically (Table 6.1) and chemically (Table 6.2) characterized. Mean diameter, uniformity coefficient, porosity and permeability were considered for the physical characterization. In addition to pH and conductivity, oxides: Cr_2O_3, Fe_2O_3, Al_2O_3, K_2O, MnO, Na_2O, MgO and SiO_2 contents were determined to characterize chemical attributes of Algerian Sahara sand dunes (western erg).

Table 6.1. Physical characterization of the sand dune.

	Equivalent diameter (mm)	Uniformity coefficient (CU)	Permeability 10^{-4} (m/s)	Mass volumic (kg/m^3)	Porosity (%)
Beni Abbes	0.22	1.39	0.976	2.63	41.39
Taghit	0.19	1.23	0.906	2.55	42.28
Timimoun	0.17	1.91	0.983	2.56	38.14

Table 6.2. Chemical characterization (%) of: SBL: Beni Abbes Sand washed; STAL: Taghit Sand washed and STIL: Timimoun Sand washed.

	Cr_2O_3	Fe_2O_3	Al_2O_3	K_2O	Na_2O	MnO	CaO	MgO	SiO_2	pH	Conductivity (µS/cm)
SBL	0.004	0.18	0.17	0.008	0.008	00	1.15	0.015	97.44	6.97	59
STAL	0.004	0.12	0.14	0.007	0.007	00	0.76	0.006	98.20	6.69	43.7
STIL	0.004	0.17	0.16	0.01	0.009	00	1.004	0.016	97.11	6.80	157.7

3. Results and discussion

3.1 Physicochemical characterization of the sand dunes

Their characterization confirmed their potential to be used as a filtration media, since the equivalent diameter, the uniformity coefficient, porosity and the permeability were in the ranges (0.17–0.22 mm), (1.91–1.39), (38–41%) and (0.906–0.983 10^{-4} m/s) respectively, the insoluble matter rate is about 98% (composed essentially of quartz), the weak organic matters and the granular analysis show that it is very uniform, meaning that (sand grain size 6) it can be characterized as small to medium. Physical (Table 6.1) and chemical (Table 6.2) characterization of the sand of dunes showed that the characteristics of sand used is close to those reported in available literature (Maazouzi et al., 2007; Touil et al., 2009).

The sand belongs to the category of fine quality as confirmed by weak porosity (38–41%). This characteristic of the sands can be quantified also by the equivalent diameter observed by the Scanning Electron Microscopy (SEM) (Fig. 6.2) and validated by the granular analysis. As noted in the Table 6.1 the uniformity coefficient (CU) is lower, down to 2, by convention, if the CU is between 1 and 2, the granulometry is said to be uniform (Gaillard, 1994; Ruban, 1996; Molle, 2003).

The results of Table 6.2 show that quartz is the most represented mineral. The presence of iron and aluminum oxides can be an indicator of the presence of FeO(OH); Fe_2O_3; $Al(OH)_3$, that are recognized as minerals of strong reactivity and as strong adsorbents of phosphorus (Molle, 2003). The calcium oxide very probably presents the concentrations due to the presence of the calcite ($CaCO_3$).

Figure 6.2. Observation to the Scanning Electron Microscopy (SEM): Sand dune of (a): Beni Abbes, (b): Taghit and (c): Timimoun).

3.1.1 X-Ray analysis of sand dunes

A good knowledge of the mineralogical composition is necessary to anticipate the difficulties that could arise in the process of valorization and specification of the mineralogical phases present in the sand. Analysis of X-rays used (Bruker D5005), has allowed one to see that the Algerian Sahara sand dunes (western erg) are composed essentially of quartz (SiO_2) Fig. 6.3.

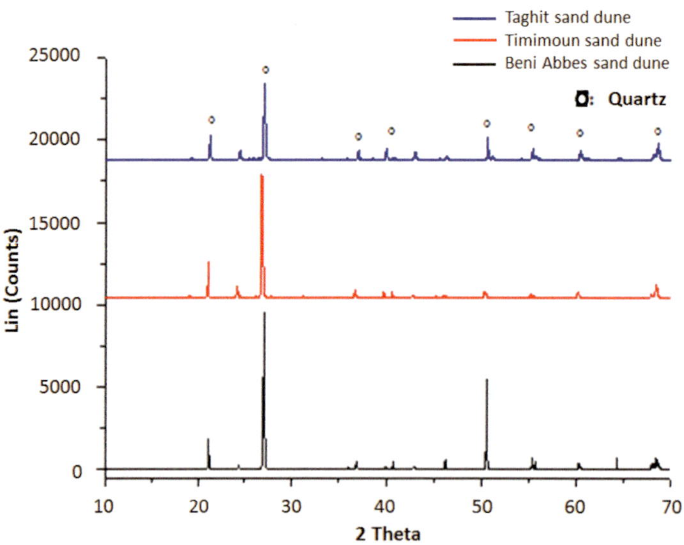

Figure 6.3. XRD patterns of samples Sand dunes [Beni Abbes (black), Taghit (blue) and Timimoun (red)].

3.1.2 Absorbance of the filtrates after passage through sand dunes

According to literature, the observations to the Scanning Electron Microscopy (SEM) of the sand show that the grains are covered with a deposit (Fig. 6.4); which is probably due to the clayey phase.

The time of washing of Beni Abbes sand lasted 65 minutes, in contrast to the ones of (Taghit and Timimoun) (90 and 95 minutes respectively). The reduction of the absorbance according to the time of washing observed (Fig. 6.5) for all wavelengths (λ) used, the maximum value of the absorbance took place in the UV domains that were equal to 0,025 for λ = 254 nm (Beni Abbes sand), a less important absorbance is observed in the domain of the visible 0,017 for λ = 400 (Beni Abbes sand) for the first filtrate (AbsF1). The same observations have been noted in the sand of Taghit and Timimoun.

The observations under a microscope (x 80) of the grain of sand (Fig. 6.6), on the top of the sand pile after the passage of washing water clearly show that a washing took place.

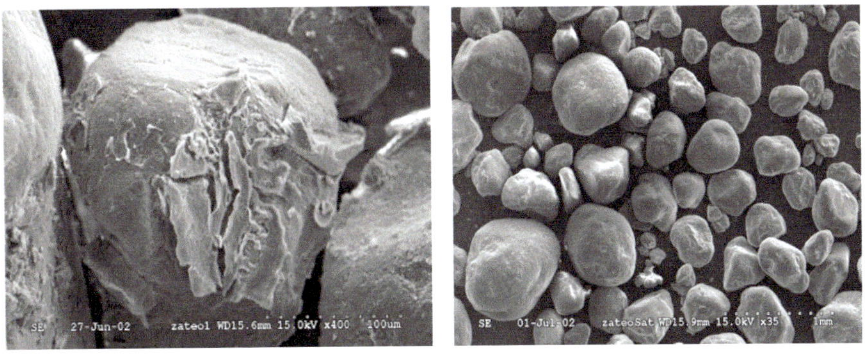

Figure 6.4. Observation to the Scanning Electron Microscopy (SEM): of ZATEO-SAT sand (Ardois et al., 2002).

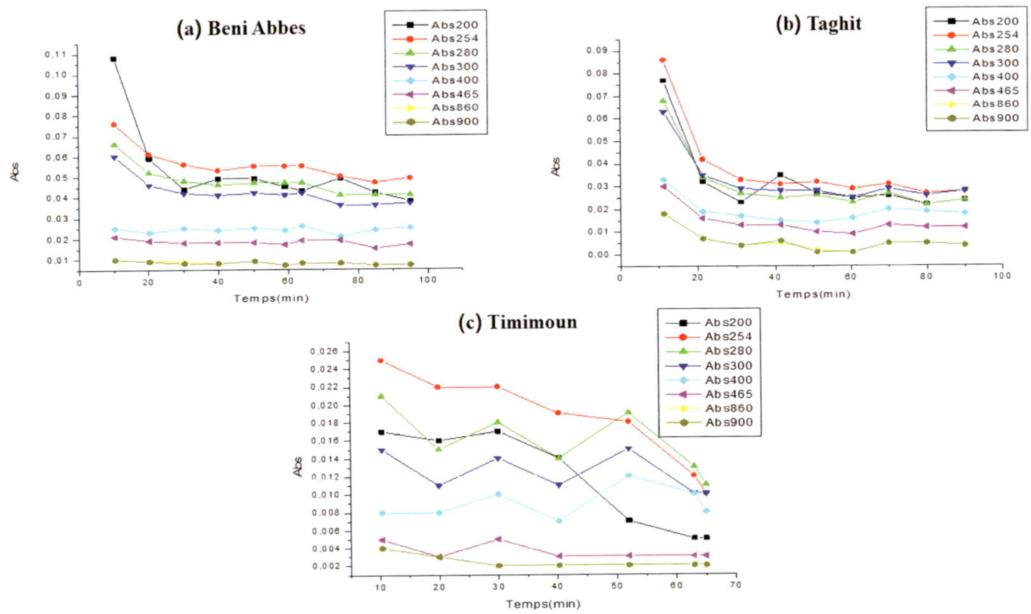

Figure 6.5. Evolution of the absorbance to different lengths of waves according to the time of washing. (a): Beni Abbes, (b): Taghit and (c): Timimoun.

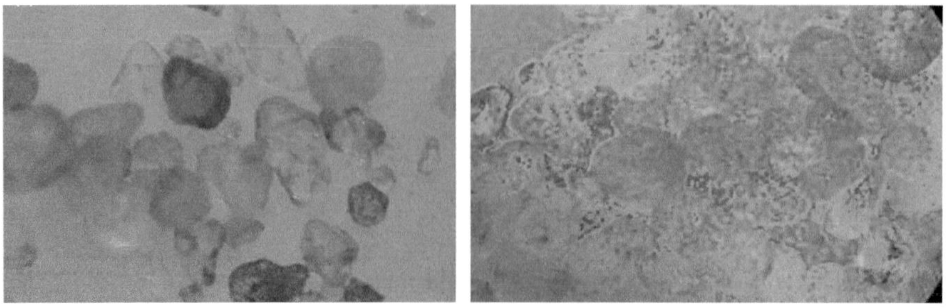

Figure 6.6. Observations under a microscope x 80 of the grain of sand, (L) washed sand and (R) unwashed sand.

3.1.3 IR spectrum of sand dunes

The IR spectrums have been recorded by means of a spectrophotometer thermo Nicolet AVATAR 320 FT- IRS. Examination of the IR spectrum of the raw sand samples and after washing (Figs. 6.7 to 6.9) shows that some strips in the specters of raw sample disappear or move after washing, probably related to the presence of the impurity related with clay. According to (Elouazzani, 2005) the strips 467, 698, 1031 cm^{-1} and 755, 789 cm^{-1} and 538 cm^{-1} are, respectively, which is characteristic to the vibration of the link Si-O and Si-O-Al of kaolinite, and the strip 875 is characteristic to the vibration of CO_3 of calcite; results from this analytic technique confirm those from chemical analysis, and the majority phases in the sand are from calcite ($CaCO_3$) and quartz (SiO_2).

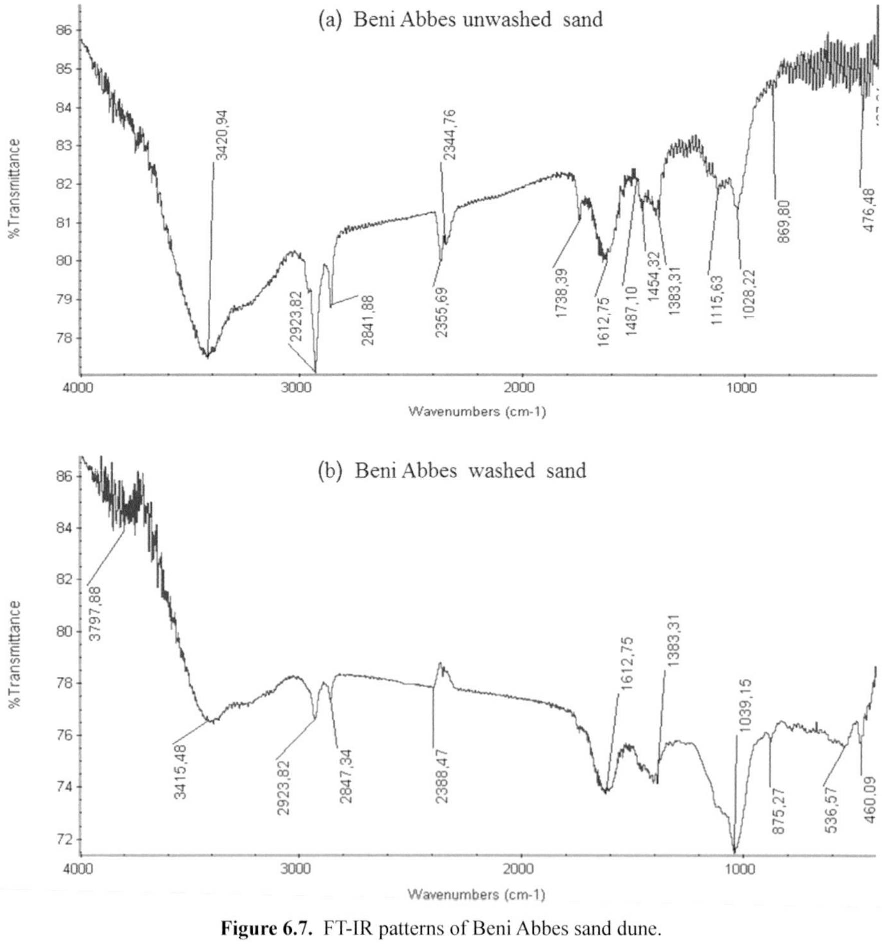

Figure 6.7. FT-IR patterns of Beni Abbes sand dune.

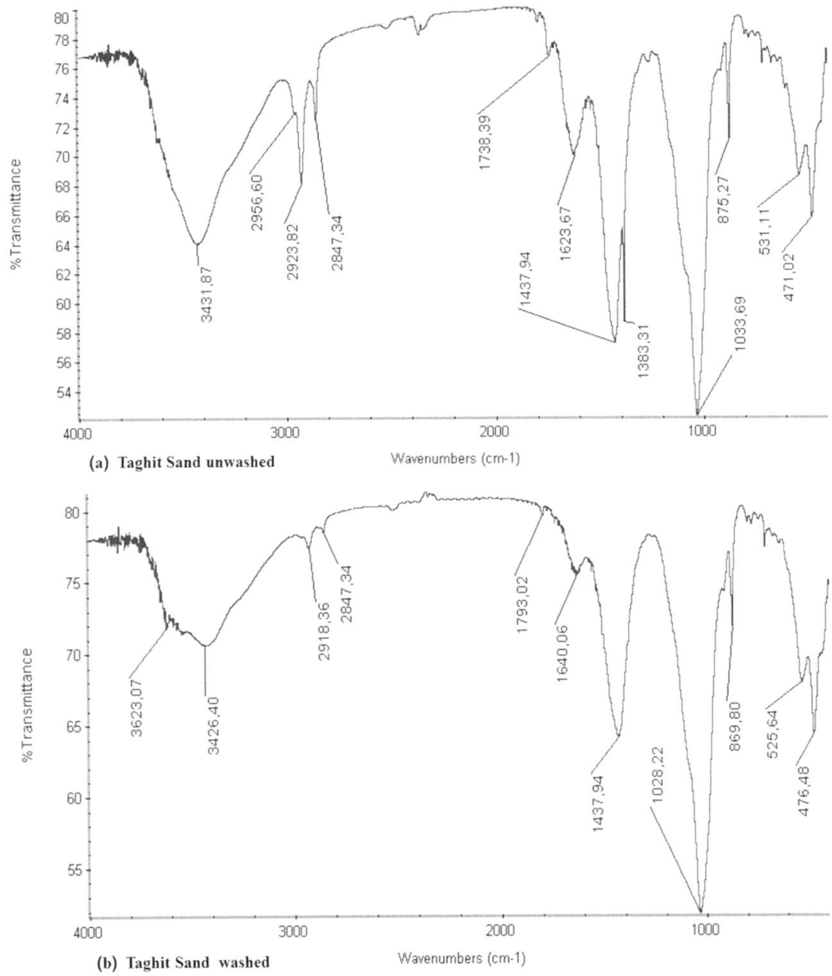

Figure 6.8. FT-IR patterns of Taghit sand dune.

4. Conclusion

The chemical analysis of the Algerian Sahara sand dunes (Beni Abbes, Taghit and Timimoun) shows that it contains 98% of SiO_2, considered as being the insoluble part. A small part of CaO, Al_2O_3 and Fe_2O_3 was confirmed by FT-IR analysis and the X-Ray analysis as well as the X fluorescence analysis. The majority phases present in the Algerian Sahara sand dunes are quartz (SiO_2) and calcite ($CaCO_3$). The erg sand dune presents favorable features for its use as being a bed filtering in pretreatment of waste water, it is believed that prior treatment (sand washing) can improve its performances taking into account the results from this work. The area has the potential for safe disposal of sewerage and other effluents but other options for utilization need to be assessed.

A comprehensive scientific survey of this vast area should be undertaken to determine the nature and extent of economic and various other potentials that the dune systems can offer.

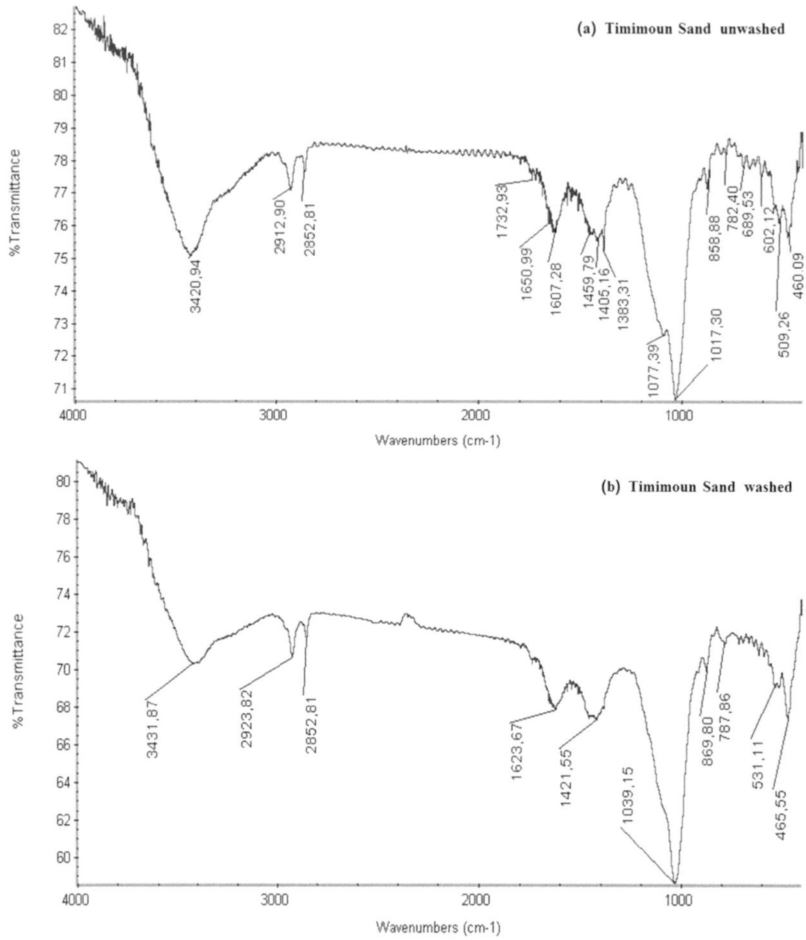

Figure 6.9. FT-IR patterns of Timimoun Sand dune. (a): Timimoun Sand unwashed; (b): Timimoun Sand washed.

Acknowledgement

The authors would like to express their gratitude to the MESRS, Algeria, for financial support.

References

Ahammed, M.M. and Chaudhuri, M. 1996. Sand-based filtration/adsorption media. Aqua – J. Water Supply: Res. Technol. 45: 67–71.

Ardois, C., Collin, J., Mallet, C. and Roujou, J.L. 2002. Caractérisation du sable éolien de Tchernobyl utilisé pour les études de sorption. Notetechnique IRSN/DPRE/SERGD n° 02-72.

Callot, Yann. 1991. Histoire d'un massif de dunes, Grand Erg Occidentale (Algérie).Géodynamique des milieux continentaux, Sécheresse 2: 26–39.

Elouazzani, D. 2005. Caractérisation physicochimique et valorisation en bâtiment et TP des cendres issues de l'incinération des boues de papeterie. Thèse de doctorat, Institut des sciences appliquées de Lyon.

Gaillard, B. 1994. Les méthodes de traçages pour l'étude des écoulements souterrains. Cours DESS Hydrogéologie, Université Grenoble I.

Hamoda, M.F., El-Ghusain, I. and Al-Mutairi, N.Z. 2004. Sand filtration of wastewater for tertiary treatment and water reuse. Desalination 164: 203–211.

Maazouzi, A., Kettab, A. and Badri, A. 2007. Etude de procédés de filtration sur sable de la région de Béchar en pré traitement de l'eau potable. Desalination 206: 358–368.

Maazouzi, A., Kettab, A. and Badri, A. 2008. Contribution à l'expérimentation de la filtration de l'eau potable en prétraitement dans quelques lits filtrants (sable) de la région de Bechar. 4ème conférence internationale sur : les Ressources en Eau dans le Bassin Méditerranéen Alger Watmed4.

Maazouzi, Kettab, A., Zahraoui, B., Khelfaoui, R. and Badri, A. 2009. Contribution à la mise en place d'un protocole de lavage de lit filtrant de sable pour les eaux potable à l'échelle de laboratoire. SéminaireMatièreOrganiquesetEnvironnement. Saint maxime (France).

Mantz, D.H. and Eng, P. 2004. New horizons for slow sand filtration, Proc. 11th Canadian National Conference and Second Policy Forum on Drinking Water and Biennial Conference of the Federal-Provincial-Territorial Committee on Drinking Water, Promoting Public Health Through Safe Drinking Water, Calgary, Alberta, 3–6 April 2004, pp. 682–692.

Molle, P. 2003. Filtre plantes des roseaux: limites Hydrauliques et rétention du phosphore. Thèse de doctorat, Université de Montpellier II.

Ruban, G. 1996. Mesure continue de la pollution par voie optique en assainissement Principe généraux et mise en œuvre, LCPC: compte rendu de synthèse, ERG2.72.10.6, 51p.

Touil, Y., Taha, S., Issaadi, R. and Amrane, A. 2009. Pilot plant for wastewater treatment involving septic pit and biological filtration on sand of dunes of the Algerian Sahara. Desalination and Water Treatment 10: 148–152.

Zri, A., Abriak, N.E. and Benzerzour, M. 2009. Valorisation des sediments dans le beton desable a base du sable de dragage. Sbeidco – 1st International Conference on Sustainable Built Environment Infrastructures in Developing Countries ENSET Oran (Algeria).

Chapter 7

Economic Heavy Minerals in the Coastal Area Between Burg EL-Burullus and Baltim, North of Nile Delta, Egypt

Mohamed S. EL-Bady

1. Introduction

Before the construction of the Aswan High Dam in 1964, the Mediterranean Sea received 134 million tons of sediments annually and about 55.5 billion m^3 of water (Abu-Zeid and ElShibini, 1997; Sharaf El Din, 1977). After the construction of the Aswan High dam, the quantity of sediments (deposited to the Mediterranean Sea) has decreased. Lesser amounts of sediments had negative effects on the agricultural croplands lands and altered most of the coastal geomorphologic units.

Many authors studied the sedimentology and mineralogy of the Nile Delta coast in relation to accretion rates and erosion of the beach (Anwar et al., 1979; El-Bousely and Frihy, 1984; Frihy and Komar, 1993; EL-Asmar, 2000; Ahmad, 2002 and others). They suggested that near the river promontories the sand is finer with the coarse sand transported to accreted areas by longshore currents. They noted that heavy minerals concentrated among the finer sediments in the eroded areas near the river mouth with decreasing amounts carried by longshore currents to the accreted areas. The most dominant minerals in the dune sand comprised of opaque minerals, amphiboles, pyroxenes, epidotes, garnet, zircon, rutile and tourmaline (Table 7.1). These heavy mineral assemblages characterize the Nile deposits (Shukri, 1950).

El-Fishawi and Molnar (1985) suggested that the normal deposition of heavy mineral occurs near the shoreline. The concentration of heavier residues such as opaques, garnet, zircon, tourmaline and rutile markedly increase, while amphiboles and pyroxenes decrease in moving from the breaker zone across the beach and backshore and up to the dune.

Erosional beaches are characterized by the concentration of high-density heavy minerals while relatively low-density heavy minerals such as amphiboles and pyroxenes are concentrated in the accretional areas. Low-density heavy minerals are winnowed by waves and currents and move offshore (Frihy and Komar, 1993). According to many authors such as (El-Fishawi and Molnar, 1985; Frihy, 1994; Frihy and Lotfy, 1994; EL-Askary and Lotfy, 1995). The finer grained and high density heavy minerals (opaques, garnet, rutile, zircon and monazite) are concentrated in eroded

Department of Geological Sciences, National Research Centre, Dokki, Cairo, Egypt.
Email: elbady_nrc@yahoo.com

Table 7.1. Locations of samples in the study area.

Samples	Location	Samples	Location
1	Shoreline EL-Burg	7	Lee side middle dune
2	Backshore, after the first sample by 50 m	8	Right side of international road toward Baltim
3	On the bottom of dune	9	Shoreline, east of Baltim resort
4	Top of dune	10	Backshore, east Baltim resort
5	Stoss side bottom dune	11	Left side of international road toward Gamasa
6	Lee side bottom dune		

areas such as the Nile Delta promontories at Damietta and Rosetta, while the areas of the coastline which suffer from accretion with shoreline advances are associated with the concentration of coarser sediments and lower density heavy minerals (augite, hornblende, tourmaline and epidote). Where the lower density heavy minerals and the light minerals (quartz and feldspars) removed by the longshore currents to the accreted areas.

The coastal area from Gamasa (31.44 N 31.52 E) and Burg EL-Burullus contain many important geomorphologic units such as coastal sand dunes, beaches, wetlands and reclamation lands. Coastal sand dunes and beaches are very important units in the area between Gamasa and Burg EL-Burullus, where the coastal sand dunes are of two kinds—old and young. The young dunes are of two types, barchans and longitudinal. The old dunes formed from the sands derived from the eroded old Sebennitic Promontory dating from about 2500–2000 B.P. (Stanley et al., 1992). The young dunes are mobile composed of fine sands, located along the shoreline and extend backshore to a maximum of 1000 m. Barchans are located on Baltim and Burg EL-Burullus beaches where plenty of sand has accumulated on the beach, but the backshore dune field is very narrow (Ahmed and El-Asmar, 2010; Ahmed et al., 2010). The longitudinal dunes are located along the Gamasa coast and southward where there is more sand and where a wide backshore dune field is present. According to El-Asmar and Haifaa (2013) the coastal sand dune area was estimated to be 165.4 km^2 in 1994. The decreasing area was monitored from 1997 (110.3 km^2) as it went to 64.3 km^2 in 2006, but had shrank to 9.3 km^2 by 2010. due to the effect of sand mining, land reclamation and the proliferation of fish farms.

2. Materials and methods

The study area is located between Burg EL-Burullus and Baltim (Fig. 7.1, Table 7.2), where the area contains belts of coastal sand dunes of barchan and longitudinal types and beaches. Two profiles perpendicular to the shoreline were studied for heavy mineral concentrations (Fig. 7.2 and

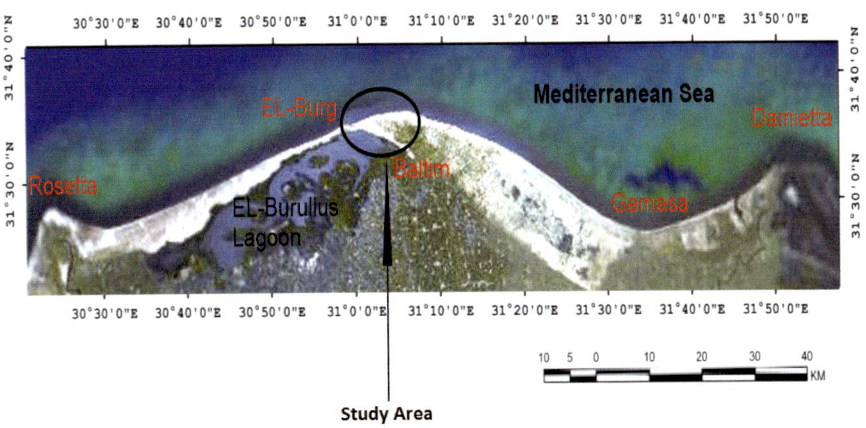

Figure 7.1. Location map for the area between Burg EL-Burullus and Baltim.

Table 7.2. Mineralogical analysis of the samples of the area between Burg EL-Burullus and Baltim.

Minerals	**Samples**										
	1	**2**	**3**	**4**	**5**	**6**	**7**	**8**	**9**	**10**	**11**
Opaque	35.9	39.6	44.1	38.8	33.2	46.3	42.7	23.4	37.6	30.9	34.3
Non Opaque	64.1	60.4	55.9	61.2	66.8	53.7	57.3	76.6	62.4	69.1	65.7
Amphipole (Horn.)	22.5	23.1	23.3	17.9	23.3	17.7	30.2	28	21.3	25.1	27.6
Pyroxene (Aug.)	14.7	13.5	13.6	13.5	15.5	8.5	17.4	25.4	15.2	17.6	13.6
Epidot	15.9	16.2	16.1	17.8	15.7	15.3	14.8	13	15.9	18.2	16.1
Zircon	19.4	19.2	22.1	22.9	18.1	21.2	19.5	14.9	19.2	16.7	18.3
Tourmaline	4.3	5.1	6	4.8	4.1	6.5	2.7	2.1	3.7	3.4	2.9
Rutile	12.9	11.2	12.2	14.6	11.7	14.1	9.4	9.9	11.8	12.6	10.7
Garnet	5.2	5.9	8.1	6.8	5.2	9.4	4.5	3.8	6.6	3.5	6.8
Monazite	3.9	4.1	4.2	5.3	3.6	5.3	2.8	2.1	4.7	2.3	3.6
Others (staulite+biotite)	1.1	1.2	1.2	1.4	1	1.2	1.2	0.9	1.1	0.6	0.4
Heavy fraction	70.2	73.8	80.2	90.4	73.5	81.7	65.8	52.3	79.7	81.2	60.9
Light fraction	29.8	26.2	19.8	9.6	26.5	18.3	45.2	47.7	20.3	18.8	39.1
Op/non-opaque	0.56	0.65	0.79	0.63	0.5	0.86	0.36	0.31	0.6	0.45	0.52

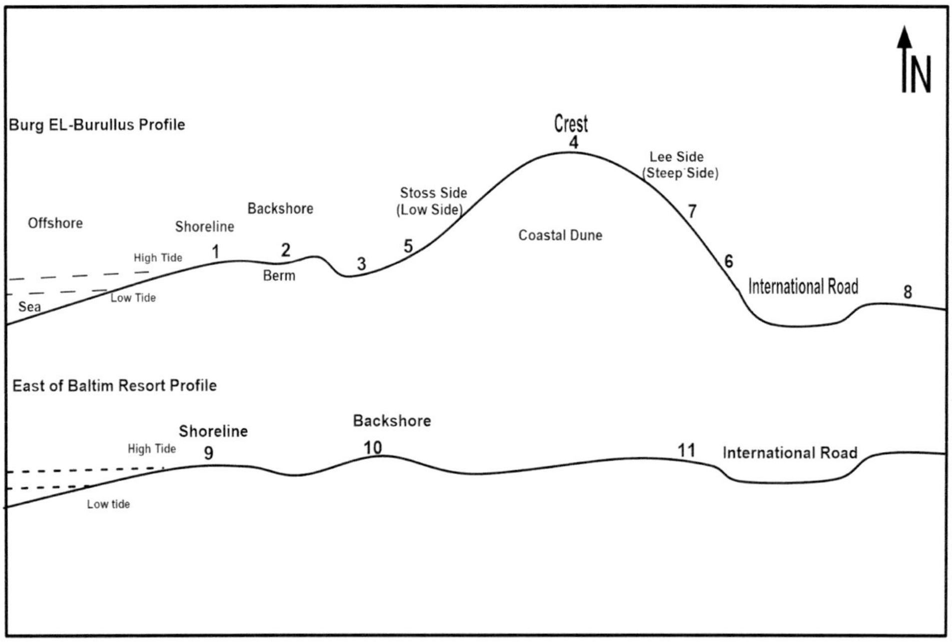

Figure 7.2. Vertical Profiles perpendicular to the beach from shoreline to dune in the area between Burg El-Burullus and Baltim.

Table 7.2). Sedimentary samples (11 in all) were collected for this study according to the method of Carver (1971), the sand fractions of 0.125 to 0.063 mm were separated using bromoform of specific gravity 2.89 as separating media into heavy and light minerals. The heavy and light minerals were individually washed and weighed. The heavy minerals were mounted in permanent slides using canada balsam (Allman and Lawrence, 1972). The opaque and non-opaque heavy minerals are calculated, and the individual non-opaque heavy minerals are calculated as a 'relative percentages of total non-opaque minerals', by counting about 400 grains per slide along randomly chosen traverses using the polarizing microscope (Dryden, 1931).

3. Results and discussion

In the study area, two profiles were studied, the first profile in the Burg EL-Burullus beach extends from the shoreline to right side of the international road (shoreline, backshore, top of dune, stoss side bottom, lee side bottom, middle lee side, right side of the international road toward Baltim). The second profile in the east of the Baltim resort and extends from the shoreline to the left side of the international road toward Gamasa (shoreline, backshore, left side of the international road) (Table 7.2, Fig. 7.2).

In the first profile, all the heavy minerals increase from the shoreline to the crest of the dune (Fig. 7.3). The total amount of heavy minerals in the EL-Burg dune increase from the base to the crest of the dune, where all heavy minerals in the lee side of the dune are more than the stoss side of the dune. After EL-Burg dune on the right side of the international road (towards Baltim), the total number of heavy minerals decrease. The overall heavy minerals are in the following order: shoreline < backshore < top of dune > stoss side bottom < lee side bottom > middle lee side > right side of the international road towards Baltim (Table 7.2 and Fig. 7.3).

Opaques, zircon, rutile, monazite, garnet and others (Staurolite + Biotite) (higher density minerals) in EL-Burg dune are more concentrated than that in the EL-Burg beach. Where, hornblende, augite and tourmaline (lower density minerals) are higher in the EL-Burg beach than that in the EL-Burg dune (Table 7.3, Fig. 7.4).

In the second profile, the total number of heavy minerals increase from the shoreline to the backshore and decrease from the backshore to left side of the international road. The total heavy minerals are in the following order: shoreline < backshore > left side of the international road. Sand samples from the shoreline and backshore of Burg EL-Burullus have a lower content of total heavy minerals compared with the shoreline and backshore samples of the east Baltim resort. Opaques, zircon, monazite, garnet, tourmaline and others (Staurolite + Biotite) of the shoreline on the east

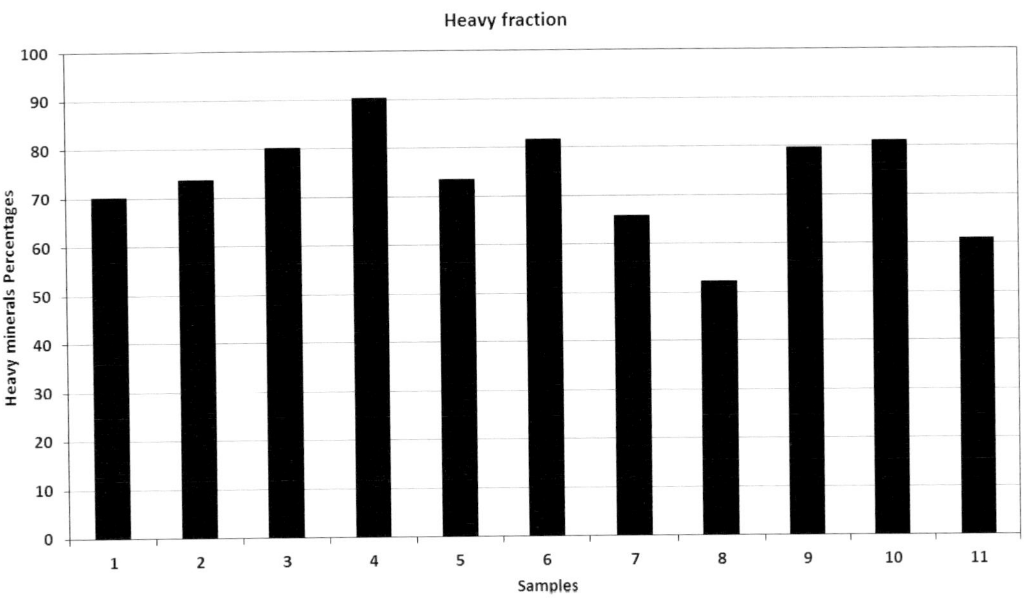

Figure 7.3. Heavy fractions percentages in the study area.

Table 7.3. Percentages of the heavy minerals species in the different locations in the study area.

Minerals	Average percentage of Burg El-Burullus eroded beach	Average percentage of Burg El-Burullus dunes	Right of international Road	Average percentage of eroded east Baltim beach	Left of international Road
Heavy percentages	74.73	77.85	52.3	80.45	60.9
Opaques	39.86	40.25	23.4	34.25	34.3
Amphibole (Hornblende)	22.96	22.27	28	23.2	27.6
Pyroxene (Augite)	13.93	13.72	25.4	16.4	13.6
Epidote	16.06	15.9	13	17.05	16.1
Zircon	20.23	20.42	14.9	17.95	18.3
Tourmaline	5.13	4.52	2.1	3.55	2.9
Rutile	12.1	12.45	9.9	12.2	10.7
Garnet	6.4	6.47	3.8	5.05	6.8
Monazite	4.06	4.25	2.1	3.5	3.6
Others (Staurolite +Biotite)	1.16	1.2	0.9	0.85	0.4

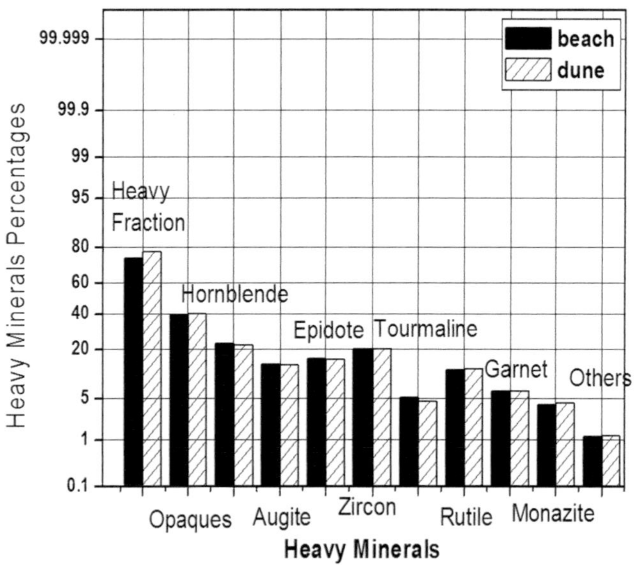

Figure 7.4. Heavy minerals in the dunes and beach of Burg EL-Burullus area, Egypt.

Baltim resort are higher than that in backshore, where hornblende, augite, epidote and rutile are lower in the shoreline than backshore of the east Baltim resort (Table 7.2).

Finally, there is a relationship between beaches and dunes in the EL-Burg area, where the higher density minerals are higher in dunes and lower in beaches, while the lower density minerals are higher in beaches and lower in dunes. The heavy mineral variations normal to the shoreline show that the opaques, garnet, rutile, zircon and monazite markedly increase while hornblende, augite, tourmaline and epidote decrease in moving from the beach to the dunes, where all heavy non opaques minerals are present in a different percentage along the two profiles due to the type of beaches and the main geomorphologic units (Figs. 7.4 and 7.5). These results are in agreement with EL-Fishawy and Molnar (1985). The heavy minerals concentration of the study area changed

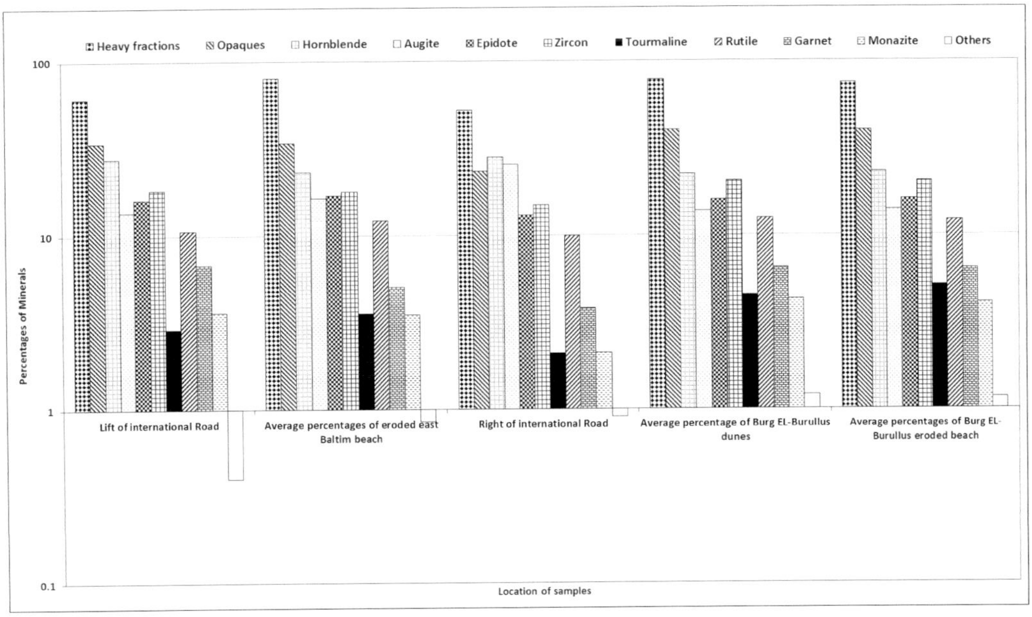

Figure 7.5. Changes of heavy minerals species in five locations between Burg EL-Burullus and Baltim, Egypt.

when the beach changed or/and by removing the coastal sand dunes. The two studied profiles are different from each other due to the type of the beach and the presence of sand dunes, where the concentrations of heavy minerals changed along each of the profiles (Fig. 7.5).

Mineralogical composition of the heavy minerals:

The optical properties of nine heavy minerals under the polarizing microscope representing the main contents of the heavy fraction of the sediments are given next:-

1. Opaque minerals

They constitute a high percentage of heavy minerals that ranges from 23.4 to 40.25%. The higher concentrations are located in the eroded beach and in the sand dunes of EL-Burg and Baltim (Table 7.3 and Figs. 7.4 and 7.5). They are studied as a single group, and are composed mainly of magnetite and ilmenite, while hematite and limonite are rare.

2. Non-opaque minerals

These constitute the main amount of heavy minerals in the study area. They are composed mainly of augite, hornblende, zircon, epidote, garnet, tourmaline, monazite and others.

Augite is the most common non-opaque heavy minerals in the study samples. It is recorded in all the investigated samples, but in high percentages on the right side of the international road and on the beach of study area (Table 7.3 and Figs. 7.4 and 7.5). Its color varies from greenish yellow to brownish violet. Most of the augite grains are short prismatic or irregular with frequent rounded grains.

Hornblende varies in color from glassy green, bluish green to rare brownish green. The grains are usually prismatic, angular to sub rounded. It is recorded in high percentages in the beach of the study area (Table 7.3 and Figs. 7.4 and 7.5).

Zircon grains are mostly colorless but vary from clear water to dusky turbid. The higher percentages of zircon are recorded in the erosional areas and in the sand dunes of El-Burg and Baltim (Table 7.3 and Figs. 7.4 and 7.5). Zircon is mostly represented by small prismatic grains with rounded edges.

Epidote is lemon yellow and greenish yellow color with rounded to sub rounded grains. Some grains exhibit weak pleochroism with high relief. It is recorded in high percentages in the beach of the study area (Table 7.3 and Figs. 7.4 and 7.5).

Garnet is represented by colorless to pink varieties and occurs in sub angular to sub rounded and rounded grains. Garnet grains have a very high relief and are isotropic under crossed nicols. The higher percentages recorded in the erosional areas and in the sand dunes of El-Borg and Baltim (Table 7.3 and Figs. 7.4 and 7.5).

Rutile is observed in reddish brown and yellowish brown kinds, where reddish brown is the most common. Rutile is observed as minute prismatic grains mostly with rounded edges. It is recorded mainly in the erosional areas and in the dunes of EL-Burg with high concentration and recorded with lower concentration in the beach (Table 7.3 and Figs. 7.4 and 7.5).

Tourmaline is observed as egg shaped grains and display different colors but the brown variation with strong pleochrism is common. It is recorded in low percentage in the erosional areas and in EL-Burg and Baltim dunes and represents a high percentage in the beach of the study area (Table 7.3 and Figs. 7.4 and 7.5).

Monazite grains are rounded to oval shape, with a lemon yellow and green color. The grains have a very high relief. Monazite grains are very rare, they are recorded in low amounts in the erosional areas and EL-Burg and Baltim dunes (Table 7.3 and Figs. 7.4 and 7.5).

4. Conclusions

The total number of heavy minerals increase from the shoreline moving to the crest of dunes, are higher on the top and on the lee side of the dune rather than the bottom and stoss side of the dune. Heavy mineral variations normal to the shoreline show that opaques, garnet, rutile, zircon and monazite markedly increase while hornblende, augite, tourmaline and epidote decrease in moving from the beach to the dunes. These results are in agreement with EL-Fishawy and Molnar (1985). The total amount of heavy minerals of the shoreline and backshore of east Baltim resort are higher than that in the EL-Burg shoreline and backshore.

References

Abu-Zeid, M.A. and El-Shibini, F.Z. 1997. Egypt's High Aswan Dam. Int. J. Water Resour. Dev. 13(2): 219–217.

Ahmad, M.H. 2002. Multi-Temporal Conflict of the Nile Delta Costal Changes, Egypt National Authority for Remote Sensing and Space Science (NARSS). Littoral 2002, The Changing Cost, EuroCoast/EUCC, Porto-Portugal, ISBN 972-8558-09-0.

Ahmed, M.H., El-Asmar, H.M. and Ali, E.M. 2010. Short term Assessment of Landuse Transformation and Shoreline Changes at the Northern Zone of the Nile Delta Coast, Egypt The 8th AARSS (The African Association of Remote Sensing of the Environment), October (25–29), 2010 Adesabeba.

Ahmed, M.H. and El-Asmar, H.M. 2010. Monitoring and assessment of coastal development along the Nile Delta coast using satellite images and GIS. 2nd Conf. on Contemporary Environmental Issues in Arid and Semi, Arid Regions (Envi 2010: Climate Change), Bibliotheca Alexandrina, Egypt, 3–5 July 2010.

Allman, M. and Lawrence, D.F. 1972. Geological Laboratory Techniques, Arco., New York.

Anwar, Y.M., Gindy, A.R., El-Askary, M.A. and El-Fishawi, N.M. 1979. Beach accretion and erosion, Brullus-Gamasa coast, Egypt. Marin Geol. 30: M1–M7.

Carver, R.E. 1971. Procedures in Sedimetary Petrology. John Wiley, New York, 653pp.

Dryden, A.I. 1931. Accuracy in percentage representation of heavy mineral frequencies. Proceedings of National Academy of Science 17: 233–238.

El-Askary, M.A. and Lotfy, M.F. 1995. The use of texture and heavy mineral properties of sand as indicators of the advancing and receding of the Nile Delta beaches, Egypt. N.jb. Geol. Palaont. Mh. 257–270 p.

El-Bouseily, A. and Frihy, O. 1984. Textural and mineralogical evidence denoting the position of the mouth of the old Canopic Nile branch on the Mediterranean coast, Egypt. Journal of African Earth Sciences 2: 103–107.

El-Fishawi, N.M. and Molnar, B. 1985. Mineralogical relationships between the Nile Delta coastal sands. Acta Mineralogica Petrographica, Szcged, XXVII: 89–100.
El-Asmar, H.M. 2000. Geoenvironmental changes along Gamassa Baltim Coast North of the Nile Delta, Egypt. Z. Geomorph. N.F. 44/1: 59–73, Berlin–Stuttgart, Germany.
El-Asmar, H.M. and Haifaa, A.A. 2013. Environmental impact assessment and change detection of the coastal desert along the central Nile Delta Coast, Egypt. International Journal of Remote Sensing Applications, Volume 3 Issue 3, September 2013.
Frihy, O. and Komar, P.D. 1993. Long-term shoreline changes and the concentration of heavy minerals in beach sands of the Nile Delta, Egypt. Mar. Geol. 115: 253–261.
Frihy, O.E. 1994. Discrimination of accreted and eroded coasts using heavy mineral composition of the Nile Delta beach sands, Egypt. Sedim. 41: 905–912.
Frihy, O. and Lotfy, M. 1994. Mineralogy and textures of beach sands in relation to erosion and accretion along the Rosetta promontory of the Nile Delta, Fgypt. J. Coast. Res. 10: 588–599.
Sharaf El Din, S. 1977. Effects of Aswan high Dam on the Nile flood and on the estuarine and coastal circulation pattern along the Mediterranean Egyptian coast. Limnol. Oceanogr. 22(2): 194–207.
Stanley, D., Warne, A., DAVIS, h., Bernasconi, M. and Chen, Z. 1992. Late Quaternary North Central Nile Delta from Manzal to Burullus lagoons. Egypt. Nation. Geograph. Res. Explor. 811: 22–51.

Part 2

Dune Landscapes of China—Origin and Evolution

The seven chapters provide a comprehensive overview of the nature and extent of sand seas and dunes both in the north and northwest of China but also describe and classify the dunes that form along the extensive coastline. Emphasis is placed on explaining the formation and development of dune field landscapes.

Chapter 8: *Lu Huayu et al.* The authors use a comprehensive approach to analyse the formation and development of the dune field landscape in north and northwest China in this chapter. They argue that the dune field landscape was established as a result of the long-term evolution of tectonic deformation and uplift, climatic drying and cooling, and intensified physical weathering in the Asian interior. The results often challenge the conclusion derived from analysis of a single site or area, and it is hoped that conclusions drawn by the authors will contribute to an improved understanding of the landform surface processes, dust emission, and the resulting climatic impact in Asia. They have investigated more than 100 depositional sequences within the dune field of the desert and the Chinese Loess Plateau, all of which were investigated in detail during the field work. Most of the sections were sampled for dating by independent optically stimulated luminescence (OSL) and magneto-stratigraphy analysis (MSA); and in addition, proxy indicators such as magnetic susceptibility, grain-size distribution and stable carbon isotopic composition were analysed, in order to reconstruct climatic changes.

Chapter 9: *Lu Qi et al.* The Kumtag Desert had remained as the last untouched natural desert in China to be fully investigated in a scientific way. From June 2007 to May 2017, teams conducted several integrated large-scale and multidisciplinary desert expeditions, professional sampling surveys and ground observations. In total, they worked for more than 500 days in the desert and travelled over 400,000 km. These expeditions gained fruitful results in many scientific research subjects, including geology, geomorphology, hydrology, pedology, meteorology, zoology and botany. In this chapter, an overview of the sand dunes in the Kumtag Desert is presented based on the two-stage expeditions.

Chapter 10: *Zhibao Dong.* The Badain Jaran Sand Sea is in Gansu province and Inner Mongolia Autonomous Region (north-western China). It features the tallest dunes on Earth and a unique mega-dune–lake landscape. It had been explored little until the 1990s, though early scientific explorations surrounding the sand sea had begun by the early 20th century. Heated debates now focus on the desert environment, and particularly how the mega-dunes and desert lakes develop and evolve. This chapter reviews the status of these debates and summarizes the supporting evidence.

Chapter 11: *Dong Xue. Nebkahs* in western and north western China play an important role in sandy areas and may be implicated in dune genesis and formation. *Nebkhas* are an important type of wind-bound landform and represents a typical biogeomorphology that forms from the interactions of plants, wind and sediments. The formation and development of *nebkhas* is closely related to climate change and human activities. The morphology, distribution and formation of the *nebkhas* on Gobi (pebble-strewn) and sandy desert in China is explained.

Chapter 12: *Dong Yuxiang*. Coastal Aeolian dunes are dynamic geomorphic systems affected by the interaction of sea, land, and atmosphere. Recently, while further exploring the evolution process of coastal Aeolian sand deposition environment, research has predominately focused on the essential characteristics that form coastal Aeolian dunes in the dynamic environment of the interaction of sea, land, and atmosphere. Moreover, comprehensive research has been performed by relating coastal Aeolian dunes to the atmosphere, waves, and beaches. This chapter aims to summarize the classification, distribution characteristics, origins, formation, and evolution of coastal Aeolian dunes along the China coastline based on a synthesis of previous research.

Chapter 13: *YuJie Niu, Victor Squires and Limin Hua.* Research on the sand dunes on the high-altitude areas of the Qinghai-Tibet Plateau, China is still at a pioneering stage. However, greater knowledge of the spatial distribution, evolutionary processes, characteristics, and driving forces of land degradation; types and geomorphology of aeolian sand dunes; and wind-driven degradation of alpine rangelands on the high altitude of Tibetan plateau, will certainly improve our understanding of the degradation processes of alpine environment. The authors try to summarize the wind-driven land degradation of alpine rangelands on the Tibetan plateau and discuss the spatial distribution, temporal variation, and driving forces of aeolian desertification; types, and geomorphology of aeolian sand dunes in detail with a view to greatly improve understanding of aeolian desertification and sand dunes on high altitude areas. In addition, they have also discussed the causes and effects of aeolian desertification, and efficient sand movement control mechanisms.

Chapter 14: *Bing Qi Zhu et al.* The focus of this chapter is on the endorheic rivers that arise on the Qinghai-Tibet plateau and run northward to end in the deserts of Inner Mongolia. Based on field and image-data survey and other analysis, it shows that only the barchans (crescent-type) dunes and the chain of barchans dunes are moving or swinging in the Hexi Corridor. The three main factors that affect the dynamic change of sand dunes are annual precipitation, annual average of wind speed and annual days of strong wind, among which the annual precipitation has the largest contribution. The influence of climatic factors on the dynamic change of sand dunes in Hexi Corridor has both positive and negative effects. Early and recent investigations about the provenance of sand dunes have shown that the sand dunes in Hexi Corridor mainly come from "sand sediments arising from local sites", with the main material sources of sand dunes being derived from alluvial plains and alluvial sediments of ancient rivers, lacustrine plains and lacustrine sediments of ancient lakes, aeolian sediments of southern and Northern denudation zones of piedmonts and alluvial sediments of modern rivers in Hexi Corridor.

Chapter 8

Formation of Dune Field Landscape in North and North-West China during Plio-Pleistocene Transition (4–2 Ma) and the Long-term Evolution

Huayu Lu,[1,*] *Zhiwei Xu,*[1] *Wenchao Zhang,*[1] *Hanzhi Zhang,*[1] *Xiaoyong Wang,*[1] *Shuangwen Yi,*[1] *Qi Lu,*[2] *Bo Wu,*[2] *Yali Zhou,*[3] *Yao Wang,*[1] *Xiaojian Zhang,*[1] *Han Feng,*[1] *Xi Chang,*[1] *Mengchun Cui,*[1] *Nana Lv,*[1] *Hanlin Wang*[1] and *Xianyan Wang*[1]

1. Introduction

Dune fields in North and North-west China (also in central and eastern Asia) encompass the Gobi and Taklimakan Deserts, other sandy deserts and sandy lands covering an area of ~ 3.8×10^6 km^2 (Fig. 8.1). The region is characterized by an arid climate with sparse vegetation and is covered by sand dunes and gravel sediments. The dune field produces a large quantity of silt particles, that are an important wind-blown dust source. The dust significantly influences the regional environment, and probably also draws down atmospheric CO_2 through its effect on biogeochemical cycles in the ocean (Martin and Fitzwater, 1988; Shoenfelt et al., 2018). Understanding the formation and development of the dune fields and deserts in North and North-west China is essential for quantitatively estimating the dust emission and its environmental impacts, and understanding the regional arid landscape evolution. However, there is still considerable debate about the timing related to the origin of the dune field landscape. The timing of the dune field landscape initiation has been variously assigned to the Eocene (Carrapa et al., 2015; Li et al., 2018), the late Oligocene–early Miocene (Guo et al., 2002; Zheng et al., 2015), the late Miocene (Sun et al., 2009; Heermance et al., 2018), the Pliocene and early Pleistocene (Ding et al., 2005; Lu et al., 2010, 2019; Sun et al., 2011), the middle Pleistocene or later (Li et al., 2014; Wang et al., 2015; Yang et al., 2015). In a recent review paper (Lu et al., 2019), we argued that the formation of the dune field landscape in

[1] Frontiers Science Center for Critical Earth Material Cycling, School of Geography and Ocean Science, Nanjing University, Nanjing 210023, China.
[2] Institute of Desertification Studies, Chinese Academy of Forestry, Beijing 100091, China.
[3] School of Geography and Tourism, Shaanxi Normal University, Xi'an 710062, China.
* Corresponding author: huayulu@nju.edu.cn

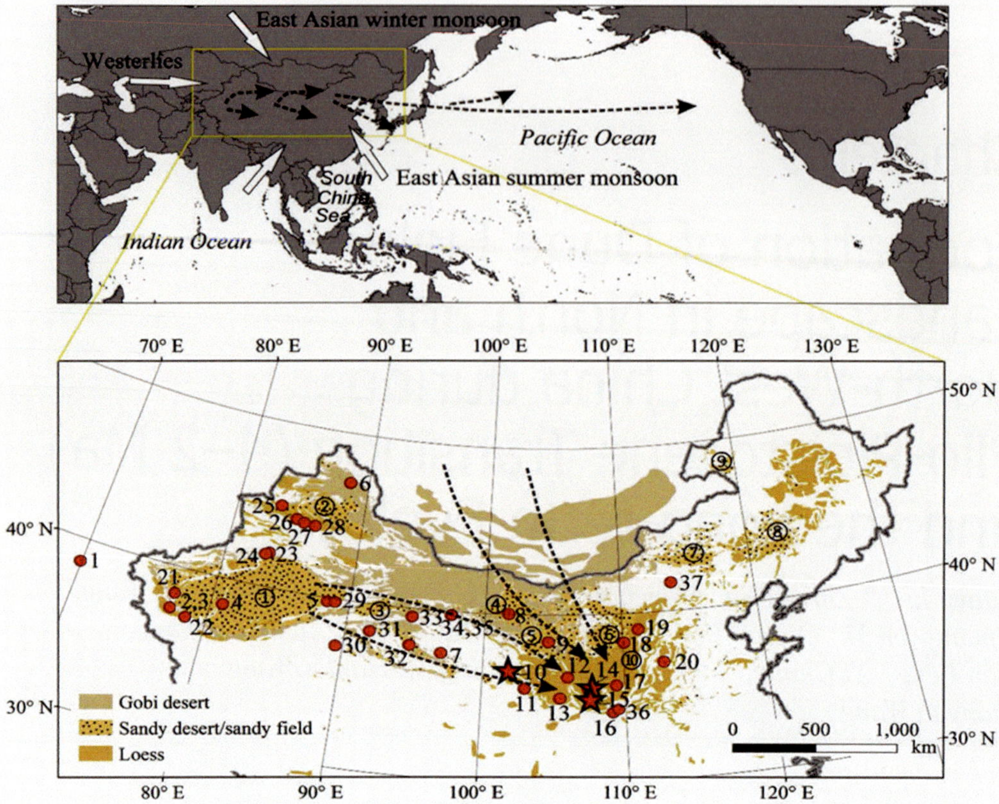

Figure 8.1. Upper figure shows the atmospheric circulation (white arrows) and dust transported pathways (black arrows) in the North and North-west China and the North Pacific Ocean. Lower figure indicates the basins, deserts, loess deposit in North and Northwest China. Red dots indicate the study sites with chronological and stratigraphic data. Red stars indicate the location of three long aeolian silt sequences (Houwan, Xifeng and Lingtai) investigated in this study.

Numbers in circles are: ① Taklimakan Desert (dune fields); ② Junggar Desert (dune fields); ③ Kumtagh Desert (dune fields); ④ Badain Jaran Desert (dune fields); ⑤ Tengger Desert (dune fields); ⑥ Mu Us sandy land (dune fields); ⑦ Otindag sandy land (dune fields); ⑧ Horqin sandy land (dune fields); ⑨ Hulun Buir sandy land (dune fields); ⑩ Chinese Loess Plateau.

Numbers without circles indicate the location of the investigated sedimentary sequences at the following sites: 1. Tajik Basin; 2. Tarim Basin 1; 3. Tarim Basin 2; 4. Tarim Basin 3; 5. Lop Nor 1; 6. Juggar Basin; 7. Qaidam Basin; 8. Badain Jaran Desert; 9. Tengger Desert; 10. Xining Basin; 11. Linxia Basin; 12. Ningxia Basin; 13. Tianshui Basin; 14. Xifeng; 15. Lingtai; 16. Guanzhong; 17. Luochuan; 18. Jiaxian; 19. Baode; 20. Yushe Basin; 21. Shache; 22. Sanju; 23. Kuche; 24. Kuchetawa; 25. Ebinur; 26. Duzishan; 27. Jingouhe; 28. Taxihe; 29. Lop Nur 2; 30. Kumukuli; 31. SG-3 & KC-1; 32. Yahu; 33. Dunhuang; 34. Laojunmiao; 35. Huoshaogou; 36. Weihe; 37. Inner Mongolia.

China occurred in the late Pliocene and was the result of ongoing changes in the topography of Asia against the background of climate change. However, the timing of the inception of the dune field landscape continues to be constrained by direct evidence.

We propose that the discrepancies between earlier studies are mainly because the sedimentary sequences that were used to indicate either local aeolian activity or the intensification of aridity within dune fields, are incomplete. These sequences are characterized by aeolian sand interbedded with lacustrine and fluvial deposits or they were interpreted as a proximal record of an unknown source. Thus, the existing evidence for dune fields formation is geographically scattered and often discontinuous due to strong wind erosion and migration of active dunes. Lakes and rivers in dune fields are always ephemeral, and consequently most of their associated lacustrine, fluvial or alluvial deposits are fragmentary in nature, so that they cannot provide complete information on dune field landscape development. Therefore, attempts to use such discontinuous records to decipher the long-term history of the dune field evolution are often controversial.

However, thick aeolian silt deposits in downwind regions of dune fields have a close affinity with the formation and development of the dune field landscape in North and North-west China. These aeolian silt depositional sequences, which are mainly located in the Chinese Loess Plateau (CLP) (Lu et al., 2010, 2019, 2021; Amit et al., 2014; Cui et al., 2015) and the North Pacific Ocean (NPO) (Rea et al., 1985; Uno et al., 2009) have proved to have originated from dune fields (Fig. 8.1), through geochemical evidence such as neodymium and strontium isotopic composition and mineral analyses (Chen and Li, 2011; He et al., 2017) and by detrital zircon age spectra (Zhang et al., 2016, 2018). Thus, the aeolian silts provide an excellent opportunity to determine the origin and long-term development of dune fields.

In the present study, a comprehensive approach to analyze the formation and development of the dune field landscape in North and North-west China has been used. It was argued that the dune field landscape was established as a result of the long-term evolution of tectonic deformation and uplift, climatic drying and cooling and intensified physical weathering and silt generation in the Asian interior. The results often challenge the conclusion derived from analysis of a single site or area, and it is hoped that our inferences will contribute to an improved understanding of the landform surface processes, dust emission and the resulting climatic impact in Asia.

2. Methods

We investigated more than 100 depositional sequences within dune fields, deserts and the Chinese Loess Plateau (CLP) (Figs. 8.1, 8.2), all of which were logged in detail in the field. Most of the sections were sampled for dating by independent optically stimulated luminescence (OSL) and magnetostratigraphy analysis; and in addition, proxy indicators such as magnetic susceptibility, grain-size distribution and stable carbon isotopic composition were analyzed, in order to reconstruct climatic changes (Lu et al., 1999, 2004a, b, 2010, 2013a, b, 2019). These published data were reinterpreted in order to investigate the development of the desert landscape. For this study, three long aeolian silt sequences from Houwan (101.78°E, 36.80°N), Xifeng (107.65°E, 35.72°N) and Lingtai (107.56°E, 34.98°N) in the Chinese Loess Plateau were sampled (Fig. 8.1), and the grain-size distributions of 1346 samples from these sequences were determined. The grain-size distributions were measured using a Malvern Mastersizer 2000 laser particle size analyzer with a measurement range of 0.02–2000 μm and an error less than 2% Pre-treatment consisted of the removal of organic matter and carbonates by the addition of 10% H_2O_2 and 10% HCl, respectively, followed by dispersal using 10 ml of 0.05 mol/L ($NaPO_3$) and treatment in an ultrasonic vibrator for 10 minutes. An 11-Ma grain size time series was obtained (Fig. 8.3; Supporting information-dataset). The grain size data for the interval of 22–11 Ma are from the Qin'an section referenced in Qiao et al. (2006).

The timescales of the three sequences were constructed by linear interpolation of the ages of each sampling level based on detailed magnetostratigraphic controls (Kukla and An, 1989; Sun et al., 1998; Lu et al., 2004b). The age of 0 ka is assigned to the top of the sequences at Lingtai; the age of 0.78 Ma was assigned to the Brunhes/Matuyama boundary determined by correlation between our magnetic susceptibility curves and the magnetic susceptibility spanning paleosol S8 at Lingtai; and the age of 6.149 Ma was assigned to the depth of 232 m in the Xifeng Red Clay sequences. The standard polarity timescale (Ogg, 2020) was used for age assignments. The sedimentation rates of the aeolian silt sequences were obtained by dividing the sedimentary thicknesses by the corresponding duration of accumulation time (Fig. 8.4).

The U-Pb ages of detrital zircon grains from nine typical loess samples from Luochuan, Xining and Jiaxian (Fig. 8.1) were measured and 2375 new ages obtained in this study (Fig. 8.5; Supporting information-dataset). More than 300 zircon grains of each sample were picked for dating with the

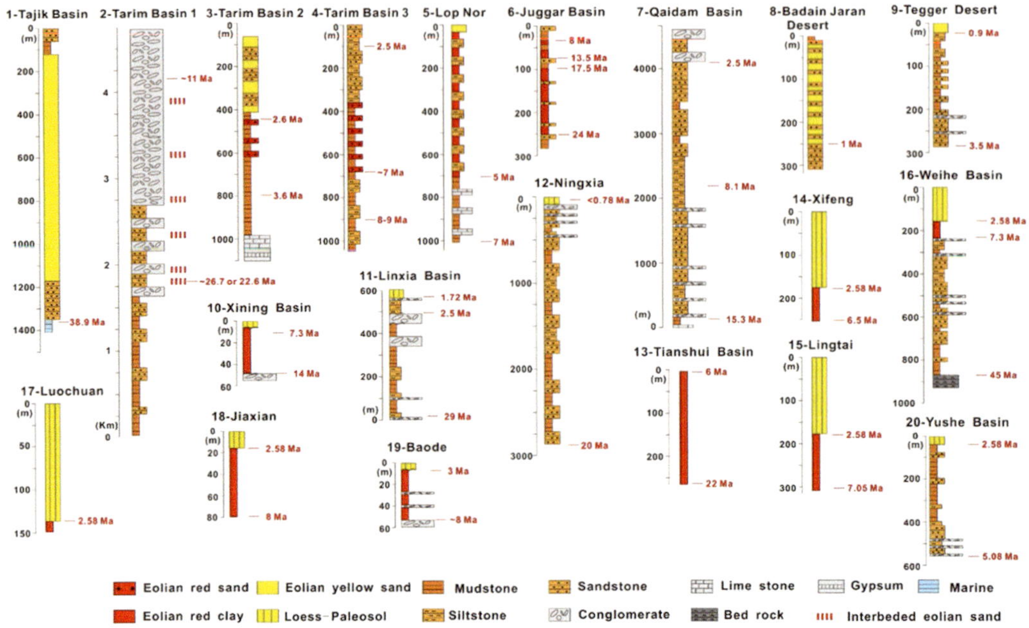

Figure 8.2. Chronological and stratigraphic data from multiple sites in the CE Asia. Site locations are shown in Fig. 8.1. The data reported are according to Carrapa et al. (2015) (1. Tajik Basin); Zheng et al. (2015) (2. Tarim Basin 1); Sun et al. (2011) (3. Tarim Basin 2); Sun et al. (2009) (4. Tarim Basin 3); Liu et al. (2014) (5. Lop Nor); Sun et al. (2010) (6. Juggar Basin); Fang et al. (2007) (7. Qaidam Basin); Wang et al. (2015) (8. Badain Jaran Desert); Li et al. (2014) (9. Tengger Desert); Lu et al. 2004b (10. Xining Basin - Houwan); Fang et al. (1997) (11. Linxia Basin); Jiang et al. (2007) (12. Ningxia Basin); Guo et al. (2002) (13. Tianshui Basin); Liu et al. (1988) and Sun et al. (1997) (14. Xifeng); Ding et al. (1998) and Sun et al. (1998) (15. Lingtai); Lu et al. (2018) (16. Guanzhong Basin); Liu (1985) (17. Luochuan); Ding et al. (1998) and Qiang et al. (2001) (18. Jiaxian); Yue et al. (2004) (19. Baode); Cao et al. (1985) (20. Yushe Basin).

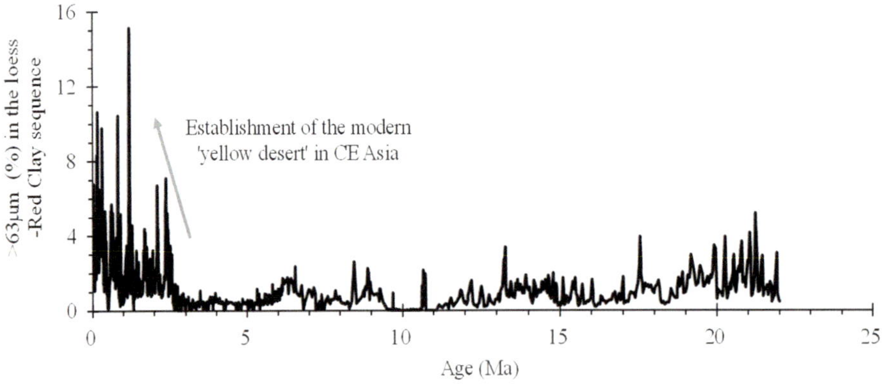

Figure 8.3. Grain-size changes in aeolian silt deposit sequences since early Miocene (data between 22–11 Ma is from Qiao et al., 2006). The significant increase in grain size at ~ 2.6 Ma indicates the formation of the desert landscape was associated with the occurring of the Northern Hemisphere Glaciations (NHG).

aim of achieving at least 120 valid ages for each sample. The zircon grains were randomly selected from each sample, while all the zircon selections were undertaken by one individual, the bias caused by individual preference can be considered the same between different samples. U-Pb age were determined using a New Wave 193 nm laser ablation system and Agilent 7700x ICP-MS in the Laboratory of Earth Surface Processes and the Environment, Nanjing University. The laser beam diameter was 25 μm, with a 10 Hz repetition rate and energy of 2–3 J/cm^2. The spot size enabled the

Figure 8.4. Dust accumulation rate in an aeolian silt sequence since the early Miocene. The major increase in dust accumulation rate at ~ 2.6 Ma indicates the formation of the present desert landscape in CE Asia. The aeolian silt deposit changes from reddish Red Clay to yellow loess sediments at ~ 2.6 Ma.

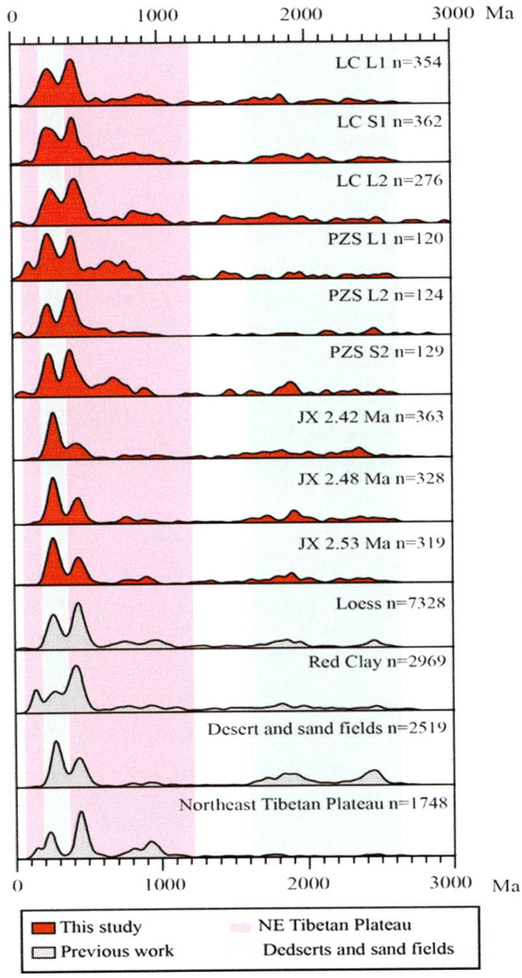

Figure 8.5. Detrital zircon U-Pb age spectra of loess, Red Clay and desert sand particles. The zircon U-Pb ages of nine loess samples (LCL1, LCS1, LCL2, PZS L1, PZS L2, PZS S2, JX 2.42 Ma, JX 2.48 Ma and JX 2.53 Ma) from this study were combined with previously published data to generate the composite probability plot. Detrital zircon U-Pb age distributions of the loess from the CLP shows peaks similar to those of the sediments from deserts; both mainly originate from the northern Qinghai–Tibet Plateau.

analysis of grains larger than 25 μm. NIST 610 was used as the standard to normalize the unknown U, Th and Pb contents. Glitter 4.4.2 was used to process the raw data, and common Pb was corrected following the method of Andersen (2002). The results are shown in Fig. 8.5.

Following pioneering work using detrital zircon U-Pb age spectra as an indicator of sediments sources of the dune fields and Chinese loess (Stevens et al., 2010), numerous zircon U-Pb ages have been generated which have served as a useful tool for pinpointing loess sources. The zircon U-Pb age data of the loess-paleosol deposits (Stevens et al., 2010; Pullen et al., 2011, 2014; Xiao et al., 2012; Che and Li, 2013; Stevens et al., 2013a, b; Nie et al., 2015; Bird et al., 2015; Licht et al., 2016; Zhang et al., 2018), desert sediments (Zhang et al., 2016, 2018; Rittner et al., 2016), Red Clay sediments (Nie et al., 2015, 2018; Shang et al., 2016; Zhang et al., 2018), and sediment and rock samples in the Northeastern Tibetan Plateau (Cao et al., 2011; Chen et al., 2004; Chen et al., 2004; Ding et al., 2010; Ding et al., 2006; Li et al., 2007; Liu et al., 2006; Luo et al., 2012; Mao et al., 1998; Pei et al., 2007a, b; Qin et al., 2005; Rui et al., 2014; Shi et al., 2004a, b; Song et al., 2004; Tseng et al., 2007; Wan et al., 2001; Wang et al., 2005, 2009; Wu et al., 2009; Xu et al., 2007; Yang et al., 2003, 2002; Yang et al., 2006; Ye et al., 2006; Zhao et al., 2012; Zhu et al., 2008, 2011) have been reported. Here we synthesized these data producing a new dataset of 14,564 ages (Fig. 8.5). The U-Pb age spectra can be regarded as a characteristic feature of typical desert sediments, loess-paleosol and Red Clay deposits. The closely match between the age spectra demonstrates a genetic linkage between the desert and loess.

Pollen analysis of 41 samples from sand land sediments from Mu Us, Otindag, Horqin, Hulunbuir of the Last Glacial Maximum (LGM) and the Holocene Optimum (HO) were obtained (Fig. 8.6). Approximately 100 g sediment of each sample was treated using standard pollen extraction procedures (Faegri et al., 1989), including immersion in HCl (10%), KOH (3%), HF (40%), followed by 7 μm sieving. All of the samples were washed with distilled water and dehydrated by sieving through a 7 μm nylon screen after each step. To calculate pollen concentration, one tablet of *Lycopodium* spores was added at the beginning of the chemical preparation (Stockmarr, 1971). Prepared samples were mounted in glycerol for storage and slide preparation. Pollen and spores were identified at magnifications of 400x or 630x and the results are presented in Fig. 8.6 (Supporting information-dataset).

In addition to these newly obtained data, we compiled a large dataset of Cenozoic vegetation records in and around the dune fields from published references. The data sources and analysis are given in the caption to Fig. 8.7.

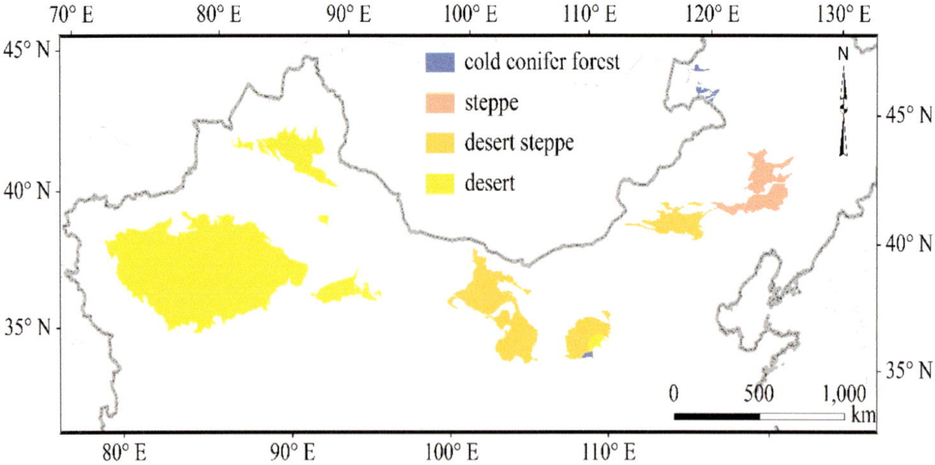

Figure 8.6. Spatial distribution of vegetation within the deserts of CE Asia during the LGM, reconstructed by newly obtained pollen data.

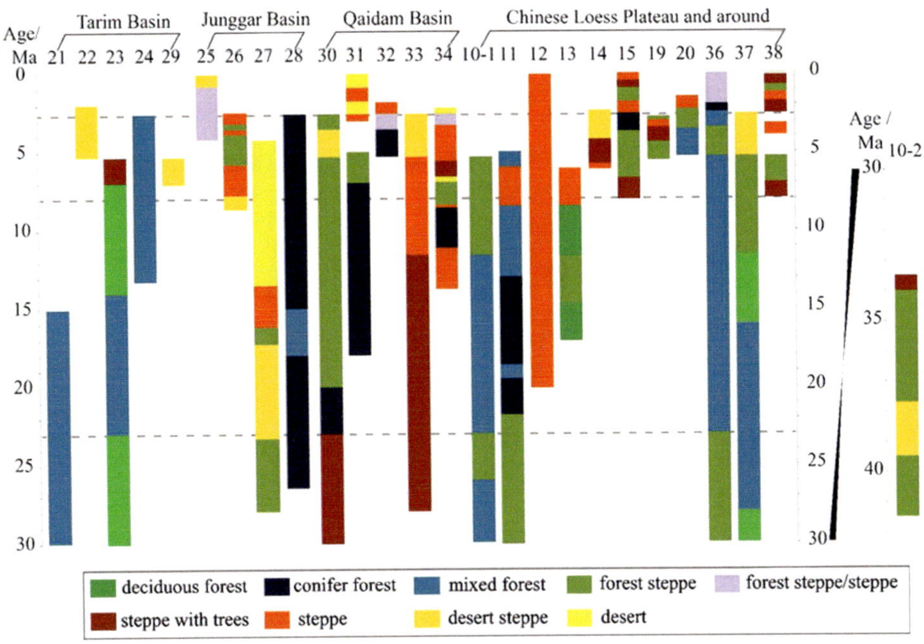

Figure 8.7. Vegetation records synthesis, as inferred from palynological and mammalian records from multiple sites in northern China. The numbers referred to are: 10. Xining Basin - Minhe (Sun et al., 1984; Wang and Deng, 2009; Wang et al., 1990a) and Shuiwan (Dupont-Nivet et al., 2008, 2010; Hoorn et al., 2012); 11. Lingxia Basin - Maogou (Ma et al., 1998); 12. Ningxia Basin - Sikouzi (Jiang and Ding, 2008); 13. Tianshui Basin - Yanwan (Hui et al., 2011); 14. Xifeng (Wang et al., 2006); 15. Lingtai - Chaona section (Ma et al., 2005b; Wu et al., 2007); 19. Baode (Li et al., 2011); 20. Yushe (Cao and Cui, 1989; Shi et al., 1993); 21. Shache (Wang et al., 1990a); 22. Sanju (Sun et al., 2008); 23. Kuche (Jin et al., 2002); 24. Kuchetawa (Zhang and Sun, 2011); 25. Ebinur (Zhang et al., 2009); 26. Duzisha (Sun et al., 2007); 27. Jingouhe (Tang et al., 2011); 28. Taxihe (Sun and Zhang, 2008); 29. Lop Nur 2 (Hao et al., 2012); 30. Kumukuli (Xiao et al., 2003); 31. SG-3 & KC-1 (Cai et al., 2012; Miao et al., 2011, 2013); 32. Yahu (Fang et al., 2008; Wu et al., 2011); 33. Dunhuang (Ma, 1991); 34. Laojunmiao (Ma et al., 2005a); 35. Huoshaogou (Miao et al., 2008); 36. Weihe (Sun et al., 1980; Tong et al., 2000); 37. Inner Mongolia (Wang, 1990; Wang et al., 1990b).

Spectral analysis of the grain-size > 63 μm% time series obtained in this study and the oxygen isotope time series of Lisiecki and Raymo (2005) was performed using the Blackman-Tukey method, and cross-spectral analysis of the two-time series was also conducted. ARAND software (Howell et al., 2006) was used for both analyses. The following parameters were used to optimize bias/variance properties of spectral estimates: number of lags = 400 (~ 1/6 length of record) and samples per analysis = 1200 (Fig. 8.8).

3. Results

The long-term depositional sequences of the aeolian sand and silt sediments are direct indicators of the dune field and desert landscape. Our field logging data and the compiled data show that North and North-west China was dry since at least the middle Eocene to late Pleistocene, with a stepwise expansion of the aeolian sand and silt deposition occurring over time (Fig. 8.2). In addition, a major expansion of the aeolian sand and silt cover occurred during 4–2 Ma. These field observations are confirmed by changes in the grain-size distributions and aeolian silt accumulation rates since the early Miocene (Figs. 8.3, 8.4). The grain size of the aeolian silt was finer and there were low amplitude fluctuations during 25–3.6 Ma, while a grain size coarsening with greater amplitude fluctuations occurred after ~ 3.6 Ma, and especially since ~ 2.6 Ma. The average sand fraction content increased from 1–2% before the Pleistocene to ~ 4% during the Pleistocene, indicates a rapid expansion of the desert cover and/or a strengthened East Asian winter monsoon circulation. In

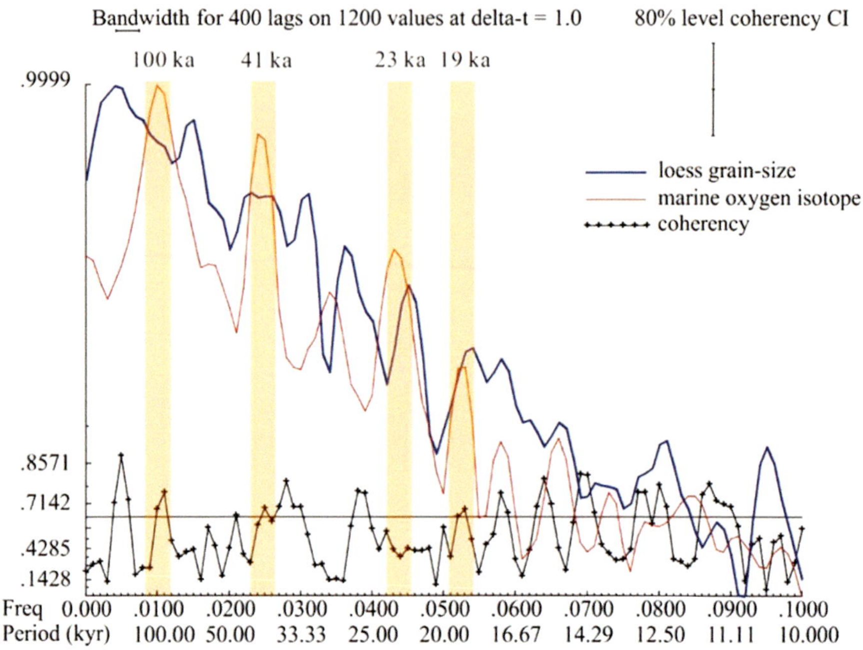

Figure 8.8. Results of cross-spectral analysis of the time series of aeolian silt grain-size and the marine oxygen isotopic record (proxy indicator of global ice volume) since the late Pliocene. Note the close coupling of cooling and drying in Asia.

parallel with the grain-size changes, the average dust accumulation rate increased from ~ 2 cm/ka before the late Pliocene, to ~ 6 cm/ka during the Pleistocene (Fig. 8.4), shows a major change in the dune field landscape in China during 4–2 Ma.

Our detrital zircon U-Pb dating results (Fig. 8.5) show that all the loess layers in the Chinese Loess Plateau (CLP) exhibit similar pronounced peaks at 250–350 Ma and 350–540 Ma, and small peaks at 1200–540 Ma and 2800–1200 Ma. These features are consistent with the detrital zircon age spectra of deserts and sand lands in North and North-west China. The Red Clay, however, shows dominant peaks at 350–540 Ma and 200–150 Ma, with smaller peaks at 350–250 Ma and other age ranges. These ages are similar to those of the Northeastern Tibetan Plateau. These findings support the widely accepted conclusion that the aeolian silt sediments of the CLP originated from the dune fields of North and North-west China.

Eighty plant taxa were observed in 41 samples from the sand lands, including 20 tree taxa such as *Pinus*, *Ulmus*, *Betula*, *Quercus* and *Celtis*. Tree pollen comprises a small proportion of the total pollen and spores. Forty-six grass and shrub taxa were observed, and they evidently dominated the vegetation. They include *Artemisia*, Chenopodiaceae, Gramineae, Aster-type, Taraxacum-type, Cyperaceae, Ranunculaceae, Leguminosae, *Thalictrum*, *Humulus*, Labiatae, Cruciferae and *Ephedra*. Seven fern spore types were found, such as *Selaginella sinensis* and Polypodiaceae; and there were also small percentages of hydrophytes such as *Zegnema* and *Concentricystes*. Pollen and spore concentrations are higher in the Holocene Optimum sediments than in the aeolian sandy deposits of the Last Glacial Maximum. Vegetation is a reliable indicator of the dune field landscape, and it typically reveals alternations of bare dune fields and vegetated conditions in arid regions. Our synthesis pollen dataset (Fig. 8.7) shows that bare dune field landscapes expanded since the late Miocene, and especially since the late Pliocene.

The cross-spectral analysis results show that fluctuations in the dune landscape were closely coupled with global temperature changes within the Earth orbital bands, with periodicities of ~ 100 kyr, ~ 40 kyr and ~ 20 kyr. They are significant at the 80% coherency level and demonstrate a statistical linkage between temperature of the Northern Hemisphere and dune fields landscape changes in North and North-west China (Fig. 8.8).

4. Discussion

4.1 Genetic linkage between dune field landscape and loess deposition

The Quaternary loess in the CLP has zircon U-Pb age distributions which closely match those of dune field surface sediments, establishing a genetic link between the two (Fig. 8.5). There are distinct U-Pb age peaks at 250–350 Ma and 400–540 Ma in the loess samples (Fig. 8.5). The zircon grains with ages of 250–350 Ma originated mainly from the Badain Jaran and Tengger dune fields, while the 400–540 Ma zircon grains came from the orogenic systems in the northern Qinghai-Tibet Plateau (Zhang et al., 2016, 2018, 2020). The Late Miocene and Pliocene Red Clay may have received greater contributions from the northern Qinghai-Tibet Plateau, with a predominant peak at 400–540 Ma (Fig. 8.5). The Quaternary loess and paleosol deposits, however, have a greater contribution from the dune fields, with a relatively larger component of zircon grains aged 250–350 Ma. In addition, the zircon grains with an age peak at 1800–2200 Ma, originating from the North China Craton, are also found in both the dune field surface sediments and loess deposits since the late Pliocene; however, the percentage is much lower for the Miocene and Pliocene Red Clay. This means that strengthened physical weathering of the North China Craton since the late Pliocene may have supplied more clastic sediments to the dune fields and loess deposits. Furthermore, the zircon age distributions in the Red Clay exhibit strong site-specific variation (Shang et al., 2016; Zhang et al., 2018), implying that prior to the late Pliocene silt/sand production and transport were limited, which excluded the existence of an extensive dune field landscape resembling that of the present day. It is well recognized that dune field sediments and the loess deposits in the Asian interior are genetically linked (Liu, 1985; Lu and Sun, 2000; Chen and Li, 2011; Stevens et al., 2010; He et al., 2017; Lu et al., 2019).

Apart from the detrital zircon U-Pb age spectra, several other lines of evidence, such as from mineral assemblages, the isotopic composition of neodymium and strontium (Chen and Li, 2011; Zhang et al., 2012; He et al., 2017), element contents and atmospheric circulation patterns (Liu, 1985; Zhang et al., 2012) support the conclusion of a genetic link between the dune fields and the aeolian silt deposits of China. Therefore, both the dune fields and loess can be regarded as a closely coupled depositional system, and that the landscape evolution of the dune fields are recorded in aeolian silt deposits.

4.2 Evidence for formation of present dune field landscape in late Pliocene

The percentage of grains of size > 63 μm in the loess-Red Clay sequences is used as an indicator of the evolution of the dune field landscape; this is because the sand particles coming from dune field sediments are transported by near-surface saltation, which is predominantly influenced by dune field extent (Ding et al., 2005; Lu et al., 2004a, 2019; Amit et al., 2014; Xu et al., 2020). The increasing content of > 63 μm grains in the loess-Red Clay sequence clearly indicates a coarsening trend in the aeolian silt since the late Pliocene (Fig. 8.3), as well as a distinct increase in the dust accumulation rate (Fig. 8.4). In addition, increased pulses of accumulation of siliceous particles in the Northern Pacific Ocean (NPO) are identified as occurring during 4–2 Ma (Rea et al., 1998; Zhang et al., 2016), which corresponds well with the increase in the dust accumulation rate on the CLP. These marked increases in the accumulation of aeolian sand and silt deposits downwind of the dune fields largely indicate that a stable dune field landscape was established during 4–2 Ma, with an abundance of silt materials produced, transported and deposited both on land and within the ocean. Before the significant increase in the aeolian silt accumulation rate, the deposition of the aeolian Red Clay underlying the loess implied that a certain amount of dune field and desert could have existed in North and North-west China and CE Asia. At that time, a broad area was covered by the Red Clay, but this deposit had a much lower accumulation rate and a smaller grain size (Wen et al., 2004; Lu et al., 2010, 2021), indicates the comparatively limited role of mountain weathering and sand and silt particles production. In addition, the sediment sources, the dune field landscape as well as the dust transport trajectory were different before and after 4–2 Ma.

It was suggested by us that in association with climatic cooling of the late Pliocene a 'Red dune field' was replaced by a 'Yellow dune field', together with the occurrence of greater mountain weathering and the production of a large amount of sand and silt which accumulated in the CLP (Figs. 8.3, 8.4). In addition to our findings, the thick and continuous aeolian silt/sand deposits in the central Taklimakan dune fields (Sun et al., 2011), at the Xining site close to the Qaidam Basin and the Badain Jaran dune fields (Lu et al., 2004b, 2019; Fang et al., 2019), the southern Mu Us dune field (Ding et al., 2005), and even loess deposits in Iran (Wang et al., 2016) and Tajikistan (Ding et al., 2002) (Fig. 8.1) support the view that the dune field landscape formation had commenced by 4–2 Ma.

4.3 Contribution of vegetation reconstruction and paleoclimatic simulation to characterizing evolution of an arid climate in North and North-west China and CE Asia

Vegetation record is an indicator of the evolution of an arid climate and dune fields and desert landscape during the Cenozoic era. Existing vegetation reconstructions in North and North-west China extended back to ~ 40 Ma (Fig. 8.7). It has been shown that there was a relatively short-lived phase of desert steppe at ~ 40 Ma in the Xining Basin, followed by a much longer period of forest steppes or forests (Dupont-Nivet et al., 2007). In the early and late Miocene, several sites close to the modern dune fields were occupied by desert steppe vegetation, indicating the early development of a desert landscape (Fig. 8.7). However, even if the desert landscape occurred sporadically at that time, widespread forests or forest steppes were still present. By ~ 8 Ma, dry climate biomes such as deserts, desert steppes and savannah became common, accompanied by forest steppes, revealing an aridification trend. The intensification of aridity occurred in the late Pliocene during 4–2 Ma when almost all vegetation types at these sites became deserts or desert steppes, indicating the final formation of a vast desert and dune field environment in North and North-west China (Fig. 8.7).

Paleoclimatic simulations show that there was a pronounced stepwise drying in interior Asia since late Miocene. Our re-analysis of the paleoclimatic simulations for the early Eocene (Li et al., 2018a), late Eocene (Li et al., 2018b), Miocene (Herold, 2011), middle Pliocene (Zhang et al., 2013), last interglacial (Otto-Bliesner et al., 2013), Last Glacial Maximum (Schmidt et al., 2014), and mid-Holocene (Schmidt et al., 2014), support the interpretation of stepwise processes of aridification. During the Eocene, the dune field region may have experienced a somewhat higher precipitation level than today, possibly due to higher temperatures and a topographic effect (Huber and Golder, 2012; Botsyun et al., 2019). During the Miocene and Pliocene, precipitation in North China decreased; however, its eastern part remained moist and may still have had relatively high precipitation (Zhang et al., 2013). During the Pleistocene, a dry climate dominated most of the dune field regions (Burke et al., 2018). In the glacial period, especially the LGM, most of the dune fields had average precipitation of less than ~ 200 mm, causing deserts and desert steppes to become the dominant surface cover in North and North-west China (Fig. 8.6). Modelling results also reveal that the drying in North China was closely associated with global cooling and ice cover expansion in the high-latitude Northern Hemisphere. These paleoclimatic modelling data show that the onset of a modern-like arid climate in the Asian interior developed in a stepwise manner and finally formed at 4–2 Ma.

As climate is one of the most important controls on the occurrence of vegetation and a desert landscape of North and North-west China and Central and Eastern Asia, the evolution and variability of the paleoclimate must be examined when deciphering the development of the desert landscape. Two aspects of climate change are clearly observed in our vegetation synthesis. One is the alternation of wet and dry conditions in North China, together with glacial-interglacial cyclicity, which have greatly influenced the development of the dune field landscape, in particular the sandlands in eastern Asia (Fig. 8.6). The intensification of glaciation in the Northern Hemisphere resulted in extremely dry conditions in the middle latitude temperate climate zone, associated with strengthened wind velocity which transported, deposited and mixed the sand sediments accumulating in the basins (Fig. 8.1).

The second aspect is the long-time evolution of an arid environment which is associated with the formation of a relatively arid climate since at least the middle Eocene in North and North-west China. These dry climatic conditions in a greenhouse world resulted in the expansion of sclerophyllous plants (Fig. 8.7). The weathering intensity was weak (such as seasonal temperature difference was small, and surface freezing and thawing weathing was weak) and therefore sediment production and the supply of sand particles was limited; a sand sea in North-west China did not yet exist until the late Pliocene, in response to global cooling and drying, when the reduction of vegetation cover and the intensification of surface weathering occurred, which produced large amounts of sand particles, resulting in the formation of the sand sea in North China, which then become a major source of aeolian silt deposits in the CLP and NPO (Figs. 8.1, 8.7).

4.4 Forcing mechanisms of the evolution of dune field landscape of China

The establishment of a modern-like dune field landscape during the late Pliocene provided a continuous and stable dust source for loess and deep-sea aeolian silt sediments. Compared with the Red Clay sediment, the Quaternary loess has a coarser grain-size and a high accumulation rate (Figs. 8.3, 8.4). Global cooling via an atmospheric bridge effect may have enhanced aridity in North-west China during 4–2 Ma. First, lower sea levels resulting from global cooling and ice sheet growth would have increased the transport distance of water vapour from the ocean to the Asian interior, intensifying drying in North-west China. Second, global cooling would also have reduced vapour circulation and weakened intensity of the East Asian summer monsoon, which would have increased aridity and the aeolian activity in the desert. In addition, cooling would have enhanced glaciation in high mountains such as the Qinghai-Tibet Plateau, which would have resulted in the production of substantially greater amounts of surface sediments, contributing to dune field and loess deposits, and increasing both of the sedimentation rate and grain-size in the dune field and loess regions (Figs. 8.3, 8.5). Thus, we concluded that global cooling was the principal underlying driver of aridification in the Asian interior during the late Pliocene.

In order to further understand the linkage between global cooling and dust accumulation, we performed cross-spectral analysis of the aeolian silt grain-size and the marine oxygen isotope time series using the Blackman-Tukey method, implemented in the ARAND software (Howell et al., 2006). The results show substantial co-variations, within the earth-orbital frequency domain, of changes in the dune field landscape (as indicated by the > 63 μm grain-size record) and the global ice-volume record (Lisiecki and Raymo, 2005) (Fig. 8.8). This further demonstrates a linkage between global cooling on the glacial-interglacial timescale and Asian aridification. The relationship can be verified by a reconstruction of the dune field landscape which shows that expansion and contraction of the dune field were closely associated with glacial-interglacial climatic fluctuations. During the last glacial period, the area of the dune field expanded and desert steppe vegetation became dominant (Fig. 8.6), while in contrast the dune field was reduced in area by 5 to 20% during the mid-Holocene (Lu et al., 2013a, 2019, 2021; Xu et al., 2015, 2020). Our analysis supports the conclusion that changes in the dune field landscape and the establishment of the modern arid environment during the late Pliocene was driven mainly by Northern Hemisphere cooling (Lu et al., 2010, 2019) rather than by tectonic deformation and uplift in the Asia interior, emphasizing that climate change plays a vital role in the formation of the desert landscape during late Pliocene.

It is widely accepted that surface uplift in the Himalayan-Tibetan Plate during the late Cenozoic caused drying in the Asian interior and a strengthening of the Asian monsoon (Kutzbach et al., 1989; Liu et al., 2019). A role for the growth of the Himalayan-Tibetean Plateau in the long-term aridification in interior Asia is not rejected, since the formation of the desert was a result of climatic drying and mountain uplift. However, it is argued that the most substantial change in the dune field landscape occurred in the late Pliocene during 4–2 Ma, which is difficult to explain by Tibetan-Himalayan uplift forcing alone, since the main production of elevated Tibetan-Himalayan topography occurred in the early or late Miocene, or even earlier (Molnar et al., 1993, 2010; Feng et al., 2021). This timing is much earlier than the major changes in desert landscape in North and North-west China, which makes it difficult to invoke topographic uplift as the primary cause of establishment of modern desert landscape Moreover, uplift in the Northeastern Tibetan Plateau, which has been the direct sand source for deserts such as the Badain Jaran and Tengger (Fig. 8.1), mainly developed in the late Miocene, ~ 8 Ma (Wang et al., 2011; Fan et al., 2019), much earlier than the timing of increase in the accumulation rate of the aeolian silt at ~ 2.6 Ma. The occurrence of major uplift several million years earlier than the formation of the desert landscape formation clearly demonstrates that uplift was not the main factor driving the establishment of the desert landscape. Therefore, we suggest that global cooling rather than regional surface deformation was primarily responsible for the establishment of the Asian interior dune fields in the late Pliocene.

4.5 A comprehensive approach to understanding formation of dune field in North and Northwest China

Asian interior dune field landscape is a product of arid climate, basin landforms and large-scale sand production (Fig. 8.9, Lu et al., 2019). The fundamental landforms and arid climate that would give rise to desert had already been established by early Miocene (Guo et al., 2002; Sun and Wang, 2005), but strong physical weathering and surface erosion in high mountains occurred in late Pliocene, along with onset of major glaciation in Northern Hemisphere (Lu et al., 2010, 2019, 2020; He et al., 2017). This makes one sceptical of earlier hypotheses of the early origin of the dune field landscape that consider sand deposition at isolated sites as an indicator of desert evolution, rather than adopting a more comprehensive approach (see the comments in Lu et al. (2019)). As sand deposit was fragmentary and localized, caution is needed before concluding that a vast desert landscape was in existence on only the basis of a discrete sand layer interbedded within other fluvial-lacustrine deposits (Fig. 8.2), in the absence of supporting evidence of climate state, topography and sand production. It is obvious that the debate on formation of Taklimakan Desert from the Eocene (Li et al., 2018) to the late Pliocene (Sun et al., 2011) is based on the issue of localized records, for which the implications of aeolian sand units in the depositional sequences may

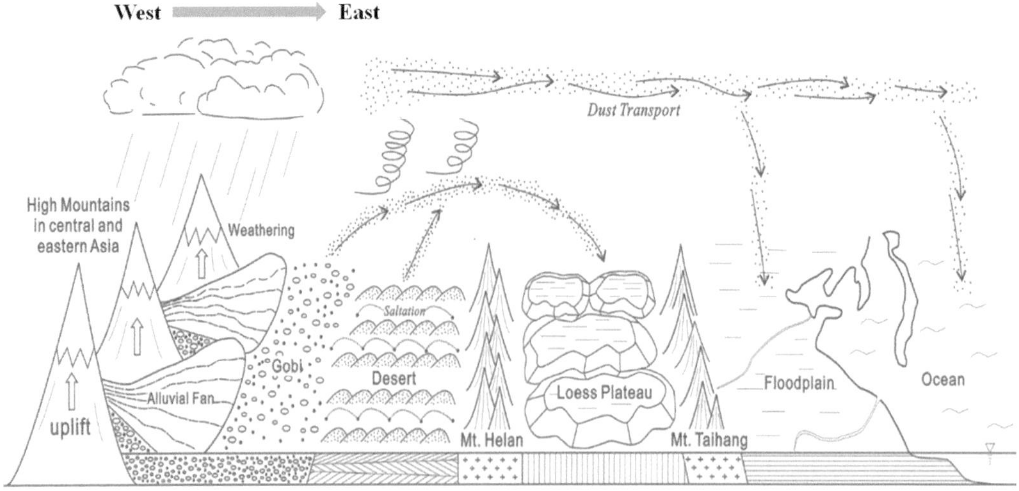

Figure 8.9. Origin, transport and deposition of dust from the Asian interior. The high- and low-level dust transport pathways are along with the trajectories of the Northern Hemisphere westerlies and the East Asian winter monsoon, respectively. The thick loess deposits of the Loess Plateau are associated with the formation and evolution of the Asian desert.

be quite different. The aeolian sand at a particular site could have originated, for example, from local fan deposits under regional dry conditions or been deposited due to strong winds but in a mild humid climate. For example, a large area of sand dunes exists in the vicinity of Poyang Lake (29°21′N, 116°10′E) in South China and the Changli coastal zone of the Yellow Sea (39°39′N, 119°18′E), even though these sand dunes occur in a humid climate with the influence of the East Asian monsoon, and their presence is due to the large-scale atmospheric circulation pattern and local wind environment. Therefore, these dune fields in isolation cannot be interpreted as indicating an arid climate.

Our conclusions do not exclude the possibility that early drying in the Asian interior was caused by growth of the Qinghai-Tibetean Plateau and the closure of Paleotethys Sea (Ramstein et al., 1997; Liu et al., 2019). The retreat of the Paleotethys Sea during middle Eocene (Carrapa et al., 2015) and global cooling since the Eocene-Oligocene transition (Dupont-Nivet et al., 2007; Zachos et al., 2008) may have resulted in reduced moisture availability and hydrological cycling, which may have commenced aridification of the Asian interior. The late Oligocene and early Miocene uplift of the Qinghai-Tibetean Plateau could subsequently have strengthened aridification (Zheng et al., 2015; Liu et al., 2019). However, these stages in the evolution of a drier climate may have been associated with a lower relief, high temperatures and weak physical weathering in the mountains of the region, thus limiting production of sand particles in Asian interior. As a result, there was still insufficient sandy supply to form dunes of the desert landscape. The most important threshold occurred during 4–2 Ma, when there was large increase in the capacity for transport of aeolian sand and silt (Fig. 8.2), as evidenced by aeolian dust accumulation on land and in the ocean (Fig. 8.4). A vast dune landscape under strong winds and dry climatic conditions, combined with a favourable topography, was able to supply abundant sand and silt to the regions downwind (Fig. 8.9).

5. Conclusions

The dune field landscape in North and North-west China is a result of the long-term evolution of mountain building, basin deepening, climatic drying and intensified physical weathering of the surface. An evolutionary tipping point in the desert and dune field formation process occurred in late Pliocene, which resulted from combination of a dry climate, abundant clastic sediments, and a favourable basin topography in Asia interior. Overall, our assessment supports the 'cooling drives drying' hypothesis (Lu et al., 2010) for the formation of the Asian desert and dune field landscape since the late Pliocene.

Since formation and evolution of dune fields in North and North-west China is an extremely complex question which needs to be further addressed, we have only analyzed the chronology of its development in this chapter. There is a general description of dune fields in Appendix 1, in which the landform, climate, ecology and present landscape of the dune fields are briefly introduced, readers who are interested in this part can go to Appendix 1 to get more information regarding dune fields in North and North-west China at end of this chapter.

6. Acknowledgements

We are grateful to Prof. Jonathan Adams and Dr. Jan Bloemendal for their careful correction of English. We thank Prof Joseph Mason, Prof. David Thomas, Dr. Thomas Stevens, Dr. Anna Bird, Prof. Michael Meadows, Ms Lingjuan Wen, Ms Quanyu Liu, Ms Cunfa Zhao and Dr. Langping Li for their assistance in the field and laboratory. This research was supported by the National Natural Science Foundation of China (awards 42021001, 41920104005, 41888101).

References

Amit, R., Enzel, Y., Mushkin, A., Gillespie, A., Batbaatar, J., Crouvi, O. et al. 2014. Linking coarse silt production in Asian sand deserts and Quaternary accretion of the Chinese Loess Plateau. Geology 42: 23–26.

Andersen, T. 2002. Correction of common lead in U-Pb analyses that do not report 204Pb. Chemical Geology 192(1): 59–79.

Bird, A., Stevens, T., Rittner, M., Vermeesch, P., Carter, A., Andò, S. et al. 2015. Quaternary dust source variation across the Chinese Loess Plateau. Palaeogeography, Palaeoclimatology, Palaeoecology 435: 254–264. https://doi.org/10.1016/j.palaeo.2015.06.024.

Botsyun, S., Sepulchre, P., Donnadieu, Y., Risi, C., Licht, A. and Caves Rugenstein, J.K. 2019. Revised paleoaltimetry data show low Tibetan Plateau elevation during the Eocene. Science 363: eaaq1436. DOI: 10.1126/science.aaq1436.

Burke, K., Williams, J., Chandler, M., Haywood, A., Lunt, D. and Otto-Bliesner, B. 2018. Pliocene and Eocene provide best analogs for near-future climates. Proceedings of the National Academy of Sciences of United States of America 115: 13288–13293.

Cai, M.T., Fang, X.M., Wu, F.L., Miao, Y.F. and Appel, E. 2012. Pliocene–Pleistocene stepwise drying of Central Asia: evidence from paleomagnetism and sporopollen record of the deep borehole SG-3 in the western Qaidam Basin, NE Tibetan Plateau. Global and Planetary Change 94: 72–81. https://doi.org/10.1016/j.gloplacha.2012.07.002.

Cao, J. and Cui, H. 1989. Research of Pliocene flora and paleoenvironment of Yushie basin on Shanxi Plateau, China. Chinese Journal of Geology 4: 369–375.

Cao, X., Lü, X., Yao, S., Mei, W., Zou, X., Chen, C. et al. 2011. LA–ICP–MS U–Pb zircon geochronology, geochemistry and kinetics of the Wenquan ore-bearing granites from West Qinling, China. Ore Geology Reviews 43(1): 120–131. http://doi.org/10.1016/j.oregeorev.2010.03.004.

Cao, Z., Xing, L. and Yu, Q. 1985. The magnetostratigraphic age and boundaries of the Yushe Formation. Bulletin of the Institute of Geomechanics 6: 144–154.

Carrapa, B., DeCelles, P.G., Wang, X., Clementz, M.T., Mancin, N., Stoica, M. et al. 2015. Tectono-climatic implications of Eocene Paratethys regression in the Tajik basin of central Asia. Earth and Planetary Science Letters 424: 168–178.

Che, X.D. and Li, G.J. 2013. Binary sources of loess on the Chinese Loess Plateau revealed by U–Pb ages of zircon. Quaternary Research 80(3): 545–551. https://doi.org/10.1016/j.yqres.2013.05.007.

Chen, D., Liu, L., Sun, Y., Zhang, A., Liu, X. and Luo, J. 2004. LA-ICP-MS zircon U-Pb dating for high-pressure basic granulite from North Qinling and its geological significance. Chinese Science Bulletin (18): 1901–1908.

Chen, J. and Li, G.J. 2011. Geochemical studies on the source region of Asian dust. Science China Series D: Earth Sciences 54(9): 1279–1301.

Chen, Z., Lu, S., Li, H., Song, B., Li, H. and Xiang, Z. 2004. The age of the Dehe biotite monzogranite gneiss in the North Qinling: TIMS and SHRIMP U-Pb zircon dating. Regional Geology of China (02): 136–141.

Cui, M.C., Lu, H.Y., Sweeney, M., Mason, J.A., Feng, H. and Xu, Z.W. 2015. PM10 emission flux in the Tengger Desert and Mu Us Sand field, northern China measured by PI-SWERL. Chinese Science Bulletin 60: 1621–1630.

Ding, L., MA, C., Li, J., Wang, L., Chen, L. and She, Z. 2010. LA-ICP-MS zircon U-Pb ages of the Lantian and Muhuguan granitoid plutons southern margin of the North China craton: Implications for tectonic setting. Geochimica (05): 401–413.

Ding, S., Pei, X., Liu, H., Li, Z., Sun, R., Liu, Z. and Meng, Y. 2006. LA-ICP-MS zircon U-Pb dating of the Xinyang Neoproterozoic granitoid gneisses in the Tianshui area, western Qinling, and its geological significance. Geology in China 33(6): 1217–1225.

Ding, Z.L., Sun, J.M., Liu, T.S., Zhu, R.X., Yang, S.L. and Guo, B. 1998. Wind-blown origin of the Pliocene red clay formation in the central Loess Plateau, China. Earth and Planetary Science Letters 161(1-4): 135–143. https://doi.org/10.1016/S0012-821X(98)00145-9.

Ding, Z.L., Ranov, V., Yang, S.L., Finaev, A., Han, J.M. and Wang, G.A. 2002. The loess record in southern Tajikistan and correlation with Chinese loess. Earth and Planetary Science Letters 200(3-4): 387–400. https://doi.org/10.1016/S0012-821X(02)00637-4.

Ding, Z.L., Derbyshire, E., Yang, S.L., Sun, J.M. and Liu, T.S. 2005. Stepwise expansion of desert environment across northern China in the past 3.5 Ma and implications for monsoon evolution. Earth and Planetary Science Letters 237(1-2): 45–55.

Dupont-Nivet, G., Krijgsman, W., Langereis, C.G., Abels, H.A., Dai, S. and Fang, X. 2007. Tibetan plateau aridification linked to global cooling at the Eocene-Oligocene transition. Nature 445(7128): 635–638.

Dupont-Nivet, G., Hoorn, C. and Konert, M. 2008. Tibetan uplift prior to the Eocene-Oligocene climate transition: Evidence from pollen analysis of the Xining Basin. Geology 36(12): 987–990.

Dupont-Nivet, G., Lippert, P.C., Van Hinsbergen, D.J., Meijers, M.J. and Kapp, P. 2010. Palaeolatitude and age of the Indo–Asia collision: Palaeomagnetic constraints. Geophysical Journal International 182: 1189–1198.

Faegri, K., Kaland, P.E. and Krzywinski, K. 1989. Textbook of Pollen Analysis. 4th Ed. John Wiley & Sons Ltd., Chichester, 328 pp.

Fan, L.G., Meng, Q.R., Wu, G.L., Wei, H.H., Du, Z.M. and Wang, E.C. 2019. Paleogene crustal extension in the eastern segment of the NE Tibetan plateau. Earth and Planetary Science Letters 514: 62–74.

Fang, X.M., Li, J.J., Zhu, J., Chen, H. and Cai, J. 1997. Absolute age determination and division of the Cenozoic stratigraphy in the Linxia Basin, Gansu Province, China. Chinese Science Bulletin 42(14): 1457–1471.

Fang, X.M., Zhang, W.M., Meng, Q.Q., Gao, J.P., Wang, X.M., King, J. et al. 2007. High-resolution magnetostratigraphy of the Neogene Huaitoutala section in the eastern Qaidam Basin on the NE Tibetan Plateau, Qinghai Province, China and its implication on tectonic uplift of the NE Tibetan Plateau. Earth and Planetary Science Letters 258(1-2): 293–306. https://doi.org/10.1016/j.epsl.2007.03.042.

Fang, X.M., Wu, F.L., Han, W.X., Wang, Y.D., Zhang, X.Z. and Zhang, W. 2008. Plio-Pleistocene drying process of Asian inland—Sporopollen and salinity records from Yahu section in the central Qaidam Basin. Quaternary Sciences 28(5): 875–882.

Fang, X.M., Fang, Y.H., Zan, J.B., Zhang, W.L., Song, C.H., Appel, E. et al. 2019. Cenozoic magnetostratigraphy of the Xining Basin, NE Tibetan Plateau, and its constraints on paleontological, sedimentological and tectonomorphological evolution. Earth-Science Reviews 190: 460–485. https://doi.org/10.1016/j.earscirev.2019.01.021.

Feng, H., Lu, H.Y., Carrapa, B., Zhang, H.Z., Chen, J., Wang, Y. and Clift, P.D. 2021. Erosion of the Himalaya-Karakoram recorded by Indus Fan deposits since the Oligocene. Geology 49(9): 1126–1131. https://doi.org/10.1130/G48445.1.

Guo, Z.T., Ruddiman, W.F., Hao, Q.Z., Wu, H.B., Qiao, Y.S., Zhu, R.X. et al. 2002. Onset of Asian desertification by 22 Myr ago inferred from loess deposits in China. Nature 416(6877): 159–163.

Han, J., Fyfe, W.F., Longstaffe, F.J., Palmer, H.C., Yan, F.H. and Mai, X.S. 1997. Pliocene-Pleistocene climatic change recorded in fluviolacustrine sediments in central China. Palaeogeography, Palaeoclimatology, Palaeoecology 135(1-4): 27–39. https://doi.org/10.1016/S0031-0182(97)00019-9.

Hao, H., Ferguson, D.K., Chang, H. and Li, C.-S. 2012. Vegetation and climate of the Lop Nur area, China, during the past 7 million years. Climatic Change 113(2): 323–338.

He, T., Liu, L.W., Chen, Y., Sheng, X.F. and Ji, J.F. 2017. A seven-million-year hornblende mineral record from the central Chinese Loess Plateau. Scientific Reports 7: 2382. https://doi.org/10.1038/s41598-017-02400-0.

Heermance, R.V., Pearson, J., Annelisa, M., Liu, L.T., Xu, J.H., Chen, J. et al. 2018. Erg deposition and development of the ancestral Taklimakan Desert (western China) between 12.2 and 7.0 Ma. Geology 46: 919–922.

Herold, N., Huber, M. and Müller, R. 2011. Modeling the Miocene climatic optimum. Part I: Land and atmosphere. Journal of Climate 24(24): 6353–6372.

Hoorn, C., Straathof, J., Abels, H.A., Xu, Y., Utescher, T. and Dupont-Nivet, G. 2012. A late Eocene palynological record of climate change and Tibetan Plateau uplift (Xining Basin, China). Palaeogeography, Palaeoclimatology, Palaeoecology 344: 16–38.

Howell, P., Pisias, N., Ballance, J., Baughman, J. and Ochs, L. 2006. ARAND Time-series Analysis Software: Brown University, Providence RI.

Huber, M. and Goldner, A. 2012. Eocene monsoons. Journal of Asian Earth Sciences 44: 3–23.

Hui, Z.C., Li, J.J., Xu, Q.H., Song, C.H., Zhang, J., Wu, F.L. and Zhao, Z.J. 2011. Miocene vegetation and climatic changes reconstructed from a sporopollen record of the Tianshui Basin, NE Tibetan Plateau. Palaeogeography, Palaeoclimatology, Palaeoecology 308(3-4): 373–382.

Jiang, H.C., Ding, Z.L. and Xiong, S.F. 2007. Magnetostratigraphy of the Neogene Sikouzi section at Guyuan, Ningxia, China. Palaeogeography, Palaeoclimatology, Palaeoecology 243(1-2): 223–234.

Jiang, H.C. and Ding, Z.L. 2008. A 20 Ma pollen record of East-Asian summer monsoon evolution from Guyuan, Ningxia, China. Palaeogeography, Palaeoclimatology, Palaeoecology 265(1-2): 30–38.

Jin, X., Wang, D., Liu, Y. and Zhang, J. 2002. Two Cenozoic palynological assemblages of the Kurha section, Kuqa, Xinjiang and their age and environmental significance. Geological Bulletin of China 21(12): 823–833.

Kukla, G. and An, Z.S. 1989. Loess stratigraphy in Central China. Palaeogeography, Palaeoclimatology, Palaeoecology 72: 203–225.

Kutzbach, J.E., Guetter, P.J., Ruddiman, W.F. and Prell, W.L. 1989. Sensitivity of climate to late Cenozoic uplift in Southern Asia and the American West: Numerical experiments. Journal of Geophysical Research: Atmospheres 94(D15): 18393–18407. https://doi.org/10.1029/JD094iD15p18393.

Li, J.X., Yue, L.P., Roberts, A.P., Hirt, A.M., Pan, F., Guo, L. et al. 2018. Global cooling and enhanced Eocene Asian mid latitude interior aridity. Nature Communications 9: 3026. https://doi.org/10.1038/s41467-018-05415-x.

Li, X.R., Fang, X.M., Wu, F.L. and Miao, Y.F. 2011. Pollen evidence from Baode of the northern Loess Plateau of China and strong East Asian summer monsoons during the Early Pliocene. Chinese Science Bulletin 56(1): 64–69.

Li, X.Y., Zhang, R., Zhang, Z.S. and Yan, Q. 2018a. Do climate simulations support the existence of East Asian monsoon climate in the Late Eocene? Palaeogeography, Palaeoclimatology, Palaeoecology 509: 47–57. doi:10.1016/j.palaeo.2017.10.029.

Li, X.Y., Zhang, R., Zhang, Z.S. and Yan, Q. 2018b. What enhanced the aridity in Eocene Asian inland: Global cooling or early Tibetan Plateau uplift? Palaeogeography, Palaeoclimatology, Palaeoecology 510: 6–14. https://doi.org/10.1016/j.palaeo.2017.10.029.

Li, Z.J., Sun, D.H., Chen, F.H., Wang, F., Zhang, Y.B., Guo, F. et al. 2014. Chronology and paleoenvironmental records of a drill core in the central Tengger Desert of China. Quaternary Science Reviews 85: 85–98. https://doi.org/10.1016/j.quascirev.2013.12.003.

Li, Z., Pei, X., Ding, S., Liu, Z., Wang, F., Li, G. et al. 2007. Zircon U-Pb dating of the Nanyili granodiorite in the Pingwu area, northwestern Sichuan, and its geological significance. Geology in China (06): 1003–1012.

Licht, A., Pullen, A., Kapp, P., Abell, J. and Giesler, N. 2016. Aeolian cannibalism: Reworked loess and fluvial sediment as the main sources of the Chinese Loess Plateau. Geological Society of America Bulletin 128(5-6): 944–956. http://doi.org/10.1130/B31375.1.

Lisiecki, L.E. and Raymo, M.E. 2005. A Pliocene-Pleistocene stack of 57 globally distributed benthic $\delta^{18}O$ records. Paleoceanography 20(2). https://doi.org/10.1029/2004PA001071.

Liu, H., Pei, X., Ding, S., Li, Z. and Sun, R. 2006. LA-ICP-MS zircon U-Pb dating of the Neoproterozoic granitic gneisses in the Yuanlong area, Tianshui City, West Qinling, China, and their geological significance. Geological Bulletin of China (11): 1315–1320.

Liu, T.S. 1985. Loess and the Environment. Science Press, Beijing.

Liu, W.G., Liu, Z.H., An, Z.S., Sun, J.M., Chang, H., Wang, N. et al. 2014. Late Miocene episodic lakes in the arid Tarim Basin, western China. Proceedings of the National Academy of Sciences 111(46): 16292–16296.

Liu, X.D., Dong, B.W., Yin, Z.Y., Smith, R.S. and Guo, Q.C. 2019. Continental drift, plateau uplift, and the evolutions of monsoon and arid regions in Asia, Africa, and Australia during the Cenozoic. Science China Earth Sciences 62: 1053–1075. https://doi.org/10.1007/s11430-018-9337-8.

Liu, X.M., Liu, T.S., Xu, T.C., Liu, C. and Chen, M.Y. 1988. The Chinese loess in Xifeng, I. The primary study on magnetostratigraphy of a loess profile in Xifeng area, Gansu province. Geophysical Journal International 92(2): 345–348. https://doi.org/10.1111/j.1365-246X.1988.tb01146.x.

Lu, H.Y. and Sun, D.H. 2000. Pathways of dust input to the Chinese Loess Plateau during the last glacial and interglacial periods. Catena 40: 251–261.

Lu, H.Y., Zhang, F.Q., Liu, X.D. and Duce, R.A. 2004a. Periodicities of palaeoclimatic variations recorded by loess-paleosol sequences in China. Quaternary Science Reviews 23(18-19): 1891–1900.

Lu, H.Y., Wang, X.Y., An, Z.S., Miao, X.D., Zhu, R.X., Ma, H.Z. et al. 2004b. Geomorphologic evidence of phased uplift of the northeastern Qinghai-Tibet Plateau since 14 million years ago. Science China Series D: Earth Sciences 47: 822–833.

Lu, H.Y., Wang, X.Y. and Li, L.P. 2010. Aaeolian sediment evidence that global cooling has driven late Cenozoic stepwise aridification in central Asia. Geological Society, London, Special Publications 342(1): 29–44.

Lu, H.Y., Yi, S.W., Xu, Z.W., Zhou, Y.L., Zeng, L., Zhu, F.Y. et al. 2013a. Chinese deserts and sand fields in Last Glacial Maximum and Holocene Optimum. Chinese Science Bulletin 58(23): 2775–2783.

Lu, H.Y., Yi, S.W., Liu, Z.Y., Mason, J.A., Jiang, D.B., Cheng, J. et al. 2013b. Variation of East Asian monsoon precipitation during the past 21 k.y. and potential CO_2 forcing. Geology 41(9): 1023–1026. https://doi.org/10.1130/G34488.1.

Lu, H.Y., Zhang, H.Z., Wang, Y.C., Zhao, L., Wang, H.L., Sun, W.F. and Zhang, H.Y. 2018. Ceonozoic depositional sequence in the Weihe Basin (Central China): A long-term record of Asian Monsoon precipitation from the greenhouse to icehouse Earth. Quaternary Sciences 38(5): 1057–1067.

Lu, H.Y., Wang, X.Y., Wang, X.Y., Chang, X., Zhang, H., Xu, Z. et al. 2019. Formation and evolution of desert-Gobi in central and eastern Asia. Earth-Science Reviews 194: 251–263. https://doi.org/10.1016/j.earscirev.2019.04.014.

Lu, H.Y., Wang, X.Y., Wang, Y., Zhang, X.J., Yi, S.W., Wang, X.Y., Stevens, T., Kurbanov, R. and Marković, S.B. 2021. Chinese loess and the Asian monsoon: What we know and what remains unknown. Quaternary International, https://doi.org/10.1016/j.quaint.2021.04.027.

Luo, B., Zhang, H. and Lü, X. 2012. U–Pb zircon dating, geochemical and Sr–Nd–Hf isotopic compositions of Early Indosinian intrusive rocks in West Qinling, central China: petrogenesis and tectonic implications. Contributions to Mineralogy and Petrology 164(4): 551–569. http://doi.org/10.1007/s00410-012-0748-2.

Ma, Y.Z. 1991. Tertiary sporo-pollen assemblages from southern Dunhuang basin, Gansu Province. Acta Micropalaeontologica Sinica 8(2): 207–225.

Ma, Y.Z., Li, J.J. and Fang, X.M. 1998. Records of the climatic variation and pollen flora from the red beds at 30.6–5.0 Ma in Linxia district. Chinese Science Bulletin 43: 301–304.

Ma, Y.Z., Fang, X.M., Li, J.J., Wu, F.L. and Zhang, J. 2005a. The vegetation and climate change during Neocene and Early Quaternary in Jiuxi Basin, China. Science in China Series D: Earth Sciences 48(5): 676–688. https://doi.org/10.1360/03yd0110.

Ma, Y.Z., Wu, F.L., Fang, X.M., Li, J.J., An, Z.S. and Wang, W. 2005b. Pollen record from red clay sequence in the central Loess Plateau between 8.10 and 2.60 Ma. Chinese Science Bulletin 50(19): 2234–2243.

Mao, J., Zhang, Z., Yang, J., Song, B., Wu, M. and Zuo, G. 1998. Single-zircon dating of Precambrian strata in the west sector of the northern Qilian Mountains and its geological significance. Chinese Science Bulletin 43(15): 1289–1294.

Martin, J.H. and Fitzwater, S.E. 1988. Iron deficiency limits phytoplankton growth in the north-east Pacific subarctic. Nature 331(6154): 341–343.

Miao, Y.F., Fang, X.M., Song, Z.C., Wu, F.L., Han, W.X., Dai, S. and Song, C.H. 2008. Late Eocene pollen records and palaeoenvironmental changes in northern Tibetan Plateau. Science in China Series D: Earth Sciences 51(8): 1089–1098.

Miao, Y.F., Fang, X.M., Herrmann, M., Wu, F.L., Zhang, Y.Z. and Liu, D.L. 2011. Miocene pollen record of KC-1 core in the Qaidam Basin, NE Tibetan Plateau and implications for evolution of the East Asian monsoon. Palaeogeography, Palaeoclimatology, Palaeoecology 299(1-2): 30–38.

Miao, Y.F., Fang, X.M., Wu, F.L., Cai, M.T., Song, C.H., Meng, Q.Q. and Xu, L. 2013. Late Cenozoic continuous aridification in the western Qaidam Basin. evidence from sporopollen records. Climate of the Past 9(4): 1863–1867.

Molnar, P., England, P. and Martinod, J. 1993. Mantle dynamics, the uplift of the Tibetan Plateau, and the Indian monsoon. Rev. Geophy. 31: 357–396.

Molnar, P., Boos, W.R. and Battisti, D.S. 2010. Orographic controls on climate and paleoclimate of Asia: thermal and mechanical roles for the Tibetan Plateau. Annu. Rev. Earth. Pl. Sc. 38: 77–102. https://doi.org/10.1146/annurev-earth-040809-152456.

Nie, J.S., Stevens, T., Rittner, M., Stockli, D., Garzanti, E., Limonta, M. et al. 2015. Loess plateau storage of northeastern Tibetan plateau-derived yellow river sediment. Nature Communications 6: 8511. https://doi.org/10.1038/ncomms9511.

Nie, J., Pullen, A., Garzione, C.N., Peng, W. and Wang, Z. 2018. Pre-Quaternary decoupling between Asian aridification and high dust accumulation rates. Science Advances 4(2): eaao6977.

Ogg, J.G. 2020. Geomagnetic polarity time scale. pp 159–192. *In*: Geologic Time Scale Elsevier.

Otto-Bliesner, B.L., Rosenbloom, N., Stone, E.J., McKay, N.P., Lunt, D.J., Brady, E.C. and Overpeck, J.T. 2013. How warm was the last interglacial? New model–data comparisons. Philosophical Transactions of the Royal Society A 371(2001), DOI: 10.1098/rsta.2013.0097.

Pei, X., Ding, S., Li, Z., Liu, Z., Li, G., Li, R. et al. 2007a. LA-ICP-MS zircon U-Pb dating of the Gabbro from the Guanzizhen ophiolite in the northern margin of the West Qinling and its geological significance. Acta Geologica Sinica 81(11): 1550–1561.

Pei, X., Ding, S., Zhang, G., Liu, H., Li, Z., Li, G. et al. 2007b. The LA-ICP-MS zircons U-Pb ages and geochemistry of the Baihua basic igneous complexes in Tianshui area of West Qinling. Science in China Series D: Earth Sciences 50(S2): 264-276. http://doi.org/10.1007/s11430-007-6028-8.

Pei, X., Ding, S., Zhang, G., Liu, H., Li, Z., Li, W. et al. 2007c. Zircon LA-ICP-MS U-Pb dating of Neoproterozoic granitoid gneisses in the north margin of West Qining and geological implication. Acta Geologica Sinica (06): 772–786.

Pullen, A., Kapp, P., McCallister, A.T., Chang, H., Gehrels, G.E., Garzione, C.N. et al. 2011. Qaidam Basin and northern Tibetan Plateau as dust sources for the Chinese Loess Plateau and paleoclimatic implications. Geology 39(11): 1031–1034. http://doi.org/10.1130/G32296.1.

Pullen, A., Ibáñez-Mejía, M., Gehrels, G.E., Ibáñez-Mejía, J.C. and Pecha, M. 2014. What happens when n – 1000? Creating large-n geochronological datasets with LA-ICP-MS for geologic investigations. Journal of Analytical Atomic Spectrometry 29(6): 971–980. http://doi.org/10.1039/c4ja00024b.

Qiang, X.K., Li, Z.X., Powell, C. and Zheng, H.B. 2001. Magnetostratigraphic record of the Late Miocene onset of the East Asian monsoon, and Pliocene uplift of northern Tibet. Earth and Planetary Science Letters 187(1): 83–93.

Qiao, Y.S., Guo, Z.T., Hao, Q.Z., Yin, Q.Z., Yuan, B. and Liu, T.S. 2006. Grain-size features of a Miocene loess-soil sequence at Qinan: Implications on its origin. Science in China Series D: Earth Sciences 49(7): 731–738.

Qin, J., Lai, S. and Li, Y. 2005. Petrogenesis and geological significance of Yangba granodiorites from Bikou area, Northern margin of Yangtze Plate. Acta Petrologica Sinica (03): 697–710.

Ramstein, G., Fluteau, F., Besse, J. and Joussaume, S. 1997. Effect of orogeny, plate motion and land–sea distribution on Eurasian climate change over the past 30 million years. Nature 386(6627): 788–795.

Rea, D.K., Leinen, M. and Janecek, T.R. 1985. Geologic approach to the long-term history of atmospheric circulation. Science 227: 721–725.

Rea, D.K., Snoeckx, H. and Joseph, L.H. 1998. Late Cenozoic Aeolian deposition in the North Pacific: Asian drying, Tibetan uplift, and cooling of the northern hemisphere. Paleoceanography 13(3): 215–224.

Rittner, M., Vermeesch, P., Carter, A., Bird, A., Stevens, T. et al. 2016. The provenance of Taklamakan desert sand. Earth and Planetary Science Letters 437: 127–137. https://doi.org/10.1016/j.epsl.2015.12.036.

Rui, L., Mi, C., Xiang-sheng, T., Heng, H.U. and De-pin, Y. 2014. Geochemical, Zircon SIMS U-Pb geochronological and Hf isotopic study on Lantian and Muhuguan Plutons in Eastern Qinling, China: petrogenesis and tectonic implications. Acta Mineralogica Sinica 34(4): 469–480.

Schmidt, G., Annan, J., Bartlein, P., Cook, B., Guilyardi, É., Hargreaves, J. et al. 2014. Using palaeo-climate comparisons to constrain future projections in CMIP5. Climate of the Past 10(1): 221–250.

Shang, Y., Beets, C.J., Tang, H., Prins, M.A., Lahaye, Y., van Elsas, R. et al. 2016. Variations in the provenance of the late Neogene Red Clay deposits in northern China. Earth and Planetary Science Letters 439: 88–100.

Shi, N., Cao, J.-X. and Königsson, L.-K. 1993. Late Cenozoic vegetational history and the Pliocene-Pleistocene boundary in the Yushe basin, SE Shanxi, China. Grana 32(4-5): 260–271.

Shi, R., Yang, J., Wu, C. and Wooden, J. 2004a. First SHRIMP dating for the formation of the Late Sinian Yushigou Ophiolite, North Qilian Mountains. Acta Geologica Sinica 78(5): 649–657. http://doi.org/10.3321/j.issn:0001-5717.2004.05.009.

Shi, R., Yang, J., Wu, C., Iizuka, T. and Hirata, T. 2004b. Island arc volcanic rocks in the North Qaidam UHP Metamorphic Belt. Acta Geologica Sinica 78(1): 52–64. http://doi.org/10.3321/j.issn:0001-5717.2004.01.007.

Shoenfelt, E.M., Winckler, G., Lamy, F., Anderson, R.F. and Bostick, B.C. 2018. Highly bioavailable dust-borne iron delivered to the Southern Ocean during glacial periods. Proceedings of the National Academy of Sciences of United States of America 115(44): 11180–11185. https://doi.org/10.1073/pnas.1809755115.

Song, S., Zhang, L., Niu, Y., Song, B., Zhang, G. and Wang, Q. 2004. Zircon U-Pb SHRIMP ages of eclogites from the North Qilian Mountains in NW China and their tectonic implication. Chinese Science Bulletin 49(8): 848–852. http://doi.org/10.1007/BF02889759.

Stevens, T., Palk, C., Carter, A., Lu, H. and Clift, P.D. 2010. Assessing the provenance of loess and desert sediments in northern China using U-Pb dating and morphology of detrital zircons. Geological Society of America Bulletin 122(7-8): 1331–1344. http://doi.org/10.1130/B30102.1.

Stevens, T., Adamiec, G., Bird, A.F. and Lu, H. 2013a. An abrupt shift in dust source on the Chinese loess plateau revealed through high sampling resolution OSL dating. Quaternary Science Reviews 82: 121–132. https://doi.org/10.1016/j.quascirev.2013.10.014.

Stevens, T., Carter, A., Watson, T.P., Vermeesch, P., Andò, S., Bird, A.F. et al. 2013b. Genetic linkage between the Yellow River, the Mu Us desert and the Chinese Loess Plateau. Quaternary Science Reviews 78(355-368). http://doi.org/10.1016/j.quascirev.2012.11.032.

Stockmarr, J. 1971. Tablets with spores used in absolute pollen analysis. Pollen et Spores 13: 614–621.

Sun, D.H., Liu, T.S., Chen, M.Y., An, Z.S. and Jone, S. 1997. Magnetostratigraphy and palaeoclimate of Red Clay sequences from Chinese Loess Plateau. Science in China Series D: Earth Sciences 40(4): 337–343.

Sun, D.H., Shaw, J., An, Z.S., Cheng, M.Y. and Yue, L.P. 1998. Magnetostratigraphy and paleoclimatic interpretation of a continuous 7.2 Ma Late Cenozoic Aeolian sediments from the Chinese Loess Plateau. Geophysical Research Letters 25(1): 85–88.

Sun, D.H., Bloemendal, J., Yi, Z.Y., Zhu, Y.H., Wang, X., Zhang, Y.B. et al. 2011. Palaeomagnetic and palaeoenvironmental study of two parallel sections of late Cenozoic strata in the central Taklimakan Desert: Implications for the desertification of the Tarim Basin. Palaeogeography, Palaeoclimatology, Palaeoecology 300(1-4): 1–10.

Sun, J.M., Xu, Q.H. and Huang, B.C. 2007. Late Cenozoic magnetochronology and paleoenvironmental changes in the northern foreland basin of the Tian Shan Mountains. Journal of Geophysical Research: Solid Earth 112: B04107. https://doi.org/10.1029/2006JB004653.

Sun, J.M. and Zhang, Z.Q. 2008. Palynological evidence for the mid-Miocene climatic optimum recorded in Cenozoic sediments of the Tian Shan Range, northwestern China. Global and Planetary Change 64: 53–68. https://doi.org/10.1016/j.gloplacha.2008.09.001.

Sun, J.M., Zhang, L.Y., Deng, C.L. and Zhu, R.X. 2008. Evidence for enhanced aridity in the Tarim Basin of China since 5.3 Ma. Quaternary Science Reviews 27: 1012–1023. https://doi.org/10.1016/j.quascirev.2008.01.011.

Sun, J.M., Zhang, Z. and Zhang, L.Y. 2009. New evidence on the age of the Taklimakan Desert. Geology 37(2): 159–162.

Sun, J.M., Ye, J., Wu, W., Ni, X.J., Bi, S.D., Zhang, Z.Q. et al. 2010. Late Oligocene–Miocene mid-latitude aridification and wind patterns in the Asian interior. Geology 38: 515–518. https://doi.org/10.1130/G30776.1.

Sun, X.J. and Wang, P.X. 2005. How old is the Asian monsoon system?—Palaeobotanical records from China. Palaeogeography, Palaeoclimatology, Palaeoecology 222(3-4): 181–222. https://doi.org/10.1016/j.palaeo.2005.03.005.

Sun, X.Y., Fan, Y.X., Deng, C.L. and Yu, Z.Q. 1980. Cenozoic sporo-pollen assemblage of the Weihe Basin, Shaanxi. Bulletin of the Institute of Geology Chinese Academy of Geological Sciences, 84–109.

Sun, X., Zhao, Y. and He, Z. 1984. The Oligocene-Miocene palynological assemblages from the Xining-Minghe basin, Qinghai Province. Geological Review 30(3): 207–216.

Tang, Z.H., Ding, Z.L., White, P.D., Dong, X.X., Ji, J.L., Jiang, H.C. et al. 2011. Late Cenozoic central Asian drying inferred from a palynological record from the northern Tian Shan. Earth and Planetary Science Letters 302: 439–447. https://doi.org/10.1016/j.epsl.2010.12.042.

Tong, G.B., Wu, X.H. and Chen, Y. 2000. Palynoevents since Late Cenozoic in Weihe Valley and Climate Changes. Journal of Geomechanics 6(4): 18–65.

Tseng, C., Yang, H., Yang, H., Liu, D., Tsai, C., Wu, H. and Zuo, G. 2007. The Dongcaohe ophiolite from the North Qilian Mountains: A fossil oceanic crust of the Paleo-Qilian ocean. Chinese Science Bulletin 52(17): 2390–2401. http://doi.org/10.1007/s11434-007-0300-3.

Uno, I., Eguchi, K., Yumimoto, K., Takemura, T., Shimizu, A., Uematsu, M. et al. 2009. Asian dust transported one full circuit around the globe. Nature Geoscience 2: 557–560. https://doi.org/10.1038/NGEO583.

Wan, Y.S., Xu, Z.Q., Yang, J.S. and Zhang, J.X. 2001. Ages and compositions of the Precambrian High-grade basement of the Qilian Terrane and its adjacent areas. Acta Geologica Sinica (English edition) 75(4): 375–384. http://doi.org/10.1111/j.1755-6724.2001.tb00055.x.

Wang, D.N., Sun, X.Y. and Zhao, Y.N. 1990. Late cretaceous to tertiary palynofloras in Xinjiang and Qinghai, China. Review of Palaeobotany and Palynology 65(1-4): 95–104.

Wang, F., Sun, D.H., Chen, F.H., Bloemendal, J., Guo, F., Li, Z.J. et al. 2015. Formation and evolution of the Badain Jaran Desert, North China, as revealed by a drill core from the desert centre and by geological survey. Palaeogeography, Palaeoclimatology, Palaeoecology 426: 139–158. https://doi.org/10.1016/j.palaeo.2015.03.011.

Wang, L., Lv, H.Y., Wu, N.Q., Li, J., Pei, Y.P., Tong, G.B. and Peng, S. 2006. Palynological evidence for Late Miocene–Pliocene vegetation evolution recorded in the red clay sequence of the central Chinese Loess Plateau and implication for palaeoenvironmental change. Palaeogeography, Palaeoclimatology, Palaeoecology 241(1): 118–128. https://doi.org/10.1016/j.palaeo.2006.06.012.

Wang, T., Wang, X., Tian, W., Zhang, C., Li, W. and Li, S. 2009. North Qinling Paleozoic granite associations and their variation in space and time: Implications for orogenic processes in the oregens of central China. Science in China Series D: Earth Sciences 52: 1359–1384.

Wang, T., Zhang, Z., Wang, X., Wang, X. and Zhang, C. 2005. Neoproterozoic collisional deformation in the core of the Qinling Orogen and its age: Constrained by Zircon SHRIMP dating of strongly deformed syn-collisional granites and weakly deformed granitic veins. Acta Geologica Sinica (02): 220–231.

Wang, W.M. 1990. Tertiary sporo-pollen assemblages from the Shangdou-Huade Basin, Inner Mongolia-with discussion on the formation of the steppe vegetation in China. Acta Micropalaeontologica Sinica 7: 239–252.

Wang, W.M., He, X.G., Shi, L., Zhang, Z.Y., Shi, G.J., Liu, S. et al. 1990. Sporo-pollen assemblage from the Miocene Tongguer Formation of Inner Mongolia and its climate. Journal of Integrative Plant Biology 32(11): 901–904.

Wang, W.T., Zhang, P.Z., Kirby, E., Wang, L.H., Zhang, G.L., Zheng, D.W. and Chai, C.Z. 2011. A revised chronology for Tertiary sedimentation in the Sikouzi basin: implications for the tectonic evolution of the northeastern corner of the Tibetan Plateau. Tectonophysics 505: 100–114. https://doi.org/10.1016/j.tecto.2011.04.006.

Wang, W. and Deng, T. 2009. Palynoflora from the stratotype section of the Neogene Xiejian Stage and its significance. Acta Palaeontologica Sinica 48(1): 1–8.

Wang, X.Y., Lu, H.Y., Zhang, H.Z., Wu, J., Hou, X.X., Fu, Y. and Geng, J. 2018. Distribution, provenance, and onset of the Xiashu Loess in Southeast China with paleoclimatic implications. Journal of Asian Earth Sciences 155: 180–187. https://doi.org/10.1016/j.jseaes.2017.11.022.

Wang, X., Wei, H.T., Taheri, M., Khormali, F., Danukalova, G. and Chen, F.H. 2016. Early Pleistocene climate in western arid central Asia inferred from loess-palaeosol sequences. Scientific Reports 6: 20560. https://doi.org/10.1038/srep20560.

Wen, L.J., Lu, H.Y. and Qiang, X.K. 2004. Changes of grain-size and sedimentation rate of the Neocene "Red Clay" deposit along Chinese Loess Plateau and its implication to palaeo-wind system. Science China Series D: Earth Sciences 48: 1452–1462.

Wu, F.L., Fang, X.M., Ma, Y.Z., Herrmann, M., Mosbrugger, V., An, Z.S. and Miao, Y.F. 2007. Plio–Quaternary stepwise drying of Asia: Evidence from a 3-Ma pollen record from the Chinese Loess Plateau. Earth and Planetary Science Letters 257(1-2): 160–169. https://doi.org/10.1016/j.epsl.2007.02.029.

Wu, F.L., Fang, X.M., Herrmann, M., Mosbrugger, V. and Miao, Y.F. 2011. Extended drought in the interior of Central Asia since the Pliocene reconstructed from sporopollen records. Global and Planetary Change 76(1-2): 16–21. https://doi.org/10.1016/j.gloplacha.2010.11.007.

Wu, M., Guo, Y. and Li, X. 2009. Zircon LA-ICP-MS U-Pb dating of the Guandiping diorite pluton from Mianxian, south Qinling, central China. Northwestern Geology (01): 73–78.

Wu, Y. 2001. Palynoflora at late Miocene-early Pliocene from leijiahe of lingtai, Gansu Province, China. Acta Botanica Sinica 43(7): 750–756.

Xiao, A., Li, D., Zhang, H. and Li, X. 2003. Climate evolution and composional features of sporopollen in the Oligocene–pliocene series of the Kumukuli Basin in Xinjiang. Geology of Shanxi 21: 36–44.

Xiao, G.Q., Zong, K.Q., Li, G.J, Hu, Z.C., Dupont-Nivet, G., Peng, S.Z. and Zhang, K.X. 2012. Spatial and glacial-interglacial variations in provenance of the chinese loess plateau. Geophysical Research Letters 39(20): L20715. https://doi.org/10.1029/2012GL053304.

Xu, W., Zhang, H. and Liu, X. 2007. U-Pb zircon dating constraints on formation time of Qilian high-grade metamorphic rock and its tectonic implications. Chinese Science Bulletin (10): 1174–1180.

Xu, Z.W., Lu, H.Y., Yi, S.W., Vandenberghe, J., Mason, J.A., Zhou, Y.L. and Wang, X.Y. 2015. Climate-driven changes to dune activity during the Last Glacial Maximum and deglaciation in the Mu Us dune field, north-central China. Earth and Planetary Science Letters 427: 149–159.

Xu, Z.W., Mason, J.A., Xu, C., Yi, S.W., Bathiany, S., Yizhaq, H., Zhou, Y.L., Cheng, J., Holmgren, M. and Lu, H.Y. 2020. Critical transitions in Chinese dunes during the past 12,000 years. Science Advances 6(9): eaay8020.

Xue, X.X., Zhang, Y.X. and Yue, L.P. 2006. Paleoenvironments indicated by the fossil mammalian assemblages from red clay-loess sequence in the Chinese Loess Plateau since 8.0 Ma BP. Science China Series D: Earth Sciences 49(5): 518–530.

Yang, J., Xu, Z., Pei, X., Shi, R., Wu, C., Zhang, J. et al. 2002. Discovery of diamond in north qinling: evidence for a giant UHPM belt across central china and recognition of paleozoic and mesozoic dual deep subduction between north China and Yangtze plates. Acta Geologica Sinica 76(4): 484–495. http://doi.org/10.3321/j.issn:0001-5717.2002.04.007.

Yang, J., Liu, F., Wu, C., Wan, Y., Zhang, J., Shi, R. and Chen, S. 2003. Two ultrahigh pressure metamorphic events recognized in the central orogenic belt of China: Evidence from the U-Pb dating of coesite-bearing zircons. Acta Geologica Sinica 77(4): 463–477. http://doi.org/10.3321/j.issn:0001-5717.2003.04.003.

Yang, X.P., Scuderi, L.A., Wang, X.L., Scuderi, L.J., Zhang, D. Guo, Li, H.W. et al. 2015. Groundwater sapping as the cause of irreversible desertification of Hunshandake Sandy Lands, Inner Mongolia, northern China. Proceedings of the National Academy of Sciences of United States of America 112(3): 702–706. https://doi.org/10.1073/pnas.1418090112.

Yang, Z., Dong, Y., Liu, X. and Zhang, J. 2006. LA-ICP-MS zircon U-Pb dating of gabbro in the Guanzizhen ophiolite Tianshui, West Qinling, China. Geological Bulletin of China 25(11): 1321–1325.

Ye, H., Mao, J., Li, Y., Guo, B., Zhang, C., Liu, J. et al. 2006. SHRIMP zircon U-Pb and molybdenite Re-Os dating for the superlarge Dongguo porphyry Mo deposit in East Qinling, China, and its geological implication. Acta Geologica Sinica (07): 1078–1088.

Yin, A. 2006. Cenozoic tectonic evolution of the Himalayan orogen as constrained by along-strike variation of structural geometry, exhumation history, and foreland sedimentation. Earth Sci. Rev. 76: 1–131. 10.1016/j.earscirev.2005.05.004.

Yue, L.P., Deng, T., Zhang, Y.X., Wang, J.Q., Zhang, R., Yang, L.,R. and Heller, F. 2004. Magnetostratigraphy of stratotype section of the Baode stage. Journal of Stratigraphy 28(1): 48–63.

Zachos, J.C., Dickens, G.R. and Zeebe, R.E. 2008. An early Cenozoic perspective on greenhouse warming and carbon-cycle dynamics. Nature 451(7176): 279–283.

Zhang, H.Y., Lu, H.Y., Jiang, S.Y., Vandenberghe, J., Wang, S.J. and Cosgrove, R. 2012. Provenance of loess deposits in the Eastern Qinling Mountains (central China) and their implications for the paleoenvironment. Quaternary Science Reviews 43: 94–102. https://doi.org/10.1016/j.quascirev.2012.04.010.

Zhang, H.Z., Lu, H.Y., Xu, X.S., Liu, X.M., Yang, T., Stevens, T. et al. 2016. Quantitative estimation of the contribution of dust sources to Chinese loess using detrital zircon U-Pb age patterns. Journal of Geophysical Research: Earth Surface 121(11): 2085–2099. https://doi.org/10.1002/2016JF003936.

Zhang, H.Z., Lu, H.Y., Stevens, T., Feng, H., Fu, Y., Geng, J.Y. and Wang, H.L. 2018. Expansion of dust provenance and aridification of Asia since ~ 7.2 Ma revealed by detrital zircon U-Pb dating. Geophysical Research Letters 45(24): 13,437–413,448. https://doi.org/10.1029/2018GL079888.

Zhang, H.Z., Lu, H.Y., Zhou, Y.L., Cui, Y.Y., He, J., Lv, H.Z., Wang, K.X. and Wang, X.Y. 2020. Heavy mineral assemblages and U-Pb detrital zircon geochronology of sediments from the Weihe and Sanmen Basins: New insights into the Pliocene-Pleistocene evolution of the Yellow River. Palaeogeography, Palaeoclimatology, Palaeoecology. https://doi.org/10.1016/j.palaeo.2020.110072.

Zhang, H., Zhang, B., Harris, N., Zhang, L., Chen, Y., Chen, N. and Zhao, Z. 2006. U–Pb zircon SHRIMP ages, geochemical and Sr–Nd–Pb isotopic compositions of intrusive rocks from the Longshan–Tianshui area in the southeast corner of the Qilian orogenic belt, China: Constraints on petrogenesis and tectonic affinity. Journal of Asian Earth Sciences 27(6): 751–764. http://doi.org/10.1016/j.jseaes.2005.07.008.

Zhang, J.Q., Liu, B., Chen, X.G., Sun, X.M. and Wang, Y.S. 2009. Environmental evolution at the drilling hole k3 in the east-south margin of Ebinur Lake, Xinjiang in Late Cenozoic. Geoscience 23(6): 1012–1021.

Zhang, R., Yan, Q., Zhang, Z.S., Jiang, D., Otto-Bliesner, B.L., Haywood, A.M. et al. 2013. Mid-Pliocene East Asian monsoon climate simulated in the PlioMIP. Climate of the Past 9: 2085–2099. https://doi.org/10.5194/cp-9-2085-2013.

Zhang, W.F., Chen, J., Ji, J.F. and Li, G.J. 2016. Evolving flux of Asian dust in the North Pacific Ocean since the late Oligocene. Aaeolian Research 23: 11–20.

Zhang, Z.Q. and Sun, J.M. 2011. Palynological evidence for Neogene environmental change in the foreland basin of the southern Tianshan range, northwestern China. Global and Planetary Change 75(1-2): 56–66.

Zhao, J., Chen, D., Tan, Q., Chen, M., Zhu, X., Guo, C. and Liu, L. 2012. Zircon LA-ICP-MS U-Pb dating of basic volcanics from Erlangping Group of the North Qinling, eastern Qinling Mountains and its geological implications. Earth Science Frontiers 19(4): 118–125.

Zheng, H.B., Wei, X.C., Tada, R., Clift, P.D., Wang, B., Jourdan, F. et al. 2015. Late oligocene–early miocene birth of the Taklimakan Desert. Proceedings of the National Academy of Sciences 112(25): 7662–7667. https://doi.org/10.1073/pnas.1424487112.

Zheng, H., An, Z. and Shaw, J. 1992. New contributions to Chinese Plio-pleistocene magnetostratigraphy. Physics of the Earth and Planetary Interiors 70(3): 146–153.

Zhu, L., Zhang, G., Guo, B. and Li, B. 2008. U-Pb (LA-ICP-MS) zircon dating for the large Jinduicheng porphyry Mo deposit in the East Qinling, China, and its metallogenetic Geodynamical setting. Acta Geologica Sinica (02): 204–220.

Zhu, L., Zhang, G., Chen, Y., Ding, Z., Guo, B., Wang, F. and Lee, B. 2011. Zircon U-Pb ages and geochemistry of the Wenquan Mo-bearing granitioids in West Qinling, China: Constraints on the geodynamic setting for the newly discovered Wenquan Mo deposit. Ore Geology Reviews 39(1-2): 46–62. http://doi.org/10.1016/j.oregeorev.2010.10.001.

Chapter 9

Formation, Evolution and Forcing Mechanism of Feathered Sand Dunes in the Kumtag Desert, China

Lu Qi,[1,2,*] *Cui Guipeng,*[1] *Gao Pan,*[1] *Cui Mengchun,*[1] *Wu Bo*[1] and *Chu Jianmin*[3]

1. Introduction

The Kumtag Desert is located at the eastern margin of the Tarim Basin in Northwest China. The desert is the fourth largest sand seas covered mainly by mobile dunes and shifting sands. It lies between 90°27′–94°48′ E, 39°00′–40°47′ N and advances across Gansu Province and the Xinjiang Uygur Autonomous Region (Fig. 9.1). The desert covers an area of 22800 km^2, stretching 280 km from east to west and 120 km from north to south. The boundary of the desert extends eastward to the Mingsha Mountain (Whispering Megadunes) in Dunhuang, westward to the Lop Nur facing the Taklimakan Sand Desert, southward to the Altun Mountains and northward to the Aqike Valley and the Beishan Mountain ranges. The natural conditions in Kumtag Desert are harsh and the climate is extremely dry. The land surface is composed of complicated landforms and diverse dunes, including the unique feathered longitudinal dunes (Wang et al., 2011).

Since the 1950s, China has sponsored and organized several large-scale desert expeditions and ground investigations in the arid zone, when the Kumtag Desert had remained as the last untouched natural desert to be fully investigated in a scientific way. The natural conditions of Kumtag Desert are extremely harsh and depopulated which constrained any scientific expedition. In December 2006, the Ministry of Science and Technology of China launched the two-stage "*Integrated Scientific Expedition of the Kumtag Desert Program*" to fill in the gap of the last unexplored desert in China (Lu, 2011; Wu, 2017). Under the leadership of the Chinese academy of Forestry, in collaboration with the Chinese Academy of Sciences, Ministry of Education, State Meteorology Administration as well as other academic and educational institutions of Gansu province, more than 150 scientists and experts of geology, geomorphology, hydrology, meteorology, zoology, botany, pedology, mapping and integration planning from 18 academic institutions were nominated and invited to form a scientific expedition team to implement this expedition program. From June 2007 to May 2017, the teams conducted several integrated large-scale and multidisciplinary desert expeditions, professional sampling surveys and ground observations. In total, they worked for more than 500 days in the desert and travelled over 400000 km. These expeditions gained fruitful results

[1] Institute of Desertification Studies, Chinese Academy of Forestry, Beijing 100091, China.
[2] Experimental Center of Desert Forestry, Chinese Academy of Forestry, Dengkou 015200, Inner Mogolia, China.
[3] Research Institute of Forestry, Chinese Academy of Forestry, Beijing 100091, China.
* Corresponding author: luqi@caf.ac.cn

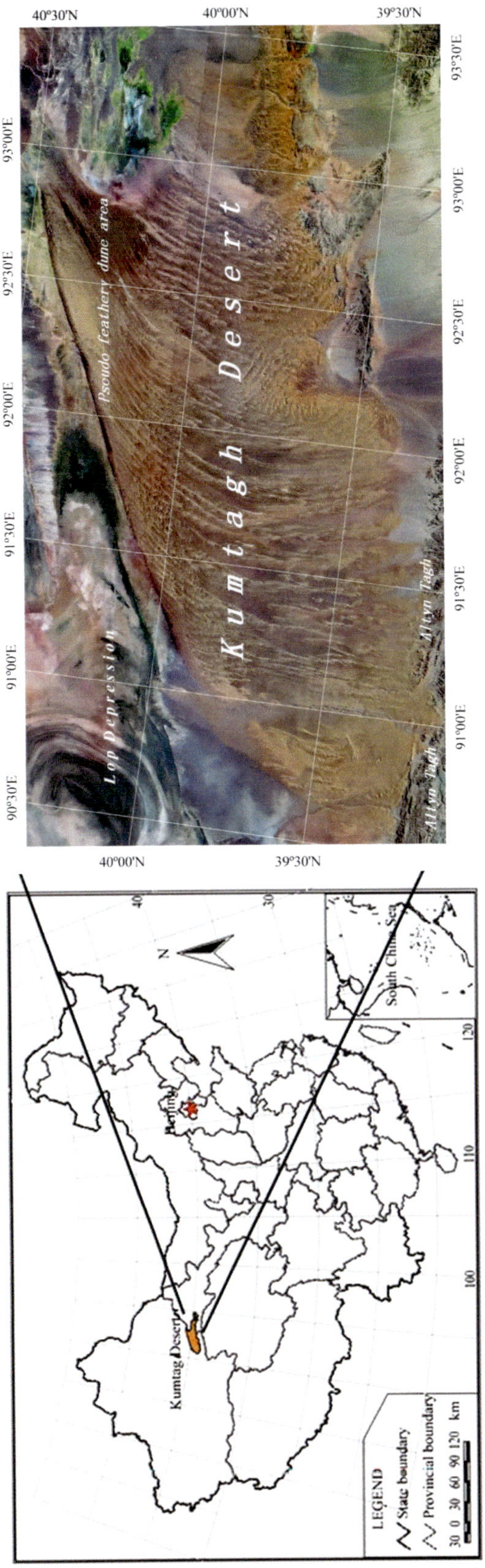

Figure 9.1. Location and map of the Kumtag Desert in China (Wang et al., 2011).

in many scientific research subjects, including geology, geomorphology, hydrology, pedology, meteorology, zoology and botany (Lu, 2011; Wu, 2017).

In this chapter, an overview of the sand dunes in the Kumtag Desert is presented based on the two-stage expeditions. The feathered longitudinal dunes are discussed from multiple points of view at various levels of morphology, characteristics of wind regime, sedimentation structure, particle gradient and mineral compositions, spectrum characteristics, particle shape and form. The formation, evolution and forcing mechanism of feathered sand dunes in the Kumtag Desert are reviewed. Natural resources and ecological environment of the desert are described to provide insights for sustainable land management in future.

2. Geological background

The Kumtag Desert lies in a middle Cenozoic faulted basin (Haranor Syneclise)[1] that is located between the Beishan Fold Belt and the Altun Fault Block, establishes a mountain-basin system with the uplifted Beishan Mountain and also forms a sloping stony plain at the north foothill of the Altun Mountains. Quaternary loose sediments are widely distributed in this basin and mainly composed of lacustrine, aeolian and fluvial deposits. Outcrops of Proterozoic metamorphic series and Precambrian-Ordovician siliceous limestone are found at the marginal regions of the basin, along with sparsely distributed Jurassic and Tertiary strata.

Based on the analysis of stratigraphic correlation, Optically Stimulated Luminescence (OSL), Electron Spin Resonance (ESR) dating and other proxy indicators, the aeolian sand deposits in the Kumtag Desert were found to have accumulated since the early Pleistocene and even the Pliocene. The formation of the Kumtag Desert can be dated back to the Pliocene, when aeolian sand deposits coexisted with alluvial deposits (Shule River Group, characterized by grey conglomerate and brick red gravelly silty mudstone) and lacustrine sediments (celadon silty mudstone). Aeolian sand deposits were widely distributed across the desert in the early Pleistocene, indicating that a desert landscape had developed at that time. After multiple expansions during the middle and late Pleistocene, the modern desert landscape was finally established. Sediments of ice melting and ice edge of the Later Quaternary Period were found for the first time, which illustrates that a cold and wet environment prevailed then. Based on the stratum chronology dating and the evidence of neotectonic movement, large areas of the denuded Yardang landform (Fig. 9.2) in the north of the Kumtag Desert were mainly formed from the later stage of the Meso-Pleistocene Period to the early

Figure 9.2. Yardang landform in the north Kumtag Desert.

[1] an intracratonic sedimentary basin.

stage of the later Pleistocene Period. The formation of the Kumtag Desert is the interactive result of the neo-tectonic movement of the unceasing uplift of the Qinghai-Tibet Plateau and long-term climate change (Ding et al., 2017; Dong et al., 2013a; Dong et al., 2010a; Li et al., 2017; Mason et al., 2009; Zhao et al., 2015).

Mineral composition of the surface samples of the Kumtag Desert is almost consistent with that of rocks of the Altun Mountains and surface sediments of the alluvial/fluvial fans of the Altun Mountains, which indicates that one major source of the Kumtag Desert sediments is the Altun Mountains. Alluvial deposits and lake sediments in the Aqik valley and lower reaches of the Shule River are prone to be eroded and transported by strong northeast winds to the hinterland of the Kumtag Desert; as a result, these places become another source of surface deposits of the Kumtag Desert. The sediments of the Kumtag Desert of the Later Cenozoic Era are mainly composed of lacustrine, aeolian silt and sand deposits with manifestations of polycycle and phase transformation variation of the stratum chronology (Stevens et al., 2010; Xu et al., 2011).

The water system distribution pattern and evolution of the Kumtag Desert are the interactive results of river hydrologic processes and land surface erosion and denudation under the impacts of climate change and human development activities in accordance with the Qinghai-Tibet Plateau uplift, particularly the neo-tectonic movement in the Quaternary Period, since the Cenozoic Era. It is found that, by conducting sample surveys of the modern water system distribution, discharge and water quality, the modern water systems of the Kumtag Desert came from the Altun Mountains, the Qilian Mountains and the Beishan Mountains ranges and the south-north oriented rivers, streams, springs, swamps and lakes, which belong to the Hexi Corridor-Alxa inland drainage system (the Shule River system) and the Tarim inland drainage system (the Lop Nur). As a result, a Modern Water System Distribution Map of the Kumtag Desert was compiled on the basis of these ground investigations. Gullies and dammed lakes are well developed in the Kumtag Desert. Meanwhile, the Evolution Map of the Drainage System of the Kumtag Desert Region and the Shule River Drainage since the late stage of the Meso-Pleistocene Period was compiled on the basis of relative geologic and historic records (E et al., 2008; Yan et al., 2011).

According to observational data from six surrounding meteorological stations for the period 2003–2012, Mean Annual Temperature (MAT) of the Kumtag Desert was 10.3°C, Mean Annual Precipitation (MAP) was 44.6 mm, mean annual wind speed was 2.4 m/s, and mean annual sunshine hours was 3213.8 hours. At the interannual scale, MAT generally increased during 2003–2012, with the maximum value of 10.9°C occurring in 2007. MAP exhibited a slightly increasing trend, reaching the maximum value of 67.7 mm in 2005 and the minimum value of 22.6 mm in 2009. By analyzing meteorological data of the Kumtag Desert and nearby districts during the last five decades, it is indicated that wind velocity is weakening and the number of occurrences of dust and sandstorms has evidently decreased. There are four main types of weather conditions affecting the Kumtag Desert and the nearby districts, namely the cold air from the mountains, cold air invading from the east, local convective airflow and prefrontal thermal depression. The formation and evolution of the Kumtag Desert and its arid climate are largely affected by local complex topography and evaporation rate (Hu et al., 2017; Huo et al., 2011; Zeng et al., 2011; Zhang et al., 2010; Zhao et al., 2010).

3. Geomorphology of the Kumtag Desert

The geomorphological types of the Kumtag Desert can be divided into five categories: aeolian landform, fluvial landform, dry denudational landform, glacial periglacial landform and other landforms, among which the aeolian landform is the dominant type. In the aeolian landform, aeolian accumulation landform (dune) prevails while the areas of wind-erosion landforms are much smaller and mainly consist of various types of yardang. The aeolian geomorphologic division of the Kumtag Desert is determined by the top-down sequential method. The first level division is separated according to the regional differentiation regularity and the genetic type of the underlying

Figure 9.3. Feathered sand dunes in the Kumtag Desert.

geomorphology, and then the second level division (sub regions) is divided by the morphology of dune. In summary, the Kumtag Desert can be separated into seven aeolian geomorphologic divisions and 23 sub regions (Dong et al., 2013b; Dong et al., 2009; Dong et al., 2010b; Dong et al., 2011a; Dong et al., 2011b; Qian et al., 2015).

A unique pattern, perceived as a 'feathered dune' (Fig. 9.3). by many researchers, has been identified for the first time in the Kumtag Desert on satellite images and aerial photographs (see below).

The morphologic characteristics of the feathered longitudinal sand dune and its formation process are proved based on the full knowledge of the formation of aeolian landforms and its development principle of the Kumtag Desert (Dong and Lü, 2014). According to ground observations and field investigations, a classification system of sand dunes suitable to the Kumtag Desert was developed; in combination with interpretation of remote sensing images, a geomorphology map of the Kumtag Desert was compiled. A unique aeolian landform unit (Fig. 9.4), i.e., sand-gravel-moraine (Dong et al., 2011b; Lu, 2011), developed in the Kumtag Desert, was defined and named for the first time and its specific geomorphologic system interlocking with nearby sand dunes and longitudinal dunes was carefully observed.

Figure 9.4. Sand-gravel-moraines in the Kumtag Desert.

The feathered patterns (pseudo-feather dunes) are distributed in the northern Kumtag Desert, between 91°24.8'–92°48.4'E and 40°1.2'–40°27.8'N, covering an area of 4000 km^2 and accounting for about 18% of the total area of the Kumtag desert (Dong et al., 2008a; Dong et al., 2008b). Through aerial photographs and satellite images, morphological characteristics of the feathered patterns in the northern Kumtag Desert were also analysed (Dong et al., 2008a; Dong et al., 2008b). The linear dunes were generally aligned from east-northeast to west-southwest and stretched over several or even several tens of kilometres, with sinuous ridges and occasional slip-faces. The linear dunes were 8 to 21 m (mostly 10 to 17 m) tall, 40 to 120 m wide, and were spaced 100 to 1000 m (mostly 100 to 400 m) apart. The dune-like drifts were almost perpendicular to the linear dunes and convex facing downwind (the prevailing wind is towards SW 205°). The dune-like drifts were relatively regularly spaced 80 to 350 m (mostly 80 to 220 m) apart and often comprised small patches of drifted sand with a spatial extent of several metres and spacing of 15 to 50 m. Coarse sand ripples (or fine gravel ridges) with a wavelength of 1 to 2 m and a height of 0.1 to 0.3 m occasionally appeared in the dark inter-dune flats (Dong et al., 2008b).

The feathered longitudinal dunes are in fact compound or complex linear sand dunes originating from barchan dunes and comprise the following elements: ① speedy wind-energy regime and sharp double-peak pattern wind regime which usually enable the formation of linear dunes; ② relatively limited sand sources from sediments; ③ high content of non-eroded coarse sands contained in sediments; and ④ flat underlying land surface.

4. Forcing mechanisms of the feathered dunes

In combination with satellite images and meteorological data, Dong (2009) summarized the morphological characteristics and formation mechanisms of typical feathered dunes that have been reported worldwide. First, feathered dunes belong to compound linear dunes, the primary ridges of which are considered to be the shafts of feathered dunes and the subsidiary ridges oblique to the primary ridges are considered to be the vanes of feathered dunes. The subsidiary ridges may appear on either end (side) or both ends (sides) of the primary ridges, elongating towards or off the primary ridges. Second, feathered dunes often have a low height and wide inter-dune area, probably due to insufficient supply of sand. Finally, the wind regime responsible for the formation of feathered dunes is complicated, with different wind energy and resultant drift direction inconsistent with the alignment of primary ridges (Dong, 2009). However, the interpretation of the formation of feathered dunes in the Kumtag Desert is disputed.

Debates on forcing mechanisms of the feathered dunes in the Kumtag Desert remained as the latest research highlight in the subject of geomorphology from multiple points of view at various levels of morphology, characteristics of wind regime, sedimentation structure, particle gradient and mineral compositions, spectrum characteristics, particle shape and form (Dong, 2009; Dong et al., 2013b; Dong et al., 2010a; Dong et al., 2008a; Dong et al., 2008b; Dong et al., 2010c; He et al., 2009a, b; Qian et al., 2009; Qu et al., 2011; Qu et al., 2007; Wang et al., 2014; Wang et al., 2017; Wang et al., 2009; Yang et al., 2016, 2018; Zeng et al., 2009). Three interpretations of forcing mechanisms of the feathered dunes were raised from three different spatial scales: ① Feathered fan for the overall Kumtag Desert from a topographical hypothesis in a desert basin or catchment scale; ② Feathered patterns (pseudo-feather dunes) from aeolian hypothesis in a dune scale; ③ Feathered texture as seif dunes/raked linear dunes from aeolian and grain colour hypothesis in a mineral scale.

4.1 Feathered fan

The Kumtag Desert is similar in general appearance to a huge feather fan that overlays the southwest-northeast inclined alluvial-diluvial plain in the north foothill of the Altun Mountains. The handle of the feather fan is situated at Shanlongsha in the northern Kumtag Desert, and the middle part is composed of linear dunes aligned almost from the northeast to southwest. Satellite images of the fantail exhibit texture features aligned almost directly from north to south, which in fact illustrate the interlacing distribution of shifting sand and diluvial plain due to the combined effect of rivers and floods coming from the Altun Mountains (Dong et al., 2008a).

It is speculated that the general appearance of the Kumtag Desert is primarily controlled by geological structures, reflecting the impact of neotectonic movement on the formation and evolution of desert as well as the development of aeolian geomorphology (Dong et al., 2008a). The Kumtag Desert lies in a middle Cenozoic faulted basin (Haranor Syneclise) that is located between the Beishan Fold Belt and the Altun Fault Block, establishing a mountain-basin system with the uplifted Altun Mountains and Beishan Mountain ranges. The Haranor Syneclise disintegrated and the Aqike Valley was formed due to neotectonic movement since Late Tertiary. A steep scarp, with a height of 50 to 100 metres, was also formed between northern Kumtag Desert and the Aqike Valley, which protects the Kumtag Desert against the impact of water mass changes in the Lop Nur Depression and thereby facilitates the preservation of aeolian geomorphology in the Kumtag desert (Dong et al., 2008a).

Is there an underlying topography of the linear dunes in Kumtag Desert? Unlike the linear dunes in other deserts in the world, the correlation between the height and spacing of linear dunes in the Kumtag Desert is relatively poor. Thus, it has been suspected that the formation of Kumtag linear dunes is related to the underlying topography, probably caused by the aeolian sand which covered the yardang landscape. The direct evidence for this is the Shanlongsha Yardang landform developed in the northeast of the desert and the yardang landform developed in the Aqike Valley (Qu et al., 2004; Qu et al., 2005). Based on the ESR dating of the yardang strata and regional geomorphological features, Qu et al. (2004) proposed that the evolution history of the Aqike Valley and the Kumtag Desert can be divide into the following four stages: (1) The big lake stage of Lop Nur. This stage is from early Pleistocene to middle Pleistocene. (2) The mountain uplift and lake retreat stage. During late middle Pleistocene (about 300 ka BP), the strong uplift of the Qinghai-Tibetan Plateau formed a barrier for maritime air mass, and the climate of Central Asia became relatively dry. Due to the left-lateral strike-slip and eastward movement of the Altun Mountains, plumetic fault valley groups occurred in the strata of old lakes. In this region, there are two groups of faults (one EW direction, the other NE direction). The water of Shule River could not feed the Lop Nur anymore and the eastern part of the lake gradually dried up. (3) The rainstorm erosion and strong wind erosion stage. During the late Middle Pleistocene and the early Late Pleistocene, many yardang groups developed at the north part of Aqik Valley. In the southern side of the valley, there are many NE direction paralleled yardang ranges and wind-eroded valleys. In the middle of the Aqike Valley, there are seasonal salt marsh and isolated yardang. (4) The expanding stage of Kumtag Desert at the end of Late Pleistocene and Holocene. At this stage, the dunes expanded northward and covered up the plumetic fault valleys. Wind erosion valleys and yardang ranges together form the most incredible plumetic[2] sand dunes (also called broom-like dunes in Qu et al., 2005) landform.

4.2 Feathered patterns

Qu et al. (2007, 2011) proposed the feather-shaped sand ridges hypothesis through field observations, aerial photograph interpretations and simulation experiments in the wind tunnel (Fig. 9.5) (Qu et al., 2011; Qu et al., 2007). Qu et al. (2007) suggested that the forming factors of feather-shaped sand ridge are as follows: (1) Extensive inclined plain. (2) Sand materials on the surface are not abundant. (3) The dynamic condition is two groups of winds in similar directions. (4) Tongue-like sand dune is the sub form of barchan sand ridge (Qu et al., 2007). The tongue-like sand dune and barchan sand ridge together form the feather-shaped sand ridge. The developing process of feather-shaped sand ridge can be summarized as the follow stages: barchan sand dune → barchan sand ridge → tongue-like sand dune → feather-shaped sand ridge.

[2] Featherlike (hence plume).

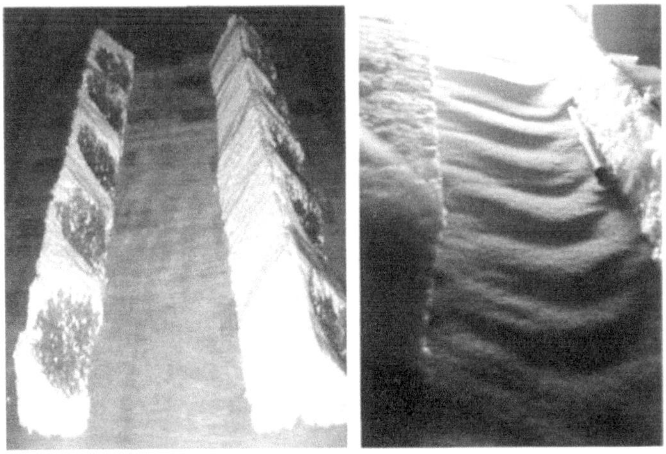

Figure 9.5. Tongue-like sand ripples in feather-shaped sand ridge in the wind tunnel (Qu et al., 2007).

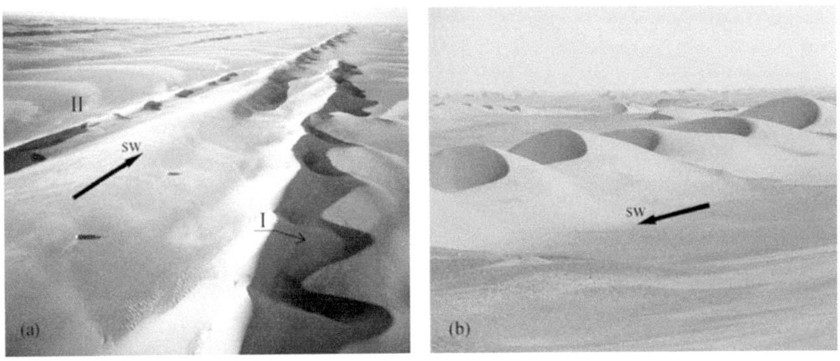

Figure 9.6. Low altitude and ground surface photograph of feathered sand ridges (Qu et al., 2011). (a) Low altitude photograph of feathered sand ridges; I, northwest horn of barchan dune left by dune ridge, II, tongue-shaped dune; location: 40°12'40.62"N, 92°06'25.98"E, photographed height 150 m, photographed time: October 2006. (b) Ground surface photograph of feathered sand ridge shows the southeast horn of barchan dune and downwind connection scene; location: 40°15'15.67"N, 92°19'20.92"E; photographed time: September 2005.

Qu et al (2011) further proposed that feathered sand ridges in the North eastern Kumtag Desert consist of crescent sand ridges and interridge tongue-shaped dunes (Fig. 9.6) (Qu et al., 2011). The differences in grain size, mineral composition and albedo between crescent sand ridges and tongue-shaped dunes, and between windward and leeward slopes of tongue-shaped dunes, result in their feathery appearance in aerial and satellite imagery (Qu et al., 2011).

Through the measurements of the sand drift potential in the region, the sand-moving wind for feathered sand ridges can be divided into three sectors; i.e., north-north easterly, easterly and east-north easterly sectors, roughly corresponding to the southeast, northwest and southwest slip faces. Qu et al. (2011) pointed out that the crescent sand ridges resulting from the connection of barchan dunes along the prevailing wind direction are longitudinal dune ridges rather than transverse ones (Fig. 9.7); and the tongue-shaped dunes and quasi-dune shapes have obvious distinctions and are new transverse dune types (Qu et al., 2011). According to McKee's dune shape classification, the feathered sand ridges are not a deformation dune type but a complex one. According to Wu's dune morphological and genetic classification, they are not dune ridges or compound dune ridges that form under the action of unidirectional winds or two winds intersecting at an acute angle, but are complex dune ridges that form under the action of three winds intersecting at an acute angle.

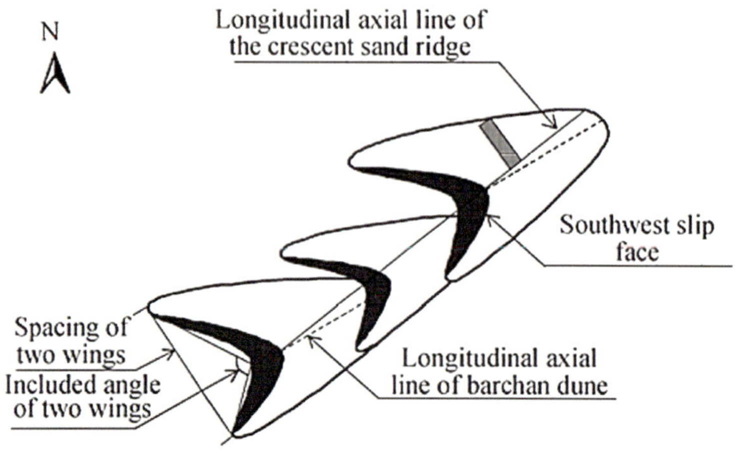

Figure 9.7. Schematic diagram of crescent sand ridge shape (Qu et al., 2011).

Figure 9.8. Quickbird satellite image and low-angle aerial photograph of the feathered patterns (pseudo-feather dunes) in the northern Kumtag Desert (Dong et al., 2008a; Dong et al., 2008b). The left is Quickbird satellite image taken in 2005 with a resolution of 0.6 m; central point location: 40°14'24.12"N, 92°26'24.53"E. The right is a low-angle aerial photo taken in 2006; location: 40°12'40.62"N, 92°06'25.98"E, photographed height 150 m.

Based on field investigations, Dong et al. (2008a, 2008b) suggested that the shafts of feathered patterns in aerial photographs and satellite images were interpreted as linear dunes composed of high albedo sand, and the vanes of feathered patterns were ascribed to albedo contrasts in sediments of bright dune-like drifts and dark inter-dune flats (Fig. 9.8).

The reasons for which vanes were not considered to be transverse dunes are as follows: (1) the height of vanes was barely distinct from surrounding areas, in discordance with the morphological features of typical dunes; (2) topographic variations were not well correlated to the patterns of bright dune-like drifts; (3) it is unlikely to produce linear and transverse dunes under the same wind regime (Dong et al., 2008a). Further analysis illustrated that the albedo contrasts are primarily associated with differences in the particle size distribution and mineral composition of sediments (Dong et al., 2008a; Dong et al., 2008b). The sediments of linear dunes and dune-like drifts were mainly composed of medium sand, fine sand and very fine sand, whereas those of inter-dune flats were mainly coarse sand and very coarse sand (Dong et al., 2008b; Liu et al., 2006). It is found that coarser sediments had lower content of minerals with high albedo such as quartz, and greater content of those with low albedo such as dacite.[3] As a result, the coarser sediments of the dark inter-dune flats had lower albedo (0.16), whereas the finer sediments of the bright linear dunes and

[3] an extrusive rock that is sometimes partly glassy and is composed of plagioclase and quartz with biotite, hornblende, or pyroxene.

dune-like drifts had higher albedo (0.31) (Dong et al., 2008b). Moreover, the coarser sediments of the dark inter-dune flats were better rounded and sorted, implying that they were more resistant to weathering and wind erosion (Dong et al., 2008b).

Dong et al. (2008a, 2008b) demonstrated that the formation and retention of vane patterns can be explained by saltation-initiated activity proposed by Bagnold (Bagnold, 1941). The finer sediments constituting the bright linear dunes and the dune-like drifts belong to the saltation category that has a propensity for self-accumulation into sand dunes even in the absence of obstacles (vegetation, etc.). Since saltating grains would bounce off a hard surface with coarse sand more effectively than they do over a bed of loose fine sand, the sand transport rate over the bright patches of finer sediments is accordingly lower than that over the surrounding dark coarser sediments, thereby resulting in the accumulation of saltating grains over bright patches and the development of bright drifts. However, the bright dune-like drifts have not evolved into transverse dunes due to insufficient supply of erodible fractions of surface sediments, which have been transported to form linear dunes (Dong et al., 2008a; Dong et al., 2008b). The initiation dynamics of original dune-like patches has not yet been well understood. According to the convex-downwind patterns of dune-like drifts, Dong et al. (2008a, 2008b) suggested that the patches might have formed during a recent period of stronger-than-normal winds, when fine sediments were sorted out and drifted over the inter-dune flats like waves under the influence of wind turbulence and were then left behind on the surface when the winds abated. Another explanation is that fine sediments constituting the dune-like drifts were blown from linear dunes by those winds that were oblique to linear dunes.

However, Wang et al. (2009) suggested that the dune 'feathers' are created by dunes of different heights, rather than by differences in material composition. Through field studies and sedimentological analyses, they found the dunes are linear dunes and zibars,[4] corresponding with the rachises and vanes respectively and appearing as feathery patterns in aerial photographs and satellite images.

4.3 Feathered texture as seif dunes/raked linear dunes

Based on field investigation and further analysis of aerial photographs and satellite images, Dong et al. (2008a, 2008b) suggested that the originally recognized feathered dunes in the northern Kumtag Desert are in fact feathered patterns. The feathered dunes in the Kumtag Desert are most likely to be seif dunes[5] with sinuous ridges, and a unique form is identified and characterized as raked linear dunes (Fig. 9.9), which covers an area of about 2400 km^2 and is mostly distributed north of 40°04' N (Dong, 2009; Dong and Lü, 2014; Dong et al., 2008a; Dong et al., 2008b; Dong et al., 2010c).

Dong et al. (2008a) provided a tentative explanation for the formation of seif dunes using the evolution model proposed by Bagnold (1941). According to Bagnold's model, the most frequent north easterly wind (the prevailing gentle wind), in the Kumtag Desert forms barchan dunes with south eastern and north western horns, and then the strongest but less frequent eastly and south-south easterly winds (sand-bearing storm winds) elongate the south eastern horns to form sief dunes, resulting in the alignment of the seif dunes from 55° to 235° (with 0° as north) (Fig. 9.10). Although the north-north eastly wind tends to elongate the north western horns of the barchan dunes, its action is limited compared with that of easterly and south-south easterly winds. As a result, the north western side of the seif dunes exhibit a clear appearance of barchan horns. However, the formation of raked linear dunes cannot be accounted for by the above explanation (Dong et al., 2010c).

On the basis of detailed wind data obtained during 2008–2010, Dong et al. (2010c) further proposed a modified version of Tsoar's (1984) model to interpret the formation of raked linear

4 are coarse-grained, low-relief, slipface-less, aeolian bedforms that occur on sand sheets and within interdune corridors of many eolian sand seas.

5 a long, narrow sand dune or chain of dunes, generally oriented in a direction parallel to the prevailing wind or in a direction resulting from two or more winds blowing at acute angles to each other.

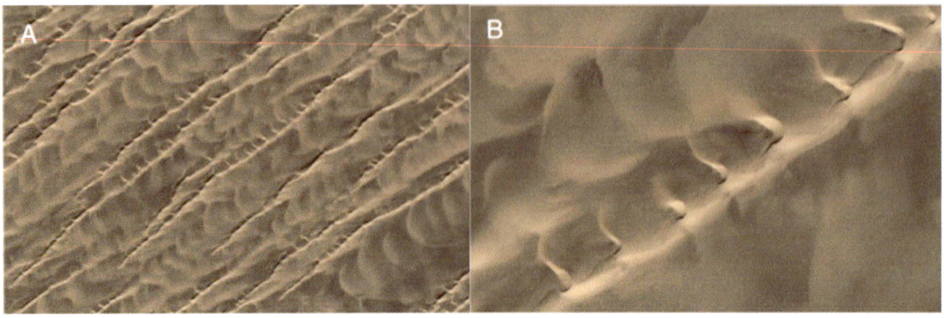

Figure 9.9. Raked linear dunes in the northern Kumtag Desert (Dong et al., 2010c). (A) An overview of the geomorphology. (B) A close-up of a single dune (central point location: 40°14'29.8"N, 92°14'6.4"E).

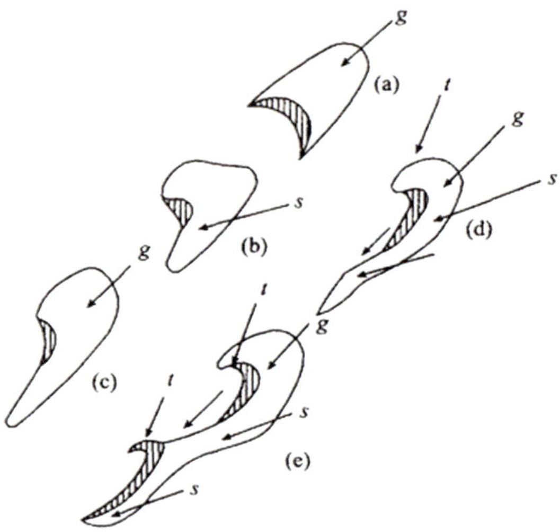

Figure 9.10. Schematic diagram of Bagnold's evolution model from barchan dunes to seif and feathered dunes (Dong et al., 2008a).

dunes (Fig. 9.11). In this model, the northern group of winds (north-north easterly and north easterly winds) dominant in spring and summer form barchans, which are then modified and reoriented by the prevailing eastern group (east-north easterly and easterly winds) in autumn that blow oblique to the barchans. As a result, the limbs of barchans closest from the eastern group of winds are eroded, while the limbs farthest from the winds extend to form the subsidiary ridges. Besides, the barchans are reoriented, with the alignment from roughly 255°–260° to 75°–80° (with 0° as north). Due to a deficiency in sediment availability, the extension of barchan limbs is very limited and therefore the formed subsidiary ridges are short with bright tailing dune-like drifts. It is found that the western group of winds (westerly and west-south westerly winds) prevail in winter in the raked linear dune area, which also plays a role in shaping the raked linear dunes by facilitating the accumulation of sediments on primary ridges. The raked linear dunes appear to be in a dynamic equilibrium with the above wind regime. Apart from the influence of the wind regime, dune collision is also likely to modify dune geomorphology; however, what is the role of dune collision in the formation of raked linear dunes in the Kumtag Desert remains unknown.

A recent study demonstrated that the raked patterns are attributed to a combination of two dune growth mechanisms (fingering mode and bed instability mode), based on numerical modelling and analysis of field data and satellite images (Lü et al., 2017). It was suggested that the raked linear dunes can be characterized as linear dunes asymmetrically indented by a line of superimposed oblique dunes under multi-directional wind regimes. The main dunes (primary ridges) are in supply-

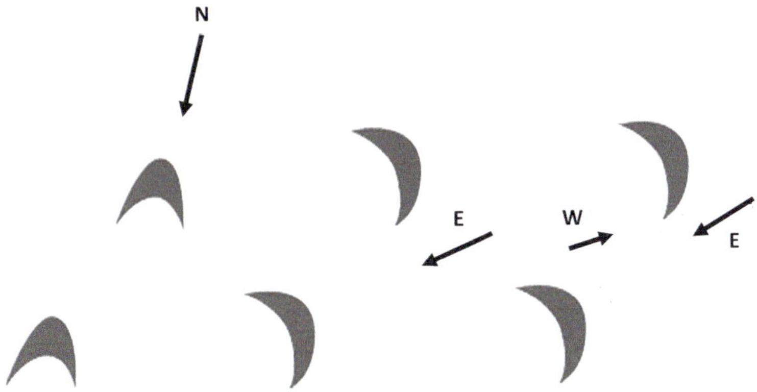

Figure 9.11. Schematic diagram of the modified Tsoar's evolution model for the formation of the raked linear dunes in the Kumtag Desert (Dong et al., 2010c).

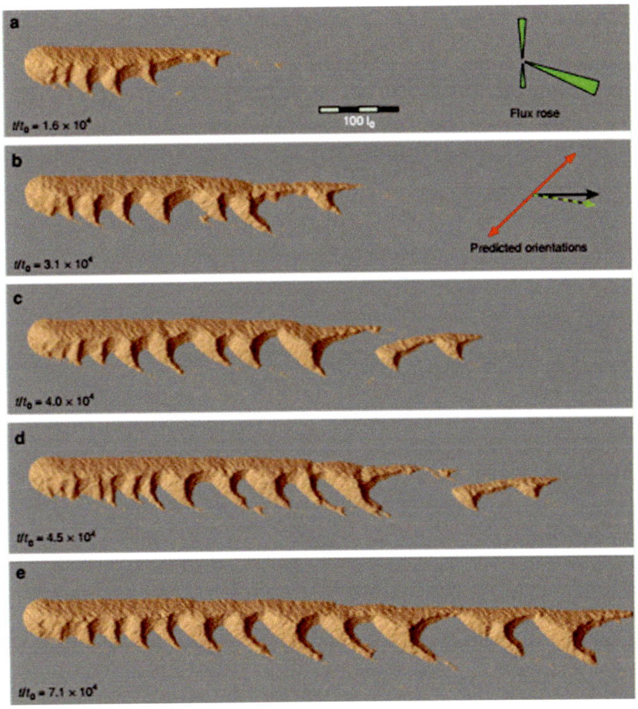

Figure 9.12. Simulations of the formation of steady-state raked linear dunes (Lü et al., 2017). (a–d) Elongating raked linear dune and development of secondary bedforms. (e) A snapshot of the steady-state raked linear dune. Insets illustrate the sand flux roses and the predicted orientations of dunes growing by extension (fingering mode, black arrow) and perpendicularly to the maximum gross bedform-normal transport (bed instability mode, red arrows). The green arrow shows the resultant sand flux at the crest of dune growing in the bed instability mode.

limited conditions and thus elongate along the resultant drift direction (fingering mode), while the superimposed dunes (subsidiary ridges) preferentially develop and grow perpendicularly to the main dunes with sufficient supply of sediments from the main dunes (bed instability mode). These two modes occur simultaneously only in the case where the angle between dune alignments and extension directions is within specific ranges. Results of numerical modelling show that the superimposed dunes can migrate along the linear dune over a long period, meanwhile the raked pattern extends and keeps a constant shape with regularly spaced subsidiary ridges of constant height and width propagating only on one side of the linear dune. An equilibrium has been reached in the mass balance between the primary and subsidiary ridges (Fig. 9.12). According to the ratio of elongation rate to the length of raked linear dune field, Lü et al. (2017) deduced that the formation of raked linear dunes in the Kumtag Desert under the present tridirectional wind regime can be dated back to at least 5500 years ago. The present wind regime may have prevailed since the Holocene Optimum, as evidenced from the presence and alignment of yardangs at the north-eastern end of the dune field.

5. Conclusion

The land surface morphology of Kumtag Desert is composed of complicated landforms and diversified dunes characterized by the unique feathered longitudinal dunes. The forcing mechanisms of the feathered dunes were raised from three different spatial scales: ① Feathered fan for the overall Kumtag Desert from a topographical hypothesis in a desert basin or catchment scale; ② Feathered patterns (pseudo-feathery dunes) from aeolian hypothesis in a dune scale; ③ Feathered texture as seif dunes/raked linear dunes from an aeolian and grain color hypothesis in a mineral scale. The so-called feathered longitudinal dunes are in fact compound or complex linear sand dunes originated from barchan dunes.

Furthermore, to understand the objectives of the future study on the feathered dunes the following research should be carried out: High resolution chronology of the sand dunes and underlying stratum to reveal the reasonability of the feather-shaped sand ridge hypothesis; Exploring the material source and erosion-deposition-transport under a catchment scale or a basin scale; forcing mechanism of the feathered dunes geomorphology under a mineral scale; sustainable management plan for feathered sand dunes in the Kumtag Desert.

References

Bagnold, R. 1941. The Physics of Blown Sand and Desert Dunes. Springer Netherlands.

Ding, F., Tang, J., Su, Z., Zhang, J. and Lu, Q. 2017. Deposited process and provenance analysis of BL sand loess section in southeastern margin of the Kumtagh Desert. Quaternary Sciences 37: 36–44.

Dong, L., Lu, H. and Wang, X. 2013a. A tentative reconstruction of the border migration of Xinjiang deserts during the Last Glacial Maximum and Holocene Optimum. Quaternary Sciences 33: 398–400.

Dong, Z., Qu, J., Lu, Q. and Qian, G. 2008a. A discussion on feathery aeolian geomorphology in the Kumtagh Desert. Journal of Desert Research 28: 1005–1010.

Dong, Z., Qu, J., Wang, X., Qian, G., Luo, W. and Wei, Z. 2008b. Pseudo-feathery dunes in the Kumtagh Desert. Geomorphology 100: 328–334.

Dong, Z. 2009. The identification of feathery dunes. Science & Technology Review 69–75.

Dong, Z., Qu, J., Lu, J., Qian, G., Luo, W., Wang, X., Zhou, Q. and Wang, X. 2009. Geomorphic Map of the Kumtagh Desert. Science Press, Beijing.

Dong, Z., Qian, G., Yan, P. and Su, Z. 2010a. Gravel bodies in the Kumtagh Desert and their geomorphological implications. Environmental Earth Sciences 59: 1771–1779.

Dong, Z., Qu, J., Lu, J., Qian, G., Wang, X. and Luo, W. 2010b. Compilation of geomorphic map of the Kumtagh Desert. Journal of Desert Research 30: 483–491.

Dong, Z., Wei, Z., Qian, G., Zhang, Z., Luo, W. and Hu, G. 2010c. "Raked" linear dunes in the Kumtagh Desert, China. Geomorphology 123: 122–128.

Dong, Z., Qu, J., Qian, G. and Zhang, Z. 2011a. Aeolian geomorphological regionalization of the Kumtagh Desert. Journal of Desert Research 31: 805–814.

Dong, Z., Su, Z., Qian, G., Luo, W., Zhang, Z. and Wu, J. 2011b. Aeolian Geomorphology of the Kumtagh Desert. Science Press, Beijing.

Dong, Z., Qian, G., Luo, W., Zhang, Z. and Lü, P. 2013b. Dune types and their distribution in the Kumtagh Sand Sea, northwestern China. Zeitschrift für Geomorphologie 57: 207–224.

Dong, Z. and Lü, P. 2014. Recent advances in research on the aeolian geomorphology of China's Kumtagh Sand Sea. Advances in Geosciences 37: 41–46.

He, Q., Yang, X., Huo, W., Wang, S., Shang, K. and Liu, H. 2009a. The characteristics of grain size from the Kumtag Desert and its environmental significance. Sciences in Cold and Arid Regions 1: 128–134.

He, Q., Yang, X., Huo, W., Wang, S., Shang, K. and Liu, H. 2009b. Characteristics of sand granularity from Kumtag Desert and its environmental significance. Journal of Desert Research 29: 18–22.

Hu, Y., Ning, G., Kang, C., Wang, S., Shang, K. and Yang, X. 2017. Temporal and spatial variability of the extreme precipitation around the Kumtag Desert. Journal of Desert Research 37: 536–545.

Huo, W., He, Q., Yang, X., Zhao, Y., Wang, S. and Li, Y. 2011. Characteristics of sand-dust weather over Kumtagh Desert during 2007–2008 and the weather styles. Journal of Desert Research, 36.

Li, L., Lu, H., Wang, X., Yi, S., Xu, Z. and Zhao, L. 2017. The deposition and environmental change of gobi Desert in North Altun Mountains (Northwestern China) in 5000 yrs. Journal of Desert Research, 9.

Liu, H., Wang, J., Liao, K., Zheng, Q., Zhang, J., Ding, F., Tang, J., E, Y., Su, Z., Zhai, X. and Zhang, G. 2006. Morphologic parameters of featherlike longitudinal ridge of Kumtag Desert. Arid Land Geography 29: 314–319.

Lü, P., Narteau, C., Dong, Z., Rozier, O. and Courrech Du Pont, S. 2017. Unravelling raked linear dunes to explain the coexistence of bedforms in complex dunefields. Nature Communications 8: 14239.

Lu, Q. 2011. A Study of Kumtag Desert. Beijing: Science Press.

Mason, J., Lu, H., Zhou, Y., Miao, X., Swinehart, J., Liu, Z., Goble, R. and Yi, S. 2009. Dune mobility and aridity at the desert margin of northern China at a time of peak monsoon strength. Geology 37: 947–950.

Qian, G., Dong, Z., Luo, W. and Wang, H. 2009. Variations of horizontal and vertical velocities over two-dimensional transverse dunes: A wind tunnel simulation of the effect of windward slope. Journal of Arid Environments 73: 1109–1116.

Qian, G., Dong, Z., Zhang, Z., Luo, W., Lu, J. and Yang, Z. 2015. Morphological and sedimentary features of oblique zibars in the kumtagh desert of northwestern china. Geomorphology 228: 714–722.

Qu, J., Zheng, B., Yu, Q. and Zhao, A. 2004. The Yardang landform of aqik valley in the east of lop-nor and its relationship with the evolution of the Kumtagh desert. Journal of Desert Research 24: 294–300.

Qu, J., Zuo, G., Zhang, K., Zu, R. and Fang, H. 2005. Relationship between the formation and evolution of the Kumtag Desert and the regional Neotectonic Movement. Arid Land Geography 28: 424–428.

Qu, J., Liao, K., Zu, R., Xia, X., Jing, Z., Dong, Z., Zhang, K., Yang, G., Wang, X. and Dong, G. 2007. Study on formation mechanism of feather-shaped sand ridge in Kumtag Desert. Journal of Desert Research 27: 349–353.

Qu, J., Liao, K., Dong, G., Niu, Q., Jing, Z. and Han, Q. 2011. Feathered sand ridges in the Kumtagh Desert and their position in the classification system. Science China Earth Sciences 54: 1215.

Stevens, T., Palk, C., Carter, A., Lu, H. and Clift, P.D. 2010. Assessing the provenance of loess and desert sediments in northern China using U-Pb dating and morphology of detrital zircons. Bulletin 122: 1331–1344.

Wang, F., Chu, J., Wang, X., Cui, X., Wu, B., Yang, W. and Lu, Q. 2011. A study of Kumtag desert: Progresses and achievements. Journal of Resources and Ecology.

Wang, L., Gong, H., Shao, Y. and Li, B. 2014. Analysis of elevation discrepancies along the Lop Nur ear-shaped stripes observed using GLAS and DGPS data. International Journal of Remote Sensing 35: 1466–1480.

Wang, Z., Sun, Q., Ren, X., Wang, T. and Chen, F. 2009. Pseudo-feathery dunes in the Kumtagh desert reclassified as linear dunes and zibars. Aeolian Research 1: 87–89.

Wang, Z., Lai, Z. and Qu, J. 2017. Inverted relief landforms in the Kumtagh Desert of northwestern China: a mechanism to estimate wind erosion rates. Geological Journal 52: 131–140.

Wu, B. 2017. Atlas of Intergraded Physical Geography of Kumtag Desert. Beijing: SinoMaps Press.

Xu, Z., Lu, H., Zhao, C., Wang, X., Su, Z., Wang, Z., Liu, H., Wang, L. and Lu, Q. 2011. Composition, origin and weathering process of surface sediment in Kumtagh Desert, Northwest China. Journal of Geographical Sciences 21: 1062.

Yan, P., Li, W., E, Y., Wang, X. and Li, X. 2011. Preliminary results of hydrological investigation in Kumtagh Desert. Journal of Desert Research 31: 242–246.

Yang, Z., Qian, G., Dong, Z., Luo, W., Zhang, Z., Lu, J. and Li, J. 2016. Grain size characteristics of the aeolian sediments from Sanlongsha Area of the Northern Kumtagh Desert. J. Desert Res. 36: 589–596.

Yang, Z., Qian, G., Dong, Z., Luo, W., Zhang, Z., Lu, J. and Li, J. 2018. Aeolian sediment transport in the Sanlongsha area to the North of Kumtagh Desert. Journal of Desert Research 38: 58–67.

Youhao, E., Wang, J., Yan, P., Gao, S., Wang, X., Su, Z., Liao, K., Liu, H. and Ding, F. 2008. Evolution of palaeo-drainage system and its relationship with the formation of desert landform in the Kumtag Desert. Acta Geographica Sinica 63: 725–734.

Zeng, S., Gong, C., Shang, K., Zeng, X. and Wang, S. 2011. Spatial and temporal variation of wind field in the Kumtagh Desert. Journal of Desert Research 31: 28–33.

Zeng, X., Wang, S., Shang, K., Yang, D. and Zeng, S. 2009. A study of wind system in Kumtag Desert. Journal of Desert Research 29: 391–396.

Zhang, Z., Dong, Z., Zhao, A. and Qian, G. 2010. The characteristic of night-day wind direction change in the Kumtagh Desert. Journal of Arid Land Resources and Environment 24: 126–130.

Zhao, L., Lu, H., Zhang, E., Wang, X., Yi, S., Chen, Y., Zhang, H. and Wu, B. 2015. Lake-level and paleoenvironment variations in Yitang Lake (Northwestern China) during the past 23ka revealed by stable carbon isotopic composition of organic matter of lacustrine sediments. Quaternary Sciences 35: 172–179.

Zhao, Y., He, Q. and Huo, W. 2010. Characteristics of climate change around Kumtage Desert. Journal of Arid Meteorology 28: 291–296.

Chapter 10

Mega-Dune—Lake System in the Badain Jaran Sand Sea

Zhibao Dong

1. Introduction

The Badain Jaran Sand Sea is situated on the Alxa Plateau in western Inner Mongolia (Fig. 10.1), between longitudes 99°48′E and 104°14′E, and between latitudes 39°20′N and 42°N, covering an area of about 50000 km^2 (Yan et al., 2001a). The sand sea is part of the Alxa Desert region, and bounded to the south by the Heli Mountains, the Beidai Mountains and the Heishantou Mountains (maximum elevation 1963 m a.s.l.) that separate it from the gobis (gravel deserts) of the Hexi Corridor. To the southeast, it is bounded by the Yabrai Mountains (maximum elevation 1957 m a.s.l.), which separates it from the Tengger Desert.

To the west and northwest, it stretches down to the flat lowlands of the Gulunai grassland (ca. 1000 m a.s.l.). To the north, it is bounded by the Guezihu wetland (ca. 1000 m a.s.l.), which merges with the black gobis and plains of Mongolia. More than 50% of the sand sea is covered by mega-dunes 200 to 300 m tall, with the tallest mega-dune exceeding 400 m. Lying among the tallest mega-dunes in the southeast, more than 100 lakes of different sizes are concentrated within an area of approximately 4000 km^2, forming unique mega-dune—lake landscape (Fig. 10.2).

The climate is an extreme continental desert type according to Köppen's climate classification (Dong et al., 2004). The mean annual precipitation is less than 90 mm, mainly falling from June through August and decreasing from the southeast to the northwest. The annual potential evaporation is more than 2500 mm, and increases from the south to the north. The mean annual air temperature ranges from 9.5 to 10.3°C, with the lowest monthly mean temperature in January (about −10°C) and the highest in July (about 25°C). Temperatures increase from the south to the north as the elevation decreases. The mean annual wind speed ranges from 2.8 to 4.6 m s^{-1}, increasing from the south to north, with the strongest winds in April and May. No runoff occurs in the sand sea, and the Heihe River to the west is the only river close by.

As a part of Alxa Desert, the Badain Jaran Sand Sea is the second largest dunefield in China. It is considered to be a mystifying place for its mega-dunes, the highest seen globally (Breed et al., 1979; Zhu et al., 1980), beautiful interdune lakes and magical booming sand. Though it has been inhabited by the Mongolian herdsmen for a long time the Badain Jaran Desert did not receive much attention until the 1980's.

Early work in the Badain Jaran Sand Sea mainly involved some expeditions in the Alxa Desert, such as those led by Scandinavian geologists (Hörner, 1936; Norin, 1941). Shortly after the reunification of China, several expeditions were organized to investigate the natural resources

School of Geography and Tourism, Shaanxi Normal University, Xi'an, Shaanxi, China.

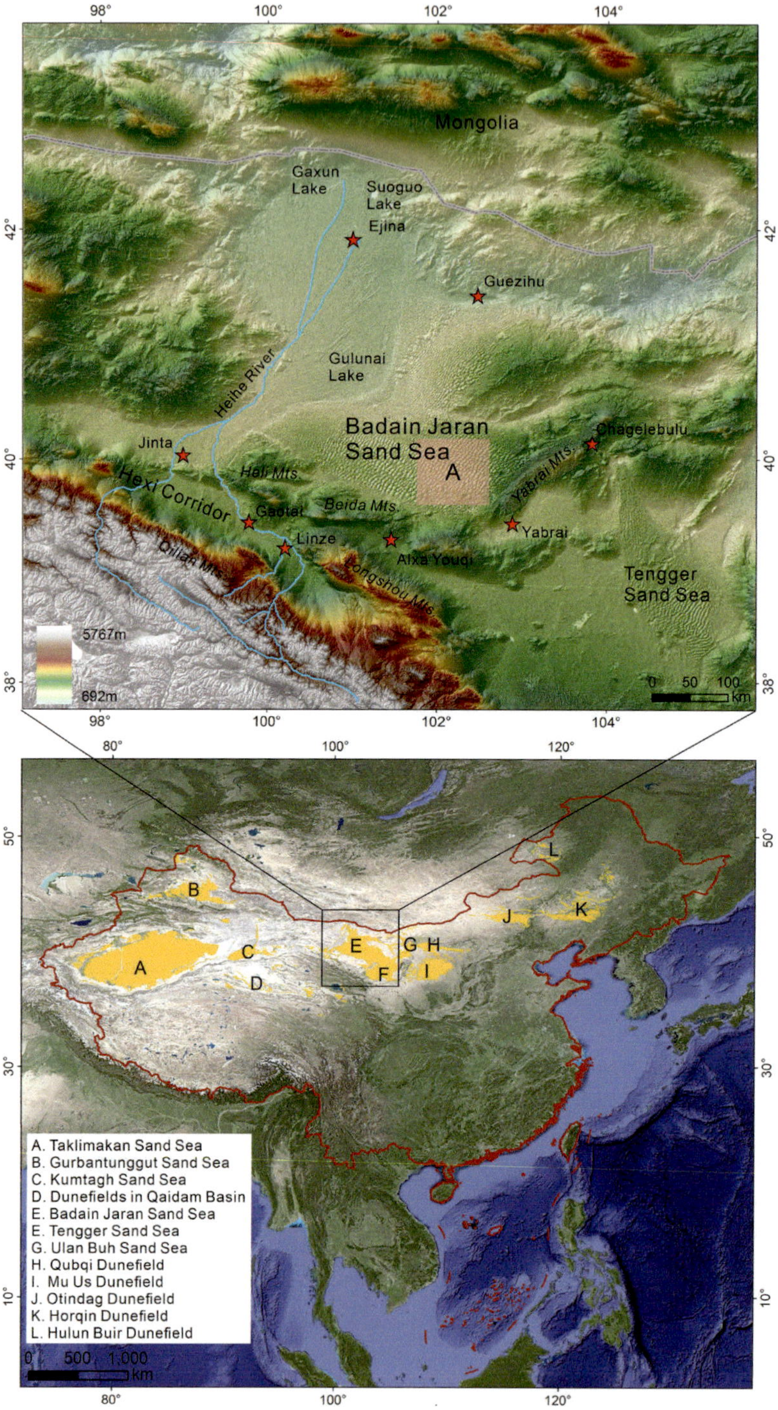

Figure 10.1. The location of the Badain Jaran Sand Sea, and main topographic features: A represents the area where the tallest mega-dunes and desert lakes are concentrated.

in the Alxa Desert. In 1957, the Chinese-Soviet Central Yellow River Complex Expedition studied the aeolian processes in the Alxa Desert. In 1958, the Qinghai-Gansu Expedition of the Chinese Academy of Sciences explored the Alxa Desert. In 1959, Chinese and Soviet investigators began a 3-year Complex Sand Expedition into the Alxa Desert. The survey results of the aeolian landforms,

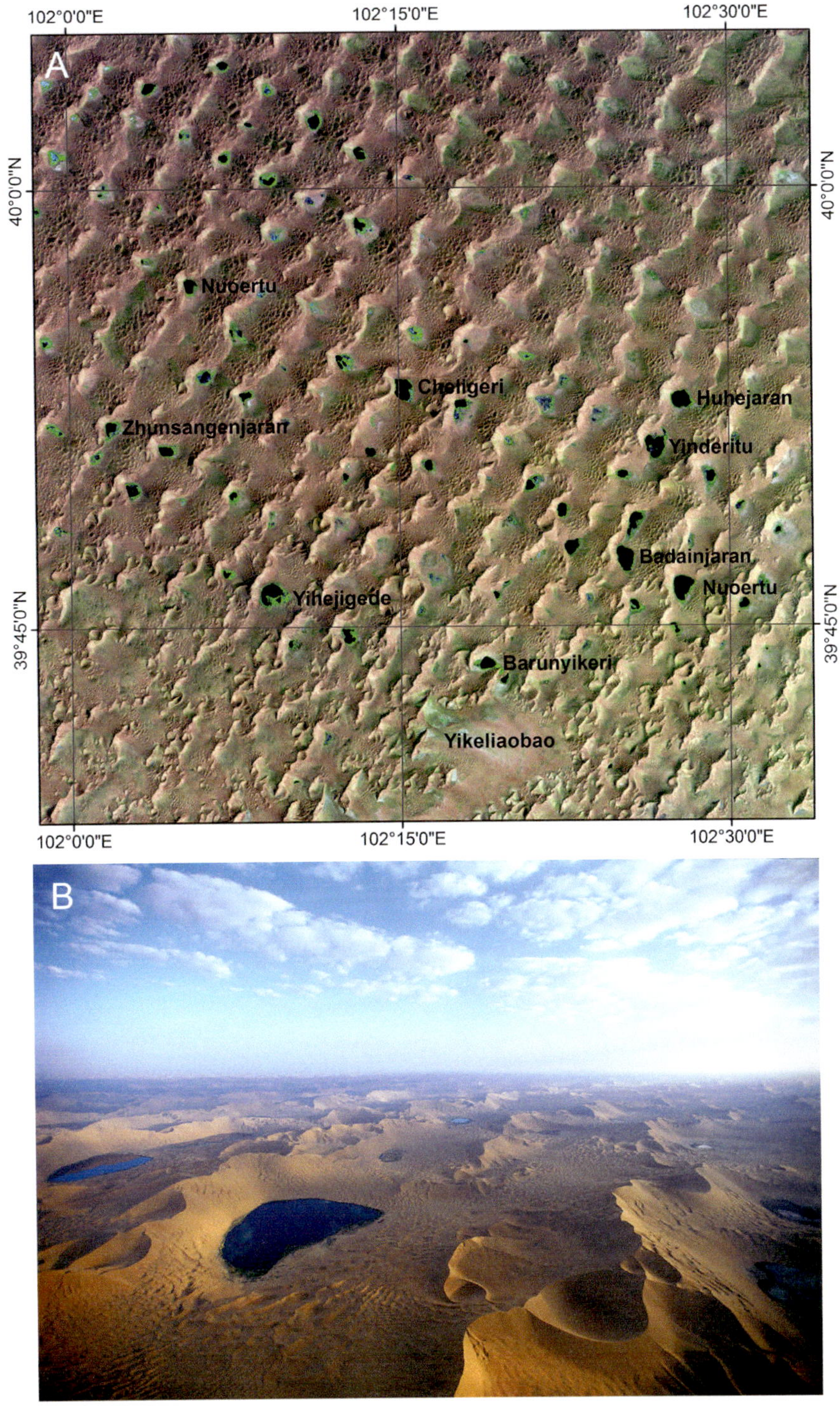

Figure 10.2. A satellite image of the area A in Fig. 10.1(A) and typical mega-dune–lake landscape of the Badain Jaran Sand Sea (B).

vegetation and water resources were reported by Li (1962), Lou (1962), Yu (1962), Sun and Sun (1964), Tan (1964). Remote sensing technology was used to recognize the general characteristics of the Badain Jaran Sand Sea after the 1970's (Breed et al., 1979; Walker et al., 1987; Zhu et al.,

1992; Guo, 2000). Breed et al. (1979) made a relatively more detailed report of the dune types, height, arrangement, spacing and their relation to the local wind regime based on the interpretation of landsat imagery. Chinese and German scientists made a comprehensive survey of the Badain Jaran Sand Sea in 1988 and compiled the 1:500,000 Map of the Aeolian Landform in Badain Jaran Sand Sea based on the landsat TM imagery (Zhu et al., 1992). The Badain Jaran Sand Sea witnessed increasing interest from scientists after the 1990's. Some researchers attempted to study the formation and evolution of the desert according to the various geological eras (Dong et al., 1995; Jakel, 1996; Yang, 2000; Yan et al., 2001b); others studied the characteristics and formation of sand dunes (Wang, 1990; Lu and Guo, 1995; Yang and Jiang, 1998; Yang et al., 1999). Hofmann (1996) studied the limnology and geochemistry of the lakes in the sand sea.

The interest of researchers in the Badain Jaran Sand Sea has been renewed in the 21st century due to advances in available research equipment and techniques. In depth scientific investigations have been conducted for the last two decades to build on earlier work, which had primarily accumulated the basic information to support new research. Researchers have focused on three key issues: the formation and evolution of the sand sea and their relation to the environment, formation of the mega-dunes and formation of the lakes. Multi-disciplinary research involving geomorphology, hydrology, geochemistry, limnology, botany, ecology, environmental sciences and other fields have increasingly been conducted. More sophisticated methods have been used. For example, isotope hydrology techniques were used to investigate the sources of lake water, high-resolution satellite images were used to analyze mega-dune geomorphology (Dong et al., 2009), ground-penetrating radar was used to study the sedimentary structures of the mega-dunes (Li et al., 2009), gravitational methods were used to detect the bedrock underlying the mega-dunes (Yang et al., 2011). The audio-magnetotelluric method was applied to study the mega-dune structure (Ren and Jin, 2011), and meteorological observations were conducted to obtain information on wind, temperature, precipitation and evaporation conditions, which are crucial factors involved in the formation and evolution of mega-dunes and lakes. Extensive sediment surveys were conducted to trace the sand sources. Dune morphology was analyzed to describe dune formation, especially for mega-dunes. Sedimentary records preserved in surface and deeper strata were obtained by drilling and were used to reconstruct the evolution and development history of the sand sea. Diverse methods were used to analyze lake water and explain lake formation and its relationship to the surrounding areas. Dune moisture contents were observed to identify their contribution to lake water, and hence the significance of atmospheric precipitation.

2. Wind regime and sediments

2.1 Wind regime

The surface wind circulation over Badain Jaran Sand Sea is controlled primarily by the central Asiatic high-pressure cell in the winter and thermal low-pressure cell that develops over the Indian subcontinent in the summer. During winter, subsidence from the central Asiatic high tends to produce northerly to northwesterly winds across the deserts. During spring the central Asiatic high weakens, its center shifts 10° westward, and a thermal low begins to develop over the Indian subcontinent. At this time, the Badain Jaran Sand Sea lies in a zone between the high and low-pressure cells, and also at the same time the highest drift potentials occur. By summer low pressure influences most of Asia, and the wind regime in the desert becomes weaker and more directionally unstable (Breed et al., 1979).

The sand Drift Potential (DP) was estimated by the method proposed by Fryberger (1979). Figure 10.3 shows the drift potentials based on the meteorological data of the four sites surrounding the Badain Jaran Sand Sea.

The available wind data for calculating DP for Dingxin, Ejin, Sharitai and Guezihu are from 1997 to 2001, 1996 to 2001, 1990 to 2001 and 1998 to 2001 respectively. The DP, Resultant

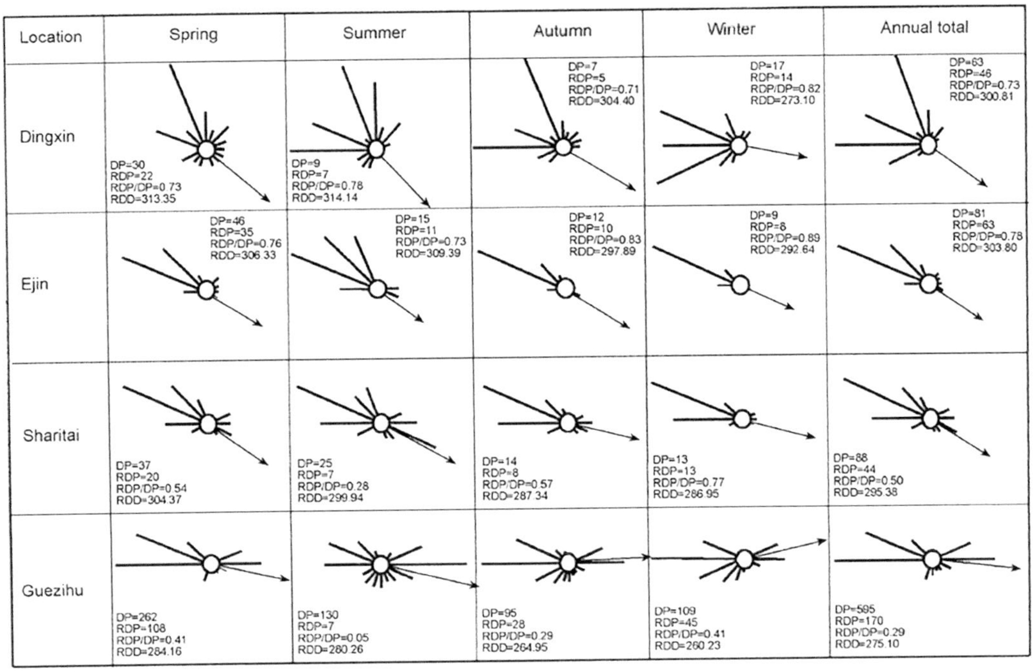

Figure 10.3. The sand drift potential in the Badain Jaran Sand Sea.

Drift Potentials (RDP), directional factor (RDP/DP) and Resultant Drift Direction (RDD) in the northern site (Guezihu) are very different from the other three sites. It is not certain what makes the differences. Except Guezihu where DP is very high (595 VU), the Badain Jaran Sand Sea generally has low wind energy environments, the annual DP ranging from 63 VU to 88 VU. Spring has the greatest DP, followed by summer. Autumn has the lowest DP. The resultant drift directions are generally towards the southeast. The directional factor varies from 0.29 to 0.78. Summer usually has the least directional factor, implying the most variable wind direction. The southwest site Dingxin has an acute bimodal wind regime, the north site Guezihu has an obtuse bimodal wind regime, while the northwest site Ejin and the southeast site Sharitai have narrow or wide unimodal wind regimes.

2.2 Sediments

Sediment, wind, and surface characteristics, including the level of vegetation cover and the presence of other obstacles, are considered to be the main factors that control dune geomorphology. The first question that appears relates to the origin of the abundant sediments constituting the mega-dunes, and particularly whether they are of local origin or have been transported from other places. Early researchers, based on analysis of the grain-size distribution and mineral composition, proposed that the sediments were mainly of local origin; that is, that they were the weathered and denuded products of the underlying sandstones, sandy conglomerates, and other clastic rocks from the Mesozoic and Cenozoic, and that the Quaternary loess deposits surrounding the sand sea contributed only a small proportion of the dune sediments (Lou, 1962; Sun and Sun, 1964). Comparison of the grain-size distribution in sediment samples from the northwestern and southeastern parts of the sand sea led Mischke (2005) to conclude that the sediments of the Badain Jaran Sand Sea mainly originated in alluvial deposits of the Heihe River west of the sand sea, and that these deposits were blown into the sand sea by prevailing winds. Evidence to support this opinion is found in the slight but distinct fining of aeolian sediments towards the southeastern part of the sand sea, which is in agreement with the SW to NE alignment of the transverse dune ridges and the presence of slope faces pointing towards the southeast. The mean particle size in dune sediments from the northwest averaged 0.21 mm, whereas the mean of sediments from the southeast averaged 0.19 mm.

Other criteria used to characterize the grain size distribution, including the 1st, 5th, 10th, and 30th percentiles of the sediment samples, had trends similar to that of the mean size (Mischke, 2005). Qian et al. (2011) recently analyzed the grain-size distribution of more than 200 samples along the dominant NW to SE wind direction, and found that the mean grain size of sediments on lower dunes surrounded by mega-dunes showed a slight downwind fining, but no downwind fining for sediments of mega-dunes. It is therefore possible that the size distribution of the sediment particles alone is not sufficient to define the sediment sources of the Badain Jaran Sand Sea, since Qian et al. (2011) found no statistically significant trend in the grain-size distribution. Sand sea developed in a rift basin that is surrounded by a series of mountains that have the potential to contribute sediments. In particular, the mountains to the south and east, where several rivers originate, might have contributed to the majority of sediments. However, the Heihe River lies upwind of the sand sea, and has possibly provided large amounts of sediments, though the present topography of the Badain Jaran Sand Sea is inclined toward the lower reaches of the Heihe River (Dong et al., 2004). This would complicate the flow of sand-laden air to some extent.

The interaction between dune morphology and wind results in spatial variations in the grain-size composition of dunes; thus, comparisons of grain-size parameters among different positions on a dune can reveal dune dynamics to some extent (Lancaster, 1995). Several grain-size studies have been conducted on dunes in recent years to understand dune geomorphology, and especially that of mega-dunes (Shao et al., 2010; Li et al., 2011; Qian et al., 2011; Zhao et al., 2011b). These comparisons were made between regions, dune types and positions on a mega-dune. Dune sediments in the Badain Jaran Sand Sea are mainly composed of fine and medium sands, with mean grain sizes ranging from 2.1 to 2.7 φ (0.15 to 0.24 mm). These grains are generally coarser than those in the dune sediments of the Taklimakan Sand Sea to the west and the Tengger Sand Sea to the east, and have a moderate size compared with dune sediments elsewhere in the world (Shao et al., 2010; Qian et al., 2011). Dune sediments generally have a unimodal size distribution, are well to very well sorted, and have a positively skewed and mesokurtic distribution. In comparison, the sediments from the spaces between the dunes have bimodal or multimodal size distributions, are coarser (with a mean size around 1.5 φ, equivalent to 0.35 mm), exhibit less sorting and have a leptokurtic distribution (Qian et al., 2011). On the windward slope of a mega-dune, sediments became finer and better sorted moving from the toe to the crest (Li et al., 2011; Qian et al., 2011; Zhao et al., 2011b). This is because finer sand grains are transported more rapidly on the slopes to upper section of a mega-dune (Dong et al., 2009; Zhao et al., 2011b). Sediments on the leeward slope were generally coarser than those on the windward slope (Shao et al., 2010).

In summary, studies of sediments in the Badain Jaran Sand Sea have mostly focused on grain-size characteristics, and future studies should include the mineralogy, element geochemistry and micro-morphology of sediment grains. Difficulties in accessibility of many regions of the desert have precluded a fully representative sampling of the sediments, which will therefore require further study.

3. Morphology of the mega-dunes

A morphological classification system that included nine main types was proposed for the Badain Jaran Sand Sea based on Landsat TM satellite images (Zhu et al., 1992): compound transverse mega-dunes, compound longitudinal mega-dunes, star mega-dunes, transverse ridges, barchanoid ridges, barchan dunes, linear dunes, star dunes and nebkhas. Compound transverse mega-dunes that are aligned from NE to SW, NNE to SSW and NEE-SWW, roughly perpendicular to the prevailing wind direction, are dominant, accounting for 97% of the mega-dune area (Liu et al., 2010), though mega-dunes in the southeast gradually become reversed mega-dunes in response to a stronger SE wind. The height of the mega-dunes increases from the northwest to the southeast. Mega-dune density, which was characterized by the total length of mega-dunes per square kilometer, increases from 80 m in the northwest to 450 m in the southeast (Liu et al., 2010).

Several researchers (Qu et al., 2003; Dong et al., 2009) have attempted to study the morphological features of mega-dunes. Based on a fractal geometry analysis, Qu et al. (2003) attempted to characterize mega-dunes in different height groups in the southeast of Badain Jaran Sand Sea using their fractal dimension. They found that the fractal dimension increased slightly with increasing mega-dune height, ranging from 1.0 to 1.3, which suggests that the mega-dunes tend to become more complex as their height increases. Dong et al. (2009) performed a detailed analysis of the geomorphological hierarchies of typical mega-dunes, and identified three statistically distinct hierarchical orders based on measurements of dune height and spacing (Fig. 10.4). Order I represents the mega-dunes as a whole, and shows typical features of reversing dunes, with typical slipfaces and horns. The top 40 to 60 m of the mega-dunes is the crest zone (the reversing zone), which is the most active part of the dune, and migrates back and forth in response to variations in wind direction and forms a reversed slipface. Superimposed on the windward slope of these mega-dunes are simple individual dunes whose size increases from the toe to the crest of the mega-dunes. The superimposed dunes are divided into two orders (Orders II and III). Order III represents dunes that are distributed on the lower quarter of the windward slope of the mega-dunes. They form in response to seasonal winds and the average wind regime over a period of several years because their height is less than 15 m.

They move quickly and are aligned in different directions at contrasting positions in response to how the local air flow is modified by the mega-dune. Order II represents dunes that are distributed between the order III dunes and the crest zone, and form in response to variations in the average wind regime over longer periods than order III dunes. They move much more slowly than order III dunes and are generally aligned in the same direction as the mega-dune, except where they lie between the horns of two neighboring mega-dunes. In the latter case, the orientation is more random, and is affected by modifications of the local airflow by the mega-dune.

Geomorphological patterns that are characterized by a height–spacing relationship are regular for the dunes of different orders on mega-dunes (Fig. 10.5).

The height–spacing relationships in the Badain Jaran Sand Sea and other deserts and sub-aqueous bedforms are similar, and for all three orders of dunes in the Badain Jaran Sand Sea, can be expressed well by a uniform linear function on a log10-log10 graph. This uniform function can be extrapolated to express the height–wavelength relationships for wind ripples that were reported by Sharp (1963), Walker (1981), and Lancaster (1995).

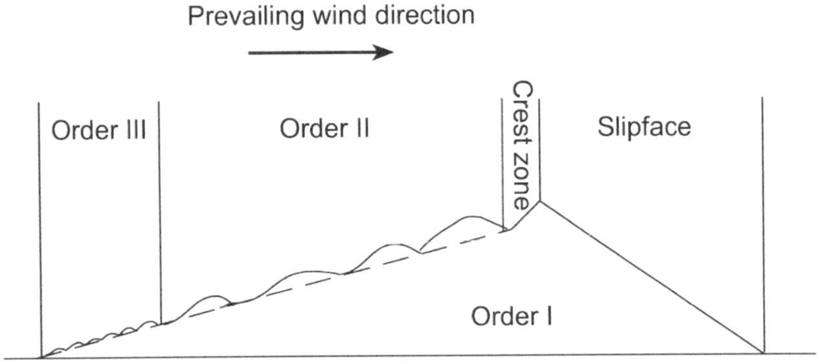

Figure 10.4. Typical hierarchical orders in a compound mega-dune system (Dong et al., 2009).

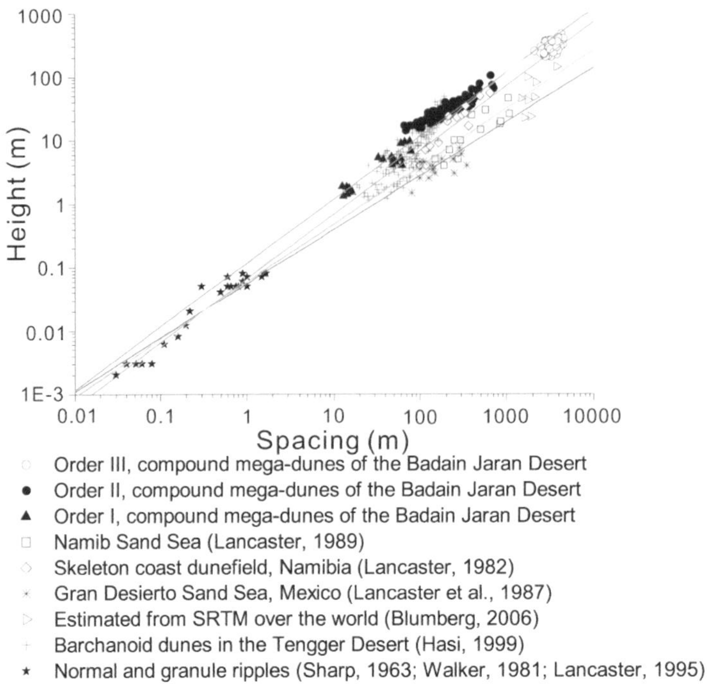

Figure 10.5. The relationship between dune height and spacing for the three hierarchical orders of dunes in the present study and in comparable studies from other areas (Dong et al., 2009).

4. Hypotheses for formation of the mega-dunes

The nature of the formation mechanisms of the mega-dunes in the Badain Jaran Sand Sea is a topic of heated debate, with several hypotheses proposed, which are summarized as the underlying topography control hypothesis, wind dominance hypothesis and groundwater maintenance hypothesis (Lou, 1962; Sun and Sun, 1964; Tan, 1964; Wang, 1990; Yan et al., 2001b; Yang et al., 2003, 2011; Chen et al., 2004a, 2006; Dong et al., 2004, 2009; Zhang and Wang, 2005; Liu et al., 2010).

4.1 Underlying topography control hypothesis

This hypothesis proposes that the mega-dunes are supported by conspicuousness that act as cores. Most early researchers accepted this hypothesis. For example, it was thought that the underlying topography controlled the formation of mega-dunes (Lou, 1962; Sun and Sun, 1964). Lou (1962) suggested two possible control mechanisms: first, that the pattern of the mega-dunes was controlled by underlying humps in the topography that were covered by drifting sand, and second, that the mega-dunes developed from sand deposits that accumulated when the drifting sand was obstructed by local stone outcrops. Sun and Sun (1964) suggested that the mega-dunes were related to the morphology of underlying formations that had been folded by tectonic movements. Several recent researchers found evidence to support the underlying topography control hypothesis (Zhang and Wang, 2005; Yang et al., 2011). Zhang and Wang (2005) thought that the underlying topography played an important role in the formation of mega-dunes, though mega-dune formation and development were affected by sand supply, wind regime, vegetation and other factors, particularly by environmental evolution. In this hypothesis, the development of sand mega-dunes was controlled by the westerlies, and the underlying landforms not only created wavy movements of the near-surface air stream but also became a barrier to the movement of blowing sand. During the humid period (the interglacial period), the dune surface was stabilized by the development of a layer of calcareous cement, thereby allowing the base of the dune to enlarge. During the arid and cold

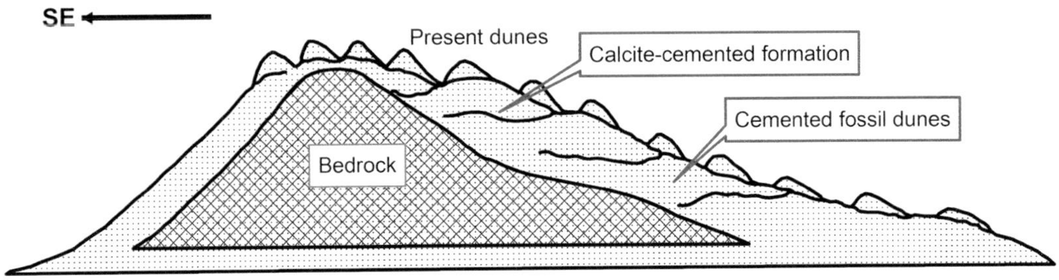

Figure 10.6. Effect of underlying terrain and of consecutive cycles of cementation by calcium compounds followed by accretion of sand and cementation on the formation and evolution of mega-dunes (Zhang and Wang, 2005).

period (the glacial period), secondary sand dunes developed above the calcareous cement layers and evolved into complex sand dunes during the course of subsequent cycles of calcareous stabilization, and dune accretion (Fig. 10.6).

Yang et al. (2011) detected the presence of the bedrock landforms hypothesized to exist by Zhang and Wang (2005) beneath the aeolian sands by applying gravitational methods. Their studies demonstrated that in addition to the average wind parameters, dune height was highly sensitive to local geology, subsurface characteristics, topography, interactions between changing climate conditions and aeolian and fluvial processes. These additional factors should be considered in any interpretation and simulation of dunes on Earth. They further anticipated that analysis of anomalous dune heights such as those seen in the Badain Jaran Sand Sea might also provide critical information on subsurface characteristics and environmental conditions on Earth and other planetary bodies.

It should be emphasized that researchers who insisted on the importance of control by the underlying topography found their evidence in areas close to the mountains in the southern and southeastern parts of the sand sea, where the underlying topography was inevitably complex. In such cases, it is highly likely that the underlying topography exerts an important impact on mega-dune development. However, it is not clear that this mechanism explains mega-dune formation farther from the mountains. For example, the stratifications underlying the mega-dunes in the southern part of the Badain Jaran Sand Sea that were examined using ground-penetrating radar showed that the underlying topography was flat, and that the mega-dunes developed on lacustrine deposits (Bai et al., 2011a, b).

4.2 Wind dominance hypothesis

In this hypothesis, wind is considered to play a key role in formation of mega-dunes. This hypothesis is proposed mainly based on mega-dune morphology and patterns because the control by the underlying topography could not produce such regular patterns as assumed by the Badain Jaran mega-dunes (Wang, 1990; Dong et al., 2009). Wang (1990) selected two 1500-km^2 sample areas to analyze the morphology of mega-dunes: one was close to the Yabrai Mountains and the other was farther away. He measured the morphometry of 102 dunes in each sample area, including their width, length, height, spacing and alignment, using aerial photographs and satellite images. He found a regular increase in the area of the mega-dune's base, its height and the spacing between the dunes with increasing proximity to the Yabrai Mountains. He attributed this to the effect of local wind regimes. The dominant northwest wind blew sand from the northwest towards the Yabrai Mountains, leading to an increase in the mega-dune size near the Yabrai Mountains. The alignment of mega-dunes from NE to SW is roughly perpendicular to the prevailing northwest wind, which strongly suggests the importance of the wind. Wang proposed a hypothesis similar to the one subsequently proposed by Zhang and Wang (2005) to explain the increased height of the mega-dunes. That is, he proposed that the mega-dunes have undergone many cycles of fixation and development in response to climate change since their formation in the Mid-Pleistocene. Shifting sand that accumulated on dunes during dry periods became fixed during wet periods, resulting in gradual increases in the height of mega-dunes after a series of dry and wet cycles. This hypothesis was supported by the work of other authors (Tan, 1964; Dong et al., 2004). Tan (1964) pointed out that lacustrine deposits from the Early Pleistocene could be found underneath the mega-dunes. Old dunes, which were fixed by calcareous cementation, initially accumulated above the lacustrine deposits. Drifting sand covered these old fixed dunes, gradually producing the current mega-dunes. Dong et al. (2004) suggested that repetitions of cycles of dune fixation and reactivation during the development process played an important role in increasing the height of the mega-dunes, and thus their overall size.

Other researchers have also explained megadune formation primarily through aeolian processes. Dong et al. (2009) measured 294 dunes and mega-dunes of different orders in the Badain Jaran Sand Sea to explain the formation of mega-dunes in terms of geomorphological patterns. Based on the height–spacing relationships (Fig. 10.6), and assuming that the dominance of wind in maintaining the evolution of mega-dunes should be characterized by the regular height–spacing relationships that have been found for simple dunes or wind ripples whose formation is dominated by the wind, they suggested that there could be few unique factors responsible for maintaining the evolution of complex mega-dunes compared with the superimposed simple dunes and dunes in other deserts, dune fields and subaqueous bedforms.

They also suggested that the tallest mega-dunes on Earth could be maintained primarily by the wind. In this hypothesis, the height–spacing relationship for mega-dunes and dunes is related to the height–wavelength relationship for wind ripples and demonstrates the purely morphometric features of the bedforms formed by wind. However, it is well known that dunes and wind ripples are formed by very different mechanisms even though they have similar height–spacing relationships. Despite this drawback, the hypothesis provides a new perspective from which to understand the formation mechanisms of mega-dunes.

4.3 Groundwater maintenance hypothesis

This hypothesis asserts that the mega-dunes are maintained by groundwater (Chen et al., 2004a, 2006), which will be discussed in more detail later. This maintenance mechanism is illustrated in Fig. 10.7.

Moisture and calcium compounds carried in the groundwater that allow cementation play important roles in forming mega-dunes. Calcium carbonate in the underlying carbonate strata is dissolved when acidic CO2-containing groundwater flows over the strata, and is carried upward by capillary forces and evaporation from the dune surface, eventually forming calcareous sinter (crystalline deposits of calcium carbonate) and calcareous cementation layers. In the Badain Jaran Sand Sea, calcareous cementations, calcareous sinters and rhizoconcretions (fossilized plant roots coated in calcium carbonate) are more abundant than in other deserts because of the abundance of carbonates in the deeper strata (possibly due to the lacustrine origin of many of these deposits; see Section 3.1 for details), which are dissolved due to the large amount of flowing groundwater (which will be discussed later). Calcareous sinters and calcareous cementation layers gradually increase in size until they eventually emerge from the surface of lake water. The inner moisture status of dunes and the huge calcareous cementation layers found at the bottom of dried lakes suggest that the calcareous sinters and calcareous cementation layers may function as kernels that induce the formation of mega-dunes. Calcareous sinters at the bottom of dry lakes also affect the flux of aeolian sands, which fall out of the air stream as its velocity decreases and are deposited above the calc-sinters to form dunes. The growing dunes further weaken the air stream that carry sand particles, leading to increased deposition, but become more severely eroded by wind because their heights intrude on layers farther from the surface where the wind speed is higher and their large volumes provide more material to be eroded. Where there is water in the sand, water molecules become adsorbed on the surfaces of sand particles to form thin films that increase the cohesion between sand particles. Enhanced cohesion between sand particles helps them to withstand the strong erosion

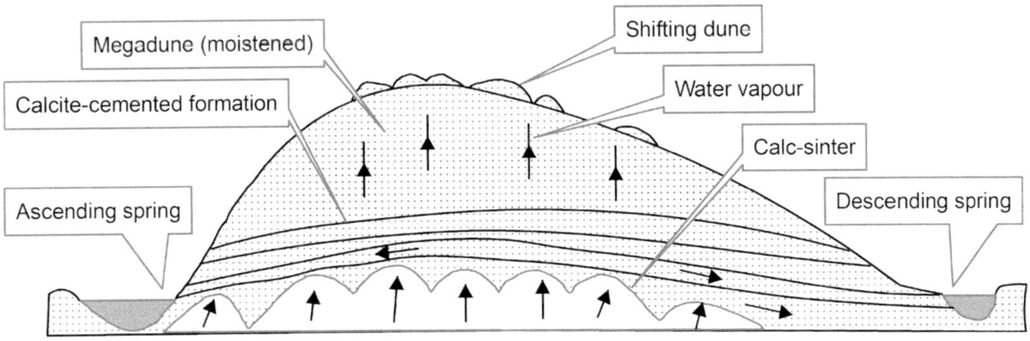

Figure 10.7. A possible formation mechanism for the mega-dunes in the Badain Jaran Sand Sea based on soil moisture dynamics (Chen et al., 2006).

process more effectively. The following evidence supports this hypothesis: (1) Mega-dunes are relatively moist below the dry surface layer, at a depth of about 0.20 m. At the top of dunes, the moisture measured to a depth of 2 m from the surface layer occupied 2 to 20% of the space between sand grains (i.e., the pore volume). (2) Calcareous sinters and rhizoconcretions are found in mega-dunes and the lakes between them. (3) Calcareous cementation layers have developed extensively in mega-dunes.

A main drawback of the groundwater maintenance hypothesis lies in the difficulty of explaining the upward movement of groundwater. It does seem possible that wet areas of the land surface would trap sand and the resulting accumulation could be the nucleus of a dune. However, this effect of groundwater cannot continue to affect dune growth after the dune has grown to a significant height. The upward movement of water from the water table cannot keep the surface of the dune moist enough to trap sand and favor more sand accumulation. Capillary rise will simply not reach far above the water table in sands. The movement of vapor by diffusion would be extremely slow, and certainly not enough to keep up with evaporation and drying of the sand dune surface.

In addition to studies of dune sediments, morphology and formation processes, researchers have attempted to study dune movement. For example, though its accuracy might be limited by resolution, a comparative analysis of Landsat TM images from 1989 and 2010 showed insignificant movement of 50–100 m tall mega-dunes in the northern part of the sand sea, significant movement towards the southeast for the 9–33 m barchanoid dunes in the eastern and southeastern parts of the sand sea, and significant movement towards the southeast and back towards the northwest by the 7–31 m barchanoid dunes in the southern part of the sand sea (Chen and Liu, 2011). Analysis of digital elevation model data for the sand sea showed that the rate of dune movement was negatively correlated with dune height (Chen and Liu, 2011).

In summary, debates about mega-dunes focus on how to understand the relative significance of wind and water versus underlying topography. Two facts should be considered in this context: the morphology of mega-dunes shows regular patterns that are usually attributed to the work of wind, and the mega-dunes are largest in the southeast, close to the Yabrai Mountains and adjacent mountain ranges. It is not yet clear whether this is due to the inherently complex underlying topography near the mountains, the blocking effect of mountains on aeolian sediment transport or the more bimodal wind in the southeast part of the sand sea because of wind flow through the low gap between the end of the Yabrai Mountains and the other mountains to the west. There is more consensus about the mechanism by which mega-dunes increase in height: this seems to result from cycles of sediment accumulation and fixation. By these means, sediments accumulated on mega-dunes have not been eroded. However, the mechanism responsible for the fixation is not yet clear: this may be due to moisture, calcareous cementation, vegetation or a combination of these factors.

5. Movement of the mega-dunes

Mega-dunes with heights as high as those in the Badain Jaran Sand Sea (i.e., more than 300 m above the bed) are believed to move at such a slow rate that movement is almost negligible. Hence, commonly used monitoring methods such as topographic surveys and comparisons of remote-sensing images have failed to reveal the movement of these mega-dunes. To resolve these problems, attempts were made to estimate the rate of advance of mega-dunes in the Badain Jaran Sand Sea by means of Optically Stimulated Luminescence (OSL) dating, which can provide accurate deposition dates of aeolian sediments (Yu and Lai, 2012). The mean rate of advance of a mega-dune can be estimated by dating sediments of different depths on the dune's slipface using the following equation:

$$D = (h_2 - h_1)\tan\alpha/(t_2 - t_1) \qquad (1)$$

where D is the mean rate of advance during the period between times t_1 and t_2, h_1 and h_2 are the depths within the dune corresponding to times t_1 and t_2 respectively, and α is the angle of the slipface with respect to the bed, which is typically about 30º. Hence, the rate of advance of a mega-dune can be estimated by dating aeolian sediments at different depths on the slipface according to the principle illustrated in Fig. 10.8. Mega-dune sediments at different depths on the slipfaces of three mega-dunes, at a position located 30% of the total slipface length from the lake surface representing the mega-dune base (Fig. 10.8).

Mega-dunes in the Badain Jaran Sand Sea are compound transverse mega-dunes with slipfaces oriented towards the southeast in response to the dominant northwestern wind; however, the slipfaces of their crests are reversed in response to the secondary southeastern wind (Dong et al., 2004). The net result is that the mega-dunes advance towards the southeast. This was also verified by the sedimentary structures of mega-dunes, which are dominated by slipface beddings inclined toward the advance direction and that lack gently inclined stoss beddings (Yang, 2000; Li et al., 2009). Advance of the mega-dunes results from deflation on the windward slope followed by deposition in the lee, as in the case of more commonly studied dunes.

At rates on the order of meters per 100 years, the mega-dunes advance much more slowly than smaller, commonly observed dunes. The Badain Jaran Sand Sea generally has a low wind energy environment according to Fryberger's (1979) classification, with an intermediate directional

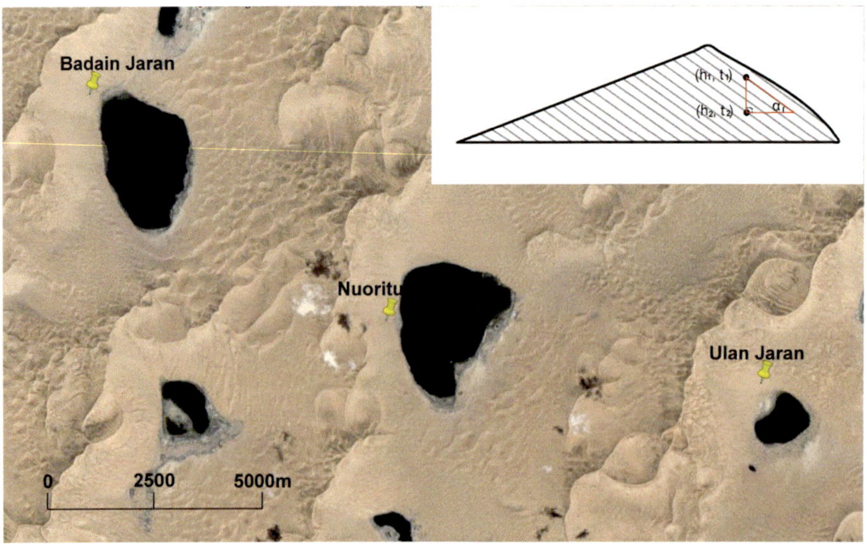

Figure 10.8. Satellite image showing the geomorphology of the three mega-dunes sampled in the present study, and (inset) graphical representation of the parameters used to estimate the rate of advance.

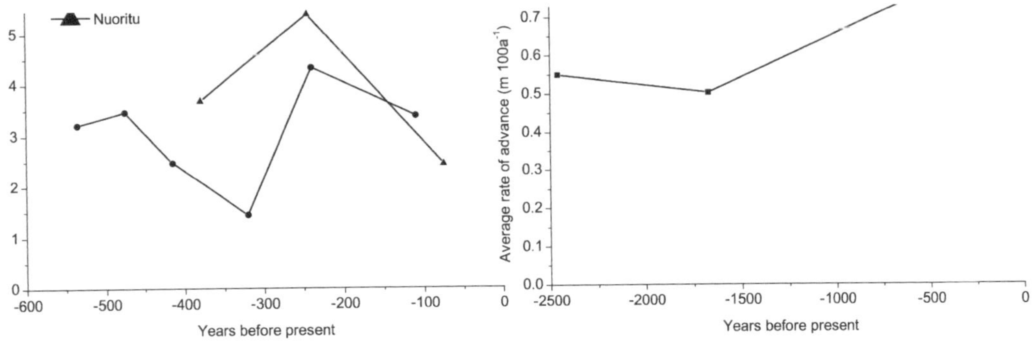

Figure 10.9. Average rates of advance of the three mega-dunes in different periods.

variability characterized by Resultant Drift Potential (RDP)/DP ratio ranging from 0.5 to 0.8 (Dong et al., 2004). Although the prevailing northwestern wind results in an overall advance toward the southeast, frequent southeastern winds, especially in the summer, results in a reversed crest zone in the top 40 to 60 m of the mega-dunes (Dong et al., 2009). This zone is the most active part of the mega-dune, and migrates back and forth in response to variations in wind direction. The reversed slipface partly offsets the mega-dune's advance, though cannot change the overall advance, because it consumes energy for the prevailing northwestern wind to adjust the slipface.

The movement of barchans is referred to understanding the advance of transverse mega-dunes in the Badain Jaran Sand Sea. The age versus depth profiles suggest that it is meaningful to describe the mega-dune rate of advance based on the average rate for each mega-dune, even though this represents different time periods for different mega-dunes. The longest time span that the rate of advance represents is less than 3 ka, which represents the Late Holocene dune reaction stage in arid regions of China (Yang, 2001), so factors that influence mega-dune advance during this period can be considered relatively stable; as a result, the mega-dune rate of advance would not have varied widely during the study period (Fig. 10.9).

The following discussion of the relationship between the rate of advance and mega-dune height, and its geomorphologic implications, is based on the average rate of advance.

A basic formula (Eq. (2)) to relate a dune's movement rate and height has been derived for transverse dunes (Simons et al., 1965).

$$V_d = q/kH\rho_p \qquad (2)$$

where V_d is the dune's rate of movement; q is the transport rate of the sand trapped by the slipface, and is approximately equal to the transport rate on the stoss slope; $k = A/LH$, where A is the two-dimensional cross-sectional area of the dune, L is its wavelength (the distance between neighboring dunes), and H is the height of the dune; and ρ_p is the bulk density of the sand in the dune. The curves in Fig. 10.10 represent a simplified function, $V_d = a + b/H$ (where a and b are regression coefficients).

Doubts, however, remain about the validity of the size versus movement curve at either extreme of the size range. Samthcin and Walger (1974) hypothesized that the relationship between the movement rate and dune height should be exponential rather than linear. However, no observations are available to support this hypothesis. The advance of the Earth's tallest mega-dunes, in the Badain Jaran Sand Sea, provides an opportunity to test their hypothesis.

Figure 10.10 shows that the relationship between the rate of advance and the height of the three mega-dunes falls on a curve that fits the linear function in Equation 2 well. This has several important geomorphologic implications. First, the tallest mega-dunes in the Badain Jaran Sand Sea experience advances similar to those of commonly observed sand dunes. Second, the possibility that the underlying relief controls mega-dune development appears to be low because no exposure of underlying bedrock is found in the stoss slopes of the mega-dunes, which have undergone a

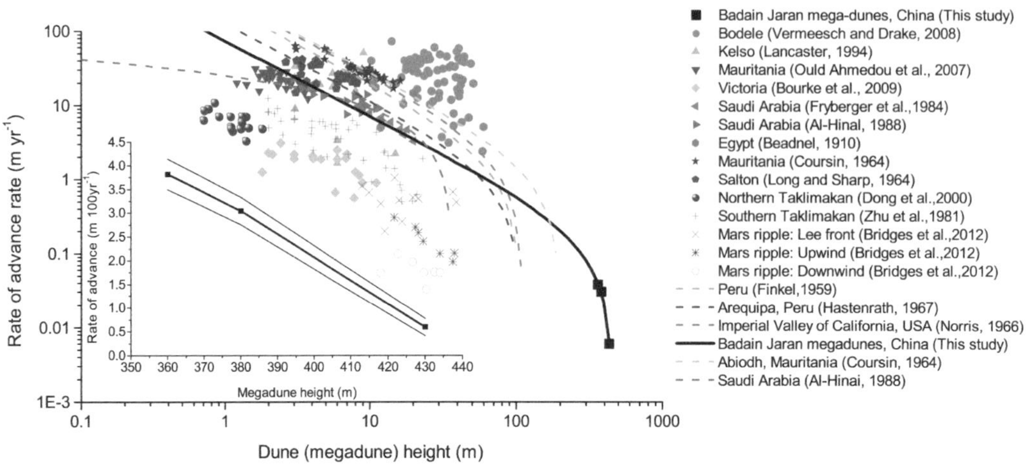

Figure 10.10. Measured rates of advance of transverse dunes from around the world and (inset graph) the estimated rate of advance of mega-dunes in the Badain Jaran Sand Sea as a function of dune height.

long history of deflation. There is increasing evidence to support our belief that no bedrock relief underlies the mega-dunes (Yang et al., 2011). Third, the present evidence for rates of mega-dune advance, combined with an earlier analysis of morphometric features (Dong et al., 2009), suggests that mega-dune development is controlled dominantly by wind rather than previously proposed factors such as the underlying relief.

6. Lakes among the mega-dunes

Formation of the lakes in the mega-dune area of the Badain Jaran Sand Sea has received continuous attention from researchers for the last two decades for three main reasons: First, lakes and mega-dunes seem to develop from a similar combination of mechanisms, because lakes are always found between the mega-dunes and larger mega-dunes are accompanied by larger lakes. Understanding the formation of lakes will therefore improve our understanding of how the mega-dunes form. Second, understanding the recharge relationship between lakes in the Badain Jaran Sand Sea and the Heihe River is of key significance for developing sustainable regional water resource utilization in such an ecologically fragile zone. The Heihe River is the second-largest inland river in China, and provides irrigation water for more than 500000 ha of farmland (Meng et al., 2004), supporting about 1.8 million people in the middle and lower reaches of the river. There is severe competition for allocation of the regional water resources. Chen et al. (2004b) suggested that utilizing the groundwater that originates in the Qilian Mountains and that recharges the lakes of the Badain Jaran Sand Sea and groundwater in the lower reaches of the Heihe River could be a practical alternative to a water diversion project. Third, isotopically anomalous groundwater has been observed in the Alxa Plateau, including in the Badain Jaran Sand Sea (Geyh et al., 1998), creating the need to define the causes of these anomalies.

Studies of lake formation in the sand sea have focused on the recharge sources. Hydrogeological surveys, hydrochemical analyses and analysis of hydrological isotopes have been conducted to analyze the recharge sources. Various hypotheses have been proposed: direct atmospheric precipitation recharge, near-source recharge, remote-source recharge and paleo-source recharge.

6.1 Direct precipitation recharge

This hypothesis proposes that the lakes are fed by atmospheric precipitation, and is based on the fact that the present lakes in the sand sea are concentrated in the southeast, where precipitation is the greatest. Precipitation feeds lakes both directly and indirectly. Wang (1990) estimated that in the mega-dune area of the sand sea with an annual precipitation of 50 mm, a 16-km^2 watershed area could directly deliver the equivalent of 800 mm of water to a 1-km^2 interdune lake. Similarly, Dong et al. (2004) suggested that the lakes were fed by springs in talus areas created by precipitation infiltration followed by downhill flow. They found that lakes were mainly limited to a west–east zone located north of the mountains, which they considered to be a zone responsible for the development of talus springs, and that the lakes collected precipitation over an area about 6500 km^2 in size to the south. The amount of water collected annually by the interdune lakes represents an estimated at 3.25×10^6 m^3, which can balance the potential evaporation (about 3300 mm) of 98 km^2 of open water, or the equivalent of 98 interdune lakes 1 km^2 in size. These, are very rough estimates that neglect the conversion efficiency of atmospheric precipitation into groundwater.

On the basis of lake and groundwater salinities Hofmann (1996) suggested the presence of two hydraulically disconnected flow systems: one lies north of an anticline located approximately 40 km from the desert's southern boundary, and is fed by direct recharge, and the other lies south of the first system and is fed by a source external to the dune field. They estimated a diffuse recharge rate of 20 mm $year^{-1}$ from the input volume needed to offset lake evaporation. Yang and Williams (2003) and Jäkel (2002) also hypothesized that direct recharge was an important groundwater source. Yang et al. (2010) thought that evaporation from lake surfaces has been overestimated, and presented new estimates of evaporation and groundwater recharge in the Badain Jaran Sand Sea based on a modified version of the Penman Equation that is more suitable for lakes. Their geochemical data and water-balance calculations suggested that local rainfall contributed significantly to groundwater recharge. They suggested that the mean annual evaporation was ~ 1040 mm from the lake surface and ~ 100 mm from land surfaces in the southeastern part of the sand sea. Both values are much lower than previously published. Consequently, about 5% of the modern precipitation could recharge the groundwater.

The direct recharge hypothesis has not been accepted by the majority of researchers (Liu, 2010). First, the annual potential evaporation, which is estimated to be about 3300 mm, is believed to be too high for the annual precipitation of about 50 mm to maintain permanent lakes, even though the lake levels seem to be slowly decreasing (Liu, 2010). In other Chinese sand seas or dune fields east of the Badain Jaran Sand Sea, where precipitation is greater, such as the Tengger Sand Sea and the Mu Us dune field, fewer permanent lakes have developed. In addition, field observations (Liu, 2010) of pore water and of post-rain infiltration in the surface sand layers of the mega-dunes indicated that a 3% moisture content accounted for more than 65% of the maximum water-holding capacity, and positive $\delta^{18}O$ values in the pore water between 0.25 m and 1 m in the soil suggested that the water had experienced several cycles of recharge and evaporation; in addition, 4 to 5.5 mm of rainfall (the most common amount for individual rain events) could only moisten a sand layer 100 mm thick. Gates et al. (2008a) estimated direct recharge in the Badain Jaran Sand Sea from mean chloride contents in 18 unsaturated-zone profiles with depths of 6 to 16 m using the chloride mass-balance method. Based on water-balance considerations, they found that direct recharge through the unsaturated zone was insufficient to support the study area's network of groundwater-fed perennial lakes. They estimated that the mean diffuse recharge was 1.4 mm $year^{-1}$, equivalent to approximately 1.7% of mean annual precipitation.

Despite these negative results, the significance of precipitation in recharging lake water in the Badain Jaran Sand Sea may have been underestimated. First, extreme weather events must be considered. Rain events with precipitation of more than 100 mm sometimes occur in the western arid regions of China, though they are rare, and lesser but still significant amounts of rainfall also occur. For example, Alxa Youqi witnessed an intense rain storm that deposited about 45 mm within 2 hours on 20 July 2012. Second, the function of mega-dunes in water conservation must be considered in more detail because little is known about this function. The fact that lakes always coexist with mega-dunes suggests that the role of these dunes may be more important than has been previously considered. The mega-dunes may conserve considerable amounts of water delivered by intense precipitation events and release this water to recharge the lakes, though as noted above, such intense precipitation is rare. In response to global climate change, precipitation in the areas south of the Badain Jaran Sand Sea has obviously increased during the last decade, and extreme weather events have become more frequent. This suggests that the role of these dunes may increase in the future.

6.2 Near-source recharge

This hypothesis asserts that lakes in the Badain Jaran Sand Sea are fed by groundwater from surrounding areas, such as the Heihe River and diluvial fans in the piedmonts south of the sand sea. Ding and Wang (2007) proposed a recharge mechanism that relates the lake water to leakage of groundwater from the Heihe River into the Jinta–Dingxin Basin (Fig. 10.11).

This was based on the estimated groundwater balance in the Badain Jaran Sand Sea and the Jinta–Dingxin Basin. They found a groundwater recharge surplus of 2.66×10^8 m^3 in the Jinta–Dingxin Basin, but a groundwater deficit of 2.55×10^8 m^3 in the sand sea. This led them to conclude that the Jinta–Dingxin Basin may feed lakes in the Badain Jaran Sand Sea. The transport of groundwater between the basin and lakes was attributed to the presence of the Altyn Fault Belt, a west to east fault belt that covers a large area of northwestern China. Ren et al. (2003) proved that the Altyn Fault Belt is composed of five faults, the longest of which extends 1700 km and passes through the Jinta–Dingxin Basin and the lake area of the Badain Jaran Sand Sea.

The near-source recharge hypothesis seems to be supported by two additional groups of evidence: First, the topography lying between the Jinta–Dingxin Basin and the lake area of the Badain Jaran Sand Sea favors an eastward flow of water. The elevations of Zhengyixia and Xiaomaying in the basin (Fig. 10.11) range between 1110 and 1228 m a.s.l., whereas the elevations of lake surfaces in the sand sea range from 1080 to 1220 m a.s.l. Second, calcareous sinters and calcareous cementation found in the lake area, combined with different ratios in $^3He/^4He$ values between river water, shallow

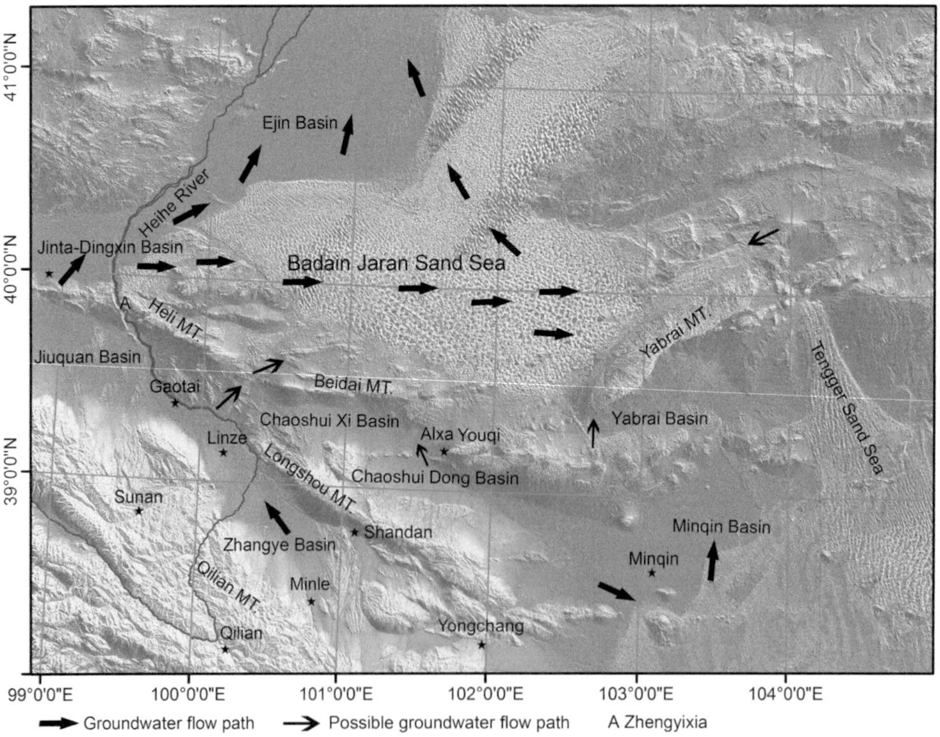

Figure 10.11. A proposed mechanism for the groundwater recharge and discharge system of the Badain Jaran Sand Sea (Ding and Wang, 2007).

groundwater and the Earth's crust, suggest that lakes in the Badain Jaran Sand Sea are recharged by deep aquifers.

Liu (2010) suggested that groundwater from other basins, such as the Chaoshui Xi Basin, Chaoshui Dong Basin and Yabrai Basin (Fig. 10.11) may feed the lakes in the Badain Jaran Sand Sea. However, Ding and Zhang (2002) estimated that the groundwater in these basins can only maintain a natural balance with evaporation, and is therefore insufficient to recharge lakes in the sand sea.

6.3 Remote-source recharge

In this hypothesis, the remote sources refer to atmospheric precipitation and lake water in the Qilian Mountains or even in the Qinghai-Tibetan Plateau. The hypothesis was first proposed to account for the isotopically anomalous groundwater of the Alxa Plateau, including the Badain Jaran Sand Sea (Geyh et al., 1998; Chen et al., 2003a, 2004b, 2005, 2006; Gu et al., 2004). Geyh et al. (1998) found that in the Gulunai area (west of the sand sea) and the Badain Jaran Sand Sea, the δD-$\delta^{18}O$ relationship for groundwater was parallel to the Global Meteoric Water Line (GMWL: $\delta D = 8\delta^{18}O + 10$), whereas the deuterium excess (d-excess) values were extremely negative, with values as low as –22.0‰. The mean d-excess values in groundwater in the southern, southeastern and northern areas of the Badain Jaran Sand Sea were about –21.0‰, but were > 0‰ in the Yabrai and Heishantou mountains to the south and southeast of the sand sea (Gates et al., 2008b). The abnormal d-excess value in groundwater is a potential tool to reveal the sources of groundwater in the sand sea using water stable isotopes ($\delta^{18}O$ and δD) and the d-excess in different water bodies. However, different alternative explanations have been proposed for the groundwater isotope data in the sand sea (Chen et al., 2003b; Zhang and Ming, 2006; Ma et al., 2007; Gates et al., 2008a, b; Liu, 2010; Zhao et al., 2011a).

Chen et al. (2004b, 2005) used the $\delta^{18}O$ and δD compositions of water samples shown in Fig. 10.12 to support the hypothesis of remote-source recharge.

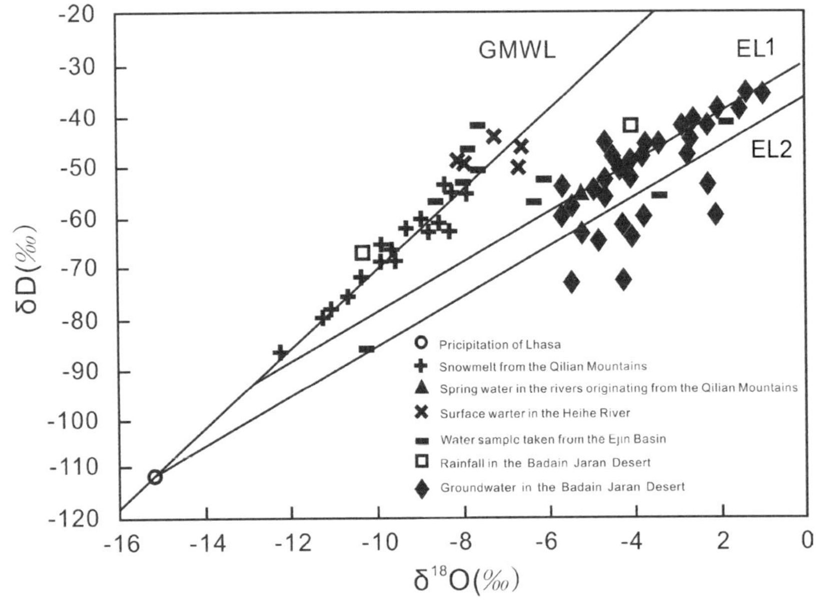

Figure 10.12. Distribution of the $\delta^{18}O$ and δD compositions measured from samples of precipitation, river water, groundwater and snowmelt in the Badain Jaran Sand Sea and its surrounding areas in the Qilian Mountains and the Qinghai–Tibetan Plateau (Chen et al., 2005). EL, evaporation lines calculated for two subsets of the data.

The average $\delta^{18}O$ and δD values in water samples from seven rivers originating from springs on the northern slopes of the Qilian Mountains fit the global meteoric water line well. The $\delta^{18}O$ and δD values of river water in the middle and lower drainage reaches of the Heihe River were also close to the global meteoric water line, despite deviation of a few points that probably resulted from the influence of the evaporation process. This was also true for one of the two precipitation samples in the Badain Jaran Sand Sea. Two evaporation lines can be identified in the remaining data. Evaporation Line I (EL1) representing samples from springs in the piedmont of the Qilian Mountains and the Longshou Mountains and several springs in the Badain Jaran Sand Sea, and this line intersects the global meteoric water line at a point with a $\delta^{18}O$ of –13.2‰, which is close to the average values for snowmelt water from glaciers in the Qilian Mountains. Evaporation line II (EL2) represents samples from springs and groundwater near Gulunai, Guezihu and the lake area of the Badain Jaran Sand Sea, and this line intersects the global meteoric water line at a point with a $\delta^{18}O$ of –15.24‰, close to the average values for precipitation measured in Lhasa.

This evidence suggests that groundwater and lakes in the Badain Jaran Sand Sea are recharged by precipitation in the Qilian Mountains and on the Qinghai-Tibetan Plateau (Chen et al., 2003b, 2004b, c, 2005). The snowmelt percolates into the mountain's Deep Big Faults system, and particularly the Qilian and Altyn faults, which in turn feeds into deep carbonate layers that extend from the mountains to the desert. The leakage through fractures in the deep carbonate layers then delivers water enriched in carbonates to the lakes and dunes in the sand sea (Fig. 10.13).

The total recharge volume was roughly estimated to be 5.0×10^8 m^3 $year^{-1}$ (Chen and Wang, 2004).

Other evidence supports the remote-source hypothesis. Calcareous sinters, rhizoconcretions and calcium cementation are found in the lakes and dunes, and are thought to be formed from carbonate-rich groundwater. The $^{87}Sr/^{86}Sr$ ratio of the water samples in the Badain Jaran Sand Sea ranges from 0.710 to 0.713, which is close to the range of 0.705 to 0.709 for carbonates created by oceanic sedimentation between 600 Ma BP and 250 Ma BP. Measurements of ^3H and three different

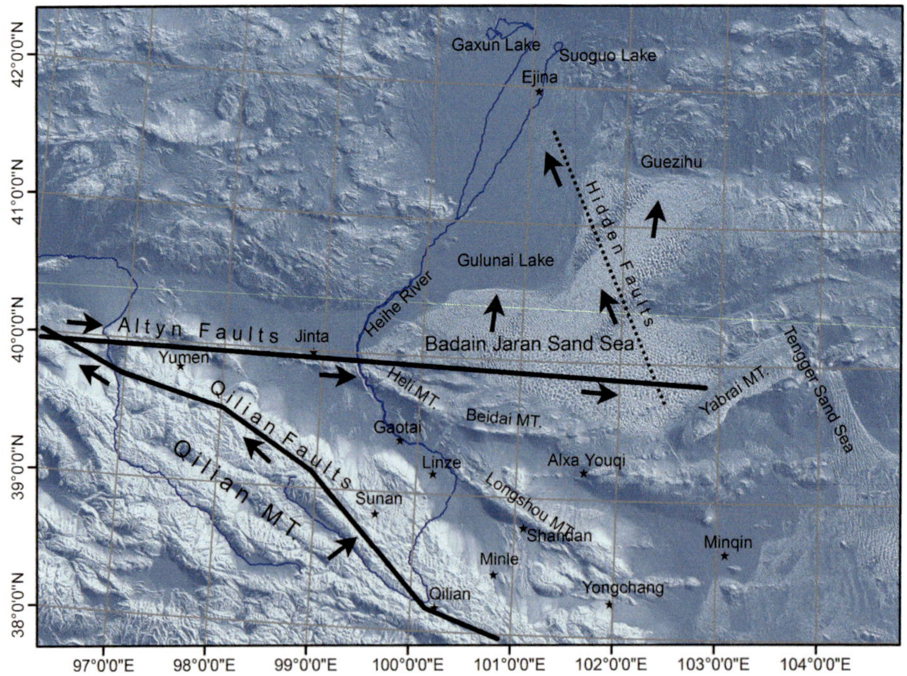

Figure 10.13. The system of large and deep faults that transports water from the Qilian Mountains and the Qinghai-Tibetan Plateau into the Badain Jaran Sand Sea. Modified after Chen et al. (2003, 2004b, 2005).

forms of chlorofluorocarbons suggested that the age of the spring water in the sand sea is between 20 and 30 years old. This relatively young age is consistent with the fast passage of the water through the faults and deep carbonate layers (Chen et al., 2005).

Several researchers have questioned the remote-source recharge hypothesis (e.g., Zhang and Ming, 2006; Ma et al., 2007, 2008; Shao et al., 2011; Zhang et al., 2011; Zhao et al., 2011a; Zhao et al., 2012). A key disputed issue is how to explain the $\delta^{18}O$ and δD compositions shown in Fig. 10.12 (Liu, 2010; Zhao et al., 2011a). Liu (2010) suggested that few researchers have paid sufficient attention to the shift in $\delta^{18}O$ values in the Badain Jaran Sand Sea, where carbonate bedrock promotes a shift of $\delta^{18}O$ of groundwater at normal temperature. Consequently, the $\delta^{18}O$ of the groundwater and lake water may have increased, and the evaporation line may have shifted to the right. Caution should also be taken in explaining the $\delta^{18}O$ and δD compositions of atmospheric precipitation because the recorded temperature increase in the study area has caused an growth of –0.2‰ to –0.52‰ in precipitation $\delta^{18}O$ in the study area (Zhang and Ming, 2006). The possibility that the Altyn fault system functions as a groundwater passage from the Qilian Mountains into the Badain Jaran Sand Sea was questioned based on an analysis of the topography of the Qilian Mountains and the elevation and discontinuities of the Altyn fault system (Zhang and Ming, 2006).

Zhang and Ming (2006) also suggested that the authors of the remote-source recharge hypothesis have insufficiently or incorrectly explained the significance of the $^{87}Sr/^{86}Sr$ ratio of the water samples; they noted that the geological significance of the $^{87}Sr/^{86}Sr$ value ranging from 0.710 to 0.713 (which is comparable to the average value of 0.7119 for the upper crust) is completely different from the range of 0.705 to 0.709 (which is comparable to the average value of 0.709198 ± 0.000020 for sea water and marine carbonates). Zhao et al. (2012) suggested that insufficient isotope samples were available to define the origin of groundwater in the Badain Jaran Sand Sea. They analyzed more than 300 samples from groundwater, river water and precipitation in the Badain Jaran Sand Sea and nearby areas in the upper, middle and lower reaches of the Heihe River and the Qilian Mountains. Their isotope results did not support the remote-source recharge hypothesis (Zhao et al., 2012). They found that: (1) There were significant δD, $\delta^{18}O$, and d-excess differences between the sand sea's groundwater and the values in other water bodies around the sand sea and in the Heihe River Basin. (2) The slope of the $\delta^{18}O$–δD regression for the Badain Jaran Sand Sea groundwater (4.509‰) was similar to those in surrounding areas (4.856‰), but was significantly lower than the slopes for groundwater in the Heihe River Basin and precipitation in the Badain Jaran Sand Sea, near Alxa Youqi, and in the upper reaches of the Heihe River Basin. (3) The d-excess value in the Badain Jaran Sand Sea groundwater was significantly lower than those in groundwater from surrounding areas, groundwater from the Heihe River Basin, precipitation in the Badain Jaran Sand Sea and near Alxa Youqi, in the upper reaches of the Heihe River Basin, and in river water in the Heihe River Basin. These results suggest that there is mixing and exchange of groundwater between the Badain Jaran Sand Sea and its nearby regions, but little or no exchange between the Badain Jaran Sand Sea groundwater and the Heihe River Basin.

6.4 Paleo-source recharge

This hypothesis proposed that lake water and groundwater in the Badain Jaran Sand Sea represented water from precipitation that fell in the distant past, when the climate was more humid. Authors who support this hypothesis believe that lakes in the Badain Jaran Sand Sea are fed by atmospheric precipitation, but that the current mean annual precipitation is far from balancing the evaporation (Edmunds et al., 2006; Ma and Edmunds, 2006; Gates et al., 2008a; Ma et al., 2007, 2008). The fact that the areas of lakes in the sand sea have been decreasing leads them to conclude that water recharge was greater in the past than at present.

Evidence of a wet paleo-climate has been found to support this hypothesis (Yang and Williams, 2003; Yang, 2002, 2006; Yang et al., 2003, 2010; Bai et al., 2011a). For example, the pattern of Holocene climatic changes led Yang et al. (2010) to emphasize the importance of paleo-precipitation in recharging lakes in the Badain Jaran Sand Sea. The chronology of lake-level changes, established by means of radiocarbon and uranium-series disequilibrium dating methods, indicated high lake levels and a wetter climate beginning ca. 10 ka BP and lasting until the late period of the Mid-Holocene. The greatest extent of lakes in the interdune depressions and surrounding areas of the sand sea indicates that the water availability was greatest during the Mid-Holocene (Demske and Mischke, 2003). Neolithic tools and pottery from the Qijia Culture (2400 to 1900 BC) suggest relatively intensive human activity in the Badain Jaran Sand Sea during the Early and Mid-Holocene, supporting an interpretation of a less harsh environment during this time (Yang et al., 2010). Ground-penetrating radar profiles and optically stimulated luminescence dating results suggested that the southern part of the sand sea had its last period of high lake levels during the MIS 5e and MIS 3 stages, and since then, alternating lacustrine deposition and calcareous cementation layers were deposited, and some shallow lakes and swamps that were present 7 ka BP were covered by aeolian sand (Bai et al., 2011a).

Based on the water chemistry and isotope profiles of groundwater and surface water samples, and those of soil moisture profiles in the Gulunai Oasis northwest of the sand sea, an attempt was made to formulate an integrated conceptual 2D hydrogeological model for the groundwater system linking the Badain Jaran Sand Sea with the Gulunai Oasis (Huang and Pang, 2007). The model suggested that the dominant recharge of groundwater below the dunes in the sand sea was from the neighboring Yabrai mountains, with the groundwater flowing northwest towards Gulunai. Groundwater emerges into lakes along its flow path and evaporates. The final discharge destination of the groundwater is the Gulunai Oasis, where it feeds a wetland ecosystem. They calculated that the vertical recharge from local precipitation that passes through the sand was small. Groundwater in the sand was probably recharged from paleo-river sources in the Qilian Mountains and from local paleo-rainfall between the Pleistocene and the Early Holocene. During the Mid- and Late Holocene, the groundwater underwent evaporation and limited recharge. Since 2 ka BP, aridity increased, and the groundwater table in the sand sea has declined slowly.

Ma and Edmunds (2006) first argued on a paleowater origin for shallow groundwater in the southern part of the sand sea using evidence from chemical and stable isotope studies from this area and analogous data from the nearby Minqin Basin (Edmunds et al., 2006). Their stable isotope relationships clearly indicated that there was no genetic relationship between the lakes and modern recharge. They also found that the shallow phreatic water and deeper bodies of confined groundwater appear to be unconnected based on chemical composition data and radiocarbon activities (Gates et al., 2008a). Ma et al. (2007) reported that the groundwater of the Badain Jaran Sand Sea was mainly found in a Pleistocene lacustrine sedimentary sandy stratum in regions near the center of the sand sea, where lake groups have formed. In the southeastern margin of the sand sea, where the lakes are widespread, the depth at the top of the aquifer layer is above 30 m, and the top of the low-permeability aquitard layer ranges from 5 to 20 m, although there is insufficient detailed data on the distribution of these depths (Ma et al., 2007).

In the region's existing lakes in the southern part of the Badain Jaran Sand Sea, the bottom of the aquitards consists of Cretaceous clastic rocks, and the aquifer layer is formed by a Holocene lacustrine sedimentary sandy, sub-sandy and aeolian sedimentary stratum. Ma et al. (2008) found that the groundwater signature from the Late Pleistocene differed markedly from that of the Holocene, as is shown clearly by the stable isotopic compositions (around –10.5% $\delta^{18}O$, versus current values of –7%). The lakes are clearly derived from a groundwater source with the same signature. This source could be paleowater from a time when the climate was much wetter than today, and other geological archive evidence suggests this would be in excess of 20 ka BP, prior to the period of aridity during the Last Glacial Maximum (Ma and Edmunds, 2006). Zhao et al. (2012) suggested that the abnormally low (negative) d-excess values in the Badain Jaran Sand Sea groundwater indicated that this water represented the results of continuous recharge during a period of cold and wet conditions, with higher relative humidity and lower temperatures than those of today.

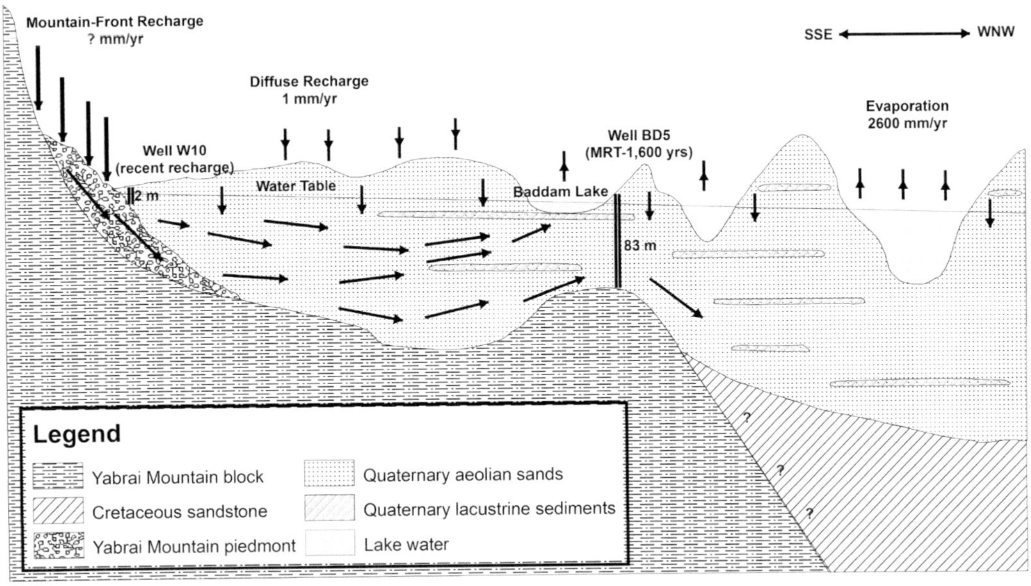

Figure 10.14. A conceptual diagram of recharge into the southeastern Badain Jaran Sand Sea based on various tracer data (Gates et al., 2008b). MRT, Mean Retention Time.

Hydrogeologic evidence suggests that a mountain range bordering the sand sea, the Yabrai Mountains that lie southeast of the sand sea, is a likely recharge zone (Gates et al., 2008b). Gates et al. (2008b) proposed a conceptual model of recharge into the Badain Jaran (Fig. 10.14) based on various tracer data.

During humid climatic phases of the Pleistocene and periods within the Early and Mid-Holocene, surface water and groundwater were more abundant in the region (Ma et al., 2004; Yang and Williams, 2003; Mischke, 2005) and the shallow Badain Jaran aquifer was probably replenished by rainfall through a number of mechanisms, including direct and mountain-front recharge. Groundwater discharged into the lowlands to the west and northwest of the desert, which are areas that also had elevated lake levels during this time, and into inter-dune lakes in the sand sea. At the end of the Mid-Holocene humid phase, approximately 6.5 ka BP, lake levels and shallow groundwater levels were 15 m higher than at present. Water tables and lake levels fell as inputs diminished because of Late Holocene aridification, and may continue to fall slowly. This conceptual model is supported by other geochemical evidence (Gates et al., 2008b; Ma and Yang, 2008; Yang et al., 2010; Shao et al., 2011). For example, Ma and Yang (2008) suggested that some kind of contact must have occurred between groundwater at the margins of the sand sea and the lakes of the southeastern Badain Jaran Sand Sea. Their isotope data shows that the groundwater in the sand sea and the Yabrai Mountains area could be phreatic water, but with a high watertable. Therefore, groundwater in the southeastern part and southern margin of the sand sea was mainly recharged by precipitation from local rainfall and by groundwater from the adjacent southern mountains. The chemical composition of the lake water revealed that the flow direction in the lake area of the Badain Jaran Sand Sea was from the southeast to the northwest, and that the lakes were recharged by both local precipitation and precipitation in the mountains to the south and southeast (Shao et al., 2011).

Both the direct-recharge hypothesis and the paleo-source recharge hypothesis recognize the importance of precipitation in recharging lakes in the Badain Jaran Sand Sea. However, the present arid climate prevents researchers from attributing these lakes primarily to precipitation recharge. The paleo-source recharge hypothesis may be promising if one accepts a combination of the past climate change and the huge water-holding capacity of the mega-dunes, which contain large amounts of water input by past precipitation. Even at present, extreme precipitation events can store large volumes of water in the mega-dunes. The juxtaposition of mega-dunes and lakes appears likely to carry meaningful information about the formation of both mega-dunes and lakes.

7. Conclusion

Mega-dunes and lakes in the Badain Jaran Sand Sea are inter-related, featuring an unique mega-dune—lake system. Mega-dunes act as water reservoir to store atmospheric precipitation, especiallty that of wetter climate periods and of extreme precipitation events in drier periods. The large volume of megadunes means huge amount of water contained. In turn, moisture held in the mega-dunes retards the mobility of sediments on the mega-dunes, also sustains vegetation to fix the shifting sand. Consequently, the mega-dunes are heightnened instead of migrating. The psooble effect of lakes on dune heightening favoured by the mega-dune—lake air circulation also needs be clarified.

References

Bai, Y., Wang, N., He, R., Li, J. and Lai, Z. 2011a. Ground penetrating radar images and optically stimulated luminescence dating for lacustrine deposition of the Badain Jaran Desert. Journal of Desert Research 31: 842–847 (In Chinese with English abstract).

Bai, Y., Wang, N., Liao, K. and Klenk, P. 2011b. Geomorphological evolution revealed by aeolian sedimentary structure in Badain Jaran Desert on Alxa Plateau, Northwest China. Chinese Geographical Science 21: 267–278 (In Chinese with English abstract).

Breed, C.S., Fryberger, S.G., Andrews, S., McCauley, C., Lennartz, D., Gebel, D. et al. 1979. Regional studies of sand seas using Landsat (ERTS) imagery. pp. 305–397. *In*: McKee, E.D. (ed.). A Study of Global Sand Seas. U.S. Geological Survey, Washington.

Chen, F., Wu, W., Holmes, J.A., Madsen, D.B., Zhu, Y., Jin, M. et al. 2003a. A mid-Holocene drought interval as evidenced by lake desiccation in the Alashan Plateau, Inner Mongolia, China. Chinese Science Bulletin 48: 1401–1410.

Chen, F. and Liu, Y. 2011. Secular annual movement of sand dunes in Badain Jaran Desert based on geographic analyses of remotely sensed imagery. Remote Sensing Technology and Application 26: 501–507 (In Chinese with English abstract).

Chen, J., Fan, Z., Wang, J., Gu, W. and Zhao, X. 2003b. Isotope methods for studying the replenishment of the lakes and downstream groundwater in the Badain Jaran Desert. Acta Geoscientica Sinica 24: 497–504 (In Chinese with English abstract).

Chen, J. and Wang, J. 2004. The impact of a new underground reservoir to be found in Badain Jaran Sand Sea in northwest China on the water transportation plan. Journal of Economics of Water Resources 22: 1–8 (in Chinese).

Chen, J., Li, L., Wang, J., Barry, D.A., Sheng, X., Gu, W. et al. 2004a. Groundwater maintains dune landscape. Nature 432: 459–460.

Chen, J., Wang, J., Zhao, X., Sheng, X., Gu, W., Chen, L. et al. 2004b. Study of groundwater supply of the confined aquifers in the Ejin Basin based on isotopic methods. Geological Review 50: 649–658 (In Chinese with English abstract).

Chen, J., Zhao, X., Wang, J., Gu, W., Sheng, X., Su, Z. et al. 2004c. Meaning of the discovery of lacustrine tufa and root-shaped nodule in Badain Jaran Desert for the study of lake recharge. Carsologica Sinica 23: 277–282 (In Chinese with English abstract).

Chen, J., Zhao, X., Sheng, X., Wang, J., Gu, W., Chen, L. et al. 2005. Geochemical information indicating the water recharge to lakes and immovable mega-dunes in the Badain Jaran Desert. Acta Geologica Sinica 79: 540–546 (In Chinese with English abstract).

Chen, J., Zhao, X., Sheng, X., Dong, H., Rao, W. and Su, Z. 2006. Formation mechanisms of mega-dunes and lakes in the Badain Jaran Desert, Inner Mongolia. Chinese Science Bulletin 51: 3026–3034.

Demske, D. and Mischke, S. 2003. Palynological investigation of a Holocene profile section from the Palaeo-Gaxun-Nur-Basin. Chinese Science Bulletin 48: 1418–1422.

Ding, H. and Zhang, H. 2002. Changes of groundwater resource in recent 50 years and their impact on ecological environment in Hexi Corridor. Journal of Natural Resources 17: 691–697 (In Chinese with English abstract).

Ding, H. and Wang, G. 2007. Study on the formation mechanism of the lakes in the Badain Jaran Desert. Arid Zone Research 24: 1–7 (In Chinese with English abstract).

Dong, G., Gao, Q. and Zou, X. 1995. Climatic changes at southern fringe of Badain Jaran Desert since Pleistocene. Chinese Science Bulletin 40: 423–427.

Dong, Z., Wang, T. and Wang, X. 2004. Geomorphology of the megadunes in the Badain Jaran Desert. Geomorphology 60: 191–203.
Dong, Z., Qian, G., Luo, W., Zhang, Z., Xiao, S. and Zhao, A. 2009. Geomorphological hierarchies for complex mega-dunes and their implications for mega-dune evolution in the Badain Jaran Desert. Geomorphology 106: 180–185.
Edmunds, W.M., Ma, J., Aeschbach-Hertig, W., Kipfer, R. and Darbyshire, D.P.F. 2006. Groundwater recharge history and hydrogeochemical evolution in the Minqin Basin, North West China. Applied Geochemistry 21: 2148–2170.
Fryberger, S.G. 1979. Dune forms and wind regime. pp. 137–160. *In*: McKee, E.D. (ed.). A Study of Global Sand Seas. Washington: U.S. Government Printing Office.
Gates, J.B., Edmunds, W.M., Ma, J. and Scanlon, B.R. 2008a. Estimating groundwater recharge in a cold desert environment in Northern China using chloride. Hydrogeology Journal 16: 893–910.
Gates, J.B., Edmunds, W.M., Darling, W.G., Ma, J., Pang, Z. and Yong, A. 2008b. Conceptual model of recharge to southeastern Badain Jaran Desert groundwater and lakes from environmental tracers. Applied Geochemistry 23: 3519–3534.
Geyh, M.A., Gu, W., Liu, Y., He, X., Deng, J. and Qiao, M. 1998. Isotopically anomalous groundwater of Alxa Plateau, Inner Mongolia. Advances in Water Science 4: 333–337 (In Chinese with English abstract).
Guo, H., Liu, H. and Wang, X. 2000. Subsurface old drainage detection and palaeoenvironment analysis using spaceborne radar images in Alashan Plateau. Science in China (Series D) 43: 439–448.
Gu, W., Chen, J., Wang, J., Zhao, X., Xie, M., Lu, J., Seiler, K.P. et al. 2004. Challenge from the appearance of vadose water within the surface layer of mega-dunes, Badain-Jaran dune desert, Inner Mongolia. Advances in Water Science 15: 695–699 (In Chinese with English abstract).
Hörner, N.G. 1936. Geomorphic Processes in Continental Basins of Central Asia, 16th International Geological Congress, Washington DC. Report, 2: 721–735.
Hofmann, J. 1996. Lakes in the SE part of Badain Jaran Shamo, their limnology and geochemistry. Geowissenschaften 14: 275–278.
Huang, T. and Pang, Z. 2007. Groundwater recharge in Badain Jaran Desert and Gurinai Oasis based on environmental tracers. Geoscience 21: 624–631 (In Chinese with English abstract).
Jäkel, D. 1996. The Badain Jaran Desert: its origin and development. Geowissenschafen 7: 272–274.
Jäkel, D. 2002. The importance of dunes for groundwater recharge and storage in China. Zeitschrift fur Geomorphologie Supplementband 126: 131–146.
Lancaster, N. 1995. Geomorphology of Desert Dunes (pp. 290). Routledge, London.
Li, B. 1962. The vegetation and its utilization in the Inner Mongolia and Northwest China. Research of Desert Control, Series 4. Science Press, Beijing. pp. 26–47 (In Chinese).
Li, E., Dong, Z. and Zhao, J. 2011. Grain size distribution of the aeolian sediments on the stoss slope of a typical mega-dune in the Badain Jaran Desert. Arid Land Geography 34: 471–478 (In Chinese with English abstract).
Li, X.Z., Wang, Z.T., Chen, F.H., Lancaster, N., Ren, X.Z. and Dong, G.R. 2009. Deposition of transverse dune mountains in Badain Jaran Desert. Quaternary Sciences 29: 797–805 (In Chinese with English abstract).
Liu, J. 2010. Recharge mechanisms of lakes and groundwater in Badain Jaran Desert. Water Resources Protection 26: 18–23 (In Chinese with English abstract).
Liu, T., Yang, X., Dong, J., Fan, X., Li, H. and Zhu, B. 2010. A preliminary study of relation between mega-dune shape and wind regime in the Badain Jaran Desert. Journal of Desert Research 30: 1285–1291 (In Chinese with English abstract).
Lou, T. 1962. The formation and utilization of the desert between Minqin and Badain Monastery. Research of Desert Control, Series 3. Science Press, Beijing, pp. 90–95 (In Chinese).
Lu, J. and Guo, Y. 1995. Study on compilation of landscape map of typical areas of high dunes in Badain Jaran Desert. Journal of Desert Research 15: 385–391.
Ma, J., Li, D., Li, X. and Edmunds, W.M. 2004. Chloride as a tracer and climatic change record from unsaturated zone of Badain Jaran Desert. Journal of Desert Research 24: 674–679 (In Chinese with English abstract).
Ma, J. and Edmunds, W. 2006. Groundwater and lake evolution in the Badain Jaran Desert ecosystem, Inner Mongolia. Hydrogeology Journal 14: 1231–1243.
Ma, J., Huang, T., Ding, Z. and Edmunds, W.M. 2007. Environmental isotopes as the indicators of the groundwater recharge in the south Badain Jaran Desert. Advances in Earth Science 22: 922–930 (In Chinese with English abstract).
Ma, J., Ding, Z., Gates, J.B. and Su, Y. 2008. Chloride and the environmental isotopes as the indicators of the groundwater recharge in the Gobi Desert, northwest China. Environmental Geology 55: 1407–1419.
Ma, N. and Yang, X. 2008. Environmental isotopes and water chemistry in the Badain Jaran Desert and in its southeastern adjacent areas, Inner Mongolia and their hydrological implications. Quaternary Sciences 28: 702–707 (In Chinese with English abstract).

Meng, J., Wu, X. and Li, Z. 2004. Land use and land cover changes in Heihe River Basin during the period of 1988–2000. Acta Scientiarum Naturalium Universitatis Pekinensis 40: 922–929 (In Chinese with English abstract).

Mischke, S. 2005. New evidence for origin of Badain Jaran Desert of Inner Mongolia from granulometry and thermoluminescence dating. Journal of Palaeogeography 7: 79–97.

Norin, E. 1941. The Tarim Basin and its border regions. Leipzig Akad. Verlagsgesellschaft, Regionale Geologie der Erde, 2, pt.4b, pp. 40.

Qian, G., Dong, Z., Luo, W., Zhang, Z., Xiao, S. and Zhao, A. 2011. Grain size characteristics and spatial variation of surface sediments in the Badain Jaran Desert. Journal of Desert Research 31: 1357–1364 (In Chinese with English abstract).

Qu, J., Chang, X., Dong, G., Wang, X., Lu, J. and Zhong, D. 2003. Fractal behavior of aeolian sand Landform in typical mega-dune area of Badain Jaran Desert. Journal of Desert Research 23: 361–365 (In Chinese with English abstract).

Ren, S., Ge, X. and Liu, Y. 2003. Progress in Altyn fault belts research. Advances in Earth Sciences 18: 386–391 (In Chinese with English abstract).

Ren, W. and Jin, S. 2011. Applying audio-magnetelluric method to the study of mega-dune structure and formation in Badain Jaran Desert, Inner Mongolia. Geoscience 25: 1167–1173 (In Chinese with English abstract).

Samthcin, M. and Walger, K. 1974. Der äolische Sandstrom aus der W-Sahara zur Atlantikkiiste. Geologische Rundschau 63: 1065–1087.

Shao, T., Zhao, J., Li, E. and Dong, Z. 2010. Grain size composition of typical mega-dune in Badain Jaran Desert. Scientia, Geographica Sinica 30: 790–795 (In Chinese with English abstract).

Shao, T., Zhao, J. and Dong, Z. 2011. Water chemistry of the lakes and groundwater in the Badain Jaran Desert. Acta Geographica Sinica 66: 662–672 (In Chinese with English abstract).

Sharp, R.P. 1963. Wind ripples. Geology 71: 617–636.

Simons, D.B., Richardson, E.Y. and Nordin, C.F. Jr. 1965. Bedload Equation for Ripples and Dunes (PP. 462, chapter H). United States Geological Survey.

Sun, P. and Sun, Q. 1964. The Hydrological Geology of the Western Inner Mongolia. Research of Desert Control, Series 6. Science Press, Beijing, pp. 121–146 (In Chinese).

Tan, J. 1964. The Types of Deserts in the Alashan Plateau, Inner Mongolia. Geographical Collections, Series 8. Science Press, Beijing, pp. 43–52 (In Chinese).

Walker, A.S. and Olsen, J.W. 1987. The Badain Jaran Desert: remote sensing investigations. Geographical Journal 153: 205–210.

Walker, D.J. 1981. An Experimental Study of Wind Ripples. M.Sc. thesis, Boston, Massachusetts Institute of Technology.

Wang, T. 1990. Formation and evolution of Badain Jaran Sandy Desert, China. Journal of Desert Research 10: 29–40 (In Chinese with English abstract).

Yan, M., Wang, G., Dong, G. and Li, B. 2001a. Study on mega dunes development and environmental change in Badain Jaran Desert. Journal of Desert Research 21: 361–366 (In Chinese with English abstract).

Yan, M., Wang, G., Li, B. and Dong, G. 2001b. Formation and growth of high mega-dunes in Badain Jaran Desert. Acta Geographica Sinica 56: 83–91 (In Chinese with English abstract).

Yang, P. and Jiang, Z. 1998. Method for measuring dune's morphological parameters and its application: an example from Badain Jaran Desert. Journal of Desert Research 18: 354–358.

Yang, P., Zou, X., Hasi and Jiang, Z. 1999. Division of Aeolian landform configuration in northern Badain Jaran Desert. Journal of Desert Research 19: 210–214.

Yang, X. 2002. Water chemistry of the lakes in the Badain Jaran Desert and their Holocene evolutions. Quaternary Sciences 22: 97–104 (In Chinese with English abstract).

Yang, X. and Williams, M.A.J. 2003. The ion chemistry of lakes and late Holocene desiccation in the Badain Jaran Desert, Inner Mongolia, China. Catena 51: 45–60.

Yang, X., Liu, T. and Xiao, H. 2003. Evolution of mega-dunes and lakes in the Badain Jaran Desert, Inner Mongolia, China during the last 31,000 years. Quaternary International 104: 99–112.

Yang, X. 2006. Chemistry and late Quaternary evolution of ground and surface waters in the area of Yabulai Mountains, western Inner Mongolia. China. Catena 66: 135–144.

Yang, X., Ma, N., Dong, J., Zhu, B., Xu, B., Ma, Z., Liu, J. et al. 2010. Recharge to the inter-dune lakes and Holocene climatic changes in the Badain Jaran Desert, western China. Quaternary Research 73: 10–19.

Yang, X.P. 2000. Landscape evolution and precipitation changes in the Badain Jaran Desert during the last 30000 years. Chinese Science Bulletin 20: 1042–1047.

Yang, X.P. 2001. Landscape evolution and palaeoclimate in the deserts of northwestern China, with a special reference to Badain Jaran and Taklamakan. Chinese Science Bulletin 46: 6–10.

Yang, X.P., Scuderi, L., Liu, T., Paillou, P., Li, H.W., Dong, J.F. et al. 2011. Formation of the highest sand dunes on Earth. Geomorphology 135: 108–116.
Yu, L.P. and Lai, Z.P. 2012. OSL chronology and palaeoclimatic implications of aeolian sediments in the eastern Qaidam Basin of the northeastern Qinghai-Tibetan Plateau. Palaeogeography, Palaeoclimatology, Palaeoecology 377-378: 120–129.
Yu, S. 1962. An Expedition to the Gobi and Badain Jaran Desert in the Western Inner Mongolia. Research of Desert Control, Series 3, 96–118. Science Press, Beijing.
Zhang, H. and Ming, Q. 2006. Hydrology and lake evolution in hyperarid northwestern China and the mystery of mega-dune formation in Badain Jaran Desert. Advances in Earth Science 21: 532–538 (In Chinese with English abstract).
Zhang, H., Wang, N., Li, Z., Dong, C., Lu, Y. and Li, G. 2011. Features of hydrogen and oxygen isotopes in lakes and groundwater in southeast Badain Jaran Desert. Journal of Desert Research 31: 1623–1629 (In Chinese with English abstract).
Zhang, W. and Wang, T. 2005. Approach to formation and evolution of mega-dunes in Badain Jaran Desert. Journal of Desert Research 25: 281–286 (In Chinese with English abstract).
Zhao, J., Shao, T., Hou, Y., Lv, X. and Dong, Z. 2011a. Moisture content of sand layer and its origin in a mega-dune area in the Badain Jaran Desert. Journal of Natural Resources 26: 694–702 (In Chinese with English abstract).
Zhao, J., Zhang, C., Dong, Z., Shao, T. and Li, E. 2011b. Particle size composition and formation of the mega-dune in the Badain Jaran Desert. Acta Geologica Sinica 85: 1389–1398 (In Chinese with English abstract).
Zhao, L., Xiao, H., Dong, Z., Xiao, S., Zhou, M., Cheng, G. et al. 2012. Origins of groundwater inferred from isotopic patterns of the Badain Jaran Desert, Northwestern China. Ground Water 50: 715–725.
Zhu, Z., Hofmann, J. and Jäkel, D. 1992. The Map of Aeolian Landform in Badain Jaran Desert (1:500,000). Xian Cartographic Publishing House, Xian (In Chinese).
Zhu, Z.D., Wu, Z., Liu, S. and Di, X.M. 1980. Deserts in China. Science Press, Beijing, pp. 107 (In Chinese).

Chapter 11

Morphology, Distribution and Formation of the Nebkhas on Gobi and Desert in China

Dong Xue[1,]* and *Yanyan Yang*[2]

1. Introduction

Nebkha is a kind of typical aeolian landforms that is commonly distributed in Gobi and Desert areas. Nebkhas as relatively high-resolution proxies in these regions, the accumulation characteristics of nebkha can be used to reconstruct environmental changes. This chapter reviewed research achievements in morphology characters, spatial distribution, type partition and sandy sediment characteristics about nebkhas. The various ecological effects of nebkhas and their response to regional environmental evolution were described. Besides, it also summarized the utilization and indicator function of nebkhas. With the development of research on nebkhas, overseas and domestic researchers are paying more and more attention to application of nebkha research in monitoring and estimating soil desertification and regional environment evolvement. Thus based on the previous studies, it was expected to help us to understand more about the interior relationship between the process of regional ecological environment change and global climate change in the future.

2. Overview of nebkhas

Nebkhas are an important kind of wind-bound landform, also known as coppices, coppice dunes, shrub-coppice dunes, coppice dunes and vegetation dunes, which are accumulations of wind–borne sediments, including sand, silt, salt grains and clay within or around the canopies of plants due to the resistance of shrubs (clumps) when aeolian sand transport flux encounter shrub obstruction in arid, semi-arid and sub-humid desert areas (Li et al., 2007). Nebkha is a type of biogeomorphology that forms from the interactions of plants, wind and sediments. Nebkhas are widespread in the world, are closely related to climate change and human activities, and play an important role in land degradation and maintaining the ecological security of oases. The study of nebkhas is not only valuable in enriching blown sand geomorphology, but also important in monitoring and evaluation of land desertification and reconstruction of regional environmental change (Yang et al., 2019; Junran Lia et al., 2018).

Numerous studies of nebkhas through field investigations and simulations of wind tunnel found that obvious interactions exist between shrubs and nebkhas of the shrub. The formation and

[1] Experimental Center of Desert Forestry, Chinese Academy of Forestry, Dengkou, Inner Mongolia, China.
[2] Institute of Ecological Conservation and Restoration, Chinese Academy of Forestry.
* Corresponding author: dongxue98765@126.com

development of nebkhas are influenced by plants through covering the ground, decomposing wind and trapping sand. The morphology of nebkhas is related to shapes, densities and structures of the plants which form them. The plants of nebkhas are able to not only trap sands through reduced wind speed, but also to protect the near surface underneath them from erosion (Wang et al., 2013). Under the same external conditions, the properties of plants are a determining factor to form nebkhas. Namely, the morphology of nebkhas is determined by the properties of nebkhas plants (Yang et al., 2012; Li et al., 2009). For example, the height of nebkhas is greatly correlated with the height of the canopy of shrubs, whereas the length of nebkhas is largely applicable with the height of the sum of shrubs and nebkhas, and there is also a good linear relationship between the volume of nebkhas and canopy of shrubs. The development of nebkhas is influenced by the growth of shrubs, and the volume and trapping capacity of nebkhas is determined by the above ground biomass of shrubs and morphology (Gao et al., 2015). Sand burial increases the scale of the scrub, which in turn increases the size of nebkhas.

The process of degradation and succession of nebkhas is the result of a combination of natural factors and anthropogenic activities to the natural ecosystem. The generation and decline of nebkhas are closely related to the positive and negative processes of land desertification. Nebkhas plays an important role in the desertification process of semi-arid grasslands and savannas, and the desertification process is actually the replacement of grasslands by nebkhas on a large scale. The invasion of plant species that form nebkhas marks the beginning of the process of desertification, which is accompanied by a significant loss of nutrients and increased soil erosion. The invasion of plant species leads to the formation of nebkhas and is a precursor to a more arid climate and appearance of desertification of the landscape. The development scale of nebkhas such as their length, width and height are controlled by plant types and growing conditions, which is very different from mobile dunes. In regions with intense sand transport activity, the formation, development and sediments of nebkhas could provide high-resolution dating and storing environment information for reconstruction of the regional environment changes.

3. Distribution and spatial pattern of nebkhas

From the point of view of the distribution, nebkhas occur in various ecosystems where there is a large amount of erodible material for their development (Qong et al., 2002; Wang et al., 2006). They develop in such sites as steppes, steppe-desert ecotones, oasis-desert ecotones, agro-pastoral transitional zones, piedmont alluvial fans, areas surrounding lakes in arid regions, floodplains or downstream areas of rivers, margins of deserts, coasts, interdune areas of slacks of active dunes and vegetated dunes. From the characteristics of climate, nebkhas are distributed from extreme arid to semi-arid regions, with annual precipitation ranging from 22 to 330 mm, water table between 1.5 and 17 m, and average annual wind speed between 2.1 and 5 m/s. The main soil types of nebkhas include wind-blown sandy soil, chestnut calcium soil and gray brown desert soil. Nebkhas are distributed not only in humid sandy coastal regions with very low elevations, but also in alpine valleys and plateau basins at elevations of 4,000 to 5,000 m (see Niu et al, this volume).

Spatial pattern analysis could reveal the quantitative characteristics of the spatial distribution of individual populations and interrelationship of various environmental factors, and reflect the heterogeneity of the spatial structure of environment and the intrinsic characteristics of plant-selected habitats. The study of the spatial distribution pattern of nebkhas at small scales could provide explanations for ecological pattern characteristics and developmental processes at large scales. The spatial pattern of nebkhas varies in different habitats. For example, the patch size and spatial distribution of *Nitraria tangutorun* nebkhas differs significantly in interdunes and fixed sandy areas. *Tamarix* nebkhas in gobi are distributed in patch, while those in deserts are distributed in dune chains. The horizontal spatial distribution, surface soil moisture and morphological characteristics of nebkhas at different evolution stages could be used as an indicator of the different developmental stages of nebkhas. Precipitation and surface runoff influence the spatial pattern of nebkhas on a large

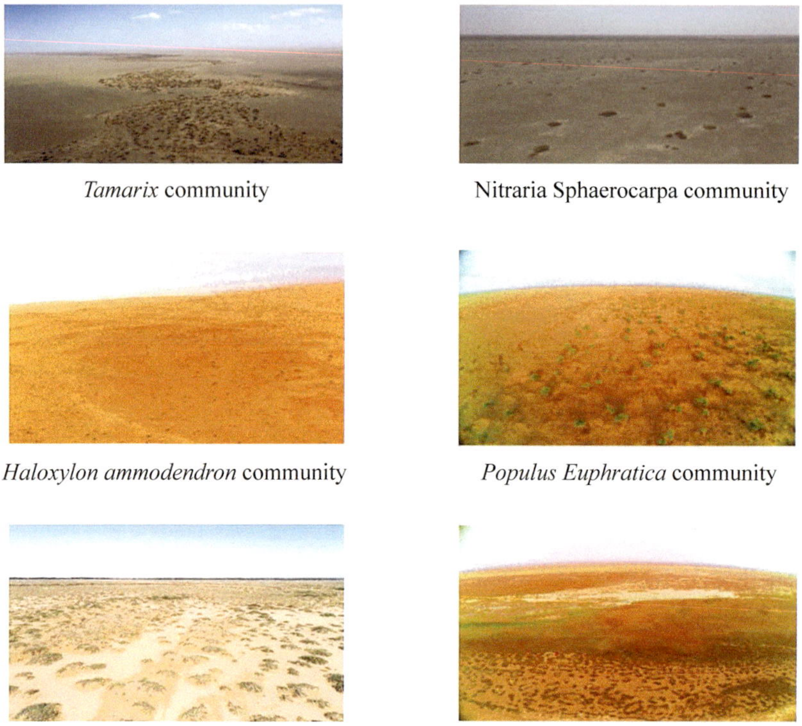

Tamarix community — Nitraria Sphaerocarpa community

Haloxylon ammodendron community — *Populus Euphratica* community

Tangutorum community — *Phragmites* and Alhagi maurorum community

Figure 11.1. Spatial pattern and distribution of different shrub types.

scale, and the environmental factors of local micro-topography also play a very important role in shaping the spatial pattern of nebkhas (Fig. 11.1).

4. Types and characteristics of nebkhas

4.1 Classification of nebkhas

Nebkhas are an important component of wind-blown dunes. In the classification according to the condition of sedimentary, wind-blown dunes were divided into loose sandy dunes and dunes under the influence of vegetation by Melton (1940) in the early period, and later the main dune types were divided into free dunes and obstacle dunes by Mabbutt. In the comprehensive classification by Cook et al., 1993, wind-blown dunes were classified as stabilized dunes, anchored dunes and mobile dunes. Nebkhas, vegetated parabolic, beach dunes and blowout are the main types of stabilized dunes, which are collectively called phytogenetic dunes.

The most common classification of nebkhas is according to the names of different plant species (Fig. 11.2). According to the types of the plants, nebkhas were classified into *Tamarix*, *Nitraria*, *Calligonum* and other three types by Tan (1962) in the earlier years, while others were classified into *Tamarix*, *Caragana*, *Nitraria* and distinct two types by Zhenda Zhu (1962). The classification system of nebkhas has improved with further research. According to plant species, plant suvirival status, development stages and material composition, nebkhas are classified according to three schemes by Mu (1995) after their studies of pattern and evolution of nebkhas in the Taklamakan Desert. (i) Nebkhas were divided into three types on the basis of plant species which were arboreal nebkhas, shrub and grass nebkhas, grass nebkhas and mixed nebkhas of various species, respectively. Arbors, shrubs and grass are always distributed in river terrace, flood plains and silt depressions and around stagnant depressions, respectively. Therefore, this classification can truly reflect the ecological environment of the development of nebkhas. (ii) Nebkhas were classified as embryonic, growing,

Figure 11.2. Nebkhas formed by different plant species.

Figure 11.3. Development processes of *Tamarix Chinensis* nebkhas species.

stable and declining forms according to their survival state and development, which could reflect the evolution of nebkhas (Fig. 11.3). (iii) Nebkhas were classified as clay nebkhas, clay-sand nebkhas and sand nebkhas on the basis of material composition and species and survival state of plants.

4.2 Morphology characteristics of nebkhas

The height of nebkhas ranges from a few centimeters to several meters, and the diameter of nebkhas could be up to tens of meters. The size of nebkhas depends on the scale and density of the plants that fix it (Fig. 11.4). The height and diameter of the most common *Caragana* nebkhas are between 1 and 1.5 m, and from 1.5 to 2.0 m, respectively. The height and diameter of *Reaumuria songarica* and *Oxytropis aciphylla* nebkhas are from 0.2–0.3 m, and less than 1 m, respectively. Nebkhas of these types are covered by short *Reaumuria songarica* and *Oxytropis Aciphylla*. *Nitraria tangutorum* nebkhas are always with larger scales, in which the diameters could be up to 3 m and the heights range between 1 and 1.5 m. Due to the late germination of *Nitraria tangutorum* in spring, sparse branches and weak trapping capacity of sand, *Nitraria tangutorum* nebkhas show semi-flowing states in spring. The height and diameter of *Tamarix chinensis* nebkhas are from 3 to 10 m, and could reach up to dozens of meters, respectively. In the shape of mounds and cones, nebkhas are distributed as isolated or in clusters in regions with lower aeolian blown sand activities. The scale of *Achnatherum splendens* nebkhas is usually small, with the height and diameter less than 0.5 m and 1 m, respectively. *Ceratoides arborescens* nebkhas are tall and dense with a diameter of up to 3 m, a height of up to 0.7–0.8 m and high sand trapping capacity, which often form a dense group of nebkhas, presenting a fixed wave-like sand landscape. In some areas, several nebkhas grow together to form aggregated dunes which present a ring or chain extending in the downwind direction.

The height of nebkhas at different development stages decreased in the order of activation stage, stabilization stage, development stage and embryonic form stage, which is mainly related

Figure 11.4. Morphology characteristics of nebkhas.

to vegetation characteristics at some stage and the balance between erosion and accumulation (Mountney et al., 2006). The morphology of nebkhas varies at different development stages, which can be broadly classified into conical and hemispherical, except for the early shield shape and late irregular shape (Wu et al., 2006). Nebkhas are usually present with relatively steep windward slopes and gentle leeward slopes (Langford, 2000). Shrubs with vertical growth, fewer branches and higher plant density form conical nebkhas with a larger height, while shrubs with creeping growth, smaller branches and lower plant density form hemispherical nebkhas with a shorter height. Compared with conic nebkhas, hemispherical nebkhas have a stronger sand fixation capacity.

Nebkhas are unstable landforms that begin to activate at the surface when erodible materials are exhausted (Seifert et al., 2009). Nebkhas may be formed by wind erosion, runoff erosion, rain splash erosion and animal activity, but the size of nebkhas which are formed by different processes may vary greatly. Nebkhas with large volume are always formed by wind erosions, while those with a smaller size are formed by animal activities and runoff erosions, although most nebkhas are likely to be the result of a combination of these dynamics in the natural field.

Zhu and Chen (1994) found the morphology of nebkhas has gone through four stages, in which the order of precedence is linear sand bars, isosceles triangle sand spouts, ovoid dunes with streamlined profiles and approximately round or oval dunes (Fig. 11.4). The external morphology of nebkhas is mechanistically a state of dynamic equilibrium, formed by the interaction of vegetation, wind and sand source in specific developmental stages. The formation and development of nebkhas is closely related to climate change and human activities. The degradation and succession process of nebkhas is mainly the comprehensive effect of natural factors and human activities on the regional ecosystem. The generation of nebkhas could have a great impact on land productivity and the ecological environment more generally.

Morphological parameters of nebkhas are considered to be the most important indicator features that can be used for regional desertification monitoring and evaluation. The formation, development and evolution of nebkhas on desert margins are inextricably linked to the surrounding environmental factors, the morphological characteristics of nebkhas in different habitats vary greatly (Xie et al., 2015; Yue et al., 2005). The correlation between the length, width, area, volume and height of nebkhas could not only reflect their stages and environmental conditions, but also are able to reflect the biomass of the plants (Bing et al., 2008). Obvious differences in basic morphology and scale of *Nitraria* exist among desert grassland, desert and gobi, in which the size of nebkhas in gobi and desert grassland is the smallest and largest, respectively. The morphological parameters, developmental trends and vegetation coverage at different development stages also differed greatly. The height of nebkhas at the development and activation stages varied differently with the horizontal scale, while the size of nebkhas grew significantly with an increase of the height and cover of plants, indicating that environmental changes can be revealed by the morphological characteristics of nebkhas. Meanwhile, the study of the morphological characteristics of nebkhas in different types in the desert can reflect the trend of desertification.

4.3 Sedimentary characteristics of nebkhas

The grain size composition of nebkhas at different positions is closely related to wind conditions and material sources, and vegetation coverage and topography also affect soil physical properties of nebkhas (Mao et al., 2015). The sediment of nebkhas is from interdunes nearby and surrounding areas, which has gone through blown erosion, preliminary sorting, polishing and deposited in and around plants due to their resistance to wind. The flow structure of plants changes during the formation of nebkhas, so that the grain size of nebkhas leads to the corresponding changes. The average grain size of surface sediment tends to be thinner from windward toes to the tops of nebkhas, and coarser from the tops to leeward toes (Zhao et al., 2007). The grain size of nebkhas can comprehensively reflect the material source, hydrodynamic environment, capacity and route of grain transport and is an important indicator for the reconstruction of paleoenvironment and paleoclimate (Zhao et al., 2007).

Regardless of the basic shape of nebkhas, there are interactions and observations among nebkhas, plants and sediments. On the one hand, the better the growth of plants, the stronger is the sand trapping capacity, and the more sediment depositions in and around the plants there is, could lead to the growth of roots and branches. On the other hand, good growth of shrubs of nebkhas increase the roughness of surface, which could lead to more sand depositions, and then increase nebkhas volume and change its morphology.

Most nebkhas contain multilayer cross bedding with a dip angle equal to or close to the angle of repose, which is generally considered to be formed by sand collapse on the leeward side.

These beddings are inclined, which are basically centered on the top of nebkhas and incline to windward slope and leeward slope respectively. The dip angle gradually increases from the bottom to the top of nebkhas. The bottom dip angles are 7° to 9°, while the top ones can reach 18° to 22°. According to sedimentary rhythms and dip angles, the bedding can be divided into several sedimentary groups, and each group can be divided into several texture layers with the same dip angle, which indicates that the development of nebkhas has experienced the periodic change of wind conditions. Meanwhile, the steep angles of the most of nebkhas show the prevailing wind direction in some regions. The layers of *Tamarix* nebkhas contain a large number of annual layers, which are composed of wind sand layers formed in spring and summer and the litter layers formed in autumn and winter. The thickness of the annual layer is larger in the middle of nebkhas near the trunk of the shrub, but smaller outside of nebkhas. The residual roots and litters inside of nebkhas are connected with each other and bonded with sand particles, forming a sticky texture of sand particles. Therefore, the nutrient content of nebkhas is higher than that of mobile sand dunes in the surrounding.

5. Ecological effects of nebkhas

Nebkhas play an important role in sandy areas and may be involved in dune genesis and formation.

5.1 The function of shrub in preventing wind and fixing sand maintains the ecological balance and stability of desert

Shrubs are the dominant vegetation type in desert ecosystems in arid areas and play an important role in maintaining the structure, function, biodiversity and the ecological service function of a desert ecosystem. Nebkhas change the geomorphic characteristics of the sand surface, increases the roughness of the near surface, and plays a role in reducing the wind speed, preventing the migration of sand dunes, intercepting litter and fixing sand. Wind speed of the downwind direction could be reduced by more than 40% by *Nitraria* nebkhas at the edge of Minqin Oasis in western China, indicating the obvious wind prevention and sand fixation effects of nebkhas. Shrub communities when composed of shrubs of different types have obvious effects on protecting and improving the desert ecological environment. The impact of shrubs on sand transport processes are very complex. From the perspective of the plants themselves, their distribution mode, porosity and plant morphological characteristics are closely related to sand transport flux, while environmental factors such as wind direction, micro-topography and hydrological characteristics are also closely related to aeolian sand transport processes (Wang, 2007). Therefore, a large number of field observations and researches need to be launched to understand the effects of plants on sand transport processes.

5.2 Fertile island effect of shrub increases soil nutrients

A fertile island of nebkhas is formed through trapping wind erosion materials (soil, dust and litter) by shrubs. A large amount of organic matter ends up in the soil in the form of rhizosphere deposition through the developed root system of plants, improving the soil fertility. The fertile islands formed by different forms of shrubs vary greatly. In arid and semi-arid areas, the fertile island effect of nebkhas formed under shrub canopy was significant (Valinte et al., 1991). The spatiotemporal dynamics of soil nutrients under and between shrubs in an arid desert ecosystem is an important basis for exploring the nutrient process and formation mechanism of fertile islands. In terms of spatial distribution, the soil under shrubs contains higher nutrient content (C, N, P, etc.) and cations than that of the soil between shrubs. On the time scale, the longer the shrub community develops, the more obvious the spatial autocorrelation of soil nutrients is (Xie et al., 2001). Especially in the desert ecosystem with water as the limiting factor, the soil nutrient elements cannot be fully absorbed and utilized by plants due to the lack of water, which allows soil nutrients to accumulate. And the spatial heterogeneity leads to adaptive spatial differentiation of shrubs, such as *Nitraria tangutorum*, leading to stronger adaptive ability to transport sand (Zhang et al., 2009; Liu et al., 2010). The effect of a fertile island gradually weakens and disappears with the distribution of nebkhas gradual transiting from an oasis to the desert, however, the fertile island also provides necessary conditions for community evolution in a region.

5.3 The shelter effect of shrub increases species diversity

As the fertile island effect of nebkhas gradually appears, its surrounding microenvironment also changes. Shelter effects of shrubs is a series of processes that are due to the enhancement of soil water heterogeneity by shrubs that not only benefit the survival and reproduction of nebkhas, but also provide good conditions for the invasion, accumulation and survival of herbaceous plants under the shrub. This is conducive to the settlement and restoration of herbaceous plants in the nebkhas. Seeds are difficult to retain on the surface sand layer because there is no barrier in areas of sand dunes, but they can stay in the shrub under the shelter effect of a shrub. Therefore, the coverage, height, density, species number and soil seed bank of plants in the shrub decrease from shrubs to the outside (Zuo et al., 2009; Maestre, 2005). The shelter effect of the shrub on herbaceous plants under their canopies not only enriches the diversity of desert plants, but also provides important species resources for the restoration of degraded vegetation, can also play an important role in maintaining the stability of desert vegetation.

5.4 Carbon sequestration capacity and carbon sink function of nebkhas

First, shrub communities forming nebkhas have strong carbon sequestration capacity, which is closely related to the structure, growth type, size and living environment of shrubs (Gratani et al., 2011; Bell et al., 2012; Erik et al., 2013; Li et al., 2014; Chen, 2014). For example, the carbon sequestration capacity of a small shrub is lower than that of medium and large shrubs. Second, shrubs and soils of nebkhas have obvious mutual feedback effects. On the one hand, wind erosion materials and litter captured, deposited and decomposed by shrubs, atmospheric dustfall, leaf and root exudates microbial activity products and other substances in the rough layer of stems enter soil through the shrub stem flow and throughfall irrigation, which causes enrichment of organic matter in the soil under nebkhas (Liu et al., 1999; Zhao et al., 2007). On the other hand, the formation of fertile islands in the soil of nebkhas will promote the growth and distribution pattern of the shrub, and further strengthen the enrichment process of organic matter in the soil of nebkhas and aid the carbon exchange process of shrubs (Maestre, 2005; Pei, 2007; Zuo et al., 2009). The soil organic carbon, total nitrogen and microbial biomass of nebkhas soil were significantly higher than those of soil in the surroundings. Evidently the carbon sink function occurs in the soil of nebkhas. The carbon content of soil is related to texture, type, particle size of the soil itself and size, species and age of shrub (Kieft et al., 1998; Jackson et al., 2002; Wheeler et al., 2007; Li et al., 2016; Niu et al., 2017). Soil organic carbon is significantly higher under *leguminous* shrubs than under *non-leguminous* shrubs.

6. Utilization and prospect of nebkhas

6.1 Environmental indication of nebkhas

Nebkhas are closely related to environmental factors. Aeolian landforms are very common due to strong sand transport activities, and the environmental evolution process is sensitive to climate change in arid and semi-arid areas. Due to the lack of high-resolution carriers, the means and results of environmental change reconstruction in these areas are relatively poor. However, nebkhas developed in these areas are aeolian sand depositaries, and their formation and evolution provide a record yielding relatively rich information about regional environmental evolution. At the same time, it is easy to observe the morphology of nebkhas. Nebkhas with different types of landforms can reflect the regional environment. With the decline of rangeland productivity, the degradation of the ecological system is accompanied by the gradual emergence of nebkhas, conversely, the appearance of nebkhas could show land degradation of the rangeland. Nebkhas are used by many scholars to monitor and evaluate desertification at a regional scale. It is helpful to define the indicative function of nebkhas scientifically and accurately by connecting the development of nebkhas with soil erosion and land desertification. The extensive development and distribution of nebkhas in the extremely arid desert environment may indicate the stability of the ecological environment in the region. In the transitional region of desert, the stable trend of sand dunes, the 'oasis-ation' of land or the better conditions of soil and water are indicated.

6.2 Conservation function of nebkhas

The nebkhas is the sink of water and nutrients in the system between nebkhas and their interdunes, while interdunes of nebkhas are the source. This source and sink system of this typical patch landform can distribute water and nutrient resources reasonably, and expand the ecosystem functions. The appearance of nebkhas can increase the surface roughness, effectively improve the water and nutrient cycle of the ecosystem by capturing water and nutrients, which is conducive to the ecosystem recovery in arid and semi-arid areas. Excessive human activities could destroy the source and sink system of nebkhas and their interdunes, cause water and nutrient loss and accelerate erosion and degradation of the ecosystem. Therefore, the conservation of nebkhas is of great significance for the stability of arid and semi-arid ecological environment. The key process of conservation is to protect the plants of nebkhas. Vegetation change has a great impact on the morphology of nebkhas, balances between soil erosion and accumulation, water balances between supply and demand, regional wind regime and sand source. Therefore, how to prevent the plants which formed nebkhas from being damaged is the key to successful conservation.

6.3 Effects of nebkhas on regional environmental change and ecological stability

The formation and development of nebkhas is closely related to climate change and human activities. The temporal and spatial variation of annual grain size of nebkhas, especially *Tamarix* nebkhas, reflects the diversity of regional aeolian environment. More objective and credible information of climate and environment change is obtained by reconstructing the spatial and temporal variation process of temperature and precipitation on the centennial scale, which further highlights the indicative role of nebkhas in change of an aeolian environment. The climate and environmental change sequences contained in the annual layer of nebkhas is relatively short (nearly 100 years). A certain period in the past can be inferred from the current year through the annual layer dating, but there is a lack of a clear (standard) year, that is, a certain baseline point in the past. This causes the definition of years to be fuzzier, and the error of chronology is large, so that it cannot clearly define a certain period of time in the past through annual layer analysis of nebkhas. There is a necessity to unite geology methods to determine the chronology of nebkhas, set up the large-scale sequences of climate and environment change and response mechanisms of nebkhas to environmental and ecological stability under conditions of global climate change. In the course of the development process of nebkhas, the function of carbon source and sink of nebkhas has been paid greater attention against the background of global warming. The study on the response of internal water and heat factors and soil carbon flux change in different types of nebkhas will be helpful to understand the internal relationship between the regional ecological process and global climate change. At the same time, the study of nebkhas will also expand from a small individual scale to large scales combined with environmental factors, climate change and ecological evolution.

7. Conclusion

A nebkha (also known as a nabkha, coppice dune, shrub-coppice dune, vegetation dune, and vegetated dune) is a vegetated, sandy mound formed when the vegetation anchors the shifting sand and traps any organic material as it blows across the landscape. The heights of these features range from a matter of centimeters to as much as 20 m. There is a considerable quantity of data on the plants that are associated with the development of nebkhas. They have to have certain qualities for this role: ability to withstand drought, tolerate salt, thrive on sand burial, have wide shallow rooting systems to obtain water from fog and occasional rain events, extract moisture from groundwater, withstand sand abrasion, and develop adventitious root. Although nebkhas are normally thought of as a type of sand dune, and most nebkhas are indeed composed of sand-sized material, largely consisting of quartz, there are many nebkhas which contain appreciable amounts of silt and clay, as well as salts such as gypsum. Although some nebkhas are ephemeral features, various dating techniques have shown that they can be decades, centuries and even millennia old.

Nebkhas are very widespread in dry and coastal environments. They have considerable importance because they can be ‘islands of fertility’ and have major hydrological and biological effects. They can, if eroded, be major sources of dust emission. The formation and development of nebkhas is closely related to the climate and human activities.

References

Bell, T.W., Menzer, O., Enrique, T.D. et al. 2012. Carbon dioxide exchange over multiple temporal scales in an arid shrub ecosystem near La Paz, Baja California Sur, Mexico. Global Change Biology 18(8): 2570–2582.

Bing, L., Wenzhi, Z. and Rong, Y. 2008. Characteristics and spatial heterogeneity of *Tamarix ramosissima* Nebkhas in desert-oasis ecotones. Acta Ecologica Sinica 28(4): 1446–1455.

Bochet, E., Poesen, J. and Rubio, J.Z. 2015. Mound development as an interaction of individual plants with soil, water erosion and sedimentation processes on slopes. Earth Surface Processes & Landforms 25(8): 847–867.

Chen, H.X. 2014. The Characters of "Fertile Island" and Carbon Sequestration of Shrub *Reaumuria Songorica* in the Desert[D]. Lanzhou: Lanzhou University.

Cooke, R., Warren, A. and Goudie, A. 1993. Desert Geomorphology. London: UCL Press.

Du, J.H., Yan, P., Zhan, X.L. et al. 2008. Surface erosion-resistance of *Nitraria tangutorun* nebkhas at different succession stages in Minqin oasis and related affecting factors. Chinese Journal of Applied Ecology 19(4): 763–768.

Erik, P., Russell, L.S., Enrique, P.S. et al. 2013. Nocturnal soil CO_2 uptake and its relationship to subsurface soil and ecosystem carbon fluxes in a Chihuahuan Desert shrubland. Journal of Geophysical Research: Biogeosciences 118: 1593–1603.

Gao, Y., Dang, X.H., YU, Y. et al. 2015. Nabkha morphological characteristics and sand fixing capacity of artemisia sphaerocphalain the southeastern edge of the Ulan Buh Desert[J]. Journal of Desert Research 35(1): 1–7.

Gratani, L., Catoni, R. and Yarone, L. 2011. *Quercus ilex* L. carbon sequestration capability related to shrub size. Environmental Monitoring & Assessment 178(1-4): 383–392.

Guoxun, X., Weicheng, L. and Wenzhi, Z. 2015. Effect of sand source and shrub features on nebkha morphological characteristics in desert steppe. Journal of Desert Research 35(3): 573–581.

Hang, F.X., Wang, M.X. and Wang, Y.E. 2002. Recent progress on the research of vegetation protection in soil erosion by wind. Acta Phytoecologica Sinica 26(5): 627–633.

Jackson, R.B., Banner, J.L., Jobbagy, E.G. et al. 2002. Ecosystem carbon loss with woody plant invasion of grasslands. [J]. Nature 418(6898): 623–626.

Jie, Z.Y., Cheng, X.X., Bao, W.F. et al. 2007. Grain-size features of sand materials in *Tamarix* cone veins and their environmental indication significance in the Lop Nur region. Arid Land Geography 30(6): 791–796.

Junran Lia and Sujith Ravi. 2018. Interactions among hydrological-aeolian processes and vegetation determine grain-size distribution of sediments in a semiarid coppice dune (nebkha) system. Journal of Arid Environments 154: 24–33.

Kieft, T.L., White, C.S., Loftin, S.R. et al. 1998. Temporal dynamics in soil carbon and nitrogen resources at a grassland-shrubland ecotone. Ecology 79: 671–683.

Langford, R.P. 2000. Nabkha (coppice dune) fields of south-central New Mexico, U.S.A. Journal of Arid Environments 46(1): 25–41.

Li, H.Q., Li, Y.N., Zhang, F.W. et al. 2014. Carbon budget of alpine *Potentilla fruticosa* shrubland based on comprehensive techniques of static chamber and biomass harvesting[J]. Acta Ecologica Sinica (4): 925–932.

Li, H., Shen, H., Chen, L. et al. 2016. Effects of shrub encroachment on soil organic carbon in global grasslands[J]. Scientific Reports 6(1): 28974.

Li, J. and Ravi, S. 2018. Interactions among hydrological-aeolian processes and vegetation determine grain-size distribution of sediments in a semi-arid coppice dune (nebkha) system. Journal of Arid Environments 154(JUL.): 24–33.

Li, J.W., Li, Z.Z., Wu, S.L. et al. 2009. The spatial heterogeneity of morphologic feature of *Nitraria* nebkhas Around Ebinur Lake, Xinjiang. Journal of Desert Research 29(4): 628–635.

Li, Z.Z., Wu, S.L., Wang, X.F. et al. 2007. Bio-geomorphologic features and growth process of *Tamarix* nabkhas in Hotan River Basin, Xinjiang. Journal of Geographical Sciences 62(5): 462–470.

Liu, F.M., Jin, Y. and Zhang, X.J. 1999. Preliminary study on fertile island effect about *Haloxylon Ammodendron*[J]. Journal of Arid Land Resources and Environment 13(3): 86–88.

Liu, G.H., Yang, Y.L., Sheng, J.D. et al. 2010. 'Fertile island' characteristics of soil nutrients in *Haloxylon Ammodendron* land in north Xinjiang. Acta Pedologica Sinica 47(3): 545–554.

Maestre, F.T. and Cortina, J. 2005. Remnant shrubs in Mediterranean semi-arid steppes: effects of shrub size, abiotic factors and species identity on understorey richness and occurrence. Acta Oecologica 27(3): 161–169.

Mao, D.L., Lei, J.Q., Li, Y.S. et al. 2015. Spatial differentiation of physico-chemical properties of surface sand materials in the oasis-desert ecotone in Cele, Xinjiang, China. Journal of Desert Research 35(1): 136–144.

Melton, F. A. 1940. A tentative classification of sand dunes and its application to dune history in the southern High Plains[J]. Journal of Geology 48(2): 113–173.

Mountney, N.P. and Russell, A.J. 2006. Coastal aeolian dune development, Sólheimasandur, southern Iceland. Sedimentary Geology 192: 167–181.

Mu, G.J. 1995. Types, genesis and evolution of shrub and grass mounds in Taklimakan Desert [J]. Arid Zone Research (supplement): 31–37.

Niu, S., Liu, B. and Wang, L. 2017. Effect of soil particle size and organic carbon characteristics on the development stages of Caragana Korshinskii Shrub Nebkhas[J]. Chinese Journal of Soil Science 48(5): 1032–1038.

Niu, Y., Squires, V.R. and Hua, L. in press. High-altitude aeolian desertification and sand dunes on the Tibetan plateau, pp. 179–194 (this volume).

Pei, S.F. 2007. Effects of Grazing and Exclosure on Soil and Vegetation in Alxa Desert Steppe. Lanzhou: Lanzhou University.

Qong, M., Takamura, H. and Hudaberdi, M. 2002. Formation and internal structure of *Tamarix* cones in the Taklimakan Desert. Journal of Arid Environments 50(1): 81–97.

Seifert, C.L., Cox, R.T., Forman, S.L. et al. 2009. Relict nebkhas (pimple mounds) record prolonged late Holocene drought in the forested region of south-central United States. Quaternary Research 71(3): 329–339.

Song-Fang, N., Bing-Ru, Li and Li-Juan, W. 2017. Effect of soil particle size and organic carbon characteristics on the development stages of *Caragana korshinskii* shrub nebkhas. Chinese Journal of Soil Science (5): 1032–1038.

Tan, J.N. 1962. Large-scale geomorphological survey and compilation of geomorphological map in aeolian sand area and its significance [A]. Sand Control Team, Chinese Academy of Sciences. Research on sand control (No. 4). Beijing: Science Press 62–75.

Valiente, B.A., Bolongaro, C.A., Briones, O. et al. 1991. Spatial relationships between cacti and nurse shrubs in a semi-arid environment in central Mexico. Journal of Vegetation Science 2(1): 15–20.

Wang, H.L., Gao, J.L., Yuan, W.J. et al. 2013. Spatially heterogeneous characteristics of surface soil particles around nebkhas in the Gobi Desert. Chinese Journal of Plant Ecology 37(5): 464–473.

Wang, S.T., Zhou, X.Y., Zhang, C.L. et al. 2007. Wind protecting effect of shrub dunes in ecotone of minqin oasis. Journal of Geographical Sciences 27(1): 104–108.

Wang, X., Wang, T., Dong, Z. et al. 2006. Nebkha development and its significance to wind erosion and land degradation in semi-arid northern China. Journal of Arid Environments 65(1): 129–141.

Wheeler, C.W., Archer, S.R., Asner, G.P. et al. 2007. Climatic/edaphic controls on soil carbon/nitrogen response to shrub encroachment in desert grassland[J]. Ecological Applications 17(7): 1911–1928.

Wu, S.L., Li, Z.Z., Xiao, C.X. et al. 2006. Research progress on nabkhas and research significance. Journal of Desert Research 26(5): 734–738.

Xie, G. and Steinberger, Y. 2001. Temporal patterns of C and N under shrub canopy in a loessial soil desert ecosystem. Soil Biology 33(10): 1371–1379.

Xie, G.X., Luo, W.C. and Zhao, W.Z. 2015. Effect of sand source and shrub features on nebkha morphological characteristics in desert steppe. Journal of Desert Research 35(3): 573–581.

Yang, F., Wang, X.Q., Yang, D.L. et al. 2012. Research on the morphological interactions between *Tamarix ramosissima* thickets and Nebkhas under different sand supply conditions: A case study in Cele oasis-desert ecotone. Journal of Ecology 32(9): 2707–2719.

Yang, Y.Y., Liu, L.Y., Shi, P. et al. 2019. Converging effects of shrubs on shadow dune formation and sand trapping. Journal of Geophysical Research: Earth Surface 124: 1835–1853.

Yong, G., Xiaohong, D., Yi, Y. et al. 2015. Nabkha morphological characteristics and sand fixing capacity of *Artemisia sphaerocphalain* the southeastern edge of the Ulan Buh Desert. Journal of Desert Research 35(1): 1–7.

Yue, X.L., Ha, S. and Zhuang, Y.M. 2005. Studies on sandy grassland nebkhas—a review. Journal of Desert Research 16(5): 738–743.

Zhang, L.Y., Shen, J.D., Wu, H.Q. et al. 2009. Soil nutrient characteristic of *Tamarix* site in Xinjiang. Scientia Silvae Sinicae 45(3): 54–60.

Zhao, H.L., Su, Y.Z., Zhang, H. et al. 2007. Multiple effects of shrub on soil properties and understory vegetation in Horqin Sand Land, Inner Mongolia. Journal of Desert Research 27(3): 385–390.

Zhu, Z.D. 1962. Classification of Dune in Arid and Semi-arid Areas of Northwest China [A]. China Sand Control Team of the Academy of Sciences. Research on sand control (No. 4)[C]. Beijing: Science Press, China, 31–47.

Zhu, Z.D. and Chen, G.T. 1994. Sandy Desertification in China. Beijing: Science Press.

Zuo, X.A., Zhao, H.L., Zhao, X.Y. et al. 2009. Changes on landscape pattern of sand dunes at different scales in horqin sandy land. Journal of Desert Research 29(5): 785–795.

Chapter 12

Type, Distribution, Formation and Evolution of Coastal Aeolian Dunes

Yuxiang Dong,[1,*] *Shuyi Fu,*[1] *Shaoyun Zhang*[1] and *Eerdun Hasi*[2]

1. Introduction

Coastal aeolian dunes are dynamic geomorphic systems affected by the interaction of the sea, land and atmosphere. Moreover, their morphological changes are the focus of coastal resource management and landscape evolution research. Therefore, they represent dynamic environmental systems that indicate the relationship between the geomorphic process and morphological response, especially the interaction between the sea and land (Dong, 2010). Researchers began studying coastal aeolian dunes in China in the 1980s, and their research focused on dune types and development processes of coastal aeolian dunes before the early 21st century. As a result, significant progress has been made with regard to dune classification, distribution, formation, evolution and sediment characteristics (Cai and Cai, 1983; Dong, 2006b; Li et al., 1987; Li et al., 1985; Wu, 1993; Wu and Wu, 1987); Subsequently, researchers concentrated on characterizing the aeolian sand transport process over beaches and dunes, which includes field investigations and observations of the wind flow over a typical dune surface, the vertical distribution pattern of sand transport, coastal aeolian dune movement and morphological changes, and look towards further clarifying the dynamic process of the evolution of coastal aeolian dunes (Dong et al., 2008; Dong and Ma, 2008; Dong et al., 2009; Han et al., 2010); Recently, while further exploring the evolution process of coastal aeolian sand deposition environment (Hu et al., 2013; Jiang et al., 2016; Jin, 2017), research has predominately focused on the essential characteristics that form coastal aeolian dunes in the dynamic environment of the interaction of the sea, land and atmosphere. Moreover, comprehensive research has been performed by relating coastal aeolian dunes to the atmosphere, waves and beaches (Dong, 2010; Dong and Du, 2014). An example of this research includes the study of typhoon response patterns of sandy landforms on the South China coast (Dong et al., 2016a, 2016b; Yang et al., 2017; Yang et al., 2017, 2018, 2020; Yang et al., 2018; Zhang et al., 2016). This chapter aims to summarize the classification, distribution characteristics, origins, formation and evolution of coastal aeolian dunes along the China coastline based on earlier research.

2. Coastal aeolian dune types

Coastal aeolian dunes in China are a typical coastal aeolian landform and are famous for their wide variety and complex morphologies. Coastal aeolian dunes vary considerably but are predominately

[1] School of Geography and Planning, Sun Yat-Sen University, Guangzhou 510006, China.
[2] School of Natural Resources, Faculty of Geographical Science, Beijing Normal University, Beijing 100082, China.
* Corresponding author: eesdyx@mail.sysu.edu.cn

Table 12.1. Types and distribution of coastal aeolian dunes in China.

ID	Types	Characteristics and distribution
1	Incipient foredunes	Incipient foredunes (Fig. 12.1(a)) are built up at the backshore or on the crest of a beach ridge of sand or shingle where dune colonization by vegetation and trap blown sand. These foredunes are several meters high, and are well developed along the northwest coast of the Liaodong Peninsula, eastern Hebei province, Shandong Peninsula, southern Fujian province, western and eastern Guangdong province and Hainan Island.
2	Transverse foredune ridges	Transverse foredune ridges (Fig. 12.1(b)) gradually grow and coalesce to form the ridges of aeolian dunes which are parallel to the coast are transverse foredune ridges, are usually less than 10 meters high. In China, they are mainly distributed along the coast of eastern Hebei province, Muping county-Beihai forest farms in Wendeng City-Shuangdao forest farm in Weihai city, the northern Shandong Peninsula, the northern Changle district in Fujian province, Haimen Bay-Jieshi Bay of eastern Guangdong, western Guangdong province, and Wenchang city in Hainan Island.
3	Barchan dunes	Barchans (Fig. 12.1(c)) are convex-facing the wind, with the horns of the crescent pointing downwind and marking the lateral advancement of sand. These dunes are markedly asymmetrical in cross-section, with a gentle windward slope and a much steeper slope (known as the slip face) leeward slope. In China, they are mainly scattered in the Shandong Peninsula, Changli coast of eastern Hebei province, northern Changle district in Fujian province, Haimen Bay-Jieshi Bay, eastern Guangdong and Hainan Island coast.
4	Nebkha dunes	In terrains with excess vegetation, coastal wind-drift sand is trapped by vegetation during transportation and forms nebkha dunes around the vegetation in the shape of domes, mounds or cones (Fig. 12.1(d)). They are generally several meters high and widely distributed along the China coastline. Nebkha dunes are largely developed in the coastal area of western Guangdong and Hainan Island.
5	Longitudinal dunes	Longitudinal dunes (Fig. 12.1(e)) are linear sand ridges, ranging from several to a dozen meters high. They rarely occur in China and are restricted to the Changli coast in eastern Hebei and the southern Fujian.
6	Climbing dunes	When the aeolian transport of sand is blocked by mountainous terrain, sand accumulates along the mountain slope and forms climbing dunes (Fig. 12.1(f)). In China, these dunes are highly dispersed in the upland coast, especially in northern China in areas such as Kongtong Island in Yantai city of Shandong province, Shicheng Island, and Dachangshan Island in the Changshan Archipelago. Climbing dunes typically do not exceed 50 m in height, and locally range between 100 m (Baisha Mountain, Liaodong Peninsula) and 140 m (Yandun Mountain, Shuangdao, Weihai, Shandong).
7	Parabolic dunes	Parabolic dunes are formed primarily by coastal plants, which anchor their trailing arms. These arms are 12–100 m high and approximately 10–20 m wide. In China, they have mainly developed on the coast between the Luanhe Estuary and Yanghe Estuary in eastern Hebei, Muping County-Beihai forest farms in Wendeng City-Shuangdao forest farm in Weihai city, the area in the northern Shandong Peninsula, eastern Changle and Gulei Peninsula coast in southern Fujian, Haimen Bay-Jieshi Bay, eastern Guangdong and Hainan Island coast.
8	Oblique dune ridges	Oblique sand ridges are aeolian in origin and intersect the coastline at an acute angle of 30°~50°. Oblique dune ridges can be up to 100 m long and approximately a few meters to a dozen meters high. They are generally fixed dunes, and in China are concentrated along the coast of the Luanhe Estuary and Dapu Estuary in Changli, eastern Changle in southern Fujian.
9	Transverse dune ridges	These long and straight dune ridges develop behind foredunes on the landward side and are perpendicular to the prevailing winds. Transverse dune ridges (Fig. 12.1(g)) are characterized by constant movement and increased accretion. They typically exceed 10 m in height, reaching approximately 50 m and are 200–300 m wide and 4–5 km long. The transverse dune ridges of the Gold Coast in Changli of Hebei province are typical transverse dune ridges and are the largest coastal aeolian dunes in China. Their relative height can reach above 40 meters.
10	Coastal sand sheets	Coastal sand sheets (Fig. 12.1(h)) are built on the vast and flat or microwave undulating sand accumulation area. They are widely distributed on the plain or mesa, which is farther inland of the trailing edge area along the coastal aeolian dune belt of China. Coastal sand sheets are concentrated on the coasts of the northwest Liaodong Peninsula, eastern Hebei province, northern Shandong Peninsula, southern Fujian province, the Haimen Bay-Jieshi Bay of eastern Guangdong, western Guangdong and Hainan Island.
11	Remnant dunes and blowouts	Common remnant dunes (Fig. 12.1(i)) and blowouts (Fig. 12.1(j)) are formed by wind erosion. However, they display various scales and shapes due to differences in the sediment sources and winds. These formations are widely distributed in aeolian geomorphology areas along China's coasts, such as the Changli coast in eastern Hebei province, the eastern Changle and Pingtan Island of southern Fujian province, Leizhou Peninsula in western Guangdong and the northeast Hainan Island.

composed of incipient foredunes, barchan foredunes, transverse foredune ridges, nebkha dunes, parabolic dunes, oblique dune ridges, barchan dunes, transverse dune ridges, longitudinal dunes, climbing dunes, coastal sand sheets, wind erosion unakas, blowouts and various erosional dune types (Table 12.1 and Fig. 12.1) (Dong, 2000; Fu et al., 1997; Li et al., 1989). Based on an amalgam of position, morphology, stability and morphological characteristics, many diverse types of

Figure 12.1a–j. Pictures of coastal aeolian dune types and landforms in China. (a) incipient foredunes on the Changle coast, Fujian; (b) transverse foredune ridges; (c) barchan dunes; (d) nebkha dunes; (e) longitudinal dunes; (f) climbing dunes, low foredunes and coastal aeolian dunefields; (g) transverse dune ridges; (h) coastal sand sheets; (i) remnant dunes; (j) blowouts (all photographs by Yuxiang Dong, (b), (d), (g), (h), (i)and (j) were taken in Piangtan, Fujian; (c), (e) and (f) were in Changli, Hebei).

Table 12.2. Coastal aeolian dunes classification (Dong, 2006a).

Position	Developmental sequence and vegetation action	Morphological type
Foredunes	Incipient dunes	Incipient foredunes, barchan foredunes, transverse foredune ridges
Backdunes	Stabilized dunes	Shadow dunes, parabolic dunes, oblique dune ridges
	Unstable dunes	Barchan dunes, transverse dune ridges, longitudinal dunes, climbing dunes, coastal sand sheets, remnant dunes and blowouts

coastal aeolian dunes mentioned above are categorized into a coastal wind-sand geomorphologic classification system consisting of two types, three sub-types and 13 basic morphological types (Table 12.1) (Dong, 2006a).

3. Distribution characteristics of coastal aeolian dunes

Macroscopically, China's coastal belts are significantly controlled by the Neocathaysian structural system, a tectonic format that developed in the Mesozoic and Cenozoic and is characterized by an NNE-NE trend. Based on the study of the geological structure, tectonic landform and regional coastal environment, Lei et al. (2014) found that the distribution of sandy beaches in China is controlled by the eastern Meso-Cenozoic tectonic belts and that sand accumulates in the uplift zones along the coast, thereby forming the genetic characteristics of segmented enrichment. Sandy beaches usually occur along low-hills and terraces with submerged valleys (headland-bay). The sediment movement offshore and sand redistribution inshore vary considerably because of the coastal geomorphology, bathymetry and wave climate variation along shore. There are three categories: (1) sand overlaps on headland-bay, (2) sediment deposition on continuous offshore bars, and (3) sand deposition on the wave-dominated delta plains.

Sandy beaches provide a significant sand source for coastal aeolian landforms. Affected by the monsoon climate, China's coastal region has an arid climate in spring and winter, and the winds are strong and sustained. These characteristics provide the necessary dynamic conditions for the development of coastal sand landforms. Coastal aeolian landforms are widely distributed but discrete due to the regional inequality of sand sources (Dong, 2006a). Thus, the dune types are diverse and relatively scattered in distribution (Fig. 12.2), and there are significant regional differences in the dune types and sizes (Dong, 2006a, 2006b). Sandy coastal landforms occur on the flat plain along the coastline (Changli, Hebei province), wide and open-platform coast (northwest part of the Shandong Peninsula), and headland-bay (northern part of the Shandong Peninsula) because of the relatively flat and less undulated regions around the Yellow Sea and Bohai Sea. Conversely, sandy coastal landforms are obstructed by the rocky headland that comprises scattered dunes with a relatively small volume and are confined to southern China. Accordingly, coastal aeolian distribution in China does not follow the zonality law. Dunes are scattered on the coastal shores from northern China's temperate sub-humid regions to southern China's sub-tropics and humid tropical regions. Overall, the distribution characteristics of coastal aeolian dunes are as follows:

3.1 Widely distributed, small, and scattered

China's coastline runs more than 18000 km, from the Yalu River in the north to the Beilun Estuary in the south (Dong, 2006a). Based on the preliminary investigation and statistical data, the coastal aeolian landforms along the coastline of China cover a total area of approximately 3.0×10^5 hm^2, with southeastern China accounting for more than 70.0% of the total dune area in China (2.338×10^5 hm^2) (Wu et al., 1995a). The remaining area is mainly distributed in the Yellow and Bohai Sea area (7.0×10^4 hm^2) (Fu et al., 1997). Coastal aeolian dunes show a zonal distribution, and the distribution area is usually a few hundred to thousands of hectares. Very few dunes can exceed 1.010^4 hm^2.

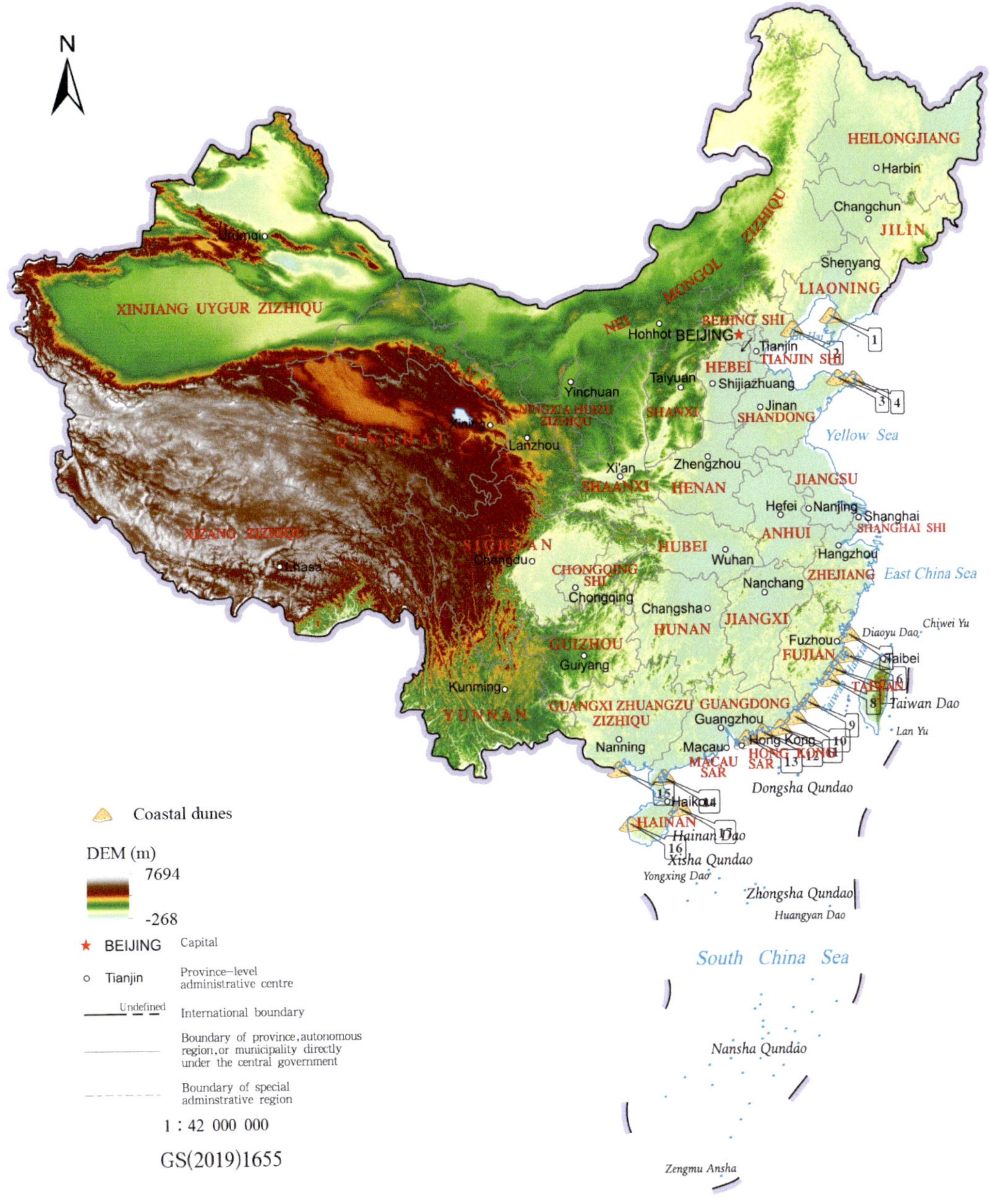

Figure 12.2. Distribution of coastal aeolian dunes in China (Dong, 2006a) 1. Xiongyue-Xianren Island on the northwest coast of Liaodong Peninsula; 2. Changli coast in Hebei province; 3. Taozi Bay in Yantai city, Shandong province; 4. Muping county-Beihai forest farm in Wendeng city-Shuangdao forest farm in Weihai city; 5. Changle coast in the Fujian province; 6. Pingtan county (Haitan Island) in the Fujian province; 7. Kuetou coast in Putian city, Fujian province; 8. Chongwu and Shenhu coast in the Fujian province; 9. Gulei Peninsula and Dongshan Island in Zhangpu County, the southern west coast of the Taiwan Strait; 10. Chaoyang district, Chenghai district in Guangdong province; 11. Shenquan Harbour-Jiazi Harbour-Jieshi Bay of eastern Guangdong, western Guangdong province; 12. Dapeng Bay on Pearl River Estuary in Guangdong province; 13. Xichong Bay in Shenzhen of Guangdong province; 14. Nansan Island, Donghai Island and Naozhou Island in Zhanjiang of Guangdong province; 15. Fengchengqisha coast of Beibu Bay in Guangxi province; 16. Dongfang coast in western Hainan Island; 17. Wenchang coast in northern Hainan Island.

3.2 Distribution topography in the estuary regions

Draping on various topographies is a key characteristic of coastal aeolian dunes. Dune terrains are predominantly divided into sandy coastal plains, which include delta estuaries and coastal plains in the bedrock headland bay (such as sand barriers, bay mouth sandbars, sand spits and tiebars), and bedrock hills and platforms composed of various rocks and other strata. Due to the combined effects of waves and currents, most coastal aeolian dunes are concentrated around the delta plains at river estuaries, and typical large-scale coastal aeolian dunes are generally distributed on the river side of the estuary. For example, the Luanhe River in Hebei province is a sandy river with an annual sediment load of 267010^5 t. It represents the largest dune coast and is 45 km long, 1–4 km wide and 10–40 m high; further, it is located north of the Luanhe and Yanghe Estuary (Dong, 2006a; Li et al., 1987; Wu and Wu, 1990). Minjiang River is the largest river along the Fujian coast, with an annual sediment load of 1019.21^5 t. South of the Changle coastline, the aeolian dune distribution has formed coastal aeolian dunes that are 10 to 20 m high, with several dunes reaching a maximum height of 40 m. In addition to the coastal plain, dunes also occur in low hills, where they reach a distribution height of more than 100 m (Dong, 2006a; Wu and Wu, 1990). And as the coastal areas in the Hanjiang Delta of Guangdong province and the Sigeng town of the southern Changhua River Estuary in western Hainan Island, which are covered by extensive aeolian zones (Wu et al., 1995b).

3.3 Distinguishing regional characteristics and diversity in dune types and sizes

China has a vast territory with twisting and varied coastlines. Owing to the topography controls, wave actions, sand beaches and longshore bars, which usually occur between the headland and bay, a sufficient number of sand sources are formed for dune development at the upper beach. However, large-scale coastal aeolian landforms typically do not show complex structures and usually take the form of incipient foredunes, longitudinal dunes and climbing dunes. The plain coasts are formed by sediment accumulation under the action of rivers, currents and waves, which are distributed in the Liaodong Bay, Bohai Bay, Laizhou Bay, northern Jiangsu, Yangtze River delta and Hainan Island. However, because of the variation in sediments distributed in these areas, including muddy and gravely-sand sediment, coastal aeolian dunes are extremely sporadic. The reasons for sporadic dune formation include the following: first, despite the fact that a significant amount of sediment supply from rivers is transported by river dynamics, currents, and wave actions and subsequently deposited along the Liaodong Bay, Bohai Bay, Laizhou Bay, the Subei and Yangzi Delta and Hainan Island, the sandy beaches contain significant amounts of fine-grained silt and coarse particles (lags, shells), which are not conducive to large-scale coastal aeolian dune formation. Therefore, coastal aeolian dunes rarely occur and are scattered in these regions. Second, coastal aeolian dune types are typical and diverse, and dune volumes are relatively large. They are incipient foredunes, transverse foredune ridges, nebkhas, parabolic dunes, oblique dune ridges, barchan dunes, transverse dune ridges, longitudinal dunes, climbing dunes, sand sheets, remnant dunes and sand blowouts (Dong, 2006a).

4. Forming conditions of coastal aeolian dunes

The aforementioned classification and distribution of coastal aeolian dunes mainly depend on the wind conditions, sediment sources and the synchronous dry season and monsoon (Li et al., 1989). Generally, coasts with an arid climate, powerful wind-force and wide and smooth terrains are favorable for coastal aeolian dune development (Dong, 2006b; Wu, 1991).

4.1 Dynamic conditions

The eastern coastal areas in China are located in the monsoon climate zone. According to the meteorological data in provinces along the coasts during 1960–2018 (Table 12.3), this area is characterized by annual mean temperatures ranging between 8.58°C and 24.06°C, annual mean

Table 12.3. Climatic characteristics of maritime provinces in China during 1960–2018.

Region	Province	Temperature (°C)	Precipitation (mm)	Mean wind speed (m/s)	Maximum wind speed (m/s)	Spring Mean wind speed (m/s)	Summer Mean wind speed (m/s)	Autumn Mean wind speed (m/s)	Winter Mean wind speed (m/s)
The Yellow Sea and the Bohai Sea	Liaoning	8.58	695.05	2.94	32.7	3.54	2.54	2.78	3.05
	Shandong	12.77	687.93	3.24	37.7	3.76	2.89	2.98	3.46
	Hebei	10.39	525.35	2.36	29.7	2.98	2.11	2.06	2.45
	Tianjin	12.78	556.39	3.25	28	3.96	3.03	2.91	3.39
	Zhejiang	17.21	1433.38	3.01	44.7	2.94	2.98	3.03	3.17
	Jiangsu	15.16	1044.49	2.87	26.3	3.25	2.85	2.58	3.06
The southern China	Fujian	18.8	1634.6	2.3	41	2.22	2.25	2.38	2.44
	Guangdong	21.86	1746.08	2.21	45	2.18	2.01	2.24	2.45
	Guangxi	20.85	1567.65	1.83	40	2	1.64	1.69	2.16
	Hainan	24.06	1807.92	2.54	33.7	2.64	2.52	2.44	2.61

precipitation between 525.35 mm to 1807.92 mm, average annual wind speeds between 1.83 m/s to 3.25 m/s and maximum wind speeds from 26.30 m/s to 45.00 m/s. The South China Coast is located in a subtropical and tropical monsoon climate zone, which is affected by cold high pressures from Mongolia and cold snaps in the winter, thereby leading to dry and cold northeast monsoons. Due to the occurrence of summer typhoons in the western Pacific and the China South Sea in the summer, and northeast onshore wind with six months (from October to March of the following year) are dominant in all coasts. This strong and sustained northeast onshore wind is a powerful force for the formation of coastal aeolian dunes in South China (Wu et al., 1995a). The Yellow Sea and the Bohai Sea are located in the temperate monsoon region. Based on an analysis of the relationship between the prevailing wind direction and ridge extend orientation of coastal aeolian dunes and ridges, researchers determined that most dunes and ridges are formed by N-NW winds, which are primarily responsible for coastal aeolian dune formation (Fu et al., 1997). In summary, southerly winds prevail along China's coasts in summer due to strong southerly winds (typhoons) that are usually accompanied by rainfall. Wet sand with moisture cannot be easily transported, resulting in billows and surges from the open sea. Therefore, the medium- and fine-grained sediments are mostly located in the wavefield with billows and surges; further, sediments are out of reach from the wind, and are thus not transported and will not be brought to the nearshore zone. The coastal areas are controlled by the continental air mass during spring, autumn and winter. The strong northeast winds bring drought and cold weather. The monsoons in winter and spring remain dry due to warming, which is conducive to the transportation of sand grains. Meanwhile, the northeast wind arrives after the prevailing westerly waves in autumn, and the amount of fine sand in the sand body is retained on most sandy shores. However, fine sand is susceptible to transport by strong winds that slant against the shore. For coasts that are perpendicular to strong winds, a short blowing opportunity is experienced. In conclusion, the heavy northeasterly winds in autumn and winter are dynamical forces for coastal aeolian activities and the formation of coastal aeolian dunes (Cai and Cai, 1983).

4.2 Foundation of sediment sources

The sediment sources of coastal aeolian dunes mainly originate from sandy materials of coastal sand beaches and sand barriers, including river sediments, coastal erosional sediments, continental shelf sediments and bioclastic sediments. River sediments are the primary source of nearshore sandy sediments. Sandy sediments with different genetic types and morphological characteristics are formed under the combined effect of runoff, tidal current and waves. Broad sandy beaches and marine deposition plains are formed in coastal areas, which provide sufficient sediment sources for dune development. Coastal erosion is the second-most important source of sand. Coastal areas are mostly headland-bay coasts and coastline bedrock and headlands are formed by debris accumulation due to wave erosion, which serves as an important coastal aeolian sand source. Continental shelf sands are residual sand left on the inner continental shelf due to the rise in sea-level and are distributed on the continental shelf around the shallow sea and transported to the shore by the combined action of tides, ocean currents and waves. These sands are an important sediment source for the development of coastal aeolian dunes. Finally, bioclastic sediment is formed by the decomposition of shells by wave action. These sediments are transported to the shore and provide sediment for dune development (Wu et al., 1995a).

The abundance of sand sources plays an important role in developing modern coastal aeolian dunes in China. The abundant sand sources in coastal areas of China have resulted in the development of various coastal aeolian dunes, such as the massive coastal aeolian dunes of Changli, Hebei province, which are formed from the continuous supply of sand by the Luanhe River. In this area, large quantities of sediment are discharged into the sea in summer and transported north under the action of southerly-wind waves. A large amount of sediment is stored in the sand ridges near the shore and is partly transported to the seaside. Especially in autumn, fine sand is transported to the shore in large quantities when winds and waves are weak (Cai and Cai, 1983). Meanwhile, the sandy beaches and longshore bar near the bedrock shores bay provide enough sediment for sand-drift activities. However, sand source and terrain conditions restrict the morphology and size of dunes, which are typically simple in type and small in this area. Currently, there is an inadequate supply of sand in coastal areas of China, and beaches can no longer be relied upon as the primary sand source. For example, the eastern coast of Laizhou Bay and the northern coast of Rongcheng in the Shandong Peninsula are characterized by fine-medium grained sands on several beaches in the summer. However, the calm conditions associated with winter minimize the transport of medium- to fine-grained sands during this season (Cai and Cai, 1983). Additionally, although the northern Jiangsu coast between the Shandong Peninsula and Yangtze River Delta is the largest plain coast in China, this is the only coastal shore without modern coastal aeolian dunes due to its powdery silt composition.

4.3 Topographic conditions

The development of large-scale coastal aeolian dunes is based on wide topographic space, which also plays an important determining role in the scale of coastal aeolian dunes (Wu et al., 1995a). Generally, in northern China (such as the western coast of Liaodong Bay) and the southern region (such as the coast along Fujian and Guangdong), the coastal aeolian dunes are larger in the coastal segments which tends to north-south or east. These characteristics are caused by strong northeasterly winds that do not co-occur to cover the wind field with billows and surges, such as the shores perpendicular to the dominant wind, and a large area within the bank levees is exposed to the wind field and is widely affected by wind. At the same time, some medium- and fine-grained sands do not enter the beach cycle but stay on the slope of the upper levee, providing a reasonable amount of sand source and wind field for wind erosion transport. Thus, these segments have become important areas for coastal aeolian dune development. For example, the Changli coast of Hebei province is located between the Liuyang Estuary and Luanhe Estuary. It contains typical morphological coastal aeolian dunes of the largest size, which vary in width from 150 to 250 m (Dong and Huang, 2013).

5. Coastal aeolian dunes evolution

5.1 Development process

Dating of coastal aeolian dunes in China appears to be difficult. ESR (Electron Spin Resonance) dating experiments indicated that most of the coastal aeolian dunes activities in China occurred since 70000a (Ye et al., 1995). During the formation of coastal aeolian dunes, a series of fluctuation changes were experienced, and the dunes reflect the characteristics of multi-period development. The development stage at 20000–12000 a B.P., subsequent decay stage at 12000–2500 a B.P., development stage from 2500 a B.P. to the current period, and the formation and development of coastal aeolian dunes are consistent with the overall changes in the global environment (Chen et al., 1991). There was a periodic change in dryness and humidity during the past 2500 years with an overall tendency toward dryness, corresponding to a decrease in humidity (Chen and Zhu, 1996). Climate change has been gradually characterized by cyclical changes in wet and dry periods since 2500 a B.P. and an overall drying trend. At the same time, several regional dry and cold periods (400–300 a B.P., 1200–1100 a B.P., 1700–1600 a B.P.) and the coupling effect of dry and cold climate has significantly contributed to the development of coastal sand and wind.

In the temperate coastal areas of northern China, such as the Changli county in Hebei province, coastal aeolian dunes were formed in a dry and cold climate in the low sea surface from 3500 to 2000 a B.P. in the Holocene (Li et al., 1987; Ying et al., 1987). After 202–907 B.C., sand dunes were aggregated and formed from integrated sand bars that had been blown by strong winds for a long period as the sea receded gradually (Gao, 1983). The dating of dunes with a surface thickness of 1 m on the Changli coast shows that the toe of the windward slope of the transverse ridge is approximately 180a, and the ages for the lower middle slope, upper-middle slope and top slope are 59a, 39a, and 13a, respectively. This decreasing trend in age from the bottom to the top indicates the landward accumulation and vertical accretion of dunes by onshore winds. The aeolian activity in the last 25a resulted in the formation of a 1m-thick barchan at the surface, with a non-significant tendency for the barchan to decrease in age from bottom to top, indicating that the barchan accreted in a southwestward direction along the ridgeline and arms with the windward slope top in an erosional environment (Jiang et al., 2016). Additionally, dating of the coastal aeolian dunes in northeastern Panjin city, Liaoning province show that they formed over the last 800 years and display the characteristics of multi-period development, with dune formation mainly influenced by dry and cold climatic conditions and winter monsoons (Wang, 2014).

The accumulation of aeolian sand in the South China Coastal Zone is the result of the cool and dry climatic period of this region since the mid-Holocene, especially in the last 3000 years during the late Holocene (Wu et al., 1995a; Wu et al., 1995b). In recent years, dating experiments of coastal aeolian dunes in the Fujian province (Chen, 1982; Chen and Zhu, 1996; Cheng et al., 2016; Hu et al., 2013; Jin et al., 2015; Wang, 2015; Wang, 1990; Zhang et al., 2007), Guangdong province (Li et al., 1987; Wan et al., 2016; Wu et al., 1995b) and Hainan Island (Wan et al., 2016; Wu et al., 1995a; Wu et al., 1995b; Zhang and Liu, 1987) revealed an earlier measured dune age of 7440 ± 1100 a B.P. and the latest age of 510 ± 90 a B.P. Most of the ages ranged between 3000–1300 a B.P. and can be divided into three phases: Phase I (8000–5000 a B.P.), Phase II (3000–1300 a B.P.), and Phase III (< 700 a B.P.). During these three periods, the development of coastal aeolian dunes corresponds to periods of relatively dry climate and low sea levels during the Holocene (Wu et al., 1995b). Hu et al. (2013) and Jin et al. (2015) established that both the Changle and Jinjiang coastal aeolian dunes formed during these three periods, and the Changle coastal aeolian dunes formed during sea-level decline, which started in 6000 a B.P. The dry and cold climate and strong winter monsoons are essential factors controlling the development of dunes. Periods of high sea levels at certain times were recorded in the Fujian-Guangdong-Taiwan region from 5600–2800 a B.P., 2800–1700 a B.P., and 1700–800 a B.P. From 5600 to 1700 a B.P., beach sediments were formed in higher sea level conditions during the seawater recession stage. A long hiatus occurred between 1700 and 800 a B.P., and between 1700 and 800 a B.P. Aeolian dunes began to develop between 800 and 200 a B.P. with an increase in the intensity of the winter monsoon in the coastal zone and strong aeolian sand activity during the Little Ice Age (LIA). Aeolian sands developed by ~ 200 a B.P. and were closely related to human activities. The coastal aeolian dunes formed 1000 years ago in the Jinjiang River Basin were characterized by strong sand activity during the LIA

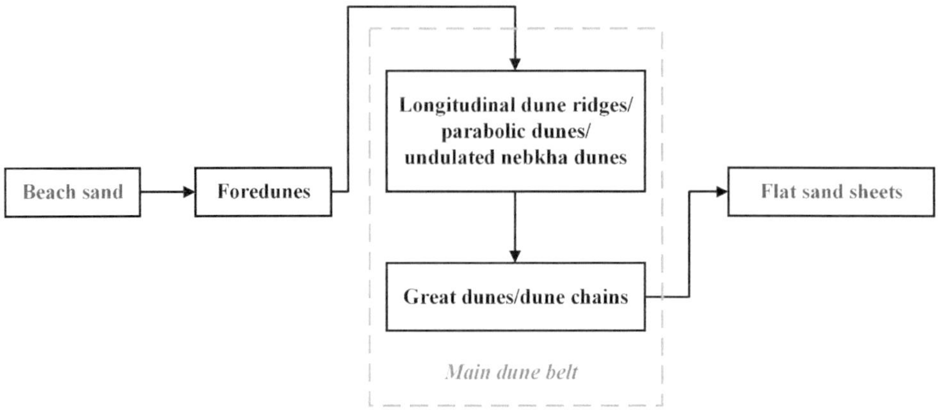

Figure 12.5. Aeolian sand geomorphic development pattern in the open tableland coast and small bay plain coast with adequate sand sources.

sand ridge to a low coppice dune and sand sheet. Therefore, the aeolian geomorphic structures are arranged from the sea to inland in the following sequence pattern: foredunes → aggraded transverse sand ridges → oblique dune ridges → nebkha dunes → flat sand sheets.

2) The aeolian sand geomorphic development pattern in the open tableland coast and small bay plain coast with adequate sand sources (Fig. 12.5). Examples of these development patterns can be seen at the northern coast, such as the platform on the northwest coast of the Shandong Peninsula, the Xiong Yue-Gui Zhou Plain in the Liaodong Peninsula and the small plains of bay or beach ridge terraces in the northern and southeastern coasts of the Shandong Peninsula. The dune development pattern is such that the beach sands are transported to the foreshore by onshore winds and accumulate to form foreshore dunes. During periods of strong wind, the foreshore dunes are eroded, and the strong onshore winds transport sands inland, such as the accumulation of various low dunes and sand mounds in the leading edge of the plateau or small plain, forming taller dune chains or transverse sand ridges at the landward edge of the dune and sand mound belt. The aeolian sands that cross the large dune belt accumulate into sand sheets on flat land. The aeolian sand geomorphic structures are arranged from the sea to inland in the following sequence pattern: foredunes → longitudinal dune ridges or parabolic dunes or undulated nebkha dunes → great dunes or dune chains → flat sand sheets.
3) The aeolian sand geomorphic development pattern on a sandy beach at headland-bay (Fig. 12.6). On the north coast of the Shandong Peninsula and the northwestern coast of the Liaodong Peninsula, the coastal plain is narrow, hills occur close to the sea and strong winds occur in the coastal zone. However, the north wind blocks and slows down in front of the hills. The development pattern of the aeolian landforms is characterized by onshore winds carrying beach sands to the shore. These sands accumulate to form dunes in front of the shore. The strong north wind carries the sands all the way to the front of the hills because the wind speed weakens when the hills block the wind, and the aeolian sand accumulates from tall dunes or dune chains in front of the hills. The resultant formations are medium-sized or longitudinal dunes or undulating aeolian sand sheets across a wider sandy land between the foreshore dunes and the large dune belt. The aeolian sands that cross the dune belt cover the seaward slopes of the dunes, forming sloping hillside wind-forming sandy lands. The aeolian sand geomorphic structures are arranged from the sea to inland as follows: foredunes → longitudinal dune ridges or undulated dune fields with small dunes → parabolic dunes or dune chains → dune ramps.

Jiang et al. (2016) specifically analyzed the sedimentary structure and dune development process of typical transverse sand ridges in the coastal aeolian dune zone of Changli, Hebei, north

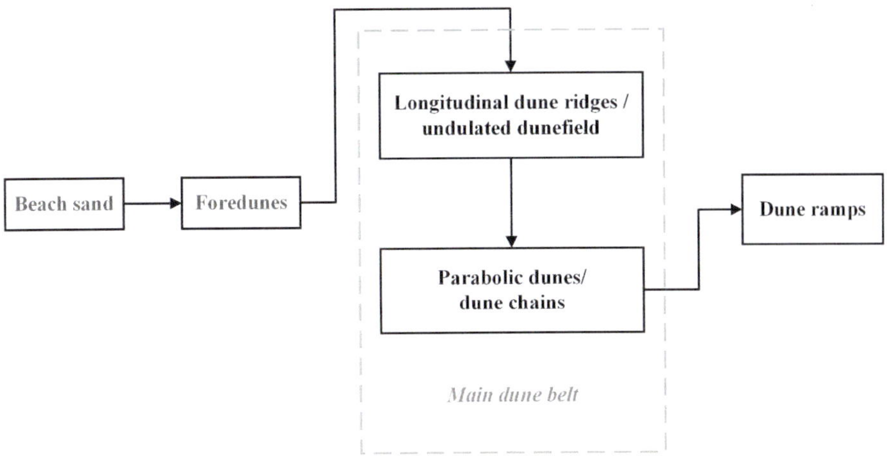

Figure 12.6. Aeolian sand geomorphic development pattern on the sand beach at headland-bay.

of the Luanhe Delta. Their research revealed an evolutionary pattern of coastal sand dam on the outer edge of 'lagoon → coastal foredune → increasing landward accretion'.

6. Coastal aeolian disaster and control in China

6.1 Hazards

Coastal aeolian disasters have existed since ancient times in China. During the late 15th century, the north coast of the Shandong Peninsula experienced severe windblown sand disasters, and large areas of farmland and houses were buried by coastal aeolian dunes (Cai, 1982). During 1850–1950, coastal areas in southern China (Wu et al., 1992; Wu and Wu, 1990; Wu et al., 1995a) and Fujian's Dongshan county experienced the burial of approximately 13 villages and 1300 hm^2 of farmland by aeolian sand. Between 1723 and 1735, 13 villages in the Pingtan county of Fujian province were buried by aeolian sand; during the 1920–50s, Guangdong province were buried by aeolian sand, leaving 2400 people homeless. During the 1929–1949 period, two villages and 70 hm^2 of farmland on Nanshan Island in the Guangdong province were buried by aeolian sand. In 200 years before the 1950s, about 13 villages and 204 hm^2 of farmland in Wenchang county were buried by aeolian sand. Modern coastal sandstorm disasters still exist, and predominately cause silted-up ports, a decline in coastal salt and aquaculture production, farmland desertification, fertility reduction and damage to crops, and the burying of villages, roads and other surface buildings. Overall, destructive characteristics associated with aeolian disasters include local sand lifting and material transportation by wind (Zhang and Yang, 1992).

6.2 Preventive measures

Planting trees is the primary measure that can combat future coastal aeolian disasters in China. Since the 1950s, a large coastal shelterbelt forest area has been established, and vast tracts of coastal moving sand have been fixed, resulting in the significant alleviation of sand damage. For example, a coastal shelterbelt forest was established on the Changli coast in Hebei province during the 1958–1962 period (Gu et al., 1992). To date, approximately 45 km^2 of sand shelterbelt forest has been established, and a large area of moving sand has been fixed, reducing the range and intensity of the dune movement. A coastal shelterbelt forest was established in Hainan's Dongfang county in the 1950s (Wu et al., 1992), and to date, over 10000 hm^2 of shelterbelt forest have been established, stabilizing large areas of coastal aeolian dunes. An investigation of the ecological benefit of coastal shelterbelt forest on Nanshan Island in Guangdong province shows that coastal shelterbelt forest reduce wind, stabilizes the sand, reduce typhoon damage, improve coastal climatic and soil nutrient conditions, increase animal and plant species and optimize the coastal environment (Xie et al., 1998). In response to the status of coastal aeolian disasters in China, essential control strategies include protecting coastal vegetation, perfecting the shelterbelt forest system, the rational development of coastal mineral resources, establishing engineering sand control installations, scientifically planning coastal tourist areas and surface building and optimum-scale development according to specified threshold capacities (Chen and Chen, 1993; Hu, 1999; Wu et al., 1992; Zhang and Yang, 1992).

7. Conclusions

On the basis of the increasing research on the type, distribution, genesis, developmental patterns and evolution of coastal aeolian dunes in China, this chapter presents the results of comprehensive research by integrating coastal aeolian dunes with atmosphere, waves and beaches, starting from the fundamental characteristics that coastal aeolian dunes are formed in the dynamic environment of sea-land-air interaction. The dynamic process of the development of coastal aeolian dunes and the evolution process of coastal aeolian sedimentary environment in China is explained, the following basic conclusions are drawn on the type, distribution, development and evolution:

1) The basic types of coastal aeolian dunes in China include incipient foredunes, transverse foredune ridges, nebkha dunes, parabolic dunes, oblique dune ridges, barchan dunes, transverse dune ridges, longitudinal dunes, climbing dunes, coastal sand sheets, remnant dunes and sand blowout/trough, that can be characterized as foreshore and backshore dunes, and as three subtypes, namely: incipient, stable, and unstable dunes.
2) Coastal sand dunes in China are characterized by the following features: widespread, but small and scattered; diverse topography, but concentrated in estuaries; and large regional differences in type and size.
3) The forming of coastal aeolian dunes in China is mainly influenced by the monsoon climate. Coastal characteristics of China include an arid climate, strong winds, abundant sand sources, flat terrain and open spaces, which are essential for dune development.
4) Modern coastal aeolian dunes in China were formed within the last few thousand years, with development patterns of sandy landscapes on the South China coast, a sand-rich open plain coast, an open tableland coast, a small plain coast of the bay and the headland bay beach shore.
5) China has achieved a certain degree of success in the prevention and control of coastal aeolian sand dune migration since the 1950s. However, because of the great value of coastal sand in terms of embankments, economic use, leisure tourism and cultural science and education, activities to exploit coastal sand have always existed. Thus, it is imperative to carry out coastal sand control and corresponding protection.

Acknowledgements

This work funded by National Natural Science Foundation of China (Grant: 41871006). The authors gratefully acknowledge all the editors for their valuable suggestions for this manuscript.

References

Cai, K. 1982. Wind sand disaster in beach of Shandong. Ocean Science Discussions (2): 68 (in Chinese).

Cai, A. and Cai, Y. 1983. A study of depositional causation and characteristics of coastal aeolian sand in coast areas in China. Journal of Desert Research 3(3): 1–10 (in Chinese).

Chen, C. 1982. Studies of holocene Geochronology in the coastal regions of southern Fujian. Taiwan Strait 1(2): 64–73. doi: CNKI:SUN:TWHX.0.1982-02-008 (in Chinese).

Chen, F. and Zhu, D. 1996. Dune formation and evolution the south coast of Minjiang Estuary. Journal of Desert Research 16: 20–26. doi: CNKI:SUN:ZGSS.0.1996-03-003 (in Chinese).

Chen, M., Chen, J. and Wang, J. 1991. The evolutional trend of swamping, salting and desertization along the coast of bohai sea and its relationship to global change. Quaternary Sciences (2): 113–122. doi: 10.1002/path.1700860204 (in Chinese).

Chen, X. and Chen, J. 1993. Wind-sand disaster in South China coast. The Chinese Journal of Geological Hazard and Control 3(3): 102–104. doi: CNKI:SUN:ZGDH.0.1993-03-015 (in Chinese).
Cheng, Y., Li, Z., Jin, J., Xu, X., Shen, J., Yu, X. and Fan, X. 2016. Luminescence character of Changle coastal dune sand: Environmental significance based on OSL dating Journal of Subtropical Resources and Environment 11(3): 61–71. doi: 10.3969/j.issn.1673-7105.2016.03.008 (in Chinese).
Dong, Y. 2000. Classification of coastal dunes in temperate zone in China. Journal of Desert Research 20(2): 159–165. doi: CNKI:SUN:ZGSS.0.2000-02-011 (in Chinese).
Dong, Y. 2006a. The coastal aeolian geomorphic types and their distribution pattern in China. Marine Geology and Quaternary Geology 26(4): 99–104. doi: CNKI:SUN:HYDZ.0.2006-04-019 (in Chinese).
Dong, Y. 2006b. Coastal aeolian research in China: Progress and prospect. Progress in Geography 25(2): 26–35. doi: 10.11820/dlkxjz.2006.02.004 (in Chinese).
Dong, Y., Hesp, P. A. and Namikas, S.L. 2008. Field measurements of structure of wind-sand flow on coastal transverse ridge in China. Scientia Geographica Sinica 28(4): 507–512. doi: 10.3969/j.issn.1000-0690.2008.04.009 (in Chinese).
Dong, Y. and Ma, J. 2008. Influence of wind velocity on the vertical distribution of different grain size sands in the wind-sand flow on the coastal dune. Acta Scientiarum Naturalium Universitatis SunYatSeni 47(5): 98–103. doi: 10.3321/j.issn:0529-6579.2008.05.021 (in Chinese).
Dong, Y., Namikas, S.L. and Hesp, P.A. 2009. Vertical distribution models of sand transport rate of different grain size group in coastal Aeolian mass flux. Geographical Research 28(5): 1179–1187. doi: 10.3321/j.issn:1000-0585.2009.05.004 (in Chinese).
Dong, Y. 2010. Progress and prospect of research on the surfzone-beach-dune interaction models. Journal of Desert Research 30(4): 796–800. doi:CNKI:SUN:ZGSS.0.2010-04-011 (in Chinese).
Dong, Y. and Huang, D. 2013. Measurements of coastal dune movement in Feicuidao of Changli, Hebei Province. Joural of Desert Research 33(2): 486–492. doi: 10.7522/j.issn.1000-694X.2013.00066 (in Chinese).
Dong, Y. and Du, J. 2014. Review on the typhoon effect on coastal aeolian landform. Journal of Desert Research 34(3): 634–638. doi: 10.7522/j.issn.1000-694X.2013.00366 (in Chinese).
Dong, Y., Huang, D. and Zhang, X. 2016a. Observation of morphological responses of a coastal climbing dune to typhoon. Journal of Desert Research 36(4): 865–870. doi: 10.7522/j.issn.1000-694X.2016.00030 (in Chinese).
Dong, Y., Huang, D. and Zhang, X. 2016b. Preliminary observation morphological responses to typhoon of coastal dunes. Geographical Research 35(10): 1925–1934. doi: 10.11821/dlyj201610011 (in Chinese).
Fu, M., Xu, X. and Xu, X. 1997. The aeolian geomorphical types in the coastal areas of the Yellow Sea and Bohai Sea, and the distribution patterns and developing models. Oceanologia and Limnologia Sinica 28(1): 56–65. doi: CNKI:SUN:HYFZ.0.1997-01-009 (in Chinese).
Gao, S. 1983. The age of the northern coastal dunes at the Luan River mouth. Transactions of Oceanology and Limnologu (3): 56–65. doi: CNKI:SUN:HYFB.0.1983-03-003 (in Chinese).
Gu, J., Sun, L. and Ge, N. 1992. The protection and influence of human activities in Changli natural protection area. Geography and Territorial Research 8(3): 23–26 (in Chinese).
Han, Q., Qu, J. and Zhang, K. 2010. Wind tunnel investigation of wind erosion modulus using tropical humid coastal sands from Southern China. Journal of Soil and Water Conservation 24(1): 37–40. doi: http://ir.casnw.net/handle/362004/11125 (in Chinese).
Hu, F.g., Li, Z. and Jin, J. 2013. Coastal environment evolution record from Anshan coastal aeolian sand of Jinjiang, Fujian Province, based on the OSL dating. Acta Geographica Sinica 68(3): 343–356. doi: 10.3969/j.issn.0375-5444.2013.03.005 (in Chinese).
Hu, J. 1999. Tourism sand dune factor and critical capacity. Geography and Territorial Research 15(1): 61–64. (in Chinese).
Jiang, F., Li, Z. and Jin, J. 2016. Sedimentary structure and development model of typical coastal dunes along the coast of Changli county, Hebei Province. Acta Oceanological Sinical 38(7): 107–116. doi: 10.3969/j.issn.0253-4193.2016.07.010 (in Chinese).
Jin, J. 2017. Chronology and Environmental Significance of Coatal Aeolian Dunes in Fujian, China. Fujian Normal University (in Chinese).
Jin, J., Li, Z. and Hu, F.g. 2015. Mid-Holocene coastal environment and human activities recorded by a coastal dune in Fujian Province, China. Acta Geographica Sinica 70(5): 751–765. doi: 10.11821/dlxb201505007 (in Chinese).
Lei, G., Cai, F., Su, X., Cao, H., Qi, H., Liu, J. and Yu, F. 2014. Tectonic geneses of regional differential distribution and accumulated morphology of Chinese sandy beaches. Journal of Applied Oceanography 33(1): 1–10. doi: 10.3969/J.ISSN.2095-4972.2014.01.001 (in Chinese).
Li, C., Chen, G. and Wang, X. 1987. Study on coastal sand sedimentation at the north bank of Luanhu River. Journal of Desert Research 7(2): 12–21 (in Chinese).
Li, C., Chen, G. and Yang, H. 1989. Eolian deposits in coastal zones of Fujian Province. Journal of Desert Research 9(4): 8–18 (in Chinese).
Li, P., Huang, Z. and Zhang, Z. 1987. Sea level changes since Late Pleistocene in eastern Guangdong. Acta Oceanologica Sinica 9(2): 216–222. doi: CNKI:SUN:SEAC.0.1987-02-010 (in Chinese).

Li, S., Liu, M. and Wang, Y. 1985. Wind-induced dunes along the coast of Shandong Peninsular. Journal of Oceanography of Huanghai and Bohai Sea 3(3): 47–56. doi: CNKI:SUN:HBHH.0.1985-03-008 (in Chinese).
Wan, Y., Zheng, Z., Xiao, Y., Tang, Y. and Chen, C. 2016. Dating and sedimentary environment reconstruction of Holocene beach profile in the Pearl River Estuary Islands. Tropical Geography/Redai Dili 36(3): 388–398. doi: 10.13284/j.cnki.rddl.002851 (in Chinese).
Wang, X. 2014. Coastal Dune in Panjin Formation Process and Its Environmental Response, China University of Geosciences (Beijing). doi: CNKI:CDMD:2.1014.238502 (in Chinese).
Wang, X. 2015. Unveiling Holocene Coastal Aeolian Environment: A Record from the Coastal Dunes at Changle, Fujian. Fujian Normal University (in Chinese).
Wang, Y. and Zhu, D. 1987. An approach on the formation causes of coastal sand dunes. Journal of Desert Research 7(3): 30–40 (in Chinese).
Wang, Y. 1990. The classification of quaternary strata in Fujian Province. Geology of Fujian 009(004): 289–306 (in Chinese).
Wu, Z. and Wu, K. 1987. Sedimentary structure and developing model of coastal dunes along the northeastern coast of Hainan Island, China. Acta Geographica Sinica/Dili Xuebao 42(2): 129–141. doi: 10.11821/xb198702004 (in Chinese).
Wu, Z. and Wu, K. 1990. Progress and problem on research of coastal aeolian erosion in China. Scientia Geographica Sinica 10(3): 230–236 (in Chinese).
Wu, Z. 1991. Research on the landforms of the wind-drift sand in south China coast. Tropical Geomorphology (S1): 95–98. doi: CNKI:SUN:RDDM.0.1991-S1-018 (in Chinese).
Wu, Z., Huang, S. and Hu, S. 1992. Coastal wind-drift sand in Hainan Island, China, and its harnessing measures. Journal of South China Normal University (Natural Science) (2): 104–107. doi: CNKI:SUN:HNSF.0.1992-02-021 (in Chinese).
Wu, Z. 1993. New progress on research of coastal aeolian geomorphology and deposits in China. Advances in Earth Science 8(2): 59–61. doi: CNKI:SUN:DXJZ.0.1993-02-009 (in Chinese).
Wu, Z., Wu, K. and Hu, S. 1995. Research on the Landforms of the Wind-Drift Sand in South China Coast. Beijing: Science Press (in Chinese).
Wu, Z., Wu, K., Huang, S., Zhu, Y., Hu, S. and Jin, Z. 1995. Research on the Holocene coastal dune in South China Coast Science in China (Series B) 25(2): 211–218. doi: CNKI:SUN:JBXK.0.1995-02-014 (in Chinese).
Xie, P., Xie, D. and Zhen, C. 1998. The ecological efficiency of coastal protect tree at Nansan island in Zhanjiang. Journal of Zhanjiang Ocean University/Zhanjiang Haiyang Daxue Xuebao 18(1): 80–82. doi: CNKI:SUN:SHDX.0.1998-01-019 (in Chinese).
Yang, L., Dong, Y. and Du, J. 2017. Research progress of coastal dunes response to storm. Advances in Earth Science 32(7): 716–722. doi: 10.11867/j.issn.1001-8166.2017.07.0716 (in Chinese).
Yang, L., Dong, Y. and Huang, D. 2017. Morphology and grain size of the near surface sediment over coastal sand sheet response to typhoon. Scientia Geographica Sinica 37(8): 1243–1250. doi: 10.13249/j.cnki.sgs.2017.08.015 (in Chinese).
Yang, L., Dong, Y., Huang, D., Huang, S. and Li, J. 2018. Morphological change of coastal transverse dune in Pingtan Island of China after Typhoon Matmo. Journal of Desert Research 38(6): 1136–1143. doi: CNKI:SUN:ZGSS.0.2018-06-003 (in Chinese).
Yang, L., Dong, Y. and Huang, D. 2018. Morphological response of coastal dunes to a group of three typhoons on Pingtan Island, China. Aeolian Research 32: 210–217. doi: 10.1016/j.aeolia.2018.03.009 (in Chinese).
Yang, L., Dong, Y. and Huang, D. 2020. Morphological change of coastal climbing dunes in Pingtan Island of China after typhoon group. Journal of Desert Research 40(2): 1–8. doi: 10.7522/j.issn.1000-694X.2019.00072 (in Chinese).
Ye, Y., Diao, S. and He, J. 1995. ESR dating of the coastal wind sand of China. Oceanologia ET Limnologia Sinica 26(5): 488–493. doi: CNKI:SUN:HYFZ.0.1995-05-005 (in Chinese).
Zhang, J., Yuan, B. and Zhou, L. 2007. Luminescence chronology of "Old Red Sand" in Jinjiang and its implications for optical dating of sediments in South China. Chinese Science Bulletin 53(4): 591–601. doi: 10.3321/j.issn:0023-074x.2007.22.011 (in Chinese).
Zhang, W., Li, Z., Jin, J., Zhang, F., Li, Z., Xu, X. and Cheng, Y. 2020. Optically stimulated luminescence dating of aeolian sand deposits in the Northeast Coast of Hainan Island. Acta Sedimentologica Sinica/Chenji Xuebao, 1–12. doi: 10.14027/j.issn.1000-0550.2020.033 (in Chinese).
Zhang, X., Dong, Y. and Yang, L. 2016. Responses of grain size of near surface layer over coastal dunes to typhoon. Journal of Desert Research 36(6): 1519–1526. doi: 10.7522/j.issn.1000-694X.2016.00072 (in Chinese).
Zhang, Z. and Liu, R. 1987. The HOLOCENE Holocene Along The Coast of Hainan Island. Scientia Geogr Aphica Sinica 7(2): 129–138. doi: CNKI:SUN:DLKX.0.1987-02-003 (in Chinese).
Zhang, Z. and Yang, Y. 1992. Features, origins and disaster reduction countermeasures of modern wind-sand hazards in the northern coast area of Jiaodong Peninsula. Journal of Desert Research 12(4): 34–40. doi: 10.1007/BF02677083 (in Chinese).

Chapter 13

High-Altitude Aeolian Desertification and Sand Dunes on the Tibetan Plateau, China

Yujie Niu,[1,2,*] *Victor R. Squires*[1] and *Limin Hua*[1,*]

1. Introduction

Aeolian Desertification (AD), also referred to as wind-induced or sandy desertification, is a type of land degradation caused by the action of windblown erosion (Wang et al., 2015). It manifests itself by the existence of various aeolian deposits and landforms typically, such as sand dunes (Pye and Tsoar, 2008). Such phenomena usually occur in arid and semiarid regions, which possess both the dry loose surface sediments and strong windblown action. In China, AD is extensively distributed in arid, semiarid and parts of sub-humid regions usually with less than 500 mm average annual rainfall (Fig. 13.1a), and has been studied in depth at low elevations in the north in the provinces of Xinjiang, Inner Mongolia, Gansu, Ningxia, Qinghai, Tibet, etc. (Wang, 2014; Yaojie et al., 2019). However, AD has rarely been studied at the Tibetan Plateau with high elevation (Dong et al., 2017), where the average altitude is about 4500 m (Fig. 13.1b), and the environment is wild, remote and sometimes inhospitable, but has global significance (See Box 13.1).

The AD and shifting sand dunes of the Tibetan Plateau have been around for a long time but are more active recently resulting in the formation of shifting dunes and frequent dust and sand storms (Dong et al., 2017; Fang et al., 2004), because of the comprehensive effect of global change and excessive human activities on the Tibetan Plateau in recent decades. The areas affected by AD and the related wind-induced land degradation (such as the formation of aeolian landforms including sand dunes, mega-ripples and yardangs, etc., and also rangeland degradation) are expanding, further threatening the survival of livestock and herdsmen living on the plateau. From 1950s to 2010s, desertified land and AD have expanded in the agropastoral regions, the semiarid lands bordering the deserts and the basins of inland river in northern China (Wang, 2014). The total area of sandy desertification in 1955 was about 0.137 million km^2 in China, and the desertified area increased over time. By 2010, it was about 0.376 million km^2, which was about 2.74 times greater than the affected area in 1955 (Fig. 13.2).

Researchers of AD began to be concerned about the significance of AD in the early 1990s on some high-altitude places of the Tibetan Plateau, due to the expansion of sand dunes (Dong et al.,

[1] College of Grassland Science, Gansu Agricultural University, Lanzhou, China.
[2] Department of Disturbance Ecology, BayCEER, University of Bayreuth, Bayreuth, Germany.
* Corresponding author: yujie.niu@uni-bayreuth.de; hualm@gsau.edu.cn

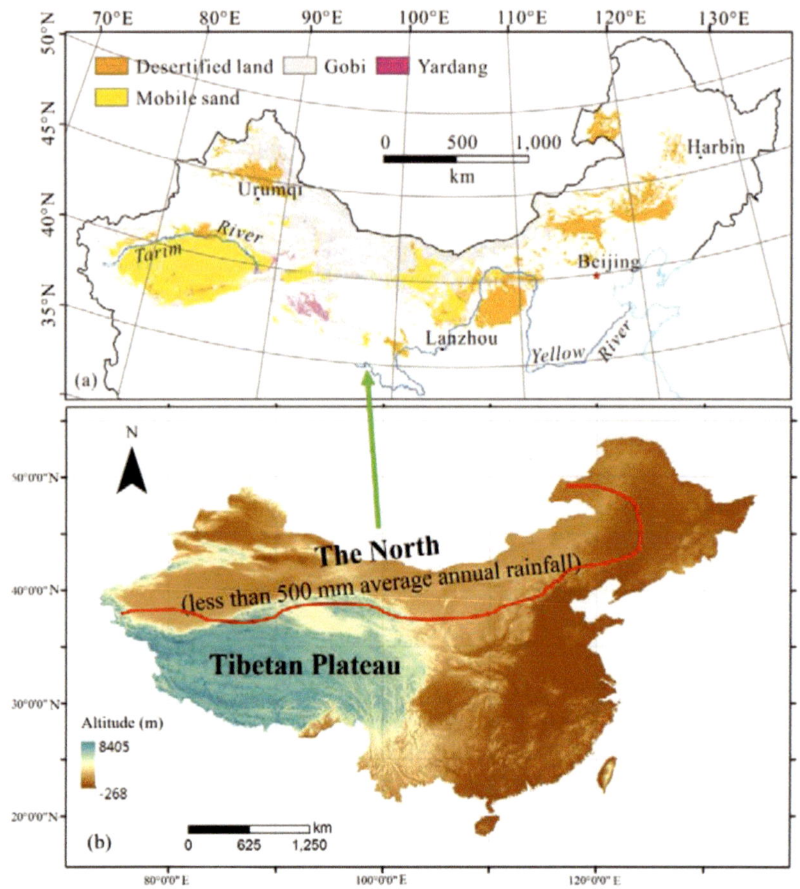

Figure 13.1. Aeolian desertification (AD) and sand dunes have been extensively studied on drylands at low elevations in the north of China, (a) Distribution of AD in northern China, based on Yaojie et al. (2019). (b) The altitudinal distribution in China with the generally low-lying areas of North China having arid and semiarid climatic regimes and the high-altitude Tibetan Plateau.

Box 13.1 The importance of the highest, coldest, and largest plateau in the world.

The Tibetan Plateau, is the world's highest and largest plateau, known as "the roof of the world" due to the high elevation and mountain systems, the "third pole" because of the cold environment and the "Asian water tower" due to the rich river systems (Immerzeel et al., 2010; You et al., 2015). From the Himalaya mountains in the south, north to the Kunlun and Qilian mountains, west to the Pamir and the Karakoram mountains, and the western Qinling mountains and the loess Plateau in east and northeast, it stretches between 26°00'–39°47'N, 73°19'–104°47'E, the area is about 2.5 million km^2. A large part of the plateau lies at elevations above 4,000 to 5,000 meters. The Tibetan Plateau's global significance on the ecology, climate and human systems at local, regional to global scales is widely recognized, providing numerous ecosystem services and functions.

Known as the "Water Tower of Asia" (Immerzeel et al., 2010), the Tibetan Plateau feeds into many famous major Asian river systems (such as the Yellow River, Yangtze River, Lancang River, Nu River, Brahmaputra River, Ganges River, Indus River, Mekong River). With numerous unique and rare animal and plant resources, it is one of the world's important biodiversity hotspots. The unique physical and geographical environment of the Tibetan Plateau also breeds unique alpine ecosystems with the largest area, the most concentrated distribution and most types in the world.

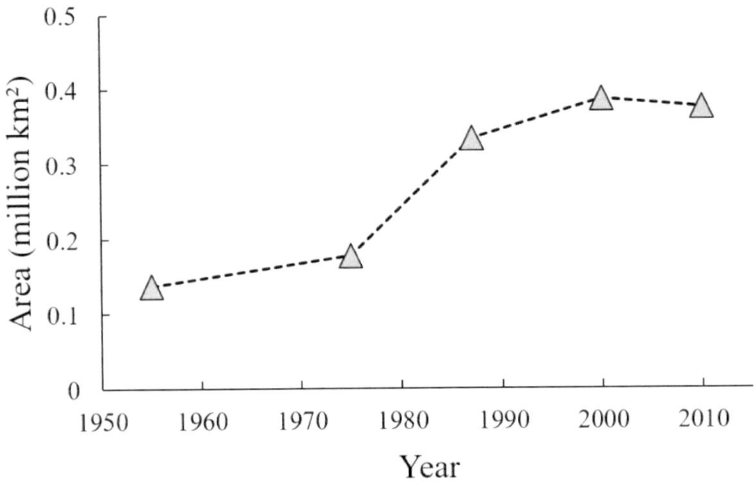

Figure 13.2. The area of AD in China from 1950s to 2010s, data based on Wang (2014).

2017) and the increasing attention to ecological environment by governments. Areas such as the Mugetan in Gonghe Basin (the earliest place to study the Tibetan Plateau deserts) (Jiong et al., 1989), Shiquanhe Basin in the province of Tibet (Li et al., 2016), low reaches of river basins of Zoige Plateau (Dong et al., 2010), and the Yarlung Zangbo River region (Shen et al., 2012) were singled out. So far, the research on the AD and sand dunes on the high-altitude areas of the plateau is still at a pioneering stage (Dong et al., 2017). Greater knowledge of the spatial distribution, evolutionary processes, characteristics and driving forces of AD; types and geomorphology of aeolian sand dunes; and wind-driven degradation of alpine rangelands on the high altitude of the Tibetan Plateau, will improve our understanding of the degradation processes of the alpine environment. It will contribute to knowledge of the high-altitude earth-surface interface, and the impacts of climate change and increased human activities on alpine ecosystems, and further provide valuable inspirations and detailed guidance for ecological restoration in global high-altitude areas.

2. AD on the Tibetan plateau

2.1 Spatial distribution of AD land on the Tibetan plateau

The Tibetan Plateau in China is one of most fragile and sensitive ecosystems in the world and is undergoing severe AD, which is widely distributed on the plateau, and is especially serious in the northwest (Fig. 13.3a). In 2015, the AD land covered an area of 392,914 km^2 (accounting for around 15.1% of the plateau), among which 28,235 km^2 was mobile sandy land (Zhang et al., 2018a). In addition, most of the AD lands were distributed on the western plateau. The distribution of AD lands, from southeast to northwest on the plateau, gradually changes from scattered small patches, to large and concentrated areas with increasingly aggregated desert patches. It is generally thought that there is little AD land in the southeast of the Tibetan Plateau, which is characterized as both high-altitude and a cold-humid climate. Nevertheless, some AD lands are distributed sporadically in the southeast of the plateau on the river flood plains such as in the source region of the Yellow River (Fig. 13.3b). As shown in Fig. 13.3c, there is a typical landscape of AD in the source region of the Yellow River. The desertification threat is becoming more and more serious for human beings living in southeastern plateau (Hu et al., 2011). There are three main distribution regions of desertification in the source area of the Yellow River, namely, the west of the source area of the Yellow River, the northern part of the river source area (Fig. 13.3b), and the eastern part of the river source area. The desertification area is about 22, 042.32 km^2 in the source area of the Yellow River, accounting for the 16.77% of the total area of the source area of the Yellow River. With the increase of land

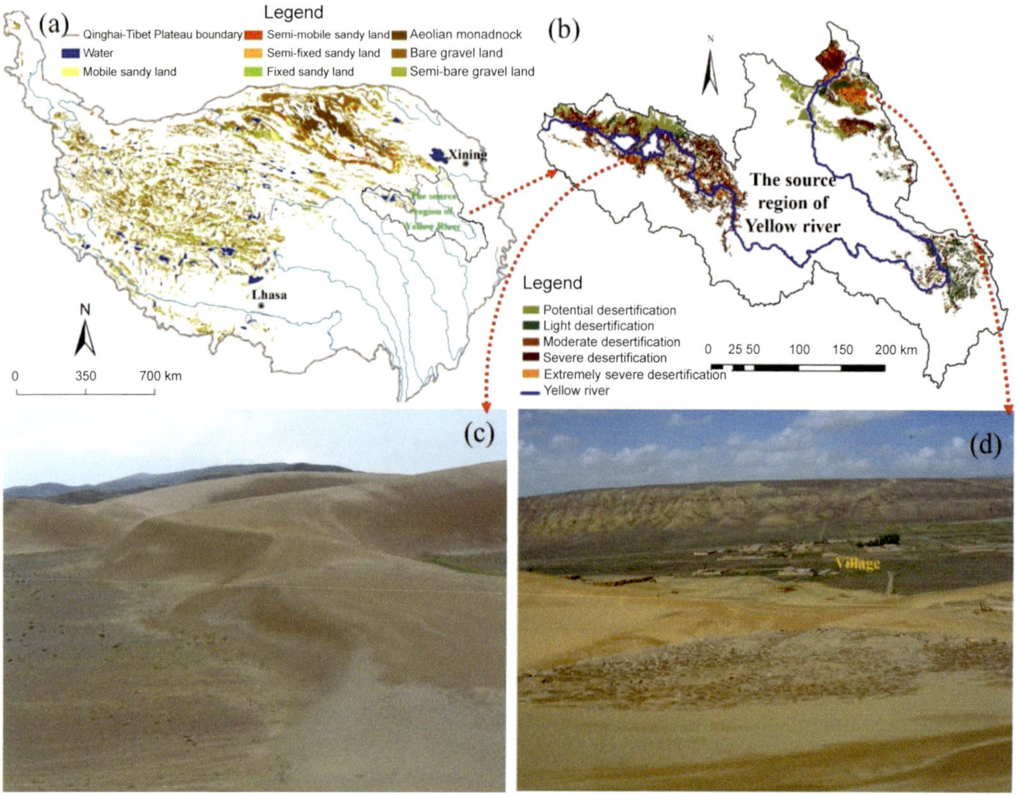

Figure 13.3. (a) Spatial distribution of AD land on the Tibetan Plateau (according Zhang et al., 2018a), and especially, (b) the distribution of desertification land in the source region of the Yellow River, modified after Hu et al. (2011). (c) Typical land scape of AD in the source region of the Yellow River on the Tibetan Plateau. (d) Spreading desertification threatens villages on Guinan county, eastern of Tibetan Plateau. The photographs by Yujie Niu.

desertification, the areas of pasture land are significantly reduced, which further intensifies the contradiction between population and cultivated land and threatens food security. With the desert expanding, many villages have also been threatened (Fig. 13.3d).

2.2 Temporal variation of AD land on the Tibetan plateau

From 1977 to 2015 (Fig. 13.4), the area of all types of AD land increased at first and then decreased, as the area of bare gravel land expanded. By 2000, the total area of AD land dominated on the Tibetan Plateau (Zhang et al., 2018a). From 1977 to 2000, the total area of desertification land (all types combined) expanded by 14948 km^2 on the Tibetan Plateau, which accounted for 3.9% of the total desertified land in 1977. That is, the desertified land area on the Tibetan Plateau increased by 650 km^2 per year from 1977 to 2000 (Fig. 13.4). Moreover, from 1977 to 1990, the AD on the Tibetan Plateau had one of the fastest growth rates with 775 km^2 added per year. However, since 2000, the area of desertification land began to decrease, and the desertified land area decreased by 579 km^2 per year from 2000 to 2015. In 2015, the total area of AD was 392,914 km^2, which accounted for around 15.7% of the Tibetan Plateau (the entire area probably is 2,500,000 km^2); around 52.8% of the desertification land on the Tibetan Plateau was made up of bare gravel land with an area about 207,431 km^2 and semi-bare gravel land was the second largest desertification land type, which accounted for around 26.3% of the desertification land. For sandy AD, the area of mobile sandy land was about 28,235 km^2 amounting for 7.2% of the total desertification land, while semi-fixed sandy land accounted for approximately 5.5% of the total desertification land with an area about 21,572 km^2, and then fixed sandy land had an area of 12,768 km^2.

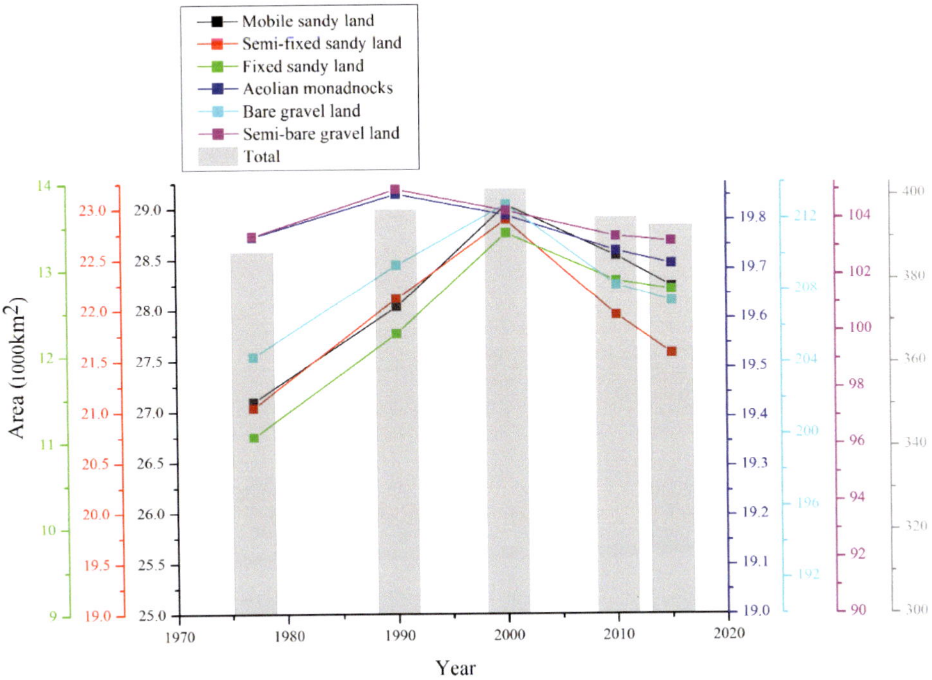

Figure 13.4. Temporal variation of AD land on the Tibetan Plateau, data from Zhang et al. (2018a).

3. Aeolian sand dunes on the Tibetan plateau

3.1 Size characteristics and microstructure of aeolian sands on the Tibetan plateau

Aeolian sand is a layer of sand that is blown and deposited by the wind, mainly in deserts and the Gobi. The aeolian transport of sand particles can be separated into four physical systems roughly according to the particle size (Nickling and McKenna, 2009), long-term suspension with a particle size about < 0.02 mm diameter, short-term suspension with a particle size about 0.02–0.07 mm diameter, saltation with a particle size about 0.07–0.5 mm diameter and reptation and creep with a particle size about > 0.5 mm diameter (Fig. 13.5a). These four aeolian transport modes will be affected by changing wind velocity, sandy soil moisture content, land berries, etc. The particle size of aeolian sand on the Tibetan Plateau mainly fall on the category of saltation (0.07–0.5 mm diameter). As Dong et al. (2017) summarily reported the mean diameter of sand grains on the Tibetan Plateau, ranging from 0.22 mm to 0.24 mm, from 0.14 mm to 0.22 mm, from 0.10 mm–0.16 mm, from 0.16 mm–0.20 mm, from 0.17 mm–0.47 mm, from 0.11 mm–0.14 mm, from 0.12 mm–0.26 mm, respectively, in the shifting dunes of headwaters region of the Yangtze River, in the shifting dunes of Gonghe Basin, in the semifixed dunes of Gonghe Basin, in the shifting dunes of Puruogangri, in the linear dunes of Qaidam Basin, in the shifting dunes of Yarlung Zangbo River valley, in the shifting dunes of the Zoige Plateau. The quartz grain surface features of aeolian sand including surface roughness, roundness and sphericity, etc. (Fig. 13.5b) are terms used to describe the shape of the corners on a particle of sediment and also evaluate the aeolian reworking history, the aeolian transport distance, the frequency and magnitude of wind-storm events in aeolian sands researches (Beal and Shepard, 1956). The Tibetan Plateau's aeolian sand grain is in a different micro-appearance compared to the grains from sand dunes in low-elevations of northern China. The aeolian sand grains on the Tibetan Plateau are poorly rounded with less spherical, and the outlines of the grains are mostly angular (Dong et al., 2017), further implying that the physical rounding effects on the aeolian sand grains of plateau are less than other grains in the deserts at lower elevations in central Asia.

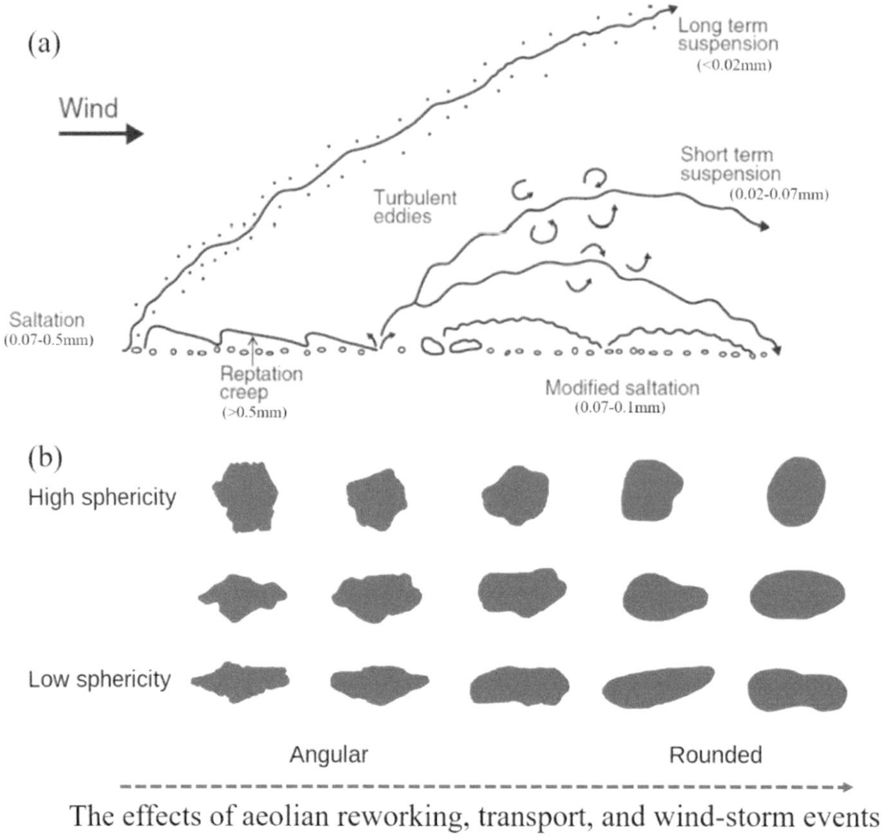

Figure 13.5. (a). Schematic of several physical methods of aeolian particles transport, from Nickling and McKenna (2009). (b). The changes of aeolian sand grains with different surface roughness, roundness and sphericity on the Tibetan Plateau under the effects of physical rounding, modified after Wikimedia Commons.

3.2 The unique aeolian climbing sand dunes on the Tibetan plateau

With the combined effects of erosion, transportation and deposition of sediment, many distinct features (such as, sand dunes, loess deposits, ventifact, yardangs, deflation hollow, desert pavement, etc.), are created by aeolian processes. Sand dunes (both active and fixed by vegetation) are major kinds of aeolian landforms (Zhenda, 1984). Traditionally research on the types and geomorphology of sand dunes tended to concentrate on the shapes and the formation process of sand dunes, that are in the area unaffected by topography. The Tibetan Plateau belongs to an enormous mountain system, which consists of contiguous mountains and hills (Niu et al., 2019a). In nature, the sand dunes are commonly influenced by topography through the changes in wind-blown sand movement. Complicated topography can change the air direction, then accelerate or decelerate the sand flow driven by winds. Aeolian sand dunes accumulated on high-relief areas due to the effects of topography are common in desert regions (Dong et al., 2018). Within the mountain system of the Tibetan Plateau there are particularly unusual sand dunes. These are usually called climbing dunes, as shown in Fig. 13.6. Additionally, as defined in ENCYCLOPEDIA OF PLANETARY LANDFORMS (Chojnacki et al., 2014), *a climbing dune is a high-order bedform with a lee face on the uphill side of a topographic slope > 8°*. Compared with basic sand dunes (See Box 13.2), climbing sand dunes are not stabilized. The study conducted by Dong et al. (2018), using the method in combination with Google Earth digitization and field examinations on the Tibetan Plateau, found that there was about 30,075 km^2 of the total area of climbing dunes; and there were the largest areas

Figure 13.6. The climbing dunes on the alpine rangelands in front of the mountains in the source region of the Yellow River on the Tibetan Plateau, the photograph by Yujie Niu.

Box 13.2 summarizing the classification of six basic sand dunes.

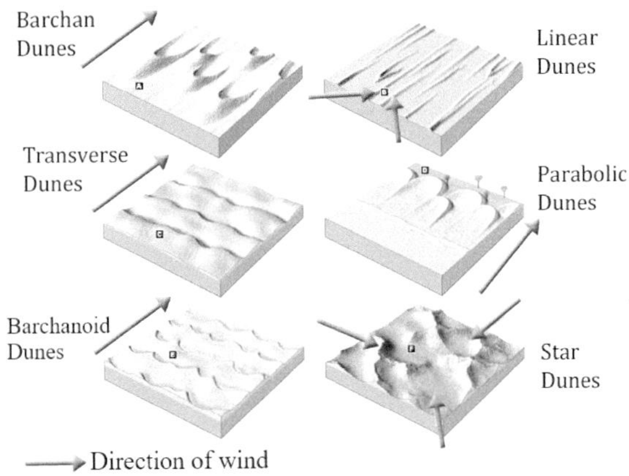

There were six basic types of sand dunes based on the shape and size according to previous research (Kate, 2017), including: barchan/crescentic dunes, linear/longitudinal dunes, transverse dunes, parabolic/blowout dunes, barchanoid dunes (transitional type between barchan dunes to transverse dunes) and star dunes (Box 13.2). Barchan dunes shaped like crescents are solitary sand dunes. They can form where there are limited sand supplies, flat and hard surfaces and lack of vegetation. Linear dunes are formed parallel to the prevailing wind, and the sand supply is moderate here. Transverse dunes shaped like long ridges, are oriented at right angles with prevailing winds. They can be formed in places with abundant sand and sparse vegetation. Parabolic dunes will form when the sands are partially covered by vegetation. Barchanoid dunes are an intermediate form between barchan and horizontal dunes. Star dunes shaped like isolated sand hills will form and develop under variable wind directions.

of climbing dunes in Qinghai, Xinjiang and Tibet. In addition, on the Tibetan Plateau, the climbing dunes make up about 7.5% of all aeolian landforms. Dong et al. (2018) also concluded (based on their research) that the maximum climbing height of the climbing dunes was 1416 m, and the average climbing height was 191 m.

4. The driving forces of aeolian desertification on the Tibetan plateau

Natural factors and human activities are the most important factors for desertification on the Tibetan Plateau (Dong et al., 2017; Xue et al., 2009). For natural factors, climate change (rising temperature, precipitation variation), permafrost degradation, rodent damage and wind play critical roles in desertification. But anthropogenic factors, such as population density, overgrazing, reclamation and road expansion, etc., are considered to accelerate soil erosion and desertification.

4.1 Natural factors

4.1.1 Climate change on the Tibetan plateau

With global warming, the temperature on the Tibetan Plateau is increasing rapidly (Fig. 13.7). From 1960 to 2012, there was an average annual temperature increase of 0.3–0.4°C per decade on the Tibetan Plateau, which was about twice the rate of global warming in the same period (Niu et al., 2019a; Stocker et al., 2013). In addition, the winter warming of the Tibetan Plateau is more obvious. Therefore, the Tibetan Plateau is often referred to as the amplifier of global warming. The variation of the warming trend on the Tibetan Plateau is relatively large, ranging from 0.09°C to 0.74°C every 10 years (Yao et al., 2015). From 1961–1990, the rate of temperature rising on the Tibetan Plateau and its surrounding areas (including China and surrounding countries) grew with the increasing elevation (Liu and Chen, 2000).

Precipitation on the Tibetan Plateau peaks in summer from June–September (Niu et al., 2019a), accounting for 60–90% of the annual precipitation (Xiao et al., 2008). From 1960 to 2012, there was an increasing trend of precipitation on the Tibetan Plateau, with an increase of 2.2% per decade (data from 95 meteorological stations on the Tibetan Plateau and surrounding areas, data source: China

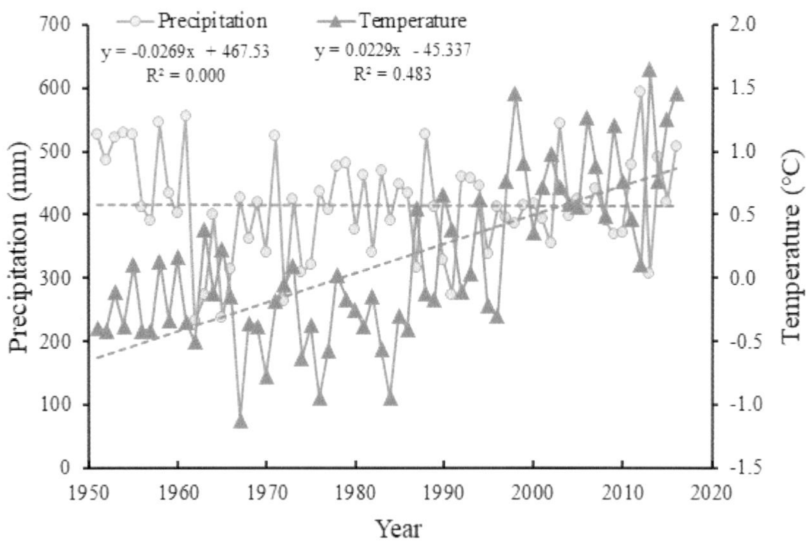

Figure 13.7. Average temperature and precipitation of Tianzhu Tibetan Autonomous County on the Tibetan Plateau from 1951 to 2016 (data from Wushaoling National Meteorological Stations on the Tibetan Plateau, data source: China Meteorological Science data sharing service network). In this alpine rangeland area, there was an increasing rate about 0.23°C per decade based on the average annual temperature from 1951–2016. Modified after Niu et al. 2019(a).

Meteorological Science data sharing service network, Yao et al., 2015). However, the difference between north and south was significant. Precipitation increased in the north and decreased in the south. This discrepancy between the north and the south is closely related to the weakening of the Indian monsoon and the strengthening of the westerlies.

4.1.2 Wind regime on the Tibetan plateau

The Tibetan Plateau is under the combined influence of four modes in space and time, namely the Indian monsoon mode, westerly mode (Yao et al., 2017), winter monsoon mode and east Asian monsoon mode. In addition, the wind regime varies greatly in different seasons affected by various atmospheric modes (Dong et al., 2017). The average wind speed of the Tibetan Plateau increases from east to west in different seasons. The average wind speeds are highest in winter and lowest in summer. In spring, autumn and winter of the Tibetan Plateau, westerly winds dominate. In summer, the northern part of the Tibetan Plateau is controlled by the north wind, and the southern part of the Tibetan Plateau by the south winds. Due to the lowest vegetation coverage in winter, the highest average wind speeds in winter (westerly winds are dominant), and the western part of the Tibetan Plateau as the main desertified area. As the soil freezes in winter, the process of desertification mainly occurs in spring and autumn, this is when the seasonally frozen ground begins to thaw.

4.1.3 Permafrost degradation on the Tibetan plateau

Permafrost is defined as any ground that stays below freezing. The entire Tibetan Plateau is potentially a permafrost distribution area, mainly including permafrost and seasonal frozen ground (Fig. 13.8a). The seasonal frozen ground is largely distributed in the northeast edge of the plateau (Li and Cheng,

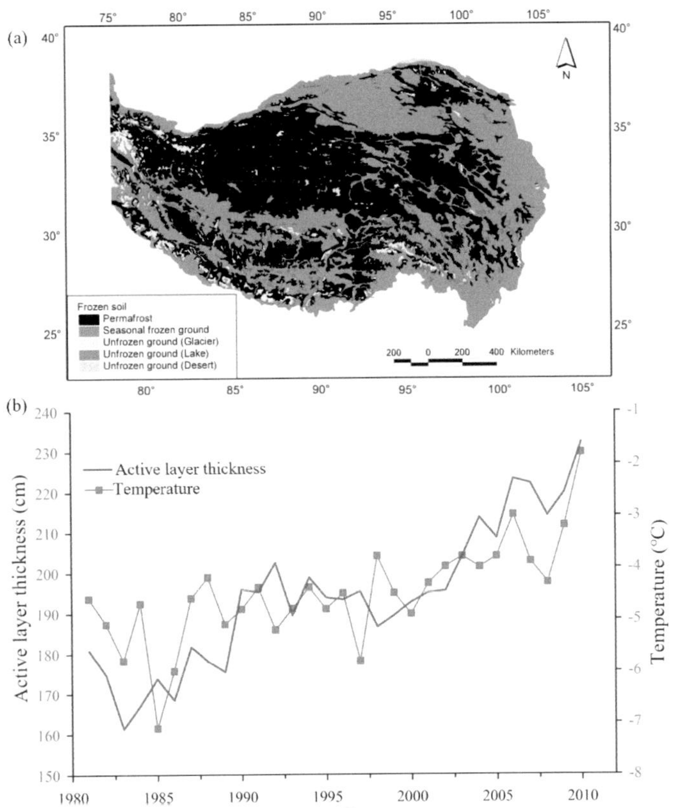

Figure 13.8. (a) The frozen soil distribution (including mainly permafrost and seasonal frozen ground) on the Tibetan Plateau (Li and Cheng 1996). (b) The changes of the ALT (active layer thickness) and temperature of the permafrost area on the Tibetan Plateau from 1981 to 2010 (from Li et al., 2012).

1996). Permafrost distribution is a typical feature of the alpine ecosystem on the Tibetan Plateau, its intact preservation is an important prerequisite to ensure the ecological health of the alpine ecosystem. The main parameters to characterize permafrost include active layer thickness (ALT) and soil temperature. Through field observations along the Qinghai-Tibet Highway on the plateau, Li et al. (2012) observed that the ALT of permafrost showed an obvious increasing trend, due to the climate changes and the effects of frequent human activities. The increase of ALT is significantly correlated with the rising of temperature. The ALT of the permafrost on the Tibetan Plateau thickens at a rate of about 1.33 cm per year from 1981–2010 (Fig. 13.8b). Permafrost degradation not only changes the vegetation ecosystem but also affects the stability of soil, aggravating AD on the Tibetan Plateau.

4.1.4 Rodent damage on the Tibetan plateau

The environmental impacts caused by rodents (plateau pika and plateau zokor) are obvious on the Tibetan Plateau (Fig. 13.9). The rodents graze on the leaves and roots of grasses on the rangelands, and compete for food with livestock, and dig holes and dump soil on the surface and expose it to wind erosion (Niu et al., 2019b). It is believed that rodents caused the rangeland degradation by assuming a direct relationship between rodents and degradation. Pika and zokor has been labeled as "rangeland pests", and the local government and herders have been fighting to control rodents for more than half a century from the 1970s. As a result, rodents have thus become an indefensible scapegoat for rangeland degradation. But there is evidence that rodents are the result of, not the cause of, the degradation of rangelands on the Tibetan Plateau, and that they are key species in maintaining the integrity of the plateau ecosystem (Niu et al., 2020). The outbreaks of rodents are often associated with environmental deterioration and habitat destruction. Rodents are prey for many plateau predator species, have a role in soil nutrient cycling and promoting vegetation succession

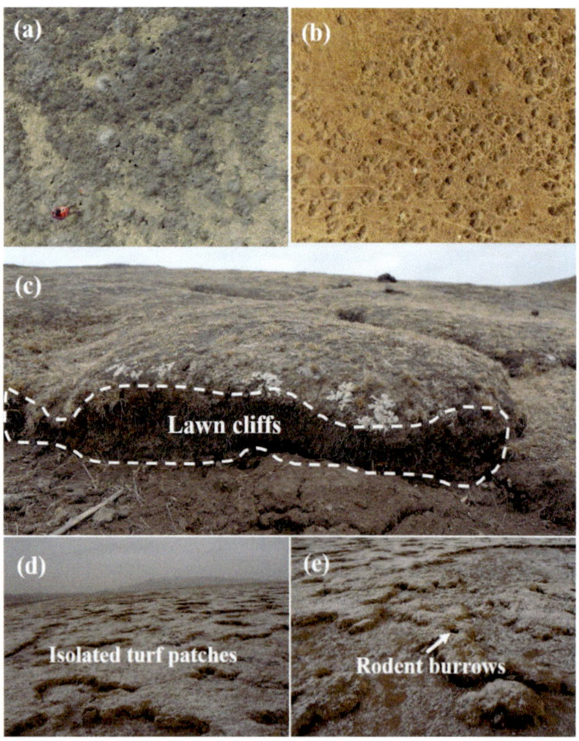

Figure 13.9. Two typical landscapes of rangeland degradation caused by burrowing rodents, which are plateau pika (a) and plateau zokor (b) on the Tibetan Plateau. Pictures (from Niu et al., 2019c) showing lawns cliffs (c) and isolated turf patches (d) and rodent burrows (e) in the degraded alpine rangeland. Photograph by Yujie Niu.

(Niu et al., 2020). However, one needs to be aware, that the tremendous increase of rodents on the degraded rangeland may trigger severe accelerated degradation by undercutting the protective turf (Fig. 13.9), thus exposing rangeland on the Tibetan Plateau to wind-erosion.

5. Anthropogenic factors

5.1 Population and economic growth on the Tibetan plateau

In the primary stage of socio-economic development, especially for developing countries, population and economic growth are usually accompanied by the expansion of the natural resource exploitation, total pollutant emission and the degradation of ecosystems (Thomas, 1956). It seems that economic development inevitably comes at the cost of environmental degradation, then environment degradation becomes the limiting factor of economic development, finally ending in a vicious circle. A large number of people are already vying for limited natural resources. In the first half of the 20th century, the total population of the Tibetan Plateau grew slowly. After the founding of 'New China' in 1949, along with the pro-natalist policy before 1980s, and a significant improvement in the quality of life and medical conditions, the population began to grow rapidly. In Tibet for example, the population had reached 3.076 million in 2012, up from 1.05 million in 1951 (Fig. 13.10a). The economic scale of Tibet is also rising very fast since 1951 (Fig. 13.10b). In 2012, Tibet's GDP reached US$11.104 billion, 142.1 times that of 1951, calculated at comparable prices. The rapid population growth and economic development, accompanied with increasing demand for biofuels are bringing enormous pressure on the natural environment throughout the Tibetan Plateau, or even the world.

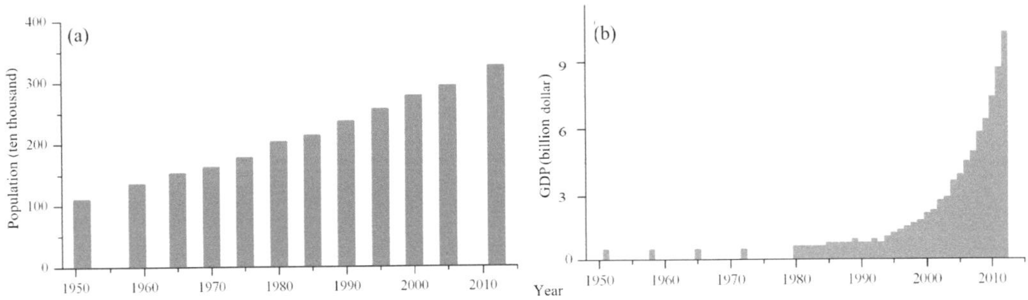

Figure 13.10. The changes of population(a) and GDP(b) in Tibet autonomous region since 1951. Modified after Yao et al. (2015).

5.2 Overgrazing on the Tibetan plateau

Grazing is the most common, convenient and economical way to utilize grasslands but more research on grassland rational utilization and cultivation improvement is required. The key problem is to control the stocking rate in the grassland. China's grasslands are being degraded at an alarming rate, largely because of overgrazing and the mismanagement of grassland resources (Niu et al., 2019c). Over the years, China's grasslands have been suffering from uncontrolled overgrazing, one-sided pursuit of livestock head count and neglect of individual livestock production performance, gradually falling into a vicious circle between grassland degradation and low livestock production performance (Squires et al., 2010). In the past few decades, researchers have shown that the average overstocking rate on rangelands across China is about 34% (Squires and Hua, 2010), and the rangelands in different regions of the Tibetan Plateau were also frequently overgrazed with overstocking rates within the range of 27–529% (Niu et al., 2019c; Zhang et al., 2014). The desertification problems caused by human overuse of natural ecosystems have attracted worldwide attention by ecologists. But until recently research about desertification on the high-altitude areas under overuse was rare. The Tibetan Plateau is a vast alpine rangeland region, where herdsmen live

Figure 13.11. Grazing on the desertified alpine rangelands next to the dune fields on the Tibetan Plateau. Photograph by Yujie Niu.

on grazing, and then livestock rely on natural pasture vegetation for survival (Fig. 13.11). Alpine areas are extremely sensitive to human disturbance due to their thin soil layers and extremely slow processes of soil formation coupled with short growing seasons (3–4 months), Overgrazing has reduced the coverage of the alpine rangelands and increased the risk of soil erosion by rain or winds. In the early stage of overgrazing, interwoven roots can still protect the soil. This layer is gradually eroded by rain, grazing and trampling by livestock, is crushed by vehicles, leaving the substructure (consisting of root and good soil) of rangelands exposed to the air, which is increasingly subject to erosion. As a result, it is difficult for herbage to grow normally and the desertified area gradually expands.

5.3 Over reclamation and road expansion on the Tibetan plateau

In the context of China's reforms and opening-up since 1980s, rapid economic development has led to wanton use of natural resources, in the fragile alpine ecosystems on the Tibetan Plateau. Activities such as conversion of rangeland to cropland, expansion of road networks and mining, etc., have exacerbated the problems (Fig. 13.12). Previous research showed that, since the 1950s, more than 193,000 km^2 of native rangeland has been cultivated, accounting for nearly 5% of the total rangeland in China (Su et al., 2005). In the eight years from 1989 to 1996, 15,460 km^2 of rangeland was still cultivated, mostly in Xinjiang and Inner Mongolia (Miao and Lu, 2008). The basic path of rangeland desertification caused by land conversion is that the best grassland is plowed first, then other vegetation cover is reduced and surface soil is exposed, when increased soil wind and water erosion occurs, finally leading to soil desertification. With the rapid development of China's national economy, the pace of roads and rail construction is increasing. In recent years, the total length of China's railways has been increasing, and the railway plays an increasingly important role in the comprehensive transportation system of the national economy. By the end of 2018, the total length of railways in operation with high-speed (over 250 km per hour) trains on China had exceeded 29,000 km, and the length of roads reached 4,846,531.68 km, ranking the first place among countries and regions in the world. On Tibetan Plateau, roads can damage previously closed and protective turfs of alpine rangelands. The turfs are eroded by water and wind slowly and exposed to overgrazing. Over time, lawns are destroyed, and rangelands develop into a gravel or sand surface slowly.

Figure 13.12. Unregulated mining encroaches into alpine rangelands on the Tibetan Plateau, exposing bare sands, that increases the risk of desertification. Photograph by Yujie Niu.

6. Sandstorms caused by aeolian desertification

Sandstorms emerge when strong winds drag a lot of dust from the ground into the air, making the air so cloudy that the furthest point at which the outline of an object can be seen is less than a kilometer away. The occurrence of sandstorms requires two conditions: first, there is dust, that is, a large area of the Gobi or desert. Second, there is a strong wind, to push the sand and dust drifting force. Extensive desertification and the windy climate on the Tibetan Plateau favor the occurrence of sandstorms that occur on the Tibetan Plateau from December to March of the following year. The center of the sandstorm occurs on the Tibetan Plateau, and the center is moving to the north over time (Ge, 2010). Compared with other areas where sandstorms occur in the north China, the plateau is more likely to lift fine particles such as dust, to an altitude of 5,500 m, and the westerly jet stream at this height is the main natural power for the long-distance transmission of dust from Asia, even to the distant north Pacific. Song (2004) concluded that most of the yellow sand in winter in Japan and South Korea could have come from the Tibetan Plateau, according to statistics analysis from 681 separate meteorological stations from 1961–2000. If desertification is left unchecked on the Tibetan Plateau, it could become one of the world's most important sources of sand in the future.

6.1 Efficiently fixing sand dunes on the Tibetan plateau

Sand dunes on the Tibetan Plateau are small hills of loose particles deposited and built by aeolian processes. The sand mass of dunes can move under the force of wind allowing the dunes to migrate. The usual method to prevent migration is to place barriers to reduce wind velocity and foster sand deposition (Heshmati and Squires, 2013; Qu et al., 2007). Sand barriers can be made with different materials such as straw, stones, clays, branches and also polyacrylamide materials etc. In North China, straws and stones are easily available locally, are often used as the materials of sand barriers. Agricultural straws from wheat, corn and other easy-to-obtain plant stalks are placed in the shape of a checkerboard with squares of various sizes (called straw checkerboard) and half of the straws is buried in the sand and the other half is exposed (Li et al., 2006; Qiu et al., 2004; Heshmati and Squires, 2013). It is effective in fixing sand dunes. Due to its constructive features (decreasing the wind velocity, preventing wind erosion and rapid results) and convenient characteristics (low cost, ease of use, environmentally friendly, etc.), the straw checkerboard technique has been used extensively to prevent desertification and control windblown sand in North China. The straw checkerboard technique was originally used in low-altitude deserts of Northwest China (Qiu et al., 2004; Zhang et al., 2018b), and because of the success in fixing sand dunes, now has become one of the most well-applied and widely-used engineering measures in desert areas throughout the country, including the high-altitude deserts of the Tibetan Plateau, and has also been extended to other places around the world. As Qiu et al. (2004) concluded that the principle of the straw checkerboard technology to fix dunes is that the straws can greatly increase the roughness of the dune surface and reduce the sand flux intensity by 99.5%, and significantly increasing the water and organic matter content of the surface soil, further promoting both the stability of the sand dune surface and the microenvironment (Fang et al., 2007). The sand barriers made by other materials, such as stones, branches etc., can also play a similar role in fixing sand dunes. Stone checkerboard is more common in controlling of desertification on the Tibetan Plateau, as stones are easily available on the plateau (Fig. 13.13). Surface cover over sand is also a common method of sand fixation. The technologies of chemical sand stabilization have also attracted much more attention due to its advantages in easy and fast construction, and improvement of plant growth conditions (Zang et al., 2015). These methods use binding materials to form a crust with sand on the surface of the moving dune to fix the sand, which can also enhance the abilities of water-keeping, and the sandy land nutrition. Nevertheless, there are some problems in application, such as the high cost and the lack of required equipment. Revegetation on the desertified area is also a preferred method worldwide due to the advantages of being aesthetically pleasing, imitates a natural system, self-repairing (Roundy and Biedenbender, 1995). It is proposed here that the checkerboard of living plants consisting of drought-tolerant native herbs or shrubs, in particular, maybe is a more innovative way to preventing desertification. However, suitable seeds of psammophytes and shrub seedlings need to be cultivated first.

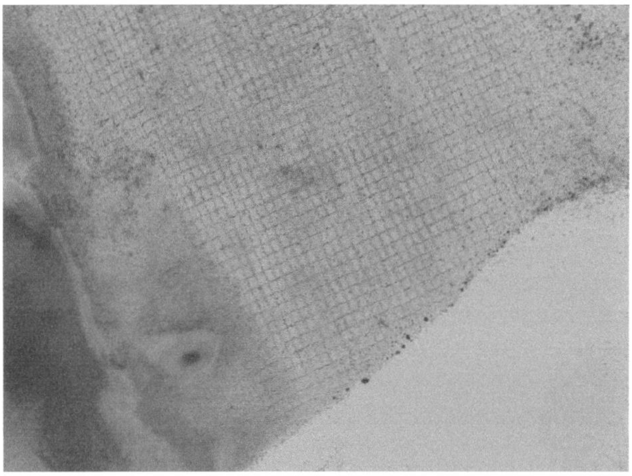

Figure 13.13. The checkerboards by straws or stones are widely used to control the movement of sand dunes in Tibetan Plateau, due to their effective and convenient features of fixing dunes on various desert environments. The photo was taken from unmanned aerial vehicle by Yujie Niu.

7. Conclusion

Aeolian desertification and shifting dunes have been widely studied on the arid and semi-arid regions of low elevations around the world but there is very little work on high elevations, especially the Tibetan plateau, where aeolian dunes have been around for a long time and become more active in recent decades. The Tibetan Plateau in China is undergoing severe AD, which is especially serious in the northwest of the plateau. Nevertheless, as turf eroded and the sandy soil was exposed in the grassland areas, desertification began to spread in eastern plateau, mainly concentrated in the degraded river source areas of alpine meadows and sandy soil steppes. Intensified human activities are the main causes to accelerate soil erosion and desertification of the alpine rangelands on the Tibetan plateau in the context of global warming. Revegetation on the desertified area by checkerboard technique with living plants consisting of drought-tolerant native herbs or shrubs maybe is a more innovative way. Without better prevention efforts, the Tibetan plateau (now a major water source in China and in neighbouring countries in SE Asia and in South Asia) in future could become one of major sand sources in the world.

References

Beal, M.A. and Shepard, F.P. 1956. A use of roundness to determine depositional environments. Journal of Sedimentary Research 26: 49–60.

Chojnacki, M., Hargitai, H. and Kereszturi, Á. 2014. Climbing dune. *In*: Encyclopedia of Planetary Landforms. Springer, New York.

Dong, M., Yan, P., Liu, B., Wu, W., Meng, X., Ji, X., Wang, Y. and Wang, Y. 2018. Distribution patterns and morphological classification of climbing dunes in the Qinghai-Tibet Plateau. Aeolian Research 35: 58–68.

Dong, Z., Hu, G., Yan, C., Wang, W. and Lu, J. 2010. Aeolian desertification and its causes in the Zoige Plateau of China's Qinghai–Tibetan plateau. Environmental Earth Sciences 59: 1731–1740.

Dong, Z., Hu, G., Qian, G., Lu, J., Zhang, Z., Luo, W. and Lyu, P. 2017. High-altitude aeolian research on the Tibetan plateau. Reviews of Geophysics 55: 864–901.

Fang, H.Y., Cai, Q.G., Chen, H. and Li, Q.Y. 2007. Mechanism of formation of physical soil crust in desert soils treated with straw checkerboards. Soil and Tillage Research 93: 222–230.

Fang, X., Han, Y., Ma, J., Song, L., Yang, S. and Zhang, X. 2004. Dust storms and loess accumulation on the Tibetan plateau: a case study of dust event on 4 March 2003 in Lhasa. Chinese Science Bulletin 49: 953–960 (in Chinese).

Fayiah, M., Dong, S.K., Khomera, S.W., Rehman, S.A.U., Yang, M. and Xiao, J. 2020. Status and challenges of qinghai–tibet plateau's grasslands: an analysis of causes, mitigation measures, and way forward. Sustainability 12.

Ge, R. 2010. Sandstorms on the plateau. Chinadialogue. Available at https:// https://www.chinadialogue.net/blog/3941-Sandstorms-on-the-plateau/en (accessed 16 November 2010).

Heshmati, G.A. and Squires, V.R. 2013. Combating Desertification in Asia, Africa and the Middle East. Springer.

Hu, G., Dong, Z., Lu, J. and Yan, C. 2011. Spatial and temporal changes of desertification land and its influence factors in source region of the Yellow River from 1975 to 2005. Journal of Desert Research 5 (in Chinese).

Immerzeel, W.W., Van Beek, L.P. and Bierkens, M.F. 2010. Climate change will affect the Asian water towers. Science 328: 1382–1385.

Jiong, J., Guangrong, D. and Shangyu, G. 1989. Reclamation experience of Mug Shoaly Land on Guinan grassland within potential desertified region of Gonghe Basin. Journal of Desert Research 9: 54–61 (in Chinese).

Kalińska, E. and Nartišs, M. 2014. Pleistocene and Holocene aeolian sediments of different location and geological history: A new insight from rounding and frosting of quartz grains. Quaternary International 328: 311–322.

Kate, M. 2017. What are the different types of sand dunes? Available at https://socratic.org/questions/what-are-the-different-types-of-sand-dunes (accessed 21 March 2016).

Li, Q., Zhang, C., Shen, Y., Jia, W. and Li, J. 2016. Developing trend of aeolian desertification in China's Tibet Autonomous Region from 1977 to 2010. Environmental Earth Sciences 75: 895.

Li, R., Zhao, L., Ding, Y., Wu, T., Xiao, Y., Du, E., Liu, G. and Qiao, Y. 2012. Temporal and spatial variations of the active layer along the Qinghai-Tibet Highway in a permafrost region. Chinese Science Bulletin 57: 4609–4616 (in Chinese).

Li, S. and Cheng, G. 1996. Map of Frozen Ground on Qinghai-Xizang Plateau[Z]. Lanzhou: Gansu Culture Press (in Chinese).

Li, X., Xiao, H., He, M. and Zhang, J. 2006. Sand barriers of straw checkerboards for habitat restoration in extremely arid desert regions. Ecological Engineering 28: 149–157.

Liu, X. and Chen, B. 2000. Climatic warming in the Tibetan plateau during recent decades. International Journal of Climatology: A Journal of the Royal Meteorological Society 20: 1729–1742.

Miao, Y. and Lu, X. 2008. Primary study on the reasons of grassland reclamation of China during the historical period. Pratacultural Science 4 (in Chinese).

Nickling, W.G. and McKenna, N.C. 2009. Aeolian sediment transport. pp. 517–555. Parsons, A. and Abrahams, A.D. (eds.). Geomorphology of Desert Environments (New York: Springer).

Niu, Y., Yang, S., Zhou, J., Chu, B., Ma, S., Zhu, H. and Hua, L. 2019a. Vegetation distribution along mountain environmental gradient predicts shifts in plant community response to climate change in alpine meadow on the Tibetan plateau. Science of The Total Environment 650: 505–514.

Niu, Y., Zhou, J., Yang, S., Chu, B., Zhu, H., Zhang, B., Fang, Q., Tang, Z. and Hua, L. 2019b. Plant diversity is closely related to the density of zokor mounds in three alpine rangelands on the Tibetan plateau. PeerJ 7: e6921.

Niu, Y., Zhu, H., Yang, S., Ma, S., Zhou, J., Chu, B., Hua, R. and Hua, L. 2019c. Overgrazing leads to soil cracking that later triggers the severe degradation of alpine meadows on the Tibetan plateau. Land Degradation & Development 30: 1243–1257.

Niu, Y., Yang, S., Zhu, H., Zhou, J., Chu, B., Ma, S., Hua, R. and Hua, L. 2020. Cyclic formation of zokor mounds promotes plant diversity and renews plant communities in alpine meadows on the Tibetan plateau. Plant and Soil 446: 65–79.

Pye, K. and Tsoar, H. 2008. Aeolian Sand and Sand Dunes. Springer Science & Business Media.

Qiu, G.Y., Lee, I.-B., Shimizu, H., Gao, Y. and Ding, G. 2004. Principles of sand dune fixation with straw checkerboard technology and its effects on the environment. Journal of Arid Environments 56: 449–464.

Qu, J., Zu, R., Zhang, K. and Fang, H. 2007. Field observations on the protective effect of semi-buried checkerboard sand barriers. Geomorphology 88: 193–200.

Roundy, B.A. and Biedenbender, S.H. 1995. Revegetation in the desert grassland. The Desert Grassland 265–303.

Shen, W., Li, H., Sun, M. and Jiang, J. 2012. Dynamics of aeolian sandy land in the Yarlung Zangbo River basin of Tibet, China from 1975 to 2008. Global and Planetary Change 86: 37–44.

Song, L. 2004. Monthly temporal-spatial distribution of sandstorms in China as well as the origin of Kosa in Japan and Korea. Chin. J. Atmos. Sci. 28: 820–827 (in Chinese).

Squires, V., Hua, L., Zhang, D. and Li, G. 2010. Towards Sustainable Use of Rangelands in North-West China. Springer Science & Business Media.

Squires, V. and Lua, L. 2010. Livestock husbandry development and agro-pastoral integration in Gansu and Xinjiang, Towards Sustainable Use of Rangelands in North-West China. Springer, pp. 19–37.

Stocker, T., Qin, D., Plattner, G., Tignor, M., Allen, S., Boschung, J., Nauels, A., Xia, Y., Bex, V. and Midgley, P. 2013. IPCC, 2013: summary for policymakers in climate change 2013: the physical science basis, contribution of working group I to the fifth assessment report of the intergovernmental panel on climate change. Camb. Univ. Press Camb. UKNY NY USA.

Su, H., Liu, G. and He, T. 2005. Reclamation of grassland and its harm. Grassland of China 27: 61–63 (in Chinese).

Thomas, W.L. 1956. Man's Role in Changing the Face of the Earth. Chicago, London, 10–13.

Wang, T. 2014. Aeolian desertification and its control in Northern China. International Soil and Water Conservation Research 2: 34–41.

Wang, T., Xue, X., Zhou, L. and Guo, J. 2015. Combating aeolian desertification in northern China. Land Degradation & Development 26: 118–132.

Xiao, C., Qin, D., Yao, T., Ding, Y., Liu, S., Zhao, L. and Liu, Y. 2008. Progress on observation of cryospheric components and climate-related studies in China. Advances in Atmospheric Sciences 25: 164–180.

Xue, X., Guo, J., Han, B., Sun, Q. and Liu, L. 2009. The effect of climate warming and permafrost thaw on desertification in the Qinghai–Tibetan plateau. Geomorphology 108: 182–190.

Yao, T., Feng, R., Chen, D. et al. 2015. Scientific assessment of environmental changes in the Tibetan plateau. Beijing: Chinese Academy of Sciences (in Chinese).

Yao, T., Piao, S., Shen, M., Gao, J., Yang, W., Zhang, G., Lei, Y., Gao, Y., Zhu, L. and Xu, B. 2017. Chained impacts on modern environment of interaction between westerlies and Indian monsoon on Tibetan plateau. Bulletin of the Chinese Academy of Sciences 32: 976–984 (in Chinese).

Yaojie, Y., Min, L., Lin, W. and A-Xing, Z. 2019. A data-mining-based approach for aeolian desertification susceptibility assessment: A case-study from Northern China. Land Degradation & Development 30: 1968–1983.

You, Q., Min, J., Zhang, W., Pepin, N. and Kang, S. 2015. Comparison of multiple datasets with gridded precipitation observations over the Tibetan plateau. Climate Dynamics 45: 791–806.

Zang, Y.-X., Gong, W., Xie, H., Liu, B.-L. and Chen, H.-L. 2015. Chemical sand stabilization: A review of material, mechanism, and problems. Environmental Technology Reviews 4: 119–132.

Zhang, C.-L., Li, Q., Shen, Y.-P., Zhou, N., Wang, X.-S., Li, J. and Jia, W.-R. 2018a. Monitoring of aeolian desertification on the Qinghai-Tibet Plateau from the 1970s to 2015 using Landsat images. Science of the Total Environment 619: 1648–1659.

Zhang, S., Ding, G., Yu, M., Gao, G., Zhao, Y., Wu, G. and Wang, L. 2018b. Effect of straw checkerboards on wind proofing, sand fixation, and ecological restoration in shifting Sandy Land. International Journal of Environmental Research and Public Health 15: 2184.

Zhang, Y., Zhang, X., Wang, X., Liu, N. and Kan, H. 2014. Establishing the carrying capacity of the grasslands of China: a review. The Rangeland Journal 36: 1–9.

Zhenda, Z. 1984. Aeolian landforms in the Taklimakan Desert, Deserts and Arid Lands. Springer, pp. 133–143.

Chapter 14

Characteristics, Dynamics, and Provenance of Sand Dunes in the Hexi Corridor, China

Bing-Qi Zhu,[1,*] *Jiaxin Zhang,*[1] *Chen Sun,*[1] *Xiaozong Ren,*[2] *Xunming Wang*[1] and *Patrick Rioual*[3]

1. Introduction

Aeolian sediments and their sedimentary strata record the environmental changes in source areas or desert areas and their responses to global climate change and human activities. They are unique and important sedimentary and geomorphological archives of dryland-landscape evolution (Goudie, 2002; Lancaster et al., 2013, 2016; Williams, 2014; Yang et al., 2019). In China, about 566,000 square kilometers of land area are covered by aeolian sand, covering a wide range of geomorphological and tectonic backgrounds ranging from 155 m below sea level to 5,000 m above sea level (Yang, 2006). Desert landscapes dominated by active sand dunes are mainly distributed in arid areas with an average annual precipitation of less than 200 mm, while the sand-land landscapes dominated by semi-active dunes and vegetated dunes mainly appear in semi-arid areas with an average annual precipitation of 200–400 mm (Zhu et al., 1980). The present geomorphology of these sand deserts is the product of long-term and short-term changes of the interaction between endogenic forces (such as tectonic movement) and external forces (such as climate) of the earth system. In turn, these deserts may directly affect the global climate system through sediment circulation (such as dust cycles) (Yang, 2006). Therefore, the understanding of desert landscape evolution will increase our understanding of the earth system.

Regarding the formation and evolution of desert landscapes in China, the loess-paleosol sedimentary sequences from the Loess Plateau indicate that deserts in Northwest China may have existed as early as 22 myr (Guo et al., 2002, 2004; Dong and Wang, this volume), Formation of the megadunes of the Badain Jaran Sand Sea (Dong Zhibao and Wang Ni'ang, this volume) but the geomorphological and sedimentological evidence found inside these deserts indicate that the modern-scale landscapes of these deserts are much younger in age (Yang et al., 2004, 2006). But up to now, due to the lack of adequate and continuous stratigraphic profiles, the geomorphological

[1] Key Laboratory of Water Cycle and Related Land Surface Processes, Institute of Geographic Sciences and Natural Resources Research, Chinese Academy of Sciences, Beijing 100101, China.
[2] School of Geography Science, TaiYuan Normal University, Jinzhong 030619, China.
[3] Key Laboratory of Cenozoic Geology and Environment, Institute of Geology and Geophysics, Chinese Academy of Sciences, Beijing 100029, China.
* Corresponding author: zhubingqi@1gsnrr.ac.cn

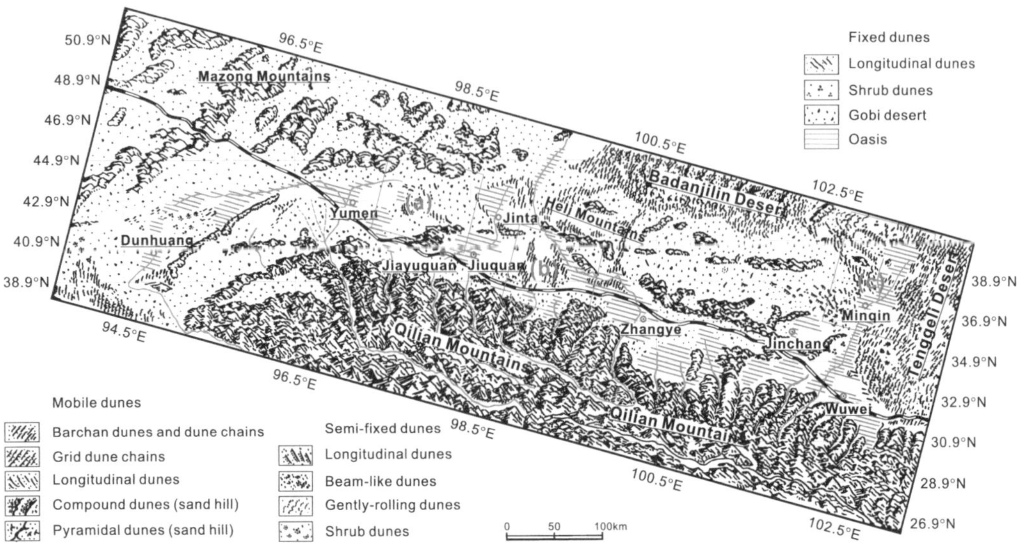

Figure 14.2. Geomorphological map of the Hexi Corridor (modified after Wang, 2003).

have occurred in the past 2 Ka (Zhu and Wang, 1992; Ren et al., 2014) and has accelerated over the past 80 years (Li and Squires, 2009).

In terms of aeolian landforms, sands dunes or dune fields in the Hexi Corridor are mainly distributed in a narrow and long belt between the Qilian Mountains and the Heli Mountains to the west of Wushaoling and the east of Paleo-Yumenguan (Fig. 14.2) (Zhu et al., 1980). Compared with the dune landforms in the adjacent areas of the Badanjilin and Tenggeli Deserts, the sand dunes in the two deserts tend to be convergent in spatial distribution, while the Hexi Corridor is different. Here the dunes are almost scattered, mainly distributed in the vicinity of oases along some rivers, in the oasis or the Gobi Desert areas (Fig. 14.2). From east to west in the dune belt, sand dunes are mainly distributed around the Minqin Oasis in the lower reaches of the Shiyang River, the Zhangye and Gaotai Oases in the middle reaches of Heihe River, the Jiuquan and Jinta Oases in the lower reaches of the Beidahe River, and the Dunhuang Oasis in the lower reaches of the Danghe River (Fig. 14.2) (Zhu et al., 1980).

The total area of dune fields in the Hexi Corridor is about 754 square kilometers, with a large number of big crescent-shaped dunes and chains of crescent-shaped dunes developed on the edges of oases. The Minqin Basin is a typical area exhibiting dune landforms development in the Hexi Corridor. It is located at the lower reaches of the Shiyang River and the western edge of the Tenggeli Desert. The annual average precipitation is about 116.4 mm and the annual average wind speed is about 2.25 m/s. A large number of crescent-shaped sand dunes are distributed on the northwestern edge of the oasis, i.e., to the windward of sand-transport winds in the oasis.

2.2 Data and analytical methods

For the study of sand dune geomorphology, the first method is to use the sample-quadrate survey procedure to measure the height and shape of typical high dunes in the field with a rangefinder, and the second is to measure the length, angle and width of the windward slope and downwind slope of each dune in the sample quadrate and between different quadrates along the local dominant wind direction by using rangefinder and remote sensing image scales (such as Google Earth scales, etc.), thus generating comprehensive geomorphic data of sand dunes in the region. In addition to the geomorphological data of sand dunes themselves, landscape researchers could also use the sample-quadrate survey method to investigate the ecological parameters of vegetation cover in the selected sampling area. For both geomorphological and ecological surveys, sub-scale sample quadrates will be selected from the upper, middle and lower parts of the windward and leeward slopes of each dune. Three quadrates can be selected from the dune slope in the windward and downwind directions of each dune along the local prevailing wind direction and the size of each quadrate can be designed as 5 m × 5 m or smaller (Chang et al., 2016a, 2017; Lang et al., 2017).

In recent years, a number of studies have been systematically carried out in different areas of the Hexi Corridor to investigate the different landform types of widespread sand dunes at a geomorphic unit scale in the field, including the crescent-shaped (barchan) dunes, chains of barchan dunes, pyramid-shaped dunes, parabolic dunes and longitudinal dunes belt. Based on field observations and satellite remote sensing image data in different periods, the geomorphological parameters and characteristics of these dunes have been obtained (Zhang and Dong, 2014; Chang et al., 2016a, 2017; Lang et al., 2017; Zhu et al., 2021, 2022; Zhu, 2022a, 2022b). The geomorphological parameters of part of these dunes in the Hexi Corridor and their comprehensive data are shown in Tables 14.1 and 14.2.

In addition to the above-mentioned field survey and measurement of geomorphic parameter of sand dunes, quantifying the structure of wind-blown sand flow and the movement rate of dunes is also the most direct and effective means to explain the dynamic change of dunes and their geomorphological evolution (Dong et al., 1998; Chen and Liu, 2011; He et al., 2012; Dong and Huang, 2013; Wang et al., 2013; Hu et al., 2016; Mao et al., 2016; Zhu, 2022a). Generally, there are two methods to study the moving velocity of sand dunes, one is early positioning observation (MDCES, 1975; Dong et al., 1998; He et al., 2012) and the other is based on remote sensing images (Chen and Liu, 2011; Dong and Huang, 2013; Mao et al., 2016). Research works based on the both methods have been carried out in dune fields of the Hexi Corridor. Accordingly, in this chapter we will integrate and organize the different observation data of dune movement measurement in the Hexi Corridor, and further discuss the geomorphological evolution of sand dunes in the Hexi Corridor.

The grain size composition and distribution of aeolian sediment is an important indicator to understand the formation and development of sand dunes. This is because the grain size parameters of sand particles can be used not only to distinguish the depositional environment (aeolian, fluvial or lacustrine), but also to identify the movement types (creep, saltation or suspension) of sediments in the transportation process. Therefore, the analysis and study on the granular sedimentology of sand dunes is a basic method to understand the genesis and evolution of dunes in the Hexi Corridor. At present, research works about the grain-size sedimentology of aeolian sediments and related aqueous sediments, such as alluvial and proluvial fans, lacustrine deposits, fluvial deposits, etc., has been widely carried out in the Hexi Corridor (Zhu and Yu, 2014; Zhu et al., 2014; Zhang and Dong, 2015; Zhang et al., 2016; Pan et al., 2019; Zhang et al., 2020; Zhu et al., 2022). We systematically collected and organized the granular evidence, which makes it possible to conduct a comprehensive and comparative study on the dunes in the Hexi Corridor from a perspective of sedimentology.

Erodible clastic sediments as material sources are the fundamental base for the formation of sedimentary landforms (Pettijohn et al., 1972; Taylor and McLennan, 1985). Therefore, identifying the source of wind-transported materials in an arid environment is a prerequisite for understanding the formation of dune landforms (Zhu et al., 1980, 1981; Yang et al., 2012). The analysis of major and trace elements, including rare earth elements, has become a reliable technique for detecting the source of desert sediments (Muhs et al., 1995, 1996; Pease et al., 1998; Honda and Shimizu, 1998; Wolfe et al., 2000; Pease and Tchakerian, 2003; Zimbelman and Williams, 2002; Muhs, 2004; Yang et al., 2007; Zhu and Yang, 2009; Jiang and Yang, 2019). The reason is that for aeolian sediments, the differences in composition and distribution of rare earth elements and other trace elements in different samples/sub-fractions are largely controlled by the parent-rock compositions, as these elements only exist in specific minerals and are difficult to be lost during transportation (Pettijohn et al., 1972; Taylor and McLennan, 1985). In the Hexi Corridor, preliminary results have been achieved in case studies of analyzing the elemental compositions of aeolian sediments using major- and trace-element geochemical methods (e.g., Ren et al., 2014; Pan et al., 2019; Zhang et al., 2020;

Table 14.1. The locations, heights, movement directions, and lengths of dunes in the Hexi Corridor.

Dune type	Dune ID	Geographic location		Height of dunes (m)	Movement direction	Length of the beaches (m)		Source
		Longitude (E)	Latitude (N)		(N-W)	Upwind	Downwind	
Barchan dunes	1	102°55’16”	38°37’52”	9.8	48°	438.5	252	Chang et al., 2016a, 2017
	2	102°55’13”	38°38’00”	11.2	48	163.3	492.7	Chang et al., 2016a, 2017
	3	102°55’05”	38°36’06”	9.5	48°	129.2	163.3	Chang et al., 2016a, 2017
	4	102°55’02”	38°37’51”	3.7	48°	304.2	484.1	Chang et al., 2016a, 2017
	5	102°56’34”	38°32’11”	7.9	45°	271.7	229.4	Chang et al., 2016a, 2017
	6	102°56’43”	38°31’59”	7.6	46°	762.3	430.1	Chang et al., 2016a, 2017
	7	102°54’37”	38°25’47”	3.9	45°	295.9	80.8	Chang et al., 2016a, 2017
	8	102°52’56”	38°25’17”	5.9	87°	42.6	52	Chang et al., 2016a, 2017
	9	98°49’44”	39°57’41”	5	51°	350.4	254.5	Chang et al., 2016a, 2017
	10	98°49’59”	39°58’07”	2.6	54°	222.8	437.9	Chang et al., 2016a, 2017
	11	98°49’18”	40°00’41”	7.2	57°	184.6	197.3	Chang et al., 2016a, 2017
Chains of barchans dunes	12	102°54’53”	38°37’46”	6.4	54°	726.9	752.8	Chang et al., 2016a, 2017
	13	102°55’55”	38°37’48”	5.8	54°	443.4	406.7	Chang et al., 2016a, 2017
	14	102°54’46”	38°37’24”	11.1	50°	794.8	658.4	Chang et al., 2016a, 2017
	15	98°51’17”	39°57’59”	13.8	53°	413.6	361.1	Chang et al., 2016a, 2017
	16	98°51’31”	39°57’31”	8.7	54°	501.8	466.2	Chang et al., 2016a, 2017
	17	98°48’04”	39°58’50”	9.6	53°	554	445	Chang et al., 2016a, 2017
Pyramid dunes	18	94°42’23”	40°05’16”	25.8	SW-NE	/	/	Chang et al., 2016a, 2017
	19	94°42’10”	40°05’14”	90.3	SW-NE	/	/	Chang et al., 2016a, 2017
	20	94°41’47”	40°05’11”	76.6	SW-NE	/	/	Chang et al., 2016a, 2017
	21	94°40’53”	40°05’11”	121.8	SW-NE	/	/	Chang et al., 2016a, 2017
	22	94°40’43”	40°05’09”	114.1	SW-NE	/	/	Chang et al., 2016a, 2017
	23	94°40’12”	40°05’24”	88.9	SW-NE	/	/	Chang et al., 2016a, 2017

Table 14.1 contd. ...

Parabolic dunes	24	102°57'15"	38°36'27"	4.6	/	286.1	35.3	Chang et al., 2016a, 2017
	25	102°57'42"	38°36'26"	4.4	/	228.9	188	Chang et al., 2016a, 2017
	26	102°58'15"	38°36'10"	3.3	/	133.3	198.5	Chang et al., 2016a, 2017
	27	93°59'40"	38°37'08"	3.7	/	396	302.2	Chang et al., 2016a, 2017
	28	98°41'36"	41°35'64"	4.4	/	59.9	0	Chang et al., 2016a, 2017
	29	98°41'20"	40°08'51"	4.1	/	15.7	17.7	Chang et al., 2016a, 2017
Accumulated sand-belts (longitudinal dunes belts)	30	103°12'36"	38°47'57"	15.2	/	70.4	farmland	Chang et al., 2016a, 2017
	31	103°13'30"	38°48'36"	17.1	/	44	farmland	Chang et al., 2016a, 2017
	32	103°32'03"	39°02'12"	18.6	/	811.7	farmland	Chang et al., 2016a, 2017
	33	103°31'29"	39°02'10"	5.6	/	707.7	farmland	Chang et al., 2016a, 2017
	34	103.29'49"	39°02'34"	12.2	/	1557.6	farmland	Chang et al., 2016a, 2017
	35	103°26'19"	39°02'20"	9.4	/	207.1	223.4	Chang et al., 2016a, 2017

Table 14.2. Morphological characteristics of barchans dunes in the Hexi Corridor.

Dune ID	Plots area	Height (m)	Thickness of camponotus (m)	Length (m)	Width (m)	Slope of leeward direction (°)	Forward direction (°)	Beach land length of upwind (m)	Beach land length of downwind (m)	Source
1	Minqin 3-1	9.8	123.3	214.6	202.5	32.6	N48°W	662.9	378	Chang et al., 2017
2	Minqin 3-2	11.2	155.7	294.4	227.4	32.2	N48°W	163.3	491.6	Chang et al., 2017
3	Minqin 3-3	9.5	12.5	197.5	209.8	32.1	N48°W	134.2	176.6	Chang et al., 2017
4	Minqin L	3.7	78.8	86.5	81.8	31.8	N48°W	251.8	367.8	Chang et al., 2017
5	Minqin 9	7.9	90.8	112.5	81.5	32.9	N45°W	214.3	253.3	Chang et al., 2017
6	Minqin 10	7.6	86.7	136.8	64.7	31.5	N46°W	137.9	430.1	Chang et al., 2017
7	Minqin 11	11.6	104.4	183.6	57.3	31.6	N46°W	179.9	243.1	Chang et al., 2017
8	Minqin 12	8.6	64.1	131.4	77.9	30.1	N46°W	135.2	697.3	Chang et al., 2017
9	Jinchang 1	8.4	130.3	197.6	123.2	31.7	N55°W	227.4	372.1	Chang et al., 2017
10	Jingta 1	6	83.7	125.3	104.8	31.5	N63°W	250.4	129.5	Chang et al., 2017
11	Jinta 2	5.3	73.5	104.8	76.6	31.7	N63°W	108.2	419.8	Chang et al., 2017

Zhu et al., 2021), which provide basic data for this study to comprehensively identify the material sources of different dunes in the study area.

3. Results and discussion: geomorphological, sedimentological and geochemical characteristics of sand dunes in the Hexi corridor

3.1 Heights and dynamics of sand dunes in the Hexi corridor

The comprehensive data on the heights of different types of sand dunes widely developed in various areas of the Hexi Corridor, as well as other geomorphic parameters of these dunes, can be found in Table 14.1 and Fig. 14.3.

It can be seen from Table 14.1, Table 14.2 and Fig. 14.3 that different types of dune landforms have a varying average height, maximum height and minimum height in the study area. In summary, (1) the average, maximum and minimum heights of the typical crescent-shape (barchan) dune in the Hexi Corridor are about 6.75 m, 11.20 m, and 2.60 m, respectively. (2) The average, maximum and minimum heights of the typical chains of barchan dunes are about 9.2 m, 13.8 m and 5.8 m, respectively. (3) The average, maximum and minimum heights of the typical pyramid-shaped dune

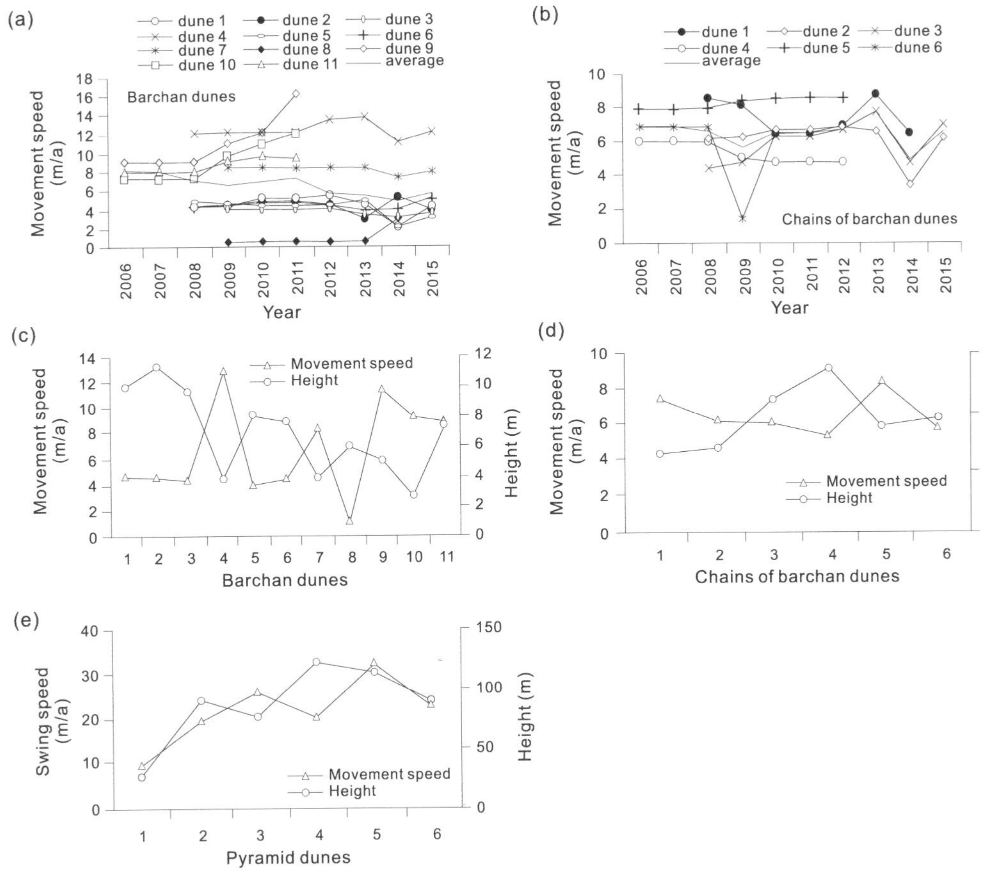

Figure 14.3. The moving speed and height of sand dunes in the Hexi Corridor (modified after Chang et al., 2016a). (a) the moving speed of crescent-shaped (barchan) dunes (see Table 14.1 for geographical locations of the corresponding dune IDs); (b) the moving speed of chains of barchan dunes (see Table 14.1 for geographical locations of the corresponding dune IDs); (c) the moving speed and height of barchans dunes (see Table 14.1 for geographical locations of the corresponding dune IDs); (d) the moving speed and height of chains of barchan dunes (see Table 14.1 for geographical locations of the corresponding dune IDs); (e) the movement speed and height of pyramidal dunes (see Table 14.1 for geographical locations of the corresponding dune IDs).

are about 86.3 m, 121.8 m and 25.8 m, respectively. (4) The average, maximum and minimum heights of the typical parabolic dune are about 4.1 m, 4.6 m and 3.4 m, respectively. (5) The average, maximum and minimum heights of the typical longitudinal dune is about 13.0 m, 18.6 m and 5.6 m, respectively.

Regarding the dynamic changes of sand dune landforms, as early as 1959–1964, the then newly established Minqin Comprehensive Experimental Station of Desertification Control (MCESDC) in China carried out the field positioning observation and research on wind-blown sand flows in the Hexi Corridor (Zhu et al., 1980; Zhu, 1994, 1999; Zhu and Wang, 1992; Wang, 2003). For example, 187 positioning observation points were set up in field along a 20 km long observation line in the Minqin Basin and a large number of observations were made to assess the structure of wind-blown sand flow, the shape of sand dunes, erosion and accumulation of aeolian sand, changes in sand ripples and sand dune movement, etc. Other researchers continued to carry out relevant research on different areas of the Hexi Corridor (MDCES, 1975; Zhu et al., 1980; Wang, 2003; Zhang et al., 2004; Qu et al., 2005; Wang et al., 2013; Yin et al., 2014, 2016; Chang et al., 2016a, 2017; Zhang et al., 2016; An et al., 2019; Chang, 2019; Hu et al., 2020; Zhu et al., 2021). Results of these studies show that in the Hexi Corridor, 80% of the sand particles of the wind-blown sand flows are moving at the height of 20 ~ 30 cm near the ground surface, of which about half of the sand particles are moving at the height of 0.3 ~ 0.5 cm near the surface. At the wind speed of 7 m/s, 75% of the sand particles are within the height of 10 cm, and only 0.035% are within the height of 76 ~ 200 cm. The movement modes of sand particles in the Hexi Corridor include creeping (wriggling/rolling), saltation (jumping/springing) and suspension (floating/levitating) and the movement modes of sand dunes include three ways: straight-forward movement, wigwagging movement and onward-wigwagging movement (Wang, 2003).

Among the different sand dunes in the Hexi Corridor, the crescent-shaped (barchan) dunes, chains of barchan dunes, pyramid-shaped dunes, parabolic dunes and longitudinal dunes have been given the most extensive attention. Based on field surveys/measurements (e.g., measurement of the electronic total station) and satellite image (e.g., Google Earth) data of different periods, researchers obtained the geomorphological parameters and moving speeds of these different dunes (e.g., Ren et al., 2010; Chang et al., 2016a), as shown in Table 14.1 and Fig. 14.3.

It can be seen that the average moving speed of the crescent-shaped dunes is about 6.6 $m^{-1}yr$, the maximum is 12.5 $m^{-1}yr$, and the minimum is only 1.0 $m^{-1}yr$ (Fig. 14.3a). The average moving speed of the chains of barchan dunes is about 6.5 $m^{-1}yr$, the maximum is 8.3 $m^{-1}yr$ and the minimum is only 5.4 $m^{-1}yr$ (Fig. 14.3b). Compared with the crescent-shaped dunes, the chains of barchan dunes move relatively slowly and the movement speed changes little. In general, the crescent-shaped dunes and the chains of barchan dunes in the Hexi Corridor move along the NW-SE direction. The direction of movement of the east part of the corridor is about NW 45°, while the movement angle of the Jinta area in the western corridor is higher (Table 14.1 and Fig. 14.3). The average swing velocity at the tops of the pyramid dunes is about 6.3 $m^{-1}yr$, the maximum is 97.4 $m^{-1}yr$ and the minimum is only 1.1 $m^{-1}yr$. The direction of movement of the pyramid dunes will also change, but its main direction of motion is SW-NE (Fig. 14.3).

The above analytical results indicate that in the dynamic process of different dunes in the Hexi Corridor, the crescent-shaped dunes move the fastest, followed by chains of barchan dunes; only the top of the pyramid dunes wigwags, while the parabolic dunes and the longitudinal dunes hardly move forward; the higher the height of the crescent-shaped dune (or the chain of barchan dunes) is, the slower their movement ; on the contrary, the higher the height of the pyramid dunes, the faster they swing. Analysis also suggests that the moving speed of sand dunes is positively correlated with the average wind speed of sandstorms (Chang et al., 2016a).

Here, the moving speed of sand dunes in the Hexi Corridor with that in other desert areas in Northwest China are compared. For example, the observation results of eight crescent-shaped dunes along the oil transportation highway in the Taklamakan Desert show that the moving speed of sand dunes in October 1991 ~ 1992 is 4.81 ~ 10.87 m, with an average of 7.29 m; the moving speed of sand dunes in October 1992 ~ 1993 is 3.33 ~ 8.89 m, with an average of 5.56 m (Dong et al., 1998). Regarding the analytical results based on high-resolution remote sensing images in 2010 ~ 2014, the average speed of dune movement in the Tenggeli Desert was 4.36 $m^{-1}yr$ in 2010 ~ 2013, and was 2.43 $m^{-1}yr$ in 2013 ~ 2014 (Hu et al., 2016).

Compared with the dynamic process of sand dunes in different deserts mentioned above, the crescent-shaped dunes, that are widely distributed on the edge of oases in the Hexi Corridor, also have obvious spatial differences in dynamic changes. For example, during the period 2006 ~ 2015, due to the influence of wind speed, strong wind days and other factors, sand dunes at the edge of the oasis in the Minqin area moved the fastest, with a moving speed of about 6.2 $m^{-1}yr$; affected by the main wind direction and other factors, sand dunes at the edge of oasis in the Dunhuang area moved the slowest, with a moving speed of about 0.8 $m^{-1}yr$; under the influence of different meteorological factors, the horizontal length of the windward slope of sand dunes in most areas of the Hexi Corridor increased in 2015 compared with 2016, while the horizontal length of the leeward slope decreased. There are three main factors influencing the dynamic changes of sand dunes in the Hexi Corridor, i.e., the annual precipitation, the annual average wind speed and the annual strong wind days, among which the annual precipitation has the largest contribution rate (Shi et al., 2018). It can be concluded that meteorological factors (or climate factors) have an important impact on the dynamic change of sand dunes in the Hexi Corridor, and this impact has both positive (promoting) and negative (inhibiting) effects.

3.2 *Granular sedimentology of sand dunes in the Hexi corridor*

Granular sedimentological analysis of aeolian and aqueous sediments (alluvial and proluvial fans, lacustrine and fluvial deposits, etc.) from the Hexi Corridor have shown that sand dunes have different grain size compositions in various regions of the Hexi Corridor (Zhu and Yu, 2014; Zhu et al., 2014; Zhang and Dong, 2015; Zhang et al., 2016; Pan et al., 2019; Zhang et al., 2020). The spatial differences in grain size compositions reflect the different sources of dune sediments and transportation modes in the formation and development of sand dunes in the Hexi Corridor.

In the Jiuquan Gaotai area in the middle and eastern Hexi Corridor (area B in Fig. 14.2), grain size parameters (including mean, standard deviation, skewness and kurtosis, etc.) of sand dunes at different geomorphological positions on the dune surface, such as the base of windward slope, slope surface, dune crest (top), base of leeward slope were determined (Zhang and Dong, 2015). The granular sedimentology shows that the grain-size frequency cumulative curves of sand dunes in this area are mostly unimodal, and a few are bimodal (Fig. 14.4); the dune surface sediments are mostly fine sand fraction and very fine sand fraction, with an average grain size of 0.07 mm ± 0.01 ~ 0.24 mm ± 0.06, which is similar to the average particle size of sand dunes around the world. The finer the dune particle, the better is the sorting degree. The mean grain size grows with the increase of the skewness values, but decreases with the increase of the kurtosis values. From upwind to downwind, the dune sediment becomes finer, the medium-sand fraction decreases, and the fine-sand fraction, very-fine-sand fraction, silt-fraction and clay-fraction increase; the source materials affect the changes in the average grain size of dune sediment from upwind to downwind (Zhang and Dong, 2015). Based on the Sahu (1964) discriminant function, the sedimentary environment in the middle and eastern part of the Hexi Corridor includes three types of deposits, i.e., aeolian deposit, lacustrine deposit and alluvial-proluvial deposit, of which aeolian deposit is the dominant (about 50%) (Zhang and Dong, 2015). In this dune field, there are three types of grain-size-distribution patterns in a dune-scale unit: the dune crest is coarser (the dune slope and inter-dune area are finer), the dune crest is finer (the dune slope and inter-dune land are coarser), and there is no significant difference between dune crest, windward and leeward slopes. Among them, the coarser-dune-crest model is the most common type, accounting for 69% of all sand dunes, while the finer-dune-crest is the second most common type (accounting for 24%) (Zhang and Dong, 2015).

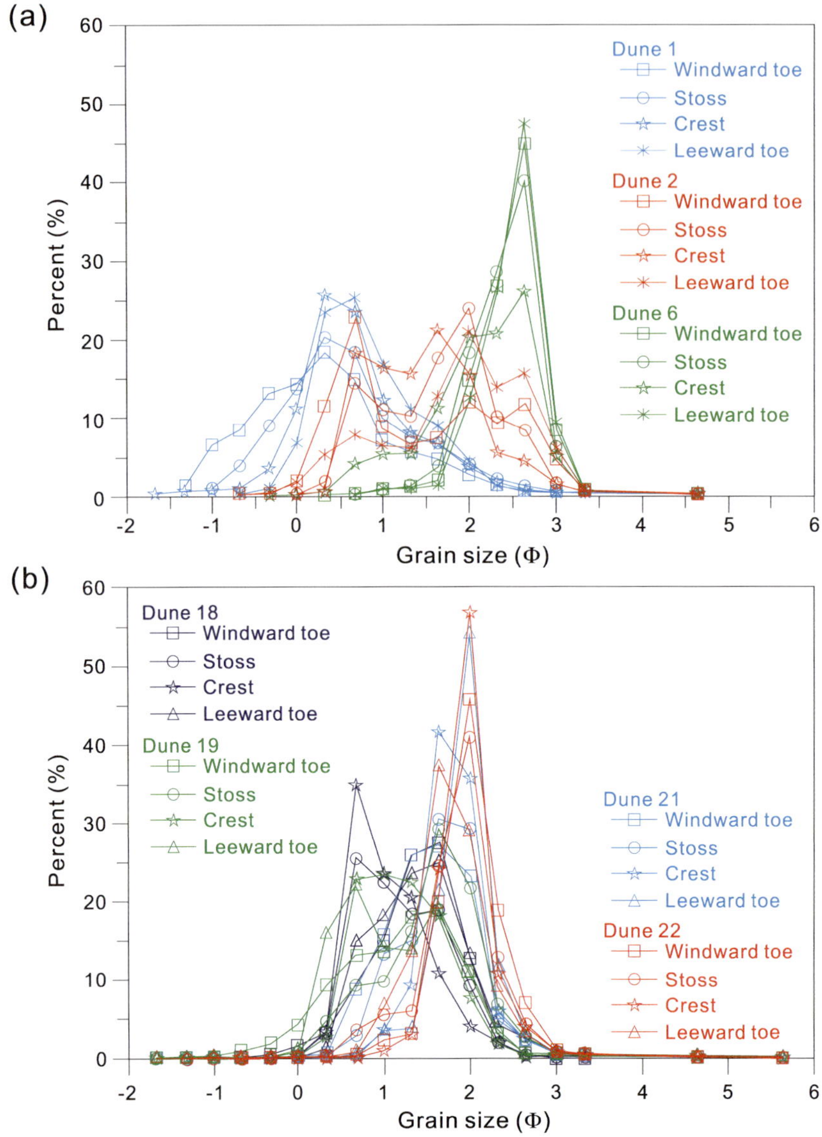

Figure 14.4. The probability cumulative distribution curves of grain size of dune sediments in the Hexi Corridor (modified from Zhang and Dong, 2015).

In the Jinta-Jiayuguan-Huahai area in the western Hexi Corridor (area A in Fig. 14.2), the crescent-shaped (barchan) dunes that developed on the margin of the Gobi Desert and ancient playas have been systematically studied to assess granular sedimentology (Pan et al., 2019). The results of this research show that the grain size of the surface aeolian sediments of crescent-shaped dunes in the western Hexi Corridor is mainly the medium-sand fraction (21.7 ~ 57.4%), followed by the fine-sand fraction (23.2 ~ 53.0%); the mean grain size ranges between 0.27 ~ 0.43 mm (while the paleolacustrine sediment ranges between 0.10 ~ 0.21 mm) (Pan et al., 2019). It is obvious that the average grain size of sand dunes in the Jiaquan-Jinta area of the western Hexi Corridor is larger than that in the Jiuquan-Gaotai area of the central-eastern Hexi Corridor, and also larger than those in other desert regions of northern China. The crescent-shaped dune sediments in this region are mainly medium to good in the sorting level. The frequency cumulative curves of dunes are mostly unimodal and nearly symmetrical, and the kurtosis is medium in level. The granular characteristics of sand dunes in this region are closely linked to their dune morphology and the properties of the underlying Gobi surface (Pan et al., 2019). For example, compared with the crescent-shaped dunes, the top of shrub dunes in this region is dominated by medium- to fine-sand fractions, not the medium-sand fraction. This is possibly because of the wide distribution of gravel around the shrub dune, which reduce the airflow velocity near the ground, making it difficult for the coarser-grained particles to move up to the surface of shrub dunes. While for crescent-shaped dunes, the length of wind flow path on the windward side of the dune is longer. During the long-distance movement of the unsaturated aeolian sand flow, the coarser-grained particles collide with the ground surface elastically, resulting in the coarser-grained particles moving to a higher position on the windward slope of crescent-shaped dune, so crescent-shaped dunes are dominated by the medium-sand-fraction particles. The clay-grain content is less in the surface layer of crescent-shaped dunes and shrub dunes, but more (up to 28.2%) in sand dunes developed on dry lake beds (playas) or ancient lake area, which is relevant to the local source of fine particles.

Generally, the probability cumulative curve of grain size of aeolian sediments is usually shown as 2 ~ 4 independent line segments, indicating that 2 ~ 4 modes of motion occur during their transportation (Visher, 1969). The probability cumulative curve of grain size of sand dunes in the Hexi Corridor basically presents a 3-stage pattern, indicating that the sand dunes here are dominated by the motion mode of saltation under the action of wind, and superimposed with small amounts of particles transported by the motion modes of suspension and creeping. This feature has been observed in sand dunes in different areas of the Hexi Corridor (Zhang and Dong, 2015; Zhang et al., 2016; Pan et al., 2019), including the sand dunes in the lower reaches of the Heihe River Basin near the Hexi Corridor (Zhu et al., 2014). But compared with sand dunes, the movement modes of Gobi sediments and lacustrine sediments are mainly creeping-saltation and saltation-suspension, respectively (Zhu and Yu, 2014; Zhang et al., 2016), indicating that the motion mode of lacustrine sediments and Gobi sediments is different from that of dune sediments. This can also be reflected from the differences in the frequency accumulation curves of dune sediments, Gobi sediments and lacustrine and fluvial sediments. Dune sediments are usually unimodal distribution, while aqueous sediments are usually bimodal or multimodal distribution (Zhu, 2007; Zhu and Yu, 2014).

3.3 Sources of dune sediments in the Hexi corridor

Regarding the source of aeolian sediments, the provenance of sand dunes in the Hexi Corridor was first investigated as early as 1959–1964 by the China's MCESDC (MDCES, 1975; Zhu et al., 1980; Chang, 2019), and work continues to this day (Ferrat et al., 2011; Ren and Wang, 2010; Ren et al., 2014; Zhang and Dong, 2015; Zhang et al., 2016, 2020; Chang, 2019). The analytical results from heavy mineral assemblages indicate that the provenance of sand dunes in the Hexi Corridor is mainly from "sands of in-situ rising" (Zhu et al., 1980; Chang, 2019). Researchers from paleogeography believe that sediments from alluvial plains and alluvial deposits of ancient rivers, lacustrine plain and lacustrine deposits of ancient lakes, aeolian deposits in the erosion zone of the southern and northern foothills of the Hexi Corridor and alluvial deposits of modern rivers are the main material sources of sand dunes in the Hexi Corridor (MDCES, 1975). Based on mineralogical analysis, Ferrat et al. (2011) observed that sand dunes in the Hexi corridor have similar concentrations of mica, quartz, illite, muscovite and albite with adjacent areas, such as the Tenggeli Desert and the Badanjilin Desert. This similar mineralogical composition may lead to the similarity in geochemical characteristics of aeolian sediments in these areas. Recently, grain-size sedimentological and geochemical studies based on detrital materials suggest that alluvial deposits in the Qilian Mountains, paleo-lacustrine deposits in low-level areas and surface aeolian deposits in the Gobi Desert areas are the main material sources for the formation and development of sand dunes in the Hexi Corridor (Wang and Wang, 2013; Zhang and Dong, 2015; Zhang et al., 2016; Zhang et al., 2020).

Due to the application of geochemical methods in recent years, it is possible to identify sediment sources more precisely (Wang, 2011; Wang and Wang, 2013; Ren et al., 2014; Pan et al., 2019; Zhang et al., 2020). In this study, based on geochemical evidences, we took the Gobi areas in the Huahai-Jiayuguan-Jinta region of the West Hexi Corridor (area A in Fig. 14.2), the Jinta-Gaotai region of the middle Hexi Corridor (area B in Fig. 14.2) and the Minqin Basin in the East Hexi Corridor (area C in Fig. 14.2) as the case examples (Ren et al., 2014; Pan et al., 2019; Zhang et al., 2020) to explore the material sources of sand dunes developed in these regions.

The Minqin Basin is dominated by an oasis landscape. It is located in the East Hexi Corridor (area C in Fig. 14.2), and the northern edge of the Loess Plateau, and is bordered by the Badanjilin Desert to the northwest and the Tenggeli Desert to the southeast (Figs. 14.1 and 14.2). The Minqin Basin is considered a natural obstacle to the convergence of the two deserts (Zhu et al., 1980). Geographically and geomorphologically, identifying the origin and transportation of aeolian sediments in the Minqin Oasis and its adjacent desert areas will help to better understand the relationship between the loess and desert in China (Liu, 1985; Sun, 2002; Yang et al., 2007; Yang et al., 2011; Ren et al., 2014, Lu et al., this volume).

The research of Ren et al. (2014; Ren, 2010; Ren and Wang, 2010) and others systematically collected aeolian sediment samples from sand dunes in the Minqin Oasis and its surrounding desert areas. Through geochemical analysis, combined with wind data and cluster analysis methods, the characteristics of compositions and spatial distributions of major and trace elements of aeolian samples from the Minqin Oasis and its adjacent deserts (the Badanjilin and Tenggeli Deserts), as well as the provenance and transportation pathways of aeolian sediments in these areas, were described. The analysis of geochemical data shows that in the bulk (whole-rock) samples of sand dunes in the Minqin Basin (M) and its surrounding areas (the Badanjilin Desert B, the Badanjilin-Minqin transition zone BM, the Tenggeli-Minqin transition zone (TM), the northeast edge of the Tenggeli Desert TNE, the southwest edge of the Tenggeli Desert TSW), the contents of major elements are higher in the content of SiO_2, reaching between 72.2 and 88.9% with an average of about 83.3%. In contrast, the contents of most trace elements are relatively low, and only the contents of Ba, Ce, Co, Mn and Sr reach > 100 ppm (Ren, 2010; Ren et al., 2014). Compared with the average composition of the Upper Continental Crust (UCC), Taylor and McLennan (1985), the concentrations of Ba, SiO_2, Rb, Sr, Al_2O_3 and K_2O in the Minqin Basin and its surrounding areas are relatively uniform (Fig. 14.5), indicating that the spatial differences of these elements in abundance are small and they are relatively homogeneous in the study area, while obvious convex and concave shapes are observed for other elements (Fig. 14.5), indicating that the spatial differences of these elements in abundance are large and they are relatively heterogeneous in the study area. The homogeneous and heterogeneous characteristics between different elements thus can be used as geochemical indicators to identify different sources of sediments in the study area. For the major elements' compositions, only SiO_2 is enriched relative to the upper continental crust (UCC), and the others are relatively depleted (Fig. 14.5). For the trace elements' abundance, most elements are depleted relative to UCC, except for Cr and Ni enriched in B and BM area and Cr enriched in TNE area. The binary and ternary diagrams of some major and trace elements and their ratios, such as Cr, Ni, Cr/V, Y/Ni, Al, V, Zr, Hf, Zr/Hf, reveal that sand dunes have different material sources between the western part of the Minqin Basin (including sub-area B, BM and TNE) and the southeast side of the Minqin Basin (TSW), while sand dunes in the Minqin Basin (M) and the Tenggeli-Minqin transition zone (TM) are related to the two big deserts, respectively (Ren et al., 2014). This geochemical evidence indicates that aeolian sediments from the Badanjilin Desert can be transported to the west side of the Minqin Oasis through mountains (Yabulai) for a long distance by the northwest wind, and aeolian sediments from the Tenggeli Desert provide sediment source to the east side of the Minqin Oasis. However, the aeolian sediments of the two deserts cannot reach or bypass the other side of the oasis (that is, not crossing the oasis barrier to achieve a confluence of the two deserts). This reveals that the Minqin Oasis is an effective barrier to prevent the migration and convergence of sand dunes between the Badanjilin and Tenggeli Deserts. However, the large number of aeolian

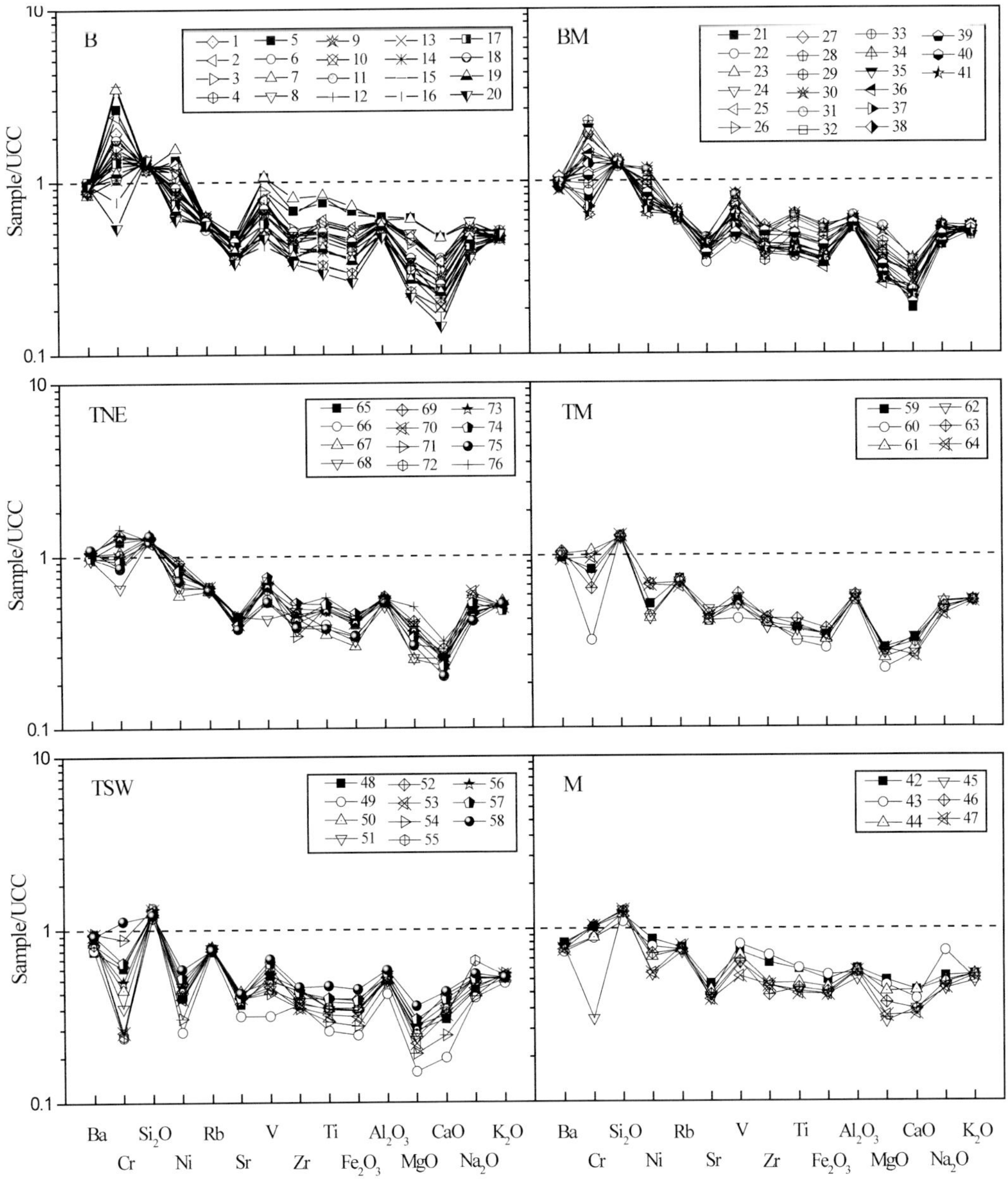

Figure 14.5. The UCC-standardized distributions and compositions of major and trace elements of sand dunes in the Minqin Basin and its surrounding areas in the western Hexi Corridor (modified after Ren et al., 2014).

dunes developed in and around the Minqin Basin also suggests that the role of an oasis in preventing aeolian erosion is limited and should not be overestimated.

Some researchers have conducted geochemical analysis of major and trace elements in aeolian sediments of barchans dunes and other sediments developed in the western Hexi Corridor (Zhang et al., 2017; Pan et al., 2019). The dunes studied are located in the Gobi area to the north and west of Jiayuguan (area A in Fig. 14.2). The dune types are mainly barchan dunes, chains of barchan dune and asymmetric barchans dunes (Zhang et al., 2017; Pan et al., 2019). The aeolian samples were mainly collected from the surface sediments of barchan dunes and asymmetric barchan dunes, including different geomorphic sites of dunes such as the crest of dune, the bottom of the windward slope, the middle of the windward slope and the bottom of the leeward slope. The analytical results show that after the standardization of UCC, the barchan dunes on the Gobi surface in the western

Hexi Corridor are significantly enriched in CaO and SiO_2 (accounting for 5.55 and 66.12% of the total rock mass, respectively). The contents of Cao, MgO and Fe_2O_3 are gradually enriched from northwest to southeast, that is, the enrichment degree increases along the dominant wind direction. The UCC-normalized concentrations of Na_2O and K_2O are both significantly < 1, indicating that alkali metal elements are significantly depleted or leached in the study area (Zhang et al., 2017; Pan et al., 2019). The contents of trace elements vary among different dune samples, reflecting a complexity of provenance of these dune sediments; however, the variations of trace elements are similar in different geomorphic positions of the one dune (Pan et al., 2019). Compared with UCC, trace elements Co, As, La, and Nd are significantly enriched, while other elements are depleted. Compared with the chemical elements in the Tenggeli and Badanjilin Deserts (Li, 2011), the Hexi Corridor has lower SiO_2 content, similar K_2O content, and lower contents of other trance elements.

Mineralogical and major- and trace-element geochemical analyses of aeolian sediments from the Jinta-Gaotai area (area B in Fig. 14.2) in the middle Hexi Corridor have been carried out (Ferrat et al., 2011; Wang and Wang, 2013; Wang et al., 2018; Zhang et al., 2020). These studies use the light/heavy mineral assemblages, the ratios of Na_2O/Al_2O_3 to K_2O/Al_2O_3 and SiO_2/Al_2O_3, Ba/Sr to Rb/Sr, Rb/Sr to Ce/Sr, and the composition of CaO and Cl to identify the provenance of aeolian sediments in the study area. Similar mineralogical compositions (mica, quartz, illite, muscovite and albite) are found in dune sediments from the Hexi Corridor and adjacent areas such as the Tenggeli and Badanjilin Deserts (Ferrat et al., 2011). This feature is also found by major and trace element analysis (Wang and Wang, 2013; Zhang et al., 2020), indicating that the geochemical characteristics of aeolian sediments in the Hexi Corridor and its adjacent areas are also similar. Compared with the composition of the upper continental crust (UCC), dune sediments in the Jinta-Gaotai area in the middle Hexi Corridor are also enriched in CaO (Fig. 14.6), which are similar to those in the fluvial and lacustrine sediments near the Taklamakan and Badanjilin Deserts and the aeolian sediments in the surrounding the Gobi Desert, but they are slightly different from those in the Kumtag and Tenggeli Deserts (Zhang et al., 2020). Through the methods of Multi-Dimensional Scaling (MDS), Principal Component Analysis (PCA) and regional topography analysis, these studies suggest that the endorheic rivers of the Hexi Corridor are not only the sediment sink of the Qilian Mountains, but also the sediment sink of the Beishan Mountains. The sand dunes from the Hexi Corridor are similar in provenance to those from the Tenggeli, Badanjilin and Kumtag Deserts (Zhang et al., 2020).

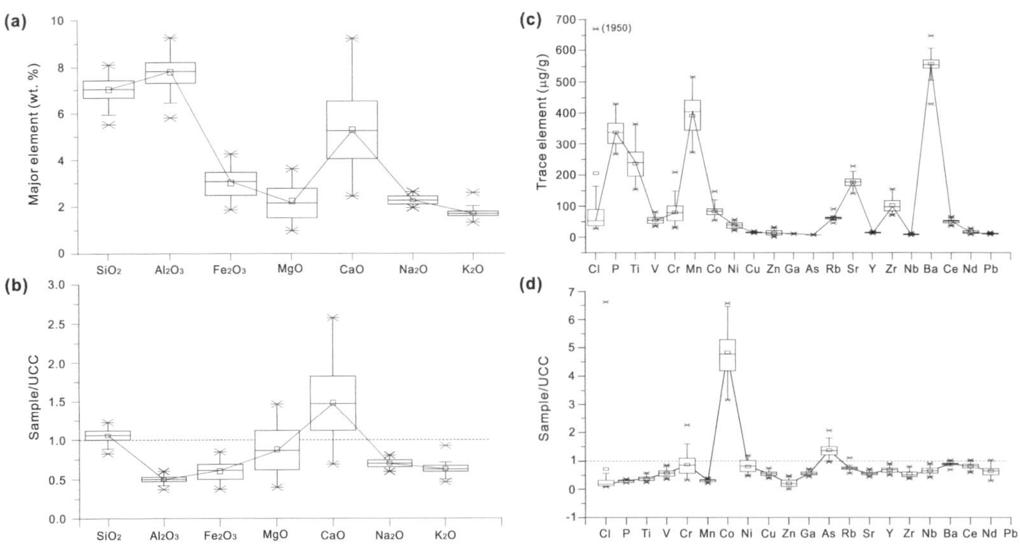

Figure 14.6. Compositions of the major and trace elements and their UCC-standardized distributions of sand dunes in the Jinta-Gaotai area of the middle Hexi Corridor (modified after Zhang et al., 2020).

Table 14.3. The average element contents (%) of sand dunes in the Hexi Corridor and other deserts and the average composition of the upper continental crust (UCC).

Regions	Fe_2O_3	CaO	MgO	SiO_2	Al_2O_3	Na_2O	K_2O	References
Hexi Corridor	3.5	5.55	2.07	66.12	9.24	2.45	2	Zhang et al., 2017; Pan et al., 2019
Badanjilin Desert	1.93	2.06	1.19	80.27	7.78	1.9	2	Zhu and Yang, 2009
Tenggeli Desert	1.96	1.3	1.12	80.94	8026	1.88	2.25	Zhu and Yang, 2009
Kumutage Desert	2.88	4.64	2.19	70.13	9.59	2.52	1.98	Dong et al., 2011
Taklamakan Desert	3.1	7.88	2.2	62.05	10.6	2.58	2.11	Zhu and Yang, 2009
Loess (CLP)	4.56	8.62	2.31	58.65	11.86	1.68	2.44	Dong et al., 2011
paleosol (CLP)	5.12	0.83	2.21	65.18	14.79	1.41	3.15	Dong et al., 2011
UCC	5	4.2	2.22	66	15.2	3.9	3.4	Taylor and McLennan, 1985
Terrestrial shale	7.22	1.3	1.2	62.8	18.9	1.2	3.7	Taylor and McLennan, 1985

Therefore, it can be concluded that dried river deposits (playas), alluvial/proluvial deposits and lacustrine deposits are the main sources of sand dunes in the Hexi Corridor, while these detrital deposits mainly come from the Qilian Mountains in the south and the Beishan Mountains in the north. Comparing the major element abundances of aeolian sediments in different deserts of China (Table 14.3), the barchan dunes in the western Hexi Corridor have some differences and similarities with deserts in Northwest and North-central China. For example, the content of Fe_2O_3 in sand dunes of the Hexi Corridor can reach 3.50%, which is generally higher than any of the other deserts, while the contents of other elements in the Hexi Corridor are in a similar range of element abundance to those of other deserts, indicating that the sand dunes in the Hexi Corridor are rich in iron element.

4. Mechanism of land degradation in the Hexi corridor and its environmental implications

Generally speaking, sand dunes in the Hexi Corridor are dominated by mobile dunes. In early research work, the formation of these mobile dunes was considered to be caused by the destruction of vegetation previously fixed on the shrub dunes in the oasis; or because the gravel Gobi Desert and related wind-erosion areas in the Hexi Corridor provided abundant sand sediments, resulting in a wide aeolian-sand transportation and accumulation to form sand dunes in the corridor area; or because that in the arid Hexi Corridor, the shallow and intermittent rivers with broad riverbeds changed their courses and the abandoned dried riverbeds were blown by wind, causing accumulation of the fluvial sediments to form source-bordering sand dunes near the river banks (Zhu et al., 1980).

4.1 Land degradation in the Hexi corridor – mechanisms and stages

The mechanisms mentioned above for the formation of sand dunes can explain the characteristics of thc sporadic or sheet-like distribution of sand dunes in the Hexi Corridor, especially the characteristics of sand dunes intermittently distributed along the dried riverbeds by winding and zigzag patterns, and scattered on gravel of the Gobi Desert at the edge of oases.

Apart from the above natural causes, it is also believed that the formation of sand dunes in the Hexi Corridor is not all controlled by the influence of natural factors, some dunes could have been formed in historical periods and are the result of human activities (Zhu et al., 1980; Zhu and Chen, 1994; Zhu, 1999; Yang et al., 2004; Chang et al., 2005; Li, 2007). For example, ruins of the Han, Tang and Ming Dynasties and sites of the Great Wall can also be found in some dune fields, such as the Shouchang ruins in the South Lake area of Dunhuang, the ruins in the dune fields of the Xicheng

Post Station to the west of Zhangye, and the sites of the Great Wall of Ming Dynasty in the dune fields in the Minqin area (Zhu et al., 1980; Zhu and Chen, 1994; Zhu, 1999), etc.

Regarding the above land-degradation mechanisms, the wind-eroded fields in the Gobi Desert and its deflation products in the western Hexi Corridor (the upper windward area) are considered to be one of the reasons for land degradation of the middle and eastern Hexi Corridor, because the former may provide the material basis for the latter to form sand dunes. However, is the existence of these western Gobi and wind erosion areas really the main cause of land degradation in the Hexi Corridor?

In order to solve this problem, research has been carried out on the Gobi soils, wind-erosion landforms and the intensity and potential of the wind erosion process in the western Hexi Corridor (e.g., Zhu et al., 1980; Zhang et al., 2004; Qu et al., 2005; Wang et al., 2013; Yin et al., 2014, 2016; Zhang et al., 2016; An et al., 2019; Hu et al., 2020). From the perspective of geomorphology, wind-eroded lands are widely distributed in the western part of the Hexi Corridor. The result of surface wind erosion in these areas is the formation of long strip-shaped aeolian monadnocks and deflation hollows (aeolian depressions), which are roughly parallel to the wind direction. Generally, these wind-eroded landforms are about 1 ~ 3 meters high, and a few are up to 5 meters (Zhu et al., 1980), which are roughly distributed along the lower reaches of the Shule River and the western part of the Dunhuang Oasis. Studies on the surface properties of gravel desert (Gobi) and its impact on wind erosion and dust emission in the west of the Hexi Corridor have shown that the gravel Gobi is very different from the sand desert (dune field) because the grain-size composition of the surface sediment and the occurrence state (erodibility) of fine granular material are completely different (Zhang et al., 2004; Yin et al., 2014, 2016; Zhang et al., 2016; Hu et al., 2020). The surface of the Gobi Desert is composed of the gravel-, sand-, fine-sand- and clay-fraction particles with coarse- and fine-grained materials mixed together, while the surface of the dune field is mainly composed of the medium-sand-fraction particles, with no or few coarser-grained gravels and finer-grained silt and clay particles (Zhang et al., 2016; An et al., 2019). A salt crust usually exists on the surface of the Gobi Desert in the West Hexi Corridor and thus the fine-grained particles are consolidated due to the presence of salt cement, leading to a weak wind erodibility in the Gobi Desert, while the surface of dunes is loose, making a strong erodibility in the dune field (Zhang et al., 2016). As the potential sediment availability (controlled by erodibility of surface fine-grained particles), the gravel coverage (surface roughness) and the average particle size of surface sediments are the main factors affecting dust emission (Pye, 1987; Gillette and Stockton, 1989; Raupach et al., 1993), three factors determine whether the Gobi Desert in the western Hexi Corridor has potential contribution to regional aridization in the Hexi Corridor.

Based on field surveys and using the ImageJ software to process high-resolution image data, Zhang et al. (2016) estimated the gravel coverage and surface salt crust status in different Gobi areas in the West Hexi Corridor, determined the proportion of total surface sediment weight occupied by gravels (diameter > 2 mm), and analyzed the sedimentological grain-size distribution of different land surface sediments. The results of this work show that: (1) in the West Hexi Corridor, the gravel coverage of the Gobi surface is moderate, with the gravel coverage mainly between 40 and 70% (average 52%, SD = ± 17%) (Zhang et al., 2016). The rate of gravel coverage in this level can produce the maximum aerodynamic roughness on the ground surface to prevent dust emission (Lyles and Tatarko, 1988; Wolfe and Nickling, 1996; Dong et al., 2002a, 2002b; Liu and Dong, 2003; Uno et al., 2006; Rostagno and Degorgue, 2011). (2) Most of the Gobi surface lands (75% of the total area) have formed a salt crust. Only the areas with high sand-transport potential of wind in the northern Hexi Corridor and the edge of sand deserts have no surface salt crust. (3) The content of erodible materials (sand, silt and clay) on the Gobi surface has a clear spatial distribution. Sediments of the Gobi surface are mostly the medium-sand and fine-sand particles (52.5 and 25.0%, respectively), while the silt and clay contents range from 9.8 to 40.1%, with most areas (about 73%) ranging from 10 to 30%. (4) In most Gobi regions, the potential transport of sand material is > 200 vector units, but 75% of these areas have solid soil crust on the ground surface (Zhang et al., 2016).

Combining the above indexes, i.e., the proportion of fine-grained dust materials, the coverage rate of salt crust and the potential transport of sand material, show that the high level of gravel coverage and surface crust rate in the West Hexi Corridor can effectively reduce the dust emission from the Gobi surface. Therefore, the Gobi area in the West Hexi Corridor is not the main source area of dust and sandstorms occurring in the middle and East of the Hexi Corridor. While the northern part of the Hexi Corridor may be the main source area of dust.

Another potential indicator to assess the degree of modern desertification in the region comes from the meteorological parameter, such as the number of sandstorm days and strong wind days (Chang et al., 2019). Sandstorm refers to a windy and sand weather phenomenon with the wind speed $\geq$ the sand-blowing wind speed and the horizontal visibility $<$ 1,000 m. The Hexi Corridor is considered to be one of the areas with the most-frequent (heavy) sandstorm in Northwest China (Zhu and Chen, 1994; Zhu, 1999). The results of meteorological data analysis from the Hexi Corridor show that since 1956, the number of local sandstorm days in the Hexi Corridor is about 11.20 day/year (Table 14.4), and the average number of strong wind days (wind speed $>$ 8 grade/day) is about 18.39 days/year (Table 14.4), but the number of local sandstorm days in the Hexi Corridor has shown a generally-decreasing trend (Fig. 14.7a), with a decline rate of 0.677 times/year (Chang et al., 2011). In addition, the frequency of sandstorms throughout northern China was also decreasing during the same period (Zhang and Ren, 2003; Li and Zhang, 2007). However, this is contrary to the situation of global sandstorms, because the number of times of global sandstorms is increasing in recent decades (Houghton et al., 2001; Ding, 2002). This indicates that the sandstorm process in the Hexi Corridor and even in the entire northern China does not respond to global changes, revealing that the cause of land degradation in the Hexi Corridor is different from other parts of the world.

As shown in Fig. 14.7a, the number of sandstorm days in the Minqin area of the Hexi Corridor has shown a decreasing trend as a whole since 1956–2008, and there are three sub-trends during this period, i.e., both the frequency and number of sandstorms decreased rapidly (1956–1969), the frequency was high and stable (1971–1987), and the frequency was low and slowly decreased (1987–2008).

However, with global warming, the temperatures in desert areas are generally increasing (Fig. 14.7b), and the frequency of sandstorms should have increased, but why has the actual situation reduced? To answer this question, one needs to analyze the climate change in arid northern China and the Hexi Corridor during this period. For nearly half a century, global warming has become a

Table 14.4. Meteorological data of sandstorm and strong wind in the Hexi Corridor (cited from Chang et al., 2015).

Weahter station	Average wind speed (m/s)	Wind speed³ 8 grade/d	Sandstorm days (d)	Maximum wind direction
Gulang	3.5	4.5	4	NW
Wuwei	1.8	9.7	4.8	NW
Minqin	2.7	25.1	27.4	NW
Jinchang	3.3	25.1	27.4	NW
Yongchang	3	24.6	4.2	NW
Zhangye	2	12.2	11.8	NNW
Linze	2.5	21.7	7.7	NW
Gaotai	2	7.8	11.1	NW
Jiuquan	2.2	16.6	10.3	NNW
Jinta	1.9	14.4	6.4	NW, WNW
Yunmen	3.8	40.6	8.1	WNW
Average	2.61	18.39	11.2	/

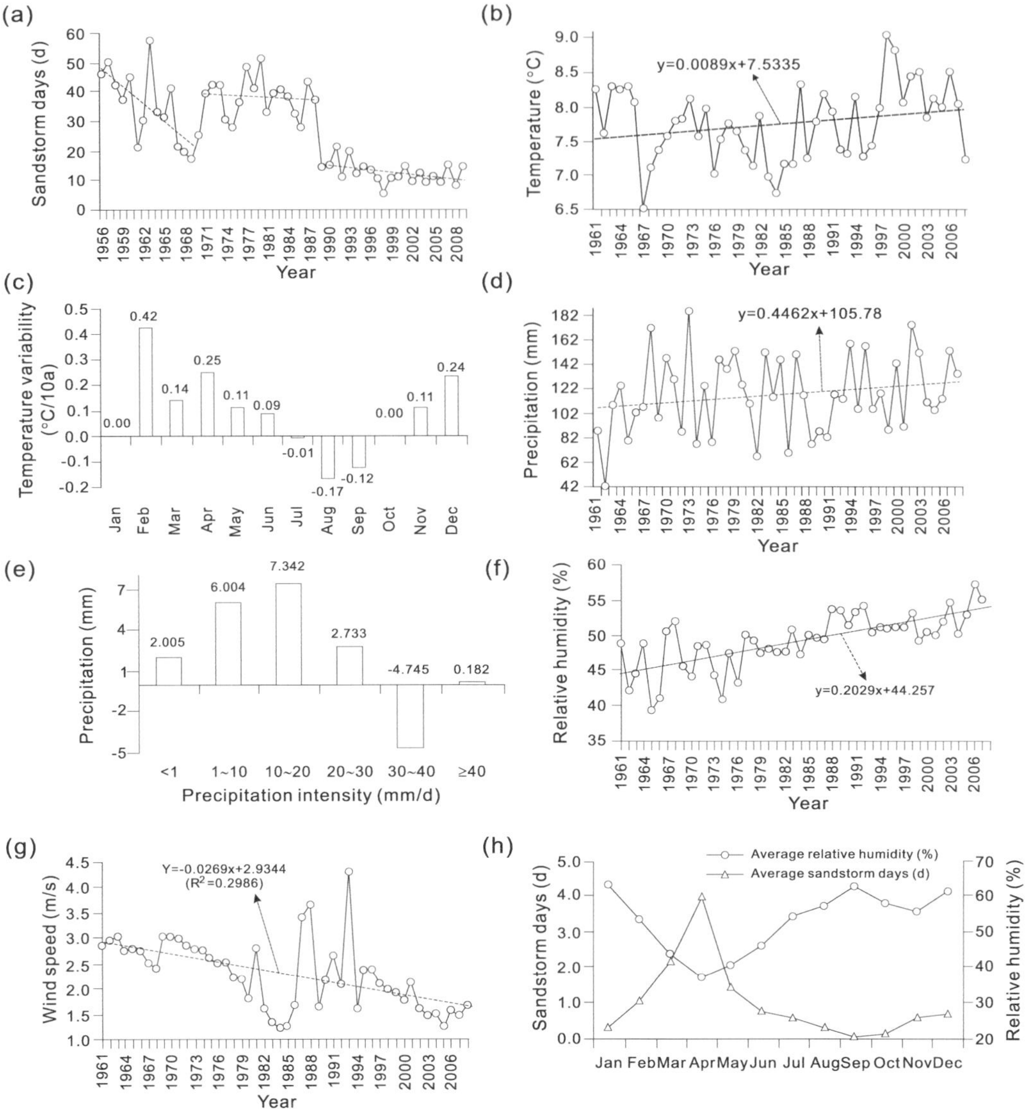

Figure 14.7. Changes of climate parameters in Hexi Corridor in recent 50 years (modified after Chang et al., 2011). (a) Variation of annual average temperature during 1961–2008 in Minqin; (b) Temperature variability during 1961–2008 in Minqin; (c) Variation of annual mean precipitation during 1961–2008 in Minqin; (d) Variation and distribution of precipitation intensity during 1961–2008 in Minqin; (e) Variation of annual relative air humidity during 1961–2008 in Minqin; (f) Variation of annual average wind speed during 1961–2008 in Minqin; (g) Variation of sandstorm days during 1956–2008 in Minqin; (h) Comparison of the monthly average sandstorm days and the monthly average relative air humidity during 1961–2008 in Minqin.

worldwide concern. In this context, how does the climate change and respond to global warming in the desert areas of northern China and the Hexi Corridor? Research on this issue have been carried out in arid regions of northern China and the Hexi Corridor (e.g., Ding, 2002; Sha et al., 2002; Chang et al., 2011, 2016b). Here, we briefly summarize the research results of climate change in the Hexi Corridor (the Minqin area) in recent decades.

1) From 1961 to 2008, the rising rate of the annual average temperature in the Minqin area was higher than that of global temperature in the 20th century and that of China in the past 100 years. Among them, the temperature increase in February was the largest, with the monthly average temperature increasing by 3.01°C (Fig. 14.7b).
2) From the 1980s to 1990s, the warmest in the world in the 20th century, the extreme maximum temperature in Minqin had increased significantly, while the extreme minimum temperature had decreased intermittently, and the instability of the extreme maximum and minimum temperatures had increased. The detailed results are as follows: the instability of the monthly average temperature in January and April increases; the isothermal date in February is 10.36 days earlier; the instability of the extreme maximum temperature in December and January increases; the variation coefficient of the extreme minimum temperature in May is as high as 287% (Fig. 14.7C).
3) During the period 1961–2008, the temperature in the Minqin area increased, the precipitation also shown an increasing trend, and the air humidity also increased significantly (Fig. 14.7d, 14.7f).
4) In the Minqin area, the instability of precipitation increased in January and the stability of annual precipitation increased (Fig. 14.7e). In general, as with the large-scale regional climate change in northern China, the problem of temperature instability should be more worthy of attention than the problem of temperature warming (Ding, 2002; Sha et al., 2002; Chang et al., 2011, 2016b).
5) The wind speed in the Minqin area continued to decrease during the period from 1961 to 2008, (Fig. 14.7g).
6) There is a significant negative correlation between the annual and seasonal distributions of sandstorms and the relative humidity of air (Fig. 14.7h).

It can be seen from the above that although the temperature in the Hexi Corridor has increased (in response to global warming), the precipitation has risen (in response to the enhancement of the Asian summer monsoon climate), the relative humidity of the air has increased, while the wind speed has decreased. Most importantly, there is a significant negative correlation between the annual and seasonal distributions of sandstorms and the relative humidity of air (Fig. 14.7h), which is an important clue to answer the above question. As a result of this environmental change, on the one hand, the dynamic force of dust-release process will be reduced (because of the decrease in wind speed), and on the other hand, the viscosity of the surface sediment particles will increase (because of the increase in humidity), which will reduce the frequency of dust storm events. Therefore, since 1961, the decreasing trend of sandstorm days in the Hexi Corridor is mainly due to the increase in the relative humidity of the air, that is, the warming and humidification of the local climate is one of the main reasons responsible for the reduction of sandstorms in the Hexi Corridor. In other words, the Hexi Corridor has a potential trend of reverse desertification so-called "land greening" in the past half century. The main influencing factor of land degradation in the Hexi Corridor is climate change, that is, controlled by natural factors.

4.2 Environmental implications of land degradation (LD) in the Hexi corridor

For more than half a century, in response to the problem of LD in the Hexi Corridor, local governments and organizations have carried out systematic engineering and theoretical researches on desertification control in the Hexi Corridor. The most significant works and researches have been done on the ecological and geomorphological control of LD, such as the study on the water physiology and ecology of desert xerophytes, the study of desert climate change and its response to global warming, the study of the phenology of desert plants, and the study of aeolian sediments accumulation on the edge of oasis, etc. (Zhu and Wang, 1992; Zhu, 1999). These ecological studies also highlight that the ecological and water environment of the Hexi Corridor has changed greatly with the above-mentioned climatic and geomorphological changes in the Hexi Corridor, such as:

1) The groundwater level in the Hexi Corridor has fallen and the regional water resources have decreased. For example, 26 motorized wells located in the inner Minqin Oasis indicate that the groundwater level in the central part of the Minqin Oasis fell at a rate of 0.54 m/y during the period 1985–2001 (Fig. 14.8); and seven motorized wells located at the edge of the Minqin Oasis indicates that the groundwater level in the marginal area of the Minqin Oasis also decreased, with a drop rate of 0.56 m/y from 1985 to 2017 (Fig. 14.8).
2) The area of natural sand-fixing forests in the Hexi Corridor has decreased as a result of die-back. Due to the drop of groundwater level, the natural sand-fixing vegetation in the Minqin area has declined on a large scale, and the desert steppe has become a sand field. For example, in 1981, the area of natural sand-fixing forest in the Minqin area was 203,951 hm^2, but by 2002, the area of natural sand-fix forest had decreased to 197,353 hm^2 (Table 14.5); in the early 1980s, there were 373 hm^2 of *Populus euphratica* forest in Minqin, but it has now disappeared (Table 14.5).

The above ecological and hydrological problems informs us that in an arid and desertified environment such as the Hexi Corridor, the reason for the continuous decrease of vegetation is not the lack of afforestation, nor the change of regional precipitation or relative humidity (although the climate in the Hexi Corridor has become more humid in the past half century, as illustrated in Fig. 14.7), but is subject to the changes of "effective water", such as groundwater and soil water (Fig. 14.8). In other words, effective moisture is the most restrictive factor for ecological sustainability in the desert environment. Therefore, the control of LD in an arid environment should be based on the principle of “centering on groundwater balance”. Deviating from the principle of "effective water balance" is a deviation from the basic law of natural environment change in sand areas. With this contradiction, it is difficult to achieve the control of LD without the overall direction of water-saving management and research on groundwater.

Therefore, it can be said that the change of water environment is one of the leading factors restricting the change of the ecological environment from an oasis landscape toward a desertified landscape in arid areas. The degradation of the ecological environment in the lower reaches of the Shiyang River Basin in the Hexi Corridor, once again illustrates this point. Nationally, from a larger spatial scale, the arid and semi-arid areas in northern China are seriously deficient in water resources. How to combat LD is the primary problem restricting the sustainable development of the region.

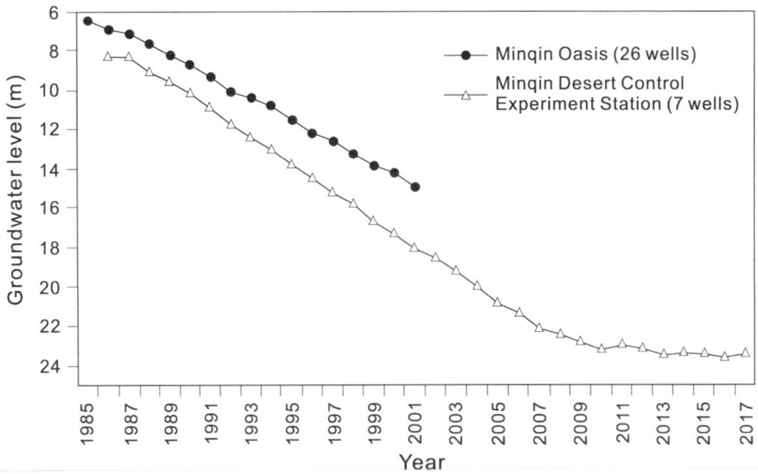

Figure 14.8. Variational trend of groundwater levels in the Minqin area (modified after Chang, 2019).

Table 14.5. Changes in the area of the sand-fixing forest in different years in the Minqin Basin (cited from Chang, 2019).

Year	**Planted forest**								**Natural forest**					
	Arbor forest		**Shrub forest**						**Nitraria tangutorum earth bags with different coverages**			**Chinese Tamarisk earth bags**	**Caragana Korshinskii**	**Populus euphratica**
	Elaeagnus angustifolia	other	Haloxylon ammodendron	Hedysarum scoparium	Caragana Korshinskii	Salix glabra	Chinese Tamarisk	other	> 0.4	< 0.3	< 0.1			
1981	16840	855.9	13044	171.8	19	690.9	227.7	4	23379	36377	12662	5238	125922	372.6
2002	16700	/	44600	100	3508	/	/	132800	19560	19560	14860	5011	125922	0

Some scholars believe that the transfer of water from the outer basins (such as the Yellow River Basin and the Yangtze River Basin) can appropriately alleviate the serious water shortage and LD in the inland area of the Hexi Corridor. However, is such a measure feasible? In view of the large-scale allocation of resources in arid areas, on the one hand, the transfer of water from the outer basin is not the fundamental solution to the problem, and it is likely to cause ecological problems in other watersheds, because almost every watershed under arid environment is short of water or potentially short of water. On the other hand, large-scale water transfer is also unrealistic. Due to the constraints of socio-economic conditions and financial resources in the Hexi Corridor, it is not feasible to transfer water from the outer basin at least in the near future. It is believed that the fundamental way to solve the problem of water resources of the inland areas of the Hexi Corridor lies in the rational utilization of existing water resources. Furthermore, some studies believe that the inland areas of the Hexi Corridor are not an absolute "resource-based water shortage", but a combination and coexistence of "resource-based water shortage" and "technics-based water shortage" (Chang and Liu, 2003). It means that the current utilization of water resources is unreasonable and the efficiency of utilization is not high. Comprehensive control is the only way to combat LD in the interior of the Hexi Corridor.

5. Summary of findings

The formation and dynamic change of sandy dune landforms in the Hexi Corridor is a typical problem of land desertification in the mid-latitudes of Northern Hemisphere, which has attracted great attention from China and the international community for nearly half a century. It is of great significance to understand the formation and evolution of aeolian dunes in this area to reveal the process and mechanism of desertification and the related environmental changes. Based on the existing high-resolution satellite image interpretations, field investigations and observations, comprehensive understanding of the evidences from geomorphological, aeolian-physical, granulometrical and geochemical analysis, this study discussed the formation of dune landform, the mechanism of desertification and their environmental implications in the Hexi Corridor. The analytical results show that 80% of the sand particles flow within a height of 20 ~ 30 cm near the surface, and about half of the sand particles flow within a height of 0.3 ~ 0.5 cm near the surface in the Hexi Corridor. The average height of the typical crescent-shaped dunes is about 6.75 m, and the minimum and maximum values are between 2.6 and 11.2 m. In terms of the inter-annual and multi-year dynamic characteristics of the sandy dunes in the Hexi Corridor, only the crescent-shaped dunes and chains of barchan dunes are moving or wigwagging, while the parabolic and longitudinal dunes did not move. Under the influence of wind speed, strong wind days and other factors, the dunes at the edge of the Minqin Oasis move the fastest, with a moving speed of about 6.2 m/a. Affected by the main wind direction and other factors, the dunes at the edge of the Dunhuang Oasis move the slowest, with a moving speed of about 0.8 m/a. The main factors affecting the dynamic changes of sandy dunes in the Hexi Corridor are the annual precipitation, the annual average wind speed and the number of annual strong wind days, of which the annual precipitation contributes the largest, indicating that the climate factors have a most important impact on the dynamic change of sand dunes. The cumulative curve of particle size frequency of dune sediments in the Hexi Corridor basically presents a three-segment model, indicating that the dune sediments are mainly in a saltation mode under the action of wind force, superimposed with a small amount of coarser and finer particles dominated by the creeping and suspension models, which is obviously different from that of the Gobi sediments with a dominant two-segment model. The early and recent investigations of dune sources and the sedimentological and geochemical evidences indicate that the sandy dunes in the Hexi Corridor are mainly derived from “locally or in-situ raised sediments”, which are mainly come from alluvial plains and ancient fluvial sediments, as well as ancient lake plains and lacustrine deposits in the Hexi Corridor, aeolian deposits in the piedmont denudation zones of the north and south mountains and modern fluvial sediments. In terms of the abundance of major and trace elements, the crescent-shaped dunes in the Hexi Corridor have certain similarities and differences to other sandy dunes in the northwest and northern deserts of China (such as the Kumtag, Badanjilin, Tenggeli, and Taklamakan Deserts, etc.) or aeolian loess in the Loess Plateau. The sandy dunes in the Hexi Corridor are relatively rich in iron and Co. Considering the proportion of fine particles on the surface, the coverage rate of surface salt crust, and the potential migration of erodible sandy materials.

6. Conclusions

it can be concluded that the Gobi in the west of the Hexi Corridor is not the main source area of sandstorms in the middle and east of the corridor, but the north is probably the main source area of dust. In the past half century, the warming and humidification of the local climate is one of the main causes of the reduction of sandstorms in the study area. During the same period, the Hexi Corridor has a potential trend of anti-desertification, which is mainly controlled by climate change, that is, an influence of natural factors. But in the oasis area of the Hexi Corridor, the effective measures to restrict desertification depend on human activities. Comprehensive management and rational utilization of the existing water resources, especially the restriction of the decline of groundwater, is the key to preventing desertification in oases, rather than water transfer from outer river basins.

Acknowledgments

The study was financially supported by the National Natural Science Foundation of China (grant nos. 41930640), the Project of the Second Comprehensive Scientific Investigation on the Qinghai Tibet Plateau (2019QZKK1003), and the Third Xinjiang Scientific Expedition Program (Grant No. 2021xjkk0803). Sincere thanks are extended to Profs. Xiaoping Yang, Zhaofeng Chang, Zhibao Dong and Zhengcai Zhang for their help in the author's (B. Zhu) research work.

References

An, F., Zhang, D., Zhao, J., Chai, C., Zhao, Y. and Sun, T. 2019. Physical and chemical properties of soils in different types in the Gobi areas of the Hexi Corridor. J. Chin. Soil Water Conserv. 6: 42–47. Doi: 10.14123/j.cnki.swcc.2019.0143 (in Chinese with English abstract).

Chang, Z. and Liu, H. 2003. Desertification control in Hexi inland river basins of Gansu Province (II). Protection Forest Science and Technology 2: 28–32 (In Chinese with English Abstract).

Chang, Z., Zhao, M., Han, F., Zhong, S. and Liang, T. 2005. Natural and artificial factors and their transfer on sand desertification of lower reaches of Shiyang River Basin. Arid Land Geography 28(2): 150–155 (in Chinese with English abstract).

Chang, Z., Han, F. and Zhong, S. 2011. Response of desert climate change to global warming in Minqin, China. Journal of Desert Research 31(2): 505–510 (In Chinese with English Abstract).

Chang, Z., Wang, Q., Zhang, J., Xi, J., Wang, Q., Zhang, D., Tang, J. and Zhang, H. 2015. Environmental conditions of barchans dune and barchans chain—a case study from the Hexi Desert area of Gansu. Animal Husbanbandry and Feed Science 7(6): 383–388.

Chang, Z., Zhu, S., Shi, X., Zhang, J., Li, Y., Wang, Q., Zhang, D. and Duan, X. 2016a. Comparisons between movement speed of main types of dunes: a case study of desert areas in Hexi region of Gansu Province. Journal of Landscape Research 8(6): 36–40.

Chang, Z., Ma, Z., Wang, D., Han, F., Duan, X., Wang, Q. and Zhang, J. 2016b. Instability of climate change in the Minqin Desert Area. Arid Zone Research 33(3): 601–608 (in Chinese with English Abstract).

Chang, Z., Li, Y., Zhang, J., Wang, Q., Zhang, D., Tang, J., Wang, Q. and Zhang, H. 2017. Stability mechanisms of barchan sand dunes: a case study in the Hexi Desert in Gansu. Acta Ecologica Sinica 37(13): 4375–4383 (in Chinese with English abstract).

Chang, Z. 2019. Problems and solutions to desertification combating in the Hexi, Gansu for 60 years. Journal of Arid Land Resources and Environment 33(9): 152–159 (in Chinese with English Abstract).

Chen, F., Yu, Z., Yang, M., Ito, E., Wang, S., Madsen, D.B., Huang, X., Zhao, Y., Sato, T., John, B., Birks, B., Boomer, I., Chen, J., An, C. and Wünnemann, B. 2008. Holocene moisture evolution in arid central Asia and its out-of-phase relationship with Asian monsoon history. Quaternary Science Reviews 27: 351–364.

Chen, F. and Liu, Y. 2011. Secular annual movement of sand dunes in Badain Jaran Desert based on geographic analyses of remotely sensed imagery. Remote Sensing Technology and Application 26(4): 501–507.
Chen, J., Li, L., Wang, J., Barry, A., Sheng, X., Gu, W., Zhao, X. and Chen, L. 2004. Groundwater maintains dune landscape. Nature 432: 459.
Chen, J., Zhao, X., Sheng, X., Dong, H., Rao, W. and Su, Z. 2006. Formation mechanisms of megadunes and lakes in the Badain Jaran Desert, Inner Mongolia. Chinese Science Bulletin 51: 3026–3034.
Ding, Y. 2002. Global climate change. World Environment 6: 9–12.
Ding, Y. 2002. Prediction of environmental changes in Western China. pp. 1–231. *In*: Qin, D. (ed.). Assessment of Environmental Evolution in Western China. Science Press, Beijing.
Dong, C., Wang, N., Chen, J., Li, Z., Chen, H. and Chen, L., Ma. 2016. New observational and experimental evidence for the recharge mechanism of the lake group in the Alxa Desert, north-central China. Journal of Arid Environments 124: 48–61.
Dong, Y. and Huang, D. 2013. Preliminary observation of movement of coastal dunes in Feicui Island in Changli, Hebei Province. Journal of Desert Research 33(2): 486–492.
Dong, Z., Chen, G., Yan, C., Han, Z. and Wang, X. 1998. Movement laws of dunes along oiltransportation highway in the Tarim Desert. Journal of Desert Research 18(4): 328–333 (in Chinese with English abstract).
Dong, Z., Qu, J., Liu, X., Zhang, W. and Wang, X. 2002a. Experimental investigation of the drag coefficients of gobi surfaces. Science in China Series D - Earth Sciences 45(7): 609–615.
Dong, Z., Liu, X. and Wang. X. 2002b. Aerodynamic roughness of gravel beds. Geomorphology 43(1-2): 17–31.
Dong, Z., Su, Z. and Qian, G. 2011. Aeolian Landforms in the Kumtag Desert. Science Press, Beijing.
Dong, Z., Qian, G., Lv, P. and Hu, G. 2013. Investigation of the sand sea with the tallest dunes on Earth: China's Badain Jaran Sand Sea. Earth-Science Reviews 120: 20–39.
Dong, Z. and Wang, N. in press. Mega-dune—Lake system in the Badain Jaran sand sea. pp. 127–150, this volume.
Ferrat, M., Weiss, D.J., Strekopytov, S., Dong, S.F., Chen, H.Y., Najorka, J., Sun, Y.B., Gupta, S., Tada, R. and Sinha, R. 2011. Improved provenance tracing of Asian dust sources using rare earth elements and selected trace elements for palaeomonsoon studies on the eastern Tibetan Plateau. Geochimic & Cosmochimic Acta 75: 6374–6399.
Gates, J.B., Edmunds, W.M., Darling, W.G., Ma, J., Pang, Z. and Young, A. 2008a. Conceptual model of recharge to southeastern Badain Jaran Desert groundwater and lakes from environmental tracers. Applied Geochemistry 23: 3519–3534.
Gates, J.B., Edmunds, W.M., Ma, J. and Scanlon, B.R. 2008b. Estimating groundwater recharge in a cold desert environment in northern China using chloride. Hydrogeology Journal 16: 893–910.
Geyh, M.A., Gu, W., Liu, Y., He, X., Deng, J. and Qiao, M. 1998. Isotopically anomalous groundwater of Alxa Plateau, Inner Mongolia. Advance in Water Science 9: 333–337 (in Chinese).
Gillette, D.A. and Stockton, P.H. 1989. The effect of nonerodible particles on wind erosion of erodible surfaces. Journal of Geophysical Research 94(12): 885–893.
Goudie, A. 2002. Great Warm Deserts of the World: Landscapes and Evolution. Oxford University Press, New York.
Guo, Z., Ruddiman, W., Hao, Q., Wu, H., Qiao, Y., Zhu, R., Peng, S., Wei, J., Yuan, B. and Liu, T. 2002. Onset of Asian desertification by 22 Myr ago inferred from loess deposits in China. Nature 416: 159–163.
Guo, Z., Peng, S., Hao, Q., Biscaye, P., An, Z. and Liu, T. 2004. Late Miocene-Pliocene development of Asian aridification as recorded in the Red-earth Formation in northern China. Global and Planetary Change 41: 135–145.
He, J., Guo, J., Xing, E., Cui, W. and Li, J. 2012. Structure of wind-sand flow and movement laws of dunes along the Yellow River in Ulan Buh Desert. Transactions of the Chinese Society of Agricultural Engineering 28(17): 71–77 (in Chinese with English abstract).
Honda, M. and Shimizu, H. 1998. Geochemical, mineralogical and sedimentological studies on the Taklimakan Desert sands. Sedimentology 45: 1125–1143.
Houghton, J.T., Ding, Y., Griggs, D.J., Noguer, M., van der Linden, P.J., Dai, X., Maskell, K. and Johnson, CA. 2001. Climate Change 2001: The Scientific Basis. Cambridge University Press, Cambridge, UK, pp. 15–108.
Hou, L., Wang, X., Hu, B.X., Shang, J. and Wan, L. 2016. Experimental and numerical investigations of soil water balance at the hinterland of the Badain Jaran Desert for groundwater recharge estimation. Journal of Hydrology 540: 386–396.
Hu, F., Zhang, K., An, Z. and Yu, Y. 2020. Composition of wind dynamic environment among desert, oasis and gobi. Journal of Desert Research 40(4): 113–119 (in Chinese with English abstract).
Hu, X., Wang, M. and Liu, Y. 2016. Analysis of movement of dunes in the Tengger Desert based on high-resolution remote sensing images. China Science and Technology Review 2: 337.
Huang, T. and Pang, Z.H. 2007. Groundwater recharge in Badain Jaran Desert and Gurinai Oasis based on environmental tracers. Geosciences 21: 624–631(in Chinese).

Jiang, Q. and Yang, X. 2019. Sedimentological and geochemical composition of aeolian sediments in the Taklamakan Desert: Implications for provenance and sediment supply mechanisms. Journal of Geophysical Research Earth Surface 124: 1217–1237.

Jin, K., Rao, W., Tan, H., Song, Y., Yong, B., Zheng, F., Chen, T. and Han, L. 2018. H-O isotopic and chemical characteristics of a precipitation-lake water-groundwater system in a desert area. Journal of Hydrology 559: 848–860.

Lancaster, N., Yang, X. and Thomas, D. 2013. Spatial and temporal complexity in Quaternary desert datasets: Implications for interpreting past dryland dynamics and understanding potential future changes. Quaternary Science Reviews 78: 301–302.

Lancaster, N., Wolfe, S., Thomas, D., Bristow, C., Bubenzer, O., Burrough, S., Duller, G., Halfen, A., Hesse, P., Roskin, J., Singhvi, A., Tsoar, H., Tripaldi, A., Yang, X. and Zarate, M. 2016. The INQUA Dunes Atlas chronologic database. Quaternary International 410: 3–10.

Lang, L., Wang, X., Zhu, B., Wang, X., Hua, T., Wang, G., Li, H. and Zhang, C. 2017. Nebkha formation and variations in sediment availability and wind-energy regime of the western Hexi Corridor over the past several dacades. Journal of Desert Research 37(4): 611–620 (in Chinese with English abstract).

Li, B. 2007. The oasis in the lower reaches of the Shiyang River has become the "the second Loulan" before the mid-Tang Dynasty. Research and Development 2: 153–157 (in Chinese).

Li, E. 2011. Comparative Study on Characteristics of Aeolian Sediments between the Badanjilin and Tengeli deserts. Ph.D doctoral dissertation, Shaanxi Normal University, Xi'an (in Chinese with English abstract).

Li, S.H. and Fan, A. 2011. OSL chronology of sand deposits and climate change of last 18 ka in Gurbantunggut Desert, northwest China. Journal of Quaternary Science 26(8): 813–818.

Li, Y. and Zhang, S. 2007. Review of the Research on the relationship between sand-dust storm and arid in China. Advances in Earth Science 22(11): 1169–1176 (in Chinese with English abstract).

Li, Y. and Squires, V. 2009. Case study 5: Hexi corridor, Gansu. pp. 151–170. *In*: Squires, V.R., Lu Xinshi,, Lu Qi, Wang Tao and Yang Youlin (eds.). Rangeland Degradation and Recovery in China's Pastoral Lands. CABI, Wallingford U.K.

Lioubimtseva, E. 2004. Climate change in arid environments: revisiting the past to understand the future. Progress in Physical Geography 28: 502–530.

Lioubimtseva, E., Cole, R., Adams, J.M. and Kapustin, G. 2005. Impacts of climate and land-cover changes in arid lands of Central Asia. Journal of Arid Environments 62: 285–308.

Liu, T. 1985. Loess and the Environment. China Ocean Press, Beijing (in Chinese).

Liu, X. and Dong, Z. 2003. Aerodynamic roughness of gravel bed. Journal of Desert Research 23(1): 40–47 (in Chinese).

Luo, X., Jiao, J.J., Wang, X. and Liu, K. 2016. Temporal 222Rn distributions to reveal groundwater discharge into desert lakes: Implication of water balance in the Badain Jaran Desert, China. Journal of Hydrology 534: 87–103.

Luo, X., Jiao, J.J., Wang, X., Liu, K., Lian, E. and Yang, S. 2017. Groundwater discharge and hydrologic partition of the lakes in desert environment: Insights from stable 18O/2H and radium isotopes. Journal of Hydrology 546: 189–203.

Lu, H. et al. in press. Formation and evolution of the dune landscape in north and northwest China, pp. 91–110, this volume.

Lyles, L. and Tatarko, J. 1988. Soil wind erodibility index in seven northern central states. Transactions of the ASAE 31(5): 1396–1399.

Ma, J., Li, D., Zhang, J., Edmunds, W.M. and Prudhomme, C. 2003. Groundwater recharge and climatic change during the last 1000 years from unsaturated zone of SE Badain Jaran Desert. Chinese Science Bulletin 48: 1469–1474.

Ma, J. and Edmunds, W.M. 2006. Groundwater and lake evolution in the Badain Jaran desert ecosystem, Inner Mongolia. Hydrogeology Journal 14: 1231–1243.

Ma, J., Ding, Z., Edmunds, W.M., Gates, J.B. and Huang, T. 2009. Limits to recharge of groundwater from Tibetan Plateau to the Gobi Desert, implications for water management in the mountain front. Journal of Hydrology 364: 128–141.

Ma, N., Wang, N., Zhao, L., Zhang, Z., Dong, C. and Shen, S. 2014. Observation of mega-dune evaporation after various rain events in the hinterland of Badain Jaran Desert, China. Chinese Science Bulletin 59: 162–170.

Machalett, B., Oches, E.A., Frechen, M., Zöller, L., Hambach, U., Mavlyanova, N.G., Marković, S.B. and Endlicher, W. 2008. Aeolian dust dynamics in central Asia during the Pleistocene: driven by the long-term migration, seasonality, and permanency of the Asiatic polar front. Geochemistry Geophysics Geosystems 9(8): Q08Q09, doi:10.1029/2007GC001938.

Mao, D., Lei, J., Zhou, J., Xue, J., wang, C. and Rehemutula, Z. 2016. Movement rules of different shifting dunes and semi-shifting dunes in Cele, Xinjiang Uygur Autonomous Region. Research of Soil and Water Conservation 23(3): 278–282.

MDCES (Minqin Desert Control Experiment Station). 1975. Deserts and Control in Gansu. Lanzhou: Gansu People's Publishing house. pp. 33–38.

Muhs, D., Bush, C., Cowherd, S. and Mahan, S. 1995. Geomorphic and geochemical evidence for the source of sand in the Algodones dunes, Colorado Desert, southeastern California. pp. 37–74. *In*: Tchakerian, V. (ed.). Desert Aeolian Processes. Chapman & Hall, London.

Muhs, D., Stafford, T., Cowherd, S., Mahan, S., Kihl, R., Maat, P., Bush, C. and Nehring, J. 1996. Origin of the Late Quaternary dune fields of northeastern Colorado. Geomorphology 17: 129–149.

Muhs, D. 2004. Mineralogical maturity in dunefields of North America, Africa and Australia. Geomorphology 59: 247–269.

Pan, K., Zhang, Z., Dong, Z., Zhang, C. and Li, X. 2019. Physicochemical characteristics of surface sediments of crescent-shaped sand dunes in the Hexi Corridor, Gansu, China. Journal of Desert Research 39(1): 44–51 (in Chinese with English abstract).

Pease, P., Tchakerian, V. and Tindale, N. 1998. Aerosols over the Arabian Sea: geochemistry and source areas for aeolian desert dust. Journal of Arid Environments 39: 477–496.

Pease, P. and Tchakerian, V. 2003. Geochemistry of sediments from Quaternary sand ramps in the southeastern Mojave Desert, California. Quaternary International 104: 19–29.

Pettijohn, F.J., Potter, P.E. and Siever, R. 1972. Sand and Sandstone. Springer-Verlag, New York.

Pu, Z. 2005. For the lost oasis—a review of studies on desertification in the Hexi Corridor during the historical period. Collections of Essays On Chinese Historical Geography 20(1): 157–158 (in Chinese).

Pye, K. 1987. Aeolian Dust and Dust Storms. Academic Press, Gainesville.

Qu, J., Huang, N., Ta, W., Lei, J. and Dong, Z. 2005. Structural characteristics of Gobi sanddrift and its significance. Advances in Earth Science 20(1): 19–23 (in Chinese with English abstract).

Raupach, M.R., Gillette, D.A. and Leys, J.F. 1993. The effect of roughness elements on wind erosion threshold. Journal of Geophysical Research 98(D2): 3023–3029.

Ren, X. 2010. Element analysis of surface sediments from active dunes in the Minqin Oasis and its adjacent deserts. Ma.D Dissertation, Lanzhou University, Lanzhou (in Chinese with English abstract).

Ren, X., Liu, T. and Wang, Z. 2010. Characters of geomorphologic parameter about barchans dunes. Research of Soil and Water Conservation 17(1): 163–166 (in Chinese with English abstract).

Ren, X. and Wang, Z. 2010. The provenance of eolian sediments in Minqin Oasis, Gansu Province. Journal of Ningxia University Natural Science Edition 31(1): 88–92 (in Chinese with English abstract).

Ren, X., Yang, X., Wang, Z., Zhu, B., Zhang, D. and Rioual, P. 2014. Geochemical evidence of the sources of aeolian sands and their transport pathways in the Minqin Oasis, northwestern China. Quaternary International 334-335: 165–178.

Rostagno, C.M. and Degorgue, G. 2011. Desert pavements as indicators of soil erosion on aridic soils in north-east Patagonia (Argentina). Geomorphology 134: 224–231.

Sahu, B.K. 1964. Depositional mechanisms from the size analysis of clastic sediments. Journal of Sedimentary Research 34(1): 73–83.

Sha, W., Shao, X. and Huang, M. 2002. Climate warming and its impact on natural regional boundaries in China since 1980s. Science China in Series D 32(4): 317–326.

Shao, T., Zhao, J., Zhou, Q., Dong, Z. and Ma, Y. 2012. Recharge sources and chemical composition types of groundwater and lake in the Badain Jaran Desert, northwestern China. Journal of Geographical Sciences 22: 479–496.

Shi, X., Li, G., Liu, S. and Wei, Y. 2018. Dynamic changes of barchans dunes and its relationship with meteorological factors along oasis fringe in Hexi Corridor. Journal of Gansu Agricultural University 2: 86–93 (In Chinese with English Abstract).

Sun, J. 2002. Provenance of loess material and formation of loess deposits on the Chinese Loess Plateau. Earth and Planetary Science Letters 203: 845e859.

Taylor, S.R. and McLennan, S.M. 1985. The Continental Crust: Its Composition and Evolution. Blackwell Scientific Publications, London.

Uno, I., Wang, Z., Chiba, M., Chun, Y., Song, S., Hara, Y., Jung, E., Lee, S.S., Liu, M., Mikami, M., Music, S., Nickovic, S., Satake, S., Shao, Y., Song, Z., Sugimoto, N., Tanaka, T. and Westphal, D.L. 2006. Dust model intercomparison (DMIP) study over Asia: overview. Journal of Geophysical Research 111: D12213, doi: 10.1029/2006JD006575.

Visher, G. 1969. Grain size distributions and depositional processes. Journal of Sedimentary Research 39(3): 1074–1106.

Wang, J., Liu, L. and Shen, L. 2013. Research of movement laws of barchan dunes in the Mu Us sandy land based on Google Earth software. Remote Sensing Technology and Application 28(6): 1094–1100.
Wang, L. 2011. Surface deposits in the Hexi Corridor and its adjacent areas and implications for provenance of Asian dust. Ma.D Dissertation, Lanzhou University, Lanzhou (in Chinese with English abstract).
Wang, L. and Wang, Q. 2013. Elemental compositions of surface deposits in the Hexi Corridor and its adjacent areas, northwestern China. Northwestern Geology 46(2): 69–80 (in Chinese with English abstract).
Wang, N., Ma, N., Chen, H., Chen, X., Dong, C. and Zhang, Z. 2013. A preliminary study of precipitation characteristics in the hinterland of Badain Jaran Desert. Advance in Water Sciences 24: 153–160 (in Chinese).
Wang, T. 2003. Desert and Desertification in China. Hebei Science and Technology Press, Shijiazhuang (in Chinese).
Wang, X., Lang, L., Hua, T., Zhang, C. and Wang, Z. 2013. Gravel cover of Gobi desert and its significance for wind erosion: an experimental study. Journal of Desert Research 33(2): 313–319 (in Chinese with English abstract).
Wang, X., Hua, T., Zhu, B., Lang, L. and Zhang, C. 2018. Geochemical characteristics of the fine-grained component of surficial deposits from dust source areas in northwestern China. Aeolian Research 34: 18–26.
Wen, J., Su, Z., Zhang, T., Tian, H., Zeng, Y., Liu, R., Kang, Y. and Velde, R. 2014. New evidence for the links between the local water cycle and the underground wet sand layer of a mega-dune in the Badain Jaran Desert, China. Journal of Arid Land 6: 371–377.
Williams, M.A.J. 2014. Climate Change in Deserts: Past, Present and Future. Cambridge University Press, New York.
Wolfe, S.A. and Nickling, W.G. 1996. Shear stress partition in sparsely vegetated desert canopies. Earth Surface Processes and Landforms 21: 607–619.
Wolfe, S., Muhs, D., David, P. and McGeehin, J. 2000. Chronology and geochemistry of Late Holocene eolian deposits in the Brandon Sand Hill, Manitoba, Canada. Quaternary International 67: 61–74.
Wu, X., Wang, X., Wang, Y. and Xu, B.X. 2017. Origin of water in the Badain Jaran Desert, China: new insight from isotopes. Hydrology and Earth System Sciences 21: 4419–4431.
Wu, Y., Wang, N., Zhao, L., Zhang, Z., Chen, L., Lu, Y., Lü, X. and Chang, J. 2014. Hydrochemical characteristics and recharge sources of Lake Nuoertu in the Badain Jaran Desert. Chinese Science Bulletin 59: 886–895.
Yang, X., Rost, K.T., Lehmkuhl, F., Zhu, Z. and Dodson, J. 2004. The evolution of dry lands in northern China and in the Republic of Mongolia since the Last Glacial Maximum. Quaternary International 118-119: 69–85.
Yang, X. 2006. Desert research in northwestern China. A brief review. Geomorphologie: Relief, Processus, Environment 4: 275–284.
Yang, X., Zhu, B. and White, P.D. 2007. Provenance of aeolian sediment in the Taklamakan Desert of western China, inferred from REE and major-elemental data. Quaternary International 175: 71–85.
Yang, X., Scuderi, L., Paillou, P., Liu, Z., Li, H. and Ren, X. 2011. Quaternary environmental changes in the drylands of China e a critical review. Quaternary Science Reviews 30: 3219–3233.
Yang, X., Li, H. and Conacher, A. 2012. Large-scale controls on the development of sand seas in northern China. Quaternary International 250: 74–83.
Yin, D., Qu, J., Zu, R., Tan, L. and An, Z. 2014. Impact of disturbing on amount of wind erosion of sandy Gobi. Journal of Desert Research 34(1): 1–8 (in Chinese with English abstract).
Yin, D., Qu, J., Zhao, S., Tan, L., An, Z. and Xiao, J. 2016. Amount of wind erosion to gravel gobi under different disturbing modes. Arid Land Geography 39(3): 495–503 (in Chinese with English abstract).
Yang, X., Ma, N., Dong, J., Zhu, B., Xu, B., Ma, Z. and Liu, J. 2010. Recharge to the inter-dune lakes and Holocene climatic changes in the Badain Jaran Desert, western China. Quaternary Research 73: 10–19.
Yang, X., Liang, P., Zhang, D., Li, H., Rioual, P., Wang, X., Xu, B., Ma, Z., Liu, Q., Ren, X., Hu, F., He, Y., Rao, G. and Chen, N. 2019. Holocene aeolian stratigraphic sequences in the eastern portion of the desert belt (sand seas and sandy lands) in northern China and their palaeoenvironmental implications. Science China Earth Sciences 62: 1302–1315.
Zhan, L., Chen, J., Li, L. and Barry, D.A. 2018. Comment on "Origin of water in the Badain Jaran Desert, China: new insight from isotopes" by Wu et al. (2017). Hydrology and Earth System Sciences 22: 4449–4454.
Zhang, H. and Ming, Q. 2006. Hydrology and lake evolution in hyperarid northwestern China and the mystery of mega-dune formation in Badain Jaran Desert. Advance in Earth Sciences 21: 532–538 (in Chinese).
Zhang, K., Qu, J., Zu, R. and Fang, H. 2004. Wind tunnel simulation about the effects of the different underlying surfaces on the features of drifting sand current. Arid Land Geography 37(3): 352–355 (in Chinese with English abstract).
Zhang, L. and Ren, G. 2003. Change in dust storm frequency and the climatic controls in northern China. Acta Meteorologica Sinica 61(6): 744–750 (in Chinese with English abstract).
Zhang, Z. and Dong, Z. 2014. Dune field patterns and wind environments in the middle reaches of the Heihe Basin. Journal of Desert Research 34(2): 332–341 (in Chinese with English abstract).
Zhang, Z. and Dong, Z. 2015. Grain size characteristics in the Hexi Corridor Desert. Aeolian Research 18: 55–67.

Zhang, Z., Dong, Z., Li, J., Qian, G. and Jiang, C. 2016. Implications of surface properties for dust emission from gravel deserts (gobis) in the Hexi Corridor. Geoderma 268: 69–77.
Zhang, Z., Dong, Z., Zhang, C., Qian, G. and Lei, C. 2017. The geochemical characteristics of dust material and dust sources identification in northwestern China. Journal of Geochemical Exploration 175: 148–155.
Zhang, Z., Pan, K., Zhang, C. and Liang, A. 2020. Geochemical characteristics and the provenance of aeolian material in the Hexi Corridor Desert, China. Catena 104483, doi: 10.1016/j.catena.2020.104483.
Zhao, J., Shao, T., Hou, Y., Lv, X. and Dong, Z. 2011. Moisture content of sand layer and its origin in a mega-dune area in the Badain Jaran Desert. Journal of Natural Resources 26: 694–702 (in Chinese).
Zhou, Y., Lu, H., Joseph, M., Miao, X., James, S. and Ronald, G. 2008. Optically stimulated luminescence dating of aeolian sand in the Otindag dune field and Holocene climate change. Science in China Series D Earth Sciences 51: 837–847.
Zhu, B. 2007. Geochemistry, hydrochemistry and sedimentology of the Taklamakan Desert in Tarim Basin, NW China. Ph.D doctoral dissertation, Institute of Geology and Geophysics Chinese Academy of Sciences, Beijing.
Zhu, B. and Yang, X. 2009. Chemical weathering of detrital sediments in the taklamakan desert, northwestern china chemical weathering of detrital sediments in the Taklamakan Desert, Northwestern China. Geographical Research 47(1): 57–70.
Zhu, B. and Yu, J. 2014. Aeolian sorting processes in the Ejina desert basin (China) and their response to depositional environment. Aeolian Research 12: 111–120.
Zhu, B., Yu, J., Rioual, P. and Ren, X. 2014. Particle size variation of aeolian dune deposits in the lower reaches of the Heihe River basin, China. Sedimentary Geology 301: 54–69.
Zhu, B., Zhang, J. and Sun, C. 2021. Physiochemical characteristics, provenance, and dynamics of sand dunes in the arid Hexi Corridor. Frontiers in Earth Science 9: 728202, doi: 10.3389/feart.2021.728202.
Zhu, B., Zhang, J. and Sun, C. 2022. Potential links of gobi, dust, and desertification: a comprehensive understanding from aeolian landforms evolution in a middle-latitude desert. Sedimentary Geology 428: 106049, doi: 10.1016/j.sedgeo.2021.106049.
Zhu, B. 2022a. Mechanism of land degradation and their environmental implications in a middle-latitude desert area of China. Land Degradation and Development 33: 145–178.
Zhu, B. 2022b. The recent evolution of dune landforms and its environmental indications in the mid-latitude desert area (Hexi Corridor). Journal of Geographical Sciences 32(4): 617–644.
Zhu, Z., Wu, Z., Liu, S. and Di, X. 1980. An Outline of Chinese Deserts. Science Press, Beijing (in Chinese).
Zhu, Z., Chen, Z., Wu, Z., Li, J., Li, B. and Wu, G. 1981. Study on the Geomorphology of Wind-drift Sands in the Taklamakan Desert. Science Press, Beijing (in Chinese).
Zhu, Z. and Wang, T. 1992. Theory and practice on sand desertification in China. Quaternary Sciences 2: 97–106 (in Chinese with English abstract).
Zhu, Z. and Chen, G. 1994. Sand Desertification in China. Science press, Beijing (in Chinese).
Zhu, Z. 1994. Development trend of sandy desertification in interland agro-pastoral region of northern China in recent ten years. Journal of Desert Research 14: 1–7 (in Chinese).
Zhu, Z. 1999. Deserts, Desertification, Land Degradation and Strategies for Rehabilitation in China. Environmental Press, Beijing (in Chinese).
Zimbelman, J. and Williams, S. 2002. Geochemical indicators of separate sources for eolian sands in the eastern Mojave Desert, California, and western Arizona. Geological Society of America Bulletin 114: 490–496.

Part 3

Morphology and Distribution of Dune Landscapes on the Indian Subcontinent

The first four chapters in this Part elaborate on many aspects related to India's large Thar Desert. In addition to details on geomorphology, sedimentology and chronology there is an examination of the impact of the Thar desert on humans (and vice versa).

Chapter 15: *Aayush Srivastava*. Geomorphology and chronology of dunes in the Thar desert is the focus here. Major sand dune types (e.g., parabolic, linear) and their orientation suggests a dominant role of the south-westerly Indian Summer Monsoon (ISM) winds in their formation and subsequent development. Luminescence dating, which provides a chronometric perspective to dune accumulation phases, has often been used to investigate and reconstruct past geomorphological changes in the Thar dunes against the backdrop of strengthening and weakening of the ISM. The chapter provides a review of over 200 published dune ages from the Thar to date, and assess their utility in palaeo reconstructions. The chapter concludes with recommendations for future chronology-based studies of aeolian features of the Thar.

Chapter 16: *Mahesh K. Gaur*, *Sudesh K. Wadhawan* and *R.P. Dhir*. The chapter deals with the geomorphological characteristics of sand dunes and the evolution of aeolian landscape of the western Rajasthan Thar Desert in India. The arid tract in Rajasthan covers an area of 19.6 million hectares, which is 62% of the total hot arid zones in India. The aeolian activity in the Thar began over > 150 ka, resolving that Thar is not of anthropogenic origin as suggested previously. The present spatial extent of the aeolian activity in the Thar is in a contracted stage compared to that in the geological past, which refutes the arguments on its rapid north-eastward expansion. The western-most belt of the Rajasthan is covered with sand dunes extending from the great Rann along the Pakistan border in Punjab. Authors have used coarse and high resolution satellite images to study temporal changes in the dunal form, size and areal extent. There is a multiplicity of types of dunes and of dune agglomerations and these are described and discussed.

Chapter 17: *Hema Achyuthan and Sudesh K. Wadhawan*. Authors report on multi- and inter-disciplinary work carried out along the eastern margin of the Thar Desert on the Late Quaternary deposits reveal that the region has been a major corridor for human migration 'Out of Africa' and occupation in a semi-arid to arid landscape since MIS 5. Stabilized dunes, arid paleosols and calcrete nodules/horizons associated with Lower to Middle, Upper and Mesolithic stone tools reveal the occupancy of early stone age man on the landscape. Detailed geoarcheological studies supported by U/Th and OSL dating methods reveal that during the last ~ 200 ka major pluvial periods along Thar desert margin were ~ 200 ka (MIS 7.3), ~ 160–150 ka, ~ 126–115 ka (Climatic Optimum), ~ 90–80 ka, ~ 55–50 ka, ~ 30 to 26 ka, and 7–6 ka, and these periods coincided with the phases of intense high summer monsoons. There were major periods of enhanced aeolian sand mobilization and accumulation.

Chapter 18: *Sudesh K. Wadhawan.* The authors attempt to elaborate on influence of the underlying geology and tectonics of both the Quaternary and pre-Quaternary regime on sediment deposition and try to analyze evolutionary trends of parabolic aeolian bedforms in the Thar Desert. The clustered parabolic dunes are characteristic aeolian landforms of the Thar Desert in NW India. These older stabilized dune forms have been worked upon by episodically accelerated aeolian dynamism that resulted in producing several evolved and superimposed complex dune forms. Based on dune morphometric analyses, it is feasible to infer a hierarchical system of aeolian bedforms. Geospatial diversity of dunes grades from poorly organized units of isolated parabolic forms lying in discernible rows astride the dominant south-westerly winds, close to the Rann of Kutch to well organized regular waves of clustered rake-like high parabolic dunes with coalescing arms forming massive, but deflated mega-barchans.
Coastal dune systems of the Indian sub-continent are a special category and in the four chapters that follow, the focus is on the factors influencing the formation and development of dunes and on the relationship between the land and the sea. People and their activities are also factors that impact on the stability of dunes. New methods of studying dunes are discussed and techniques are explained.

Chapter 19: *Neelratan Singh et al.* focus on the cold arid and tectonically active, Ladakh Himalayan Zone. The objective of the present study is to understand morphology and analyze migration rate of dunes in the selected part of Shyok Valley, Ladakh Himalaya. The authors endeavor to explain the processes behind the formation of these barchanoid dunes and the reasons responsible for their migration. The morphometric analysis of the dunes has been carried out. This is a first of its kind of attempt made to calculate the migration rate of the dunes of Leh-Ladakh of India.

Chapter 20: *Linto Alappat et al.* India is characterized by a long coastline measuring over 7500 km. Each dune system in the coast is genetically associated with change in prevailing climatic conditions or sea level. The chapter will review the dunes as a geomorphological feature, paleo-climate proxy to differentiate their characteristics at different geographical regions along the coast, categorize them based on their chronology of formation and evolutionary process. The new methods of studying dunes, including chronology will be discussed based on the available luminescence ages and its methodological advancement will be briefly explained.

Chapter 21: *Ashis K. Paul et al.* The chapter highlights the diversity of coastal sand dunes in different coastal sections of India with categories of primary, mature and ancient dunes. A scheme of dune classification is prepared for the entire coastal zones of India based on the physiographic settings and morphologic types of the dunes in the present work. A hypothesis is built up regarding the evolution of coastal sand dunes and other aeolian deposits which is tested in the field and laboratory for the identification of geomorphological diversities of dune landscapes. The study reveals that various anthropogenic pressures have degraded the heritage landscapes along the coastal belts. Some remedial actions are mentioned and novel ones are proposed.

Chapter 22: *U. de S. Jayawardena.* This chapter highlights some important features along the coastal zone of Sri Lanka with the special attention on the dune sand deposits. There are many coastal habitats and geomorphological features. Pleistocene dune deposits occur with old raised beach and Holocene dunes occur with younger raised beach. The formation and shapes of dune deposits in Sri Lanka depends on the rainfall distribution, wind strength, wind direction and sand supply. No accurate estimate of the rate of annual deposition of dune sands has been calculated. The laboratory investigations indicate that the dune sand can be mixed with river sand, cement and water for the building construction industry. There are many environmental problems within the coastal zone due to weak law and order implementation. In the future these problems may be increased due to the sea level rise within next 100 years. Therefore, some new policies required to protect the people, society, environment and resources in the coastal zone of Sri Lanka.

Chapter 23: *M.S. Islam*. This chapter is an overview of coastal sand dune system of Bangladesh based on available documents, and field observation. Due to adequate supply of sediment, prevailing wind system, tidal range and wave activities, the coastal belt of Bangladesh, particularly the eastern belt, is ideal for the growth and development of sand dune landscape. Dunes are widely distributed along the on-shore coastal belt and also in off-shore islands. The existing scientific knowledge on morphological and ecological characteristics of coastal dune system of Bangladesh is poor. Despite morphological and ecological importance, this unique habitat is being seriously degraded due to natural hazards and also by unwise human interventions. The chapter highlights the importance of P^3C doctrine (Protection, Preservation, Prevention and Control) of dune management practice in Bangladesh.

Chapter 15

Dune Accumulation Chronologies from the Thar Desert, India: A Review

Aayush Srivastava[1,2]

1. Introduction

Often considered as the easternmost extension of the vast Saharo-Arabian mid-latitude desert belt, the Thar is a subtropical dryland in South Asia, covering an area of ~ 200,000 km^2 and supporting a large and dense human population (Singhvi and Kar, 2004; Roy et al., 2009). Geographically, the Thar Desert is mainly located in the state of Rajasthan in North-west India, bordering the irrigated Indus River plains in Pakistan to the west, the Punjab plains in the north and north-east, the Aravalli range in the south-east and the Rann of Kucch in the south. The Luni (~ 500 km long) is the only river system in the desert, which is dotted by various saltwater lakes including Didwana, Sambhar, Lunkaransar, Bap-Malar, etc. The Thar is unique with distinct dunefields of different morphometric types present throughout the desert, extending beyond the extent of occurrence of dune activity today (Wadhawan, this volume). Several types of sand dunes have been mapped in the Thar by extensive fieldwork in the past (Goudie et al., 1973; Wasson et al., 1983; Kar, 1987; Kar, 1993) and more recently by digital interpretation of remotely sensed images (Moharana et al., 2013). These dunes include parabolic, linear, transverse, barchans, etc., which vary in dimensions and cover ~ 60% of the total landscape in the Thar. Moharana et al. (2013) defined and mapped two more types, the dune complex (in areas where the form is a result of complex wind patterns) and the dissected dune complex (where the dune slope has numerous gullies). The orientation of these dunes suggests a dominant role of the south-westerly Indian Summer Monsoon (ISM) winds in their formation (Fig. 15.1). Based on meteorological observations, the ISM winds have been shown to exceed the threshold for sand movement, before the onset of rainfall, in the pre-monsoon period (Wasson et al., 1983; Kar, 1993). Dune activity is more pronounced during this period, and only after this increased wind intensity that rainfall is delivered to the region. The amount of annual rainfall ranges from 100 mm or less in the west to about 500 mm in the east, about ~ 90% of which is received during the southwest monsoon season between July and September. Due to monsoonal rainfall, large expanses of dunes and interdune plains are vegetated, comparable to the dryland landscapes in the Southern Hemisphere, notably south-west Kalahari, inland Australia and central Argentina (Bullard et al., 1995; Hesse, 2010; Zárate and Tripaldi, 2012).

[1] School of Earth and Environmental Sciences, University of St Andrews, UK.
[2] School of Geography and the Environment, University of Oxford, UK.
Emails: as537@st-andrews.ac.uk, aayush.srivastava@ouce.ox.ac.uk

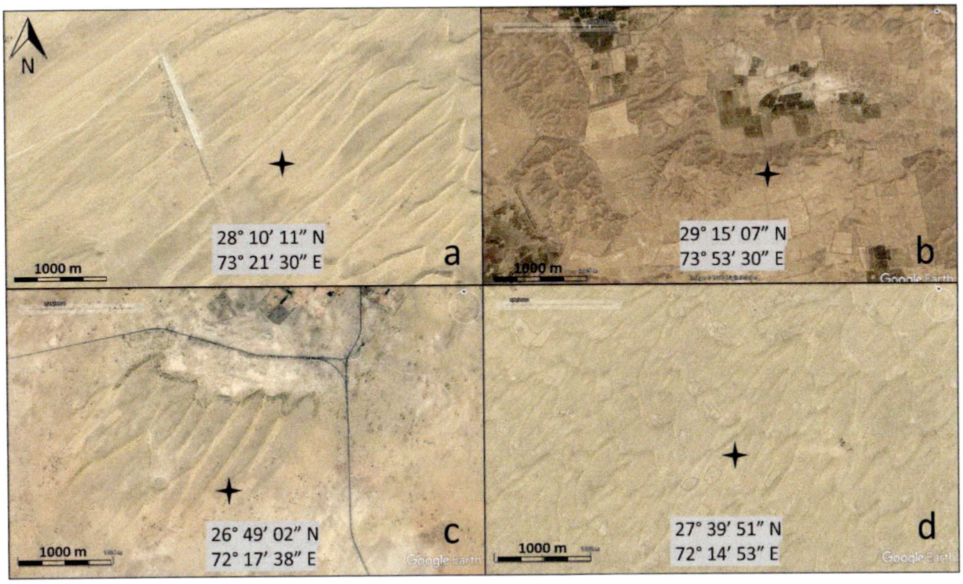

Figure 15.1. Examples of different dune types present in the Thar: (a) linear dunes in the central Thar (b) dune complex in the northeastern Thar (c) parabolic dunes in the central Thar, and (d) transverse dunes in the western Thar. Note, the orientation of these dunes is in the SW direction controlled by the dominant ISM winds.

Earlier studies by Goudie et al. (1973) and Wasson et al. (1984), based on the presence of sand dunes in the areas where modern precipitation is several times higher than what would allow dune activity, hypothesized that the Thar experienced expansion and contraction during the late Quaternary. The location of the Thar in the monsoonal system, which controls wind seasonality, precipitation and vegetation cover, implies that minor perturbations in monsoon dynamics may impact on the geomorphic processes, including aeolian dynamics, in this seasonally semi-arid to arid region. The extensive dune systems in the Thar, which are strongly aligned in the dominant direction of SW monsoon winds, therefore provide an opportunity to understand temporal and spatial changes in aeolian dynamics in the late Quaternary.

Additionally, the density of human population in the Thar is very high today, at more than 110 persons per km^2 (in sharp contrast with other deserts which have population densities below 10 persons per km^2; Census of India, 2011) and consequently, the Thar is undergoing intense human pressure (Kar, 2013; Dhir, 2018). The Central Arid Zone Research Institute, India reports that most of the dune systems in the Thar, which occupy 45–50% of the total area of the desert, are semi-stabilized, potentially making them sensitive to both activation through natural processes, and those enhanced by human activities. It, therefore, becomes necessary to understand the past thresholds, time scales and amplitudes of the variability of geomorphic processes shaping the dunes to make meaningful comparison with changes caused in modern times (Singhvi and Kar, 2004; Singhvi and Porat, 2008).

2. Generating dune accumulation chronologies

A major limit on paleoenvironmental and paleoclimate reconstructions in deserts is the scarcity of traditionally dateable material such as organic matter and carbonates that could serve as proxy of past changes (Lancaster, 2008). Even when present, these materials often come with complications. Radiocarbon ages from drylands are often erroneous because of contamination from the incorporation of old and modern carbon and the dating system (mostly in carbonates) not being closed. Earlier attempts to quantify dune records and develop a chronology (e.g., Goudie et al., 1973; Raghavan and Courty, 1987) were based on radiocarbon dating and proved difficult because of scarcity of dateable material and contamination of carbonate nodules. Other methods such as uranium series disequilibrium dating require pure carbonates that are rarely found in drylands (Singhvi et al., 1996). Due to extreme changes in temperature and lack of organic matter, amino acid racemisation methods are also difficult to apply in deserts (Owen et al., 2007). The use of tephra as time markers has also remained limited in deserts because preservation of the tephra layer is rare in dunefields owing to high aeolian activity (Lowe, 2011).

However, the development and applications of luminescence dating to sands (Singhvi et al., 1982) have indubitably transformed the analysis of geomorphic and sedimentary environments in drylands by adding a new dimension of chronometric control to the interpretation of sedimentary records and stratigraphic correlations. Luminescence dating is the ideal means to directly date periods of aeolian sand deposition, as it determines the time elapsed since burial of sediment, and hence, from multiple vertical samples phases of dune accumulation can be established (Duller, 2004). Lancaster (2008) noted that the advances in luminescence dating have occurred during a period in which significant conceptual progressions have also been made in aeolian geomorphology. Consequently, several luminescence dating techniques have been applied in a range of aeolian contexts around the world and over 4000 ages have been generated to reconstruct the timing and rate of dune accumulation (Lancaster et al., 2016).

Box 15.1 Luminescence dating of sand dunes

Natural minerals in the environment, such as quartz and feldspars, act as dosimeters, producing luminescence in response to naturally occurring radioactivity in the environment (e.g., Aitken, 1985; Aitken, 1998; Duller, 2008). This luminescence is produced on contact with heat (thermoluminescence; TL) or light (Optically Stimulated Luminescence; OSL), either in the environment or the laboratory. For dating purposes, it is the magnitude of the luminescence signal, accumulated since the last exposure of sediment to light which is of interest. The equivalent dose (D_e) is the amount of laboratory radiation (in Grays) required to replicate the natural luminescence signal, and the D_e is the numerator of the age calculation equation. To calculate an age, it is also necessary to measure the rate of radioactivity delivered to the sample from the surrounding sediment matrix, and this is called the environmental Dose rate ($\dot{D}$). A luminescence age is derived using the equation below,

$$\text{Age (ka)} = \textit{Equivalent dose (Gray)/Total environmental dose rate } (\textit{Gray.ka}^{-1})$$

Generally, for dating dune sands, the OSL signal from quartz is preferred (Lancaster, 2008; Singhvi and Porat, 2008) because it bleaches rapidly in nature (e.g., Godfrey-Smith et al., 1988; Duller, 2008), with near complete bleaching of the quartz OSL signal after 100 seconds exposure to daylight.

In deserts, cycles of erosion, transportation and deposition of sediments are observed over numerous temporal and spatial scales. During deposition, mineral grains present in sediments are exposed to surrounding ionizing radiation and electrons are trapped in their crystal defects. The electron populations build up and the luminescence signal is acquired over time, proportional to the radiation exposure (the environmental dose rate) (Fig. 15.2a). The luminescence is zeroed by exposure to sunlight with erosion and transportation (Fig. 15.2b). With another cycle of burial and exposure to ionizing radiation, free electrons are again stored in charge defects within grains' crystal lattices (Fig. 15.2c), and further light exposure of grains to light during erosion and transportation zeroes the luminescence (Fig. 15.2d). The grains are buried again, and luminescence is acquired with exposure to ionizing radiation (Fig. 15.2e). Following sampling without exposure to light, the natural luminescence signal can be measured, and from this, the luminescence age is calculated (Fig. 15.2f).

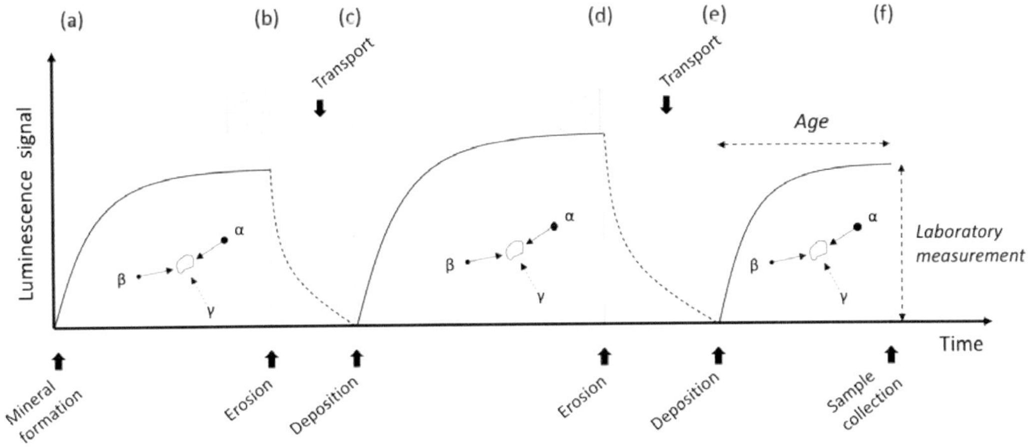

Figure 15.2. Schematic representation of luminescence signal accumulation and resetting in nature.

3. Dune accumulation chronologies from the Thar Desert

The Thar witnessed the first application of luminescence dating to sand dunes anywhere in the world nearly four decades ago by Singhvi et al. (1982) who used TL dating to date eight dune samples from archaeological contexts at Amarpura, Langhnaj and Bagor (Fig. 15.3). This seminal study added a quantitative dimension to the study of palaeo-desert dynamics and inspired several investigations later. There are now over 200 age records from the Thar, most of which have been summarized by Singhvi and Kar (2004) and key results further assessed by Singhvi and Porat (2008). In 2016, for the International Union for Quaternary Research (INQUA) dune atlas chronologic database, Lancaster et al. (2016) collated all the Thar dune ages (n = 146) published by Singhvi and colleagues (e.g., Andrews et al., 1998; Kar et al., 1998; Thomas et al., 1999; Kar et al., 2001; Juyal et al., 2003; Dhir et al., 2010). Three key points need to be considered here (i) there are multiple occurrence of the same data in different publications and in such cases, chronological data reported from their first accessible source are included here; (ii) some data are available in wider literature, but not included in the database (e.g., Chawla et al., 1992; Singhvi et al., 1994), (iii) recent studies have now expanded the available luminescence age data set by providing new suites of dune ages (Shitaoka et al., 2012; Durcan et al., 2019; Srivastava et al., 2019a; Srivastava et al., 2019b; Srivastava et al., 2020).

The two dune sequences that provide the oldest evidence of dune activity in the Thar Desert are in central Thar and include ages of ~ 187 ka at 16R dune (Singhvi et al., 2010) and ~ 155 ka at Chamu dune (Fig. 15.4; Dhir et al., 2010). These two dune sites remain the only dated evidence of dune activity beyond 100 ka in the Thar, with accumulation events occurring both before and between ~ 105–81 ka at Chamu (Dhir et al., 2010) and at ~ 130–109 ka and ~ 80 ka at 16R dune (Singhvi et al., 2010). Recently investigated, two linear dune samples provide the first reported occurrence of pre-Holocene dune activity from the northern Thar and demonstrate that the dunes have been present in the region since at least ~ 58 ka (Srivastava et al., 2019b). These dated records extended dune activity in the region back to the late Pleistocene, and attest to the antiquity of the Thar. Other reported older phases of dune accumulation include ~ 64 and ~ 60 ka at Shergarh Tri-junction (Andrews et al., 1998), ~ 55 ka and ~ 43 ka at Shergarh Tri-junction (Andrews et al., 1998), ~ 58 ka at Chamu (Dhir et al., 2010), ~ 40 ka 48 at 16R dune (Singhvi et al., 2010) (all in the east-central Thar) and ~ 40–30 ka at Awai (western Thar) (Chawla et al., 1992). After ~ 30 ka, the majority of evidence of dune activity comes from the southern margin of the Thar, with only one record dating to ~ 27 ka from Chamu in the central Thar (Dhir et al., 2010) and another dating to ~ 21 ka at 16R dune in the eastern Thar (Singhvi et al., 2010). In the southern Thar, dune activity has been recorded at ~ 25 ka at Dharoi and ~ 26 ka at Akhaj (Juyal et al., 2003). Post ~ 18 ka, evidence

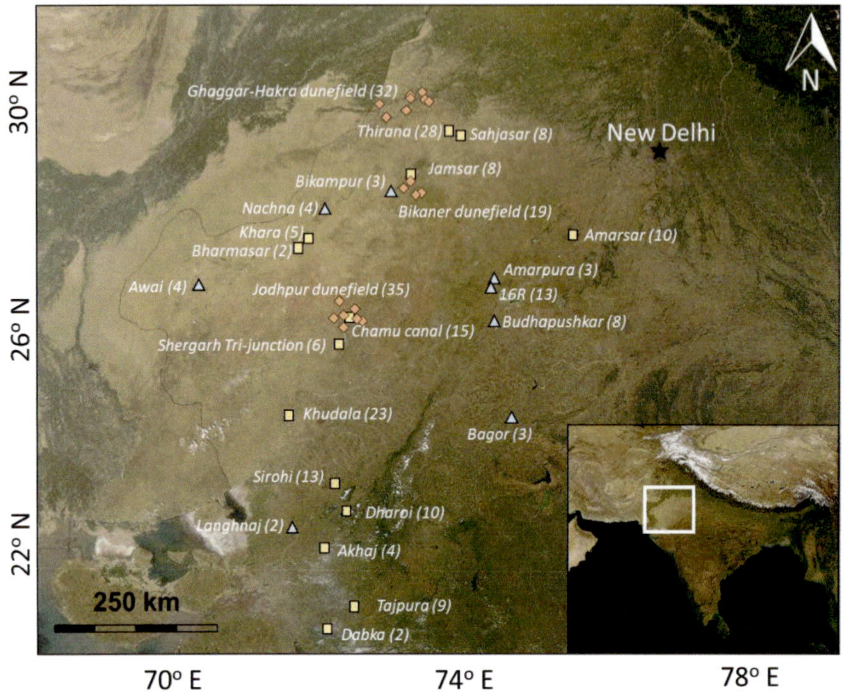

Figure 15.3. All dated dune sites in the Thar (with number of published ages). Yellow square symbols denote sites listed in Singhvi and Kar (2004) and the INQUA dunes atlas chronologic database (Lancaster et al., 2016). Blue triangle symbols indicate sites reported elsewhere, while orange diamonds represent dunefield-based investigations conducted recently in the desert.

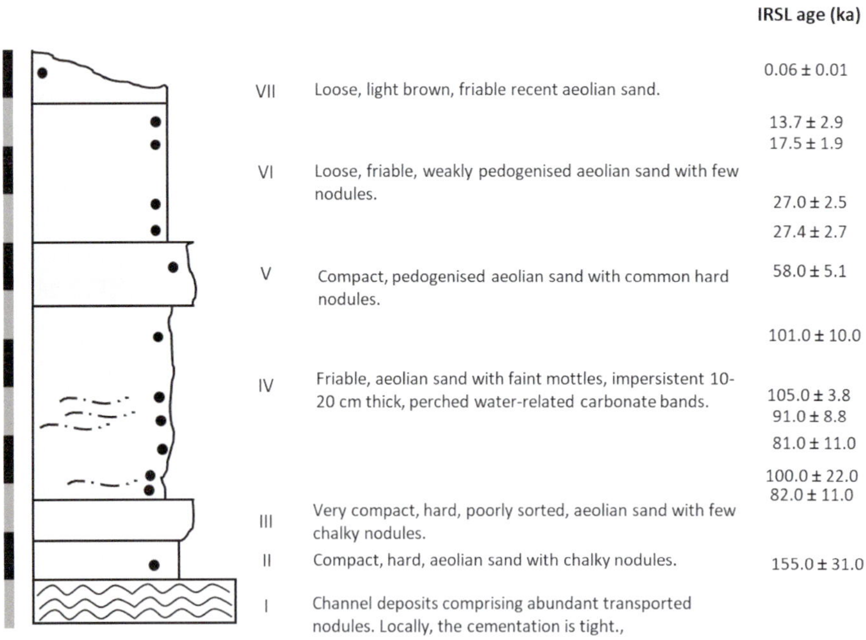

Figure 15.4. Field log and some analytical data of 12 m thick Chamu dune section demonstrate the antiquity of the Thar dune accumulation (reproduced from Dhir et al. (2010)). Sampling points for dating are represented by solid circles.

for dune activity is more widespread in the Thar. In the east and central regions, dune accumulations occurred at between ~ 17–13 ka at Chamu in the central Thar (Dhir et al., 2010), between ~ 15–13 ka at Nachna in the west (Chawla et al., 1992), ~ 13.5–11.5 ka at Bikampur in the north (Chawla et al., 1992) and ~ 13–11 ka at 16R Dune in the east (Singhvi et al., 2010).

With the inclusion of new records published in the last few years, there are multiple new Holocene records of dunes from the Thar. From the available data, the pattern of luminescence ages displays two primary characteristics: (1) phases of dune accumulation throughout the Holocene, and (2) two intense periods of accumulation around the early Holocene and in the last millennium. The records of early Holocene dune accumulation have been reported from various dune sites notably from the central and northern Thar (Srivastava et al., 2019a, 2019b). Interestingly, these records were seldom reported in earlier studies, despite this period being marked by strengthened ISM winds. Prasad and Enzel (2006) attributed this absence to seemingly large uncertainties associated with previously published ages, while Kar et al. (2004) described the presence of this phase of accumulation from dune records from the southern margin of the Thar. These early Holocene accumulation periods match with other dune records from monsoon dominated dunefields in the north-eastern Rub'al Khali (Atkinson et al., 2011; Atkinson et al., 2012) as well as an array of other geochemical and palaeoecological records from other terrestrial and marine archives (Srivastava et al., 2020). Multiple spatially heterogeneous middle and late Holocene records of dune activity have been reported at ~ 4–3, 2–1 and 0.6–0.2 ka from linear, parabolic and source-bordering dunes, which are attributed to more regional factors including sediment supply facilitated by drying river channels and alluvial plains of regional rivers. The latest phase of dune activity, marked by modern ages (< 100 a), display the role of anthropogenic pressure on the landscape as another important control on dune development and evolution (Singhvi and Kar, 2004; Srivastava et al., 2019b).

3.1 Sampling strategies

The identification of key sampling regions, which include spatial coverage that covers the geographic variability in the Thar, is an important objective in the review of published literature. So far, the application of intensive dunefield sampling programs has been less prevalent in the Thar compared to many other deserts worldwide such as the Kalahari in Africa (Telfer and Thomas, 2007; Stone and Thomas, 2008), the Rub' al Khali in Arabia (Leighton et al., 2013; Atkinson et al., 2011), the Negev in Israel (Roskin et al., 2011) and the Strzelecki and Tirari in Australia (Fitzsimmons et al., 2007). The published luminescence chronologies (and the subsequent interpretation of ages) mainly stem from scattered and sporadic records with variable spatial coverage and different ranges of sampling depths. Further, often missing information about the location and depth of dated sediments is another concern, which makes robust interrogation of existing luminescence age data sets difficult (e.g., Thomas and Bailey, 2019). In some studies, there is a bias towards focusing on deep trenches (often associated with the expansion of irrigation canals into the region) to establish the antiquity of dune accumulation phases in the desert (Kar, 2013; Dhir et al., 2010). As a result, the Holocene histories are sometimes not recorded. Hesse (2016) observed this issue in the context of Australian dunes and highlighted that under-representation of upper deposits might withhold useful insight into zones of reworking, erosion and depositional structures. For a comprehensive paleoenvironmental perspective, sampling should be methodical (spatially over a dunefield and vertically through a dune profile), as Thomas and Burrough (2016) noted, to capture dune system response to external forcing, and not localized dune activity 'noise'. Notably nearly all the dune investigations in the Thar made use of naturally exposed dune sections or artificially dug them, which allowed visibility of a clear stratigraphy and identification of any distinct change in sedimentology.

Furthermore, sensitivity to climatic and forcing factors (e.g., vegetation, sediment supply, wind direction) which determine sediment transportation, accumulation and preservation varies between different morphometric types and within the same type spatially and temporally (Thomas, 2013; Thomas and Burrough, 2016). Therefore, the geomorphologic understanding of dune type and

Table 15.1. Luminescence ages from the Thar dunes included in the INQUA dunes atlas chronologic database and published elsewhere (Lancaster et al., 2016) [Y/N = data available/not available].

Sample location	Ages (#)	Dune type	Depth data	Protocol	Luminescence type	Mineral	Age model	Reference
Amarpura	3	*N*	*Y*	MAAD	TL	Quartz	*N*	Singhvi et al. (1982)
Bagor	3	*N*	*Y*	MAAD	TL	Quartz	*N*	Singhvi et al. (1982)
Langhnaj	2	*N*	*Y*	MAAD	TL	Quartz	*N*	Singhvi et al. (1982)
Nachna	3	*Interdune*	*Y*	MAAD	TL	Quartz	*N*	Chawla et al. (1992)
Bikampur	4	*N*	*Y*	MAAD	TL	Quartz	*N*	Chawla et al. (1992)
Awai	4	*N*	*Y*	MAAD	TL	Quartz	*N*	Chawla et al. (1992)
Budhapushkar	8	*N*	*Y*	MAAD	TL	Quartz	*N*	Singhvi et al. (1994)
Thirana I	16	parabolic	*Y*	MAAD	TL	Quartz	*N*	Thomas et al. (1999)
Thirana II	12	parabolic	*N*	MAAD	OSL	Quartz	*N*	Singhvi and Kar (2004)
Sahjasar	8	parabolic	*N*	MAAD	BGSL	Quartz	*N*	Thomas et al. (1999)
Sirohi	13	undifferentiated	*N*	MAAD	IRSL	Feldspars	*N*	Thomas et al. (1999)
Chamu	15	sand sheet	*Y*	MAAD	IRSL	Feldspars	*N*	Dhir et al. (2010)
Shergarh	6	not mentioned	*N*	MAAD	IRSL/OSL/TL	Mixed	*N*	Andrews et al. (1998)
Jamsar	8	linear	*Y*	MAAD	GLSL	Quartz	*N*	Thomas et al. (1999)
Amarsar I	6	parabolic	*Y*	MAAD	TL	Quartz	*N*	Singhvi and Kar (2004)
Amarsar II	4	parabolic	*Y*	MAAD	TL	Quartz	*N*	Singhvi and Kar (2004)
Ghughalia	3	not mentioned	*Y*	MAAD	IRSL	Feldspars	*N*	Thomas et al. (1999)
Khara	5	transverse	*N*	MAAD	IRSL	Feldspars	*N*	Kar et al. (1998)
Bharmasar	2	transverse	*N*	MAAD	IRSL	Feldspars	*N*	Kar et al. (1998)
Dharoi I	7	obstacle	*Y*	MAAD	BGSL	Quartz	*N*	Juyal et al. (2003)
Dharoi II	3	obstacle	*Y*	MAAD	BGSL	Quartz	*N*	Juyal et al. (2003)

Akhaj	4	poorly organised	*Y*	MAAD	BGSL	Quartz	*N*	Juyal et al. (2003)
Tajpura I	4	obstacle	*Y*	MAAD	BGSL	Quartz	*N*	Juyal et al. (2003)
Tajpura II	5	obstacle	*Y*	MAAD	BGSL	Quartz	*N*	Juyal et al. (2003)
Dabka	2	poorly organised	*Y*	MAAD	BGSL	Quartz	*N*	Juyal et al. (2003)
Khudala	23	parabolic	*Y*	MAAD	TL/BGSL/IRSL	Mixed	*N*	Kar et al. (2001)
16R	13	*N*	*Y*	MAAD/SAR	TL/OSL	Quartz	*N*	Singhvi et al. (2010)
G-H dune field	13	source-bordering	*Y*	SAR	OSL	Quartz	*N*	Shitaoka et al. (2012)
G-H dune field	13	source-bordering	*Y*	SAR	OSL	Quartz	*CAM*	Durcan et al. (2019)
Jodhpur dune field	31	parabolic	*Y*	SAR	OSL	Quartz	*CAM*	Srivastava et al. (2019a)
Bikaner dune field	19	linear	*Y*	SAR	OSL	Quartz	*CAM*	Srivastava et al. (2019b)

development is critical for interpreting palaeoenvironments (Lancaster, 2008; Thomas, 2013). For instance, whilst parabolic dunes in Jodhpur dunefields have recorded rapid accumulation in the last 70 years, with rates up to 5 m/year, such rapid and modern signatures of dune accumulation are absent from the records from the investigated linear dunefield in the Bikaner region. It is interesting because both dunefields, only ~ 200 km apart, experience comparable rainfall and support equally dense human and livestock populations. This further highlights the importance to consider the different sensitivities of different dune types to the same external controls. However, in many studies the type and morphological description of dunes is often missing or ambiguous (Table 15.1). There are instances of contradictory descriptions as well between original citations and later reviews. For example, the INQUA dunes atlas database (Lancaster et al., 2016), the sampled dune site Jamsar in northern Thar is a parabolic dune, while Thomas et al. (1999) described it as a linear dune site. The inability to procure aerial photographs due to government restrictions, the poor resolution in old Google Earth images and an altered modern landscape, owing to agricultural activity imply that it is difficult to pin-point the exact dune form and sampling location.

Sampling strategy and location within a dune may also impact on the chronology produced and it is important to know the details of sampled dune to understand the process based of dune systems and their diverse response to drivers of accumulation (e.g., Telfer and Thomas, 2007; Stone and Thomas, 2008; Telfer, 2011). Srivastava et al. (2019b), in their investigation of a parabolic dunefield in central Thar, demonstrated that samples collected at comparable depths from a dune's nose and arm may result in significantly different chronology. Since luminescence ages only reveal a part of sedimentation history, the incorporation of all available evidence including stratigraphic and sedimentological data is important when interpreting chronologies. It allows for recognition of a more complex system and a thorough assessment of the meaning of the dates in terms of dune development (Fitzsimmons and Telfer, 2008; Singhvi and Porat, 2008).

3.2 Dune types

Parabolic dunes, being the most widespread in arid and semi-arid landscape of the Thar, are the most widely studied dune type in terms of chronological investigations (nearly one-third of available records; Lancaster et al., 2016). Not only have parabolic dunes provided evidence of the oldest known phases of dune activity in the Thar, the presence of sparse calcified nodules further attested to the greater stability of these dunes (Wasson et al., 1983; Singhvi and Kar, 2004). Though in the last century, due to the anthropogenic disturbances, parabolic dunes in central Thar have reactivated and shown modern accumulation of up to ~ 5 m per year (Srivastava et al., 2019b). Linear dunes are the most studied dune type worldwide, but have received very limited attention in the Thar. An analysis of a linear dunefield in the Bikaner region in the northern Thar has shown these dunes preserved evidence of older and multiple phases of past periods of dune activity (Srivastava et al., 2019a), just like linear dunes elsewhere (Thomas, 2013). More mobile forms transverse dunes have also been investigated to constrain dune migration rates due to environmental changes and anthropogenic pressure. To date, there is just one study on these dunes, conducted by Kar et al. (1998) who sampled a 7 m high transverse dune in western Thar and based on five luminescence ages provided evidence of high variability in dune accumulation in the last two millennia. Source-bordering dunes and associated fluvial deposits, along the Ghaggar-Hakra palaeochannel forming the northern margin of the Thar, have been a subject of recent studies, due to the potential link between fluvial activity in this system and urban collapse/demise of the Indus Civilisation (Saini et al., 2009; Saini and Mujtaba, 2012; Shitaoka et al., 2012; Durcan et al., 2019). These studies reveal that Ghaggar-Hakra was a highly dynamic geomorphic system, particularly in the middle to late Holocene, and further research has been recommended by the authors for an enhanced understanding of changing environmental conditions in the region.

3.3 Luminescence dating methods

The Thar dune ages have been generated using a wide array of laboratory methodologies, though the majority of ages were derived from analyses in two luminescence laboratories: at the Physical Research Laboratory in India and more recently at University of Oxford. Following Singhvi et al. (1982)'s seminal study, earlier studies primarily used thermoluminescence (TL) (e.g., Chawla et al., 1992; Singhvi et al., 1994; Singhvi et al., 2010), while later studies employed Infra-Red Stimulated Luminescence (IRSL) and blue/green light stimulated luminescence (BLSL/GLSL) signals for age determination (e.g., Thomas et al., 1999; Kar et al., 2001; Singhvi and Kar, 2004). It is notable that the majority of published Thar dune ages used Multiple Aliquot Additive Dose (MAAD) methods to determine D_e (as summarized in Table 15.1), which is required to calculate a luminescence age. Though these were widely used in 1980s and 1990s, MAAD methods are now largely redundant issues since they often add a relatively large uncertainty to a determined age. Consequently, often published Thar dune ages have 15–60% uncertainty associated with them, which makes the pulling apart of distinct phases of dune accumulation and their interrogation with other proxies difficult. The more useful Single Aliquot Regeneration (SAR) method (Murray and Wintle, 2000; Wintle and Murray, 2006), which corrects for sensitivity changes and generates D_e by interpolation and has replaced MAAD since 2000, has been adopted in the recent investigations (e.g., Srivastava et al., 2019a; Srivastava et al., 2019b; Durcan et al., 2019). These instrumental and methodological advances in luminescence dating make it possible to establish the timing of dune accumulation with a better chronological precision (< 10%) compatible with the resolution of late Quaternary environmental changes. In another distinctive study, Durcan et al. (2019) examined single grains of Thar quartz using the SAR protocol and showed that only 2–4% of quartz grains provide a measurable OSL signal after a set laboratory dose and pass the satisfy or the rejection criteria (Wintle and Murray, 2006). Multiple-grain studies by Srivastava et al. (2019b) also revealed a large number of measured aliquots failed on one or more quality checks required for the application of the SAR protocol (Murray and Wintle, 2000). The collective observation confirms that the use of the Thar quartz to generate a robust luminescence age is time intensive.

Several studies in the past used the IRSL signal from feldspar for luminescence age calculation (e.g., Kar et al., 1998; Kar et al., 2001; Dhir et al., 2010; Table 15.1). Although IRSL is favorable for dating of older sediments, some feldspar signals are athermally unstable (termed as anomalous fading), which if not corrected, results in underestimation of ages (Wintle, 1973; Wallinga et al., 2001). Several correction methods have been developed based on laboratory measurements of the fading rate in the sample (e.g., Huntley and Lamothe, 2001; Kars and Wallinga, 2009) and to obtain a stable luminescence signal in feldspars (e.g., Thomsen et al., 2008; Buylaert et al., 2009). For the available published IRSL ages from the Thar (Table 15.1), fading corrections were not made. However, when dating samples from a parabolic dune in central Thar, Andrews et al. (1998) suggested that feldspars did not show any age underestimation or anomalous fading by showing concordance of the IRSL ages with OSL ages; albeit the OSL ages had large uncertainties. Age uncertainties, which seldom fall below 1 σ errors of 5%, can also appear due to multiple reasons other than choice of dating protocol, and are listed in Table 15.2.

3.4 Representation of luminescence data

A suite of dune luminescence ages, which comprises point age data with one sigma errors, requires meaningful representation which is vital for analysis and interpretation, and integration with other proxies of paleoclimatic and paleoenvironmental changes. The Thar dune ages have seen diverse representation of records and probability density plots remain more commonly used (e.g., Thomas et al., 1999; Singhvi and Kar, 2004). However, this method is not preferred for representing geoproxy data as peaks within such plots are determined by the error distribution rather than true values and may produce a deceptive picture (Galbraith, 1998; Galbraith and Roberts, 2012). Singhvi and Porat (2008) followed a new approach and displayed Thar dune records as horizontal bars corresponding to

Table 15.2. Potential luminescence dating challenges related to the Thar dune sediments.

Challenge	Explanation	Consequence	Possible solution
Quartz's luminescence properties	Thar dune quartz is not very bright (compared to quartz from other drylands) and a large number of measured aliquots failed on one or more quality checks (e.g., signal levels; recycling ratio; recuperation).	Generation of a robust luminescence age becomes time intensive.	The fast ratio can be applied to test the dominance of preferable fast component in quartz OSL signal, and a large number of aliquots per sample can be assessed to generate a robust dataset.
Signal saturation	The upper limit of quartz OSL signal is limited by a saturation of signal at high doses.	Interpolation of the natural signal onto the dose response curve at higher doses is likely to result in large and asymmetric uncertainties in D_e.	A minimum D_e can still be calculated from saturated samples, and therefore minimum age can be obtained.
Partial bleaching	Sediments have incomplete resetting of the luminescence signal before burial, resulting in residual signals.	Rarely a problem in aeolian setting like the Thar, but the impact of partial bleaching is most significant for young samples which have small doses.	Its effects can be detected by analyzing the single grain D_e distributions or comparison of quartz and feldspar age estimates; and reduced by statistical techniques using the appropriate age model, e.g., minimum age model.
Post-depositional mixing	Sediments deposited at different times undergo post-depositional mixing due to bioturbation (disturbance of the sediment by plants and animals) or geomorphic processes (e.g., sediment slumping into gullies).	Sample may exhibit multi-modal D_e distributions but unlike poorly bleached sediments this distribution may have modes both higher and lower than the D_e distribution associated with the true burial age.	Its effects can be detected by analyzing the single grain D_e distributions reduced by statistical techniques using the appropriate age model, e.g., finite fixture model.
Beta heterogeneity	Variation in sedimentary matrix which causes grains deposited at the same time to receive different radiation depending upon the range of distances from 'hotspots' or areas of high beta radiation and 'coldspots' or areas of low beta radiation.	Sample exhibits multi-modal D_e distributions, which is difficult to differentiate from bioturbation.	Its effect can be reduced by statistical techniques using the appropriate age model, e.g., minimum age model, the lowest 5% model, etc.
Dose rate change over time	In their radioactive decay, ^{238}U and ^{232}Th radioactive decay are in secular equilibrium. This is because the nature is an open system, atoms are gained and lost during the burial history of a sample resulting in disequilibrium which can make dose rate calculations erroneous.	The problem is more pronounced in hydrologically variable environments where there is greater probability for daughter products to get mobilized in/out of sediment system.	The complete identification and characterization of disequilibrium require lengthy spectrometry measurements, and its correction is often problematic. The uncertainties associated with radionuclides can be widened to account for the disequilibrium.
Anomalous fading	Some feldspar signals are athermally unstable, resulting in signal loss during the burial period.	It hampers the reliability of luminescence dating with feldspars by underestimating the ages.	Application of the post-IR IRSL signal of K-feldspar stimulated at a higher temperature after a prior low-temperature IR stimulation has significantly lower fading rate.

each sampled site. One difficulty in such representation is the uncertainty whether ages include one sigma errors or are errorless means. The other difficulty is the issue of inclusion of sites which have a single age, making the bar (widths) narrower and biased towards an age with wide errors. Thomas and Burrough (2012), in the context of the Kalahari, suggested placing records in 1 ka time slices as binary data simplifying whether dune activity (index) was recorded for a given time window. This approach also does not consider uncertainties associated with individual ages, and remains prone to sampling biases. In the case of the Thar, such binary bin representation will result in the index being on meaning consistent dune activity throughout the Holocene, under-representing periods of most intense accumulation. More recently, Srivastava et al. (2019a, 2019b) plotted calculated Thar ages as a simple function of depth, but this method does not allow an effective comparison and integration with other proxies. As dune systems do not offer continuous and complete record of accumulation, Srivastava et al. (2020) used a new model called Accumulation Intensity (*AI*) model (Thomas and Bailey, 2017) to provide an improved interpretations of dune geochronological data. Whilst their resultant *AI* curves provide a nuanced assessment of Thar dune accumulation histories, temporally these were restricted only to the Holocene. Nevertheless, with more data becoming available as more studies are conducted, alternative methods for data representation should be considered and evaluated.

3.5 *Interpretation of the Thar dune accumulation chronologies*

Dunes, like many other geomorphic features, have complex accumulation histories involving multiple cycles of erosion and subsequent re-deposition that may result in the removal of the onset (Nanson et al., 1992), end phases (Munyikwa, 2005), or the total of aeolian events. This results in 'filtering' of signal between initial deposition and the interpretation of the luminescence ages (Bailey and Thomas, 2014). The extraction of information about the past from dune sequences, therefore, requires an enhanced understanding of factors including sediment supply, transport capacity of the wind and sediment availability (Preusser, 2009), sediment preservation (Munyikwa, 2005; Stone and Thomas, 2008), dune formation, migration and stabilization (Thomas and Burrough, 2012). A recommended way to minimize this issue and capture as complete as possible record of preserved history is to sample at multiple locations in the dunefield, which Thomas and Burrough (2016) suggested as the primary depositional unit. Some Thar dune studies based their interpretation on a single dune yet detailed investigations (e.g., Andrews et al., 1998; Dhir et al., 2010). However, as suggested theoretically by Bailey and Thomas (2014) and demonstrated in the context of linear dunes in the northern Thar by Srivastava et al. (2019a), a single dune profile does not capture a representative record for a dunefield or desert as a whole. An integrated assessment of Thar dune ages derived from multiple sites had been attempted notably by Singhvi and Kar (2004), Singhvi and Porat (2008) and more recently Srivastava et al. (2020).

Singhvi and Kar (2004), who first synthesized all available Thar dune records, suggested that major phases of pre-LGM (Last Glacial Maximum) dune accumulation during the late Pleistocene between ~ 100 and ~ 115 ka, ~ 75 ka, ~ 55 ka and between ~ 30 and ~ 25 ka. The dune accumulation in the Thar was episodic as conditions encouraging dune accumulation were reached for limited durations (Singhvi and Porat, 2008). Singhvi and Kar (2004) postulated that during the LGM, the Thar could have extended beyond its present boundary due to prevalent arid conditions, but wind strength was not sufficient to result in widespread large-scale accumulation. In fact, the maximum mobility and accumulation occurred during the transition period from an arid to wetter phase and when the south-westerly ISM winds strengthened (Kar et al., 2004). Therefore, this interpretation was used to explain why the peak of aridity at ~ 18 ka was marked by very little dune activity. The re-establishment of the monsoon winds in the region, hence, caused the peak of dune accumulation after the LGM. It is noteworthy that dry periods are expected to be more active (in both erosion and accumulation) than wetter periods, therefore, dated dune sediment ages are most likely to reflect periods of drier and/or windier conditions (Bailey and Thomas, 2014).

Recently, Srivastava et al. (2020) applied the new quantitative approach of *AI* methodology (Thomas and Bailey, 2017) to new sets of chronologies generated as well as old dates, and analyzed generated *AI* peaks alongside other paleoenvironmental records (variability in the ISM and associated aridity) to allow an enhanced recognition of the complexity of the Thar landscape development and reconstruct regional paleoenvironmental changes. The widespread early Holocene dune accumulation records generate a broad distinctive *AI* peak and show an in-phase relationship with increased productivity records and strengthened ISM winds deduced from the Arabian Sea sediments, high precipitation records inferred from oxygen isotope ratios from speleothems in Arabia and central India, and high lake levels along the northern Thar margin based on sedimentological and geochemical observations. Whilst there is a paleoclimatic evidence of an enhanced ISM during the early Holocene (e.g., Gupta et al., 2003; Gill et al., 2017), the correlations are equivocal in the middle and late Holocene. During the middle to late Holocene, against the backdrop of a weakening ISM, the phases of dune accumulation have been associated with enhanced sediment supply from drying river channels in northern Thar, and both drying river channels and a lowered (Arabian) sea level in the southern Thar. This local role of changing sediment supply has been shown as an important control on dune development in the eastern Rub'al Khali (Preusser, 2009) and central Australia (Maroulis et al., 2007). In the last century, another important control on dune development in the Thar that has emerged is anthropogenic pressure on landscape. Srivastava et al. (2019b) presented a history of dune activity in the twentieth century indicating the sensitivity of dunes to landscape alterations. There are several gaps in the aeolian history of the Thar dunes, notably during the LGM and mid-Holocene. These gaps in the chronological records, which are important for interpretation, could represent an artefact of sampling or periods of relatively wetter and/or calmer conditions with little dune activity. Confidence in environmental interpretations of gaps, however, increases when the differences remain open in the face of additional new data, but such interpretations remain provisional in the absence of independent verification (Bailey and Thomas, 2014).

4. Conclusions and future considerations

Despite a relatively large published luminescence age data set, the investigation of the Thar dune records in the context of understanding past landscape dynamics remains challenging. This chapter has collated the published dune accumulation records from the Thar and reviewed them in the context of the developments and improvements that have occurred in dune sampling methods and luminescence dating protocols over the years. The chapter briefly describes the relationship between OSL age clusters in dunefield chronologies and past environmental or climatic change, suggesting the relationship may not be straightforward just as in drylands elsewhere (Thomas and Burrough, 2012; Telfer and Hesse, 2013; Thomas, 2013; Lancaster et al., 2013). Nevertheless, the ongoing development and application of new field-based and laboratory techniques allow for future expansion of the research conducted in the Thar, which will be critical in providing a more temporally-complete assessment of the region's dune accumulation history.

- Results from recent studies (e.g., Durcan et al., 2019; Srivastava et al., 2020) confirm that intensive fieldwork strategies and modern dating protocols have the potential to elucidate complex dune chronologies which significantly enhance analyses of dune development. However, despite once being at the forefront of application of luminescence dating to sand dunes, the accumulation history of Thar dunes and their implications for geomorphological and paleoenvironmental significance remain under-studied when compared with other deserts (e.g., the Negev, the Rub' al Khali, the Mallee; Roskin et al., 2011; Leighton et al., 2013; Lomax et al., 2011) which are well dated, both spatially and in terms of full-dune profile records. As dunefield based studies from the Thar are not yet present, more intensive and systematic dunefield-based chronometric studies are recommended to help further identify spatial and temporal variability across the Thar dunefields.

- The post-infrared signals have been demonstrated to offer improved signal stability over IRSL signals (e.g., Zhang and Li, 2020 and references therein), and have been used to date sediments in the regions to the north (Singh et al., 2017) and south of the Thar (Blinkhorn et al., 2019). Therefore, since the Thar quartz has shown saturation after ~ 120 Gy (Srivastava et al., 2019a), future studies could consider exploring feldspars. As current feldspar ages from Thar dunes have not been corrected for anomalous fading (Lancaster et al., 2016), investigations of the IRSL signals will provide the added benefit of checking the robustness of already published records.
- Whilst luminescence ages from dune sediments are increasingly being used in paleoenvironmental and geomorphological studies, it has often been noted that publications presenting luminescence ages from the Thar dunes do not mention and describe vital information like stratigraphic and geographical location, sample preparation, number of aliquots studied, over-dispersion in calculated values as well as the age models used (Srivastava, 2019). Collectively, this opacity of luminescence data undermines the robustness and reproducibility of ages, and affects subsequent interpretation. As noted earlier, due to under reporting of sample information more than 50% of Thar ages from the INQUA dune atlas database were not eligible for *AI* analysis. Future studies and papers are recommended to be more thorough in data inclusion and presentation while publishing.
- In the Thar, there remains a marked unevenness in spatial extent and type of dune investigated. Over 50% of dune records came from parabolic dunes, ~ 20% from undifferentiated dunes, ~ 4 from transverse dunes and none from linear dunes (Lancaster et al., 2016). Less dominant dune types, especially linear and transverse dunes, require further investigation. Whilst linear dunes have seen first dunefield based investigations only recently, there still remains more potential to investigate the vast linear dunefields in the Pakistani Thar, from where no luminescence dating based study of dunes has been reported to date. More mobile forms such as transverse dunes should also be investigated to constrain dune migration rates due to environmental changes and anthropogenic pressure. Sand ramps in the southern Thar are another untapped aeolian feature in the Thar which should be further explored. Investigations of sand ramps from southern Africa have established the potential of their records in proxy-poor environments (e.g., Telfer et al., 2014; Rowell et al., 2018).
- Dhir et al. (2004) described widespread and morphologically well-evolved calcretes in the Thar and interpreted these in the context of past aggradation episodes of sediments. The ages of these deposits which lie in the range of 350–1650 ka (determined using electron spin resonance technique), allow for further assessment of Quaternary paleoenvironmental and paleoclimatic changes as the oldest Thar dune record extends back to ~ 200 ka. Furthermore, future studies may look beyond the stratigraphic aspect of the aeolian bedforms formation in the Thar Desert. Researchers may use advanced biotechnological tools to assess the impact of microbial diversity on the process of stabilization of dunes, elemental cycles in the deserts, management of desert vegetation, microfloral diversity, etc.

References

Aitken, M.J. 1985. Thermoluminescence Dating. Academic Press, London.

Aitken, M.J. 1998. Introduction to Optical Dating: The Dating of Quaternary Sediments by the Use of Photon-Stimulated Luminescence. Oxford University Press, Oxford.

Allchin, B., Goudie, A. and Hegde, K. 1978. Prehistory and palaeogeography of the Great Indian Desert.

Andrews, J.E., Singhvi, A.K., Kailath, A.J., Kuhn, R., Dennis, P.F., Tandon, S.K. and Dhir, R.P. 1998. Do stable isotope data from calcrete record late Pleistocene monsoonal climate variation in the Thar Desert of India? Quaternary Research 50(3): 240–251.

Atkinson, O.A., Thomas, D.S.G., Goudie, A.S. and Bailey, R.M. 2011. Late Quaternary chronology of major dune ridge development in the northeast Rub'al-Khali, United Arab Emirates. Quaternary Research 76(1): 93–105.

Atkinson, O.A., Thomas, D.S.G., Goudie, A.S. and Parker, A.G. 2012. Holocene development of multiple dune generations in the northeast Rub'al-Khali, United Arab Emirates. The Holocene 22(2): 179–189.
Bailey, R.M. and Thomas, D.S.G. 2014. A quantitative approach to understanding dated dune stratigraphies. Earth Surface Processes and Landforms 39(5): 614–631.
Blinkhorn, J., Ajithprasad, P., Mukherjee, A., Kumar, P., Durcan, J.A. and Roberts, P. 2019. The first directly dated evidence for Palaeolithic occupation on the Indian coast at Sandhav, Kachchh. Quaternary Science Reviews 224: 105975.
Bullard, J.E., Thomas, D.S.G., Livingstone, I. and Wiggs, G.F.S. 1995. Analysis of linear sand dune morphological variability, southwestern Kalahari Desert. Geomorphology 11(3): 189–203.
Buylaert, J.P., Murray, A.S., Thomsen, K.J. and Jain, M. 2009. Testing the potential of an elevated temperature IRSL signal from K-feldspar. Radiation Measurements 44(5-6): 560–565.
Census of India. 2011. Provisional Population Totals. New Delhi: Government of India. http://censusindia.gov.in/2011-prov-results/
Chawla, S., Dhir, R.P. and Singhvi, A.K. 1992. Thermoluminescence chronology of sand profiles in the Thar Desert and their implications. Quaternary Science Reviews 11(1-2): 25–32.
Deotare, B.C., Kajale, M.D., Kshirsagar, A.A. and Rajaguru, S.N. 1998. Geoarchaeological and palaeoenvironmental studies around Bap–Malar playa, district Jodhpur, Rajasthan. Current Science pp. 316–320.
Dhir, R.P., Tandon, S.K., Sareen, B.K., Ramesh, R., Rao, T.K.G., Kailath, A.J. and Sharma, N. 2004. Calcretes in the Thar desert: Genesis, chronology and palaeoenvironment. Journal of Earth System Science 113(3): 473–515. doi:10.1007/bf02716737.
Dhir, R.P., Singhvi, A.K., Andrews, J.E., Kar, A., Sareen, B.K., Tandon, S.K., Kailath, A. and Thomas, J.V. 2010. Multiple episodes of aggradation and calcrete formation in Late Quaternary aeolian sands, Central Thar Desert, Rajasthan, India. Journal of Asian Earth Sciences 37(1): 10–16.
Dhir, R.P., Joshi, D.C. and Kathju, S. 2018. Thar Desert in Retrospect and Prospect. Scientific Publishers.
Duller, G.A.T. 2004. Luminescence dating of Quaternary sediments: recent advances. Journal of Quaternary Science 19(2): 183–192.
Duller, G.A.T. 2008. Single-grain optical dating of Quaternary sediments: why aliquot size matters in luminescence dating. Boreas 37(4): 589–612.
Durcan, J.A., Thomas, D.S.G., Gupta, S., Pawar, V., Singh, R.N. and Petrie, C.A. 2019. Holocene landscape dynamics in the Ghaggar-Hakra palaeochannel region at the northern edge of the Thar Desert, northwest India. Quaternary International 501: 317–327.
Fitzsimmons, K.E., Rhodes, E.J., Magee, J.W. and Barrows, T.T. 2007. The timing of linear dune activity in the Strzelecki and Tirari Deserts, Australia. Quaternary Science Reviews 26(19-21): 2598–2616.
Fitzsimmons, K.E. and Telfer, M.W. 2008. Complexities in the preservation and interpretation of Late Quaternary dune records: examples from the Tirari Desert, Australia and the southwestern Kalahari, South Africa. Chungará 30: 295–308.
Galbraith, R.F. 1988. Graphical display of estimates having differing standard errors. Technometrics 30(3): 271–281.
Galbraith, R.F. and Roberts, R.G. 2012. Statistical aspects of equivalent dose and error calculation and display in OSL dating: an overview and some recommendations. Quaternary Geochronology 11: 1–27.
Gill, E.C., Rajagopalan, B., Molnar, P.H., Kushnir, Y. and Marchitto, T.M. 2017. Reconstruction of Indian summer monsoon winds and precipitation over the past 10,000 years using equatorial pacific SST proxy records. Paleoceanography 32(2): 195–216.
Godfrey-Smith, D.I., Huntley, D.J. and Chen, W.H. 1988. Optical dating studies of quartz and feldspar sediment extracts. Quaternary Science Reviews 7(3-4): 373–380.
Goudie, A.S., Allchin, B. and Hegde, K.T.M. 1973. The former extensions of the great Indian sand desert. Geographical Journal pp. 243–257.
Gupta, A.K., Anderson, D.M. and Overpeck, J.T. 2003. Abrupt changes in the Asian southwest monsoon during the Holocene and their links to the North Atlantic Ocean. Nature 421(6921): 354–357.
Hesse, P.P. 2010. The Australian desert dunefields: formation and evolution in an old, flat, dry continent. Geological Society, London, Special Publications 346(1): 141–164.
Hesse, P.P. 2016. How do longitudinal dunes respond to climate forcing? Insights from 25 years of luminescence dating of the Australian desert dunefields. Quaternary International 410: 11–29.
Huntley, D.J. and Lamothe, M. 2001. Ubiquity of anomalous fading in K-feldspars and the measurement and correction for it in optical dating. Canadian Journal of Earth Sciences 38(7): 1093–1106.
Juyal, N., Kar, A., Rajaguru, S.N. and Singhvi, A.K. 2003. Luminescence chronology of aeolian deposition during the Late Quaternary on the southern margin of Thar Desert, India. Quaternary International 104(1): 87–98.
Kar, A. 1987. Origin and transformation of longitudinal sand dunes in the Indian desert. Zeitschrift fur Geomorphologie 31: 311–337.

Kar, A. 1993. Aeolian processes and bedforms in the Thar Desert. Journal of Arid Environments 25(1): 83–96.

Kar, A., Felix, C., Rajaguru, S.N. and Singhvi, A.K. 1998. Late Holocene growth and mobility of a transverse dune in the Thar Desert. Journal of Arid Environments 38(2): 175–185.

Kar, A., Singhvi, A.K., Rajaguru, S.N., Juyal, N., Thomas, J.V., Banerjee, D. and Dhir, R.P. 2001. Reconstruction of the late Quaternary environment of the lower Luni plains, Thar Desert, India. Journal of Quaternary Science 16(1): 61–68.

Kar, A., Singhvi, A.K., Juyal, N. and Rajaguru, S.N. 2004. Late Quaternary aeolian sedimentation history of the Thar Desert. Geomorphology and Environment, ACB Publications, Kolkata, pp. 105–122.

Kar, A. 2013. Quaternary geomorphic processes and landform development in the Thar Desert of Rajasthan. Landforms Process. Environ. Manag, pp. 225–256.

Kars, R.H., Wallinga, J. and Cohen, K.M. 2008. A new approach towards anomalous fading correction for feldspar IRSL dating—tests on samples in field saturation. Radiation Measurements 43(2-6): 786–790.

Lancaster, N. 2008. Desert dune dynamics and development: insights from luminescence dating. Boreas 37(4): 559–573.

Lancaster, N., Yang, X. and Thomas, D.S.G. 2013. Spatial and temporal complexity in Quaternary desert datasets: implications for interpreting past dryland dynamics and understanding potential future changes. Quat. Sci. Rev. 78: 301–302.

Lancaster, N., Wolfe, S., Thomas, D.S.G., Bristow, C., Bubenzer, O., Burrough, S.L., Duller, G.A.T., Halfen, A., Hesse, P.P., Roskin, J., Singhvi, A.K., Tsoar, H., Tripaldi, A., Yang, X. and Zarate, M. 2016. The INQUA dunes atlas chronologic database. Quaternary International 410: 3–10.

Leighton, C.L., Bailey, R.M. and Thomas, D.S. 2013. The utility of desert sand dunes as Quaternary chronostratigraphic archives: evidence from the northeast Rub'al Khali. Quaternary Science Reviews 78: 303–318.

Lomax, J., Hilgers, A. and Radtke, U. 2011. Palaeoenvironmental change recorded in the palaeodunefields of the western Murray Basin, South Australia–New data from single grain OSL-dating. Quaternary Science Reviews 30(5-6): 723–736.

Lowe, D.J. 2011. Tephrochronology and its application: a review. Quaternary Geochronology 6(2): 107–153.

Maroulis, J.C., Nanson, G.C., Price, D.M. and Pietsch, T. 2007. Aeolian–fluvial interaction and climate change: source-bordering dune development over the past ~ 100 ka on Cooper Creek, central Australia. Quaternary Science Reviews 26(3-4): 386–404.

Moharana, P.C., Gaur, M.K., Choudhary, C., Chauhan, J.S. and Rajpurohit, R.S. 2013. A system of geomorphological mapping for western Rajasthan with relevance for agricultural land use. Annals of Arid Zone 52(3&4): 163–180.

Munyikwa, K. 2005. The role of dune morphogenetic history in the interpretation of linear dune luminescence chronologies: a review of linear dune dynamics. Progress in Physical Geography 29(3): 317–336.

Murray, A.S. and Wintle, A.G. 2000. Luminescence dating of quartz using an improved single-aliquot regenerative-dose protocol. Radiation Measurements 32(1): 57–73.

Nanson, G.C., Chen, X.Y. and Price, D.M. 1992. Lateral migration, thermoluminescence chronology and colour variation of longitudinal dunes near Birdsville in the Simpson Desert, central Australia. Earth Surface Processes and Landforms 17(8): 807–819.

Owen, L.A., Bright, J., Finkel, R.C., Jaiswal, M.K., Kaufman, D.S., Mahan, S., Radtke, U., Schneider, J.S., Sharp, W., Singhvi, A.K. and Warren, C.N. 2007. Numerical dating of a Late Quaternary spit-shoreline complex at the northern end of Silver Lake playa, Mojave Desert, California: A comparison of the applicability of radiocarbon, luminescence, terrestrial cosmogenic nuclide, electron spin resonance, U-series and amino acid racemization methods. Quaternary International 166(1): 87–110.

Prasad, S. and Enzel, Y. 2006. Holocene paleoclimates of India. Quaternary Research 66(3): 442–453.

Preusser, F. 2009. Chronology of the impact of Quaternary climate change on continental environments in the Arabian Peninsula. Competus Rendus Geoscience 341(8-9): 621–632.

Raghavan, H. and Courty, M.A. 1987. Holocene and Pleistocene pedo-sedimentary environments in the Thar desert (Didwana, Inde). pp. 639–646. *In*: Micromorphologie des sols. Réunion internationale de micromorphologie des sols. 7.

Roskin, J., Porat, N., Tsoar, H., Blumberg, D.G. and Zander, A.M. 2011. Age, origin and climatic controls on vegetated linear dunes in the northwestern Negev Desert (Israel). Quaternary Science Reviews 30(13-14): 1649–1674.

Rowell, A.L., Thomas, D.S.G., Bailey, R.M. and Holmes, P.J. 2018. Sand ramps as palaeoenvironmental archives: Integrating general principles and regional contexts through reanalysis of the Klipkraal Sands, South Africa. Geomorphology 311: 103–113.

Roy, P.D., Nagar, Y.C., Juyal, N., Smykatz-Kloss, W. and Singhvi, A.K. 2009. Geochemical signatures of Late Holocene paleo-hydrological changes from Phulera and Pokharan saline playas near the eastern and western margins of the Thar Desert, India. Journal of Asian Earth Sciences 34(3): 275–286.

Saini, H. and Mujtaba, S. 2010. Luminescence dating of the sediments from a buried channel loop in Fatehabad area, Haryana: insight into Vedic Saraswati River and its environment. Geochronometria 37(1): 29–35.
Saini, H.S., Tandon, S.K., Mujtaba, S.A.I., Pant, N.C. and Khorana, R.K. 2009. Reconstruction of buried channel-floodplain systems of the northwestern Haryana Plains and their relation to the 'Vedic' Saraswati. Current Science 97(11): 1634–1643.
Shitaoka, Y., Maemoku, H. and Nagatomo, T. 2012. Quartz OSL dating of sand dunes in Ghaggar Basin, northwestern India. Geochronometria 39(3): 221–226.
Singh, A., Thomsen, K.J., Sinha, R., Buylaert, J.P., Carter, A., Mark, D.F., Mason, P.J., Densmore, A.L., Murray, A.S., Jain, M. and Paul, D. 2017. Counter-intuitive influence of Himalayan river morphodynamics on Indus Civilisation urban settlements. Nature Communications 8(1): 1–14.
Singhvi, A.K., Sharma, Y.P. and Agrawal, D.P. 1982. Thermoluminescence dating of sand dunes in Rajasthan, India. Nature 295(5847): 313.
Singhvi, A.K., Banerjee, D., Rajaguru, S.N. and Kumar, V.K. 1994. Luminescence chronology of a fossil dune at Budha Pushkar, Thar Desert: Palaeoenvironmental and archaeological implications. Current Science, pp. 770–773.
Singhvi, A.K., Banerjee, D., Ramesh, R., Rajaguru, S.N. and Gogte, V. 1996. A luminescence method for dating 'dirty' pedogenic carbonates for paleoenvironmental reconstruction. Earth and Planetary Science Letters 139(1-2): 321–332.
Singhvi, A.K. and Kar, A. 2004. The aeolian sedimentation record of the Thar Desert. Journal of Earth System Science 113(3): 371–401.
Singhvi, A.K. and Porat, N. 2008. Impact of luminescence dating on geomorphological and palaeoclimate research in drylands. Boreas 37(4): 536–558.
Singhvi, A.K., Williams, M.A.J., Rajaguru, S.N., Misra, V.N., Chawla, S., Stokes, S., Chauhan, N., Francis, T., Ganjoo, R.K. and Humphreys, G.S. 2010. A ~ 200 ka record of climatic change and dune activity in the Thar Desert, India. Quaternary Science Reviews 29(23-24): 3095–3105.
Srivastava, A. 2019. Late Quaternary dune activity in the Thar Desert and its implications for palaeoenvironmental reconstructions: an insight from optically stimulated luminescence dating. Unpublished PhD thesis. University of Oxford.
Srivastava, A., Durcan, J.A. and Thomas, D.S.G. 2019a. Analysis of late Quaternary linear dune development in the Thar Desert, India. Geomorphology 344: 90–98.
Srivastava, A., Thomas, D.S.G. and Durcan, J.A. 2019b. Holocene dune activity in the Thar Desert, India. Earth Surface Processes and Landforms 44(7): 1407–1418.
Srivastava, A., Thomas, D.S.G., Durcan, J.A. and Bailey, R.M. 2020. Holocene palaeoenvironmental changes in the Thar Desert: An integrated assessment incorporating new insights from aeolian systems. Quaternary Science Reviews 233: 106214.
Stone, A.E.C. and Thomas, D.S.G. 2008. Linear dune accumulation chronologies from the southwest Kalahari, Namibia: challenges of reconstructing late Quaternary palaeoenvironments from aeolian landforms. Quaternary Science Reviews 27(17-18): 1667–1681.
Telfer, M.W. and Thomas, D.S.G. 2007. Late Quaternary linear dune accumulation and chronostratigraphy of the southwestern Kalahari: implications for aeolian palaeoclimatic reconstructions and predictions of future dynamics. Quaternary Science Reviews 26(19-21): 2617–2630.
Telfer, M.W. 2011. Growth by extension, and reworking, of a south-western Kalahari linear dune. Earth Surface Processes and Landforms 36(8): 1125–1135.
Telfer, M.W. and Hesse, P.P. 2013. Palaeoenvironmental reconstructions from linear dunefields: recent progress, current challenges and future directions. Quaternary Science Reviews 78: 1–21.
Telfer, M.W., Mills, S.C. and Mather, A.E. 2014. Extensive quaternary aeolian deposits in the Drakensberg foothills, Rooiberge, South Africa. Geomorphology 219: 161–175.
Thomas, D.S.G. and Burrough, S.L. 2012. Interpreting geoproxies of late Quaternary climate change in African drylands: implications for understanding environmental change and early human behaviour. Quat. Int. 253: 5–17. http://dx.doi.org/10.1016/j.quaint.2010.11.001.
Thomas, D.S.G. and Burrough, S.L. 2012. Interpreting geoproxies of late Quaternary climate change in African drylands: implications for understanding environmental change and early human behaviour. Quaternary International 253: 5–17.
Thomas, D.S.G. 2013. Reconstructing paleoenvironments and palaeoclimates in drylands: what can landform analysis contribute?. Earth Surface Processes and Landforms 38(1): 3–16.
Thomas, D.S.G. and Burrough, S.L. 2016. Luminescence-based dune chronologies in southern Africa: analysis and interpretation of dune database records across the subcontinent. Quaternary International 410: 30–45.

Thomas, D.S.G. and Bailey, R.M. 2017. Is there evidence for global-scale forcing of Southern Hemisphere Quaternary desert dune accumulation? A quantitative method for testing hypotheses of dune system development. Earth Surface Processes and Landforms 42(14): 2280–2294.

Thomas, D.S.G. and Bailey, R.M. 2019. Analysis of late Quaternary dunefield development in Asia using the accumulation intensity model. Aeolian Research 39: 33–46.

Thomas, J.V., Kar, A., Kailath, A.J., Juyal, N., Rajaguru, S.N. and Singhvi, A.K. 1999. Late Pleistocene-Holocene-history of aeolian accumulation in the Thar Desert, India. Zeitschrift fur Geomorphologie Supplementband, pp. 181–194.

Thomsen, K.J., Murray, A.S., Jain, M. and Bøtter-Jensen, L. 2008. Laboratory fading rates of various luminescence signals from feldspar-rich sediment extracts. Radiation Measurements 43(9-10): 1474–1486.

Wadhawan, S.K. in press. Geospatial Diversity, Geologic Significance and Evolutionary Trends of Parabolic Dunes in Thar Desert of Indian Subcontinent. pp. xx-xxx (this volume).

Wallinga, J., Murray, A.S., Duller, G.A. and Törnqvist, T.E. 2001. Testing optically stimulated luminescence dating of sand-sized quartz and feldspar from fluvial deposits. Earth and Planetary Science Letters 193(3-4): 617–630.

Wasson, R.J., Rajaguru, S.N., Misra, V.N., Agrawal, D.P. and Dhir, R.P. 1983. Geomorphology, late Quaternary stratigraphy and paleoclimatology of the Thar dunefield. Zeitschrift für Geomorphologie. Supplementband 45: 117–151.

Wasson, R.J., Smith, G.I. and Agrawal, D.P. 1984. Late Quaternary sediments, minerals, and inferred geochemical history of Didwana Lake, Thar Desert, India. Palaeogeography, Palaeoclimatology, Palaeoecology 46(4): 345–372.

Wintle, A.G. 1973. Anomalous fading of thermo-luminescence in mineral samples. Nature 245(5421): 143.

Wintle, A.G. and Murray, A.S. 2006. A review of quartz optically stimulated luminescence characteristics and their relevance in single-aliquot regeneration dating protocols. Radiation Measurements 41(4): 369–391.

Wintle, A.G. 2008. Luminescence dating: where it has been and where it is going. Boreas 37(4): 471–482.

Zárate, M.A. and Tripaldi, A. 2012. The aeolian system of central Argentina. Aeolian Research 3(4): 401–417.

Zhang, J. and Li, S.H. 2020. Review of the Post-IR IRSL dating protocols of K-Feldspar. Methods and Protocols 3(1): 7.

Chapter 16

Geomorphological Characteristics of Sand Dunes and Evolution of Aeolian Landscape for Sustainable Future in Western Thar Desert, Rajasthan, India

Mahesh K. Gaur,[1,*] *Sudesh K. Wadhawan*[2] and *R.P. Dhir*[3]

1. Introduction

Landscapes around the world are dynamic and constantly changing. Various weathering, erosion and deposition processes on the earth's surface are constantly giving new shapes to the existing geomorphological features. Under the influence of these persistent natural phenomena, old structures are destroyed and transformed into new topographical features. However, the landscape also preserves distinct signatures of tectonics, driven by internal geological processes of the earth and counteracted and modified by surface processes that are essentially controlled by the climate regime. It is recognized that the most important geomorphological features in today's desert areas are attributed to large-scale Paleogene-Neogene crustal movements of up- or down-warping, folding or faulting and acted on by the earth's surface processes leading to weathering, physical disintegration, and fluvial and/or aeolian erosion of detrit products and deposits in aggrading basins as geo-patterns that characterize the evolutionary development of the landscape (McKee, 1979; Wasson and Hyde, 1983; Pye and Tsoar, 1990; Wadhawan, 1996, 2005, 2018). In the case of the Thar Desert, it is fairly well documented that the development of the Thar Desert in western India is closely related to the litho-tectonic architecture of the continental sedimentation basins that occurred during the Paleogene-Neogene continental drift and the docking of the Indian plate with the origin of Eurasia. Here the Paleogene-Neogene tectonic domain in the Thar was exposed to tensile stress fields that manifested themselves in a series of normal faults and grabens/sequence of half graben structures and step-faults that influenced the architectural development of continental sedimentation basins (Kar, 2020; Wadhawan, 2010, 2018). Therefore, both tectonic and climatic causes are found to be responsible for the aggradations of continental deposits in the Thar Desert (Valdiya, 2002; Wadhawan, 2010, 2018; Singhvi et al., 2012).

[1] ICAR-Central Arid Zone Research Institute, Jodhpur-342003, India.
[2] Amrita Vishwa Vidyapeetham University, Kollam-690525 and Former Director General, Geological Survey of India.
[3] 498, Defense Colony, Jodhpur-342008, India.
* Corresponding author: maheshjeegaur@gmail.com

The Thar Desert is characterized by arid-semiarid ecologically fragile environment and occupies a unique tectonic-sedimentary domain in the North-West (NW) of India. Winds are dominant factors with sufficient scope in the arid regions to play a significant role in shaping land surfaces and depending on the wind regime and sand supply. They strongly contribute to the development of distinctive geomorphic features including diverse sand dunes types. The Thar Desert in north-western India is a natural museum and model for various wind activities as it consists largely of a sandy plain and repeated patterns of sand dunes that overlie sand sheets and Hamada (a kind of barren bedrock plains and rocky ridges) as the highlands have been eroded at base of the Precambrian, Mesozoic and Tertiary rocks.

The bedrock with a sand-covered surface of varying thicknesses in the Thar Desert occasionally includes low rock outcrops and ridges that encompass the Precambrian Aravalli range of hills along the eastern edge of the desert and over the Neo-Proterozoic-Early Paleozoic volcanic and sedimentary rocks in the central parts to the fluvial and shallow marine sedimentary rocks on the western extremities covered by thick Quaternary sediments (Wadhawan in this volume). Various studies have been carried out on sand dunes over the past few decades in order to understand their transport mechanisms and provenance. Desert dunes are most widely studied as they are important indicators of aeolian dynamics and evidence of evolutionary aspects of sand recycling, ancient and modern environments (McKee, 1979; Gabris et al., 2012; Lancaster, 1988a, b).

It is observed that the Indus, which flows along the southwestern edge of the Great Indian Sand Desert, Thar, appears to be deeply ingrained, like the Ganges in the fore-deeper regions of the Himalayas in northern India. The geological evolution of this Himalayan perennial river and its tributaries, as the major sources of sediments for southwesterly summer monsoonal winds, have effectively contributed to the exposed lower reaches of the river-delta and coastal confluence region of the vast Indus River in the Arabian Sea (East et al., 2015; Clift et al., 2012).

However, due to human interference and rapid land-use changes, major modifications are occurring in the arid and semi-arid areas. The main transformations are induced by increasing human population and pressures of various kinds in both the rural and urban contexts (Gaur et al., 2012). The development of natural oil and gas fields, the increasing exploitation of aquifers, and a number of ambitious surface water utilization schemes through canals and a network of irrigation distributors, for both drinking, household and irrigation purposes, have continued to expand. These have exerted extra stress on the fragile desert ecosystem. In many areas, the rural population may reach the limits of cultivable land and have caused irreversible and negative effects on soils and vegetation (Warren et al., 1996; Dhir et al., 2018; Wadhawan, 2015, 2019; Kar and Kumar, 2020).

This chapter synthesizes information based on the geomorphological properties and distribution of sand dunes, their genetic importance in the developing aeolian landscape in the western Thar Desert of Rajasthan (India). An attempt is also made to characterize and understand the effects of human interventions and interactions between landscapes, climate and tectonics for a sustainable future of desert societies.

2. Thar Desert in Indian subcontinent: setting and context

The Thar or the Great Indian Sand Desert extends between the foothills of the denuded Precambrian Aravalli hill ranges on the eastern desert margin in the Indian state of Rajasthan and the Indus River in Pakistan (Fig. 16.1). Since it is located on the eastern extremities of vast arid land along the Tropic of Cancer, which includes the hyperarid parts in North Africa—the Sahara, Rub-al Khali and Lut pf Arab, the Thar desert experiences lesser aridity, due mainly to its location at the transition of two major rain-bearing climate systems, the Indian Summer Monsoon (ISM) that brings about 100–500 mm rainfall, and a Mediterranean westerly system which, sometimes brings showers of < 100 mm in winter and spring. Most of the landforms in the Thar Desert, currently have retained strong imprints of the fluctuating arid-semiarid climate during the late Quaternary period, which over the time experienced relative dominance of fluvial and aeolian processes. However, some

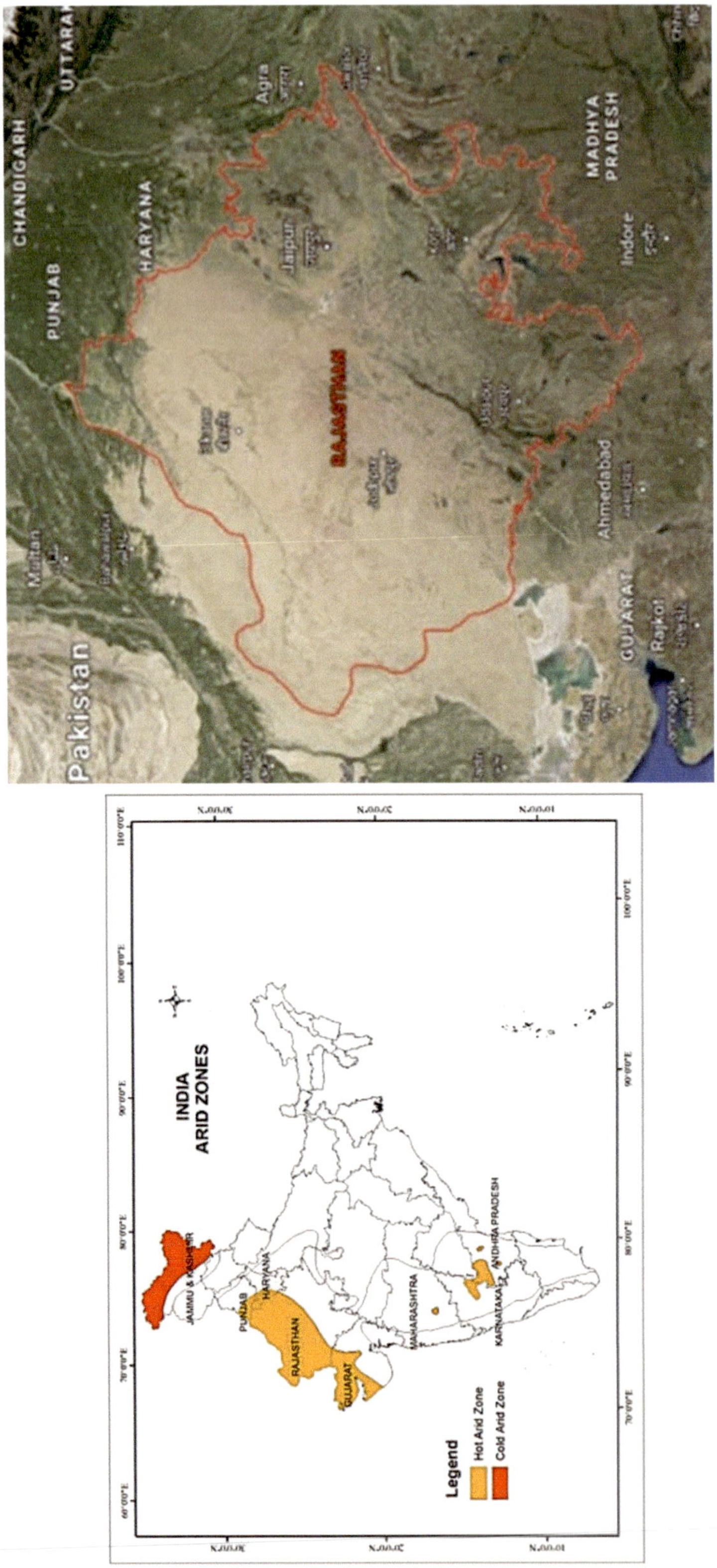

Figure 16.1. Extent of Arid Zone in India and location map of Thar Desert in Rajasthan as viewed in the satellite image [Source: Google Earth].

landforms also show the evidence of neotectonic movements, whereas the ominous signatures of human influences also exist.

Since the current climate is arid-semiarid, the weathering processes in the region are very slow. However, prolonged exposures to harsh and extreme conditions have prepared materials for transportation and subaerial denudation. Physical weathering is dominant throughout the region and occurs through diurnal and annual thermal expansion and contraction. Taking into account, the dominant geomorphological processes, the resulting construction of landforms can be broadly classified into fluvial, aeolian and lacustrine. It is noteworthy that the neotectonics efficacies of the operative earth processes correspond to the prevailing climatic zones, which fluctuated between warmer wet and cooler dry phases on different time scales and geospatial zonation that have left their traces in the Quaternary sedimentary layers (Dhir et al., 1992, 2018; Wadhawan, 2018; Kar, 1994, 2020).

The occurrence of dunes outside the boundaries of modern deserts and in areas where dunes are stabilized and partially gullied is clear evidence of previous arid environmental extensions. Today such dunes are mostly vegetated and consolidated. The westernmost belt of Rajasthan is covered with sand dunes that stretch from the Great Rann of Kutch in the southwest and to the north along the Pakistan border with Punjab. There is a variety of dune types and dune agglomerations in different parts of the *Marusthali.* Due to their shape, size, wind direction and vegetation cover, different types of dunes have been recognized in this region, namely, (i) Longitudinal dunes (*Seif* of Sahara or *erg* of Arabia); (ii) Crescent-shaped dunes (*Barchans*) of Turkmenistan; and (iii) Transverse dunes (Misra, 1968).

The longitudinal dunes are aligned from southwest to northeast, parallel to the prevailing winds and are mostly sword-shaped. These dunes are found in the western and southern parts of the region. The wind speed of the region is 1–7 km $hour^{-1}$ (up to 2.4 m^{-sec}). The wind direction is generally from southwest to northeast and is responsible for the formation and orientation of these longitudinal dunes. Crescent shaped dunes (barchans and transversely oriented dunes) have a gently sloping convex windward side and steep leeward side, which vary in width from 100 to 200 m and in height from 10 to 20 m. These dunes are mobile. These can be found isolated or longitudinally downwards continuation or occasionally next-to-one another in overlapping clusters. The transverse dunes are formed against the wind and are often found in the western and northern parts of the *Marusthali.* The windward side is long and gently sloping, while the leeward side is steep and abrupt. Such dunes are stabilized in semi-arid parts. Trees are usually found on the leeward side (Ghosh, 1952). The obstacle, parabolic, longitudinal and transverse dunes are stabilized and petrified and have well-developed nodular or finger type calcareous concretions of 1 to 100 mm size. They are 10 to 80 m high. The windward slopes are gentler and the leeward slopes are steep. These dunes could be developed into good grasslands. The diameter of sand grains in these dunes varies from 0.06 to 0.25 mm.

The wind becomes an effective geomorphological agent due to three important conditions: - (i) a sparse coverage of vegetation; (ii) the presence of dry, loose sand material on the surface; and (iii) higher wind speed to push and move these surface materials. The current landscape of the Thar Desert is primarily influenced by the action of wind, and the wind erodes surface materials through two main processes, namely deflation and attrition. Deflation lifts away the loose sand particles of varying grain sizes. Deflation occurs when winds blow fast enough or with enough turbulence over an area of loose sediments and are able to pick up and remove small boulders. The finest particles, clays and silts, transported by winds, are moved in suspension, carried by vertical currents. Such particles essentially consist of fine dust that remains in suspension as long as the strength of the upward air currents exceeds the tendency of the particles to settle down to the ground due to the force of gravity. Lancaster (1981) concluded that the progressive loss of coarse grains in the inland dunes is mainly related to the direction of sand transport. The sediments carried in suspension by the wind form its suspended load. When the wind speed exceeds 16 kilometers per hour (4.4 m^{-sec}), the grains of surface sand are often set in motion. The speed needed for the wind to pick up and move a grain is greater than what it takes to keep it moving, and this is mainly due to surface factors such as roughness and cohesion.

Another way that the wind erodes is through abrasion. Once the load of fine sediments is in the air, the impact of these wind-driven solid particles is more effective than the wind alone at removing and entraining other grains with it or breaking off new boulders. Where there are loose dust particles on the land surface, strong winds pick them up and carry them in suspensed forms. The result is a thick, dark, fast-moving cloud of dust that swirls over the land. Dust storms and dust-whirling winds, often blowing at speeds of 140 to 150 km/hour (over 40 m^{-sec}), are prevalent in the Thar Desert from April to June. Such dust storms are so strong that visibility drops to almost zero, thus blocking almost all sunlight (Yang et al., 2000). They can also be very destructive, removing layers of surface materials and depositing them elsewhere, sometimes in thick, suffocating new layers, all within a few hours. The Thar Desert is known for *sandstorms*, these usually occur in Bikaner, Churu, Jaisalmer and north-west of Jodhpur where loose sand on the surface is abundant.

But, the junction between hills and rocky plains in the Thar Desert is extremely abrupt in many places, so that it stands astride the highland-lowland boundary (Siwana-Jalore area). The wind action is very strong and shows itself in the wind-abraded surfaces, grooved surfaces, incision marks on rocky surfaces and in the alignment of dunes of various types.

Transportation, erosion and deposition are important factors of these prevailing winds in the entire Thar Desert as the vegetation cover is sparse or non-existent and soil particles are loose. The wind acts on the finer material and leaves behind gravel and stones (e.g., Barmer and Jaisalmer). The desert pavement itself is susceptible to wind abrasion. Wind erosion is responsible for most of the erosion processes in this hot arid zone. Around 28% area of the Thar Desert is covered by sheets of sand, but the sand dune is the most typical shape created by wind deposition and molding. This is one of the main features of the accumulation of desert landforms. An uninterrupted field of dunes stretches here over a large area, although the monotony of this vast sandy desert is interrupted in many places by the rocky outcrops. As such, all of arid western Rajasthan is covered with sand dunes and sand sheets of various types and shapes.

According to a classical and detailed study of the global sand seas (Breed et al. in McKee (ed.) 1979), based on the Landsat (ERTS) image interpretation, the distribution and geomorpholgy of the aeolian sand dunes in the Thar Desert were mapped at a regional scale. The average annual rainfall isohyets and the prevailing wind patterns are also displayed (Fig. 16.2).

The map (Fig. 16.2) shows diferent types of dune fields with crescent-shaped/transverse, linear and parabolic arrangement, which lie next to each other on the west and northeast of the flood-plain of Indus River and Indus and Hakra-Ghaghar Delta region ultimately meeting the Rann of Kutch and the Arabian Sea. The orientation of the dune pattern indicates that the wind direction has dominantly been from southwest to northeast. Therefore, the readily available abundant source of sand through the majestic Indus River and its associated, but currently non-existent tributaries of Ghaghar-Hakra (the lost Sarasvati River) and others. It is reasonable to assume that the humid winds from the Arabian Sea must have supported sufficient vegetation cover, even during semi-arid-arid climate, to develop anchored clusters of parabolic dunes immediately northeast downwind of the parched Rann of Kutch (see Wadhawan this volume). The frequent revisit of source-sink cycles due to fluctuating climatically controlled dominance of aeolian and fluvial forces during the Quaternary/Pleistocene led to such massive accumulation of sediments in the dune fields that occupy the Neogene continental basins or the depocentres in the Thar Desert. The transverse dune field on the extreme west and its continuation wind direction to areas west of Bikaner and Ganganagar indicate an abundant supply of fine sand-sized sediments close to the source to the prevailing southwesterly winds. This dune field also rests over a thicker column of Quaternary sediments, as confirmed by deep drilling by GSI and paleochannels and isotope studies by ISRO and BARC (Wadhawan and Kumar, 1996; Rajavat et al., 1999; Nair et al., 1999). Field studies have shown that the formation of linear dunes was supported by greater aridity and intensity, acute bimodal winds that not only independently developed linear dunes, but also effectively drain the nosal closures of the parabolic

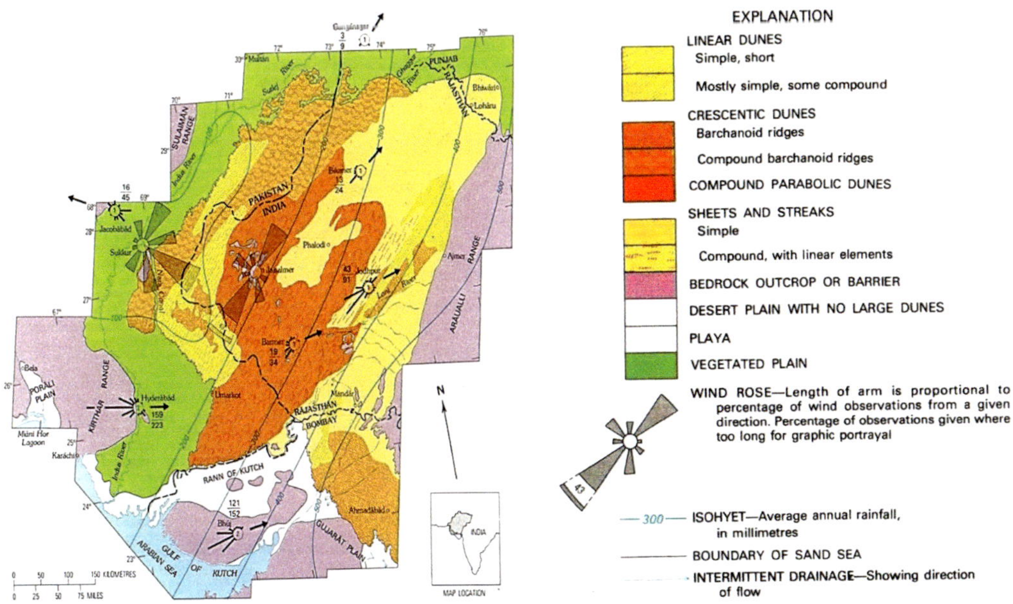

Figure 16.2. Distribution and morphology of the aeolian sand deposit in the Great Thar Desert [After Carol S. Breed et al., 1979, in USGS Professional Paper#1052; Ed. Mc Kee, 1979].

dunes formed earlier. However, in absence of exposed internal structures, it is difficult to distinguish such detached linear dunes, which have morphologically appeared as parallel arms of parabolic dunes, from those which develop independently. Both variants can be found in the Thar Desert, as linear dunes and poarabolic dunes often merge (Wadhawan, this volume).

3. Regional geological setting of Thar Desert

The desert tract of Rajasthan is characterized by low and erratic rainfall, high atmospheric aridity, abundant solar radiation, strong dust-laden winds, sparse vegetation cover and grand dunes. It occupies an area of about 208,228 km^2 or 61.0 and 6.35% of Rajasthan state and India respectively. It extends between latitudes 24°37'N to 30°10'N N and longitudes 69°29'E to 76°5'E.

The Aravalli hills run along the eastern margin, separating them from the eastern part of the state. The Thar Desert extends over 12 districts of the state of Rajasthan namely, Barmer, Bikaner, Churu, Ganganagar, Hanumangarh, Jaisalmer, Jalore, Jhunjhunun, Jodhpur, Nagaur, Pali and Sikar. It has a population of 27.1 million people with an average population density of over 130 persons per square kilometer. This hot arid tract is the most densely populated area of all the regions of similar aridity in the world. It supports a total domestic livestock population (2012 census) of nearly 30.18 million head—an increase of 14.63 million heads (94.72%) from the 1956 census figures. The desert region has about 52.3% of the state's livestock (Gaur and Chauhan, 2016). The human population density varies from 17/km^{-1} in Jaisalmer district to 361/km^{-1} in Jhunjhunun district and that of livestock from 83/km^{-1} in Jaisalmer to 274/km^{-1} in Sikar (Gaur et al., 2018).

3.1 Climate variability

The climate of the Thar desert is characterized by extreme temperature; erratic, irregular and insufficient rainfall; and a pronounced degree of aridity. The monsoons show greater temporal and spatial variability. Droughts and famines are common and similar with the name of the region. The arid conditions have been around for a long time since the 3rd century B.C. Kautilya's *Arthashastra* (Economics) had discussed the aridity and the precarious nature of rainfed agriculture during that time. From Ashoka's time (240 B.C.) to the Skanda Gupta (457 A.D.) period, agriculture was only

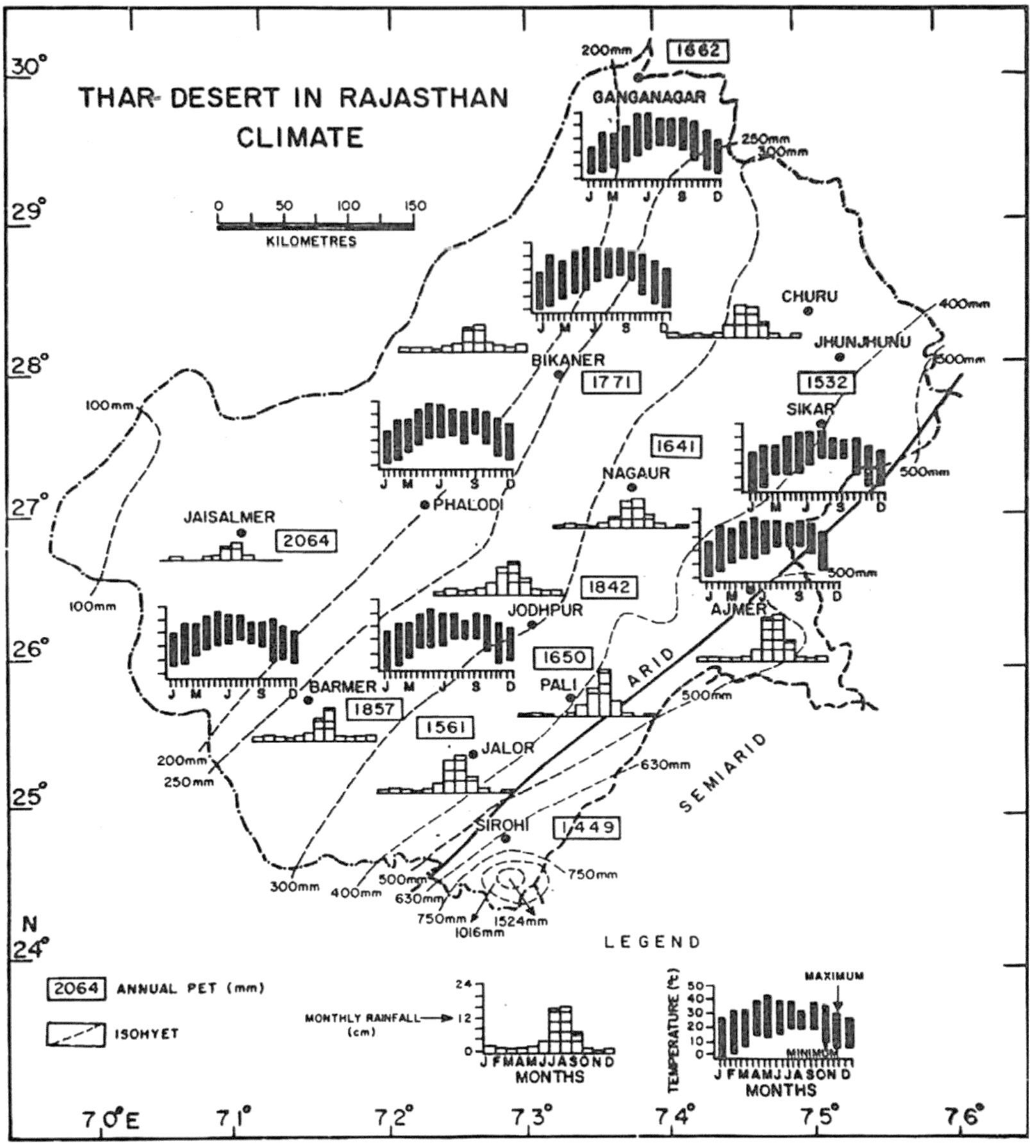

Figure 16.3. Climate variability of Thar Desert in Rajasthan.

possible wherever irrigation was available. Beal (1881) referred to the visit of Chinese traveler Hiuen Tsang in 7th Century A.D. who had to cross the vast tract of desert to reach Sind from Gurjadesh (present-day Barmer). Based on archaeological resaecrh, Ghosh (1952) determined that "… the Rajputana desert is at least two millennia old and its nucleus must be older still. But as yet practically no date to trace its history any further."

Extremes of temperature are due to location, low relative humidity, the sandy nature of soil, the lack of vegetation cover and dry winds. The maximum temperature reaches up to 45–50°C during summer months and the minimum temperature remains at 3–10°C during winter months. At its eastern margin, the mean annual rainfall is > 450 mm and around 100 mm on the westernmost border where the potential evapotranspiration ranges from 1800 to 1600 mm. About 80 to 90% of total precipitation occurs during the monsoon season (first week of July to the mid-September), with frequent annual fluctuations. The coefficient of variation is 40% in the east and 70% in the west (Fig. 16.3).

The region has a strong wind regime, the mean wind speed in June, the windiest month, is 14 to 27 km hr^{-1}. The average wind direction in Bikaner, Phalodi, Jodhpur and Barmer is from the south-west while the wind in Ganganagar is from the north-west (Gaur and Squires, 2016). The extent of rainfed cropped area varies from year to year and depends on the amount, distribution and intensity of precipitation. Therefore, during moderate and severe drought years, mobility of sand dunes increases. In the period 2018–19, the gross irrigated area was 30.4% of the total geographic area of the arid tract. As a result, a vast sandy tract and other peripheral area have been used for agricultural activities, particularly in the last 2–3 decades, which helped in combating sand movement.

The wind velocity and direction have a direct influence on the intensity of wind abrasion and the formation of dunes in the Rajasthan desert. It is observed that the orientation of the dunes in Barmer, Bikaner, Churu and Jhunjhunun is related to the direction of the wind. The ventifacts in Jaisalmer area are the product of high velocity winds of (Sharma, 1987).

The region has some spectacular and peculiar landforms and the sand dunes cover 48.86% of the region, but their density and configuration vary. Large stretches of light textured sandy plains dominated by the Torripsamments and coarse loamy Camborthids are associated with these dunes. Paleorthids or hard pan soils occupy a sizeable area. Medium and fine textured Camborthids and Calciorthids are found in the central and southern parts. Precipitation has an important influence on geomorphological processes such as chemical weathering, fluvial erosion, accelerated soil erosion, etc. Catastrophic rain (i.e., rains of unusual intensity), especially in arid and semi-arid areas, leads to considerable changes in the surface form (Gaur and Squires, 2016).

Lancaster (1988a) suggested that the spatial variability of particle size and other structural parameters of sand dunes is mostly controlled within dunes profile (dune-interdune) and also regional factors (e.g., wind direction; inland activity). The four main factors that normally control sorting on mainland dunes have been reviewed by Livingstone et al. (1999); and Liu et al. (2014). These are: (1) Particle size sorting is preferred in the inland process, usually for saltation in the range of 125 to 250 mm (2ϕ); (2) the variability (expansion and compression) of air flow streamlines which can cause a changing pattern of shear stress across the dunes; (3) the effect of gravity, either by causing sliding and falling on a dune slip face (> 300) or by transporting the grains; and (4) geological processes that can affect the sediment source and particle size distribution. The combination of these controlling factors leads to a number of different scenarios of variation in sorting and the particle size distribution in the inland dunes.

4. Tectonic set-up in Thar Desert region

Most of the area covered by the existing Thar Desert appears to have remained low-lying peneplain during the geological period as it was submerged under shallow transgressions of the sea during the Mesozoic and Early Tertiary. It is essentially confined to the West Rajasthan Shelf (WRS). The tectonic disposition and basement configuration of intra-cratonic basins and sedimentary formations ranges from the vastly deformed Precambrian Aravalli and Delhi Supergroups in the east, undeformed sedimentary rocks of Late Proterozoic to Early Paleozoic Marwar Supergroup of sedimentary rocks in the middle, to Mesozoic and Cainozoic cover sedimentary formations on the western and north-western fringes in western parts of Rajasthan and Haryana in NW India. It is noteworthy that to the west, successive geological young age of sedimentary formations was mapped from the Aravalli Hill ranges (average 500 m above msl) on the east to the recent sand covered nearly horizontally disposed Tertiary rocky plains [hamadas and isolated low ridges—mesas, outcropping at about 80 m above msl] in the west of the Thar Desert. Distinctive geomorphic expressions and relative lowering of relief are also deciphered and recorded in an east-west transect along northern parts of the Thar Desert (Wadhawan, 1996, 2010, 2018).

Major geological sedimentary basins that have been delineated in the Thar Desert region include: (1) the Nagaur - Ganganagar basin (Marwar Supergroup of rocks) bound by the Aravalli hill ranges in the east and the Delhi - Lahore - Sargoda subsurface ridge/high on the north and northwest;

(2) the Jaisalmer basin in the west dominated by the Mesozoic and intermittent Tertiary strata; (3) the Palaeogene Barmer - Sanchor basin and (4) the contiguous Cambay (Khambhat) graben and (5) the Kachchh basin in Gujarat. The paleogeographical extent of the shallow Tertiary (Mid-Eocene) sea is inferred from the occurrence of marine fossiliferous strata and lignite deposits in the marshy/deltaic environment in the western parts of the Thar region, as it has an impact on the source of evaporite minerals and persistent salinity aspects of today's inland lakes (Wadhawan and Sharma, 1997; Roy, 1999).

Periods of epiirogenic movements in extra-peninsular India are derived from the fractures caused by the processes of peneplanation that led to the formation of the erosion level or the planation surfaces. Three planation surfaces had been identified and assigned as: 1st during post-Mesozoic era; 2nd during Early and Middle Tertiary and 3rd as Early Pleistocene synonymous with the aggradation of the Older Alluvium in the Gangetic plains (Heron, 1953). A significant rise in the Himalayan barrier (the Tibetan Plateau) during the Mid-Miocene could have established monsoon conditions over the Thar Desert (Ahmed, 1969).

The rapid uplift in the Delhi area during the Middle to Late Pleistocene rejuvenated a consequent Himalayan stream—the Yamuna. It continued to destroy the mighty Drishadvati by forward erosion. This, coupled with the westward deviation of the courses of the Sutlej and Beas, withdrew the perennial water flow from the tributaries of the Vedic Sarasvati which ultimately led to its drying out in the huge sandy tracts of the encroaching Thar Desert (Yash Pal et al., 1980; Bakliwal and Grover, 1988; Roy, 2003).

Although climate change is responsible for aeolian episodes, field studies and interpretation of subsurface geological data have now revealed that tectonically active Paleogene - Neogene domains influence the successive burial of different sedimentation units of fluvial/fluvio-lacustrine and aeolian generation, particularly in the northern and western parts of the Thar Desert. The isopach map of the sub-terranean Quaternary and Tertiary formations and configuration of the paleo-basin show that the present day Ghaggar Basin (Nagaur basin) in northern Rajasthan deepened to an unusual depth and hosted a thick (over 350 m) pile of continental sediments in its northern and north-western parts. A north-flowing structurally controlled paleodrainage pattern for the early Quaternary fluvial regime is deduced.

The interpretation of various geological, geophysical and seismo-tectonic data in the Thar Desert helps to recognize that the terrain is traversed by multiple block faults, which manifested as large lineaments that were neotectonically active and shaped the geomorphic and geological evolution of the dryland environment (Fig. 16.4; Bakliwal and Ramasamy, 1987; Ramasamy, 1999; Kar et al., 2004; Kar, 2020).

5. Sand dunes: Aeolian dynamism in Western Thar Desert

A sand dune is an accumulation of sand grains that are piled up by the wind under the influence of gravity into a mound or ridge. Dunes are found wherever loose sand is blown up and deposited by wind (Bagnold, 1941). A sand dune formation in dry and semi-dry lands is the result of earth surface processes whereby the land is transformed into sand covered otherwise fertile areas which can consequently lead to desertification (Fadhil, 2013). Dunes, which are piles of unconsolidated sand sculpted by the wind to form bedforms, respond rapidly to changes in wind conditions and the resulting changes in surface dune morphology are internally represented by bounding surfaces, which are a common feature of aeolian sandstones (Bristow and Mountney, 2013). Sand dunes formed by the wind erosion process are shaped in arid areas when winds regularly blow over poorly vegetated areas. Sand dunes, that are not covered by vegetation, move in the direction of dominant winds, endangering urban areas, agricultural crops, forest plantations, irrigation canals, and road networks (Fadhil, 2013). Sand dunes are the most spectacular features of the hot Indian dry zone. These are of variable shape, size and size and their frequency increases from east to west (Sharma, 1981). Wilson (1972) categorized three universally recognized and mutually discontinuous groups of

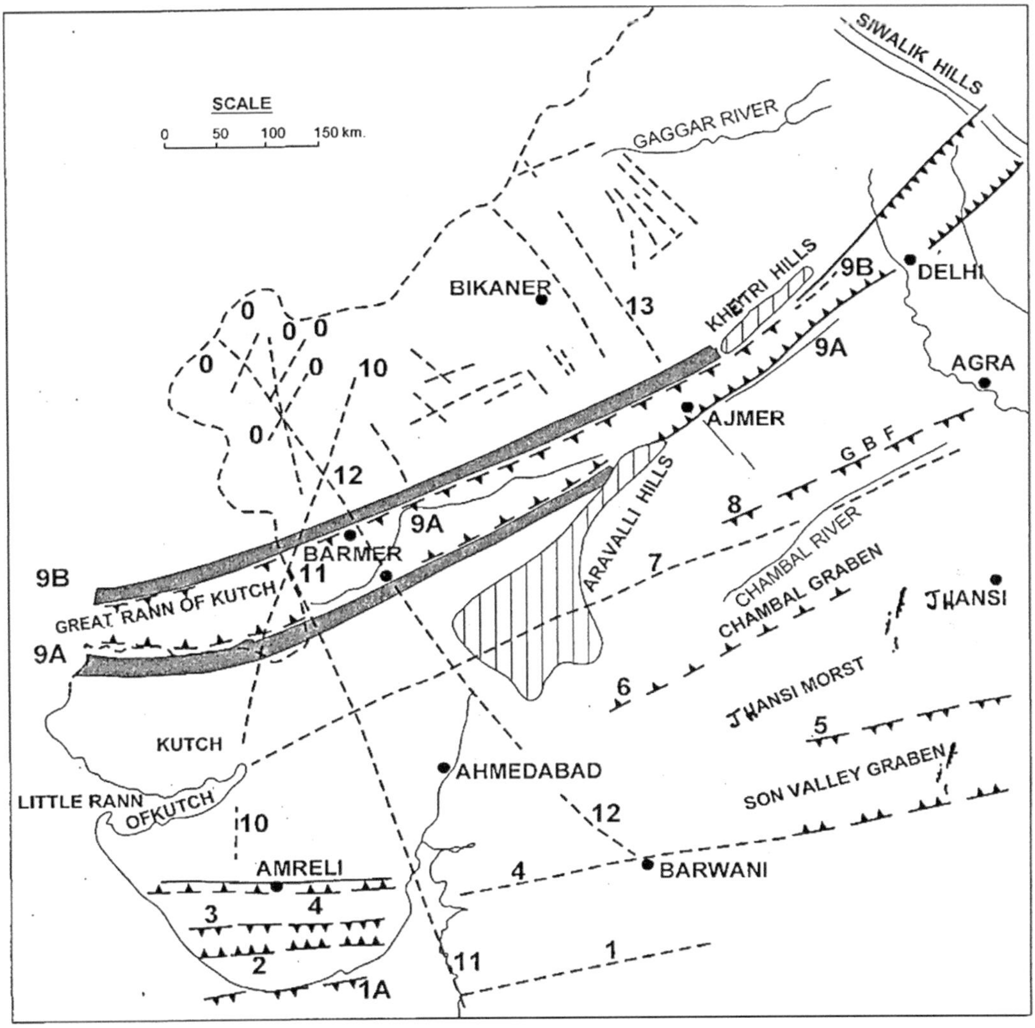

Figure 16.4. Major Inferred Tectonic Features and Lineaments with horst and graben structure in the Thar Desert in Rajasthan and Gujarat (Source: published literature cited in the text). Legend: 1 = Tapti Lineament, 2 = Sawar–Kundla Lineament, 3 = Amreli-Junagarh Lineament, 4 = Bhawnagar-Gondal Lineament, 5 = Asmara Lineament, southern boundary of the Jhansi Horst, 6 = Jalawar Lineament, southern boundary of the Chambal graben, 7 = Chambal-Jamnagar Lineament, 8 = Great Boundary Fault, 9 = Luni-Sukri Lineament, defining a major graben, 10 = Lathi–Rajkot Lineament, 11 = West Coast Lineament, 12 = Jaisalmer–Barwani Lineament, 13 = Raisinghnagar–Tonk Lineament.

aeolian bedforms. These can be distinguished by their wave-length alone, viz., draas (macroforms) spaced 0.5 to 5.0 km apart; dunes (mesoforms) having a distance of 50 to 500 m; and wind ripples (microforms) spacing 0.01 to 1.0 m. It is remarkable that there are no transitions between these three groups of aeolian bedforms and that the forms in one group do not grow into those of another. Distinct groups of bedform elements have also been derived, each indicating the corresponding possible modes of their origin respectively (Table 16.1; Wilson, 1972).

An interdependence of bedform spacing and grain-size has also been suggested by Wilson (1972), i.e., larger widely spaced dunes comprised coarser sand, as the mobilization of larger grains required greater sheer velocities inherently associated with larger scale atmospheric circulations. Subsequently, it was demonstrated that grain-size alone cannot distinguish between dunes and draas and there existed a continuum of scale between them (Wasson and Hyde, 1983a, 1983b).

Table 16.1. Aeolian bedform elements (After Wilson, 1972).

Order	Wavelength	Height	Orientation	Suggested Name
1st	300–5500 m	20–450 m	Longitudinal/Transverse	Draas
2nd	3–600 m	0.1–100 m	-do-	Dunes
3rd	15–250cm	0.2–5 cm	-do-	Aerodynamic ripples
4th	0.5–2000 cm	0.05–100 cm	Transverse	Impact ripples
------	1–3000 cm	0.05–100 cmk	Longitudinal	Secondary ripple sinuousity

(1st to 3rd order aeolian bedforms owe their origin to primary aerodynamic instability whereas the 4th order bedforms are formed either by impact mechanism or secondary Taylor – Gortler vertices).

5.1 Inland dunes

These dunes are aeolian sand dunes that are found on the mainland. These are located quite far from the coastal areas. The shape of these inland dunes varies depending on the prevailing wind direction and its intensity. The majority of them are very irregular dunes or shifting sand belts. However, there are still very well-developed parabolic dunes and longitudinal dunes. Strong air currents loosen the light, fine-grained soil particles, particularly of silt and sand, often transporting them for kilometers and depositing them elsewhere. As the silt is lighter and finer than sand, it is transported quite rapidly and this over-time resulted in the formation of aeolian sand and sand dunes, whilst the silt was transported much further and redeposited on the scantly vegetated areas. The dunes tend to *'wander'* when the wind is strong. At this time, most of the pre-existing inland dunes were formed. With the end of the Ice Age, due to availability of moisture and vegetative coverage, the mobility of the dunes rapidly came to a halt. Anthropogenic activity has been the key to all recent episodes of inland dune development. It induced changes in the nature of the dunes from stable to semi-stable and unstable.

6. Sand dunes of Thar Desert: origin and geomorphology

Geological evidence suggests that it was only after the last glacial period (also known as Last Glacial Maximum [LGM]), that desiccation and aeolian dynamics set in western Rajasthan after a long period of alternating wet and dry conditions. The aeolian deposits in the Thar desert belong to the Pleistocene and recent times. It is assumed that the large amount of sand in this area comes in various forms from prolonged aridity and sand drifts by the south-west monsoon (Sharma, 1987). The wind blows with considerable force and the sand develops from atmospheric weathering of rocky outcrops. The sand thus obtained consists of quartz with grains of felspar and hornblende. The sand particles are evenly rounded by attrition (Misra, 1968). Such sands also have their origin in high mountain ranges and were later transported and deposited by rivers in the past times (Pithawala, 1952). In the dune-free areas of Barmer, Jaisalmer and Bikaner, marine deposits of Jurassic and Eocene periods are exposed, showing an anomaly in the nature of the rock deposits in the region (Misra, 1968). Wind-deposited sands were dissected and re-deposited by torrential rains on the old alluvium and piedmont deposit skirting these hills and later wind-worked to form high hummocks and micro-dunes (Abhichandani and Roy, 1966).

The geomorphology of the area is quite old and the landforms from the Archaean to the Quaternary period occur in this region. After the Tertiary period, this region has been subjected to different climatic fluctuations, sub-aerial denudation and deposition. The existing major landforms of this region were formed during the Neogene-Quaternary period by fluvial and aeolian processes operating on various lithological formations and under different climatic phases (Ghose, 1965; Singh, 1977a). The major humid phase, which was responsible for the formation of etensive alluvial plains and sand-covered pediments, was followed by a major arid phase, probably in the pre-Holocene period. The major arid phase, which was responsible for the formation of three lakes in the Rajasthan desert, terminated around 10,000 years B.P. (Singh et al., 1975). During this arid phase, the highly active sand dunes, the sandy undulating aggraded older alluvial plains and the interdunal plains formed. The barriers created by the formation of sand dunes astride the major drainage channels segmented many valleys into sandy inland basins (Ghose et al., 1975). The major arid phase responsible for the formation of aeolian landforms, was probably replaced by a major humid phase, in the early Holocene period. From 10,000 to 3800 years B.P., the climate became humid again but from 9500 to 5000 years B.P. it was less humid (Singh et al., 1974). The fluvial activity intensified and the highly active dunes were stabilized and dissected. This major humid phase was replaced by a second arid phase, probably in the late Holocene period. The second arid phase set in around 3800 years B.P. (Singh et al., 1974). Due to renewed aeolian activity, the younger aeolian landforms viz., barchan dunes, shrub-coppice dunes and sand ripples were formed on top of the old fluvial and aeolian landforms (Singh, 1977b). The aeolian history of the Thar Desert and its eastern fringe areas is at least 150,000 years (ka) old. The south-west monsoon was responsible for these aeolian activities. A strong phase of aeolian sand accumulation occurred during 10 and 14 ka when the south-west monsoon began to build up after its weakest phase during the Last Glacial Maximum (Kar et al., 1998).

However, the origin of sand dunes has been a controversial subject. Wadia (1961) reported that sand dunes of the Indian arid zone were transported by south-west winds from the Rann of Kutch and partly from the lower Indus, while Ghosh (1952) observed that sand from the dunes of Rajasthan tended to be fine grained and rounder than beach sand. The roundness of the sand grain was caused during wind transportation by abrasion such as by bouncing off obstacles and wearing away their irregularities, rather than by the capacity of winds to pick up the rounded particles, as reported by Pettijohn and Lundahl (1943). The analysis of the size and shape of dune sand grains from different parts of the arid zone has indicated that dune sands had originated due to the reworking of the alluvial sediments by the aeolian action under prolonged dry climatic conditions. Sand dunes of different types and magnitude are comparatively more degraded and reactivated. Wind erosion and depositional activity in the Shergarh-Sai-Suwalia-Sointra sector; Phalodi-Jelasar-Chadi-Kelensar-Chaku sector and Osian-Samrau-Bhikamkor-Dabri-Khetasar sector of Jodhpur district have further desertified the existing stabilized sand dunes. The fresh sands of 1 to 2 m, 2 to 3 m and 3 to 4 m in the form of sand sheets, transverse ridges and barchan dunes were deposited, causing deterioration in the potential of these dunes.

According to Williams (2014) "in Rajasthan, a stable isotope analysis of pedogenic carbonate was used within a 70 ka sequence of aeolian sands (Andrews et al., 1998; Singhvi et al., 2010). They concluded that atmospheric CO_2 concentration were lower during glacial periods. Carbon isotopes measured in organic matter within the sand profiles consistently showed values close to –21.6, indicating that aeolian sands were deposited during a transitional climatic regime. The earlier studies implicit assumption that sand depositions occurred during glacial maxima, was thus replaced with a more nuanced interpretation of maximum dune activity during the transition from weak to stronger monsoon in the immediate postglacial time. In this case, wind strength was more important than aridity in promoting sand movement".

6.1 Zones of sand dunes in Thar Desert

Given the applied environmental factors, it possible to divide the Thar Desert into two distinct groups based on the relative dominance of aeolian and fluvial processes. The 300 mm isohyet serves the dividing line between these two (Singh, 1977a). The aeolian process has become more intense, re-activating and degrading the stabilized dunes in the west of 300 mm isohyet, resulting in the formation of new dunes. The region contains a vast expanse of sand that extends across entire western Rajasthan. The sandy deserts , most desert-like areas are known as 'thul or thal', while tracts that are relatively less sandy are known as '*thali*'. The satellite imageries (Landsat and IRS) and topographical sheets reveals that sand dunes and sheets are discontinuous in nature. Western Rajasthan has two major and two minor sand dunes and sand sheet zones (Sharma, 1987).

Miajlar Thal, Jaisalmer, Shahgarh, Ramgarh, Tanot (in Jaisalmer district), Pugal, and Mahajan are all part of the first major zone of sand dunes (in Bikaner district). A complex system of sand dunes characterizes the area west and south of Jaisalmer. Longitudinal or siefs dunes predominate in this area, with transverse dunes of barchan types developing at times. The dunes around Barmer are mostly semi-stable to stable barchan sand dunes. Beyond Jaisalmer, the thals of Pugal and Birsilpur have barchan sand dunes with a steep concave northern face reaching a height of 50 m (Sharma, 1981). The second major zone of medium to high and very high dunes passes through Shri Dungargarh, Sardarsahar, Nagaur and ends near Rajgarh and Suratgarh. The northwestern limit of this zone is formed by the old course of the Ghaggar River. The third small dune zone passes through Sanchore, Jalor, Saila, Didiwana, Sikar and Jhunjhunun before joining the two major zones near Surajagarh. The fourth patch of dunes runs through the Pushkar valley, Sambhar, and continues as isolated obstacle dunes all the way to Bharatpur. These four zones are made up of dunes of various types and origins that are unevenly distributed (Sharma, 1987).

In the Rajasthan desert, investigations on the dynamics of sand dunes and temporal changes of dunes have been carried out by Pandey et al. (1964), Vats et al. (1976), Saxena and Singh (1976), Singh (1977a, 1982), Bedi (1982), Allachin and Goudie (1971), Allachin et al. (1978), Kar (1987, 1990, 1993, 1994, 1996, 1998), Kar et al. (2004), Juyal et al. (2003), Singhvi (2004), Singhvi et al. (2012), Wasson and Hyder (1983), Pye and Tsoar (1990), Wasson et al. (1983), Wadhawan (1994, 1996, 1998, 2005, 2010), Yang et al. (2002), and many others. Sand dunes cover more than 80% of the area in the districts of Barmer and Jaisalmer. Dunes cover 40–60% of the land area in Jaisalmer, primarily in the Barmer, Bikaner, and Mohangarh tehsils (settlements), as well as the Suratgarh and Anupgarh tehsils of Ganganagar and Bikaner tehsils. In the desert region, the degree and extent of sand dunes decreases towards the east and north.

6.2 *Wind regime and dune orientation*

The orientation of sand dunes in India's Thar Desert is determined by the south-western monsoonal winds. Long narrow sand ridges of the sief type can be seen running parallel from south-west to north-west. During the summer, the wind blows from the south-west to the north-east, whereas in the winter, the wind blows from the north-east to the south-west. However, the winter winds have an impact on its appearance. The western part of the Jaisalmer district has the highest percentage of land covered by sand dunes. In the Luni basin, the percentage of land covered by sand dunes gradually decreases towards the south-eastern part of the desert. It clearly demonstrates the existence of a close relationship between wind direction and dunes orientation. The axial direction of the old and new dunes systems is nearly identical, implying that the wind regime has not changed in recent times.

Aside from the wind's direction, the velocity of the wind is also important in the movement of sand particles. At the surface, fine-grained sands (0.25 to 0.01 mm) can move with a wind velocity of 3.5–5 m sec^{-1}. Table 16.2 shows the velocity required for the movement of sand particles of various sizes, as proposed by Singh (1977a).

The average size of sand grains in the Thar Desert ranges from 0.12 mm to 0.14 mm, with three classes of roundness identified (Singh, 1977b): (i) angular to sub-angular, (ii) sub-angular to sub-

Table 16.2. Grain size of sand dunes.

Types of sand grains	Diameter (mm)	Wind velocity (m sec^{-1})
Fine grained	0.1 to 0.25	4.5 to 6.7
Medium grained	0.25 to 0.50	6.7 to 8.4
Coarse grained	0.50 to 1.0	9.8 to 11.4
Very coarse grained	1.0 to 2.0	11.4 to 13.0

rounded, and (iii) rounded to well-rounded. Abrasion during wind-borne transportation causes this roundness in sand particles.

7. Dynamics of sand dunes

Sand dunes are the most spectacular features of the Rajasthan desert, and cover approximately 48.86% of the region in various forms, magnitude and orientation. Longitudinal and transverse dunes cover 21.74 and 25.60% of the land area, respectively. Barchan dunes can be found in a small area with a geographical spread of 2.99% in Rajasthan's Jaisalmer and Barmer districts. From east to west, the intensity rises. Due to intense biotic activity, stabilized dunes are reactivated, and sands are carried away by strong southwest winds and deposited on agricultural lands in the form of active dunes and sand sheets, affecting food and fodder production. Sand dunes are the most spectacular features of the Rajasthan desert, and they cover approximately 48.86% of the region in various forms, magnitude and orientation. Longitudinal and transverse dunes cover 21.74 and 25.60% of the land area, respectively. Barchan dunes can be found in a small area with a geographical spread of 2.99% in Rajasthan's Jaisalmer and Barmer districts. From east to west, the intensity rises. Due to intense biotic activity, stabilized dunes are reactivated, and sands are carried away by strong southwest winds and deposited on agricultural lands in the form of active dunes and sand sheets, affecting food and fodder production. The majority of the old dunes were formed between 13 and 18 ka BP, when the desert extended beyond the Aravalli hill ranges in the east (Goudie et al., 1973; Wasson et al., 1983; Singhvi and Kar, 1992). Goudie et al. (1973) observed that current dune formation is largely limited to areas receiving 250 mm or less of annual rainfall, and suggested that dune-building activities have now shrunk westward by 350 kilometres from an earlier drier climatic phase. This was dubbed as a 'dune shift', and it presented an intriguing paradigm of periodic spatial shifts in dune forming climates.

When one dune migrates over another, superposition surfaces form. This is most likely to happen where small, fast-moving dunes migrate over larger, slower dunes or are superimposed on the flanks of larger dunes. It could also occur when transverse dunes migrate across larger dune bedforms (known as megadunes or draa) (Kar, 1994; Wadhawan, 1998). The easternmost sand dunes are vegetated or cultivated almost to their crests and are partly gullied. New crescentic dunes are scarce, and high dunes are mostly reactivated in areas close to major habitations. Westward, natural vegetation becomes increasingly sparse, dune slope cultivation becomes less regular, and high dunes reactivate more frequently. The grasses and small shrubs begin to take over the non-crescentic dune slopes. The larger dunes are encased in a thick smock of loose sand. The development of luminescence dating, which allowed for determining the time of sand deposition, provided a way to examine aeolian dynamism from a chronometric standpoint (Singhvi et al., 1982, 1994).

Sand dunes in arid western Rajasthan were mapped at 1:50,000 scale using Landsat and Resourcesat LISS-III data from 2015–16. The mapping process began with a digital interpretation of the False Color Composite (FCC) of 2015–16 images, followed by ground truthing of the mapped units. The mapping involved a digital interpretation of the false color composite (FCC) of 2015–16 images, followed by ground truthing of the mapped units. The mapping shows that aeolian type of topography is dominant in the area (~ 74% of the total area). The classification system was developed by the Indian Space Research Organization and the Geological Survey of India and it served as the basis for the mapping. Dominant sand dunes were mapped on the basis of their shape. The types of sand dunes and their extent are described in detail below.

7.1 Formation and types of sand dunes in Thar Desert

The formation of dunes is governed by two important principles: (a) a sand mound or hill is formed from a free sand patch that is larger (than a few meters) and is likely to grow in size and thickness, and (b) a low velocity air deposit sand to form a cliff-like dune. When the sand mound reaches a certain size, it develops a classic dune asymmetry with a gentle windward slope and a steep angle of repose face to the leeward side. Although the sand on the windward slope is firm and compact, it blows over the crest and cascades down to the slip face, where it may become loose. In India's Thar Desert, dunes come in a wide variety of shapes and sizes.

The potential of integrating Remote Sensing (RS), Spatial Analysis (SA), and modeling to predict future changes in real-world dune systems is becoming more substantial. Remote sensing images have been used to document and understand the evolution of aeolian sand dunes for more than five decades. Remote sensing (RS) has revealed a rich diversity of dune field patterns from the early 1970s. Recent high resolution satellite data have enabled the exploration of many sand regions that were previously inaccessible due to vast sandy terrains. However, the synoptic perspective provided by remote sensing imagery has only recently stimulated a shift away from single dune studies and toward dunefield-scale studies that include spatial analysis (SA) of boundary conditions, dune activity, dune patterns and hierarchies and dune–dune relations. Singh (1982) distinguished three types of dunes: (a) simple, (b) compound, and (c) complex. Pandey et al. (1964) discovered that major and minor obstacles cause the formation of simple dune types such as parabolic, longitudinal, transverse and barchan. In satellite imagery, the barchan and transverse dunes form parallel wavy lines. Transverse sand ridges connect two or more dunes of the same type in compound dune types. The coalesced barchan and parabolic dunes result in compound dune types. Barchans, barchanoids and megabarchanoids are crescentic in shape and lack vegetative coverage. Complex dunes form when two or more basic types of dunes are formed in some areas by winds from multiple directions. The variations in shape and morphology of the dunes are reflected in the types. Thomas (1988) elucidated the complexities of sand ridges in relation to wind regimes.

The following major types of dunes have been identified and mapped based on field observations and interpretation of satellite images:

a. Obstacle dunes	:	old dunes systems
b. Parabolic sand dunes	:	old dunes systems
c. Longitudinal dunes	:	old dunes
d. Transverse dunes	:	old dunes and new dunes
e. Compound and complex dunes	:	old dunes and new dunes
f. Barchan & Barchanoids dunes	:	new dunes

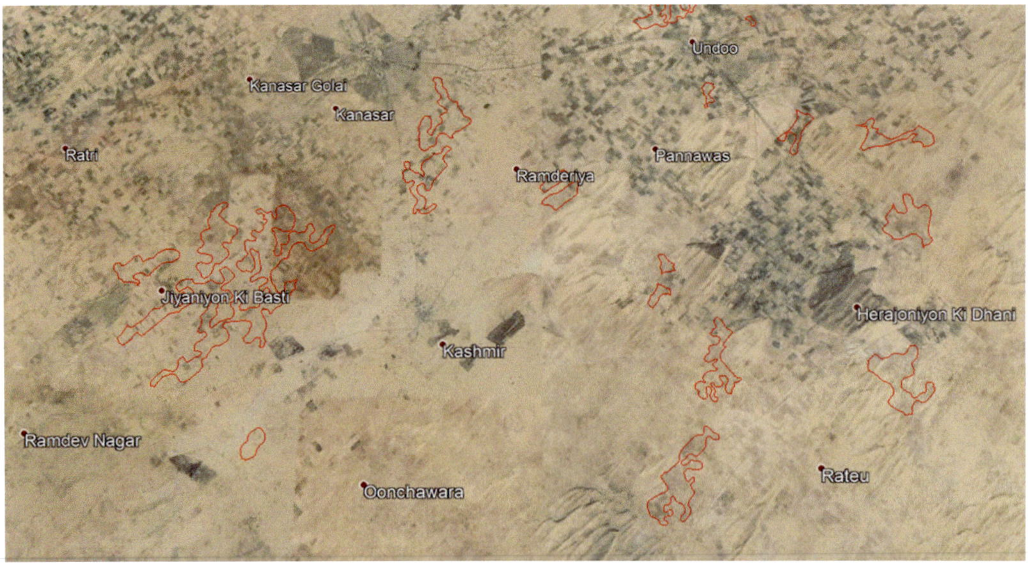

Figure 16.5. Aeolian Plain (Kotra Village, Sheo Block, Barmer District, India).

7.2 Obstacle sand dunes

These dunes are scattered, occur in discontinuous patches, mostly in the central and north-eastern parts of the Rajasthan desert. The windward and leeward obstacle dunes against the isolated rhyolitic and granitic hills, and a chain of hills of sand stone formations mostly occur in Barmer, Jodhpur, Nagaur, Jaisalmer and Jalore (Siwana) districts. These range in height from 10 to 40 meters, and the sands are compact and highly calcareous due to the presence of lime nodules. Their sand grains have an average diameter of 0.14 mm. These dunes are severely eroded and have been separated from the hills by gullies ranging in depth from 25 to 50 meters and width from 50 to 100 meters. The lime nodules of 8 to 10 cm length and 2 to 5 cm width are exposed at crests and flanks. The grain size of these dunes varies from 0.03 mm to 0.25 mm and sand grains of 0.06 mm to 0.12 mm are prevails in the region. As a result, these dunes have become well cemented and immobile. The windward sides and flanks have slopes of 2° and 10°, respectively (Singh, 1977b).

These sand dunes, known as windward and leeward obstacle dunes, are formed on the windward and/or leeward sides of the isolated and continuous chains of hills due to the obstruction in the path of sand-laden winds during the prolonged arid phase. Due to the fixed eddy, decreased wind velocity and a calm zone on the upwind side of the hills, windward obstacle dunes of flat semi-cone shape were formed. The prevailing southwest winds swept the sand over the hills and deposited it on the leeward side from the obstruction's slopes up to a few kilometers in the direction of the downward wind, forming the leeward obstacle dunes. At certain places, leeward obstacle dunes had also been formed by the north-east winds which are gently upsloping towards the hill slopes. This indicates that these dunes were active and stabilized in a prolonged humid pre-Holocene period, around 10,000 years ago, at the time of formation in the Holocene period (Singh, 1977b). Currently, these dunes are badly dissected.

7.3 Parabolic sand dunes

Parabolic sand dunes are spread over a larger area than any other type of dunes. The term "parabolic dunes" refers to sandy concave crescents with a low inclining slope towards the windward side. Simple parabolic dunes are distinctive U or V shaped dunes that are aligned parallel to the dominant wind direction, with the dune's nose pointing downwind and two arms trailing in the upwind direction. These dunes can be found in almost every district of the Thar Desert. Parabolic dunes are most commonly found near the margins of desert dune fields with a single dominant wind direction. They are oriented with the crescent's horns pointing upwind. Parabolic dunes have a long gentle windward slope and a short leeward sand-fall face. Another distinguishing feature of parabolic dunes is the presence of vegetation, roots and gently warped or a hummocky strata where sand is trapped in bunches of vegetation, and/or scoured from between vegetation stands. These dunes are common in the Thar Desert (Mainguet, 1984) and were formed by depositional or erosional processes. These are primarily concentrated in Jodhpur, Barmer, Jaisalmer, Bikaner, Nagaur, Jalore, Shri Ganganagar, Churu and Jhunjhunu districts of the arid zone (Fig. 16.6).

These dunes are dominant and accounting for roughly 65 to 70% of the total area. These dunes vary in height from 10 to 25 meters and are partially-stabilized, immobile and vegetated. These dunes have an average length and width of 1238 m and 226 m respectively. Fresh sands of 3 to 4 m thickness cover the coalesced parabolic dunes. The diameter of the sand grains ranges from 0.03 mm to 0.25 mm where 60–70% of the sand grains being between 0.06 mm to 0.12 mm in size. Barchan dunes are superimposed on the flanks of these dunes. This indicates the presence of renewed aeolian activities. The occurrence of vegetation supports in anchoring the dune arms and the trapping of wind-blown sand (Singh, 1977b).

Breed et al. (1979) and Wadhawan (1998) observed parabolic dunes in the Thar Desert as rake-like clusters that formed dominant and characteristic dunes. These dunes were formed as a result of erosional and depositional processes during the prolonged arid phase (Wadhawan, this volume). The sands were eroded away by winds, resulting in the long spoon-shaped hollow or

Figure 16.6. Parabolic dunes in Thar Desert.

parabolas with flanks tapering up to several kilometers towards the gentle windward slopes. Typical elongate clustered parabolic dune forms show remarkable parallelism of northeasterly oriented arms or the longitudinal elements, both in older stable dunes and in younger semi-stable and actively accreting dunes. It implies that the prevalent formative wind regime has remained highly persistent unidirectional since the formation of clustered parabolic megadunes (Wadhawan, 1998). It is remarkable that inter-arm spaces in a parabolic cluster enclose elongated oval shaped basins that generally slope southwesterly towards a larger upwind deflationary opening. However, in some places the tailend/windward components of parabolic clusters appear to have been choked by stabilized sandy accumulations with varying degrees of gullying and ephemeral incised flow-collapse features. Chronologically these degraded aeolian deposits on the upwind margins may be the remnants of older putative dunes that were scoured and deflated into parabolic shapes by younger aeolian dynamism (cf. Wadhawan, 1994, 1998 and this volume). It is worth reiterating that deflation of larger stable dunes and deposition of mobile sand in the form of multiples of small-scale crescentic dunes/linear arrays of low barchans perched on the flanks of parabolic dune complexes or downwind extensions on to the vast sandy interdunes reflect on the activities commensurate with the prevalent milder wind regime. If such landscape changes with milder winds continue indefinitely, large dunes may degrade and eventually disappearance in order to "remain in dynamic equilibrium with the constant flow conditions" (Wilson, 1972).

7.4 Longitudinal dunes

In North Africa and Saudi Arabia, these are known as seif dunes, while in Australia are known as sand ridges. They appear as straight parallel ridges in dark gray to dark tones on satellite imagery. On the Landsat imageries of 1980s and mid-90s, these sand dunes reach hundreds of kilometers in length, more than a kilometer in width and hundreds of meters in height. Longitudinal sand dunes named for their supposed parallelism to prevailing winds, but it now appears that they are formed by alternate convergent winds blowing from south-west to north-east. These are elongated sand ridges that usually form a series of parallel or gently curving and branching dunes. Longitudinal dunes thrive in areas with an abundant supply of sand. These dunes are mostly found in the south-western parts of Jodhpur district. They are well vegetated and stabilized, with compact sand grains. They range in height from 6 meter to 60 meters and have an orientation ranging from North 40° East to North 45° East. They are well-stabilized with sand grains ranging in size from 0.07 mm to 0.15 mm in diameter, with an average diameter of 0.10 mm (Singh, 1977b).

Dune sands are highly calcareous and compact, as they are mixed with lime nodules. The dunes are in an alluvial phase, but they have been reactivated due to renewed aeolian activity, and the transported sands often pile up on leeward sides and crests. The flanks of these dunes are superimposed by barchan dunes ranging in height from of 2.5 to 4.5 meters. These dunes are found in Phalodi, Jodhpur and Shergarh tehsils of Jodhpur district as well as in the western parts of Jaisalmer and Barmer districts (Fig. 16.7). The development of these dunes appears to be the result of human-induced reactivation and breakdown of a transverse sand ridge that coalesce the trailing arms of the parabolic dunes that is followed by erosion caused by strong winds. This process, which separated the parabolic dunes into two parallel sand ridges, forming a gully in between

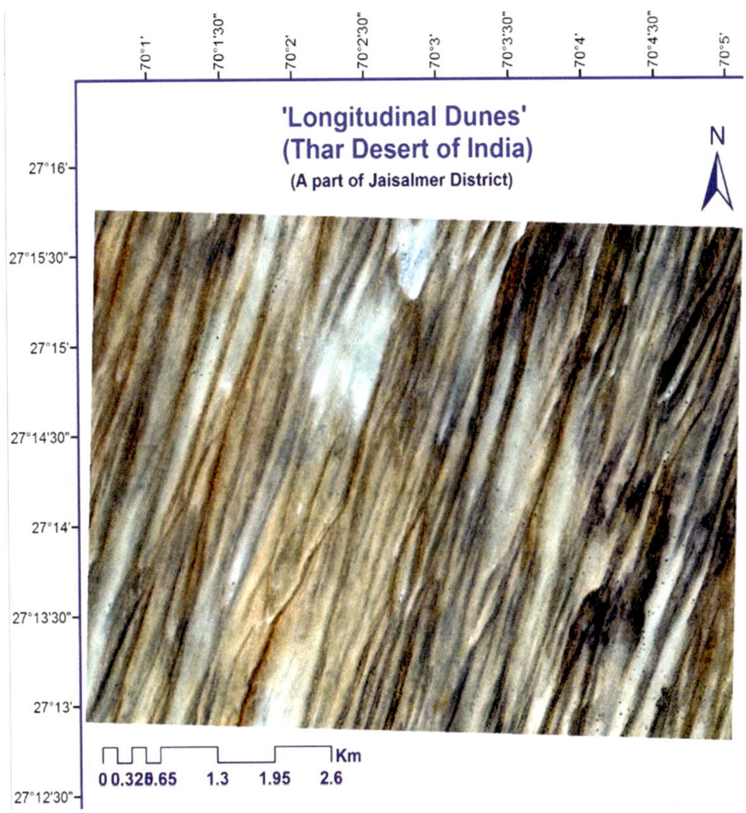

Figure 16.7. Longitudinal dunes in Thar desert.

them, and subsequent sand accumulation enhanced their length. These newly developed dunes are surrounded primarily by parabolic dunes indicating their origin from these dunes. McKee (1979), Singh (1982), Verstappen (1968) and King (1960) reported that longitudinal dunes originated from parabolic dunes while Bagnold (1941) considered that they originated from barchan dunes. Breed et al. (1979) described a field of large linear dunes with crescentric dunes in the western Thar Desert near Jaisalmer and east of Jodhpur that receives less than 100 mm annual rainfall.

According to Glennie (1970), a high wind velocity environment is favorable for the development of linear or *seif* dunes. According to Bagnold (1941), longitudinal dunes are a modification of the basic barchan form caused by strong cross winds that transverse to the prevailing wind direction. According to Verstappen (1968) and Singh (1977a and b), the longitudinal dunes are the result of the parabolic dunes rather than barchans dunes. The central curved parts came into existence due to a funneling effect where strong winds removed the top soil layers and stretched the parallel ridges in the direction of the prevailing wind.

7.5 Transverse sand dunes

Transverse dunes are similar to barchan dunes in that they are the result of a single dominant wind direction with a steeper leeward side. They may result from the coalescence of barchan dunes, as sand supply increases, to form a barchanoid ridge and then a transverse dune. They are associated with massive quantities of sand and relatively ineffective winds. The prevailing winds generated these dunes due to prodigious sand cover on the ground, abundant sand supply, scarce vegetative cover and persistent and widespread wind action under dry climatic conditions created the wave-like ridges at right angles. Their crests are normal to the prevalent wind direction, they have a gentle windward slope, a steep sand-fall slope on the lee and move slowly downwind. When the sand supply diminishes, transverse dunes break up into barchans, but when the sand supply is abundant, they aggregate into dune complexes. These are sporadic and occupy only a small area of the arid zone of Rajasthan. These are most common in the south-western part of the Pugal area of Bikaner district, southern Hanumangarh district around Rawatsar and the south-eastern Jodhpur district. Small patches of transverse dunes can also be found in the districts of Churu and Jhunjhunun. Many of these dunes have been leveled and converted to agricultural fields as the Indira Gandhi Nahar Pariyojna (canal project) has been extended to Rawatsar (Hanumangarh district), Pugal (Bikaner district) and northern part of Jodhpur district (Fig. 16.8).

These dunes range in height from 10 to 40 meters and have an average length and width of 750 m and 150 m respectively. The sand grains range in size from 0.03 mm to 0.18 mm, with an average diameter of 0.11 mm. The sands are dense and highly calcareous. Due to renewed aeolian activity, these dunes have undergone several changes, resulting in the deposition of fresh sands of 1 to 3 m thickness on the flanks, crests and the leeward side. The presence of reactivation surfaces near the crest of the transverse dune indicated that it was subject to reworking at the crest (Singh, 1977b). However, there appear to be fewer reactivation surfaces within the dune's main body. This could be because the transverse dune's main slipface was less prone due to shrub-coppice coverage and the presence of moisture at the tail end. When viewed normal to the dune migration direction, the transverse dune strata are more continuous and sub-horizontal (Mckee, 1966).

7.6 Linear dunes

Linear dunes are widespread and are the most common dune form on earth. Linear dunes are formed in areas with two wind directions with the dune crest lines aligned close to the resultant wind direction, in agreement with the gross bedform-normal rule of Rubin and Hunter (1987). Although opposing winds transport sand back and forth across the crests of linear dunes, bedforms appear to propagate by deposition at the end of the dune, thus advancing across the desert while possibly

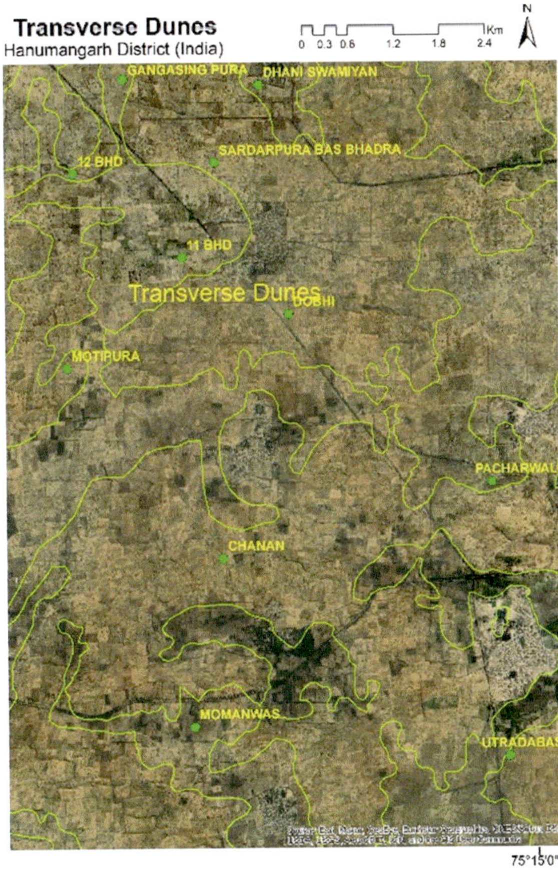

Figure 16.8. Transverse dunes in Thar Desert.

migrating laterally or back and forth in a sinuous manner. Several models for the structure of linear dunes have been proposed, starting with Bagnold (1941) who proposed that linear dunes formed under equal and opposing winds would build up by vertical aggradation with strata dipping in opposite directions on either side of the dune, forming a tent-like structure. McKee and Tibbits (1964) dug shallow trenches in a Libyan linear dune and used their observations to back Bagnold's model. However, Rubin and Hunter (1985) disagreed with this interpretation that in nature winds are almost never equal and opposite, and that as a result there should be net lateral migration of linear dunes.

In addition, Tsoar (1982) observed that some linear dunes have sinuous crest lines, which could influence the sedimentary structures if strata were preserved within the sinuous curves of linear dunes. The models for linear dune development were tested using Ground Penetrating Radar (GPR) (Bristow et al., 2000). It was proved that linear dunes can migrate laterally, forming unimodal cross-strata that can be mistaken for transverse dune deposits, as proposed by Rubin and Hunter (1985). With a sinuous crest-line, sand is deposited on the outside of the bends rather than the inside and this can result in bimodal dips where sinuous curves propagate along the crest of the dune. Furthermore, superimposed dunes form trough cross-strata on the flanks of larger dunes (Fig. 16.9). As a result, linear dunes can show a variety of patterns of aggradation (Singh, 1977b).

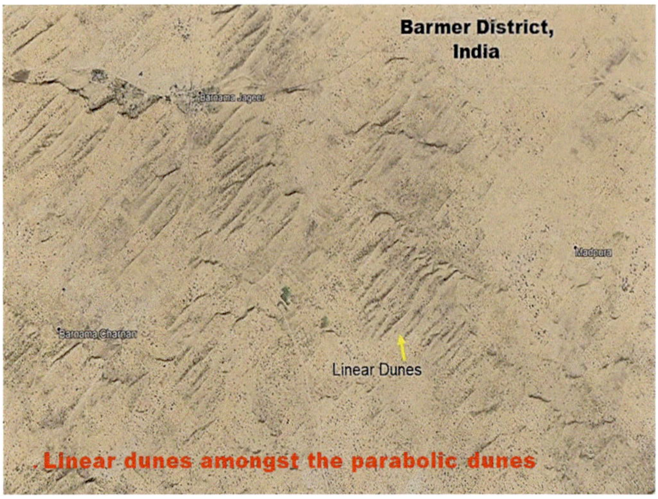

Figure 16.9. Linear dunes amongst the parabolic dunes in Thar Desert.

7.7 Barchan dunes

Barchan and barchanoids dunes are crescentic dunes with a slipface and two horns that point downwind. These are the classic aeolian landforms and one of the simplest types of sand dune. Barchan and parabolic dunes are two forms of crescent dunes of complex type. These were formed by unidirectional uniform winds on the hard-bare surfaces with scanty or devoid of vegetation and an inadequate sand supply. These are formed when prevailing winds deposit sand on bare surfaces, forming a crescent-shaped barchan dune with a concave leeward side and a convex windward side. At the same time, sand is swept around the main body of the dune and into the horns. Though shape evolution is also possible because the shape of barchan dunes is not aerodynamically stable as described by Zhu et al. (2014). These mechanisms cause the dune to move downwind over time while maintaining its shape. The sand in the barchan is often very well sorted, as constant reworking is in progress while the dune moves. They are isolated but mobile in nature. Individual and coalesced barchan dunes are scattered in the central, western and south-western parts of the Rajasthan arid zone. They are associated with areas where there is a single dominant wind-direction and limited sand supply so that they advance over a bedrock floor. A fairly constant wind direction is necessary for the formation of the most perfect barchan–side winds tend to make sickle-shaped or irregular dunes. These dunes can be differentiated from parabolic dunes on the basis of lack of four long arms of equal length. In barchan dunes most of the sand is accumulated in the central convex part and gradually decreases on both ridges (see also Embabi, this volume).

These dunes can be found in the central, western and south-western part of the arid Rajasthan. They are well-developed in the districts of Bikaner, Churu, Jhunjhunun and Jodhpur and are usually found near settlements on older dunes and in the river beds (Fig. 16.10).

They represent a remarkable balance of accumulation, transportation and erosion. These dunes are active, have no vegetative coverage and are composed of non-calcareous, loose and single grained sands. The dunes range in height from 2 m to 15 m and are also found in 'dune chains'. The sand grain size ranges from 0.06 mm to 0.12 mm, with 0.12 mm being the average size. The sand grains in these dunes are finer than in all other types of dunes with an average diameter 0.11 mm (Singh, 1977b).

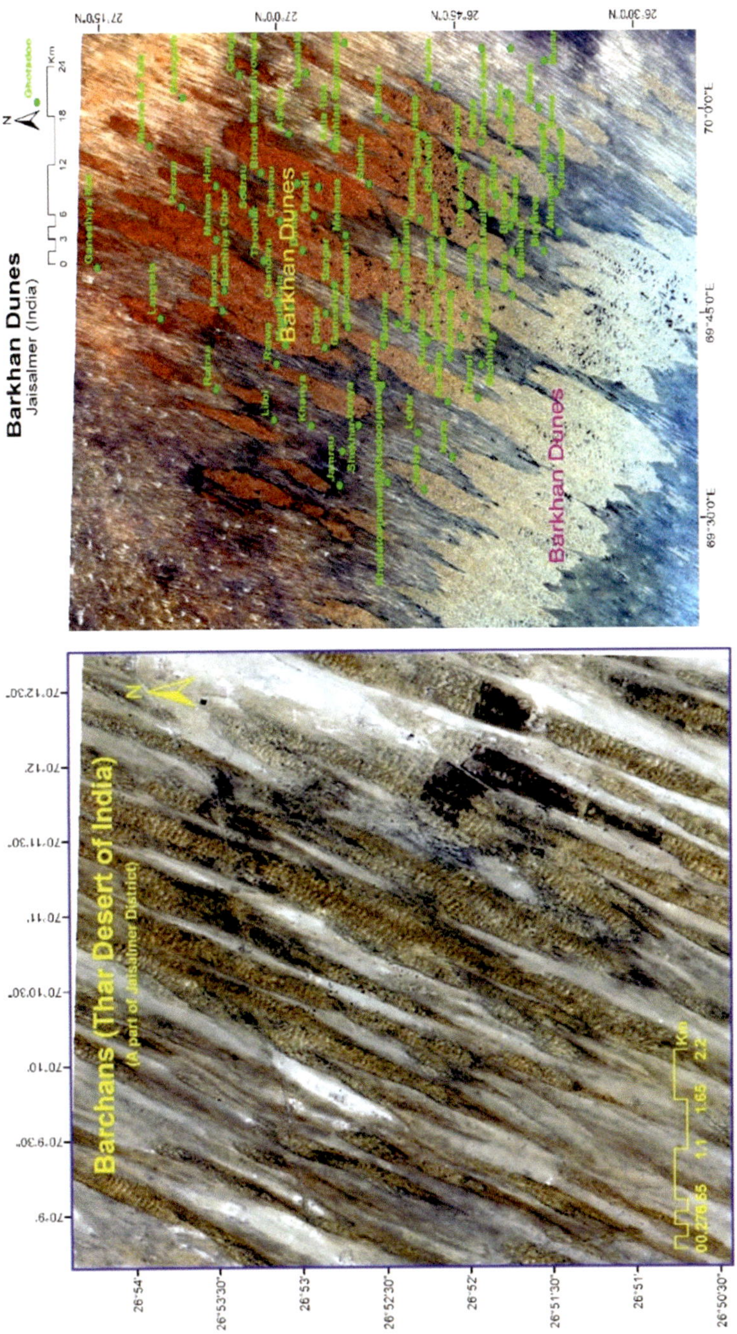

Figure 16.10. Barchans and Barchanoid dunes over-riding the older linear dunes.

7.8 Dune complexes

Very large dunes, known as megadunes or *draa*, are almost always compound or complex bedforms characterized by the development of superimposed dunes. The strata include multiple superposition and reactivation surfaces. Dune complexes are made up of two or more types of dunes that are superimposed or coalesced. It has been suggested that compound dunes are a consequence of climate change, with larger dunes attributed to the LGM when the desert areas were more extensive, with increased sediment supply, and stronger winds capable of constructing larger dunes (Glennie, 1970). These dunes are very common in the major area of vast sand spread and the presence of different types of dunes with varying orientations suggesting that they were produced under different wind regimes. They are sand masses made up of components that differ in form and/or discordant in their relationships to one another or to the major form (Fig. 16.11). They include a great variety of pyramidal, peaked, ridged, pitted, branching, radiating and intersecting assemblages. The use of GPR to image the stratigraphy within megadunes, combined with optical dating to determine the time of sand deposition and burial it is now possible to test the LGM relict-dune hypothesis.

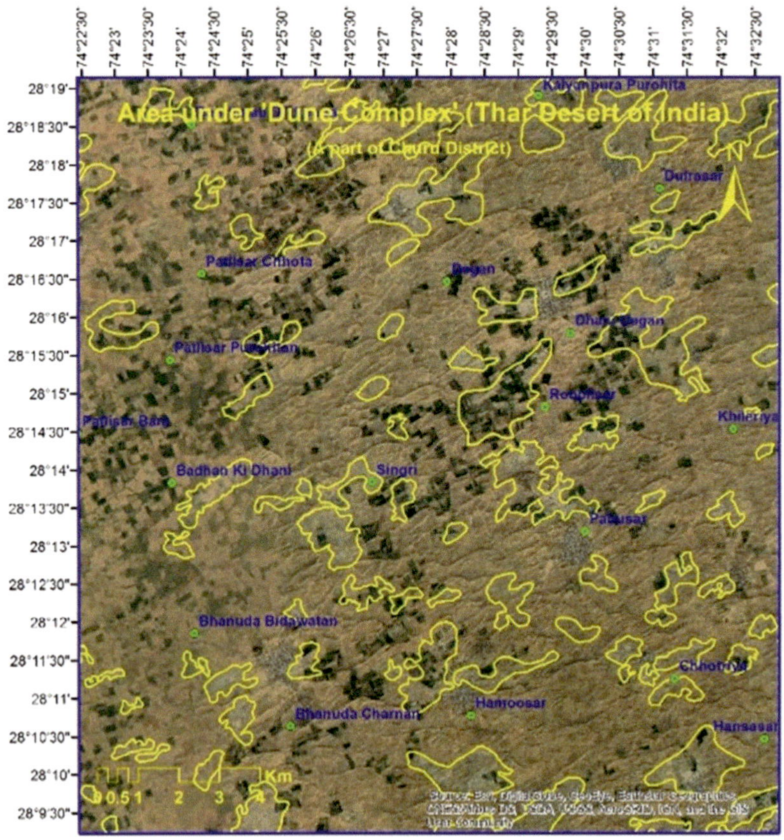

Figure 16.11. Dune complexes.

7.9 Vulnerability of sand dunes

The vulnerability classification of different types of sand dunes was prepared based on weathering resistant capacity (bearing capacity), wind erosion severity, vulnerable wind erosion classes and other collateral data such as soil erosion, slope and so on. The classes *viz*., high, moderate and low were made on this basis. Sand dunes and other sandy landforms are the most vulnerable landforms to wind erosion/deposition activity. However, sandy landforms in the east are somewhat more stable

Table 16.3. The relationship between bearing capacity of the sand dunes and vulnerability level.

S.No.	Sand Dune Unit	Erosion Resistance Capacity	Vulnerability Level
1	Interdunal Depressions	Low	Low
2	Sand Sheets	Low	Moderate
3	Aeolian Plain	Low	High
4	Desert Pavement	Low to Moderate	High
5	Deflation Plains	Low to Moderate	High
6	Sand Dunes	Moderate to High	High
7	Playa	Low to Moderate	Moderate
8	Dissected Dune Complex	Moderate	Moderate
9	Pediplain	Low	Moderate

than those lying in the west. Table 16.2 shows that sand dunes, aeolian plain, desert pavements and deflation plains having slopes, show the high vulnerability to the land degradation and desertification in the Thar Desert (Kalwar et al., 2005; Gaur et al., 2020). All the sand dune units show a moderate to high category because of high resistance capacity like-pediment volcanic, pediment granite, etc.

Table 16.3 also shows the reverse relationship between the bearing capacity of sand dunes and vulnerability level. The probability of aeolian erosion activity is entirely based on the binding capacity of sand particles. The vulnerability level is higher in dune fields with low bearing capacity. Though erosivity is a major factor for spatial pattern of sandy landforms, the erodibility of sediments is also significant.

8. Land use change and temporal changes in dune forms, size and area

The arid areas of the Rajasthan have a very dynamic environment wherein gradual changes have taken place during the last 60 years, but they have accelerated since 1995 for a number of reasons. Life of the inhabitants is governed by the vast sea of sand and it is one of the most mobile elements here. Its generation, mobility and accumulation are controlled by the moisture regime and xerophytic vegetation. Any change in these factors engenders a corresponding change in sand movement and deposition. Therefore, the behavior of sand is very important. The availability of coarse and high-resolution satellite images provides an opportunity to easily monitor temporal changes in the dunal form, size and areal extent. Another important aspect that has brought changes in the dunal forms is development of infrastructures, the expansion of agricultural activities, urbanization, increase in irrigated areas, growth of industrial and mining activities, etc. (Gaur and Squires, 2016). The regional pattern of population growth rate in western Rajasthan shows that extreme hot arid region district (Barmer) has the highest, 32.55% growth rate whereas Indira Gandhi Canal Region (Ganganagar and Hanumangarh) district has the least population growth of 10.06% only. High population growth rate of Barmer may be attributed to the increased economic and employment opportunity in the district due to oil and gas exploration (Gaur and Squires, 2016). This increase is a result of developmental activities including the construction of border roads and trans-border railway line, laying of petroleum pipelines, allotment of land, adaption of improved agro-horticultural practices, large scale sale and purchase of land, mining activities, setting of ancillary industrial units and hotels for energy officials and staff, etc. The development of pasture lands, and vocational training activities for sustenance and drinking water facilities through government and corporate efforts are other reasons. These have resulted into inward migration of population to the Barmer district (Gaur and Sharma, 2013).

Land use and land cover changes are mainly driven by three factors: (1) natural processes such as climate and atmospheric changes, wildfire and pest infestation; (2) direct anthropogenic activity such as deforestation, industrialization, urbanization and road-building; and (3) indirect effects of human activity such as water extraction leading to the lowering of the water table. Natural processes and anthropogenic activities can both improve or degrade the state of the land, so it is essential to distinguish beneficial from detrimental changes (Turner and Meyer, 1991; Wadhawan, 2015, 2019).

8.1 Modeling aeolian landscapes

In aeolian geomorphology, modeling the formation and growth of dune ecosystems is a crucial part of the extension and exploration of fundamental research questions. Models incorporate and reflect our current understanding of the several processes and conditions that operate in an aeolian environment, including saltation process physics, boundary layer flow dynamics over topography, surface conditions, climate controls the and effects of vegetation over a number of multiple temporal and spatial scales. Aeolian landscape development models incorporate and portray contemporary knowledge of the physical processes and fundamental concepts involved, and are often the only way to investigate long-term dune development and examine dune fields on other planets and moons (Baas, 2013). A number of complex conceptual or semi-quantitative models have been developed to characterize key types of sand dunes, beginning from the initiation of a standardized bedform. A fundamental relationship between the local gradient of transport movement, the local slope and the rate of migration was established by Bagnold (1941). The gentle wind was predicted by Tsoar's (1984) model as the one oblique to the barchan, which then elongated the far-sided horn. Livingstone (1988, 1989) suggested a different conceptual model of their formation, based on the diagonal back-and-forth motion of ephemeral bedforms on the active crest of the larger-scale underlying dune ridge and much larger and broader linear dunes in the Namib Sand Sea. Wasson and Hyde (1983) developed a synthesis model for the occurrence of the key dune forms based on a sample inventory of dune fields across major deserts. This approach was further integrated by Yizhaq et al. (2007, 2009) into a vegetation cover hysteresis model in relation to wind power. A biogeomorphic model of dune mobilization and stabilization was created by Hugenholtz and Wolfe (2005). Modeling of dune field activation and stabilization remains an important area of research, as sedimentary dune deposits and their dating are frequently used to infer records of aeolian activity. To solve the average Navier-Stokes equations for the turbulent wind area over dunes, Computational Fluid Dynamics (CFD) simulations can be used (Wiggs, 2001) (Fig. 16.12).

Jackson and Hunt (1975) developed the most commonly used method in dune modeling for airflow over a low hill, it was subsequently modified and revised by Hunt et al. (1988). Fowler (2001, 2002) and Ellis and Fowler (2010) examined the evolution of the dune profile as a more pressing problem of growth of bedforms, shifting to higher-order nonlinear evolution equations beyond simple bed stability analysis that can be applied not only to dunes but also to other sedimentary deformations under fluid flows such as subglacial drumlins.

The AEOLUS-II model was developed by Namikas and Sherman (1998). However, it calculates aeolian sedimentation/deflation along a transect because of (negative/positive) changes in transport flux between one gridpoint and the next. The transect models require uniform or user-defined wind forcing along the profile and do not dynamically evaluate the near-surface airflow and shear velocity over the developing topography (Baas, 2013).

Howard et al. (1978) developed the first application of a sediment continuity model over a Digital Elevation Model (DEM) of an existing (measured) barchan dune. Herrmann's (2002) model was used to simulate evolution of dune fields and interactions between migrating dunes, including a putative 'solitary wave behavior' of barchans (Schwammle and Herrmann, 2003). However, during the 1990s the modeling of aeolian landscapes turned from detailed reductionist analysis of airflow forcing and sand transport response, to explore instead the capabilities of simple Cellular Automata (CA) to simulate the nonlinear dynamics of dune field development as emerging self-organizing patterns. Nishimori and Ouchi (1993), and Nishimori et al. (1998), established a similar CA model, but based on a more mathematical treatment of the lattice algorithm. The CA technique for modeling dunes and bedforms continues to produce fruitful research. Discrete ECogeomorphic Aeolian

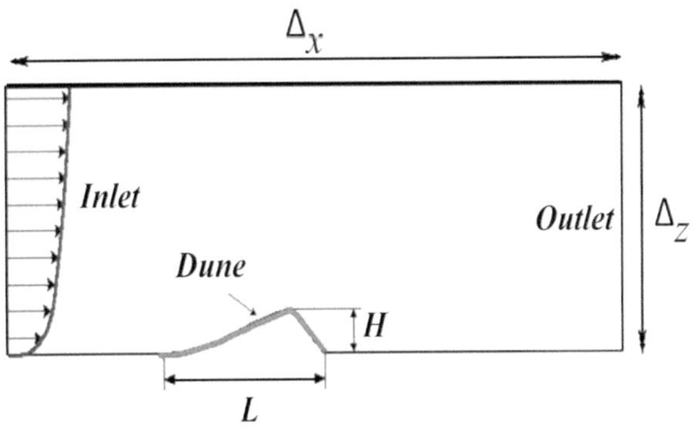

Figure 16.12. Schematic diagram of the computational setup. The simulation box has width Δ_x = 100 m and height Δ_z = 20 m, while the dune has height H = 3.26 m and total along-wind width L = 25 m. The wind blows from left to right and the wind velocity increases with the height (Source: Araùjo et al., 2013).

Landscape (DECAL) is a cellular automaton-based model which incorporated mutual feedback processes between geomorphic forcing and ecological growth to investigate fundamental controls, self-organizing and non-linear behavior in semi-arid aeolian dune environments (Nishimori and Tanaka, 2001).

Long time-scale processes associated with the dynamics of dunes and dune fields can be efficiently simulated on a computer, making numerical modeling a useful tool for dune research (Kroy et al., 2002; Sauermann et al., 2001). In order to model dune morphodynamics, an analytical description of the average turbulent wind flow over the sand terrain is required. Looking ahead, there may be perhaps two areas where aeolian landscape modeling could improve in the near future. First, one needs to look at the role of boundary conditions and the effects of external and internal sediment supply (spatially and temporally) on the formation and dynamics of dune fields. A second area for potentially novel development in dune modeling is the application of advanced CFD to a sand transport system with a deformable sedimentary bed. The next logical step is to directly incorporate sediment transport and dynamic topography into these CFD models, so that one can surmount the limitations and assumptions of the linear approximations (Baas, 2013).

9. Conclusion

Aeolian erosion develops through two principal processes: deflation (removal of loosened material and its transport as fine grains in atmospheric suspension) and abrasion (mechanical wear of coherent material, through creep and saltation). The relative significance of each of these processes is a function of surface material properties, the availability of resistant angular abrasive particles and the climatic regime. A number of dunes exist in the Thar Desert. Chronometric dating has confirmed that the aeolian activity in the Thar began over > 200 ka, resolving that the Thar is not of anthropogenic origin as suggested earlier. The present spatial extent of the aeolian activity in the Thar is in a contracted stage compared to that in the geological past, which refutes the arguments for its rapid north-eastward expansion. The region is dominated by aeolian bedforms of different dimensions in well demarcated dune fields. The shape of dunes depends fundamentally on wind-flow directionality and availability of mobile sediments. The thickness of aeolian cover can range from 1 m to 100 m. The current dune migration rates in areas of significant human-induced disturbances are much higher than during the geological past. Several older dunes are stabilized and only reactivated sands from their crests or flanks show a significantly low rate of movement, while some are so stable that these are called fossil dunes. There are, however, some active dunes which have a higher rate of movement and pose the maximum threat to human activities. Minor changes in the atmospheric circulation patterns result in amplified changes in rainfall, the winds and the aeolian dynamism. In the Thar Desert the aeolian activity is usually restricted to the period of strong summer south-west monsoon winds. Sand and dust carrying winds start gusting from March onwards when the surface is adequately dry after the preceding year's monsoon rainfall (July–September). The Thar sand dunes have mean sand grain sizes of 0.15 mm (2.74 phi). Dust particles have a platy structure with increased inter-particle cohesion. The sand particles are granular and cohesionless. The dunes of the Thar Desert in Rajasthan have been classified on the basis of their morphology, age and stability. The major dune types identified are: parabolic dunes, linear dunes, transverse dunes, obstacle dunes (along the hill slopes), barchans (and associated low dunes). In areas of constant wind-flow direction, crescent-shaped barchans are the characteristic dune type, provided the amount of sand is sufficiently low, such that the dunes occur on top of bedrock and well separated from each other. Most of the recently conducted studies in the aeolian spheres, has, however, indicated that improved understanding of un-stabilized structures and their effect on the transport of sediments could prove to be a fruitful way forward in the near future. The geomorphic characteristics of the sand dunes provides understanding about different aeolian landscape elements and their response mechanism under different climate-driven processes. For several decades, the paradigm for desert dune geomorphology studies has been that good quality empirical data on wind flow and sand flux would enable one to understand how dunes are formed and retain their shape. The quantitative study of aeolian dunes with RS and SA is still in its infancy and needs to be taken up vigorously to demonstrate the potential of this approach for addressing fundamental research questions about the morphometry, dynamics, organization and evolution of dunes.

Some of the biggest uncertainties in climate change predictions come from our lack of understanding and comprehension of the impact of some of the smallest airborne dust particles. These dust particles are transported between continents from some of the hottest and most unpopulated regions on the planet, and potentially end up fertilizing tropical rain forests or ocean algal blooms. Yet, even in the harshest of desert environments, vegetation or bacteria may still exist, and help to shape the windy landscapes that they reside in. Landscapes where wind-blown sand forms some of the most exquisite, self-organized patterns visible on satellite images of earth and recognizable on other planets. Wind-tunnel experiments and numerical modeling of dune dynamics has a long history, but now focus has turned to modeling methods, and very significant progress has been made in developing dune development models. Even though, CFD and other models have also been applied to modeling flow over aeolian dunes. However, the main factors controlling the shape of the separating streamline, which affects the shear stress on the windward side, are still uncertain. Advances in remote sensing and GIS technologies have the potential to alleviate our limited view of the landscape, but temporal data availability remains an issue to the timescales of landscape change that field scientists have described. Hugenholtz et al. (2012), however, argued that it would soon be possible to use remotely sensed data to develop models that integrate topography, climate, sediment and ecosystem components to develop well-constrained estimates of landscape change in response to different climate change scenarios (e.g., Pelletier et al., 2009) and ultimately integrated into GCMs – the Global Climate models (Thomas et al., 2005), airflow models (Liu et al., 2011), and ecosystem models (Mangan et al., 2004). Furthermore, in order to gain a better understanding of spatiotemporal patterns of dune field activity, and to validate and refine numerical models based on sediment transport equations that give rise to spatial variations of erosion and deposition, greater focus on three-dimensional topographical changes in dunefields is important.

References

Abhichandani, C.T. and Roy, B.B. 1966. Rajasthan Desert—its origin and amelioration, IGJ. XLI (3&4): 37.
Ahmed Enayat. 1969. Origin and geomorphology of the Thar Desert. Annals of Arid Zone 8(2): 171–180.
Allchin, B. and Goudie, A. 1971. Dunes, aridity and early man in Gujarat, Western India. Man 6: 248–65.
Allchin, B., Goudie, A. and Hegde, K.T.M. 1978. The Prehistory and Palaeogeography of the Great Indian Desert. Academic Press, London, 370 p.

Andrews, J.E., Singhvi, A.K., Kailath, A.J., Kuhn, R. Dennis, P.F. Tandon, S.K. and Dhir, R.P. 1998. Do stable isoptope data from calcrete record late Pleistocene monsoonal climate variation in Thar Desert of India? Quaternary Research 50: 240–251.
Araùjo, A.D., Parteli, E.J.R., Pöschel, T., Andrade Jr., J.S. and Herrmann, H.J. 2013. Numerical modeling of the wind flow over a transverse dune. Sci. Rep. 3: 2858; DOI:10.1038/srep02858.
Baas, A.C.W. 2013. Modelling aeolian landscapes. pp. 313–327. *In*: Shroder, J. (Editor in Chief), Lancaster, N., Sherman, D.J. and Baas, A.C.W. (eds.). Treatise on Geomorphology, Academic Press, San Diego, CA, vol. 11, Aeolian Geomorphology.
Bagnold, R.A. 1941. The Physics of Blown Sand and Desert Dunes. Methuen & Co., London.
Bakliwal, P.C. and Ramasamy, S.M. 1987. Lineament fabric of Rajasthan and Gujarat, India. Records Geological Survey of India, Jaipur. 113(7): 54–64.
Bakliwal, P.C. and Grover, A.K. 1988. Signatures and migration of Saraswati River in Thar desert, western India. Records Geological Survey of India 116(3-8): 77–88.
Beal, S. 1881. The Life of Hiuen Tsaing. Trubner and Co. Ltd., London.
Bedi, N. 1982. Applied geomorphic studies on Landsat imagery. *In*: Sharma, H.S. (ed.). Perspectives in Geomorphology, Vol 3. Concept, New Delhi.
Breed, C.S., Steven, G.F., Sarah, A., Camilla, M., Frances, L., Dana, G. and Kevin, H. 1979, Regional studies of sand seas using Landsat (ERTS) imagery. pp. 305–421. *In*: McKee, E.D. (ed.). A Study of Global Sand Seas. Geological Survey Professional Paper 1052 (Washington D.C.: Castle House Publications Ltd).
Bristow, C. and Mountney, N.P. 2013. Aeolian stratigraphy. pp. 246–268. *In*: John F. Shroder (ed.). Treatise on Geomorphology, Volume 11. San Diego: Academic Press.
Bristow, C.S., Bailey, S.D. and Lancaster, N. 2000. The sedimentary structure of linear sand dunes. Nature 406: 56–59.
Clift, P.D., Carter, A., Giosan, L., Durcan, J., Duller, G.A.T., Mclin, M.G., Alizai, A., Tabrez, A.R., Danish, M., VanLaningham, S. and Fuller, D.Q. 2012. U-Pb zircon dating evidence for Pleistocene Saravati river and capture of the Yamuna river. Geology 40: 211–214.
Dhir, R.P., Kar, A., Wadhawan, S.K., Rajaguru, S.N., Misra, V.N., Singhvi, A.K. and Sharma, S.B. 1992. Thar Desert in Rajasthan - Land, Man and Environment. Bangalore: Geological Society of India, 191 p.
Dhir, R.P., Singhvi, A.K., Andrews, J.E., Kar, A., Sareen, B.K., Tandon, S.K., Kailath, A. and Thomas, J.V. 2010. Multiple episodes of aggradation and calcrete formation in Late Quaternary aeolian sands, central Thar Desert, Rajasthan, India. Journal of Asian Earth Science 37: 10–16.
Dhir, R.P., Joshi, D.C. and Kathju, S. 2018. Thar Desert in Retrospect and Prospect. Scientific Publishers India, Jodhpur, 405pp.
East Amy, E., Clift Peter, D., Carter, A., Alizai, A. and Sam VanLaningham. 2015. Fluvial-Eolian interactions in sediment routing and sedimentary signal buffering: an example from the Indus Basin nd Thar Desert. Journal of Sedimentary Research 85: 715–726.
Ellis, A.S. and Fowler, A.C. 2010. On an Evolution Equation for Sand Dunes. Oxford Centre for Collaborative Applied Mathematics, Oxford.
Embabi, N.S. in press. Barchan Dunes in Egypt (current volume).
Fadhil, A.M. 2013. Sand dunes monitoring using remote sensing and GIS techniques for some sites in Iraq. In: Proc. SPIE 8762, PIAGENG 2013: Intelligent Information, Control, and Communication Technology for Agricultural Engineering 876206 (19 March 2013); https://doi.org/10.1117/12.2019735.
Fowler, A.C. 2001. Dunes and drumlins. pp. 430–454. *In*: Balmforth, N.J.P.A. (ed.). Geomorphological Fluid Mechanics. Lecture Notes in Physics. Springer, Berlin, Heidelberg.
Fowler, A.C. 2002. Evolution equations for dunes and drumlins. Revista de la Real Academia de Ciencias Exactas, Fı´sicas y Naturales. Serie A, Matema´ticas 96(3:, 377–387.
Gabris, Gy., Horvath, E., Novothny, A. and Ruszkiczay-Rudiger, Zs. 2012. Fluvial and aeolian landscape evolution in Hungary – results of the last 20 years research. Netherlands Journal of Geosciences—Geologie en Mijnbouw 91: 111–128.
Gaur, Mahesh K., Pandey, C.B. and Roy, M.M. 2012. Urbanization in Arid Zone of Western Rajasthan: Planning for Future. National Conf.: Sustainable Cities in Arid Region: Planning, Design and Management. Organizer: MNIT, Jaipur & Rajasthan State Commission on Urbanization, Jaipur. 26–28 Dec., 2012.
Gaur, Mahesh Kumar and Sharma, N. 2013. Demography and occupational structure of Thar Desert. *In*: Mahesh Gaur (ed.). Remote Sensing Applications in Dryland Natural Resource Management. pp. NIPA, New Delhi.
Gaur, Mahesh K. and Squires, Victor R. 2016. Geographic extent and characteristics of the world's arid zones and their peoples. pp. 3–20. Mahesh K. Gaur and Victor R. Squires (eds.). Climate Variability Impacts on Land Use and Livelihoods in Drylands. Springer International Publishing.

Gaur, Mahesh Kumar and Chouhan, J.S. 2016. Multi-temporal satellite data for wastelands monitoring. pp. 309–328. Mahesh Kumar Gaur et al. (eds.). Remote Sensing & GIS Application for Natural Resources Mapping & Monitoring. Scientific Publishers, Jodhpur.

Gaur, Mahesh K., Mathur, B.K. and Chauhan, J.S. 2018. Livestock population dynamics in arid zone of india. internat. J. Agriculture Sciences 10(16): 6920–6929.

Gaur, Mahesh K., Kumar, M., Chauhan, J.S., Mishra, A., Sheikh, S.I., Mehrishi, Pragya and Goyal, Prachi. 2020. Spatio-Temporal Dynamics of Land Degradation in Rural Areas: A Geospatial Analysis of the Arid Environment in Thar Desert, India. Internat. Journ. Current Micro. & App. Sc. 912: 1553–1568. doi.org/10.20546/ijcmas.2020.912.184.

Gaur, Mahesh K. and Squires, Victor R. 2020. Changes in agricultural land use and food security: Challenges and opportunities for western drylands of India. pp. 257–280. *In*: Squires, V.R. and Mahesh Kumar Gaur (eds.). Food Security and Land Use Change under Conditions of Climate Variability: A Multidimensional Perspective. Springer.

Ghose, B. 1965. The genesis of the desert plains in the central Luni basin of western Rajasthan. J. Ind. Soc. Soil Sci. 13(2): 123–126.

Ghose, B., Singh, S., Vats, P.C. and Kaith, D.S. 1975. Aerial photogrammetry and photogeomorphology applied to quantitative analysis of drainage basins for engineering planning. Ind. Geog. J. 50(2): 27–31.

Ghosh, A. 1952. The Rajputana Desert: Its Archaeological Aspect, Proc. Symp. Rajputana Desert (Bull. 1, Nat. Inst. Sc. India), 37–42.

Glennie, K.W. 1970. Desert Sedimentary Environments: Developments in Sedimentology (Vol. 14), Elsevier Publishing Co., NY, p. 221.

Goudie, A.S., Allchin, B. and Hegde, K.T.M. 1973. The former extensions of the Great Indian Sand Desert; The Geographical Journal 139: 243–257.

Heron, A.M. 1953. Geology of Central Rajputana. Memoir Geological Survey of India 79: 339.

Herrmann, H.J. 2002. Evolution and shapes of dunes. Comptes Rendus Physique 3(2): 197–206.

Howard, A.D., Morton, J.B., Gad-El-Hak, M. and Pierce, D.B. 1978. Sand transport model of barchan dune equilibrium. Sedimentology 25(3): 307–338. doi:10.1111/j.1365-3091.1978.tb00316.x.

Hugenholtz, C.H. and Wolfe, S.A. 2005. Biogeomorphic model of dunefield activation and stabilization on the northern Great Plains. Geomorphology 70(1–2): 53–70.

Hugenholtz, C.H., Levin, N., Barchyn, T.E. and Baddock, M.C. 2012. Remote sensing and spatial analysis of aeolian sand dunes: a review and outlook. Earth Sci. Rev. 111: 319–334.

Hunt, J.C.R., Leibovich, S. and Richards, K.J. 1988. Turbulent shear flows over low hills. Quarterly Journal of the Royal Meteorological Society 114(484): 1435–1470.

Jackson, P.S. and Hunt, J.C.R. 1975. Turbulent wind flow over a low hill. Quarterly Journal of the Royal Meteorological Society 101(430): 929–955.

Juyal, N., Kar, A., Rajaguru, S.N. and Singhvi, A.K. 2003. Luminescence chronology of aeolian deposition during the late Quaternary on the southern margin of Thar Desert, India. Quaternary International 104: 87–98.

Kalwar, S.C., Sharma, M.L., Gurjar, R.D., Khandelwal, M.K. and Wadhawan, S.K. (eds.). 2005. Geomorphology and Environmental Sustainability: Felicitation Volume in the Honour of Prof. H.S. Sharma. Concept, New Delhi. P. 372.

Kar, Amal. 2020. Quaternary Aeolian Landscape Development in Thar Desert and Its drivers. Proc Indian Natn Sci Acad 86 No. 1 March 2020 pp. 405–417. DOI: 10.16943/ptinsa/2020/49778.

Kar, Amal and Anil Kumar. 2020. Evolution of arid landscape in India and likely impact of future climate change. Episodes, 43: 511–523. DOI: 10.18814/epiiugs/2020/020033.

Kar, A. 1987. Origin and transformation of longitudinal sand dunes in the Indian desert. Zeitschrift fur Geomorphologie 31: 311–337.

Kar, A. 1990. The megabarchanoids of the Thar: Their environment, morphology and relationship with longitudinal dunes. The Geographical Journal 156: 51–61.

Kar, A. 1993. Aeolian processes and bedforms in the Thar Desert. Journal of Arid Environments 25: 83–96.

Kar, A. 1994. Sand dunes and their mobility in Jaisalmer district. pp. 395–418. *In*: Dikshit, K.R., Kale, V.S. and Kaul, M.N. (eds.). India-Geomorphological Diversity. Rawat, Jaipur.

Kar, A. 1996. Morphology and evolution of sand dunes in the Thar Desert as key to sand control measures. Indian Journal of Geomorphology 1(2): 177–206.

Kar, A., Felix, C., Rajaguru, S.N. and Singhvi, A.K. 1998. Late Holocene growth and mobility of a transverse dune in the Thar desert. J. Arid Environment 38(2): 175–185.

Kar, A. 1998. Present-day mobile crescentic dunes in the Thar desert, India. pp. 155–164. *In*: Alsharhan, A.S., Glennie, K.W., Whittle, G.L. and Kendall, C.G.St.C. (eds.). Quaternary Deserts and Climatic Change. A.A. Balkema, Rotterdam (Netherlands) and Brookfield (USA).

Kar, A., Singhvi, A.K., Juyal, N. and Rajaguru, S.N. 2004. Late Quaternary Aeolian Sedimentation History of the Thar Desert. pp. 107–124. *In*: Sharma, Singh and De (eds.). Geomorphology and Environment. ACB Publications, Kolkatta.

King, D. 1960. The sand ridge deserts of south Australia and related aeolian landforms Quaternary arid cycles. Royal Society of South Australia Transactions 83: 99–108.

Kroy, K., Sauermann, G. and Herrmann, H.J. 2002. Minimal model for aeolian sand dunes. Phys. Rev. E66: 031302.

Lancaster, N. 1981. Grain size characteristics of Namib Desert linear dunes. Sedimentology 28(1): 115–122.

Lancaster, Nicholas. 1988a. Development of linear dunes in the southwestern Kalahari, southern Africa. J. Arid Environ. 14(no. 3): 233–244.

Lancaster, N. 1988b. Grain-size characteristics of linear dunes in the southwestern Kalahari. J. Sediment. Petrol. 56395–56400.

Liu, B., Qu, J., Zhang, W. and Qian, G. 2011. Numerical simulation of wind flow over transverse and pyramid dunes. J. Wind Eng. Indus. Aerodyn. 99: 879–888.

Liu, B., Qu, J., Ning, D., Gao, Y., Zu, R. and An, Z. 2014. Grain-size study of inland sediments found east of Kumtagh Desert. Inland Res. 13: 1–6.

Livingstone, I. 1988. New models for the formation of linear sand dunes. Geography 73(319): 105–115.

Livingstone, I. 1989. Monitoring surface change on a Namib linear dune. Earth Surface Processes and Landforms 14(4): 317–332.

Livingstone, I., Bullard, J.E., Wiggs, G.F.S. and Thomas, D.S.G. 1999. Grain-size variation on dunes in the Southwest Kalahari, Southern Africa. J. Sediment. Res. 69: 546–552.

Mainguet, M. 1984. A classification of dunes based on aeolian dynamics and the sand budget. pp. 31–58. *In*: Deserts and Arid Lands, edited by El-Baz (The Hague: Martinus Nijhoff Publishers).

Mangan, J.M., Overpeck, J.T., Webb, R.S., Wessman, C. and Goetz, A.F.H. 2004. Response of Nebraska sand hills natural vegetation to drought, fire, grazing, and plant functional type shifts as simulated by the century model. Climatic Change 63: 49–90.

McKee, E.D. and Tibbits, Jr. G.C. 1964. Primary structures of a seif dune and associated deposits in Libya. Journal of Sedimentary Petrology 34: 5–17.

McKee, E.D. 1966. Structures of dunes at White Sands National Monument, New Mexico (and a comparison with structures of dunes from other selected areas). Sedimentology 7: 3–69.

McKee, E.D. 1979. A Study of Global Sand Seas (Washington, DC: U.S. Government Printing Office).

Misra, V.C. 1968. "The Marusthali" in R.L. Singh (ed.). India: Regional Studies (New Delhi, 21st IGU Publication).

Nair, A.R., Navada, S.V. and Rao, S.M. 1999. Isotopic study to investigate origin and age of groundwater along palaeochannels in Jaisalmer and Ganganagar districts of Rajasthan. pp. 315–319. *In*: Radhakrishna, B.P. and Merh, S.S. (eds.). Vedic Sarasvati, Geol. Soc. India Memoir No. 42.

Namikas, S.L. and Sherman, D.J. 1998. AEOLUS II: an interactive program for the simulation of aeolian sedimentation. Geomorphology 22(2): 135–149.

Nishimori, H. and Ouchi, N. 1993. Formation of ripple patterns and dunes by windblown sand. Physical Review Letters 71(1): 197–200.

Nishimori, H., Yamasaki, M. and Andersen, K.H. 1998. A simple model for the various pattern dynamics of dunes. International Journal of Modern Physics B 12(3): 257–272.

Nishimori, H. and Tanaka, H. 2001. A simple model for the formation of vegetated dunes. Earth Surface Processes and Landforms 26(10): 1143–1150.

Pandey, S., Singh, S. and Ghose, B. 1964. Orientation, distribution and origin of sand dunes in the central Luni basin. Proceedings, Symposium on Problems of Indian Arid Zone. CAZRI, Jodhpur, pp. 84–91.

Pelletier, J.D., Mitasova, H., Harmon, R.S. and Overton, M. 2009. The effects of interdune vegetation changes on eolian dune field evolution: a numerical-modeling case study at Jockey's Ridge, North Carolina, USA. Earth Surf. Process. Land. 34: 1245–1254.

Pettijohn, F.J. and Lundahl, A.C. 1943. Shape and roundness of Lake Erie beach sands. J. Sediment. Petrol. 13: 69–78.

Pithawala, M.B. 1952. The Great Indian Desert: A Geographical Study. Proc. Symp. Rajputana Desert (Bull. 1, Nat. Inst. Sc. India), 141.

Pye, K. and Tsoar, H. 1990. Aeolian Sand and Sand Dunes. Unwin and Hyman, London, 396p.

Rajawat, A.S., Narain, A., Navalgund, R.R., Pathak, S., Sharma, J.R., Soni, V., Babel, M.K., Srivastava, K.S. and Sharma, D.C. 1999. Potentials of Radarr (ERS-1/2 SAR) and high-resolution IRS-1-C data in reconstructing palaeodrainage network of western Rajasthan. pp. 245–258. *In*: Radhakrishna, B.P. and Merh, S.S. (eds.). Vedic Sarasvati, Geol. Soc. India Memoir No. 42.

Ramasamy, S.M. 1999. Neotectonic controls on the migration of Sarasvati river of the Great Indian Desert. Memoir Geological Society of India. No. 42: 153–162.

Roy, A.B. 1999. Evolution of saline lakes of Rajasthan. Current Science 76(3): 290–295.

Roy, A.B. 2003. Late Quaternary neotectonic deformations as cause of drainage disorganisation and extinction of the Vedic Sarawati. GSI Proceedings of GEOSAS-IV. 182–191.

Rubin, D.M. and Hunter, R.E. 1985. Why deposits of longitudinal dunes are rarely recognised in the geologic record. Sedimentology 32: 147–157.

Rubin, D.R. and Hunter, R.E. 1987. Bedform alignment in directionally varying flows. Science 237: 276–278.

Sauermann, G., Kroy, K. and Herrmann, H.J. 2001. A continuum saltation model for sand dunes. Phys. Rev. E64: 31305.

Saxena, S.K. and Singh, S. 1976. Some observations on the sand dunes and vegetation and Bikaner district in western Rajasthan. Annals of Arid Zone 15(4): 313–322.

Schwammle, V. and Herrmann, H.J. 2003. Solitary wave behaviour of sand dunes – colliding dunes appear to traverse through one another and emerge unscathed. Nature 426(6967): 619–620.

Sharma, H.S. (ed.). 1981. Perspectives in Geomorphology, Vol. 1, Concept, New Delhi.

Sharma, H.S. 1987. Tropical Geomorphology. Concept, New Delhi. P. 385.

Singh, G., Joshi, R.D., Chopra, S.R. and Singh, A.B. 1975. Late quarternary history of vegetation and climate of Rajasthan Desert, India. Phil.Trans.Soc., London 267(B): 467–501.

Singh, S., Kolarkar, A.S. and Saxena, S.K. 1974. Relationship between morphology, soil and vegetation of the sand dunes in Jodhpur district, Annual Report, CAZRI, Jodhpur, pp. 31–33.

Singh, S. 1977a. Sand dunes and palaeoclimate in Jodhpur district, western Rajasthan, Man and Environment 1: 7–15.

Singh, S. 1977b. Geomorphological Investigation of the Rajasthan Desert, CAZRI Monograph No. 7.

Singh, S. 1982. Types and formations of sand dunes in the Rajasthan Desert. pp. 165–183. *In*: Perspectives in Geomorphology, edited by H. . Sharma (New Delhi; Concept Publishing Company).

Singhvi, A.K., Sharma, Y.P. and Agrawal, D.P. 1982. Thermoluminescence dating of sand dunes in Rajasthan, India; Nature 295: 313–315.

Singhvi, A.K. and Kar, A. (eds.). 1992. Thar Desert in Rajasthan (Bangalore: Geological Society of India), 191 p.

Singhvi, A.K., Banerjee, D., Rajaguru, S.N. and Kishan Kumar, V.S. 1994. Luminescence chronology of a fossil dune at Budha Pushkar, Thar desert: Palaeoenvironmental and archaeological implications; Current Science 66: 770–773.

Singhvi, A.K. 2004. Quaternary history and palaeoenvironmental record of the Thar Desert in India-Preface. Proceedings of the Indian Academy of Sciences-Earth and Planetary Sciences 113(3): 367–368.

Singhvi, A.K. and Kar, A. 2004. The aeolian sedimentation record of the Thar desert. Proceedings of the Indian Academy of Science 113: 371–401, Bangalore

Singhvi, A.K., Williams, M.A.J., Ragaguru, S.N., Mishra, V.N., Chawla, S., Stokes, S., Chauhan, N., Francis, T., Gajoo, R.K. and Humphries, G.S. 2010. A ˜ 200 ka record of climate change and dune activity the Thar Desert, India. Quaternary Science Reviews 29: 3095–3105.

Singhvi, A.K., Bhatt, N., Glennie, K.W. and Srivastava, P. 2012. India Arabia and adjacent regions. pp. 151–206. *In*: Metcalfe, S.E. and Nash, D.J. (eds.). Quaternary Environmental Change in the Tropics. Chichester, Wiley.

Thomas, D.S.G. 1988. Analysis linear dune sediment-form Relat. Kalahari Dune Desert. Earth Surf. Process. Landforms 13: 545–553.

Thomas, D.S.G., Knight, M. and Wiggs, G.F.S. 2005. Remobilization of southern African desert dune systems by twenty-first century global warming. Nature 435: 1218–1221.

Tsoar, H. 1982. Internal structure and surface geometry of longitudinal seif dunes. Journal of Sedimentary Petrology 52: 823–831.

Tsoar, H. 1984. The formation of seif dunes from barchans—a discussion. Zeitschrift Fur Geomorphologie 28(1): 99–103.

Turner, B.L. II and Meyer, W.B. 1991. Land use and land cover in global environmental change: Considerations for study, International Social Science Journal 130: 669–677.

Vats, P.C., Singh, S., Ghose, B. and Kaith, D.S. 1976. Types, orientation and distribution of sand dunes of Bikaner district. Geographical Observer 12: 67–75.

Valdiya, K.S. 2002. Saraswati: The River that Disappeared. University Press, Hyderabad. 116p.

Verstappen, H.Th. 1968. On the origin of longitudinal (Seif) Dunes. Zeitseh. Fur Gomorph. 12(2): 200–220.

Wadhawan, S.K. 1994. Dune dynamics and evolution of aeolian landforms in parts of Jaisalmer District, Rajasthan, India. Journal of the Indian Society of Remote Sensing 22(2): 65–77.

Wadhawan, S.K. and Kumar, V. 1996. Subsurface Quaternary aeolian stratigraphy in the Ghaggar basin of Thar Desert, India. Jour. Arid Environment 32(1): 37–51.

Wadhawan, S.K. 1996. Textural attributes of recent aeolian deposits in different sub-basins of the Thar Desert, India. J. Arid Environment 32(1): 59–74.
Wadhawan, S.K. and Sharma, H.S. 1997. Quaternary stratigraphy and morphology of Desert Ranns and evaporite pans in central Rajasthan, India. Man and Environ. 22(2): 1–10.
Wadhawan, S.K. 1998. Late Quaternary evolution of clustered parabolic megadunes in Thar Desert, India. pp. 185–195. *In*: Alsharhan, A.S., Glennie, K.W., Whittle, G.L. and Kendall, C.G.St.C. (eds.). Quaternary Deserts and Climatic Change. A.A. Balkema, Rotterdam (Netherlands) and Brookfield (USA).
Wadhawan, S.K. 1999. Quaternary Geology and Stratigraphy of the Shergarh – Dechu sub-basin along the eastern margin of the Thar Desert, Rajasthan, India. 1999. Man and Environ. 24(1): 63–75.
Wadhawan, S.K. 2005. Geological evolution of Thar Desert in Rajasthan and Gujarat, India. GSI Final Report, Jaipur. Records of Geological Survey of India 137(7): 56–58.
Wadhawan, S.K. 2010. Quaternary continental sedimentation basins and dune fields in Thar Desert of NW India. Gondwana Geological Magazine (presently J. Geoscience Research) Spl. Vol. No. 12: 265–270.
Wadhawan, Sudesh Kumar. 2015. Geotechnical applications for geohazards mitigation in dryland environment of Thar Desert, India. Journal of Engineering Geology Vol. XL No. 2, 48–60 (Proceedings International Conference on Engineering Geology in New Millennium. EGNM-2015), IIT Delhi, New Delhi.
Wadhawan, S.K. 2018. Palaeogene–Neogene tectonics and continental aggradational basins in north-western India: Implications for geological evolution of Thar Desert. The Indian Paleogene, 151–166.
Wadhawan, S.K. 2019. Desert Landforms. GeoGraphy and you. March 1–15, 2019. 19(17): 23–27.
Wadhawan, S.K. in press. Geospatial Diversity, Geologic Significance and Evolutionary Trends of Parabolic Dunes in Thar Desert of Indian Subcontinent (current volume).
Wadia, D.N. 1961. Geology of India. Macmillan.
Warren, A., Sud, Y.C. and Rozanov, B. 1996. The future of deserts. Jour of Arid Environments 32(1): 75–89.
Wasson, R.J. and Hyde, R. 1983. Factors determining desert dune type, Nature 304(5924): 337–339.
Wasson, R.J., Rajaguru, S.N., Misra, V.N., Agrawal, D.P., Dhir, R.P., Singhvi, A.K. and Rao, K.K. 1983. Geomorphology, late Quaternary stratigraphy and palaeoclimatology of the Thar dunefield. Zeitschrift fur Geomorphologie, Supplememtband 45: 117–151.
Wiggs, G.F.S. 2001. Desert dune processes and dynamics. Progr. Phys. Geog. 25: 53–79.
Williams, M.A.J. 2014. Climate Change in Deserts: Past Present and Future. Cambridge Univ. Press, Cambridge.
Wilson, I.G. 1972. Aeolian bedforms, their development and origins. Sedimentology 19: 173–210.
Yang, Y., Squires, V. and Lu, Qi. 2002. Global alarm: Dust and Sand storms from the World's drylands. United Nations, Beijing ISBN 92-1-1201140-4.
Yash Pal, Baldev Sahai, Sood, R.K. and Agrawal, D.P. 1980. Remote sensing of the 'Lost' Saraswati River. Proceedings of Indian Academy of Science (Earth and Planetary Sciences) 89(3): 317–331.
Yizhaq, H., Ashkenazy, Y. and Tsoar, H. 2007. Why do active and stabilized dunes coexist under the same climatic conditions? Physical Review Letters 98(18).
Yizhaq, H., Ashkenazy, Y. and Tsoar, H. 2009. Sand dune dynamics and climate change: a modeling approach. Journal of Geophysical Research-Earth Surface, 114.
Zhu, B.Q., Yu, J.J., Rioual, P. and Ren, X.Z. 2014. Particle size variation of inland dune deposits in the lower reaches of the Heihe River basin, China. Sediment. Geol. 301: 54–69.

Chapter 17

Late Quaternary Paleoenvironments of the Eastern Margins of the Thar Desert, India

A Corridor for Human Migration and Occupation

Hema Achyuthan[1,*] and *Sudesh K. Wadhawan*[2]

1. Introduction

The period particularly from Marine Isotope Stage (MIS) 8 to 1 has witnessed frequent climate shifts in time, frequency, duration and space. This period is important as *Homo sapiens* migrated 'Out of Africa' especially since MIS 5 and populated the rest of the world. The migration gateway and the path of migration to different regions of the world can be traced applying multi and interdisciplinary tools including integrated studies on Quaternary geology and sedimetology of the host deposits, geomorphological context and landscape evolution, desert soils and calcretes, application of isotopic analyses, chronmetric dating (Carbon dating/Thermoluminescence/OSL/IRSL dating); palaeoclimatological interpretations for the reconstruction of past conducive climate for adaptation of fauna and flora and their environmental linkages, availability of resistant rocks needed for innovative uses as stone tools, and evolving technology for making of palaeolithic tools in the vicinity of geoachaelogical sites and along possible migratory routes of homonin, etc. (cf. Rapp and Hill, 1998; Fouache et al., 2010). Understanding the past man and land relationship and prevailing paleoclimate demands an integration of such tools from Quaternary sciences. Thus, multidisciplinary teams of geologists, geomorphologists, geochronologists, botanists, zoologists, chemists, climate modellers, remote sensing experts, GIS experts and other technical experts often come together bringing their expertize and work as a team integrating their data to reconstruct the past 2.6 million year in particular and explain 'Man' and the landscape relationship in terms of landscape evolution and influence of paleoclimate shifts. For instance, the Jawalapuram site in South India where in the Middle Paleolithic Assemblages from the Indian Subcontinent, both before and after the Toba Super-Eruption were studied in detail by Petraglia et al. (2007), applying an inter and multidisciplinary approaches.

[1] Professor, Institute for Ocean Management, Anna University, Chennai 600025, Tamil Nadu.
[2] Adjunct Professor, Amrita Vishwa Vidyapeetham Kollam-690525, Kerala and DG, Geological Survey of India (Retd.).
* Corresponding author: hachyuthan@gmail.com

MIS 5 represents a geologic temperature record, spanning 130 to 80 ka and the last interglacial period is dated around 124 ka–119 ka (MIS 5e), when the temperatures were at least 2°C warmer than the present compared with the global mean surface temperatures (Rohling et al., 2007; Carlson, 2008; Alley et al., 2010; Dansgaard et al., 1993; Dubey et al., 2011; Blinkhorn, 2013; Blinkhorn et al., 2019). During this period the mean sea level was 4–6 m higher than at present, following reductions of the Greenland ice sheet. Stable oxygen isotopes of planktonic foraminifera and age constraints data obtained from corals suggest average rates of sea-level rise of 1.6 cm per century (Rohling et al., 2007; Carlson, 2008; Alley et al., 2010; Chappell and Shackleton, 1986; Berger, 2008). These findings are important to understand current climate change, because global mean temperatures during MIS 5e particularly were comparable to the expected climate change today (Heinrich, 1988; Shackleton et al., 2003; Hansen et al., 2015; Hearty and Tormay, 2017). MIS 5 and 3 are considered to be similar in past climate conditions, arid and warmer with shorter interludes of wetter settings. MIS 3 represents a period between 57 ka and 29 ka and is often well thought-out to be a relatively warmer period during the last glacial cycle. During MIS 3 several abrupt climate warm phases known as the Dansgaard-Oeschger (D-O) events occurred and the warm phases were followed by cold stages, so-called the Heinrich events. The D-O and the Heinrich events effected both floral as well as faunal life and these have been registered in the Greenland ice core oxygen isotope record that reveal transitions from cold, stadial climate conditions to mild, interstadial conditions, eventually followed by a return to cold stadial conditions (Dansgaard et al., 1993). It is well established now that during this period the Upper Paleolithic tool technology commenced (Antonio, 1996), Neanderthals disappeared (Higam et al., 2014) and far-off continents such as Australia got populated (Clarkson et al., 2017, 2020). MIS 5 and 3 periods are important stages to reconstruct the past climate when undergoing rapid climatic change and when 'Man' was migrating and occupying several territories and populating (Clarkson et al., 2017; Blinkhorn and Petraglia, 2017).

In order to understand the regional context of paleo-precipitation, paleotemperature and paleo-vegetation type and its evolution process for the last interglacial-glacial cycle on global deserts and their margins, it is imperative to consider interpretations made by a number of researchers from well documented sections studied, notably in China, Saudi Arabia, etc. (Cronin, 2010; Petraglia et al., 2011; Wang et al., 2018). These findings indicate that the paleo-vegetation in the Chinese Loess Plateau (CLP) experienced semi-humid forest-grassland or semi-arid grassland that kept pace with the paleoclimate although the paleoclimate underwent multiple cycles of shifts (Wunsch et al., 2013; Bol'shakov, 2014). However, geological and historical evidences supported by climate models reveal that in future global cooling might take place and this will have a significant impact on agriculture and natural ecosystem (Stothers, 1984; Brovkin et al., 2009). However, the gaps in knowledge especially in relation to climate oscillations and their influence on homonins, during the shift from Marine Isotope Stage MIS 5 to 4 and MIS 3 to 2 have been bridged from studies on the CLP deposits which recorded these variations in detail along with evolution of vegetation (Liu, 1985). The transitional boundary of MIS 5 to 4 in the loess-paleosol sequence was calculated to be around 73.9 ka (Porter and An, 1996; Lu et al., 2007). During the MIS 5 to 4 transition, in the central CLP, woody plants and C4 grasses decreased markedly, and this area was overwhelmingly dominated by arid steppes with few trees (Zhang et al., 2006).

Desert boundaries are remarkably sensitive margins to climate shifts. These dryland margins expand to cover larger areas when aridity sets-in, while it shrinks during prolonged wetter conditions. Thus, late Quaternary sediment sequences from the eastern margin of the Thar Desert hold records of the past regional hydroclimate signals and archaeological evidences that are essential to assess the paleoenvironmental shifts. The eastern margins of the Thar Desert have recorded several of such events in the sucessive sequences of continental deposits mostly along the Aravalli hill ranges and have immense bearing on occurrence of geoacrchalogical sites and migratory aspects of homonins (Dhir et al., 1992; Raghav et al., 1995; Raghav, 1999; Raghvan, 1987). It is important to understand as to what type of climate was prevailing in the Indian subcontinent when the northwestern region of India was already experiencing arid and semi-arid conditions (Juyal et al., 2006; Wadhawan and Kumar, 1996; Achyuthan et al., 2007a, b; Wadhawan et al., 2009; Saini et al., 2009). Some pertinent issues that need to be adressed are: What was the spatial extent of the desertic conditions and intensity of aridity during MIS 5-3? How often did the desert margins shift (in expansion and/or in shortening)? Did 'Early Man' cross the deserts or did he take other routes. What were the pathways and what are their signatures? Was the desert vast in expanse as it is today or even larger? Whether it had experienced wetter interludes that shortened the desert area. Did 'Man' experience arid climate during migration? Notwithstanding some initial and discrete efforts in pursuing directed investigations in the Thar Desert, Rajasthan, Gujarat and parts in southern India (Dhir et al., 1992; 2010, 2018; Achyuthan et al., 2007a, b; Singhvi et al., 2010; Nagar et al., 2010; Petraglia et al., 2012a, b; Wadhawan [this volume]), some of these questions have still remained inadequately answered as more multidisciplinary work needs to be carried out in India to comprehensively understand the paleoclimatic variations particularly since MIS 7.3 and its influence on human occupation of the terrain.

This contribution synthesizes multidisciplinary data based on key paleolithic sites discovered along the eastern margins of the Thar Desert such as Singi Talav, 16 R dune section-Didwana, Dayalpura, Amarpura, Indola ki Dhani, Pir Baba Masjid, Ambali-Rol-Genana, Chhajoli-Kataoti, Jayal, Amarpura and Budha Pushkar (Raghavan, 1987; Blinkhorn, 2018; Blinkhorn et al., 2017, 2019) dune sections and several sites around salt lakes (Wadhawan and Sharma, 1997; Saini et al., 2005; Achyuthan et al., 2007a, b). It is also attempted to provide robust and near continuous proxy evidences for paleoenvironmental changes using archaeological evidences along the eastern margin of the Thar Desert that acted as a gateway to a conducive regional corridor for human migration waves to sites 'Out of Africa'.

2. Thar Desert: regional context, climate and bio-geodiversity

Occurring as an eastern extension of the Saharo-Arabian Desert, the regional setting of the Thar Desert in western India (Fig. 17.1) has been elaborated in earlier chapters [e.g., Wadhawan, this volume], it is pertinent to note that the continental collision and subduction of the Indian plate is responsible for uplift of the Himalayan mountainous ranges to the northwest and north of the Thar Desert, standing in stark contrast to the more muted relief of the sub-continent elsewhere (Hu et al., 2016).

The Thar Desert sands overlie the Archean basement (Early Precambrian) gneisses (metamorphic rocks formed between 4 billion and 2.5 billion years ago), Neo-Proterozoic (Late Precambrian)

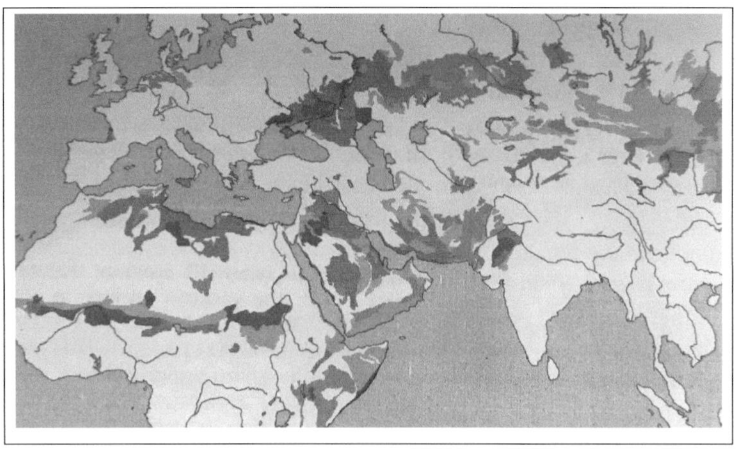

Figure 17.1. Map showing extensive Saharo-Arabian desert regions in parts of eastern Africa, Middle East, Arabia, extending through the Thar Desert and the Chinese desert region.

volcanic and sedimentary rocks (formed about 780 to 541 million years ago), sedimentary rocks of Jurassic and early Cenozoic age. These consolidated rocks were later covered by Quaternary deposits of aeolian dune sands and fluvio-aeolian sand sheets (material deposited by rivers and reworked by winds). The vast diversity of sand dunes has been formed in multiple successions in different sub-basins in the Thar Desert and occur at relatively lower elevations in the north and northwestern region and attain higher general ground elevations towards the eastern and northeastern margins defined dominantly by the Aravalli hill ranges (Wadhawan, 1996, 2010, 2018; Raghav et al., 1995; Wadhawan and Kumar, 1996; Saini and Muztaba, 2012). The Orsang, Mahi, Sabarmati and Indus Rivers remain perennial in the Thar region, whereas the Ghaggar-Hakra and Luni Rivers flow intermittently along the desert margins (Fig. 17.2) during the monsoon season, eventually forming saline pools during the dry season. There are several buried paleochannels that can be traced joining the Luni from the NNW and west that are now covered by the sand dunes, while the

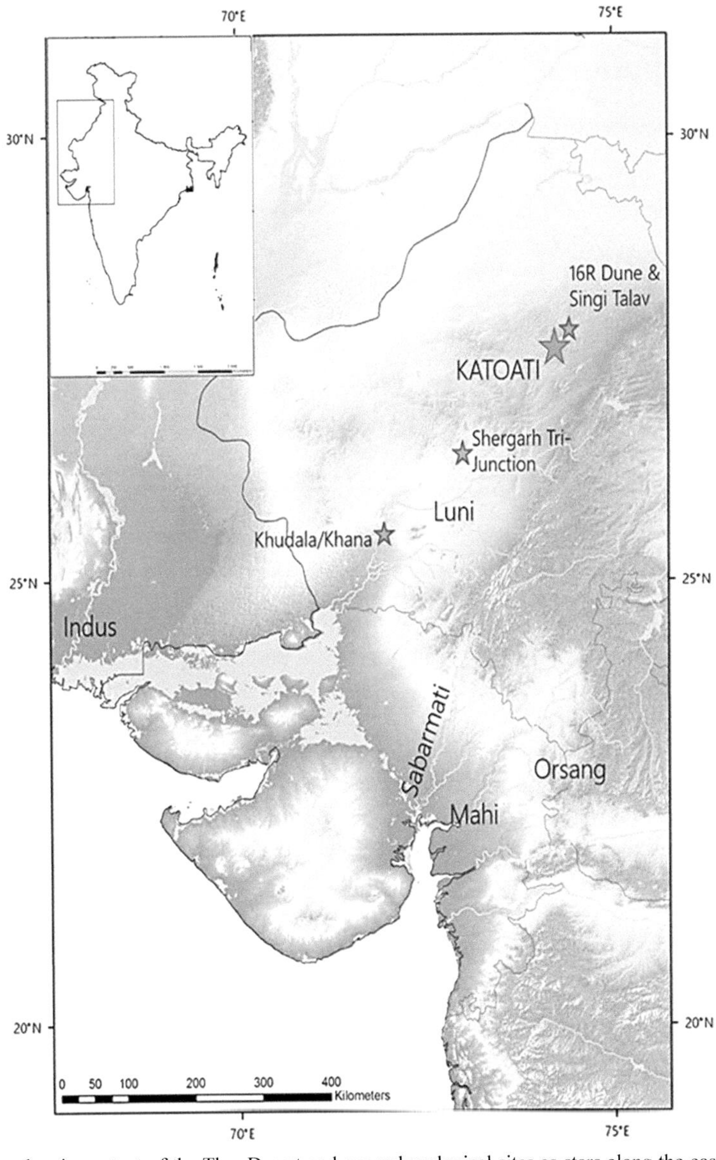

Figure 17.2. Map showing extent of the Thar Desert and geoarchaeological sites as stars along the eastern margins (after Blinkhorn et al., 2019).

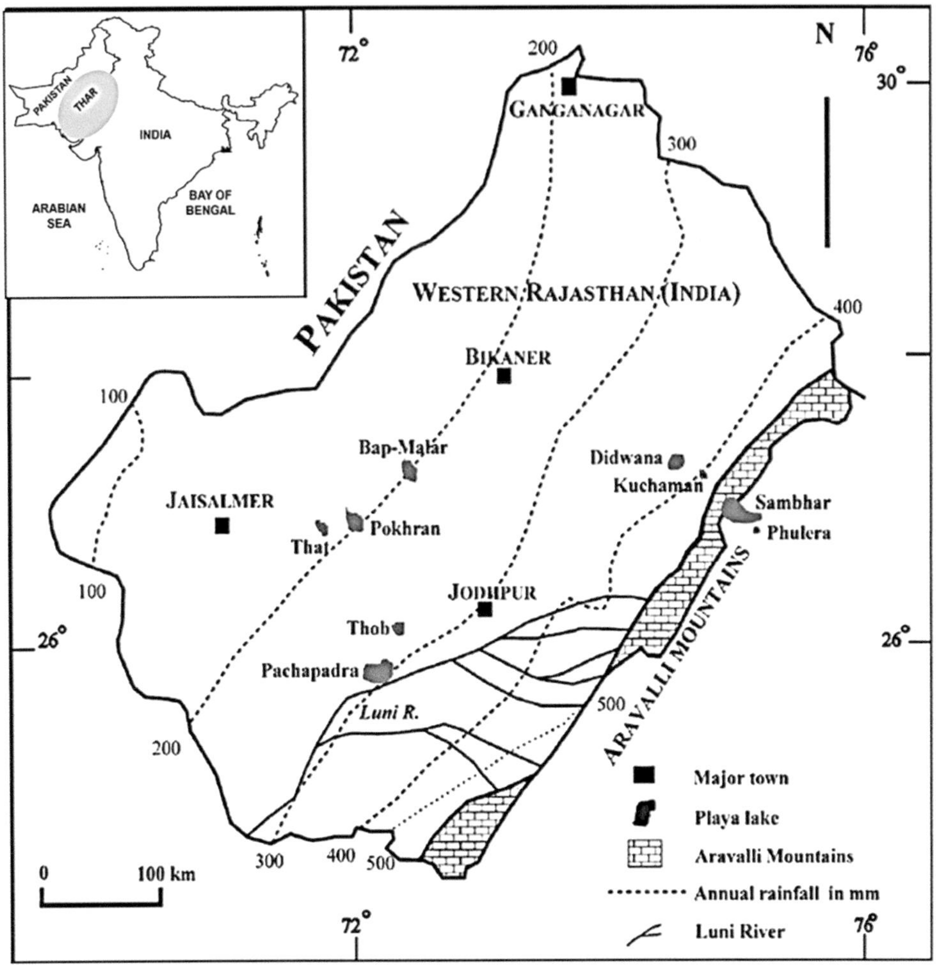

Figure 17.3. Map of the Luni River draining the margin of the Thar Desert and receiving tributaries only from the east and southeast (after Roy and Smykatz-Klossa, 2007).

Luni receives tributaries that are active from the NE, E and SE regions (Raghav, 1999; Ramasamy, 1999) (Fig. 17.3).

Several playas (saline lake beds), locally known as *dhand*s or *sars* and *ranns* occur scattered throughout the Thar Desert region. The saline lakes carry water with lower salinity owing to the summer monsoonal rains and these localized water pools in the disrupted channels get evaporated during peak summer forming a saline crust on the surface. The uplift of the Tibetan Plateau and the surrounding regions played a key factor driving the Asian Monsoon systems (Tada et al., 2016). Located at the margins of monsoonal precipitation, the eastern margin of the Thar Desert has been particularly sensitive to the shifts in monsoonal intensity and aridity conditions. It is along the Thar Desert margin that monsoonal precipitation sharply increases as one moves from west to east, accompanied by the change in seasonality in rainfall from winter to summer. The average amount of annual rainfall in the desert is generally low, ranging from about 4 inches (100 mm) or less in the west to about 20 inches (500 mm) in the east. Precipitation amounts fluctuate widely from year to year. About 90% of the total annual rainfall occurs during the season of the southwest monsoon, from July to September (*see also* Indian monsoon. The Thar Desert in India presently receives ~ 80% of its annual precipitation from the southwest summer monsoon and the eastward migration of mid-latitude winter cyclones (originating over the West Asia and Mediterranean terrains) contributes < 20% of the annual rainfall. The eastern margin of the desert receives an average annual precipitation of ~ 600 mm and the western margin receives ≤ 100 mm. The low and variable rainfall in the region is related to lack of atmospheric moisture condensation. Instrumental records from 1901–2010 AD indicate that the annual precipitations of the Thar Desert decreased in the amount of summer precipitation during the El Niño years. May and June are the hottest months of the year, with temperatures rising to 122°F (50°C). In this region, the maximum temperature reaches up to 45–50°C during summer months and the minimum temperature remains at 3–10°C during winter months (source: Indian Institute of Tropical Meteorology, Pune). Annual evapo-transpiration is ~ 1500 mm in the eastern margin and more than ~ 2500 mm in the western margin. During January, the coldest month, the mean minimum temperature ranges between 41 and 50°F (5 and 10°C), and frost is frequent. Dust storms and dust-raising winds, often blowing with velocities of 87 to 93 miles (140 to 150 km) per hour, are prevalent from April to June.

Water is scarce and is drawn from deep borewell sometimes more than 500 m and is often saline or brackish. Some sweet water aquifers have been delineated in the central part of the desert, e.g., the Late Proterozoic Marwar Supergroup of rocks – the Jodhpur and Nagaur Groups of sandstones and the Lower Jurassic Lathi Formation of coarse sandstone. Elsewhere the groundwater is generally brackish. Seasonal rain fall is collected in tanks and reservoirs constructed in depressions receiving inland local drainage and is used for drinking and domestic purposes.

Arid and semi-arid regions, covering most of the Thar Desert and eastern margins, are dominated by grasslands associations of *Dicanthium-Lasiurus-Cenchrus,* with patchy xeromorphic woodland including stunted, thorny or prickly shrubs and perennial herbs which are resistant to drought (Negi, 1996; Sharma and Mehra, 2009). Stabilized sand dunes are colonized by a range of psammophytic grasses, such as *Aristida adescensionis* and shrubs such as *Calotropis procera*, *Acacia senegal* and *Prosopis cineraria*. Aquatic habitats include the salt lakes, which are often seasonally inundated, and the lake margins are dominated by a range of grasses and sedges, such as *Eleusine compressa*, *Eragrostis ciliaris* and *Dactyloctenium aegyptium* (Sharma and Mehra, 2009). The Thar Desert exhibits a high degree of biodiversity amongst its native fauna, due in part to the location at the juxtaposition between different biogeographical zones, and heavily influenced by the Thar Desert lakes that offer suitable habitats for migratory avifauna from northern Asia (Sharma and Mehra, 2009; Dhir et al., 2018). Halophytic invertebrate populations thrive in the salt lakes of the Thar Desert, feeding on similarly salt-tolerant algae and bacteria, and providing a key nutritional resource for both migratory and non-migratory fauna (Kumar, 2005).

3. Quaternary stratigraphy along eastern margin of Thar Desert

The Quaternary stratigraphic units in the desert margin include the hardpan calcretes of Late Neogene to Early Quaternary age forming the pediment surface. The pediment surfaces of the Precambrian basement rocks and the hills of Aravalli – Delhi Supergroup Fold Belts comprise the granite gneiss around the Sambhar Lake, Kuchaman-Nawa Lake, areas, hornblende gneiss around the Talchappar-Parihara Lakes area and micaceous quartzites around Ladnun and carbonaceous phyllites/phyllitic quartzites and slates around Didwana, rhyolites around Malani near Jodhpur are concealed by geosols of hardpan calcretes. The hardpan calcretes here are nearly a meter and a half thick with a sharp contact with aeolian deposits above (Achyuthan, 2003; Rahgav et al., 1995). The widespread calcretes form a distinct element of the Quaternary landscape of eastern margin of the Thar Desert. Micromorphology included pedogenetic and groundwater features within the hardpan calcretes. Thickening of calcite laminae downward and tapering at the sideward edges around the fresh, unaltered minerals of quartz and feldspars indicated cumulative and compound pedogenesis, which probably occurred locally and downward movement of carbonate solution and pore waters. Carbonate rich solutions were possibly derived from the upper horizons or surfaces bringing about the process of dissolution and re-cementation of individual laminae. Occurrence of fibrous palygorskite as coatings around the detrital grains and siderite (oolitic and pisolitic in shape) points to a subalkaline–subacidic process of pedogenesis in a semi-arid to extremely arid climate (Dhir et al., 2010, 2018). Stable $\delta^{13}C$ and $\delta^{18}O$ isotope data of the hardpan calcrete laminae vary between 0.3 to 1.5 and 5.9 and 1.5%, respectively indicating their formation at or near surface (capillary fringe), most likely supporting a thin column of soil (Achyuthan, 2007a and b). The source of most of the calcite is groundwater; however, calcite nodule formation was largely dependent on pedogenic processes associated with evaporation, evapotranspiration and/or microenvironmental changes in pH and CO_2 partial pressure. Dust is also a major source for carbonate enrichment in soils. Although it is usually assumed that the powdery calcretes are younger in age compared to the harder multi-phase crystalline and complex forms, the occurrence of Middle Palaeolithic tools below the hardpan calcrete at Roopangarh and Dayalpura, and above the hardpan calcretes at Mitri, Genana and Rol, suggests that the morphology and maturity of calcretes may not be a reliable indicator of its age (Fig. 17.4; Raghavan, 1987; Misra et al., 1982; Achyuthan et al., 2007a, b; Blinkhorn, 2013).

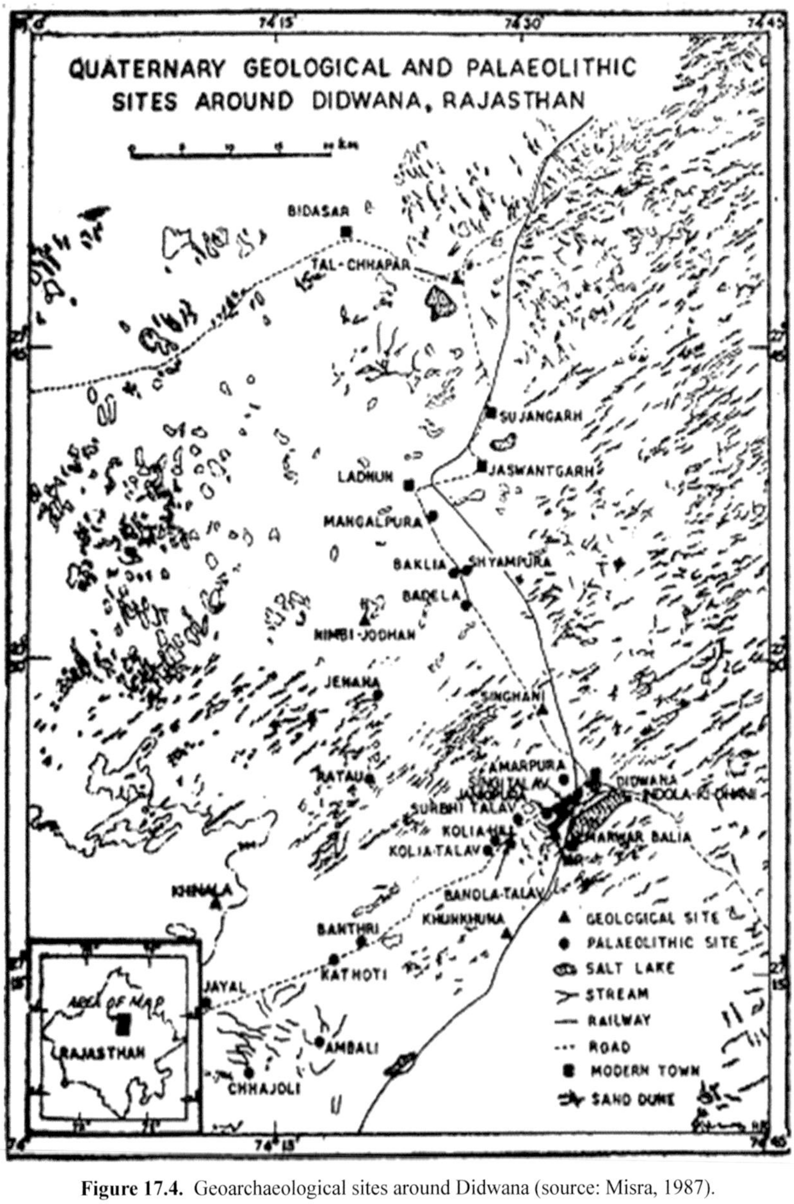

Figure 17.4. Geoarchaeological sites around Didwana (source: Misra, 1987).

3.1 Alluvium and aeolian deposits

The ephemeral River Luni or the Lavanyavati, flows principally during and directly after the monsoon period. Thereafter, the residual surface channel waters form isolated ponds and cess-pools of brackish to saline. It rises in the Aravalli Hills along the eastern margin of the Thar Desert, near Pushkar in Ajmer district, Rajasthan and passes through the pediments buried under obstacle aeolian sand sheet and alluvial plains before debouching in the tectonically structured Gulf of Cambay [Khambhat]. The river incised, during the Holocene wet phase and presented a large river cut bank litho-sections at a number of sites. Jain et al. (2005) studied exposed litho-sections from the middle-reaches of the Luni valley, and had defined two distinct phases of alluvial successions, dating depositional environments from Marine Isotope Stages (MIS) 5-1. The humid phases occurred during MIS 5 (dating to > 86 ka) at Khudala and 79 ka at Karna (Juyal et al., 2006) by forming aggraded and braided stream channel deposits. A period of pedogenesis is chronologically constrained to the period ca. 70–30 ka, and ephemeral stream deposits are evident at the transition between MIS 3-2 (Jain et al., 2005). Aeolian horizons at 27 ka indicated significant decline in fluvial activity, followed by reactivated incision around 22–14 ka and the deposition of gravel braided, mixed load meandering and ephemeral sand bed channel deposits between around 14 ka and 8 ka suggested onsiderable fluvial flux with the transition from MIS 2-1 (Jain et al., 2005). Thereafter, there has been a period of younger aeolian dynamism and deposition that caps these fluvial sequences in the area.

The sand dunes in the eastern margin of the Thar Desert are nearly 40 to 50 m thick, linear or occur as obstructed dunes abutting against the denuded Aravalli hill ranges or over undulating pediplain developed over the basement bed rock or their hardpan calcretes encrustations. Currently the sand dunes are stabilized owing to consolidation and afforestation, however the crests of the dunes are active and seasonally mobile. The source of such enormous sediments for dune-building activity and the age of the dune-building episodes in the Thar Desert have been debated, with some suggestions that recent, anthropogenic factors were responsible (see Allchin et al., 1978). However, the absolute dating of aeolian sediments into the Mid-Pleistocene clearly proves the existence of fluctuating aridity within the region over a period of over 200 ka years (Raghavan et al., 1989; Dhir et al., 2010; Singhvi et al., 2010). Two distinctive sand dune sequences investigated in central and eastern margin of the Thar Desert provide evidence for Mid-Pleistocene activity, dating to 187 ka and 140 ka at 16R Dune (Singhvi et al., 2010) and 155 ka at Chamu (Dhir et al., 2010; Fig. 17.5).

These sites also provide the only dated evidence for aeolian activity during MIS 5, with aggradational events occurring both before (Unit III) and between (Unit IV) 81–105 ka at Central Thar Desert (Dhir et al., 2010) and at 130–109 ka and 80 ka at 16R Dune, eastern margin of the Thar desert (Achyuthan et al., 2007; Singhvi et al., 2010). There is inadequate record on dating of aeolian activity to MIS 4, with the exception of aeolian horizons at Shergarh Tri-Junction dating to 60 ka and 64 ka (Fig. 17.6; Dhir et al., 1992; Wadhawan, 1999; Andrews et al., 1998; Blinkhorn, 2017). As these dates, derived from felspar aliquotes protocol and were not corrected for fading, they do not substantiate a stratigraphic order in the field and their error ranges overlap considerably with MIS 3 (Blinkhorn and Petraglia, 2017).

Aeolian activity during MIS 3 and aggradation events are observed around 55 ± 6 ka and 43 ± 5 ka at Shergarh Tri-Junction (Andrews et al., 1998); 58 ka at Chamu (Dhir et al., 2010); 40 ka at 16R Dune (Achyuthan et al., 2007; Singhvi et al., 2010) and 40–36 ka at Awai, western Thar Desert (Chawla et al., 1992). All aeolian deposits from MIS 6-3 show considerable pedocalcic pedogenesis and calcrete formations, as nodular calcretes that also occur at places as laterally extensive bands (Raghavan, 1989).

Chronometrically dated evidence for aeolian activity during the transition of MIS 3-2 is dated 27 ka from Chamu in central Thar Desert, Rajasthan (Dhir et al., 2010) and at the LGM the upper palaeolithic horizon is dated 26 ka at 16R Dune (Fig. 17.7) and at the railway cutting dune section near Didwana (Raghavan et al., 1989). Within the Thar Desert aeolian aggradations occurred at 17–13 ka at Chamu (Dhir et al., 2010) between 15–13 ka at Nachna (Chawla et al., 1992); 10–9 ka at Awai (Chawla et al., 1992); at 13 ka, 12 ka, 11 ka and 9 ka at 16R Dune and Amarpura (Achyuthan

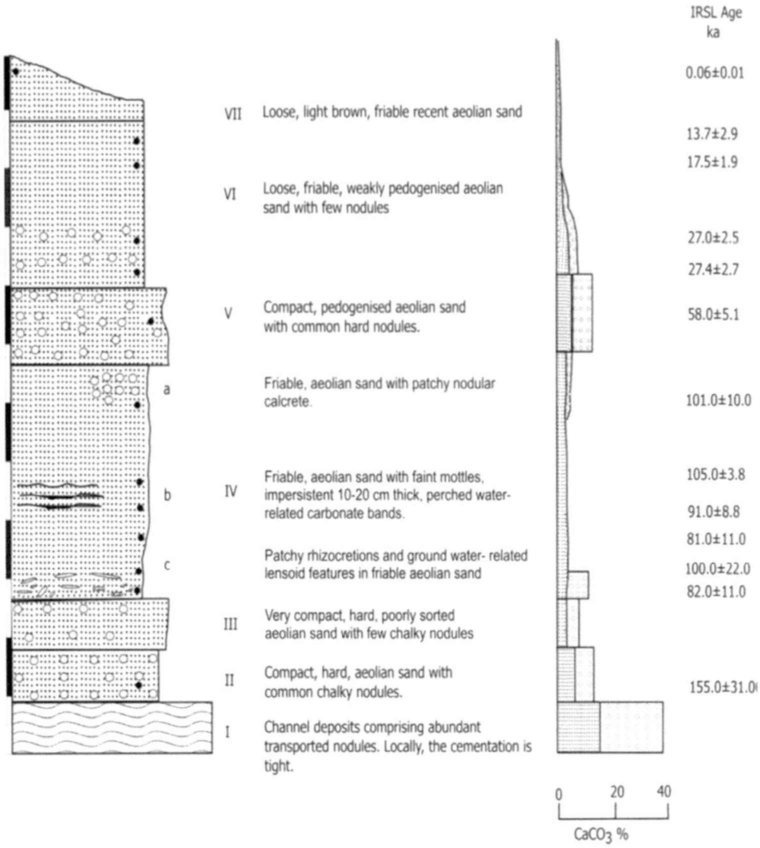

Figure 17.5. Field log and some analytical data of Chamu Section, Central Thar Desert. Symbols used in the log: aeolian aggradations (fine dots), calcrete nodules (hollow circles), sampling site for dating (solid circle). The scale (in meters appears on the left). Other features in the log are described in the text alongside. The $CaCO_3$ carbonate profile shows its content in percent in finely dispersed form (fine dots). The hollow circles in carbonate profile indicate nodules and secondary carbonate. Also shown are IRSL ages of host sands [after Dhir et al. 2010].

Figure 17.6. Alternating sequence of calcretised colluvium debris - aeolian sand deposits, section exposed at Shergarh Tri-junction, Central Thar Desert (after Dhir et al., 1992).

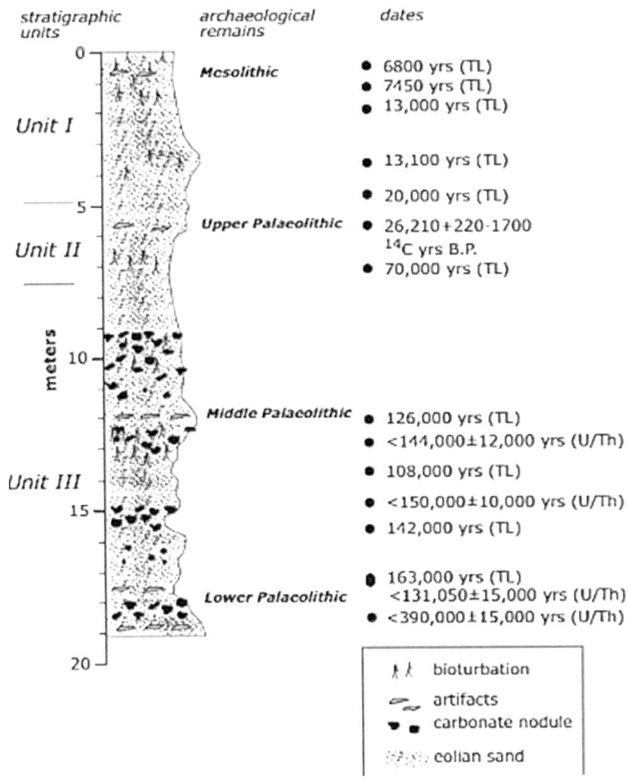

Figure 17.7. 16 R dune section with palaeolithic tools and absolute dates (after Achyuthan et al., 2007b).

et al., 2007); 13.5–11.5 ka at Bikampur (Chawla et al., 1992); 9 ka at Shergarh Tri-Junction (Andrews et al., 1998). It is important to note that Middle Paleolithic and Upper Paleolithic and Mesolithic remains occurred at sections. Didwana 16 R dune are also the singular aeolian section preserving in situ lithic tools that is well dated representing the eastern margins of the Thar Desert (Achyuthan et al., 2007b; Singhvi et al., 2010).

3.2 Lake deposits

The majority of lakes occurring along the Thar Desert margin such as the Sambhar, Didwana, Talchappar-Parihara, Nawa, Pachpadra, Kuchaman, etc., are essentially playas. These mostly become dry and shrink in size during hot summers. The location of these lakes/playas is also controlled by over-all tectonic disposition, as they occur along pull-apart basins and tectonic sags or depressions (Sinha-Roy, 1986; Wadhawan, 2018) or in drainage channels that got disrupted by blocking due to aeolian sand accumulations in the terminal Pleistocene (Achyuthan et al., 2007b; Kar, 2020). Most lake margins are also associated with historical sites of the Mughal period such as those around Talchappar and Parihara salt lakes as these sites used to be suitable grounds for hunting wildlife (Achyuthan, 2007b).

Sinha and Sakar (2009) provided dated evidence of the Late Pleistocene evolution of the lake sediments through the recovery of a 23 m lake core. Three major litho-units have been described, including (Unit III) an upper organic rich mud ^{14}C dated between 6.8 ka to modern overlying a carbonate mud dating to 7.5–9.5 ka, sand and silt dominated Unit II dating to 19–22 ka, and Unit I, a carbonate mud covering a thick carbonate sand (Sinha et al. 2006). Although evidence for sharp aridity is evident during the LGM and between 7.5–6.8 ka, this research suggests that Sambhar Lake has not been completely desiccated throughout this observed history (Sinha et al., 2006). Significantly, Sambhar Salt Lake is the only lake along the eastern margin of the Thar Desert that was clearly present even before the LGM, and has yielded Middle palaeolithic and Acheulean tools on its eastern periphery and with ongoing research by the authors suggest its water-full high lake levels were present much earlier in the Late Pleistocene.

4. Geoarchaeology: human migration and occupation

Paleoenvironmental records and climate model simulations indicate that increased humidity occurred during insolation maxima, in particular during interglacial and interstadials, such as those during Marine Isotope Stage 7 (MIS 7, ~ 240–190 ka) and the interstadials of MIS 5 (~ 130–75 ka) (Scerri et al., 2018). During these periods, the landscape of Arabia and Asia featured extensive and active river networks, lakes and wetlands with widespread vegetation cover (Scerri et al., 2018). Archaeological evidences suggest that around 200 ka, Homo sapiens emerged in Africa and by 40 ka, *Homo sapiens* had spread throughout Eurasia, and the Neanderthals, became extinct. However, it is notable in some regions of the subcontinent that distinctly Late Acheulean technologies continued to the end of MIS 6 ~ 130 ka (Haslam et al., 2011). A review on homonin migration "Out of Africa" suggest that warm and wet conditions from (MIS 5) 120 to 90 ka could have facilitated early waves of human migration toward the Levant and Arabia, as supported by fossil and lithic evidence (Blinkhorn and Petraglia, 2017). However, the primary out-of-Africa event, is constrained by genetic studies to around 65–55 ka that occurred during a cold and dry time. This complicates the climate and human migration relationship, suggesting that both "push" and "pull" factors may have prompted *Homo sapiens* to colonize Eurasia (Scerri et al., 2018). The earliest modern human populations that dispersed into South Asia used Middle Palaeolithic technologies (Petraglia et al., 2007; Blinkhorn and Petraglia, 2017; Clarkson et al., 2020). Scerri et al. (2018) presented the first chronometrically dated Acheulean sites from the Arabian Peninsula, a vast and poorly known region that forms more than half of Southwest Asia and the results show that Acheulean hominin occupation expanded along hydrological networks into the heart of Arabia from Marine Isotope Stage (MIS) 7 until at least ~ 190 ka. Paleolithic assemblages between Arabia and Pakistan have been found as surface finds and these sites have not been chronologically constrained and yet are records of dispersal from 'Out of Africa'.

Acheulian tools-archaeological evidence from Singi Talav (Fig. 17.8a & b) are associated with calcareous pan lacustrine deposits (Gaillard et al. 1993; Raghavan et al., 1993), the edge of the Didwana Lake. The Didwana Lake shrank, and subsequently shifted to its current position owing to aridity setting-in and forming linear dunes running parallel to the lake margin and causing a change in the catchment of the basin. This situation is also observed at Budha Pushkar and several other sites along the Thar Desert margin. The inhabitation of the Thar Desert throughout the Palaeolithic appears closely associated with enhanced monsoonal precipitation (Blinkhorn, 2017). However, in spite of some rare reporting of lithic tools amongst the winds deflated lag deposits or gravel spread (serir) near Bhojka (Fig. 17.9a & b; Wadhawan, 2022, this volume) on the way to Jaisalmer and geoarchaeological sites near Bap playa, the core of the Thar Desert is largely devoid of lower Palaeolithic sites and therefore it is open to conjecture as to what scale of precipitation may have been necessary to enable Late Acheulean occupations of the Thar Desert.

Acheulean tools of Singi Talav (Fig. 17.8b) have not yet been chronologically constrained. However, in India, the middle Son valley, Central India, contains a number of late Acheulian sites preserved in aggraded and incised fluvial deposits of the late Pleistocene to Holocene that has both fossils and lithic artefacts containing Lower and Middle Paleolithc ruins (Misra, 1997; Jones and Pal, 2009; Petraglia et al., 2012a, b). The type sites of Sihawal and Patpara here have been dated by OSL method to 119+–12 ka; and 140+–11 ka respectively. In the absence of an elaborate detailed geoarchaeological investigations work, considering similarities in lithic tools and the dates of associated sediments, it is reasonable to infer and piece together the available evidences and suggest that the late Acheulean of Singi Talav and other Middle Paleolithic assemblages in the eastern margin of the Thar Desert could be contemporaneous or could be slightly older than the evidences from the Son valley with the Thar Desert being the migratory corridor. The Middle Palaeolithic

Figure 17.8 a & b. Singi Talav: a clay pan site presently filled with water. 17.4b represents the late Acheulean tools from Singi Talav, Didwana (after Gaillard et al., 1993).

Figure 17.9 a & b. Several wind-abraded and iron-mineral stained gravels, Palaeolithic implements and associated waste debris indicative of patterned controlled anthropogenic cobble percussion was found at several wind-deflated gravel-pavement (Serir) locations at Bhojka, Lawan, Bap, etc., in the Thar Desert.

occupations of the Thar Desert are known from mid-MIS 5 and extending into mid-MIS 3 from sites such as Budha Pushkar, Amarpura, 16 R-Didwana, Kataoti, Jayal, Ambali, Mangalpura (Misra et al., 1989; Achyuthan, 2007; Blinkhorn et al., 2017) and the Earlier Middle Palaeolithic occupations of the Thar Desert are coincident with phases of enhanced monsoonal intensity, resulting in notable fluvial activity, whereas the younger Middle Palaeolithic assemblages appear associated with some humid interludes amidst the otherwise dry environment and sand-dunes reconstruction. The known Late Palaeolithic occupations of the Thar Desert are currently restricted to MIS 2 and can be broadly confined to a phase of increasing aridity before the Last Glacial Maximum, associated with aeolian dynamism and dune construction.

However, it is worth noting that evolving landscape provided resistant rocky raw material for Palaeolithic tools and the availability of fresh water in lakes that attracted a variety of game for the hunting and gathering migratory *Homo sapiens*. Besides, *in situ* shattering of rock boulders due to characteristic and prolonged diurnal temperature variations in arid-semiarid dryland environment of desert around Jayal – Kataoti gravely boulder ridges would make easy availability of sharp-edged hunting and cutting lithic tools Fig. 17.10 a & b.

5. Discussion

Over the past decade the chronological framework of the Quaternary events in the Thar Desert is well worked. However relatively few numbers of archaeological sites and their restricted spatial range limits understanding of the pattern of continuity of occupation (Blinkhorn, 2017; Blinkhorn and Petraglia, 2017). Before the application of chronometric dating methods, archaeological and environmental data presented by Allchin et al. (1978) recognized notable behavioural continuity in the Thar Desert. It is now evident that at least a group of hominin populations occurred in South

Figure 17.10a & b. *In situ* shattering of rock boulders due to characteristic and prolonged diurnal temperature variations in arid-semiarid dryland around Jayal – Kataoti Ridges [17.10b] showing dislodged gravel and boulders from Neogene conglomerate, would make easy availability of sharp-edged hunting and cutting lithic tools to Palaeolithic hominins.

Asia, with the arrival of *Homo sapiens* and possible disappearance or further dispersal of a poorly known local population (Blinkhorn, 2017; Blinkhorn and Petraglia, 2017).

The geospatial extent of the Thar Desert was greatly reduced during MIS 5 owing to climate amelioration and thus passage for homonins entering the Indian subcontinent would have been relatively easier when compared to an entry in during MIS 6 or MIS 4 (Petraglia et al., 2012b; Boivin et al., 2014). The desert corridor including the Sahel in North Africa and that extended along the southern Arabia and the Makran coasts, reduced the Thar Desert extent which were supporting probably savanna woodlands (Bovine et al., 2014). Palaeoenvironmental evidence from the desert during MIS 5 indicates a high percentage of C4 plants with incursions of C3 vegetation, increased summer rains and a strong monsoon (Achyuthan et al., 2007). The region would have been drained by network of streams and rivers, and perhaps playa lakes were beginning to form. The discovery of numerous Middle Palaeolithic sites in many basins of peninsular India, together with stratified occurrences post-dating Late Acheulean assemblages (Misra, 2001), indicates that hominins were present in the region for a considerable period of time (Boivin et al., 2014). Archaeological excavations at Jwalapuram, in the Jurreru Valley of southern India demonstrate that Middle Palaeolithic hominins were present by c. 78 ka (Petraglia et al., 2007, 2012b; Boivin et al., 2014). Analysis of cores from the Jwalapuram localities indicates affinities with sub-Saharan African MSA assemblages produced by *Homo sapiens* (Clarkson et al., 2012). Though the tool technology remains poorly described, the potential presence of Middle Palaeolithic assemblages dated 130 to 109 ka in 16R Dune in the Thar Desert (Achyuthan et al., 2007; Singhvi et al., 2010) and c. 125 to 75 ka in the Iranamadu Formation of Sri Lanka (Deraniyagala, 1992; Abeyratne, 1997) also suggested MIS 5 occupations that may represent *Homo sapiens*, although an archaic species cannot be ruled out (Petraglia 2010). Based on palaeo-environmental records, India appears to have been amenable to human habitation in MIS 5 and into MIS 4 (Fig. 17.2). India had favorable environments, such as broad areas of savannah and dry tropical woodlands, along with moist tropical woodland and grassland mosaic vegetation in the Ganges region.

Achyuthan identified sites that were certainly present in the early to middle stages of the Middle Pleistocene, indicating the long-term presence in the region of one or more archaic species (Petraglia, 2010; Boivin et al., 2014). Recent dating of Late Acheulean sites in the Son valley of northern India dated to c. 140 to 130 ka (Haslam et al., 2010) suggested that at least one archaic species of Homo sapiens continued to inhabit India at the MIS 6/5 transition (Petraglia et al., 2010). The coincidence with the earliest potential dates for Middle Palaeolithic sites in the Thar Desert at 130 to 109 ka is obviously suggestive of an early human presence. Late Acheulean sites reported from the Thar Desert from sediment deposits have been OSL dated to ~ 100 ka (Juyal et al., 2000). The appearance of Middle Palaeolithic assemblages between 114–96 ka in the Thar Desert have been directly attributed to the arrival of modern humans (Blinkhorn et al., 2013, 2017, 2019). Detailed geoarchaeological studies supported by U/Th and OSL dating reveal that during the last ~ 200 ka major pluvial periods along Thar Desert margin were ~ 200 ka (MIS 7.3), ~ 160–150 ka, ~ 126–115 ka (Climatic Optimum), ~ 90–80 ka, ~ 55–50 ka, ~ 30 to 26 ka and 7–6 ka, and these periods coincided with the phases of intense high summer monsoons. The major periods of enhanced aeolian sand mobilization and accumulation, as well as high dune-building activities, were evidenced during the transition to/from the peaks of vigorous monsoon, especially at ~190 ka (after the MIS 7.3), ~ 160–~ 140 ka, 125–100 ka, 75–65 ka, 24–17 ka, 14–12 ka, 5–3 ka and 2–1 ka (Kar, 2020). It has been established through detailed chronometric assessment of the Quaternary continental sediments in central and eastern parts of the Thar Desert that the dune sand accumulations and accretions lagged major peaks in aridity for instance the Last Glacial Maximum (LGM) due mainly to the reduction of sand moving winds (Singhvi et al., 2010). The current zone of sand dune mobility is very heavily impacted by pastoral and agricultural activity to offer clear information with regards to how similar activity could be inferred in the past (Dhir et al., 2018). However, the presence of sand dunes far beyond the modern arid zone indicate that sufficient wind speeds (> 6 m/s) and sediment supply were, at some remote period, available across the Thar region (Allchin et al., 1978). Paleoclimate shifts were short/abrupt, gradual or long lasting, and are apparent as shifts in the intensity of summer monsoons first and the major climate change resulted in ecological composition on either margin of the Thar Desert (Holt et al., 2013). When these factors are combined, it becomes clear that expansion across boundaries would have involved more appreciable differences at a human scale than are apparent across the physical boundary between Africa and Eurasia (Blinkhorn, 2017). Climatic events appear to have played a part in migration especially in to Asia and the Thar Desert wetter during large part of MIS 5 and 3 was a major corridor for human settlements across parts of India. However, there is a need to search for larger stone tool assemblages to comprehensively evaluate patterns of technological variability, and a greater number of assemblages older than 100 ka are required to fill a broad chronological gap in regional records.

6. Summary

In this chapter a synthesized review of the multidisciplinary geoarchaeological work carried out in Thar Desert with a focus on the palaeolithic sites investigated along its semi-arid eastern margin. The paleoenvironments of the Thar Desert during the late Quaternary period and its implications on the human occupation in a semi-arid to arid landscape is elucidated and described. The rocky pediments of the Aravalli Hill ranges in the eastern margin of the Thar Desert have been covered by episodic obstructed and stabilized sand dunes and variously dissected sand sheets bordering the prominent inland salt lakes. The aeolian sequences display multiple phases of sand accretion interspersed with palaeosols and development of calcrete horizons that indicate wet interludes in overall semi-arid environment. Periods of expansion of the desert in the past dryland environments in the Thar region were observed in the poorly consolidated aeolian deposits, while the palaeosols and calcrete formations indicate ameliorated conditions and contraction of the desert margins. The oldest preserved aeolian sequences dated so far in the Thar Desert are from the base of the north-easterly oriented stabilized obstacle dune abutting against a denudational hill near Didwana in eastern margin of the Thar Desert. This site and several archaeological sites around Chhajoli, Kataoti, Jayal and Budha Pushkar were excavated for improved understanding of the man-landscape relationship and past climate changes (Singhvi et al., 1994; Blinkhorn et al., 2017; Blinkhorn, 2018).

The wetter periods aided landscape stability by increased vegetation resulting in aeolian sediment pedogenesis and calcrete formation. It is noted that in the Thar Desert their truncated profiles have been left exposed owing to the process of succeeding aeolian deflation and aggradations, wherein the pre-existing soil layers got eroded away resulting in truncated soil profiles (Dhir and Singhvi, 2012). Detailed investigations of several dune sections, palaeosols, calcrete horizons/bands within aeolian layers aided with absolute age determinations enabled elaborate interpretations of the different phases of dune-building activities separated by relative stability in the landscape around Didwana and other parts along the Thar Desert margin. However, the periods of human occupancy were marked by *in situ* occurrence of different tools of Lower and Middle Palaeolithic periods. The beginning of long sequences of aeolian sand deposition is dated to > 300 ka thus marking the onset of the dunes and aeolian sand dunes and sheet deposition along the Thar Desert margin. Detailed geoarchaeological studies supported by U/Th and OSL and IRSL dating reveal that during the last ~ 200 ka major pluvial periods along Thar Desert margin were recorded at ~ 200 ka (MIS 7.3), ~ 160–150, ka, ~ 126–115 ka (Climatic Optimum), ~ 90–80 ka, ~ 55–50 ka, ~ 30 to 26 ka, and 7–6 ka, and these periods coincided with the phases of intense high summer monsoons. It is further corroborated that wetter conditions and the stabilization of landscape offered a conducive pathway for 'Out of Africa' migration of humonins since MIS 5 and 3. The major periods of enhanced aeolian sand mobilization and accumulation, as well as high dune-building activities, were evidenced during the transition to/from the peaks of vigorous monsoon, especially at ~ 190 ka (after the MIS 7.3), ~ 140 ka, 125–100 ka, 75–65 ka, 24–17 ka, 14–12 ka, 5–3 ka and 2–1 ka.

Largest in situ collection of cultural artefacts assemblage from the Didwana and Sambhar Salt lake areas along eastern margin of the Thar Desert have revealed significant evidences for Lower, Middle to Upper Palaeolithic hominin occupation and possible linkages for their migration and dispersal routes. This corresponds and correlates with similar geoarchaeological sites in the Son valley in the farther east and Jwalapuram in South India are amply supported by several similarities on evolving palaeolithic tool making technology, particularly continued production of the biface, choppers and numerous small lithic tools (Petraglia et al., 2007; Clarkson et al., 2012, 2017; Blinkhorn et al., 2017, 2019).

It is remarkable that the Last Glacial Maximum (LGM; end MIS 3 to early MIS 2; ~ 24–17 ka) was the driest period with intense aeolian and widespread activity and dune development, expanding the desert boundary well beyond its present eastern limit, across the Aravalli hill ranges. Subsequent to the LGM, since ~ 17 to 15 ka, the monsoons gradually strengthened in its intensity and amount of rainfall. However, this rainfall lagged the beginning and onslaught of higher wind speed by years to centuries (Singhvi et al., 2010; Kar et al., 2002). The beginning of this strong aeolian phase dated to ~ 16–14 ka, probably caused large-scale re-mobilization of the older dunes expanding the desert margin far beyond the present arid boundary. By early Holocene at ~ 11.7 ka, the monsoons and the aeolian processes decreased thus stabilizing the landscape with the spread of desert vegetation and consolidation of sand dunes.

During 11.7 to 6.5 ka the SW monsoon intensified again, enhancing freshwater intake into the lakes with expanded high lake margins and receiving higher influx of silt and clay. Lake and aeolian sediments stratigraphy and available OSL dates indicate that the aeolian activities intensified from ~ 5.0 ka to 3.5 ka due to inferred decline in the strength of the westerlies. Short interludes of intense monsoons between, the period ~ 4.0 and 2.7 ka again witnessed the rise and decline of the lake margins. Such variations in landscapes and easy availability of raw material for Palaeolithic tools have greatly influenced human occupation and migration along the Thar Desert margin.

References

Abeyratne, M., Spooner, N.A., Grün, R. and Head, J. 1997. Multidating Studies of Batadomba Cave, Sri Lanka, Quaternary Science Reviews.

Achyuthan, H. 2003. Petrologic analysis and geochemistry of the Late Neogene-Early Quaternary hardpan calcretes of Western Rajasthan, India. Quaternary International 106: 3–10. DOI.10.1016/S1040-6182(02)00158-1.

Achyuthan, H., Kar, A. and Eastoe, C. 2007a. Late Quaternary-Holocene lake-level changes in the eastern margin of the Thar Desert, India. J. Paleolimnol. 38: 493–507. https://doi.org/10.1007/s10933-006-9086-6.

Achyuthan, H., Quade, J., Roe, L. and Placzek, C. 2007b. Stable isotopic composition of pedogenic carbonates from the eastern margin of the Thar Desert, Rajasthan, India. Quat. Int. 162: 50–60. https://doi.org/10.1016/j.quaint.2006.10.031.

Allchin, B., Goudie, A. and Hegde, K.T.M. 1978. The Prehistory and Palaeogeography of the Great Indian Desert. Academic Press, London.

Alley, R.B., Andrews, J.T., Brigham-Grette, J., Clarke, G.K.C., Cuffey, K.M., Fitzpatrick, J.J., Funder, S., Marshall, S.J., Miller, G.H., Mitrovica, J.X., Muhs, D.R., Otto-Bliesner, B.L., Polyak, L. and White, J.W.C. 2010. History of the Greenland ice sheet: Paleoclimatic insights. Quaternary Science Reviews 29: 1728–1756.

Andrews, J.E., Singhvi, A.K., Kailath, A.J., Kuhn, R., Dennis, P.F., Tandon, S.K. and Dhir, R.P. 1998. Do stable isotope data from calcrete record late pleistocene monsoonal climate variation in the Thar Desert of India? Quat. Res. 50: 240–251. https://doi.org/10.1006/qres.

Andrews, J.E., Singhvi, A.K., Kailath, A.J., Kuhn, R., Dennis, P.F., Tandon, S.K. and Dhir, R.P. 1998. Do stable isotope data from calcrete record late pleistocene monsoonal climate 727 variation in the Thar Desert of India? Quaternary Research 50: 240–251.

Antonio, G. 1996. Explaining the upper palaeolithic revolution. pp. 220–39 (Chap. 8) in Contemporary Archaeology in Theory: A Reader. Cambridge, MA: Blackwell.

Berger, W.H. 2008. Sea level in the late quaternary: patterns of variation and implications. International Journal of Earth Sciences 97: 1–21.

Blinkhorn, J. 2013. A new synthesis of evidence for the Upper Pleistocene occupation of 16R Dune and its southern Asian context. Quaternary Interntional 300: 282–291. https://doi.org/10.1016/j.quaint.2013.01.023.

Blinkhorn, J., Achyuthan, H., Petraglia, M. and Ditchfield, P. 2013. Middle palaeolithic occupation in the thar desert during the upper pleistocene: The signature of a modern human exit out of Africa? Quat. Sci. Rev. 77. https://doi.org/10.1016/j.quascirev.2013.06.012.

Blinkhorn, J. 2014. Late Middle Palaeolithic surface sites occurring on dated sediment formations in the Thar Desert. Quaternary International 350. https://doi.org/10.1016/j.quaint.2014.01.027.

Blinkhorn, J. and Petraglia, M.D. 2017. Environments and cultural change in the Indian subcontinent: Implications for the dispersal of Homo sapiens in the late Pleistocene. Current-Anthropology 58(S17): 1537–5382. https://doi.org/10.1086/693462.

Blinkhorn, J., Achyuthan, H., Ditchfield, P. and Petraglia, M. 2017. Palaeoenvironmental dynamics and Palaeolithic occupation at Katoati, Thar Desert, India. Quat. Res. 87: 298–313. https://doi.org/10.1017/qua.2017.7.

Blinkhorn, J. 2018. Buddha Pushkar revisited: Technological variability in Late Palaeolithic stone tools at the Thar Desert margin, India. J. Archaeol. Sci. Reports 20. https://doi.org/10.1016/j.jasrep.2018.04.020.

Blinkhorn, J., Ajithprasad, P., Mukherjee, A., Kumar, P., Durcan, J.A. and Roberts, P. 2019. The first directly dated evidence for Palaeolithic occupation on the Indian coast at Sandhav, Kachchh. Quat. Sci. Rev. 224: 105975. https://doi.org/10.1016/j.quascirev.2019.105975.

Bol'shakov, V.A.A. 2014. Link between global climate variability in the Pleistocene and variations in the Earth's orbital parameters. Stratigr. Geol. Correl. 22: 538–551.

Boivin, N. Crowther, A. and Prendergast, M. 2014. Indian Ocean food globalization and Africa. African Archeological Review (31) 4547–581.

Bradák, B., Csonka, D., Novothny, A., Szeberényi, J., Medveďová, A., Rostinsky, P., Fehér, K., Végh, T. and Megyeri, M. 2011. Late Pleistocene paleosol formation in a dynamic aggradational microenvironment—A case study from the Malá nad Hronom loess succession Slovakia. Catena 199: 87–105.

Brovkin, V., Raddatz, T., Reick, C., Claussen, M. and Gayler, V. 2009. Global biogeophysical interactions between forest and climate. Geophysics. Research 36: L07405.

Carlson, A.E. 2008 Response of the southern Greenland ice sheet during the last two deglaciations. Geology 36: 359–362.

Chawla, S., Dhir, R. and Singhvi, A. 1992. Thermoluminescence chronology of sand profiles in the Thar desert and their implications. Quat. Sci. Rev. 11: 25–32.

Chappell, J. and Shackleton, N.J. 1986. Oxygen isotopes and sea level. Nature 324: 137–140.

Clarkson, C., Jones, S. and Harris, C. 2012. Continuity and change in the lithic industries of the Jurreru Valley, India, before and after the Toba eruption. Quaternary International 258: 165–179.

Clarkson, C., Jacobs, Z., Marwick, B., Fullagar, R., Wallis, L., Smith, M., McNeil, J., Cox, D., Arnold, L.J., Hua, Q., Huntley, J., Brand, H.E.A., Manne, T., Fairbairn, A., Shulmeister, J., Lyle, L., Salinas, M., Page, M., Connell, K., Park, G., Norman, K., Murphy, T. and Pardoe, C. 2017. Human occupation of northern Australia by 65,000 years ago. Nature 547: 306–310.

Clarkson, C., Harris, C., Li, B., Neudorf, C.M., Roberts, R.G., Lane, C., Norman, K., Pal, J., Jones, S., Shipton, C.J., Gupta, M.C., Mishra, D.P., Dubey, A.K., Boivin, N. and Petraglia, M.D. 2020. Human occupation of northern India spans the Toba super-eruption ~ 74,000 years ago. Nat. Commun. 11: 961. https://doi.org/10.1038/s41467-020-14668-4.

Cronin, T.M. 2010. Paleoclimates: Understanding Climate Change Past and Present, Columbia University Press, ISBN 0-231-14494-6, ISBN 978-0-231-14494-0, google books.

Dansgaard, W., Johnsen, S.J., Clausen, H.B., Dahl-Jensen, D., Gundestrup, N.S., Hammer, C.U., Hvidberg, C.S., Steffensen, J.P., Sveinbjörnsdottir, A.E., Jouzel, J. and Bond, G. 1993. Evidence for general instability of past climate from a 250-kyr ice-core record Nature 364: 218–220.

Deraniyagala, S.U. 1992. The prehistory of Sri Lanka: An ecological perspective (Memoir). Dept. of Archaeological Survey, Govt. of Sri Lanka.

Dhir, R.P., Kar, A., Wadhawan, S.K., Rajaguru, S.N., Misra, V.N. and Singhvi, A.K. 1992. Thar Desert in Rajasthan – Land, Man and Environment. Geological Society of India, Bangalore. 191p.

Dhir, R.P., Singhvi, A.K., Andrews, J.E., Kar, A., Sareen, B.K., Tandon, S.K., Kailath, A. and Thomas, J.V. 2010. Multiple episodes of aggradation and calcrete formation in Late Quaternary aeolian sands, Central Thar Desert, Rajasthan, India. J. Asian Earth Sci. 37: 10–16. doi:10.1016/j.jseaes.

Dhir, R.P. and Singhvi, A.K. 2012. The Thar Desert and its antiquity. Current Science 102(7): 1001–1008.

Dhir, R.P., Josi, D.C. and Kathju, S. 2018. Thar Desert in Retropspect and Prospect. Scientific Publishers, Jodhpur. 405p.

Dubey, J., Ghosh, R., Agrawal, S., Quamar, M.F., Morthekai, P., Sharma, R.K., Sharma, A., Pandey, P., Srivastava, V. and Ali, S.N. 2018. Characteristics of modern biotic data and their relationship to vegetation of the Alpine zone of Chopta valley, North Sikkim, India: Implications for palaeovegetation reconstruction. The Holocene 28: 363–376.

Fouache Eric, Pavlopoulos, K. and Stiros Stathis. 2010. Landscape evolution and geoarchaeology. Quaternary International 216: 1–2.

Fleitmann, D., Burns, S.J., Pekala, M., Mangini, A., Al-Subbary, A., Al-Aowah, M., Kramers, I. and Matter. A. 2011. Holocene and Pleistocene pluvial periods in Yemen, southern Arabia. Quaternary Science Reviews 30: 783–787.

Gaillard, C., Raju, D.R., Misra, V.N. and Rajaguru, S.N. 1983. Acheulian occupation at Singi Talav in the Thar desert: A preliminary report on 1982 excavation. Man and Environment 7: 141–152.

Gaillard, C. 1993. Contribution a`la connaissance du Pale´olithique infe´rieur-moyenen Inde. The`se de doctorat d'E´tat, Universite´de Provence, p. 653 (unpublished thesis).

Hansen, J., Sato, M., Hearty, P., Ruedy, R., Kelley, M., Masson-Delmotte, V., Russell, G., Tselioudis, G., Cao, J., Rignot, E., Velicogna, I., Kandiano, E., von Schuckmann, K., Kharecha, P., Legrande, A.N., Bauer, M. and Lo, K.-W. 2015. Ice melt, sea level rise and superstorms: Evidence from paleoclimate data, climate modeling, and modern observations that 2°C global warming is highly dangerous. Atmospheric Chemistry and Physics Discussions 15(14): 20059–20179. doi:10.5194/acpd-15-20059-2015.

Hearty, P.J. and Tormey, B.R. 2017. Sea-level change and superstorms; geologic evidence from the last interglacial (MIS 5e) in the Bahamas and Bermuda offers ominous prospects for a warming.

Haslam, M., Clarkson, C., Petraglia, M.D., Korisettar, R., Jones, S., Shipton, C., Ditchfield, P. and Ambrose, S.H. 2010.The 74 ka Toba super-eruption and southern Indian hominins: Archaeology, lithic technology and environments at Jwalapuram Locality 3. Journal of Archaeological Science 37(12): 3370–3384. DOI: 10.1016/j.jas.2010.07.034.

Heinrich, H. 1988. Origin and consequences of cyclic ice rafting in the Northeast Atlantic Ocean during the past 130, 000 years. Quaternary Res. 29: 143–152.

Higham, T. et al. 2014. The timing and spatiotemporal patterning of Neanderthal disappearance. Nature 512: 306–309.

Holt, B.G., Lessard, J.-P., Borregaard, M.K., Fritz, S. a, Araújo, M.B., Dimitrov, D., Fabre, P.-H., Graham, C.H., Graves, G.R., Jønsson, K. a, Nogués-Bravo, D., Wang, Z., Whittaker, R.J., Fjeldså, J. and Rahbek, C. 2013. An update of Wallace's zoogeographic regions of the world. Science 339: 74–8. https://doi.org/10.1126/science.1228282.

Hu, Y., Ruan, Q., Lui, J. and Marwick, B. 2020. Luminescence chronology and lithic technology of Tianhuadong Cave, an early Upper Pleistocene Paleolithic site in southwest China. Quaternary Research 94: 121–136. DOI: https://doi.org/10.1017/qua.2019.67.

Hu, Z.B., Pan, B.T., Guo, L., Vandenberghe, J., Liu, X., Wang, J., Fan, Y., Mao, J., Gao, H. and Hu, X.F. 2016. Rapid fluvial incision and headward erosion by the Yellow River along the Jinshaan gorge during the past 1.2 Ma as a result of tectonic extension. Quaternary Science Reviews 133: 1–14.

Jain, M., Tandon, S., Singhvi, A., Mishra, S. and Bhatt, S. 2005. Quaternary alluvial stratigraphical development in a desert setting: a case study from the Luni River basin, Thar Desert of western India. Fluv. Sedimentol. VII: 349–371.

Jones, S.C. and Pal, J.N. 2009. The Palaeolithic of the Middle Son valley, north-central India: Changes in hominin lithic technology and behaviour during the Upper Pleistocene. Journal of Anthropological Archaeology 28(3): 323–341. DOI: 10.1016/j.jaa.2009.05.003.

Juyal, N., Raj, R., Maurya, D.M., Chamyal, L.S. and Singhvi, A.K. 2000. Chronology of late Pleistocene environmental changes in the lower Mahi basin, western India. Journal of Quaternary Science 15: 501–508.

Juyal, N., Chamyal, L.S., Bhandari, S., Bhushan, R. and Singhvi, A.K. 2006. Continental record of the southwest monsoon during the last 130 ka: evidence from the southern margin of the Thar Desert, India. Quat. Sci. Rev. 25: 2632–2650.

Kar, A. 2020. Quaternary Aeolian Landscape Development in Thar Desert and Its Drivers. Proc Indian Natn Sci Acad 86 No. 1 March 2020 pp. 405–417. DOI: 10.16943/ptinsa/2020/49778.

Kar, A., Ranhotra, P.S., Bhattacharyya, A. and Sekar, B. 2002. Vegetation vis-à-vis climate and glacial fluctuations of the Gangotri Glacier since the last 2000 years. Current Science 82(3): 347–351.

Kumar, S. 2005. Fauna of Sambhar Lake. Wetland Ecosystem Series. Edited and Published by the Director, Zool. Surv. India, Kolkata. 6: 1–200.

Liu, T. 1985. Loess and the Environment; China Ocean Press: Beijing, China, pp. 60–65.

Lu, Y., Wang, X. and Wintle, A. 2007. A new OSL chronology for dust accumulation in the last 130,000 yr for the Chinese Loess Plateau. Quaternary Research 67: 152–160.

Misra, T.N. 1997. Balathal: A Chalcolithic settlement in Mewar, Rajasthan, India: Results of first three seasons' excavations. South Asian Studics 13: 251–73.
Misra, V.N. 1987. Middle Pleistocene adaptations in India. pp. 99–119. *In*: The Pleistocene Old World: Regional Perspectives (ed.) Olga Soffer (New York: Plenum Press).
Misra, V.N. 2001. Archaeological Evidence of Early Modern Human Occupation in South Asia Firenze University Press.
Misra, V.N., Rajaguru, S.N. and Raghavan, H. 1988/1989. Late Middle Pleistocene environment and Acheulian culture around Didwana, Rajasthan. Proceedings of the Indian National Science Academy 54A 3: 428–435.
Misra, V.N., Rajaguru, S.N., Raju, D.R., Raghavan, H. and Gaillard, C. 1982. Acheullan occupation and evolving landscape around Didwana in the Thar Desert, India. Man and Environment VI: 72–86.
Misra, V.N. 1995. Geoarchaeology of the Thar desert, northwest India. Mem. Soc. India 210–230.
Nagar, Y.C., Sastry, M.D., Bhushan, B., Kumar, A., Mishra, K.P., Shastri, A., Wadhawan, S.K. and Singhvi, A.K. 2010. Chronometry and formation pathways of gypsum using Electron Spin Resonance and Fourier Transform Infrared Spectroscopy. Quaternary Geochronology 5(6): 691–704. DOI: 10.1016/j.quageo.2010.05.001.
Negi, S.S. 1996. Biosphere Reserves in India: Landuse, Biodiversity and Conservation. Indus Publishing Company, Delhi.
Petraglia, M.D. 2010. The Early Paleolithic of the Indian subcontinent: Hominin colonization, dispersals and occupational history. pp. 165–179. *In*: Fleagle, J.G., Shea, J., Grine, E.E. and Baden, A.L. (eds.). Out of Africa1: The first Hominin Colonization of Eurasia. Springer, Dordrecht.
Petraglia, M.D., Korisettar, R., Boivin, N., Clarkson, C., Ditchfield, P., Jones, S., Koshy, J., Lahr, M. M., Oppenheimer, C., Pyle, D., Roberts, R., Schwenninger, J.L., Arnold, L. and White, K. 2007. Middle Paleolithic assemblages from the Indian subcontinent before and after the Toba super-eruption. Science 317(5834): 114–116. doi: 10.1126/science.1141564.
Petraglia, M.D., Alsharekh, A.M., Crassard, R., Drake, N.A., Groucutt, H.S., Parker, A.G. and Roberts, R.G. 2011. Middle Palaeolithic occupation on Marine Isotopic Stage 5 lakeshore in Nafud Desert, Sauudi Atrabia. Qauetrnary Science Reviews 30: 1555–1559.
Petraglia, M.D., Alsharekh, A., Breeze, P., Clarkson, C., Crassard, R., Drake, N.A., Groucutt, H.S., Jennings, R., Parker, A.D., Parton, A., Roberts, R.G., Shipton, C., Matheson, C. Abdulaziz al-Omari and Veall, M.A. 2012a. Hominin Dispersal into the Nefud Desert and Middle Palaeolithic Settlement along the Jubbah Palaeolake, Northern Arabia. PLoS ONE 7(11): e49840. https://doi.org/10.1371/journal.pone.0049840.
Petraglia, M.D., Ditchfield, P., Jones, S., Korisettar, R. and Pal, J.N. 2012b. The Toba volcanic super-eruption, environmental change, and hominin occupation history in India over the last 140,000 years. Quaternary International 258: 119–134.
Porter, S. and An, Z. 1996.Correlation between climate events in the North Atlantic and China during the last glaciation. Nature 375: 305–308.
Raghav, K.S., Wadhawan, S.K. and Sareen, B. 1995. Quaternary stratigreaphy of Jaipur upland along eartern margin of Rajasthan Desert and its palaeoenvironmental implications—an overview. pp. 1–10. *In*: First Indian Workshop on Desert Margins and Palaeomonssoon. Geological Survey of India. Jaipir.
Raghav, K.S. 1999. Evolution of Drainage Basins in parts of Northern and Western Rajasthan, Thar Desert, India. Memoir Geol. Soc. India No. 42: 175–185.
Raghavan, H., Rajaguru, S.N. and Misra, V.N. 1989. Radiometric Dating of a Quaternary Dune Section, Didwana, Rajasthan. Man and Environment 13: 19–22.
Raghavan, H. 1987. Quaternary Geology of Nagaur District, Rajasthan. Thesis submitted to Poona University, Pune, (unpublished thesis).
Raghavan, H., Gaillard, C. and Rajaguru, S.N. 1993. Genesis of calcretes from the calc-pan site of singi talav near didwana, Rajasthan, India—a micromorphological approach. Geoarchaeology 6(2): 151–168. Https://Doi.Org/10.1002/Gea.3340060203.
Ramasamy, S.M. 1999. Neotectonic controls on the migration of Sarasvati. River of the Great Indian Desert. Memoir Geol. Soc. India No. 42: 155–162.
Rapp, G. and Hill, C. 1998. Geoarchaeology: The Earth Science Approach to Archaeological interpretation. Yale University Press, London, 274pp.
Rohling, E.J., Grant, K., Hemleben, C., Siddall, M., Hoogakker, B., Bolshaw, M. and Kucera, M. 2007. High Rates of Sea-Level Rise During the Last Interglacial Period. Nature Geoscience 1(1): 38–42. Doi: 10.1038/Ngeo.2007.28.
Roy, A.B. 1999. Evolution of saline lakes of Rajasthan. Current Science 76(3): 290–295.
Roy, P.D. and Smykatz-Klossa, W. 2007. REE geochemistry of the recent playa sediments from the Thar Desert, India: An implication to playa sediment provenance. Chemie der Erde 67: 55–68.
Saini, H.S., Tandon, S.K., Mujtaba, S.A.I. and Pant, N.C. 2005. Lake deposits of the northern margin of Thar Desert: Holocene paleoclimatic implications. Current Science 88(12): 1994–2000.
Saini, H.S., Tandon, S.K., Mujtaba, S.A.I., Pant, N.C. and Khurana, R.K. 2009. Reconstruction of buried channel-floodplain systems of north-western Haryana plains and their relation to the 'Vedic' Saraswati. Current Science 97(11): 1634–1643.

Saini, H.S. and Mujtaba, S.A.I. 2012. Depositional history and paleoclimatic variations at the north-eastern fringe of Thar Desert, Haryana plains, India. Quaternary International 250: 37–48. Doi: 10.1016/j.quaint.2011.06.002.

Scerri, E.M.L., Shipton, C., Clark-Balzan, L., Frouin, M., Schwenninger, J.L., Groucutt, H.S., Breeze, P.S., Parton, A., Blinkhorn J., Drake, N.A., Jennings, R., Cuthbertson, P., Omari A A., Alsharekh, A.M. and Petraglia, M.D. 2018. The expansion of later Acheulean hominins into the Arabian Peninsula. Scientific Reports, vol. 8, pp. 1-9. DOI:10.1038/s41598-018-35242-5.

Shackleton Nicholas J., Maria Fernanda Sa´nchez-Gon, Delphine Pailler and Yves Lancelot. 2003. Marine Isotope Substage 5e and the Eemian Interglacial. Global and Planetary Change 36: 151–155.

Sharma, K.K. and Mehra, S.P. 2009. The Thar of Rajasthan (India): Ecology and conservation of a desert ecosystem. pp. 1–12. *In*: Sivaperuman, C., Baqri, Q.H., Ramaswamy, G. and Naseema, M. (eds.). Faunal Ecology and Conservation of the Great Indian Desert. Springer-Verlag Berlin Heidelberg. DOI: 10.1007/978-3-540-87409-6_1.

Singhvi, A.K., Banerjee, D., Rajaguru, S.N. and Kishan Kumar, V.S. 1994. Luminescence chronology of a fossil dune at Budha Pushkar, Thar Desert : Paleoenvironmental and archaeological implications. Curr. Sci. 66: 770–773.

Singhvi, A.K., Williams, M., Rajaguru, S.N., Misra, V.N., Chawla, S., Stokes, S., Chauhan, N., Francis, T., Ganjoo, R.K. and Humphreys. 2010. A ~ 200 ka record of climatic change and dune activity in the Thar Desert, India. Quaternary Science Reviews 29(23): 3095–3105. DOI: 10.1016/j.quascirev.2010.08.003.

Sinha, R. and Sakar, S. 2009 Climate-induced variability in the Late Pleistocene–Holocene fluvial and fluvio-deltaic successions in the Ganga plains, India: A synthesis. Geomorphology 113(3-4): 173–188.

Sinha, R., Smykatz-Kloss, W., Stuben, D., Harrison, S.P., Berner, Z. and Kramar, U. 2006. Palaeogeography Palaeoclimatology Palaeoecology 203: 252–270.

Sinha-Roy, S. 1986. Himalayan collision and indentation of the Aravalli orogen by Bundelkhand wedge: implications for neotectonics in Rajasthan. pp. 18–21. *In*: Proceedings of International Symposium on Neotectonics in South Asia. Survey of India, Dehra Dun.

Stothers, R. 1984. The great Tambora eruption in 1815 and its aftermath. Science 224: 1191–1198.

Tada, R., Zheng, H. and Clift, P.D. 2016. Evolution and variability of the Asian monsoon and its potential linkage with uplift of the Himalaya and Tibetan Plateau. Prog. Earth Planet. Sci. 3: 4, 10.1186/s40645-016-0080-y.

Torre, I., Mora, R., Arroyo, A. and Benito-Calvo, A. 2014. Acheulean technological behaviour in the Middle Pleistocene landscape of Mieso (East-Central Ethiopia). Journal of Human Evolution 76: 1–25.

Vriend, M., Prins, M.A., Buylaert, J.-P., Vandenberghe, J. and Lu, H. 2011. Contrasting dust supply patterns across the north-western Chinese Loess Plateau during the last glacial-interglacial cycle. Quaternary International 240(1): 167–180. DOI: 10.1016/j.quaint.2010.11.009.

Wadhawan, S. 2022. Geospatial Diversity, Geologic Significance and Evolutionary Trends of Parabolic Dunes in Thar Desert of the Indian Subcontinent (this volume).

Wadhawan, S.K. and Kumar, V. 1996. Subsurface Quaternary aeolian stratigraphy in the Ghaggar basin of Thar Desert, India. Journal of Arid Environments 32(1): 37–51.

Wadhawan, S.K. 1996. Textural attributes of recent aeolian deposits in different sub-basins of the Thar Desert, India. Journal of Arid Environments 32(1): 59–74.

Wadhawan, S.K. and Sharma, H.S. 1997. Quaternary stratigraphy and morphology of Desert Ranns and evaporite pans in central Rajasthan, India. Man and Environment XXII(2): 1–10.

Wadhawan, S.K. 1999. Quaternary Geology and Stratigraphy of the Shergarh – Dechu sub-basin along the eastern margin of the Thar Desert, Rajasthan, India. 1999. Man and Environment XXIV(1): 63–75.

Wadhawan, S.K., Nagar, Y.C., Juyal, N. and Singhvi, A.K. 2009. Chronology of past desertification events inferred from continental deposits along eastern margins of Thar desert–implication for urban development in Jaipur drylands, India. International Asia-Pacific Conference on Luminescence and Spin Resonance PRL Ahmedabad, India

Wadhawan, S.K. 2010. Quaternary Continental Sedimentation Basins and Dune Fields in Thar Desert of NW India. Gondwana Geological Magazine 12: 265–270.

Wadhawan, S.K. 2018. Palaeogene–Neogene tectonics and continental aggradational basins in north-western India—implications for geological evolution of Thar Desert. pp. 151–166. *In*: Bajpai, S., Tripathi, S.C. and Prasad, V. (eds.). The Indian Paleogene. Springer Nature, Dordrecht.

Wang, X., Cui, L., Yanng, S., Zhai, J. and Ding, Z. 2018. Stable carbon isotope records of black carbon on Chinese Loess Plateau since last glacial maximum: An evaluation on their usefulness for paleorainfall and paleovegetation reconstruction. Palaeogeogr. Palaeoclimatol. Palaeoecol. 509: 98–104.

Wunsch, C., Schmitt, R.W. and Baker, D. 2013. Climate change as an intergenerational problem. Proc. Natl. Acad. Sci. 110: 4435–4436.

Zech, R., Zech, M., Marković, S., Hambach, U. and Huang, Y. 2013. Humid glacials, arid interglacials? Critical thoughts on pedogenesis and paleoclimate based on multi-proxy analyses of the loess–paleosol sequence Crvenka, Northern Serbia. Palaeogeography, Palaeoclimatology, Palaeoecology 387: 165–175.

Zhang, J., Ru, W. and Li, B. 2006. Relationships between vegetation and climate on the Loess Plateau in China. Folia Geobotanica 41: 151–163.

Chapter 18

Geospatial Diversity, Geologic Significance and Evolutionary Trends of Parabolic Dunes in Thar Desert of the Indian Subcontinent

Sudesh K. Wadhawan

1. Introduction

The Thar Desert, the world's seventh largest desert in the subtropics, is characterized by arid-semiarid ecologically fragile environment and occupies a unique tectonic-sedimentary domain in the north-western parts of the Indian subcontinent. The Thar Desert lies at the eastern most extensive part of the vast Sahara-Arabian Desert on a global aridity and deserts distribution pattern in the Northern Hemisphere. It is dominated by aeolian bedforms: sand dunes of different types and dimensions. The Thar Desert in India is largely located in Rajasthan, state of India, especially between the foothills of the Aravalli Hill ranges in the east and the international borders with Pakistan in the west. Further west within Pakistan, the desert extends up to the fertile alluvial plains of the Indus River (Fig. 18.1a and b). The Indian side of the hot and arid Thar Desert is spread over 19.6 Mha in Rajasthan; 6.6 Mha in Gujarat; 1.5 Mha in Punjab and 1.3 Mha in Haryana (Dhir et al., 1992). Geologic disposition of dryland spread and landforms of the Thar Desert are inferred to be confined essentially to the West Rajasthan Shelf (Sinha Roy et al., 1998; Wadhawan, 2010, 2018). Geological evolution of the Thar is closely linked to the litho-tectonic architecture of the continental sedimentation basins that were created during the Paleogene-Neogene continental drift and docking of the Indian plate with that of Eurasia (Valdiya, 2002, 2013; Singhvi et al., 2012). Paleogene-Neogene tectonic domain in the Thar has been subject to extensional stress fields manifested in a series of normal faults and grabens/succession of half graben structures and step faults that influenced architectural evolution of continental sedimentation basins. Therefore, both tectonic and climatic causes are considered responsible for the aggradations of continental deposits in the Thar Desert (Wadhawan and Sural, 1992; Wadhawan, 2005, 2018). The Indus River flowing along the southwestern margin of the Great Indian Sand Desert – the Thar, appears to be deeply entrenched like the River Ganges in the Himalayan fore-deep region in the northern parts of India. The geological evolution of this Himalayan perennial river and its tributaries have effectively controlled and contributed to the

Adjunct Professor, Amrita Vishwa Vidyapeetham University, Kerala-690525; Former Director General, Geological Survey of India.
Email: wadhawansk.leo@gmail.com

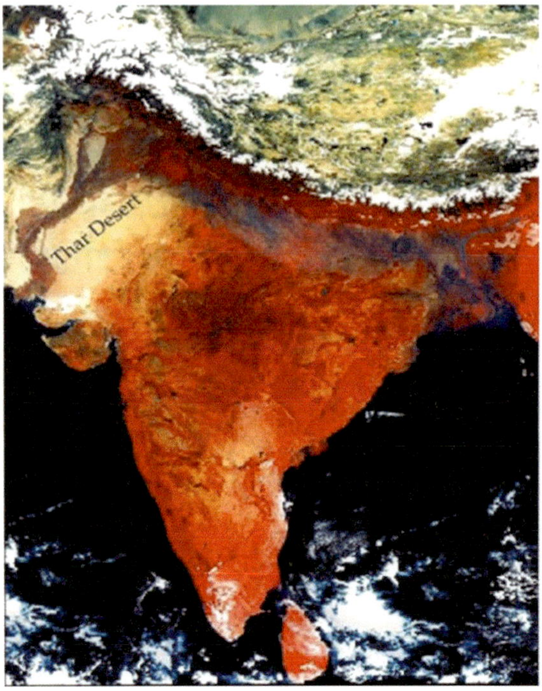

(a)

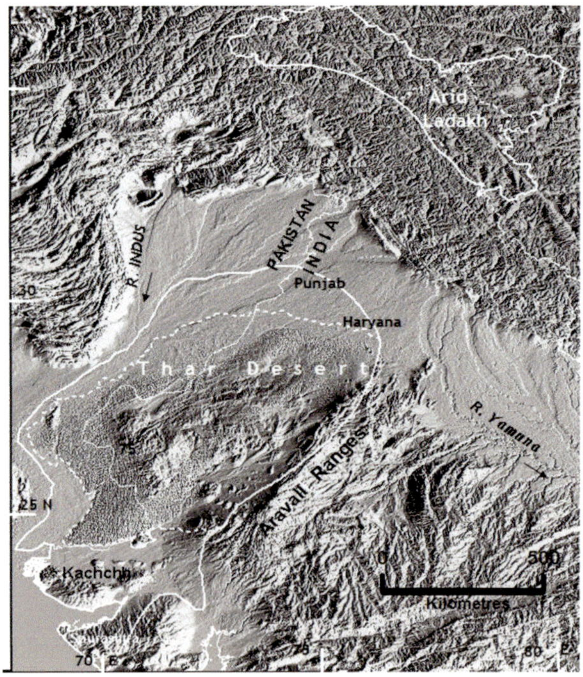

(b)

Figure 18.1. (a) IRS satellite image of Thar Desert in Indian subcontinent. (b) DEM generated Image [on the right] is based on SRTM 30-GTOPO data at ~ 1 km resolution, with annotated boundaries [Image curtsey: Dr. Amal Kar, personal communication].

major source of sediments for southwesterly summer monsoonal winds to work on the exposed lower reaches of the fluvial-deltaic and coastal confluence region of the mighty Indus River in the Arabian Sea.

Observation of dune alignment using remotely sensed imagery has shown that regionally, these dunes are closely aligned with the prevailing wind direction of the southwest monsoon system. Therefore, the Thar dune systems potentially provide a rich archive of past climatic and geomorphological change. Quaternary deposits and landform assemblages that provide a spectacular landscape include a diverse variety of dune forms and playa – saline lake deposits in the hot Thar. Studies on saline lakes have provided a high-resolution paleoclimate record, but only up to the past ~ 15 ka or so. However, strong aeolian activity in the Thar Desert has been a characteristic of the latest 250–300 ka period (Singhvi and Kar, 2004; Dhir et al., 2010). It is marked by several episodes of greater aridity, stronger wind regime and sand dynamism followed by periods of stability implying climatic amelioration resulting in consolidation and formation of soils. Wetter episodes accumulated fluvial deposits and coastal sediments that acted as provenance or the source for material for winds to work on during prolonged dry spells and generate a variety of aeolian bed-forms and dunes in different sedimentation sub-basins with distinctive substrata in the Thar (Dhir et al., 2018; Singhvi et al., 2012; Wadhawan, 2010, 2018).

Available field evidence in arid areas of the Thar Desert confirmed the existence of four major Late Paleogene-Neogene sediments depocenters or continental basins that were filled with different episodic deposits of fluvial and aeolian origin (Fig. 18.2; Wadhawan, 2018). Although the Quaternary sedimentation history in each of these continental basins is unique,it is broadly comparable and multi-cyclic varying in degree or intensity of development: starting generally with coarse colluvium/ill-sorted polymictic fluvial clastic deposits, through silt-clays of playa/lacustrine deposits and culminating with sandy aeolian deposits. Large scale drainage disruption/segmentation in the Thar Desert is inferred either due to neotectonics or obstruction of rivers channels by extending dune fields driven by the southwest winds, for instance the present day desiccated Ghaggar River along the northern periphery of the Thar Desert. It was therefore, established that the fluvial and fluvio-lacustrine deposits formed the base for preferred aeolian deposition and constrained the overall configuration and geospatial disposition of older aeolian deposits (dune fields) in the Thar Desert (Wadhawan, 2010; Singhvi et al., 2012).

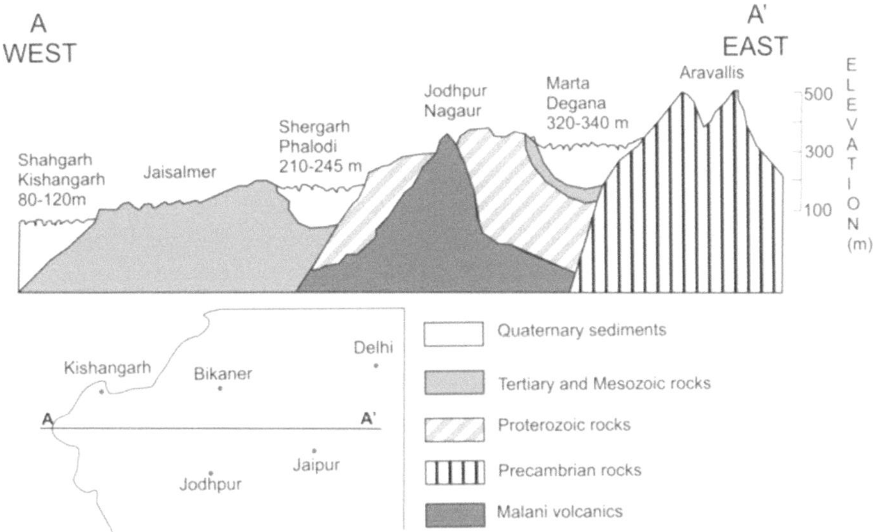

Figure 18.2. Schematic profile along 27° North showing Neogene/Quaternary Sub-basins in the Thar Desert [After Wadhawan, 1996, 2010].

It is notable that the Shergarh-Dechu – Phalodi Basin in central Thar Desert is dominated by the occurrence of characteristic stabilized parabolic dunes of clustered or rake-like disposition and derivative linear dunes comprising fine well-sorted sand and consolidated with pedocalcification. Several segments in the dune field show superimposed multi-tiered arrangement of complex dunes with higher relief thereby representing different episodes of aeolian dynamism and dune morphologic evolution. As the parabolic dunes characteristically occur as dominant aeolian landforms in the central sector of the Thar Desert, the present contribution attempts to characterize and elucidate various facets of parabolic dune diversity, its distribution patterns and morphometry, their inter-relationships, genetic significance and geological evolution.

1.1 Methodology

Preparatory works for this research included procurement, followed by study of relevant maps, satellite images, geocoded sheets and aerial photo coverage on different scales. Satellite images were sourced from ISRO and other agencies, LANDSAT TM FCC and IRS B&W for different resolutions. Geoscientific literature survey and rigorous study and interpretation on lighted ground-glass top tracing tables. These interpretations were geo-referenced and corroborated with detailed study and analyses from Survey of India modern topographical sheets on 1:50,000 scale. Aerial photo-coverage for the selected areas of interest was studied on 1:25,000 scale to interpret greater details on the dune patterns and dune morphometry. Field surveys and ground truth validations were carried out along with detailed strata-logging of exposed/excavated sections and sampling for sedimentological studies and OSL dating at laboratories. Various dune patterns were inferred and delineated/plotted manually for a clearer understanding of the sequence of events and morphologic impacts of episodic aeolian dynamics. Areas of interest were selected and listed for further detailed field surveys and photography. The collected field and laboratory data sets were synthesized for integrated outcomes and improved understanding of the genesis and evolutionary history of the parabolic dunes in the Thar Desert.

2. Geological and tectonic setup

The tectonic architecture and basement configuration of intra-cratonic basins and sedimentary formations in NW India range from the highly deformed Precambrian Aravalli (3.5–2.5 Ga) and Delhi Supergroups (~ 1 Ga), as part of the Great Indian Proterozoic Fold Belts along the western edge of the ancient Rodinia Supercontinent (Ashwal et al., 2013). A series of igneous events between 1700 Ma and 750 Ma along the south-western margin of the Aravalli Hills produced Sendra Granite, Erinpura Granite, Malani Igneous Suite, Mt. Abu Granitoids, etc., in the eastern parts and up to the Jaipur-Ajmer uplands and along the eastern margins of the Thar Desert (Meert et al., 2010). These basement rocks are covered unconformably by undeformed sedimentary rocks of Late Proterozoic to Early Paleozoic - Marwar Supergroup of sedimentary rocks in the middle, to Mesozoic and Cainozoic cover sedimentary formations on the western and north-western fringe areas in the western parts of Rajasthan and Haryana in NW India. A schematic latitudinal profile, along 27° N, also depicts a general fall in average elevation from 500 m high accordant summits of the Aravalli hill ranges on the east to 80–120 m amsl on the western part of the Thar that forms the prevailing core of the active sands covered desert (Fig. 18.2). This appears to be in direct correspondence to the prevailing climatic gradient - the conditions of extreme aridity on the west that grades gradually and becomes semi-arid with increasing rainfall towards the east.

Geoscientific research has revealed that geological evolution of the Thar Desert is intimately linked to and influenced by the plate tectonic movements - the northward drifting Indian Plate (Valdiya, 2013; Sinha-Roy, 1986; Bajpai et al., 2018); creation of continental pull-apart basins, succession of stepped graben and development of Paleogene-Neogene/Quaternary valley fills (Wadhawan, 2010, 2018). Quaternary sedimentation started in northern and western parts of the Indian subcontinent after the closure of Tethys Sea and partial uplift of the Siwalik foreland in front

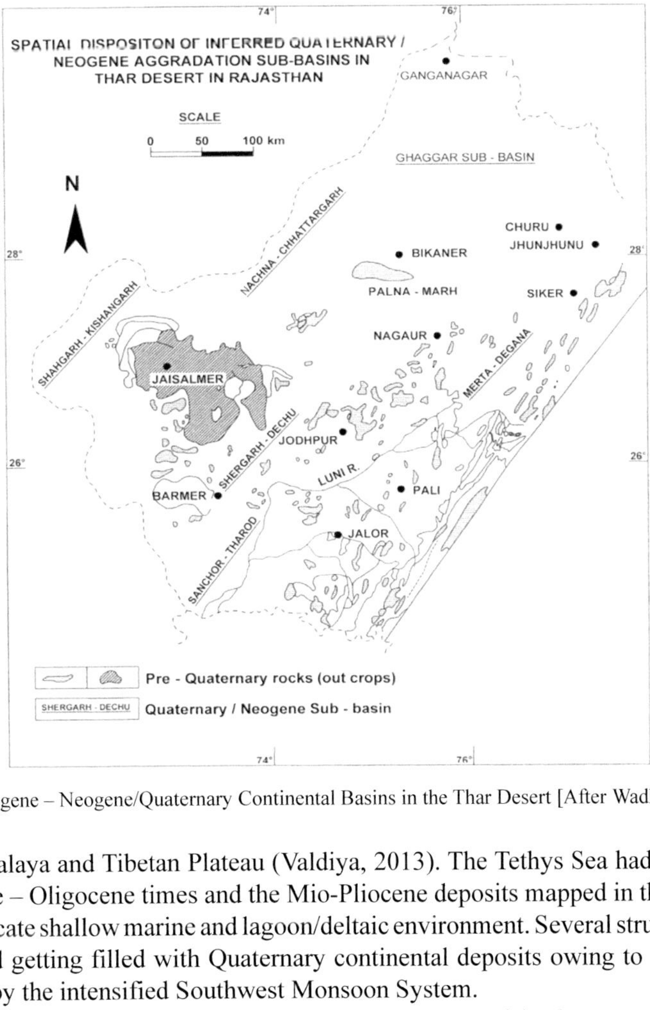

Figure 18.3. Paleogene – Neogene/Quaternary Continental Basins in the Thar Desert [After Wadhawan, 1996, 2018].

of the rising Himalaya and Tibetan Plateau (Valdiya, 2013). The Tethys Sea had started regressing since Mid-Eocene – Oligocene times and the Mio-Pliocene deposits mapped in these north-western parts of India indicate shallow marine and lagoon/deltaic environment. Several structurally controlled sub-basins started getting filled with Quaternary continental deposits owing to accelerated fluvial activity induced by the intensified Southwest Monsoon System.

Further, based on detailed field surveys, geological mapping of the Quaternary formations, their nature of surface disposition and inferred sub-surface configuration in arid-semiarid western parts of India, it is feasible to delineate a few large and broadly linear Neogene sub-basins or tectonic depressions filled with aggradational continental Quaternary sediments and separated, respectively by rocky relief features of denudation. A series of north-easterly trending linear stepped graben structures have been deciphered showing successive westerly downthrown blocks respectively. (Figs. 18.2 & 18.3; Dhir et al., 1992; Wadhawan, 1996, 2005, 2010).

2.1 Geospatial dune diversity

Analyses of such geospatial distribution of dune fields along a profile across the middle of the Thar Desert reveal the presence of three delineated Quaternary continental sedimentation basins (Fig. 18.2) and the existence of the fourth one—a major dune field occupying vast areas in the north, around Bikaner – Churu - Ganganagar, in the northern parts of the Rajasthan (Figs. 18.3 & 18.4). These continental basins or the discrete dune fields in the Thar are briefly enumerated here:

1) **Merta - Degana – Sambhar Salt Lake Basin**, west of Jaipur uplands (Aravalli hill ranges) occupying tracts along the eastern margin of the Thar Desert; it is dominated by (i) thick consolidated sand sheets preserved with multiple aeolian deposition episodes characterized and impregnated by the formation of different types of calcretes, (ii) obstacle dunes abutting against

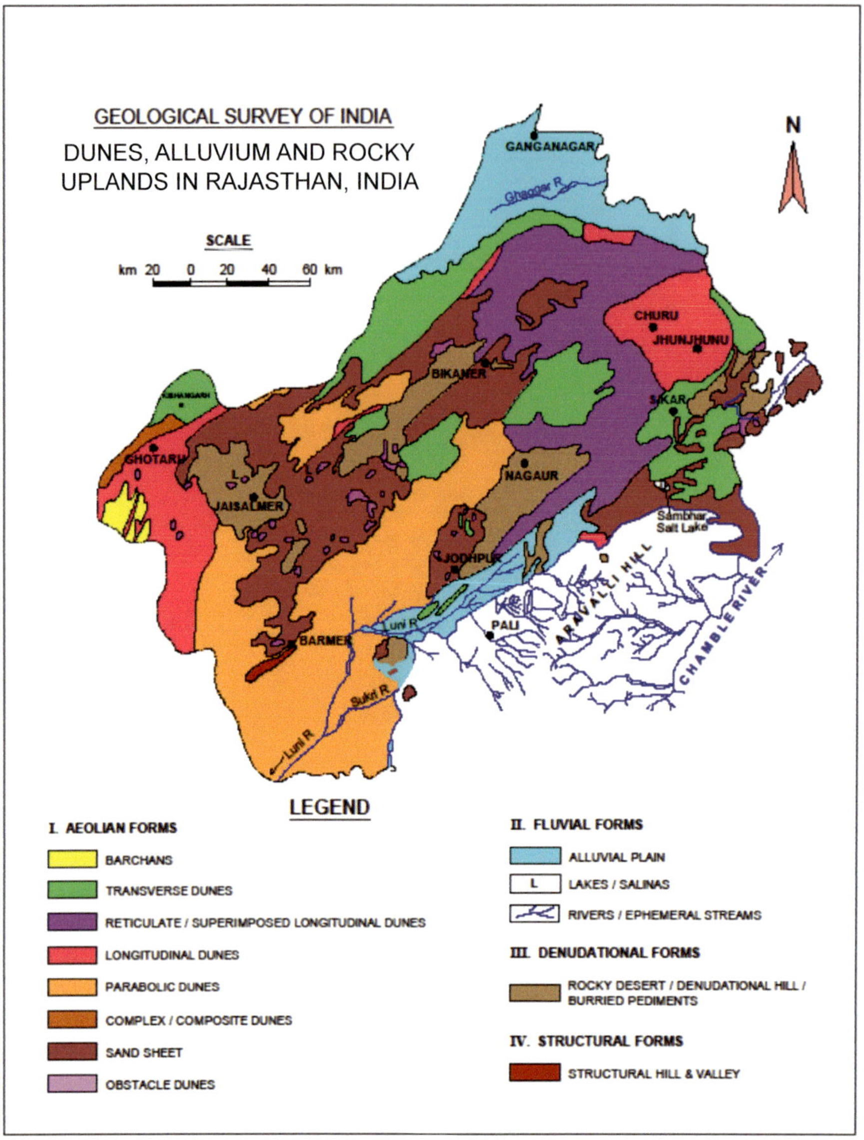

Figure 18.4. Distribution and morphology of the aeolian sand deposits/Dunes in the Thar Desert, Indian subcontinent [After GSI, 2001].

the Aravalli hill ranges. These ramp-like stabilized dunes are at present intensely dissected into a network of deep gullies. (iii) semi-stable and seasonally active network of low reticulate and linear dunes also abound in undulating plains covered with aeolian or fluvio-aeolian sand sheets. (iv) inland draining, dry ephemeral channels and 'river bank' low crescentic dunes are very common in these areas and (v) several saline lakes are also present, such as Sambhar, Kuchaman, Didwana and Tal Chhappar, etc., along with associated geo-archaeological evidences of Paleolithic human occupations (Blinkhorn, 2013; Singhvi et al., 2010).

2) **Shergarh - Dechu - Phalodi Basin** and its south-western extensions along Barmer – Sanchor -Tharad and further south-westwards up to the Rann of Kutch sub-basins occupying the central parts of the Thar Desert. This basin is dominated by characteristic stabilized parabolic dunes of coalescing/clustered or rake-like disposition and few obstacle or anchored dunes on the lee and stoss of residual hills. The dune ridges show varying relief ranging from 20 to 50 m. Several

segments in the dune field show superimposed multi-tiered arrangement of complex dunes representing different episodes of aeolian dynamism and dune morphologic evolution, and (ii) a subordinate number of linear dunes derived from deflation of parabolic dunes comprising well-sorted fine sand and consolidated with pedocalcification; (iii) vast areas covered with interdune sand sheets impregnated with calcrete formation. (iv) older lacustrine and playa deposits rich in gypsum deposits such as gypsite and calcareous gypcrete. Major playas located in the basin include Phalsund, Kwas, Thob, Bap, etc.

3) **Shahgarh - Kishangarh Basin** on the west and northwest of Jaisalmer dissected rocky uplands and Hamada/desert pavements. This tract is characterized by (i) clusters of parabolic dunes on the east that grade laterally as well as down-wind to well-defined, sharp crested linear dunes. (ii) The high and large transverse dunes occur further down-wind tracts west of Tanot and Kishangarh areas. The dune relief is generally high, 40–70 m. (iii) Active barchans: as this basin falls under extreme aridity and strong windy conditions, there is spontaneous sand dune movement and frequent formation of low migratory and active barchans that occur in successive bands over-riding the older stabilized or the fossil dunes. Larger aeolian tracts of such coalescing active barchans form wide-spread megabarchanoids in the southwest upwind area in the basin (Kar, 1993). These clusters of active sand megabarchanoids are exclusively formed here and their persistent occurrences show that the supply of sandy sediments and their aeolian mobility has been inherently high. (iv) lacustrine and playa deposits occur on the peripheral parts mostly with rocky sub-strata or along the dried-up paleo-channels, e.g., Mohangarh gypsum deposits, Kanod Rann, Mitha and Khara Ranns, Pokrann and Lawan Rann, etc., that represent the climatic fluctuations during late Pleistocene and Holocene periods.

   Synthesis of dune trends in all these above sedimentation basins, particularly the longitudinal elements of the dunes suggest that the overall sand shifting winds remained moving from SW to NE implying that the causative monsoon atmospheric circulation patterns have remained persistent and fairly stable since at least the late Quaternary period (Singhvi et al., 2012). These Quaternary continental basins support distinctive aeolian landforms that have evolved as per the prevailing climate gradient of decreasing aridity from west to east and a corresponding increase in rainfall.

4) **Bikaner-Churu-Ganganagar Basin** in northern parts of the Thar Desert shows remarkably distinctive dune morphologies of a younger relative age. The area also supports over 300 m thick column of Quaternary deposits (Wadhawan and Virendra Kumar, 1996; Saini and Mujtaba, 2012), as has been confirmed by a sub-surface drilling program during potash mineral investigations in the area. The desiccated Ghaggar River basin here occupies a buried part of tectonically sagged or sinking fore-deep that abuts against the Himalayan Mountains further northwards. The dunes are low relief semi-consolidated transverse, network of linear dunes with transverse segments, barchans and star dunes along the northern margin of the Thar Desert with the alluvial plains in Punjab and Haryana. The area is dotted with shallow playa and gypsum deposits in the inland depressions surrounded by dunes or laid along the dried-up paleo-channels in the area, Pallu, Lunkaransar, Jamsar, etc.

3. Distribution of parabolic dunes

The Study of geospatial distribution patterns of different types of dunes in the Thar Desert have been mapped by various departments (Fig. 18.4, GSI, 2001). With the advent of satellite data and remote sensing techniques, it became of interest to study synoptic views and accurately map the surface geomorphic expressions and dune patterns in different dune fields of the world, including India (e.g., Mc Kee, 1979; Singh, 1977; Wadhawan et al., 1999; Kar, 2011, etc.).

The most striking feature of the Thar Desert and particularly the central parts of the Thar in Jodhpur district (Shergarh - Dechu – Phalodi basin of Quaternary continental sedimentation) is a wide-spread occurrence of stabilized clustered parabolic dunes of the past aeolian dynamism.

Two forms of compound parabolic dunes can be recognized from space imageries. These are V-shaped and complex U-shaped or rake-shaped with several coalesced secondary trailing arms. These unique en-echelon clustered parabolic dunes in the Thar Desert have been variously described as the 'compound parabolic', 'nested' or 'over-riding parabolic', 'multi-tiered clustered' aeolian landforms (Verstappen, 1970, 1983; Singh, 1977; Breed et al., 1979; Wasson et al., 1983; Wadhawan, 1988, 1994, 2010; Kar, 2011).

The study of space imageries (LANDSAT TM-FCC and IRS black and white and FCC) covering the western parts of India reveal that the stabilized parabolic dunes in the Thar Desert occur and spread from near the northern margin of the Great Rann of Kutch in the southwest to far-inland in broadly linear NE oriented tract covering over 500 km in length and over 150 km wide in the south, tapering to about 50 km in the northeast. It is remarkable that the parabolic dunes formed in down-wind areas close to the coastal sabkha tidal environment of the Rann of Kutch/Arabian Sea occur mostly in poorly organized clusters of single units lying in discernible rows astride the dominant south-westerly winds and separated by arrays of transverse sandy flats or inter-dune depressions (Fig. 18.5a & b).

Figure 18.5a shows the occurrence of parabolic dune fields in the downwind close proximity to the higher tidal flat of the Great Rann of Kutch. The north-easterly oriented, 2 to 4 km long arms of individual parabola are clearly demarcated. The deflationary tracts show a width ranging from 500 m to 1 km. The linear arms are also seen as the deflated segments or the remnants of parabolic dunes whose apical closures appear to have been blown-off by strong winds. Most of the parabolic dune closures here are well developed and the narrow to broad 'U' shaped closures occur towards the NE, i.e., the down-wind end of the aeolian dynamism indicating a complex mix of deflationary erosion and deposition of fine sediments. At places, the closures of one parabolic dune appear to have been advancing and merging with the next succeeding parabola. Some transverse interdune depressions are inundated with coastal water. The coastal tracts of these parabolic dunes are close to the international border, inhospitable and inaccessible and therefore, no impact of human interventions is noticed. It is noteworthy that the role of vegetation in carving of a parabolic dune is not distinct or totally absent.

However, a spectacular natural transformation in dune construction is observed in the satellite imagery scene covering parabolic dunes on further north-easterly tracts (Fig. 18.5b), just down-wind of areas seen in Fig. 18.5a, shows excellent transformation of individual scattered parabolic dunes into organized patterns of coalescing arms and creating groups of regularly spaced transverse clustered parabolic dunes. It is reasonable to visualize that these megadunes acquire their aeolian mega-landforms ranging from 'ribbed mega-barchans' to prominently sinuous transverse dunes with well carved series of numerous NE-SW oriented linear arms linked by sandy transverse closer towards the downwind. The NW-SE span of these transversals range from 3 to over 15 km whereas the average length of arms is 1.2 km. The parabolic closures on the NE of these coalesced dunes are clearly discernible. These short armed clustered parabolic dunes also occur with well-defined inter-megadunes space that ranges from less than 2 to 4 km apart.

The plan view of various parabolic dune patterns as studied on space imageries (LANDSAT TM/FCC) and field photos are shown in Figs. 18.6, 18.7, 18.8.

A small sampled area of compound parabolic dunes in the Thar Desert shows an average of seven arms, with a mean length of 2.6 km, mean width of 2.4 km, length: width ratio of 1.1 and an average interdune space of 2.3 km (Breed et al., 1979). Parabolic dune 'chains' of 7 to 12 km long with dune-relief of 10 to 85 m are reported with an average length of 1288 m and mean width of 226 m for individual parabolic dunes from Shergarh area in Jodhpur district (Singh, 1977).

Comparatively smaller sized crescentic and linear dunes of a younger generation are present in the area. They represent aeolian dynamics of lesser intensity. However, their super-imposed nature of occurrence has resulted in a multi-tiered arrangement of composite dune morphologies. These

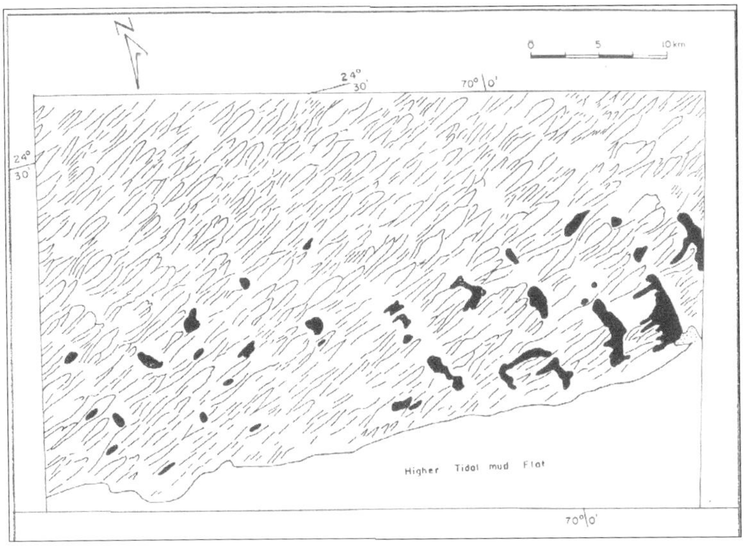

(a)

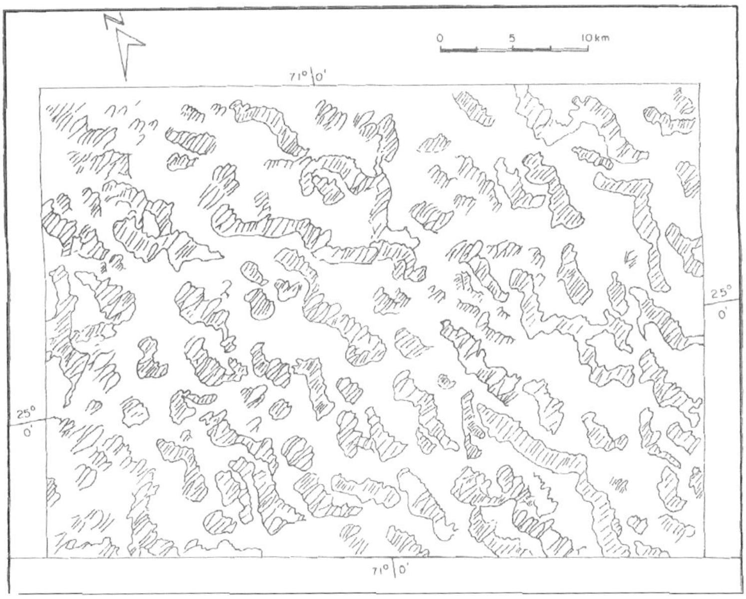

(b)

Figure 18.5. (a) LANDSAT satellite images showing distribution and morphological organization of the parabolic dunes bordering high tidal flats of Rann of Kachchh/Arabian Sea and further north-easterly tracts in the central Thar Desert, Indian subcontinent. (b) LANDSAT satellite images showing distribution and morphological organization of the parabolic dunes bordering high tidal flats of Rann of Kachchh/Arabian Sea and further north-easterly tracts in the central Thar Desert, Indian subcontinent. Note the organized segregation of mega-parabolic into transverse configurations with well-defined spaces between thee succeeding clusters covered with aeolian sand sheets.

aspects and quantitative dune morphometric parameters are described in the following sections, to infer dimensional relationships vis-a-vis aeolian bedform hierarchies. Results of aeolian geomorphic studies on anchored and stabilized dunes of older generation and the semi-stable to mobile dunes of younger aeolian episodes are presented and the evolutionary history of aeolian landforms discussed in this chapter.

(a)

(b)

Figure 18.6. (a) Synoptic view of the stabilized parabolic dunes photographed from top of a residual hill near Shergarh. Golden yellow bright sand is recently mobilized accumulation as obstacle –lee extension from the high convex apical parts of the parabolic dune. (b) Synoptic view of the stabilized obstacle dune anchored by residual hills, with sub-recent remobilized yellow sand accumulations.

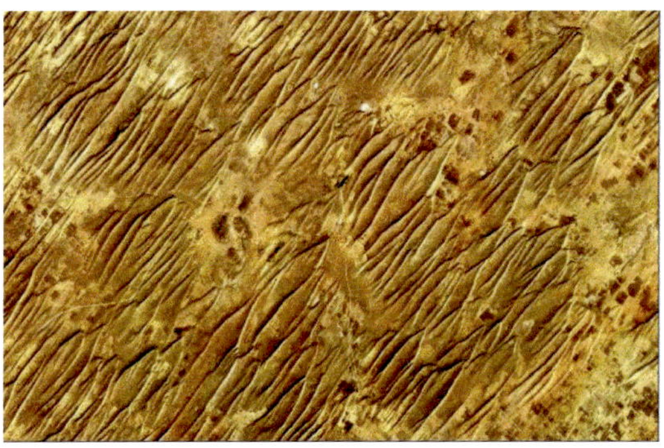

Figure 18.7. Plan view of various parabolic dune patterns as studied on space imagery, Shergarh – Dechu – Phalodi in central tract of Thar Desert.

(a)

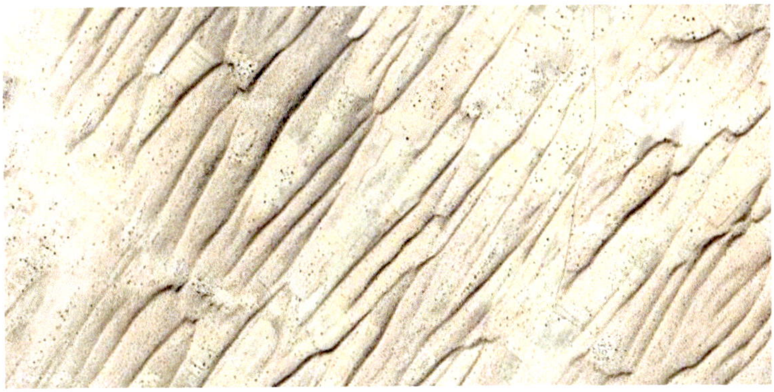

(b)

Figure 18.8. (a) Clustered Parabolic Dunes with coalesced arms and vast interdune spaces covered with sand sheet, in Shergarh – Dechu central Thar Desert; (b) A zoomed – up view to show sharply formed linear arms of the parabolic dunes and inter arm sand spaces (LANDSAT TM/FCC images courtesy: Dr. M.N. Mishra).

Dune morphometry - clustered parabolic megadunes

Parabolic dunes in the area occur in distinctive clusters whereby each parabola is linked to an adjacent dune so that they appear to share arms in planimetric configuration. The arms are oriented along NE-SW (N50°–55°E, dominant). They show remarkable parallelism though their lengths very considerably, being larger in the middle of the parabolic cluster and tapering and/or shortening gradually on either margin. A cluster may comprise up to 20 such individual parabolic arms. At places the sub-parallel arms of the clustered parabolics appear to be shaped as symmetrically disposed petals with a large central deflation hollow, followed on either side by several arms spaced by successively lesser transverse distances (Figs. 18.6 & 18.7).

Parabolic dune clusters in the area are presently observed to constitute recurring configurations of megabarchanoids on a higher scale dune hierarchy. The clustered parabolic dunes are closed by transverse closures that show varied orientations changing from NW-SE to nearly N-S on one extreme and nearly E-W on the other. Such patterns appear to have formed because of different rates of migration of the convex nosal parts of individual parabolics as governed by large scale atmospheric circulation patterns (Warren and Allison, 1998; Pye, 1993; Wilson, 1973). Different stages of dune development are observed in a well-developed complex parabolic, the central blown-out hollow opens with a narrow curvilinear passage on the windward south-westerly direction, gradually flaring laterally and reaching a maximum width span in the middle before narrowing again towards the north-easterly downwind parts. Broadly rounded (convex) to angular (hair-pin) nosal closures are noticed spatially disposed in parabolics and intense deflation at places has resulted in completely blowing out of the nosal parts, thus forming a funnel-like opening (Fig. 18.5a). Such thoroughly deflated down-wind ends of parabolics have given rise to longitudinal dunes (cf. Verstappen, 1970, 1983; Cook and Warren, 1973). Although such occurrences of complete blow-out are rare and far between in the area, yet it may be emphasized that by analogy each arm of a clustered parabolic is in essence a longitudinal dune element being aligned parallel to the resultant wind flow vector, whereas a transverse/oblique element forms a convex closure.

Sequential succession of these clustered parabolics is separated from each other by vast interdune sandy plains. The wave-length or the down-wind inter-dune spaces between clustered parabolics show variable dimensions ranging from about 550 m to about 5 km. Seventy five percent measurements of wave-lengths of mega-dunes (draas) fall between 1.0 and 2.0 km. The sand dune construction is observed to average of about 40 m above ground level in the nosal closure to the NE.

Interdune spaces tend to be of smaller dimensions frequently tend to form over-riding parabolics, while resting over the sandy alluvium, in marked contrast to well defined, larger interdune spaces over hard rock substrata or the duricrusted pediments occurring on the either sides of the dune field. Faster movement of sand over harder rock substrata has also crafted hair-pin, sharp-nosed inverted V-shaped parabolics in the area.

Morphometric measurements carried over these megadunes showed a wide range of dimensions (Wadhawan, 1998). The width, horn to horn straight distance, varies from 0.491 to 3.735 km; the longitudinal length of constituent parabolics ranges from 0.88 to 3.023 km, and the inter-arm transverse spacing from 139 m to 611 m. Table 18.1 depicts the representative interrelated morphometric measurements of megadunes in the area. In order to characterize the megadunes in the area, it is further attempted to delineate and make planimetric measurements of a few representative megadunes. The features measured include maximum width, length, area covered, circumference and number of arms present in the coalesced parabolics or the stabilized megabarchanoids (Table 18.2). Dune morphometry of clustered parabolic dune complexes showed coalescence of 15 to as many as 35 well-formed arms and the length of these arms ranges from 0.55 km to 3.380 km with modal dimension ranging from 1.0 to 1.5 km. The inter-arm spacing varies from 139 m to 665 m. However, the highest frequency of occurrence is confined to 200 m–300 m (37%) followed by 300–400 m. With the next frequent support occurrence of 100–200 m and 400–500 m (12% each) respectively.

Measurements were also made on linear distances between the successive parabolic dune clusters along NE-SW direction thus recording the inter-megadunes spaces, as well as the interdune arm spaces within the individual parabolics in a cluster. The latter measurements were made across the dune-arm alignment in a cluster with a view to know the width of aeolian deflationary path of eddies (Table 18.2). The interdune spacing within clustered parabolics varies from 139 to 611 m with highest frequency of occurrence in 200 to 400 m range. The inter-megadunes spaces measured showed a range of 500 m to over 4 km with maximum number occurrences confined to 1.5 to 2.0 km. Based on dune morphometric analyses the present study area in the Thar Desert, four dune size categories can be deciphered and grouped in a scheme of aeolian bedform hierarchy (Wadhawan, 1998). The smallest dunes are a series of low ephemeral migratory barchans with a wave length of about 50 m occurring in plain sand or as falling and climbing dunes over the older stabilized higher order dunes. Next in higher order are the meso-scale transverse/barchanoid and linear dunes occurring in diverse combinations with wave lengths of 200–300 m. They occur dominantly over the vast sandy interdune rolling plains carved by larger and partly degraded megadunes of clustered parabolics configuration. These third higher order megadunes show 200–400 m interdune spaces amongst their north-easterly oriented linear arms and 1 to 1.5 km distances between the transverse/parabolic closures of successive down-wind clusters. The fourth and highest order dune

Table 18.1 Representative morphometric parameters of clustered parabolic megadunes in Shergarh – Dechu - Phalodi Blocks Central Thar Desert, India (SoI Toposheet Nos. 45A, B, C and 40 N).

Sr. No.	**Cumulative width of Megadunes (Km)**	**Length of arms NE wind vector wise (Km)**	**Inter-arm spacing (Km)**
Shergarh - Dechu – Phalodi Block (Satellite Imagery, 1991:D-149-042 LANDSAT TM 2,3,4)			
1.	2.079	2.583 2.137 2.074	0.405 0.245 0.344
2.	1.517	1.580 1.181	0.329 0.303
3.	3.520	1.328 1.013 1.122	0.181 0.331 0.225
4.	1.240	1.477 1.365	0.479 0.324
5.	1.500	1.129 1.205	0.262 0.219
6.	3.373	3.223 2.333 1.834	0.539 0.665 0.459
7.	3.735	1.051 1.088	0.286 0.238
8.	3.331	0.910 0.937	0.290 0.269
9.	1.555	1.152 0.943	0.235 0.178
Phalodi –Chadi Block (Satellite Imagery, 1991:D-149-041 LANDSAT TM 2,3,4)			
10.	2.858	1.461 1.109 1.203	0.339 0.372 0.187
11.	0.928	1.565 1.324	0.414 0.139
12.	1.077	1.008 0.990	0.163 0.253
13.	0.491	1.201	0.491
14.	2.399	1.214 1.098	0.231 0.205
Range	0.491 to 3.735	0.876 to 3.223	0.139 to 0.665

morphology here is inferred to be that of crescentic mega-barchans with wave lengths of 0.5 to 3.5 km and cumulative lengths varying from 1.5 to 8.25 km and width from 1.25 to 8.6 km respectively (Wadhawan, 1998). The dune relief amongst these megadunes also shows variation from 50–80 m in the upwind sectors that decreases to a modal height of 25 m in the downwind sectors of the central basin in the Thar Desert.

Table 18.2. Morphometric parameters of clustered parabolic megadunes (Mega-Barchanoids) in Shergarh – Dechu – Phalodi Basin, Central Thar Desert, India.

Sl. No.	Locality (Toposheet No.)	Maximum length (Km)	Maximum width (km)	Planimetric area (sq km)	Circumference of megadune (km)	No. of arms (longitudinal elements)	Dominant Dune relief (m)
1.	East of Matora (45 B/13)	6.3	1.25	3.352	12.584	25	42
2.	North of Pindji Ki Dhani (45 B/13)	8.25	2.0	9.460	21.731	35	38
3.	Southeast of Bhikamkor (45 B/13)	4.25	1.25	2.934	9.536	21	24
4.	Barthal Ka Dhora (45 B/13)	2.105	3.305	4.852	10.545	22	37
5.	East of Sekhala	1.647	4.475	5.599	17.434	18	35
6.	Sekhala South (45 B/6)	1.972	6.290	9.917	28.660	33	44
7.	Sekhala East (45 B/6)	2.783	3.557	6.593	27.663	29	32
8.	Sekhala (45 B/6)	1.974	3.999	6.459	13.863	25	40
9.	Chamu West (45 B/10)	3.587	8.680	15.860	39.532	72	50
10.	Solankiya Tala (45 B/6)	4..301	6.102	19.689	32.357	52	50
11.	Southeast of Jalora (45 B/5)	2.010	3.850	5.837	15.536	35	42
12.	Northwest of Dechu (45 B/5)	2.255	3.010	4.174	11.642	21	30
13.	Hathundi (45 F/1)	2.550	3.000	4.728	1.976	17	31
14.	Samrau Northeast (45 B/9)	1.510	3.410	3.239	11.016	21	43
15.	Samrau South (45 B/9)	2.000	2.805	3.389	11.598	13	51
16.	Kalakondi Ka Dhora (45 B/9)	3.300	7.500	13.340	26.774	39	50

4. Parabolic dunes and evolution of the aeolian bedform hierarchy

Following salient field observation and inferences drawn from remote sensing and aerial photo-interpretation and follow-up field surveys in central parts of Rajasthan desert (Shertgarh – Dechu – Phalodi Basin) that have a direct bearing on the evolutionary trends of aeolian landforms we concluded that:

1. In a parabolic dune cluster, the longitudinal ridge elements generally show a bulbous profile, a gentle rise in relief beginning from the tail end to a crestal hump close to the middle reaches followed by a saddle in the down-wind direction, before the two adjacent longitudinal segments converge to form the transverse apical component or the closure of parabola.

2. Typical elongated clustered parabolic dune forms show remarkable parallelism of north-easterly oriented arms or the longitudinal elements, both amongst the older stable as well as younger semi-stable and actively accreting dunes. It implies the continuance of highly persistent unidirectional nature of the prevalent formative wind regime, though of milder intensity, since the formation of clustered parabolic megadunes.
3. Inter-arm spaces in a parabolic cluster enclose elongated oval shaped basins that generally slope south-westerly, towards a wide upwind deflationary opening. However, at places the tail-end/windward components of parabolic clusters appear to have been choked by stable sandy accumulation that also show a varying degree of gullying and incised by ephemeral fluvial flow-collapse features. An assemblage of such wash-out morphologies also resembles inverted 'Y' junctioned bifurcations near the windward end of longitudinal arms. Chronologically these degraded aeolian deposits on the up-wind margins may represent remnants of older putative dunes, scoured and deflated into parabolic forms by younger aeolian dynamism (Wadhawan, 1994).
4. The majority of the large stabilized compound and complex parabolic dunes in the Thar Desert are relict features that are subject to a varying degree of in situ degradation and reworking. Interdune sandy plains downwind of the clustered parabolic chain in less than 300 mm annual rainfall zone show near parallel sand streaks and transverse accumulations resting astride the path of the dominant summer wind. The growth and integration of these dune elements results in a new parabolic form. However, it is notable that their morphologic dimensions are smaller and they have their own characteristic set of smaller interdune spaces. In a frequency – magnitude framework of natural events, it needs to be emphasized that for the creation of large bedforms stronger winds and a large supply of sand are essential rather than milder winds acting over a longer duration. The later aeolian dispensation has been observed presently to result in dissipation of megaforms and spread of sand cover in linear multiples of low barchans in more arid core of the Thar Desert (Figs. 18.7 & 18.8).
5. Intervening spaces between relatively older and stabilized clustered parabolic megadunes host a series of low barchanoids and longitudinal dunes of smaller dimensions (meso-scale). The longitudinal dune elements occur superimposed over the windward slopes of the barchanoids. These low relief, 5 to 20 m high, younger dunes have shorter interdune spaces, 225 to 300 m in case of transverse and 150 to 350 m for the longitudinal dune elements. They are observed to occur in different stages of integrations resulting into low relief clustered parabolic megadunes.
6. Present day wind regime and vegetation cover appear to be favoring the formation of clustered parabolic dunes through a mechanism of integrating barchanoid/transverse and longitudinal dune elements though in characteristically smaller dimensions, but more frequently occurring patterns. These dunes ominously follow a proportionate relationship between the dune space and size, being smaller for low relief dunes and larger interdune spaces for the higher and longer clustered older megadunes respectively.
7. These observations in the Thar Desert tend to support Wilson's (1972) contention that the hierarchy of dune development is compatible with wind strength such that when strong persistent winds produce a set of large ridges that are then superseded by much gentler winds, a set of smaller ridges may develop on top or flanks of them or in the intervening interdune spaces filled with sandy sheet deposits.

5. Discussion and conclusion

It is generally recognized that changes in intensity and duration of monsoon winds have affected periodic aridity conditions resulting in contraction and expansion of the Thar Desert in the Indian subcontinent particularly during the late Quaternary. This phenomenon has been observed to be impacting on transportation and accumulation of sand deposits and construction of aeolian landforms in characteristic source-sink systems operative under overall dryland environment. Therefore, it is imperative to study and evaluate the evolutionary growth trends in delineated tectonically controlled geological basins. Accordingly, four such major basins were demarcated and characterized (Wadhawan, 1996, 2010, 2018).

The stabilized and multi-tiered complex parabolic dunes occurring typically in the north-easterly oriented central tract (Shergarh-Dechu-Phalodi basin) of the Thar Desert in the Indian subcontinent are presently deciphered to constitute patterns of larger crescentic megadunes of higher order in the aeolian bedform hierarchy. These megadunes show effects of intense deflation and scooping up of sands by exceptionally high-speed winds of the past extreme and prolonged aridity, OSL dated at an excavated section of consolidated sand sheet in the Chamu area, as ca. 115–90 ka (Dhir et al., 2010).

Complex evolutionary morphologic modifications are recorded owing to integration of degradation, mobility and deposition of abundant sand supply into successively superimposed dunes or free crescentic and linear dunes of low relief of younger generations, dated to ca. 60–52, 30–24, 16–11, 4–5.5 ka (Singhvi and Kar, 2004; Dhir et al., 2010).

The complex parabolic dune morphology may consist of super-imposition of a variety of architectural elements: longitudinal, transverse/oblique, in different dimensional hierarchies of aeolian bedforms. It is observed in the Thar Desert that such complex parabolics constitute megadunes that have experienced a multifaceted evolutionary history, incorporating numerous episodes of aeolian deposition, stabilization and partial reactivation/degradation or reworking by successively younger transgressive aeolian dynamism (Wadhawan, 1994, 1998). Besides, the consolidated nature of quartzose sand-sized sediments coupled with post-depositional modifications such as pervasive pedocalcic pedogenesis seen as lumpy kankars or calc-cemented sand aggregate impregnation, calcified root-casts (pedotubels); selective gully erosion of complex parabolic and obstacle dunes and wind deflation induced features such as lag-deposits over the dune surfaces together support inferences that these aeolian deposits are paleo-environmental indicators of the earlier aridity and intense aeolian dynamism.

Detailed studies by multi-institutional teams aided by absolute chronometry suggested that during the last ~ 200 ka, major periods of enhanced aeolian sand mobilization and accumulation, as well as prolonged dune-building activities largely coincided with and distinctly became more pronounced during the periods of transition from a drier to a wetter phase, rather than during the peak of dry period or the glacial maxima. Such peaks of aeolian sand accumulation are reported especially at ~ 190 ka (after the MIS 7.3), ~ 140 ka, 115–100 ka [after the ~ 126 ka (Climatic Optimum, when heavy and widespread monsoonal rains were experienced)]; 75–65 ka, ~ 55–50 ka, 30–25 ka, 14–7 ka, 5–3 ka and 2–1 ka (Singhvi and Kar, 2004; Dhir et al., 2010; Srivastava et al., 2019; Kar, 2020).

Dune morphometric data also reveal the dominance of 50–80 m high dunes in upwind sectors which decrease in modal height to 20–30 m in the down-wind sectors. The dunes over harder and rocky substrate grow into better defined larger sized dunes because of upwind roughness and greater sand transmission through aeolian saltation and near surface rolling of sediments. These dune patterns are also accompanied with greater interdune spaces compared to those dunes developed in the thick sandy tracts. Successive organization of parabolic dunes from singular dune patterns to aggregations into large transverse dunes with scooped up shorter linear arms clustered parabolic dunes and well defined interdune sandy corridors. Further down-wind these clustered parabolic dunes show longer lengths of linear and sub-parallel arms. Transformation of presently stabilized clustered parabolic dunes into linear dunes appear to be contemporaneous.

Genetic significance of these aeolian deposits is related to confinement of stabilized clustered parabolic dunes to a geologic - tectonic architecture—an aggrading rift basin (a major graben) formed along the Kuch – Luni-Sukri tectonic lineament (Ramasamy, 1999; Bakliwal and Wadhawan, 2003) that hosts present day ephemeral River Luni and its tributaries draining the catchment on the southwest of the Aravalli hill ranges, but its tributary channels got buried and lost under advancing dune field sediments. Some paleochannels and defunct drainage are inferred and present that ostensibly made confluence near the locality called Pachhpadra salinity affected depressions (Wadhawan, 1988; Wadhawan et al., 1999).

As the sediments within the basin get subjected to frequent reworking by the changing aeolian and fluvial regimes as a reflection of climate changes, consequently the landforms also suffer degradation and transformation. However, sand sheets that occur in the intervening spaces between the successive dunes will have a greater potential to preserve the records of sediments provenance and climate dependent geomorphic processes bringing about evolutionary changes. Therefore, interdune deposits provide a vast arena for detailed studies of properly logged and sampled sections in the area in terms of geochemistry and isotopic investigations, sedimentology, mineralogy-including heavy mineral characterization and clay mineralogy, and OSL dating will be rewarding research efforts for the future.

References

Ashwal, L.D., Solanki, A.M., Pandit, M.K., Corfu, F., Hendriks, B.W.H., Burke, K. and Torsvik, T.H. 2013, Geochronology and geochemistry of Neoproterozoic Mt. Abu granitoid, NW India: Regional correlation and implications for Rodina paleogeography. Precambrian Research 236: 265–281.

Bajpai Sunil, Tripathi Satish C. and Prasad Vandana (eds.). 2018. The Indian Paleogene, Springer Nature. 327p. https://doi.org/10.1007/978-3-319-77443-5_1.

Bakliwal, P.C. and Wadhawan, S.K. 2003. Geological evolution of Thar Desert in India—issues and prospects. Proceedings of the Indian Science Academy (INSA) 69A-2: 51–166.

Beveridge Carrie, Kocurek Gary, Ewing Ryan C., Lancaster N., Morthekai P., Singhvi Ashok K. and Mahan Shanon, A. 2006. Development of spatially diverse and complex dune-field patterns: Gran Desierto Dune Field, Sonoran, Mexico. Sedimentology 53: 1391–1409.

Blinkhorn James. 2013. A new synthesis of evidence for the Upper Pleistocene occupation of 16R Dune and its southern Asian context. Quaternary International, doi: 10.1016/j.quaint.2013.01.023.

Breed, C.S., Fryberger, S.G., Andrews, S., McCauley, C., Lennartz, F., Gebel, D. and Horstman, K. 1979. Regional studies of sand seas, using LANDSAT (ERTS) imagery. pp. 305–397. *In*: Mc Kee Edvin, D. (ed.). A Study of Global Sand Seas. USGS Professional Paper 1052. United States of America, Washington.

Cook, R.U. and Warren Andrew. 1973. Geomorphology in Deserts: Los Angeles, California University. Press. 374 p.

Dhir, R.P., Kar, A., Wadhawan, S.K., Rajaguru, S.N., Misra, V.N., Singhvi, A.K. and Sharma, S.B. 1992. Thar Desert in Rajasthan - Land, Man and Environment. Bangalore: Geological Society of India, 191 p.

Dhir, R.P., Singhvi, A.K., Andrews, J.E., Kar, A., Sareen, B.K., Tandon, S.K., Kailath, A. and Thomas, J.V. 2010. Multiple episodes of aggradation and calcrete formation in Late Quaternary aeolian sands, central Thar Desert, Rajasthan, India. Journal of Asian Earth Science 37: 10–16.

Dhir, R.P., Joshi, D.C. and Kathju, S. 2018. Thar Desert in Retrospect and Prospect. Scientific Publishers India, Jodhpur, 405pp.

Geological Survey of India. 2001. Geomorphological Map of India (scale 1: 5,00,000 scale), Government of India, Hyderabad.

Kar, A. 1993. Aeolian processes and landforms in the Thar Desert. Jour. Arid Environments 25: 83–96.

Kar Amal. 2011. Quaternary geomorphic processes and landform development in the Thar Desert of Rajasthan. pp. 223–254. *In*: Bandyopadhyay et al. (eds.). Landforms Processes & Environment Management. ACB Publications Kolkata.

Kar Amal. 2020. Quaternary Aeolian Landscape Development in Thar Desert and Its Drivers. Proc Indian Natn Sci Acad 86 No. 1 March 2020 pp. 405–417. doi: 10.16943/ptinsa/2020/49778.

Mc Kee Edvin, D. (ed.). 1979. A Study of Global Sand Seas. USGS Professional Paper 1052. United States of America, Washington. 429 p.

Meert, J.G., Pandit, M.K., Pradhan, V.R., Banks, J., Sirianni, R., Stoud, M., Newstead, B. and Gifford, J. 2010, Precambrian crustal evolution of Peninsular India: A 3.0-billion-year odyssey. Journal of Asian Earth Sciences 39: 483–515.

Pye, K. 1993. Late Quaternary development of coastal parabolic megadunes complex in north-eastern Australia. Spec. Publs. Int. Association of Sedimentologists 16: 23–44.

Ramasamy, S.M. 1999. Neotectonic controls on the migration of Sarasvati River of the great Indian Desert. Mem. Geol. Soc. 42: 153–162.

Saini, H.S. and Mujtaba, S.A.I. 2012. Depositional history and paleoclimatic variations at the north-eastern fringe of Thar Desert, Haryana plains, India. Quaternary International 250: 37–48. doi: 10.1016/j.quaint.2011.06.002.

Singh, S. (ed.). 1977. Geomorphological investigations of Rajasthan Desert, India. Central Arid Zone Research Institute [CAZRI] Monograph, Jodhpur, 93p.

Singhvi, A.K. and Kar, A. 2004. The aeolian sedimentation record of the Thar Desert. Proceedings Indian Academy of Sciences (Earth & Planetary Science) 113: 371–401.

Singhvi, A.K., Williams, M.A.J., Rajaguru, S.N., Misra, V.N., Chawla, S., Stokes, S., Chauhan, N., Francis, T., Ganjoo, R.K. and Humphreys, G.S. 2010. A ~ 200 ka record of climatic change and dune activity in the Thar Desert, India. Quaternary Science Review 29: 3095–3105.

Singhvi, Ashok K., Bhat Nilesh, Glennie Ken W. and Srivastava Pradeep. 2012. India Arabia and adjacent regions. pp. 151–206. Sarah E. Metcalf and David J. Nash (eds.). Chapter 5 in Quaternary Environmental Change in the Tropics. John Willey & Sons Ltd.

Sinha-Roy, S. 1986. Himalayan collision and indentation of the Aravalli orogeny by Bundelkhand wedge: implications for neotectonics in Rajasthan. pp. 18–21. *In*: Proceedings of International Symposium on Neotectonics in South Asia. Survey of India, Deha Dun.

Sinha-Roy, S., Malhotra, G. and Mohanty, M. 1998. Geology of Rajasthan. Geological Society of India, 278p.

Srivastava, A., Thomas, D.S.G. and Durcan, J. 2019. Holocene dune activity in the Thar Desert Earth Surface Process Landform, doi: 10.1002/esp.4583.

Valdiya, K.S. 2002. Saraswati: The River that Disappeared. Universities Press, Hyderabad, 116p.

Valdiya, K.S. 2013. The River Saraswati was a Himalayan-born River. Current Science 104(1): 42–54.3.

Verstappen, H.TH. 1970. Aeolian geomorphology of the Thar desert and Paleoclimate. Z. Geomorph. Suppl. 10: 104–120.

Verstappen, H.TH. 1983. Applied Geomorphology – Geomorphological surveys for environmental development. Elseviers, Amsterdam, 437 p.

Wadhawan, S.K. 1988. Evolution of Quaternary aeolian deposits of Jodhpur and Barmer districts, Rajasthan, India. pp. 64–78. *In*: Patel, M.P. and Desai, N. (eds.). Proceedings of the National Seminar on Recent Quaternary Studies in India. M.S. University, Baroda.

Wadhawan, S.K. and Sural, Basudev. 1992. Quaternary aeolian stratigraphy and neotectonics in Rajasthan and Gujarat. Records of Geological Survey of India 125(7): 18–19.

Wadhawan, S.K. 1994. Dune dynamics and evolution of aeolian landforms in parts of Jaisalmer district, Rajasthan. Jour. Ind. Soc. Rem. Sens. 22(2): 65–77.

Wadhawan, S.K. 1996. Textural attribute of recent aeolian deposits in different sub-basins of the Thar Desert India. J. Arid Environment 32: 59–74.

Wadhawan, S.K. and Virendra Kumar. 1996. Subsurface Quaternary aeolian stratigraphy in the Ghaggar basin of Thar Desert. J. Arid Environment 32: 37–51.

Wadhawan, S.K. 1998. Late Quaternary evolution of clustered parabolic megadunes in Thar Desert, India. pp. 185–195. *In*: Alsharhan, A.S., Glennie, K.W., Whittle, G.L. and Kendal, C.G.St.C. (eds.). Quaternary Deserts and Climate Change. AA Balkema Rotterdam.

Wadhawan, S.K., Sareen, B.K., Pal, N.K. and Raghav, K.S. 1999. Final repost on geological and environmental evolution of the Thar Desert in Rajasthan and Gujarat. Records of Geological Survey of India 132(7): 22–27.

Wadhawan, S.K. 2005. Geological evolution of Thar Desert in Rajasthan and Gujarat, India. GSI Final Report, Jaipur. Records of Geological Survey of India 137(7): 56–58.

Wadhawan, S.K. 2010. Quaternary continental sedimentation basins and dune fields in Thar Desert of NW India. Gondwana Geological Magazine (presently J. Geoscience Research) Spl. 12: 265–270.

Wadhawan, S.K. 2018. Paleogene – Neogene tectonics and continental aggradational basins in north-western India: Implications for geological evolution of Thar Desert. pp. 151–166. *In*: Sunil Bajpai, S.C. Tripathi and Vandana Prasad (eds.). The Indian Paleogene. Springer Nature.

Warren Andrew and Allison David. 1998. The Paleoenvironmental significance of dune size hierarchies. Paleogeography, Paleoclimatology, Paleoecology 137: 289–303.

Wasson, R.J., Rajaguru, S.N., Misra, V.N., Agrawal, D.P., Dhir, R.P., Singhvi, A.K. and Rao, K.K. 1983. Geomorphology, late Quaternary stratigraphy and paleoclimatology of the Thar dune field. Zeitschrift Geomorphology Supplement 45: 117–151.

Wilson, I.G. 1972. Aeolian bedforms, their development and origins. Sedimentology 19: 173–210.

Chapter 19

Sand Dunes Morphology and Migration in Shyok Valley, Ladakh Himalaya, India

Neelratan Singh,[1,9,*] *Elora Chakraborty,*[2] *Shubhendu Shekhar,*[3] *Mahesh K. Gaur,*[4] *Atul Kumar Singh,*[1] *Pankaj Kumar,*[1] *Rajesh Kumar,*[2,10] *Shukla Acharjee,*[5] *K. Rajendra,*[6] *B. Deshmukh,*[7] *Milap C. Sharma,*[2] *Sundeep Chopra*[1] and *Yamuna Singh*[8]

1. Introduction

The Shyok River is a tributary of the Indus River that flows through northern Ladakh and enters into Gilgit–Baltistan, covering a distance of ~ 550 km. The literal meaning of the word *Shyok* is sorrow and the river is known as the 'river of death' (Kapadia, 1999). The river was blocked several times by the advancing Chong Kumdan Glacier until 1929 and the dammed river caused severe floods whenever the dam was breached (Kapadia, 1999). The Shyok River originates from the Rimo Glacier, one of the tongues of the Siachen Glacier. The alignment of the Shyok River is not usual, after its origin from the Rimo Glacier it flows in the south-easterly direction; and from the Pangong range, it takes a north-western turn, and there after flows parallel to its previous path. The Shyok Valley widens at the confluence with the Nubra River but suddenly turns into a narrow gorge near Yagulung (34.77°N 77.14°E), flowing through Bogdang, Turtuk and Tyakshi before crossing into Baltistan (Ahmed and Rogers, 2015). The river joins the River Indus at Keris, east of the town of Skardu. The geomorphology of the Shyok-Nubra Valleys occupies the discontinuous steeply thrusted suture (or paleao-subduction) zones with ophiolithic melanges. The geomorphological typology and evaluation of the terrain describes the region as an ectropic zone with retarded valley cutting. The region presents zones with perched, barren moonland which is mostly devoid of vegetation due to

[1] Geochronology Group, Inter University Accelerator Centre, New Delhi, India.
[2] Centres for Study of Regional Development, Jawaharlal Nehru University, New Delhi, India.
[3] Department of Geology, Center for Advanced Studies, University of Delhi, Delhi, India.
[4] ICAR- Central Arid Zone Research Institute, Jodhpur, India.
[5] Centre for Studies in Geography, Dibrugarh University, Dibrugarh, India.
[6] Amity School of Earth & Environmental Science, Amity University, Haryana, India.
[7] Discipline of Geology, School of Sciences, IGNOU New Delhi, India.
[8] Centre for Earth Sciences, University of Hyderabad, Hyderabad, India.
[9] CSIR-CIMFR Nagpur Research Centre, Nagpur (India).
[10] Department of Geography - Sikkim University, Gangtok, India.
* Corresponding author: neelratan.geol@gmail.com

cloud and air-mass barrier ranges of the greater Himalayas. The region is tectonically active due to the presence of the Karakoram Fault in the valley and is responsible for valley widening (Searle et al., 1989). Active tectonics control the dynamics of the river and thus control the sedimentation rate and pattern (Singh et al., 2016, 2017; Gaur et al., 2019). The dynamics of the Shyok River is controlled by the Karakoram fault resulting in the widening of the narrow valley. The valley widens near Hunder and Sumur, and narrow downs near Udmaru and Skuru. The Shyok River channel with gentle slopes at the valley bed and fine-grained bed load shows the same evidences of braiding. The migration of dunes depends on the weather conditions, topography and wind speeds (Moreno-Casasola, 2008). Due to the shape of the Shyok Valley, dunes are found to be moving from narrower regions to wider regions.

Earlier studies by Pant et al., 2005; and Singh et al., 2019, mainly focused on the shapes and types of dunes in Shyok Valley, however no detailed study on the process of dune formation and migration in the Shyok Valley were conducted. Although at different locations the morphology of sand dunes were studied by Bullard and Livingstone (2002); Bullard and McTainsh (2003); and Liu et al., 2005; and Liu and Coulthard (2015), and migration of sand dunes by Gay (1999) and snow cover distribution by Gaur et al. (2019, 2022).

Both depositional (moraines, terraces, alluvial-colluvial fans, bajada, dunes) and erosional (channels and terraces) landforms are preserved in this area (Pant et al., 2005; Sant and Wadhawan, 2011a, b; Blöthe et al., 2014). The valley is partially filled with alluvial fans with patches of lake sediments on the left side and the fluvio-lacustrine and alluvial fan deposits on the right side (Lal et al., 2019). Mainly barchanoid type of dunes have been reported in the Shyok Valley by Singh et al., 2019; which also happen to be one of the most commonly occurring types of dunes in desert environment (Yang et al., 2019). Barchans dunes are crescent-shaped, formed due to scarce supply of sand (Khosronejad and Sotiropoulos, 2017).

The Shyok and Nubra Rivers transport high sediment load due to their proximity to the glaciers and alluvial fans, as they lose their ability to incise rapidly due to continuous aggradation. Various types of sediment/deposit exist such as fluviatile, glacial, fluvio-glacial, slush-flowed or solifluction debris. Predominantly, the coalescence of fans form valleys sides and are products of an environmental transitional from glacial to fluvial (Fort, 1983).

The objective of the present study is to understand morphology and analyze migration rate of dunes in the selected part of Shyok Valley, Ladakh Himalaya.

2. Study area

The study area lies in the downstream of Shyok Valley. Large impressive aeolian dune fields are present at the confluence of the Nubra and Shyok Rivers in Ladakh. Colluviated aeolian deposits are present on many landforms, sometimes containing buried soils such, as in the case of Suru Valley on the north side of the Nun Kun Massif (Lee et al., 2014). The sand dunes of the Shyok watershed lying between 34°41’18.N to 34°46N latitudes and 77°7’13.E to 77°18’30E, longitudes has been studied (Fig. 19.1). These dunes patches lie in and around Udmaru and Skuru near Hunder in Shyok Valley, Ladakh Himalaya. Three major dune complexes have been identified as dunes complex 1, 2, and 3 in the valley within the stretch on the valley sides amongst narrow and steep gorges (Fig. 19.2). The area has active and dormant sand dunes covered by vegetation. Due to vegetation on the dune surfaces, the dunes are dormant. Annual movements in individual dunes are visible and based upon data the direction of movement is SE.

Dune Complex 1:

The Shyok River bends at acute angles at the site of Dune complex 1 because of the presence of faults. Due to the bend of the river, the sediments are deposited in the manner of a point bar largely due to structural controls. Within this dune complex, a total of 19 sand dunes have been identified and measured for their dimensions and movement.

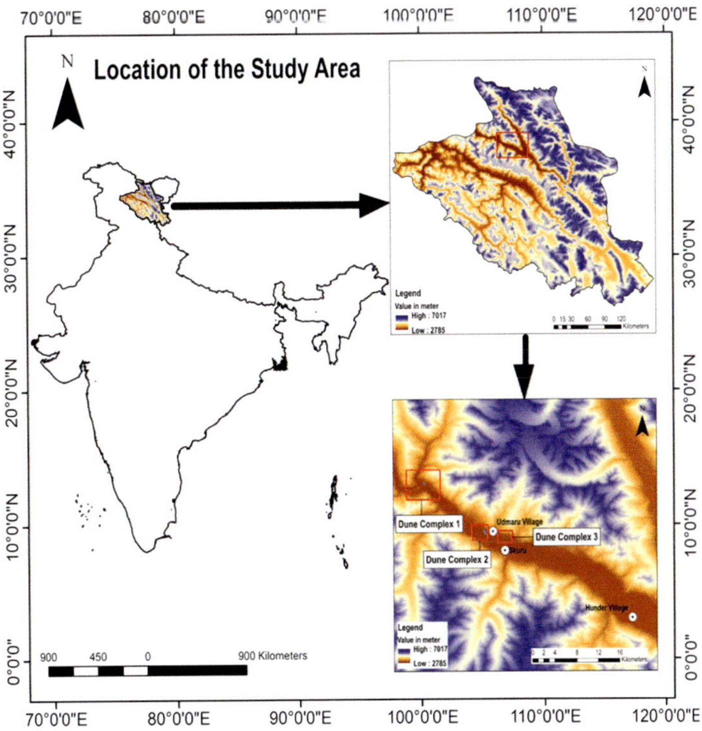

Figure 19.1. Location map of study area.

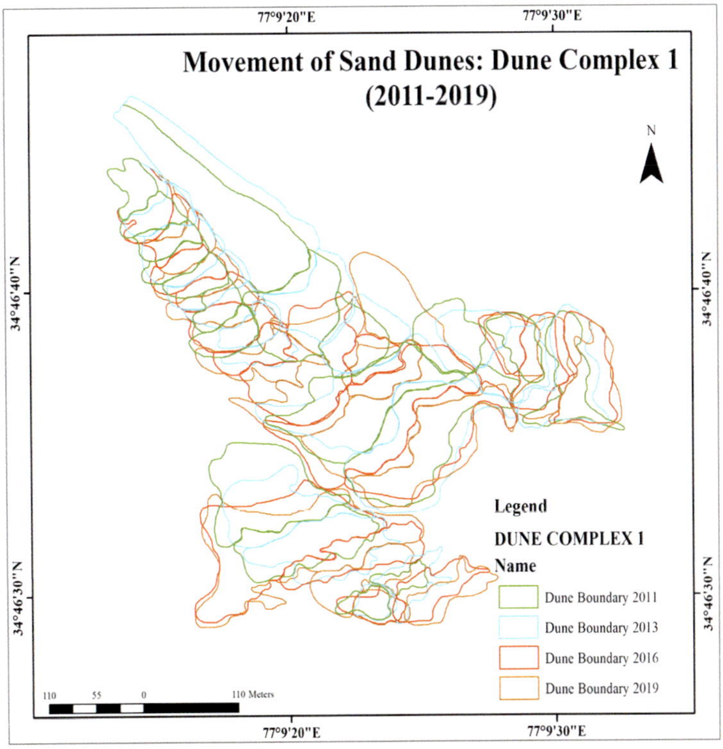

Figure 19.2. Movement of Sand Dunes: Dune Complex 1 (2011–2019).

3. Methodology

High-resolution satellite imageries from Google Earth sourced from LANDSAT Copernicus mission and MAXAR technologies have been assessed to ascertain the dimensions and movement of the dunes. The images have been duly geo-rectified using Ground Control Points collected from the field and using Sentinel 2A optical satellite imageries. Ground control points help in ascertaining the true locations of certain permanent features on the ground on respective satellite imageries. This forms a very important aspect of dealing with the mapping of any feature concerning changes over a certain timescale. The morphology has been studied using images and most of the dunes are barchans. It is also visible in figures.

The analysis was rendered through the help of Google Earth Engine QGIS Plugin as well as ArcGIS.

Sentinel 2A Optical satellite imagery Number L1C_T43SFU_A007753_20180831T054522 captured on 2018/08/31 with tile number T43SFU and L1C_T43SGU_A007710_20180828T053642 captured on 2018/08/28 with tile number T43SGU have been used to geo-reference the Google earth images to correct any undesirable shifts on the QGIS software (Fig. 19.3).

The dimensions of the barchanoid dunes were measured after vectorization of the perimeter of the dunes for each period of observation. The lengths across the axes of the sand dunes were also measured along with their areas. Total displacements have been ascertained within the mentioned timescales to assess the rate of displacement for every year. This process helped in reaching the rate of dune migration over the timescale of the study. All 19 dunes of the dune complex were studied for migration.

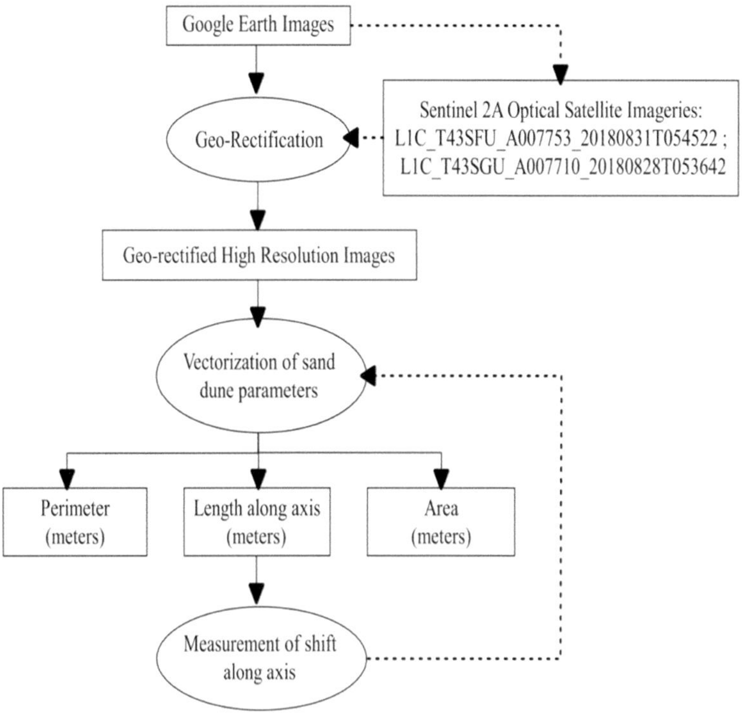

Figure 19.3. Flow chart showing procedures to use remotely sensed data.

4. Results

Dune Complex 1:

SD001:

The first dune of the dune train has been identified as a barchanoid type dune. The dune has shown a gradual migration in the south-easterly direction. It has been tracked since 2011 till 2019 for its dimensions and movements. The perimeter of the dune in 2011 was recorded at 196 meters with a length of 39.1 meters along the axis and an area of 1545 m^2, respectively. In 2013, the length of the dune along its axis increased to 44.4 m, increasing the area of the dune to 2030 m^2. The displacement noted between 2011–2013 was placed at 8.06 m. In 2016, however, the length altered further to 66 m and the perimeter and area to 284 m and 3350 m^2 respectively. The displacement noted from 2013–2016 was 21.81 m. In 2019, the length of the barchanoid type dune was recorded at 85 meters, while the perimeter stood at 309 meters and area at 4071 m^2 respectively. The displacement between 2016–2019 was 36.64 bringing up the total displacement to 66.51 meters with a rate of movement of 8.31 m yr^{-1}. There is a gradual increase in the dimensions of SD001 over the period of study (2011–2019) with the maximum displacement recorded in the years 2016–2019.

SD002:

Sand dune SD002, a part of the same dune train has recorded a perimeter of 181 meters in 2011 with the length at 29.6 m along the axis and area at 1526 m^2, respectively. The perimeter of the dune altered to 202 meters and length along the axis reduced to 24.7 meters while the area stood at 1831 m^2 respectively. The displacement in 2011–2013, was noted to be 5.79 m in the south-easterly direction. In 2016, the perimeter further increased to 214 meters while the length also increased to 32.2 m which altered the area marginally to 1881 m^2 respectively. The displacement noted in 2013–2016 was placed at 28.91 meters. In 2019, the length of the dune stood at 30.6 m along the axis while the perimeter increased was recorded at 219 meters, while the area was calculated at 2288 m^2 respectively. The displacement in 2016–2019 was recorded at 32.42 which brought up the total displacement of the dune from 2011–2019 to 67.12 m with a rate of displacement at 8.39 m y^{-1}.

SD003:

The dimensions of the sand dune have been measured from 2011–2019. In 2011, the perimeter of the dune was recorded at 180 meters with a length of 22.6 m and an area of 1542 m^2 respectively. In 2013, the perimeter of the dune was measured at 256 meters while the length along the axis was 57 meters and the area was 3878 m^2 respectively. The displacement of the dune in 2011–2013 was recorded at 34.17 m. In 2016, the perimeter reduced to 201 m with the reduction of the length along the axis to 32.5 m. The area of the dune recorded in that year was 2131 m indicating the lack of supply accretion of sand to the dune in that period 2013–2016. The displacement however was placed at a mere 3. 49 m. In 2019, the dune perimeter increased to 217 m, while the length along axis changed to 32.2 m and area to 2413 m^2 respectively. The displacement of the dune from 2016–2019 was recorded at 33.72 m bringing up the total displacement from 2011–2019 to 71.38 m with a rate of 8.92 m yr^{-1} in the south-south easterly direction.

SD004:

The fourth dune in the dune complex 1 has also been assessed for its dimensions and displacement in 2011–2019. In 2011, the perimeter of the dune was recorded at 176 m and the length along the axis at 25.6 m and area at 1501 m^2 respectively. In 2013, the perimeter of the dune altered to 219 m and the length to 30.5 m and area to 2332 m^2 respectively. In 2016, the perimeter was further altered to 201 meters and length remained 30.5 m while the area reduced to 2037 m^2. The displacement of the dune measure was put at a mere 3.61 m. Both SD003 and SD004 showed minimal movement in the years 2013–2016. In 2019, SD004 recorded a perimeter of 226 m and a length of 36.4 m along the axis. The area of the dune was calculated at 2502 m^2 in that year. The displacement in 2016–2019 was measured at 49.23 m bringing up the total displacement of the dune to 88.07 m. The rate move dune migration was calculated to be 11.08 m yr^{-1} in the south-easterly direction. The dimensions of the dune and magnitude of movement greatly differed in the mentioned timeframe of 2011–2019.

SD005:

SD005 is the fifth dune in the dune complex, which, in 2011, reflected a length of 23.2 m while the perimeter was noted at 171 m. The area was calculated at 1573 m^2. The dimensions of the sand dune increased to 211 m in its perimeter, 31.9 m in length, and 1831 m^2 in the area in 2013. The displacement was noted at 44.02 m from its initial position in 2011–2013. In 2016, the perimeter was noted at 259 m and length at 48.3 m, while the area altered to 3192 m^2 respectively. The displacement was recorded at 20.14 m in 2013–2016. In the last year of the probe, 2019, the length of the sand dune altered to 36.8 m and perimeter at 239 m, while the area was recorded at 2393 m^2 respectively. The displacement of the dune during 2016–2019 was recorded at 27.42 m bringing up the total displacement to 91.58 m over 8 years (2011–2019). The rate of movement was calculated at 11.44 m yr^{-1}.

SD006:

The sixth dune in dune complex 1 was also assessed for its dimensions and displacement in 2011–2019. In 2011, the perimeter of the dune was recorded at 166 m and the length along the axis at 33.0 m and area at 1467 m^2 respectively. In 2013, the perimeter of the dune altered to 184 m and the length to 25.6 m and area to 1661 m^2 respectively. The displacement recorded in 2011–2013 was recorded at 36.7 m. In 2016, the perimeter was further altered to 235 meters and length remained 28.7 m while the area reduced to 2084 m^2. The displacement of the dune measure was put at 23.61 m. In 2019, SD006 recorded a perimeter of 366 m and a length of 52.1 m along the axis. The area of the dune was calculated at 3356 m^2 in that year. The displacement in 2016–2019 was measured at 39.68 m bringing up the total displacement of the dune to 98.99 m. The rate move dune migration was calculated to be 12.37 m yr^{-1} in the south-easterly direction. The dimensions of the dune and magnitude of movement greatly differed in the mentioned timeframe of 2011–2019.

SD007:

SD007 and SD010 had been individual dunes in 2011 and 2013, but post-2013 both the dunes have been observed to coalesce into a single dune of greater dimensions. Therefore, both the dunes have been addressed together in this section of the study. In 2011, dune SD007 had a length of 32.1 m, while the perimeter was recorded at 189 m and area at 3350 m^2 respectively. SD010 on the other hand had a length of 59 m along its axis, the perimeter of 354 m and area of 6339 m^2 respectively. In 2013, the length of SD007 altered to 24.6 along the axis, and the perimeter and the area altered to 183 m and 1323 m^2 respectively. The dune showed a displacement of 27.69 m during 2011–2013 m. SD 010 on the other hand had a length of 50.7 m in 2013, while the perimeter and the area were recorded at 297 m and 4291 m^2 respectively. The next year of the probe, 2016 observed a coalesced dune of a greater magnitude with a perimeter 646 m and length 72.2 m along the axis respectively. The area in this year was noted at 8582 m^2 and displacement from 2013 at 75.2 m at an average from both the dunes. The perimeter of the coalesced dune in 2019 was 603 m, while the length along its axis was 46.3 m respectively. The area was recorded at 7681 m^2 and displacement from 2016 at 9.26 m respectively. The total displacement from 2011 to 2019, therefore, was 112.15 m and the rate of displacement was calculated at 14.01 m yr^{-1} respectively.

SD008:

The largest dune of the dune complex with a perimeter of 557 m and a length of 238 m in 2011 ceased to exist after 2013. The initial area of the long dune was 10039 m^2 which reduced marginally in 2013 before completely vanishing thereafter. The dune seems to provide sediments for the other dunes in the complex post-2013 due to the action of wind. The initial displacement was noted at 13.2 m.

SD009:

The dimensions of the sand dune SD009 were measured from 2011–2019. In 2011, the perimeter of the dune was recorded at 245 meters with a length of 57.9 m and an area of 3363 m^2 respectively. In 2013, the perimeter of the dune was measured at 255 meters, while the length along the axis was 56 meters and the area was 3993 m^2 respectively. The displacement of the dune in 2011–2013 was recorded at 6.59 m. In 2016, the perimeter reduced to 179 m with the reduction of the length along the axis to 54.8 m. The area of the dune recorded in that year was 1834 m^2 indicating the lack of supply accretion of sand to the dune in this period 2013–2016. The displacement however was placed at 9 m. In 2019, the dune perimeter increased to 238 m, while the length along axis changed to 63 m and area to 3101 m^2 respectively. The displacement of the dune in 2016–2019 was recorded at 20.56 m bringing up the total displacement from 2011–2019 to 36.15 m with a rate of 4.51 m yr^{-1} in the south-easterly direction.

SD011:

Sand dune SD011, a part of the same dune train recorded a perimeter of 507 meters in 2011 with the length at 54.2 m along the axis and area at 9176 m^2 respectively. The perimeter of the dune altered to 543 meters and length along the axis reduced to 47.9 meters, while the area stood at 9046 m^2 respectively. The displacement in 2011–2013, was noted to be 36.78 m in the south-easterly direction. In 2016, the perimeter further increased to 610 meters, while the length also increased to 58 m which altered the area to 11906 m^2 respectively. The displacement noted in 2013–2016 was placed at 24.05 meters. In 2019, the length of the dune stood at 72 m along the axis while the perimeter increased was recorded at 786 meters, while the area was calculated at 14568 m^2 respectively. The displacement between 2016–2019 was recorded at 10 m which brought up the total displacement of the dune from 2011–2019 to 70.83 m with a rate of displacement at 8.85m yr^{-1}.

SD012:

The 12th dune of the dune complex, also identified as a barchanoid type dune, has shown a gradual migration in the south-easterly direction. The dune was tracked from 2011 till 2019 for its dimensions and movements. The perimeter of the dune in 2011 was recorded at 516 meters with a length of 72.3 meters along the axis and an area of 10357 m^2 respectively. In 2013, the length of the dune along its axis increased to 63.2 m, increasing the area of the dune to 11018 m^2. The displacement noted between 2011–2013 was placed at 8.06 m. In 2016, however, the length altered further to 102 m and the perimeter and area to 603 and 12200 m^2 respectively. The displacement noted from 2013–2016 was 15.15 m. In 2019, the length of the barchanoid type dune was recorded at 86 meters while the perimeter stood at 744 meters and area at 14386 m^2 respectively. The displacement from 2016–2019 was 25.2 bringing the total displacement to 48.41 meters with a rate of movement of 6.05 m yr^{-1}. There was a gradual increase in the dimensions of SD012 over the period of probe (2011–2019) with the maximum displacement recorded in the years 2016–2019.

SD013:

The dimensions of sand dune SD013 were measured between 2011–2019. In 2011, the perimeter of the dune was recorded at 240 meters with a length of 69.6 m and an area of 3697 m^2 respectively. In 2013, the perimeter of the dune was measured at 209 meters, while the length along the axis was 68.6 meters and the area was 2872 m^2 respectively. The displacement of the dune in 2011–2013 was recorded at 22.93 m. In 2016, the perimeter increased to 386 m with the reduction of the length along the axis to 40 m. The area of the dune recorded in that year was 4251 m^2 however, the displacement was placed at 14.2 m. In 2019, the dune perimeter increased to 306 m while the length along axis changed to 35 m and area to 3533 m^2 respectively. The displacement of the dune in 2016–2019 was recorded at 27.5 m bringing up the total displacement from 2011–2019 to 64.63 m with a rate of 8.07 m yr^{-1} in the south-easterly direction.

SD014:

The dimensions of sand dune SD014 were measured between 2011–2019. In 2011, the perimeter of the dune was recorded at 210 meters with a length of 30.7 m and an area of 2105 m^2 respectively. In 2013, the perimeter of the dune was measured at 272 meters, while the length along the axis was 43.5 meters and the area was 3607 m^2 respectively. The displacement of the dune from 2011–2013 was recorded at 14.08 m. In 2016, the perimeter reduced to 172 m with the reduction of the length along the axis to 35.8 m. The area of the dune recorded in that year was 1494 m^2 indicating the lack of supply accretion of sand to the dune between 2013–2016. The displacement however was placed at 20.37 m. In 2019, the dune perimeter further reduced to 150 m while the length along axis changed to 48.9 m and area to 1452 m^2 respectively. The displacement of the dune in 2016–2019 was recorded at 9.5 m bringing up the total displacement from 2011–2019 to 43.95 m with a rate of 5.49 m yr^{-1} in the South-easterly direction at first, followed by an easterly direction post-2013.

SD015:

SD015 is the 15th dune in the dune complex, which, in 2011, reflected a length of 30 m while the perimeter was 411 m. The area was calculated at 3914 m^2. The dimensions of the sand dune decreased to 396 m in its perimeter, 37 m in length, and 5429 m^2 in area in 2013. The displacement was noted at 8.05 m from its initial position in 2011–2013. In 2016, the perimeter was noted at 274 m and length at 40.4 m, while the area altered to 2811 m^2 respectively. The displacement was recorded at 17.11 m in 2013–2016. In the last year of the probe, 2019, the length of the sand dune altered to 30.9 m and perimeter at 278 m, while the area was recorded at 4431 m^2 respectively. The displacement of the dune during 2016–2019 was recorded at 21.17 m bringing up the total displacement to 46.33 m over 8 years (2011–2019). The rate of movement was calculated at 5.79 m yr^{-1}.

SD016:

The 16th dune in the dune complex 1 has also been assessed for its dimensions and displacement in 2011–2019. In 2011, the perimeter of the dune was recorded at 420 m and the length along the axis at 86.2 m and area at 8390 m^2 respectively. In 2013, the perimeter of the dune altered to 405 m and the length to 64 m and area to 5621 m^2 respectively. The displacement recorded in 2011–2013 was 12.56 m. In 2016, the perimeter was further altered to 530 meters and length remained at 54.9 m, while the area increased to 7238 m^2. the displacement of the dune measure was put at 40.66 m. In 2019, SD016 recorded a perimeter of 534 m and a length of 59.5 m along the axis. The area of the dune was calculated at 6798 m^2 in that year. The displacement from 2016–2019 was measured at 10.15 m bringing the total displacement of the dune to 63.37 m. The rate move dune migration was calculated to be 7.92 m yr^{-1} in the south-easterly direction. The dimensions of the dune and magnitude of movement greatly differed in the mentioned time frame of 2011–2019.

SD017:

The dimensions of sand dune SD017 were measured from 2011–2019. In 2011, the perimeter of the dune was recorded at 379 meters with a length of 48.9 m and an area of 5836 m^2 respectively. In 2013, the perimeter of the dune was measured at 415 meters, while the length along the axis was 52.2 meters and the area was 5999 m^2 respectively. The displacement of the dune in 2011–2013 was recorded at 12.8 m. In 2016, the perimeter decreased to 180 m with the reduction of the length along the axis to 29.2 m. The area of the dune recorded in that year was 1837 m^2 however, the displacement was placed at 28.09 m. In 2019, the dune perimeter increased to 200 m, while the length along axis changed to 36.8 m and area to 2396 m^2 respectively. The displacement of the dune in 2016–2019 was recorded at 17.28 m bringing up the total displacement from 2011–2019 to 58.17 m with a rate of 7.27 m yr^{-1} in the south-easterly direction.

SD018:

Sand dune SD018, a part of the same dune train recorded a perimeter of 152 meters in 2011 with the length at 32.3 m along the axis and area at 1159 m^2 respectively. The perimeter of the dune altered to 127 meters and length along the axis increased to 35.7 meters, while the area stood at 1095 m^2 respectively. The displacement between 2011–2013, was noted to be nil. In 2016, the perimeter increased to 263 meters, while the length also increased to 36.2 m which altered the area to 3279 m^2 respectively. The displacement noted in 2013–2016 was placed at 13 meters. In 2019, the length of the dune stood at 34.8 m along the axis, while the perimeter increase was recorded at 282 meters, the area was calculated at 3206 m^2 respectively. The displacement from 2016–2019 was recorded at 42.8 m which brought up the total displacement of the dune from 2011–2019 to 55.8 m with a rate of displacement at 6.97 m yr^{-1}.

SD019:

Sand dune SD019 recorded a perimeter of 159 meters in 2011 with the length at 21.5 m along the axis and area at 908 m^2 respectively. The perimeter of the dune altered to 282 meters and the length along the axis increased to 30.2 meters, while the area stood at 2254 m^2 respectively. The displacement from 2011–2013, was noted to be nil. In 2016, the perimeter increased to 327 meters, while the length also decreased to 7 m which altered the area to 4502 m^2 respectively. The displacement noted from 2013–2016 was placed at 14.79 meters. In 2019, the length of the dune stood at 47.3 m along the axis, while the perimeter increase was recorded at 332 meters, the area was calculated at 4199 m^2 respectively. The displacement during 2016–2019 was recorded at 43.7 m which brought the total displacement of the dune from 2011–2019 to 58.49 m with a rate of displacement at 7.31 m yr^{-1}.

5. Discussion and conclusion

The Shyok and Nubra Valley regions are tectonically active with the presence of Karakoram Fault in the valley and it is widening (Searle et al., 1989; Imsong et al., 2017). The flow mechanism is created here due to the widening of the narrow valley, the wider valley towards Hunder and Sumur, and a narrow valley towards Udmaru and Skru. The region experiences cold and arid conditions throughout the year also known as the cold desert of India.

The occurrence of barchans type of sand dunes suggests that wind was predominantly unidirectional during formation of barchans dunes in the area (Hatano and Hatano, 2001; Singh et al., 2019). Barchan types of sand dunes are created where there is a limited supply of sand, having a flat ground, in which wind is unidirectional (Pidwirny, 2015; Singh et al., 2019).

The results of present study suggest that the formation of sand dunes in the valley is controlled by lithology and geomorphic agents, with the dominance of aeolian processes compared to fluvial and glacial processes (Singh et al., 2019). The valleys are unusual due to the appearance of barchanoids dunes whereas these dunes are absent in other river valleys of Ladakh Himalaya. In the case of the cold desert landsacpe of Ladakh, these crescent-shaped barchan dune are a direct result of scarce supply of sand (Khosronejad and Sotiropoulos, 2017).

Three geomorphic processes, viz. glacial, fluvial and aeolian, appear to have played key roles in the formation of sand dunes in river beds of the Ladakh area (Singh et al., 2019). Some of the dunes are stable whereas some are migratory. The dunes migrate with little or no change in their shape. The direction of migration of majority of the dunes is from NW to SE while some dunes move from E of SE, which is in the upstream direction of the Shyok River.

Acknowledgement

The First author (NS) thankful to Science & Engineering Research Board (SERB), DST, New Delhi for funding through the Young Scientist Scheme (YSS/2015/001185) a start-up Grant.

References

Ahmed, M.F. and Rogers, J.D. 2015. Regional level landslide inventory maps of the Shyok River watershed, Northern Pakistan. Bull Eng. Geol. Environ. DOI 10.1007/s10064-015-0773-2.

Bagnold, R.A. 1941. The Physics of Blown Sand and Desert Dunes. Methuen, London.

Bagnold, R.A. 2012. The Physics of Blown Sand and Desert Dunes. (Revised) Courier Corporation.
Blöthe, J.H., Munack, H., Korup, O., Fülling, A., Garzanti, E., Resentini, A. and Kubik, P.W. 2014. Late quaternary valley infill and dissection in the Indus River, western Tibetan Plateau margin. Quaternary Science Reviews 94: 102e119.
Bullard, J.E. and Livingstone, I. 2002. Interactions between aeolian and fluvial systems in dryland environments. Area 34(1): 8–16.
Bullard, J.E. and McTainsh, G.H. 2003. Aeolian-fluvial interactions in dryland environments: examples, concepts and Australia case study. Progress in Physical Geography 27(4): 471–501.
Fort, Montique. 1983. Geomorphological Observations in the Ladakh area (Himalaya). Quarterly Evolution and Present Dynamics. Hindustan Pub., Delhi.
Gaur, Mahesh Kumar, Goyal, R.K., Raghuvanshi, M.S., Bhatt, R.K., Pandian, M., Mishra, A. and Sheikh, S.I. 2019. Geospatially extracting snow and ice cover distribution in the cold arid zone of India, Internat. J. Syst. Assurance Engineering and Management, pp. 1–16. https://doi.org/10.1007/s13198-019-00883-w.
Gaur, Mahesh K., Goyal, R.K., Saha, D., Neelratan Singh, Shubhendu Shekhar, Ajai and Chauhan, J.S. 2022. The Estimation of snow cover distribution using satellite data in the cold arid leh region of Indian Himalaya. Manuscript no. Polish Journal of Environmental Studies 31(1): 1–11. DOI: 10.15244/pjoes/135606.
Gay Jr, S.P. 1999. Observations regarding the movement of barchan sand dunes in the Nazca to Tanaca area of southern Peru. Geomorphology 27(3-4): 279–293.
Hatano, Y. and Hatano, N. 2001. Dune morphology and sand transport. Forma 16: 65–75.
Imsong, W., Bhattacharya, F., Mishra, R.L. and Phukan, S. 2017. Karakoram Fault in Nubra and Shyok valleys, Ladakh Himalaya. Current Science 112(11): 2295.
Kapadia, H. 1999. Across Peaks & Passes in Ladakh, Zanskar & East Karakoram. Indus Publishing.
Khosronejad, A. and Sotiropoulos, F. 2017. On the genesis and evolution of barchan dunes: morphodynamics. Journal of Fluid Mechanics 815: 117–148.
Lal, Ravish, Saini, H.S., Pant, N.C. and Mujtaba, S.A.I. 2019. Tectonics induced switching of provenance during the Late Quaternary aggradation of the Indus River Valley, Ladakh, India. Geoscience Frontiers 10(1): 285–297. DOI: 10.1016/j.gsf.2017.12.016.
Lee, S.Y., Seong, Y.B., Owen, L.A., Murari, M.K., Lim, H.S., Yoon, H.I. and You, K.C. 2014. Late Quaternary glaciation in the Nun-Kun massif, Northwestern India. Boreas 43: 67–89.
Liu, B. and Coulthard, T.J. 2015. Mapping the interactions between rivers and sand dunes: implications for fluvial and aeolian geomorphology. Geomorphology 231: 246–257.
Liu, L.Y., Skidmore, E., Hasi, E., Wagner, L. and Tatarko, J. 2005. Dune sand transport as influenced by wind directions, speed and frequencies in the Ordos Plateau, China. Geomorphology 67(3-4): 283–297.
Moreno-Casasola, P. 2008. A case study of conservation and management of tropical sand dune systems: La Mancha–El Llano. In Coastal Dunes (pp. 319–333). Springer, Berlin, Heidelberg.
Pant, R.K., Phadtare, N.R., Chamyal, L.S. and Juyal, N. 2005. Quaternary deposits in Ladakh and Karakoram Himalaya: a treasure trove of the palaeoclimate records. Current Science 88(11): 1789–1798.
Pidwirny, M. 2015. Understanding Physical Geography, 1st Edition. Date Viewed. http://www.physical geography.net/understanding/contents.html.
Sant, D.A. and Wadhawan, S.K. 2011a. Linkage of paraglacial processes from last glacial to recent inferred from Spituk Sequence, Leh Valley, Ladakh Himalayas, India. Geological Society of India 78: 147–156.
Sant, D.A. and Wadhwan, S.K. 2011b. Morphostratigraphy and palaeoclimate appraisal of the Leh valey, Ladakh Himalayas, India. Journal of the Geological Society of India 77: 499–510.
Searle, M.P., Rex, A.J., Tirrul, R., Rex, D.C., Barnicoat, A. and Windley, B.F. 1989. Metamorphic, magmatic and tectonic evolution of the central Karakoram in the Biafo-BaltoroHushe regions of north Pakistan, Geol. Soc. Am., Spec. Pap. 232: 47–73.
Singh, A.K., Jaiswal, M.K., Pattanaik, J.K. and Dev, M. 2016. Luminescence chronology of alluvial fan in North Bengal, India: Implications to tectonics and climate. Geochronometria 43(1): 102–112.
Singh, A.K., Pattanaik, J.K. and Jaiswal, M.K. 2017. Late Quaternary evolution of Tista River terraces in Darjeeling-Sikkim-Tibet wedge: Implications to climate and tectonics. Quaternary International 443: 132–142.
Singh, N., Ali, S., Raju, N., Singh, Y., Rajendra, K., Kumar, M., Deshmukh, B., Jhariya, D., Kumar, V., Sharma, M. and Chakraborty, E. 2019. Dataset on Mapping and Morphometry of Sand Dunes in Nubra and Shyok Valleys, Ladakh Himalaya, India. Indian J. Sci. Res. 10(1): 11–19.
Slattery, M.C. 1990. Barchan migration on the Kuiseb river delta, Namibia. South African Geographical Journal 72(1): 5–10.
Yang, J., Dong, Z., Liu, Z., Shi, W., Chen, G., Shao, T. and Zeng, H. 2019. Migration of barchan dunes in the western Quruq Desert, north-western China. Earth Surface Processes and Landforms 44(10): 2016–2029.

Chapter 20

Coastal Sand Dunes of Southern India

Linto Alappat,[1,*] *Srinivasalu S.*[2] and *Ashok Kumar Singhvi*[3]

1. Introduction

Aeolian processes occur in, both arid and coastal regions (Lancaster, 2002, 2015; Speirs et al., 2008) and have been examined extensively to understand their evolution through time. Particle transport by wind depends on its size, shape, density along with the atmospheric density, wind velocity and atmospheric turbulence (Greely, 1984). These forcings are location dependent. Surface winds create landforms like sand sheets, sand dunes or loess cover. Often this process is assisted by fluvial, glacial or littoral processes that facilitate the sediment supply.

Formation of sandy landforms requires persistent wind, adequate supply of sand-sized clastic sediments and conditions conducive for the deposition of sediments that are being transported, like a topographic barrier or similar obstacles and their eventual preservation. Surface roughness created by plants, rocks or other materials modify the wind shear stress acting on land surface and attenuates the rate of mobility of sediment grains by wind, due to increased surface roughness (Grilliot et al., 2018; Lancaster, 2004). An additional factor that governs the movement of sediment particles in coastal regions is soil moisture and the salt content in the soil crust. Inter-granular movement of silt or clay-sized particles in the pore spaces indurates the sediments and restrict the mobility of detrital grains. Large scale accumulation of dunes during low sea levels associated with cooler and drier climates is facilitated by increased sediment supply due to reduced soil moisture, extended bathymetric surface and increased flood plains of rivers.

Coasts with sandy beaches and consistent onshore winds are favorable locales to form dunes. Generally, dunes along the modern coastline are linear/longitudinal due to the bi-directionality of wind regimes. Directional variability of wind and sand supply are important factors that determine the dune types with grain size, vegetation cover and sediment supply playing a minor role (Lancaster, 2004; Wasson and Hyde, 1983). As in the case of the loess deposits (Novothny et al., 2011; Singhvi et al., 2001; Wacha et al., 2011; Zhang et al., 2018), sand deposits in the form of sand sheets or dunes apprise one on the contemporary paleo-climates (Glennie and Singhvi, 2002; Preusser et al., 2005; Reimann et al., 2011; Singhvi and Kar, 2004; Tsoar et al., 2009).

[1] Department of Geology and Environmental Science, Christ College Autonomous, Irinjalakuda, Thrissur, Kerala, 680 125 India.
[2] Department of Geology, Anna University, Chennai, Tamil Nadu, 600 025 India.
[3] AMOPH Division, Physical Research Laboratory, Ahmedabad, 380 009, India.
* Corresponding author: alappatlinto@christcollegeijk.edu.in

In arid regions dune aggradation is episodic, due to specificity of environmental conditions in which the accumulation and preservation occurred. It is not the originally created, but the preserved fraction of the landform record that remains and are available for study (Kucerek and Lancaster, 1999; Singhvi and Porat, 2008). Diagenetic alteration and soil formation in buried aeolian deposits provide information on periods of landscape stability and ameliorated climates. During such periods, iron bearing minerals are altered to show a distinct red coloration for dunes and have been studied extensively (Alappat et al., 2013, 2017; Folk, 1976; Gardner, 1981; Jayangondaperumal et al., 2012; Joseph and Thrivikramji, 2005; Rao et al., 2006; Roskin et al., 2012; Singhvi et al., 1986; Singhvi and Porat, 2008; Srivastava et al., 2016; Zhang et al., 2008; Zheng and Wei, 1998).

1.1 Coastal dunes

Widespread occurrence of coastal aeolianites have been reported around the world, mostly along the mid-latitude regions and many of these are of Quaternary age (Brooke, 2001). Coastal dunes in the tropical and temperate regions have been described by Hesp (2008). The morphological development of dunes depends on plant density, their distribution, height and cover; wind velocity and rates of sand transport. These led to distinct differences in the occurrence of dune fields and types (Hesp, 2008). Coastal dunes migrate under conditions of consistent winds, sporadic storm events and human intervention through land-use changes including agriculture that is facilitated due to increase in sediment supply. Gao et al. (2020) described the mobility of coastal dunes under varied bio-geomorphological factors under given boundary conditions of sand supply, wind, vegetation and anthropogenic activities.

Occurrence of aeolianites and ages of their formation suggest that both the glacial and interglacial periods provided conditions conducive for their creation. Aeolianites that formed during the interglacials are attributed to sand deposition during higher sea level stands and those formed during glacial periods of lower sea-level, are associated with the reworking of shelf sediments (Brooke, 2001). The mobility of coastal dunes has also adversely affected crops and agriculture, (Ogura and Yura, 2008). However at times, measures to keep the dunes and sand sheets mobile to conserve the biological diversity of sand dwelling species characteristic of such active dunes, become necessary (e.g., Bar, 2013).

Globally, large scale stabilization of coastal dune fields has been a consequence of land-use changes, variation in climate through decrease in windiness or increase in temperature and rainfall and hence vegetation, (Gao et al., 2020). The sandy habitat of coastal plains host numerous endemic plants and animal species suggesting ecological uniqueness of coastal dunes (e.g., Bar, 2013; Rodrigues et al., 2011). In India, the western coast shows more diversity with 267 species of sand dune flora in Karnataka compared to 163 species on the East coast of Tamil Nadu (Rodrigues et al., 2011). This chapter reviews the present understanding of coastal dunes of South India, in terms of its geological and geomorphological settings, sediment characteristics and paleoclimatic significance.

1.2 Coastal dunes of India

On Indian coasts, beach ridges and associated coastal dunes occur as discontinuous, narrow patches of longitudinal dunes, parallel/sub-parallel to the coast (Kunte and Wagle, 2005) (Fig. 20.1).

Geomorphological features along the coastal plain shows varied depositional settings such as marine, aeolian, marginal marine, fluvial, etc. Sedimentary archives along the coast have been used to inform one on contemporary climates, sea level and/or tectonics (Alappat et al., 2015; Mauz et al., 2013; Minoura and Nakamura, 2020; Roskin et al., 2011; Szkornik et al., 2008; Telfer and Thomas, 2007; Thomas, 2009). Beach ridge complexes are largely associated with river mouths and delta plains. These occur as continuous deposits in the east coast and as narrow and discontinuous patches in the west (Kunte and Wagle, 2005), Fig. 20.1. Miliolites at the Kutch and Saurashtra coasts are the earliest known coastal aeolinite deposits of India with the depositional age extending to Marine Isotopic Stage 6 (Acharya and Bhatt, 2019; Baskaran et al., 1989; Sharma et al., 2017). Parabolic

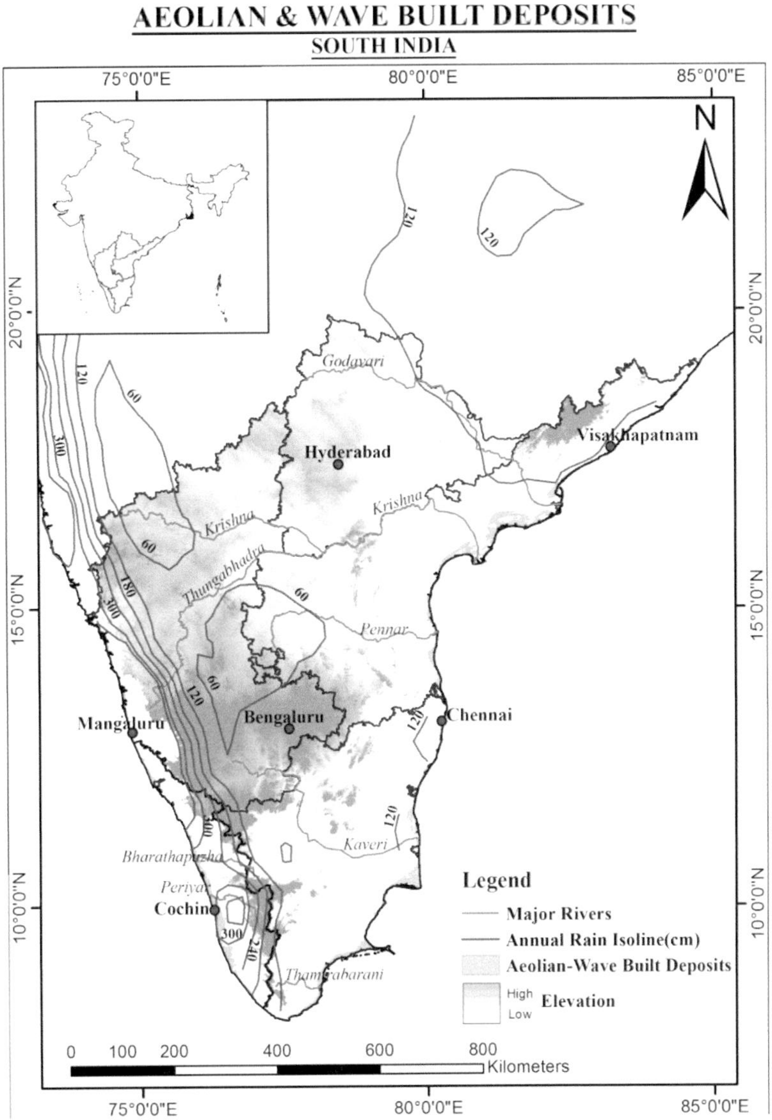

Figure 20.1. The map showing the distribution of Aeolian and wave built deposits of South India and precipitation contours in the study area. Occurrence of sand deposits are associated with river mouth, deltas and sedimentary basins. Data obtained from the Bhukosh portal of Geological Survey of India at http://bhukosh.gsi.gov.in/Bhukosh/Public. The rainfall data are generated using the gpcp precipitation data (https://psl.noaa.gov/data/gridded/data.gpcc.html (0.5 × 0.5 version, with long term mean 1981/01/01 - 2010/12/31).

dunes in the semi-arid eastern coast are the blowout of sand from the vegetated patches sandy surface.

2. Coastal dunes of South India

Sand dunes are considered as an important geomorphological element of the landscape as they mitigate the effects of natural disasters like cyclones and tsunami wave run-ups and regulate the hydrological balance through ground water recharge and irrigation. The dunes in Kerala and Karnataka also act as natural barriers in stopping the rivers draining their water to the sea, help recharging the ground water and impede intrusion of saline water. Along the western coast, coastal dunes occur as narrow disconnected patches along the beach. An exception is the South Kerala

Figure 20.2. Field photographs showing various Aeolian dune formation at the coasts of Tamil Nadu. a. Indurated dune sand occurring at the coasts of southern Tamil Nadu. b. Massive sand dune deposition at the coast of Tamil Nadu in South of Cuddalore region. c. Sand dunes overlying the beach ridges of Central Kerala located near Kodungallur. d. Photograph of linear dune and interdune area showing intense agriculture in the fertile lands of inter-dunal depressions at the coast of Tamil Nadu near Vedaranyam.

Sedimentary Basin (SKSB), where coastal dunes are associated with paleo-beach ridges and, in the near coastal regions, reworked sand invariably caps the wave-built forms. Inland dunes form due to the migration of sand from back shore regions. The sediments for dune building are largely brought by rivers to the coastal region, and then redistributed by littoral processes.

Two types of loose or indurated dunes occur at the coasts and are differentiated by their color and mineral content. Dunes along the coasts of Kerala and Tamil Nadu occur a few hundred meters to few kilometers inland from the present coast and are associated with beach ridges of mid- to late Holocene period as reworked deposits (Alappat et al., 2015). Increase in human activity has led to a large-scale leveling of the dunes and sandy areas are being artificially stabilized. Generally, the older dunes of Pleistocene to Holocene ages occur inland and modern linear and fore dunes occur at the backshore region of the present coast. The aeolian deposits at the coast of Kerala and at the near coastal region to the north of Point Calimer in Tamil Nadu are white, massive deposits and sand sheets (Fig. 20.2).

The dunes in southern Tamil Nadu near Ramanathapuram- Thoothukudi- Tirunelveli-Kanyakumari region are red with a narrow strip of white dunes closer to the present day coast (Figs. 20.3, 20.4). Red sands in Kerala are an extension of Teri sands of Tamilnadu and belong to the Vembanad formation, comprising of fine to medium grain sandy loam (Vaidyanadhan and Ramakrishnan, 2008).

Different names have been given to the red colored sands. In Tamil Nadu, these are referred to as 'Teri sands' or simply 'Teris' (Jayangondaperumal et al., 2012), which in the Tamil language means waste sands. The red sands of Andhra Pradesh that occur between Visakhapatnam and Bheemunipatnam are locally known as *Erra Matti Dibbalu* meaning 'Red Sand Hills' (Srivastava et al., 2016).

Coastal dunes have been studied for, (a) their economic significance as placer sands or industrial raw materials (Chandrasekharan and Murugan, 2001), (b) sediment characteristics (Joseph et al., 1997, 1999, 2002), (c) source of coloration and geochemical changes (e.g., Gardner, 1981,

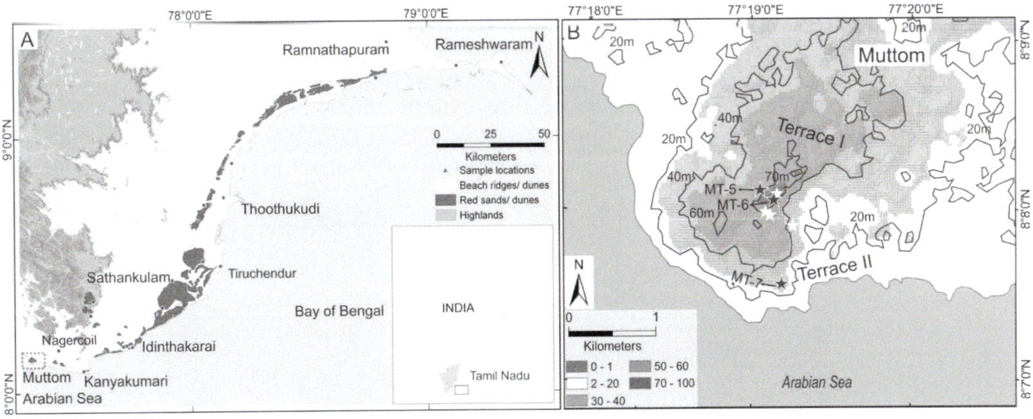

Figure 20.3. Map showing the distribution of the red sands of south India. The elevation details and locations of various section discussed in the text at the Muttom area at the south west coast of Tamil Nadu are shown in the Map (Alappat et al., 2017).

Figure 20.4. a. Gully formation at red dunes of Muttom in the west coast. b. Relict mound/hummock found in the coastal dunes between Thoothukudi and Tirunelveli in the east coast showing aeolian dune with calcrete nodule formation. The hardened Pleistocene stabilized dune surface is exposed due to alluvial and Aeolian erosion in the area. c. ~ 12 m aeolian sand section at Muttom in the west coast.

Gardner and Pye, 1981 and references therein), (d) paleo-climate and paleo monsoon reconstruction (Jayangondaperumal et al., 2012; Joseph and Thrivikramji, 2005; Thrivikramji et al., 2008; Rao et al., 2006; Srivastava et al., 2016; Sathiyaseelan et al., 2021), (e) human migration and inhabitation (Gardner and Martingell, 1990) and (f) ecological characteristics and functions of dunes in supporting flora and fauna (Rodrigues et al., 2011; Sridhar, 2009). Figure 20.5 provides a generalized stratigraphy of coastal deposits overlain by red sands.

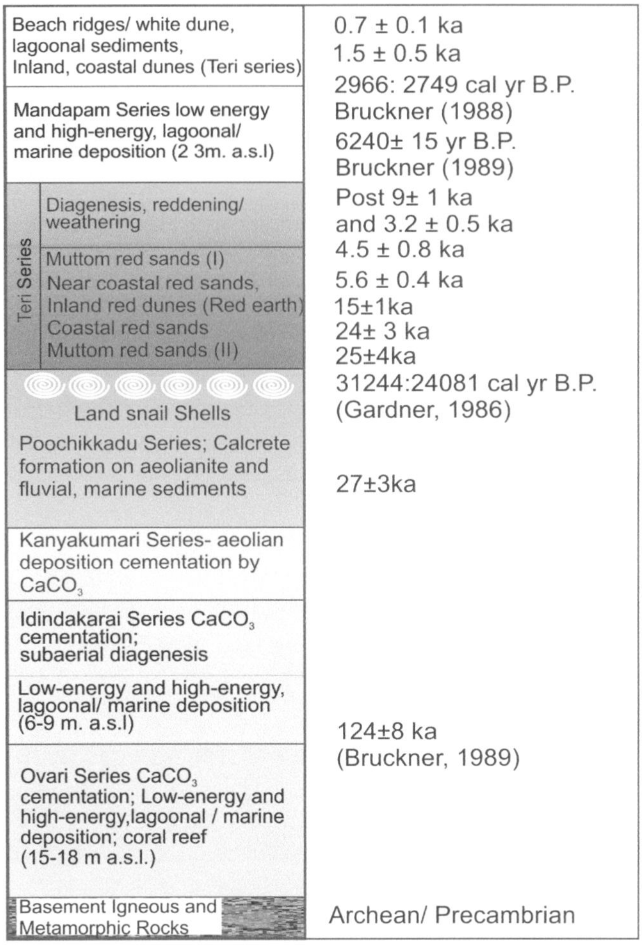

Figure 20.5. Generalized regional stratigraphy of the area (modified from Jayangondaperumal et al., 2012) with additional OSL ages from Alappat et al., 2013, 2017.

The reddening of sands in the coasts of southern Tamil Nadu is due to iron oxides formed by *in-situ* weathering of garnets (Gardner and Pye, 1981), whereas the detrital origin of pedogenic haematites was attributed to the presence of a similar color in sand dunes of Andhra Pradesh (Srivastava et al., 2016). The weathering was *in-situ*, as inferred by the presence of well-formed secondary minerals. The reddening of the dune sands occurs due to progressive loss or leaching of ferromagnesian minerals through hydrolysis (Berner, 1969; Van Houten, 1968; Walker, 1967). Places that experience cyclic precipitation and temperature changes provide beneficial locales, (Folk, 1976; Gardner and Pye, 1981; Pye and Tsoar, 2009) and this can take from a few tens of years to few centuries to form (Alappat et al., 2013; Singhvi et al., 1986). Studies show that the reddening and sand accretion are sensitive to changes in, (i) meteorological parameters, (ii) sea level, (iii) fluvial activity, (iv) agricultural and, (v) anthropogenic activities (Alappat et al., 2013, 2017; Jayangondaperumal et al., 2012; Kunz et al., 2010a, b).

Based on their occurrences, the Teri sands were earlier categorized as inland (Teris), coastal (Teris) and near coastal (Teris) dunes by Joseph et al. (1997, 1999). These categories primarily imply the distance from the coast, age of formation, potential sources of sand and associated genetic process such as sea-level positions and mass wasting. Near coastal dunes are near continuous coast parallel deposits from Kanyakumari to South Rameshwaram of Tamil Nadu. Coastal dunes occur as discontinuous ridges further inland and the thickest deposits are at Kudiramozhi and Sattankulam

(SK 1) in Tamil Nadu and Bhimuni in Andhra Pradesh. Inland dunes occur at the foothills of the Western Ghats as random outcrops (Fig. 20.2).

Based on chronology, heavy mineral content, sediment character and geomorphology, coastal dunes also may be classified as; (i) Near Coastal Dunes (NCD) associated with present shoreline/ littoral and aeolian process, (ii) Coastal Dunes (CD); associated with late-Quaternary coastal aeolian and mid-Holocene shoreline process and, (iii) Inland Dunes (ID); associated with weathering and a mass wasting process and source proximal deposition of sands due to fluvial and aeolian reworking and deposition. Distinct geomorphological features separate the three dune types. The dunes in the east coast are dissected by rivers and streams, debouching into the Bay of Bengal. Discrete separation of near coastal and coastal dunes is visible through discontinuous coast parallel paleo-lagoons and swales. The ancient landscape is bordered by modern coastal dunes in the eastern side of the coast, with a relatively young age of deposition.

2.1 Climate

As per Thornthwaite's climate classification, the study area comprises distinct climate zones. The west coast comprising Kerala, parts of coastal Karnataka, has a humid and moist-humid climate. The east coast at Tamil Nadu and Andhra Pradesh has a dry sub-humid and semi-arid climate. The coastal segment has three seasons, viz., SW monsoon (June to September), NE monsoon (October to January) and fair weather during February to May (Sanil Kumar et al., 2006). Orographic influence of the Western Ghats on SW monsoon winds results in high annual rainfall on the west coast (> 250 cm) and a rain-shadow region on its lee side (< 100 cm) (Fig. 20.1; Ramakrishnan and Vaidyanadhan, 2008).

Kerala and Andhra Pradesh receive high precipitation from the SW monsoon during June to September, and Tamil Nadu receives the most of its rain from NE monsoon during October to December. Cyclonic storms that form in the Bay of Bengal during October-November, also provide considerable rain to the east coast (Ramakrishnan and Vaidyanadhan, 2008).

Coastal areas in Kanyakumari region with geomorphological settings similar to that of Kerala receive an average annual rainfall of 106 cm (average from 2014–2018). In the west coast high wave activity are associated with southerly winds occurs during SW monsoon. Wave activity is high both during NE and SW monsoon along the east coast, mainly due to cyclones. The average wind speed during SW monsoon is 35 km/h (9.7 m^{-sec}) and during the NE monsoon 20 km/h. In the east coast wind speed exceeding 100 km/h (28 m^{-sec}) occur during cyclones (Sanil Kumar et al., 2006). The Rameshwaram coast has a micro-tidal regime with wave height in the range of 0.2–0.8 m during spring and neap tides (Stoddart and Pillai, 1972).

2.2 Geomorphology

The Western and Eastern Ghats are continuous hill ranges along the coast. The Western Ghats act as a topographic barrier and modulate the winds, precipitation and the drainage of rivers. The west coast does not have deltas, but has headlands, bays and lagoons at irregular intervals. Rivers are the major source of sediments to the coasts and these sediments are stored in the gulf, tidal marsh, bays, beaches and dunes (SanilKumar et al., 2006). The coast of Karnataka has a long linear coastline, with stable beaches, estuaries, spits, mudflats, shallow lagoons, islands and sand dunes. Along the coast of Kerala, laterite cliffs, rocky promontories, beaches, estuaries, lagoons, spits, beach ridges, dunes and bars are found (Alappat et al., 2015; Nayak, 2005). Parts of the Kerala and Karnataka coasts also experience high erosion (SanilKumar et al., 2006).

The Eastern Ghats occur as detached hills in Tamil Nadu and as a continuous stretch in Andhra Pradesh (Ramakrishnan and Vaidyanadhan, 2008). The east coast has numerous well-developed deltas with paleo river channels, ancient beach ridges and strandlines (Nayak, 2005). The shoreline of Tamil Nadu is linear in its northern part and is irregular with rocky promontories and cliffs south of Vedaranyam. The average width of the coastal zone is about 50 km, with widest swaths of

125–160 km in the Cauvery delta and narrowest swath of 2 km at Kanyakumari. The coast has two embayments namely the Palk Bay and Gulf of Mannar. River Tambaraparni deposits sediments to the coastal segment between Thoothukudi and Kanyakumari. River Cauvery transports sediment and forms a non-protruding delta between Vedaranyam and Chidambaram. Numerous small and medium rivers supply sediments to the coastal plain.

The deltaic coasts of Andhra Pradesh comprise the Krishna-Godavari Deltas along with lagoons, bays, tidal mudflats, creeks, spits, barrier bars, marshes, swamps, beach ridges, dunes, coastal alluvial plains, etc. (Nayak, 2005). The landscape comprises highlands, fluvial/marine terraces, vegetated and un-vegetated dunes, spits, beach ridges/rocks, barrier islands swales beaches and lagoons, salt marshes, coral reefs, mudflats, wetlands, flood plains, estuaries, rivers, paleo-river channels, delta plains and deltas. Ephemeral rivers, Champavathi and Gosthani that flow through the red sands of Bhimunipatnam originate in the Eastern Ghats and carry large amounts of sediments to the Bay of Bengal. The red sand areas in southern Tamil Nadu and Andhra Pradesh are classified as having badland topography where over 15 m thick sand sections stand exposed due to weathering and gully erosion.

2.3 Geology

The northern part of South peninsular India comprises of archaean Dharwar craton and the southern part has Precambrian charnockites, gneisses, and granites (Chacko et al., 1987, 1996). The Eastern Ghats comprises, charnockites, khondalites and migmatitc gneiss of the Eastern Ghats mobile belt and Proterozoic sedimentary sequences of the Cuddappah basin (Ramakrishnan and Vaidyanadhan, 2008). North east Tamil Nadu has exposures of Mio-Pliocene Cuddalore formation.

The area along the east and west coasts has bed rocks which are overlain by fluvial, fluvio-marine and marine deposits. The red coastal dunes overlie the basement sandstone or the weathered Archaean basement rocks, where as in the near coastal dunes these rest on a discrete calcrete horizon with land-snail shells embedded in them (Fig. 20.3). A continuum of quaternary sediments comprising fluvial, fluvial-marine to marine sedimentary facies are exposed at the coast of Tamil Nadu (Goswami et al., 2019).

The substrate for dunes in some areas are marine terrace formations associated with the last interglacial sea level high stand (MIS-5e, ~ 125 ka) which overlie a crystalline basement (Brückner, 1989). Horizontal to sea-ward dipping grits containing abundant mollusc shells and bivalves occur under the Teri/red dunes in Mandapam and Rameshwaram regions of the east coast (Stoddart and Pillai, 1972).

The beach ridge succession separating ancient lagoons from the modern beach has been reported at Tiruchendur in Tamilnadu (N8°25' E78°03'). Brückener (1989) ascribed it to Holocene transgression dating to 7266–7147 cal yr BP (Tuti-5A). The radiocarbon dates were recalibrated using IntCal20 calibration with OxCal v.4.3.2 (Ramsey, 2009). Coastal dunes reach a height of ~ +25 m above sea level at Rameshwaram relative to the average surface elevation of +3 m in the coastal plains (Stoddart and Pillai, 1972). Raised reefs of the Mandapam and Rameshwaram Islands were radiocarbon dated to 2926–2125 cal year BP and attributed to local emergence (Fig. 20.5; Stoddart and Pillai, 1972).

3. Mineralogy and sediment characteristics

Red dune sands are a rich source of heavy minerals such as ilmenite, magnetite, almandine garnet, hornblende, epidote, sillimanite, leucoxene, kyanite, zircon, rutile, monazite, pyroxenes, amphiboles, etc. (Chandrasekharan and Murugan, 2001; Krishna et al., 2016; Nayak et al., 2012) which are mined. Heavy mineral deposits at Kolachal-Manavalakurichi were the first placer deposits identified in India and have been mined for over a century. The heavy mineral placers are derived through erosion of the rocks in drainage basins of major rivers (Krishna et al., 2016; Mallik et al., 1987; Roonwal, 1997).

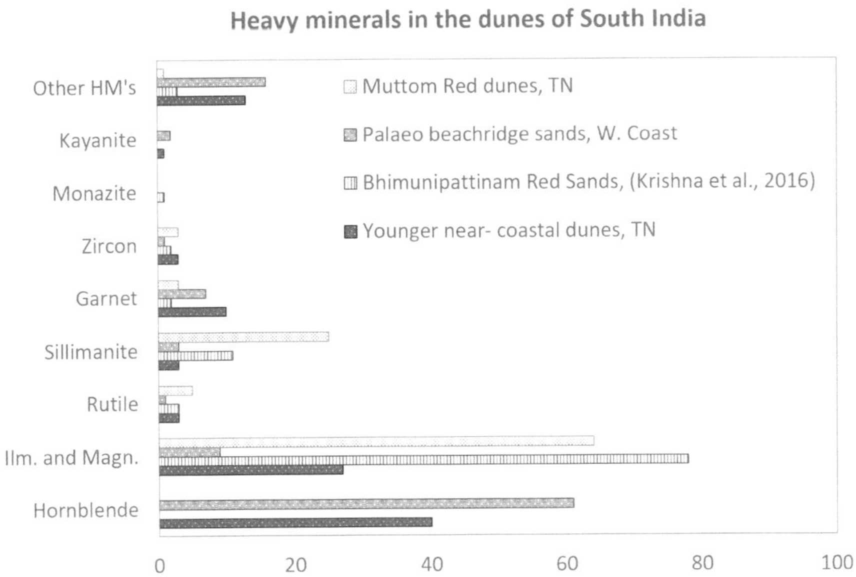

Figure 20.6. Distribution of heavy minerals in the dunes of south India as seen in the near coastal dunes (dunes of Tamil Nadu and beach ridge-dune sequences of Kerala) and red sands of Tamil Nadu and Andhra Pradesh.

The heavy-mineral assemblage of near coastal dune sands (Fig. 20.6) in Tamil Nadu and Kerala comprises amphiboles, (ilmenite, magnetite, clinopyroxenes (augite and diopside), epidote (clinozoisite and zoisite), garnet, accessory amounts of aluminosilicates (sillimanite and kyanite), zircon, rutile, staurolite, monazite, tourmaline and titanite (sphene). The lighter minerals comprise of quartz, feldspar and mica. The red dunes are rich in ilmenite, magnetite, rutile, sillimanite, garnet, zircon, monazite, kayanite, etc. This accumulation suggests that the heavy minerals are sourced from the drainage basin of rivers draining to the coasts and sand grains originate from low grade metamorphic rocks, granites, granite gneiss, charnockites and alluvium are sourced from the Cuddalore sandstone in the east coast (Sundararajan et al., 2009).

Variations in the relative concentrations of heavy minerals in the near coastal dune sands and coastal dunes with small amounts of ilmenite, near complete absence of high specific gravity minerals like monazite and magnetite are attributed to the preferential sorting of sediments by the wind along with a varied provenance. The presence of hornblende in near coastal dunes and its absence in the red coastal dunes is attributed to variations in the source rock and their chemical weathering leading to removal of unstable minerals.

Major sources of sediments in central Kerala and the east coast of Tamil Nadu are hornblende-bearing charnockites and quartzo-feldspathic gneisses with enclaves of mafic granulites, contributing hornblende to the sands of these regions. Red dunes of Tamil Nadu and Andhra Pradesh are mostly derived from Tonalite-Trondhjemite-Granodiorite (TTG) and khondalite (garnet-sillimanite-graphite gneiss) rocks, which are devoid of hornblende (Kar et al., 2003; Ravindra Kumar and Sreejith, 2016; Sreejith and Ravindra Kumar, 2013; Rajesh, 2007). These rocks are a rich source of ilmenite, magnetite and sillimanite. Garnets are mainly removed from the red sands due to chemical weathering leading to its depletion in the heavy mineral suite. Higher concentration of augite, diopside and epidote is attributed to the hornblende-bearing charnockites and quartzo-feldspathic gneisses. A similar dispersal of heavy minerals and preferential segregation was reported north of Nagapattinam between Vellar and Karaikkal (Chandrasekar and Rajamanickam, 2001; Mohan, 1995). Heavy mineral analyses of sediments from dune sand shows that the hinterlands, where the rivers drain their waters are a major source of these sediments. They were later reworked by wind from floodplains or from the beach by the onshore winds.

The mineralogical maturity of sands is related to the chemical weathering of the sediment. X-ray diffraction studies and IR spectral studies on the red dunes of Tamil Nadu at Muttom suggest the dominance of quartz. Older indurated units contain kaolinite and hematite with traces of ilmenite. Un-consolidated sands contained ilmenite and lower amounts of kaolinite. Change in the concentration of ilmenite reflected a preferential sorting during aeolian transport. The presence of quartz as a major fraction in sands indicates a high chemical maturity.

Chemical analyses of sands suggested abundance (in decreasing order) of SiO_2, Al_2O_3, Fe_2O_3 and TiO_2. The SiO_2 concentration is between 77–94%, Fe_2O_3 and TiO_2 content ranges between 1–6% and Al_2O_3 is within 2–11%. Dune sands have a high textural maturity and SiO_2/Al_2O_3 ratio between 7 and 47. Near complete depletion of Mg, Ca, Na and K was also observed. The Chemical Index of Alteration (CIA) of unaltered bedrocks and that of the red dunes were, 38 (mafic rocks) to 59 (granites and gneiss) and between 96 and 98 for white dunes, indicating higher concentration of secondary aluminous clay minerals. These correlate with XRD and IR data and indicates presence of kaolinite in red sands due to *in-situ* weathering of feldspars leading to incorporation of clay into dune sands.

Grain size analyses of sand samples from different sections from coastal and near coastal dunes show no systematic variation in the sand size. The grain size ranges between 275 to 313 μm (medium sand) for near coastal sand in the east coast. The eastern most dunes comprise moderately well-sorted fine-grained sands (182–215 μm). The west coast has moderately to well-sorted medium sand, with mean grain size between ~ 373 and 468 μm. The grain size for red dunes at different locations has a varied distribution of coarse-to-fine sand ranging between 700 to 215 μm with moderate sorting. Varied distribution of grain size indicates contemporary energy conditions during their deposition, irrespective of location and geographic settings.

Scanning electron microscopy of the surface texture of quartz grain indicated that the older coastal and inland dune sands comprised angular to sub-angular grains, with low abrasion. Younger near coastal sands from the fore dunes closer to the coast had a larger number of well-rounded grains and higher abrasion indicating a long-distance transport. The surface textures of quartz grains from near coastal sands had crescentic-shaped marks, drag marks and v-shaped pits, possibly due to shore transport under sub-aerial or subaqueous conditions. The original mechanical surface of sand grains had silica precipitation and solution etching pits. The drag marks/scratches on the grains of near coastal dune suggest grain to grain or, grain to bed collision (Williams and Thomas, 1985), confirming the input of sand from the onshore region through landward migration. Crescentic marks and moderately abraded edges suggested multiple cycle of aeolian reworking and transportation.

3.1 Chronology and paleo-environmental significance

Reddening: The process of reddening occurs over short times under favorable conditions of rainfall and temperature facilitating *in-situ* weathering and alteration of iron-bearing minerals (Gardner, 1981, 1983; Gardner and Pye, 1981; Alappat et al., 2017). Thin section studies and the presence of well-formed clay minerals suggested that the reddening occurred *in-situ* during post depositional digenetic alteration (e.g., Berner, 1969; Gardner, 1981) or by the presence of detrital sand and clay during deposition (e.g., Bensing et al., 2005; Besler, 2008; Srivastava et al., 2016). Mineral magnetism of red sands of Andhra Pradesh, suggest that reddening of sands was due to the deposition of paedogenic crystalline hematite and dissolved iron on sand and clay from detrital source transported from red soils of the hinterland by fluvial processes (Srivastava et al., 2016).

Occurrence of red sands in the coastal plain and in the foot hills of the Eastern Ghats in the hinterland (inland dunes and coastal dunes) show that the formation of red dunes is controlled by the geology of the source area, sand supply and climate, rather than their geographic position in the coastal plain. Movements of the coastline during the Quaternary due to sea level changes were considered as a reason for occurrence of coastal aeolianite deposits around the world (Brooke, 2001). *In-situ* weathering of feldspar, amphibole, pyroxene and iron rich garnet led to the formation of clay-sized aggregates and lateritic coating on the surface of quartz sand grains. The formation of coastal and near coastal dunes were initially controlled by the sea-level, as the littoral area ensures steady sand supply and strong, consistent onshore winds facilitates the migration of foredunes in the backshore region to inland. When the shoreline is pro-graded, the reworking of earlier formed beach deposits acted as the source for sand in coastal dunes.

Chronology: The chronology of coastal dunes was developed using radiocarbon and OSL dating (Alappat et al., 2013, 2017; Gardner, 1981; Jayagondaperumal et al., 2012; Joseph and Thrivikramji, 2005; Kunz et al., 2010 a,b; Srivastava et al., 2016; Sathiyaseelan et al., 2021). Dunes of southern India show two distinct occurrences viz., (a) stratigraphically older aeolian deposits as red sands at the coasts of southern Tamil Nadu at Thoothukudi, Tirunelveli and Kanyakumari region and at the coasts of Andhra Pradesh, between Visakhapatnam and Bheemunipatnam and, (b) sand deposits between the mid-Holocene paleoshore line inland and present-day shoreline, largely as light yellow, gray aeolian sand deposits. The reported OSL ages of red dunes and near coastal dunes are collated in Table 20.1.

Initially, the intensity of color of reddened sands (Redness index) was correlated with age (Norris, 1969; Turner, 1980; Walker, 1979), but it was later suggested that such a correlation did not exist as reddening occurred rather rapidly, (e.g., Roskin et al., 2012). The presence of well-formed clay minerals, led Gardner to suggest that the reddening was *in-situ* and this was used in developing the methodology for luminescence dating, where it was assumed that the grains were transported as uncoated grains were totally bleached at the time of burial (Singhvi et al., 1986) and that reddening occurred later. It was also argued that co-precipitation of uranium with iron during reddening did not affect the dosimetry to any significant extent. The bulk of the chronological data are now accrued from luminescence ages. In addition, radiocarbon ages of two aragonite land snail shells collected below the red sands unit from a section in the east coast had given ages between 31–29 ka Cal BP and 26–24 ka Cal BP (Gardner, 1981). However, apart from the inherent difficulties of radiocarbon dating, lack of well-preserved organic materials in the stratigraphic context makes it challenging to apply radiocarbon dating for finding the stages of accumulation at larger parts of the coastal dunes.

Coastal sand dunes are reported along the coast of South India (Rodrigues et al., 2011), however, the chronology of formation from the coasts of Karnataka is not available. Dune formation episodes from the coasts of Kerala, Tamil Nadu and Andhra Pradesh are reviewed below (Fig. 20.9).

Dune sands at Tamil Nadu accreted distinct periods of deposition; (a) the weathered reddened dune sands at inland, coastal and near coastal settings during 19 ± 2 ka to 0.5 ± 0.1 ka; and (b) the un-weathered near coastal longitudinal dunes during 5.3 ± 0.4 ka to 0.02 ± 0.001 ka (Table 20.1a, b; Fig. 20.5). A layer of coarser calcareous sand unit was reported from Sathiyaseelan et al., 2021, having a higher antiquity of 27 ± 2.3 ka. Since the proximity of the section to the River Vaiga and distinct hiatus with differences in the lithology of the unit, the depositional facies of the sand unit need to be ascribed. Along the coast of Tamil Nadu, at an elevated promontory at Muttom, NW of Kanyakumari, a stacked record of sediments has been preserved (Fig. 20.7; Alappat et al., 2017). Section MT-7 close to the modern coast had an age of 4.5 ± 0.4 ka (Fig. 20.7). Section MT-4 further inland to MT-7 accreted around 14 ± 1 ka. This sample was collected from the bed rock-aeolian sand contact. Samples further inland had an age of 19 ± 2 ka (Vidyasakar et al., 2017). The OSL ages at Muttom (Fig. 20.7) show a spatio-temporal variation in the deposition of sand, indicating a topographically limited supply and laterally discontinuous record of sediment accumulation. Un-weathered near coastal dunes are separated from the rest of the coastal plain by buried paleo beach ridges and lagoons. Un-published results from buried beach deposits showed an age of 5–6 ka (Fig. 20.8).

The dunes of Kerala were inactive during the last millennium and a widespread aeolian deposition occurred during 4.7 ± 0.6 ka and 2.3 ± 0.3 ka (Table 20.1a; Fig. 20.9). The dunes were associated with the progradation of the coast since the mid-Holocene transgression and were formed

Table 20.1a. List of OSL ages, sample code, description and reference of ages reported from the coastal dunes of South India (the details of near coastal dunes).

OSL ages of Near Coastal Dunes, from the coasts of Kerala and Tamil Nadu				
Sl. No.	**Sample ID**	**Age (yrs)**	**Depositional Settings**	**Reference**
1	CUD 3-A	55 ± 5	Longitudinal dunes of 6 m high, located 300 m inland, between late Holocene strandlines and present days coastline.	Kunz et al., 2010b
2	CUD 3-B	94 ± 6		Kunz et al., 2010b
3	CUD 3-C	105 ± 8		Kunz et al., 2010b
4	CUD 3-D	128 ± 11		Kunz et al., 2010b
5	CUD 3-E	171 ± 15		Kunz et al., 2010b
6	CUD 4-A	16 ± 1	Isolated single elongated dunes having length of 35 m and height of 7 m, located 2 km inland, between inland dunes and present days coastline.	Kunz et al., 2010b
7	CUD 4-B	27 ± 2		Kunz et al., 2010b
8	CUD 4-C	26 ± 2		Kunz et al., 2010b
9	CUD 4-D	36 ± 3		Kunz et al., 2010b
10	CUD 4-E	32 ± 3		Kunz et al., 2010b
11	CUD 4-F	99 ± 7		Kunz et al., 2010b
12	CUD 4-G	144 ± 8		Kunz et al., 2010b
13	CUD 4-H	172 ± 12		Kunz et al., 2010b
14	CUD 5-A	134 ± 9	Elongated dunes of 5m high, located 2 km inland, between inland dunes and present days coastline.	Kunz et al., 2010b
15	CUD 5-B	177 ± 13		Kunz et al., 2010b
16	CUD 5-C	160 ± 12		Kunz et al., 2010b
17	CUD 5-D	160 ± 12		Kunz et al., 2010b
18	CUD 5-E	144 ± 11		Kunz et al., 2010b
19	CUD 5-F	163 ± 13		Kunz et al., 2010b
20	CUD 1-E	61 ± 5	Longitudinal dunes of 13m high, located 5 km inland, between late Holocene strandlines and present days coastline.	Kunz et al., 2010a
21	CUD 1-D	206 ± 16		Kunz et al., 2010a
22	CUD 1-C	198 ± 18		Kunz et al., 2010a
23	CUD 1-B	233 ± 21		Kunz et al., 2010a
24	CUD 1-A	772 ± 72		Kunz et al., 2010a
25	CUD 2 P1-D	73 ± 5	Longitudinal dunes located 5 km inland having 200 cm thick soil-like horizon with potsherds and brick fragments.	Kunz et al., 2010a
26	CUD 2 P1-C	339 ± 27		Kunz et al., 2010a
27	CUD 2 P1-B	707 ± 60		Kunz et al., 2010a
28	CUD 2 P1-A	1244 ± 113		Kunz et al., 2010a
29	CUD 2 P1-AA	3291 ± 303		Kunz et al., 2010a
30	CUD 2 P2-B	336 ± 24		Kunz et al., 2010a
31	CUD 2 P2-A	717 ± 62		Kunz et al., 2010a
32	VDB-1A	4 ± 1	Close to modern coast from the backshore region at Vedaranniyam, Tamil Nadu.	Alappat et al., 2011
33	VDC-1 A	5 ± 1	Fore dune located 300 m away from the modern shoreline at the southern part of Cauvery delta, Tamil Nadu.	Alappat et al., 2011
34	VDC-1 B	6 ± 1		Alappat et al., 2011
35	VDC-1 C	9 ± 1		Alappat et al., 2011
36	VDC-1 D	9 ± 1		Alappat et al., 2011
37	VDC-1 E	17 ± 4		Alappat et al., 2011

Table 20.1a contd. ...

...Table 20.1a contd.

OSL ages of Near Coastal Dunes, from the coasts of Kerala and Tamil Nadu				
Sl. No.	**Sample ID**	**Age (yrs)**	**Depositional Settings**	**Reference**
38	TKN-1A	73 ± 8	Linear, massive dune, 3 km inland, southern part of Cauvery delta, Tamil Nadu.	Alappat et al., 2011
39	TKN-1B	73 ± 8		Alappat et al., 2011
40	TKN-1C	1182 ± 132		Alappat et al., 2011
41	TKN-2 A	94 ± 18		Alappat et al., 2011
42	TKN-2 B	194 ± 28		Alappat et al., 2011
43	TKN-2 C	44 ± 5		Alappat et al., 2011
44	TKN-2 D	55 ± 6		Alappat et al., 2011
45	TKN-2 E	298 ± 32		Alappat et al., 2011
46	SBD-1 A	77 ± 8		Alappat et al., 2011
47	SBD-1 B	89 ± 10		Alappat et al., 2011
48	SBD-2 A	43 ± 5		Alappat et al., 2011
49	SBD-2 B	39 ± 4		Alappat et al., 2011
50	SBD-2 C	79 ± 8		Alappat et al., 2011
51	SBD-3 A	49 ± 5		Alappat et al., 2011
52	SBD-3 B	47 ± 5		Alappat et al., 2011
53	SBD-3 C	80 ± 8		Alappat et al., 2011
54	KD-1 A	624 ± 65	Kodiyakkadu, Tamil Nadu from a curvilinear dune/ridge aligned parallel to the coastline	Alappat et al., 2011
55	PV-1 A	227 ± 28	Sand sheet, 2 km inland, southern part of Cauvery delta, Tamil Nadu.	Alappat et al., 2011
56	PV-1 B	32 ± 3		Alappat et al., 2011
57	PP-1 A	62 ± 6	Linear dune, 1 km inland, southern part of Cauvery delta, Tamil Nadu with sparse or no vegetation.	Alappat et al., 2011
58	PP-1 B	97 ± 10		Alappat et al., 2011
59	PP-1 C	95 ± 10		Alappat et al., 2011
60	KNP-1A	1.3 ± 0.2	Near coastal dunes. Brownish gray aeolian sandy ridge located ~ 3 km inland, adjacent to beach ridge.	Alappat et al., 2015
61	KNP-1B	4.1 ± 0.5		Alappat et al., 2015
62	CHK-1A	3.2 ± 0.3	Near coastal dunes. Massive, light yellow fine to medium sand sheet located ~ 3 km inland.	Alappat et al., 2015
63	EDT-1A	3.3 ± 0.4	Near coastal dunes. Gray fine to medium sand dune located ~ 3 km inland	Alappat et al., 2015
64	EDT-1B	4.7 ± 0.6		Alappat et al., 2015
65	CKP(N)-1A	0.6 ± 0.1	Near coastal dunes. Sand dune located ~ 3 km inland	Alappat et al., 2015
66	CKP(N)-1B	2.3 ± 0.3		Alappat et al., 2015
67	CKP(N)-1C	3.1 ± 0.3		Alappat et al., 2015
68	CKP(N)-1D	3.2 ± 0.3		Alappat et al., 2015
69	SNC-1A	4.5 ± 0.5	Near coastal dunes, adjacent to beach ridge, located 2.7 km inland	Alappat et al., 2015
70	KHL-1A	4.2 ± 0.5	Near coastal dunes, located near Paravoor, western side of Vembanadu Lake.	Alappat et al., 2015

Table 20.1b. List of OSL ages, sample code, description and reference of ages reported from the coastal dunes of South India (the details of red sands).

Red Dunes of Tamil Nadu and Andhra Pradesh.				
Sl. No.	**Sample ID**	**Age (ka)**	**Depositional Settings**	**Reference**
1	MP1	15.4 ± 1.0	Inland fluvial Teri, foothills of Western Ghats, Tamil Nadu.	Jayangondaperumal et al., 2012
2	TT1	11.4 ± 0.9	Red coastal Teri deposits, Tamil Nadu, moderate development of soil with rootlets.	Jayangondaperumal et al., 2012
3	P1	4.1 ± 0.4	Red coastal Teri deposits, Tamil Nadu. NE of Karamaniyar River, compact sand.	Jayangondaperumal et al., 2012
4	P2	2.3 ± 0.2		Jayangondaperumal et al., 2012
5	KV1	0.1±0.03	Red coastal Teri deposits, inter-dunal depression Kudiramoli Teri, Tamil Nadu.	Jayangondaperumal et al., 2012
6	KM1	5.6 ± 0.4	Near-coastal Teri dunes, Tamil Nadu, immediately above the calcrete unit that forms the base.	Jayangondaperumal et al., 2012
7	KLS1	4.2 ± 0.4	Near-coastal Teri dunes, Tamil Nadu, in a matrix of aragonite land snail shells horizon, to compare the 14C age and the OSL age.	Jayangondaperumal et al., 2012
8	G1	0.032 ± 0.02	Near-coastal Teri dunes, Tamil Nadu, located close to the modern coast.	Jayangondaperumal et al., 2012
9	B1	0.13 ± 0.009		Jayangondaperumal et al., 2012
10	B2	0.22 ± 0.02		Jayangondaperumal et al., 2012
11	MUT-1	14 ± 2	Red sands. MT-I, II at the NW side of the dune on Terrace I at ~ 30–75 m above msl and one km inland, at Muttam, Tamil Nadu	Vidyasakar et al. 2017
12	MUT-2	17 ± 2		Vidyasakar et al. 2017
13	MUT-3	19 ± 2		Vidyasakar et al. 2017
14	LST-2	16 ± 2		Alappat et al., 2013
15	LST-3	9 ± 1		Alappat et al., 2013
16	LST-4	12 ± 1		Alappat et al., 2013
17	LST-14	501 ± 51 a	Red sands. MT-III, upper part, loose massive orange red fine sand, at Muttam, Tamil Nadu	Alappat et al., 2013
18	LST-15	12 ± 1	Red sands. MT-III, a 8 m section of indurated sand with clay, at Muttam, Tamil Nadu.	Alappat et al., 2013
19	LST-16	15 ± 2		Alappat et al., 2013
20	LST-17	12 ± 1	Red sands. MT- IV, a ~ 4 m thick unit of brownish red, consolidated medium sand at the SE of the upper terrace underlain by weathered country rock at the base, at Muttam, Tamil Nadu.	Alappat et al., 2013
21	LST-18	14 ± 1		Alappat et al., 2013
22	LST-5	104 ± 16 a	Red sands. Inland dunes (VK-1, PG-1) at the foot hills of Western Ghats at Tamil Nadu	Alappat et al., 2013
23	LST-6	170 ± 24 a		Alappat et al., 2013
24	LST-7	282 ± 38 a		Alappat et al., 2013
25	LST-8	2 ± 0		Alappat et al., 2013
26	LST-9	6 ± 1		Alappat et al., 2013
27	LST-10	233 ± 34 a		Alappat et al., 2013
28	LST-11	119 ± 18 a		Alappat et al., 2013
29	LST-12	697 ± 89 a		Alappat et al., 2013

Table 20.1b contd. ...

Table 20.1b contd.

Red Dunes of Tamil Nadu and Andhra Pradesh.				
Sl. No.	**Sample ID**	**Age (ka)**	**Depositional Settings**	**Reference**
30	LST-20	6 ± 1	Red sands. Coastal dunes from the Sattankulam Teri (SK-1, 2), Tamil Nadu	Alappat et al., 2013
31	LST-21	8 ± 1		Alappat et al., 2013
32	LST-22	8 ± 1		Alappat et al., 2013
33	LST-23	9 ± 1		Alappat et al., 2013
34	LST-24	3 ± 0		Alappat et al., 2013
35	LST-25	6 ± 1		Alappat et al., 2013
36	MT-5A	0.12 ± 0.01	Red sands. Terrace I (MT-5 and 6) 30–75 m above MSL), 1 km from the coast at Muttam, Tamil Nadu	Alappat et al., 2017
37	MT-5B	13.6 ± 1.6		Alappat et al., 2017
38	MT-6A	13.2 ± 1.4		Alappat et al., 2017
39	MT-6B	15.9 ± 2.1		Alappat et al., 2017
40	MT-7A	0.14 ± 0.02	Red sands. Terrace II, ~ 300 m to the coast and located at 20–30 m above MSL, at Muttam, Tamil Nadu	Alappat et al., 2017
41	MT-7B	0.19 ± 0.02		Alappat et al., 2017
42	MT-7C	0.36 ± 0.04		Alappat et al., 2017
43	MT-7D	4.5 ± 0.5		Alappat et al., 2017
44	IMD/OSL/1	0.85 ± 0.1	Bheemuni red sand deposits, Andhra Pradesh, ~ 2 to 5 km inland from shoreline, > 15 m in height from MSL, dominated by dune sand.	Srivastava et. al., 2016
45	IMD/OSL/2	29.3 ± 3.5		Srivastava et. al., 2016
46	IMD/OSL/3	2.2 ± 0.1		Srivastava et. al., 2016
47	SOS/OSL/1	12.1 ± 0.3		Srivastava et. al., 2016
48	SOS/OSL/2	48.9 ± 1.7		Srivastava et. al., 2016
49	SOS/OSL/3	2.02 ± 0.02		Srivastava et. al., 2016

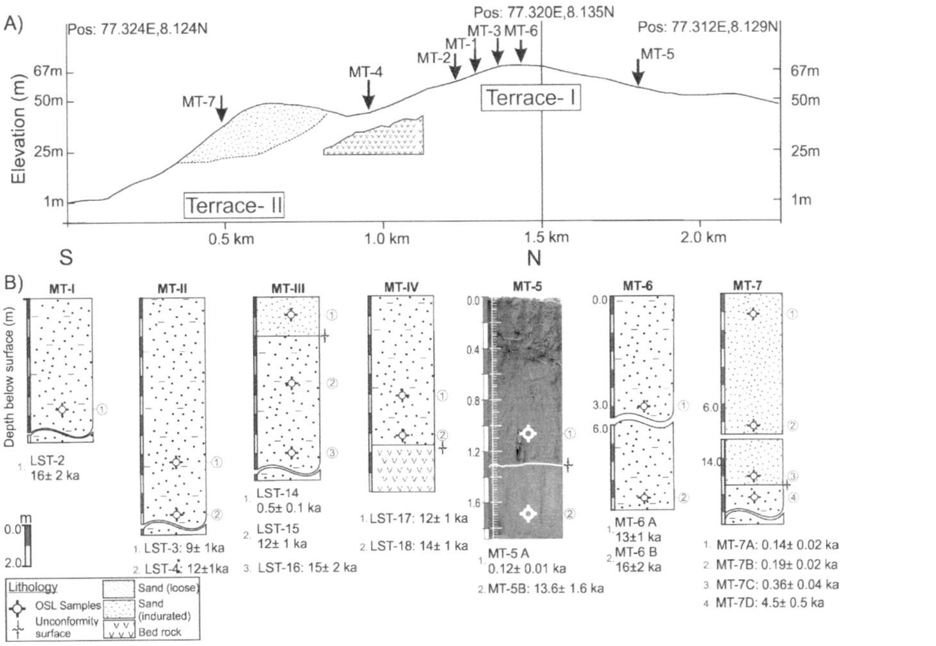

Figure 20.7. a. Illustrations showing the disposition and elevation of the dune field in the SW coast of Tamil Nadu at Muttam. b. Shows the enlarged view of the central part of the dune field with stratigraphic position of the samples and ages from sections MT-1 to 4 (Alappat et al., 2013 and 2017).

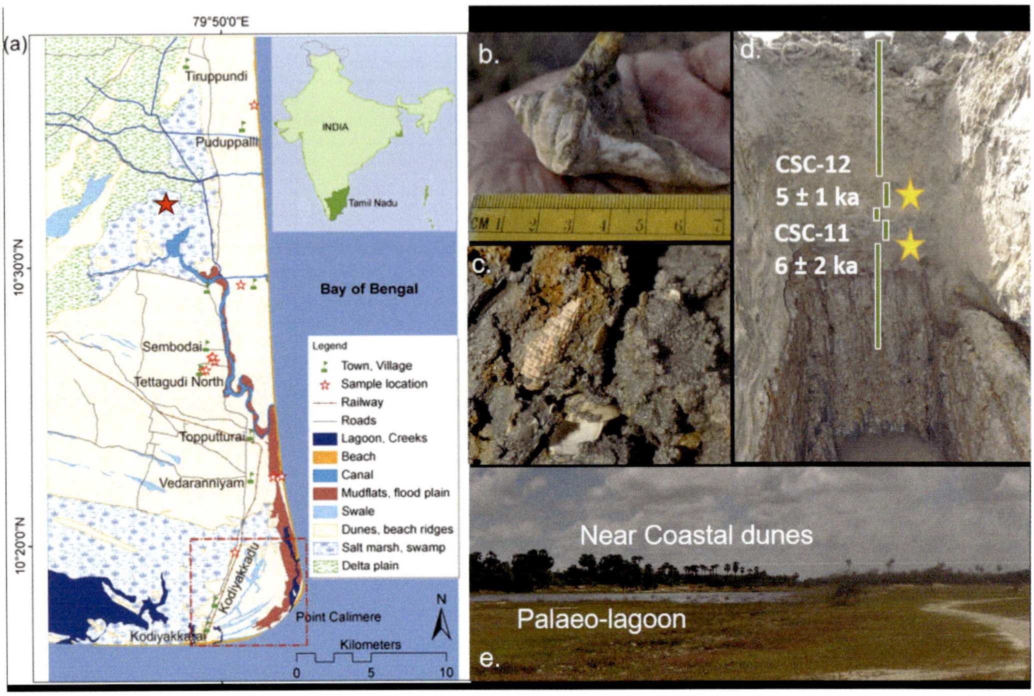

Figure 20.8. Figure showing the details of the paleo-lagoon at Kiranoor Lake, near Thirupoondi at the coast of Tamil Nadu. Fig. 20.8.a. show the map with sample locations of near coastal dunes and trench location at the paleo lagoon at Kiranoor Lake. Fig. 20.8b, c shows the marginal marine mollusc shells present in the sediments from the excavated trench, Fig. 20.8d. Photograph showing the different stratigraphic layers of excavated trench at Kiranoor Lake. Fig. 20.8e. Photograph showing the positions of near coastal dunes adjacent to the paleo-lagoon. The dunes are located at the eastern side of the lake.

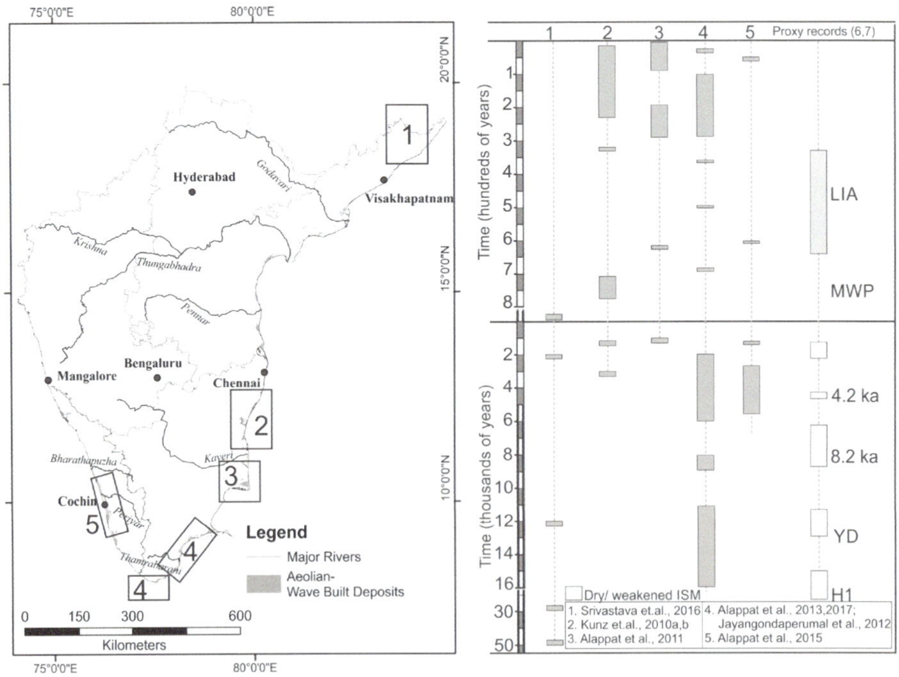

Figure 20.9. Distribution of various sand dune sequences studied in south India with their geographic position and deposition ages are shown. The important climatic events reported by (Dutt et al., 2015; Rawat et al., 2015) are shown in the diagram to correlate with periods of reduced Indian summer monsoon and sand dune accretion.

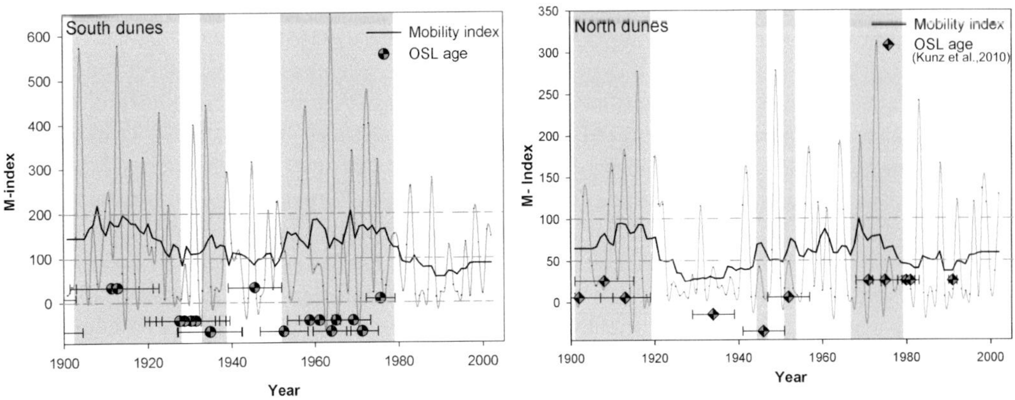

Figure 20.10. Lancaster's sand mobility index (M-index) calculated for the last century for the coast of Tamil Nadu along with the measured OSL ages (Alappat et al., 2011). Darker line shows the running average of mobility values and the lighter line in the background depicts the annual variation in the mobility index. The observed correlation with enhanced mobility values and OSL ages are highlighted with grey background. (a) Sand mobility plot for the southern dunes (Nagapattinam) and (b) for Northern dunes (Cuddalore region) Kunz et al. (2010a,b).

by aeolian reworking of beach ridges. The depositional chronology of sand deposition at the coast of Andhra Pradesh ranges from 48.9 ± 1.7 ka to 0.85 ± 0.1 ka (Fig. 20.9).

Older records of activity of sand mobilization were reported from the reddened sands of Tamil Nadu and Andhra Pradesh. The studies show that the east coast dunes of Tamil Nadu were largely mobilized/active during the last millennium (Figs. 20.9, 20.10).

Red Sands

Stratigraphically, the red dunes in the region are older with depositional ages ranging from 48.9 ± 1.7 ka to the present. At the coasts of Andhra Pradesh, basal age of red sands is 48.9 ± 1.7 ka and the aeolian accumulation continued till 12.1 ± 0.3 ka (Srivastava et al., 2016). The TL dates of red sands of Sri Lanka in similar geomorphological conditions were 74.2 ka (Singhvi et al., 1986).

Given the stratigraphic relations of red sands having accreted during periods of low sea level stands and the geomorphologic similarities of deposition at Andhra Pradesh and Sri Lanka, the red sands of Tamil Nadu may also possess sands of comparable antiquity. It is also important to distinguish facies associations of the red beds of these areas, as often the weathering obscures the stratigraphic signatures of depositional process. For the paleoclimatic interpretation of the depositional ages of the red sands, the environment of deposition is critical for correlation with other records.

The OSL ages on the upper part of the red sands at the Muttom section shows recent mobilization of the reworked sand in the area during the last 500 years (Fig. 20.7). Such dune reactivation was also observed at Bheemuni red sand deposits of Andhra Pradesh with age of deposition of 0.85 ± 0.1 ka (Srivastava et al., 2016). The current badland topography that the area possesses could have begun after 9 ka at the coasts of Tamil Nadu. Mineral magnetic susceptibility studies on the red dune sands of the period suggest enhanced precipitation and digenesis around 14 ka (Srivastava et al., 2016; Vidyasagar et al., 2017)

It has been suggested that the preservation of aeolian record in dry land occurs during the transitional climate from dry to wet (Singhvi and Porat, 2008). The mineral magnetic susceptibility studies on red sands of Bhimunipattanm, Andhra Pradesh show higher amounts of pedogenic magnetic minerals, suggesting warm and wet climatic conditions during the Holocene (Srivastava et al., 2016). The impact of monsoon precipitation on sand accretion in the area was described using earlier proxy studies which suggest an intensified precipitation during 14 ka to 7 ka (Bera et al., 1995, 1997; Tiwari et al., 2011). Kumaran et al. (2014) used the presence of buried fossil woods and swamp-marsh along the coast of Kerala and suggested a phase of heavy rainfall and warm and humid climatic condition during late Pleistocene to early Holocene.

Enhanced precipitation during this period facilitated the chemical weathering of the mineral constituents, leading to reddening and induration of sands and thereby their stabilization. Similar observations were reported from the red sands of Andhra Pradesh by Srivastava et al. (2016). Evidence for higher sea-level stands during the Holocene between 5–9 ka was reported both along the east and west coasts (Alappat et al., 2015; Banerjee, 2000; Bruckner, 1989; Goswami et al., 2019; Maya et al., 2017; Padmalal et al., 2014). It is likely that the sand supply was attenuated by increasing sea-level. The formation of near coastal dunes at the lower level closer to the present-day shoreline is dated at 4.5 ka and this was correlated with transgression (Alappat et al., 2017). Comparable ages were obtained for the dune formation in the east and west coasts (Alappat et al., 2015; Kunz et al., 2010a, b).

The coastal dune sands between Kanyakumari and Tiruchendur had depositional ages in the range 8 ± 1 to 3 ± 0.4 ka (Alappat et al., 2013). The near coastal red dune at Muttom region has a massive 15 m thick sand deposit formed during the last 400 years (Alappat et al., 2017). The ages are comparable with the ages reported by (Jayangondaperumal et al., 2012) showing a wide spread, discontinuous dune activity in the east coast, an area which is dominated by the NE monsoon. The dunes at the foot hills of the Western Ghats were formed by source proximal sands of mass wasting (Joseph et al., 1997) and were deposited fluvially around ~ 6 ka. A distinct stratigraphic contact in the section separating the middle Holocene sands from the late Holocene younger deposits can be correlated with the weakened SW monsoon in this region (Alappat et al., 2013). Dune accretion which started at 1.524 ± 0.18 ka continued till the last century with breaks in deposition. Fluvial red sand unit overlying the sand stone basement was dated to 15.4 ± 1.0 ka by Jayangondaperumal et al. (2012) displayed the onset of dune formation in this area at least during this period.

3.1.1 Near coastal dunes

This category of dunes largely comprise formation during the Holocene epoch, largely within a geomorphic zone between early-mid Holocene to the present-day day shoreline. The source of sands for the dune formation was attributed to the aeolian reworking of paleo-strand lines/sediment associated with Holocene sea-level high stand along the coasts (Rao, 1982). A section at Kiranur Lake (10°38'5.4N 79°49'45.0E; Fig. 20.3), north Vedaranniyam forms a cuspate foreland with a rapid rate of coastline progradation in this area (Fig. 20.8). The sediments in this paleo-lake ~ 3 km inland to the present-day coast have a Holocene antiquity comprising marginal marine to lacustrine sediment units with preserved fossil shells. Early-mid Holocene transgressive, estuarine deposits are overlain by shallow marine clay, followed by a succession of beach barriers and lagoonal deposits of a higher sea level. The upper part of the sediments shows weakly pedogenized sandy clay horizon. The section also comprised numerous well-preserved marginal marine mollusc shells (Fig. 20.8b,c). Two OSL samples collected from the littoral facies in the upper unit of the sediments provided an age of deposition during 5.0 ± 1 ka and 6 ± 2 ka (Fig. 20.8d). The ages of sand dunes east of this lake (Fig. 20.8e) will be described later. A sand dune section sampled from the inner side of the beach ridge complex at the south of Vaigai River, near Ramanathapuram had given an age of sand accretion above the calcareous coarse sand unit between 5.3 ± 0.4 and 0.92 ± 0.1 ka (Sathiyaseelan et al., 2021). Radiocarbon age of samples collected near the Vedaranyam region provided radiocarbon ages of 3570 ± 205 years cal BP and 1020 ± 80 years Cal B.P. (Ramasamy et al., 1998). Sand beach ridges about 1.5 km in the Godavari Delta region with marine shells was reported by (Banerjee, 2000). This along with the OSL dating of beach rocks from the coast of Tamil Nadu between Kanyakumari and Rameswaram showed depositional age of 4.2 to 2.4 ka indicating a higher sea-level stand during that time (Thomas, 2009).

The ages of littoral sediments of paleo-lagoon, along with other reported ages provide evolutionary chronology of landforms in the area. This lacustrine/littoral sequence of sedimentation was overlain by modern dune deposits in the coasts of Tamil Nadu. Similarly, the wetland succession in the west coast that borders the coastal dunes at central Kerala had shown age of deposition of 7 ± 1 ka (Alappat et al., submitted). Littoral sands of paleo-beach ridges of central Kerala were also found to be overlain by aeolian deposits (Alappat et al., 2015). The distribution of sand dunes in the coastal area indicates a spatial variation of near coastal dune deposits in association with the early-mid Holocene coastal progradation.

The chronology of dunes ranges between 4.7 ± 0.6 ka to 4 ± 1 years (Alappat et al., 2015, 2011). Near coastal dunes at Kerala and Tamil Nadu coasts suggest reactivation of near coastal dunes during the late Holocene to the present (Alappat et al., 2011, 2013, 2105, 2017; Kunz et al., 2010a,b) (Figs. 20.9, 20.10). The oldest dune sands of eastern coastal dunes are 3.3 ka (Kunz et al., 2010) at Cuddalore and 1.2 ka at Nagapattinam region (Alappat et al., 2011). The reactivation of dunes during the past two centuries has been attributed to change in meteorological conditions and land use associated with anthropogenic activities, which enhanced sand mobility. Rapid accumulation of sand up to ~ 3 m within a decade or so was reported in the coastal dunes of Nagapattinam and Cuddalore (Alappat et al., 2011; Kunz et al., 2010a, b; Fig. 20.10). Apart from the sand supply and favorable meteorological conditions, the beach morphology like a wider shelf, shallow sea and onshore winds towards the land contribute to the formation of coastal dunes (Hesp, 2012; Sherman and Lyons, 1994). The influence of human inhabitation along the coast of the Mediterranean Sea at Israel over the past 200 years was attributed to enhanced sand mobility due to the removal of vegetation (Bar, 2013). The fore dunes close to the modern beach provided a depositional age of ~ 20 years, indicating active dune formation in the coastal front, controlled by high energy wave events and later migrating landward due to onshore winds.

The dune mobility index (Lancaster, 1988) for coastal dunes of Tamil Nadu suggested that the periods of sand mobility in the area was correlated with periods of lower precipitation (Fig. 20.10; Alappat et al., 2011). Climate and anthropogenic activities in the form of agriculture, grazing and leveling control the sand mobility and dune reactivation. Control of climate is reflected through vegetation cover, which is in turn related to precipitation and evaporation. Wind energy also causes changes in dune stability through the Bagnold's threshold velocity for the wind to move a grain. Lancaster's mobility index (M-index; Lancaster, 1988), used the ratio of time when wind blowing above the threshold velocity to the ratio of precipitation to potential evapotranspiration. Lancaster (1988) suggested that, dunes are fully active for $M > 200$, stabilized inter dunes and active flanks occur for values of $200 < M > 100$, crests are active under $100 < M > 50$ and dunes are inactive and vegetated when $M < 50$. The mobility index includes the effect of wind strength and dune surface moisture on sand grain movement. Deposition of modern dunes of Tamil Nadu, show stages of partially (M-value between 100 and 200) or fully active phase (M values > 200) in the Nagapattinam region and dune crests were active during the last century in the Cuddalore region (Alappat et al., 2011). The area is used for cultivation of paddy, groundnut and other vegetables in the inter-dune depressions and has been leveled for settlements or plantations such as coconut, cashew and casuarina, etc. (Fig. 20.2).

The chronology of accumulation of sand in different sections described in this chapter along with published paleoclimatic reconstruction (Bera et al., 1995, 1997; Dutt et al., 2015; Rajagopalan et al., 1997; Rawat et al., 2015; Sukumar et al., 1993; Tiwari et al., 2011) is shown in Fig. 20.9. No systematic coherence of sand mobilization/accretion at different sites described was observed in correlation with reported proxy records of aridity from other parts of the country (Fig. 20.9). This indicates the local control of dune formation, such as sand availability, dune surface erodibility along with global climatic changes. Large scale accumulation of fluvial deposits inland during flash floods and later dune formation due to aeolian reworking in the Hagari River, Anantapur, district of Andhra Pradesh was reported by (Kale et al., 2020), suggesting the control of sediment availability in the formation of sand dunes. Anthropogenic large-scale extraction of ground water led to the lowering of the water table and also to increased sediment supply. Control of regional precipitation may exist but this would need larger data for statistically significant inferences. The sensitivity of sand dunes to climatic episodes can be varied based on the site-specific parameters like availability of sands for aeolian reworking, preservation potential at the site and the presence of hardened duricrusts/paleosols (which may be controlled by the mineralogy of the sands leading to the ease of chemical weathering, hydrological parameters like water saturation or rate of surface runoff, etc.). Further the response time of a sediment record in dunes as a proxy to a climatic event recorded in other terrestrial proxies like speleothems may be different due to the lag in response for creating conducive conditions for sand mobility and preservation. These make it difficult to correlate events between proxies with different response times and thresholds.

4. Conclusions

The studies on the coastal dunes show stages of sand accretion during the late Pleistocene to early Holocene (~ 49–9 ka) and a second stage at the middle Holocene to recent times. Based on the chronology, heavy mineral constituents, sedimentary characteristics and geomorphological association, the dunes are classified as near coastal dunes, coastal dunes and inland dunes.

Red sand occurs in all three category dunes, whereas the light-yellow gray dune sands occur as near coastal dunes, associated with mid-Holocene littoral sands (as aeolian reworked deposits). There is evidence of in-situ reddening of sand existing in dunes of Tamil Nadu and Andhra Pradesh. Reddening occurs due to the chemical weathering of iron containing heavy minerals and can occur rapidly and therefore cannot be used as an indicator of antiquity of sands.

Detrital origin of pedogenic crystalline haematite due to the oxidation of ferromagnesian minerals under tropical warm and wet climatic conditions was given as the cause of formation of red sands in Andhra Pradesh. The redness is due to haematite coating on sand grains along with kaolinite and other secondary clay minerals. While sand accretion occurs during transition from arid to semi-arid conditions, the reddening of sands may occur even during the semi-arid conditions. The presence of un-weathered late Holocene coastal dunes along the coast associated with modern shoreline indicated that the reddening process occurred prior to this period. The chronology of dune aggradation phases shows that the process was controlled by availability of sand during lower sea level and coastal progradation after the early Holocene sea level high stand. This is inferred from the association of dunes with other landforms like paleo-lagoons, floodplains, beach ridges and deltas. The dunes are a rich source of placer heavy minerals and their origin is linked to the drainage basins of major rivers in the hinterland. Variation in sediment source for different parts of the coasts and is observed to be related to the coloration of the sands. The recent aggradation of sands in the coastal plains shows increased sand mobility associated with human influenced land-use changes and drought-induced decline in the water table in regions along the coast. The mobility index shows that the dunes were active during the period of deposition as either a whole or partial reactivation on the dune crest. Dune formation along the east coast of Tamil Nadu is controlled by wind speed, vegetation content and sand moisture content as the area had plenty of sand available from past shorelines and flood plains of rivers. The textural and geochemical analyses of the dune sands show high textural maturity for sediments with higher proportion of stable mineral phases like quartz in the sand and fine-grained fraction.

Acknowledgements

Assistance in the preparation of Figure 1 by Mr. Ayyappadas C.S. and Dr. Midhun M. for the preparation of precipitation contours using the gpcp precipitation data are greatly acknowledged.

References

Acharya, B. and Bhatt, N. 2019. Sedimentological analysis of a late quaternary coastal dune system: an example from gopnath, South-East Saurashtra, Western India. Ed. Nick Lancaster. Sedimentology 66(2)(February 1): 435–458. http://doi.wiley.com/10.1111/sed.12481.

Alappat, L., Frechen, M., Ramesh, R., Tsukamoto, S. and Srinivasalu, S. 2011. Evolution of Late Holocene Coastal Dunes in the Cauvery Delta Region of Tamil Nadu, India. Journal of Asian Earth Sciences 42(3): 381–397.

Alappat, L., Frechen, M., Tsukamoto, S., Anupama, K., Prasad, S., Gopakumar, P.G. and Sree Kumar S. (Submitted) Evidences of Mid-Holocene land-sea interactions and formation of Wetlands of Central Kerala in the south west coast of India. Quaternary International.

Alappat, Linto, Frechen, M., Sree Kumar, S., Suresh Babu, R., Ravur, R. and Tsukamoto, S. 2015. Evidence of late holocene shoreline progradation in the coast of kerala, south india obtained from osl dating of paleo-beach ridges. Geomorphology 245: 73–86.

Alappat, Linto, P. Seralathan, A.D. Shukla, K., Pillai Thrivikramji and Singhvi, A.K. 2013. Chronology of red dune aggradations of south india and its paleo-environmental significance. In Geochronometria 40: 274–282. Springer. https://link.springer.com/article/10.2478/s13386-013-0118-5.

Alappat, Linto, Joseph, S., Tsukamoto, S. Kaufhold, S. and Frechen, M. 2017. Chronology and Weathering History of Red Dunes (Teri Sands) in the Southwest Coast of Tamil Nadu, India. Zeitschrift Der Deutschen Gesellschaft Fur Geowissenschaften 168(1)(March 1): 183–198.

Banerjee, P.K. 2000. Holocene and Late Pleistocene Relative Sea Level Fluctuations along the East Coast of India. Marine Geology 167(3–4)(July 15): 243–260.

Bar, P. 2013. Restoration of Coastal Sand Dunes for Conservation of Biodiversity: The Israeli Experience. In, 173–185. Springer, Berlin, Heidelberg. https://link.springer.com/chapter/10.1007/978-3-642-33445-0_11.

Baskaran, M., Shrikant Deshpande, Rajaguru, S.N. and Somayajulu, B.L.K. 1989. Geochronology of Miliolite Rocks of Kutch, Western India. Journal Geological Society of India 33(June): 588–593.

Bensing, J.P., Mozley, P.S. and Dunbar, N.W. 2005. Importance of clay in iron transport and sediment reddening: evidence from reduction features of the Abo Formation, New Mexico, USA. – J. Sediment. Res. 75(4): 562–571.

Bera, S.K., Gupta, H.P. and Farooqui, A. 1995. Berijam Lake. 20,000 yrs. sequence of paleofloristics and paleoenvironment in Palni Hills, South India. Geophytology 26: 99–104.

Bera, S.K., Farooqui, A. and Gupta, H.P. 1997. Late Pleistocene–Holocene vegetation and environment in and around Marian Shola, Palni Hills, Tamil Nadu. Paleobotanist 46: 191–195.

Berner, R.A. 1969. Goethite stability and the origin of red beds. Geochim. Cosmochim. Acta 33: 267–273.

Besler, H. 2008. Chapter Six: Quartz sand as an indicator of humid periods. *In*: Besler, H., Bolten, A., Bubenzer, O., Hilgers, A. and Van Loon, A.J. (ed.). The Great Sand Sea in Egypt – formation, dynamics and environmental change—a sediment-analytical approach. Dev. Sedimentol. 59: 129–144.

Brooke, B. 2001. The distribution of carbonate eolianite. Earth-Science Reviews 55(1–2)(October 1): 135–164.

Brückner, H. 1989. Late Quaternary Shorelines in India. pp. 169–194. *In*: Late Quaternary Sea-Level Correlation and Applications. Springer Netherlands.

Chacko, T., Ravindra Kumar, G.R.A. and Newton, R.C. 1987. Metamorphic P-T conditions of the Kerala (south India) Khondalite Belt, a granulite facies supracrustal terrain. J. Geol. 95: 343–358.

Chacko, T., Lamb, M. and Farquhar, J. 1996. Ultra-high temperature metamorphism in the Kerala Khondalite Belt. – Gondwana Research Group Memoir 3: 157–165.

Chandrasekharan, S. and Murugan, C. 2001. Heavy minerals in the beach and the coastal red sands (Teris) of Tamilnadu. *In*: Dhana Raju, R., Ali, M.A. and Krisnan, S. (ed.). Beach and inland heavy mineral sand deposits of India. Exploration and Research for Atomic Minerals 13: 87–109.

Craddock, R.A. 2012. Aeolian Processes on the Terrestrial Planets. Progress in Physical Geography: Earth and Environment 36(1)(February 10): 110–124. http://journals.sagepub.com/doi/10.1177/0309133311425399.

Dutt, S., Gupta, A.K., Clemens, S.C., Cheng, H., Singh, R.K., Kathayat, G. and Edwards, R.L. 2015. Abrupt Changes in Indian Summer Monsoon Strength during 33,800 to 5500 Years B.P. Geophysical Research Letters 42(13) (July 16): 5526–5532. http://doi.wiley.com/10.1002/2015GL064015.

Folk, R.L. 1976. Reddening of desert sands: Simpson Desert, N. T., Australia. J. Sediment. Res. 46: 604–615.

Gao, J., Kennedy, D.M. and Konlechner. T.M. 2020. Coastal dune mobility over the past century: a global review. Progress in Physical Geography: Earth and Environment (May 22): 030913332091961. http://journals.sagepub.com/doi/10.1177/0309133320919612.

Gardner, R.A.M. and Pye, K. 1981. Nature, origin and paleoenvironmental significance of red coastal and desert dune sands. Progr. Phys. Geogr. 5: 514–534.

Gardner, R.A.M. 1981. Reddening of Dune Sands—Evidence from Southeast India. Earth Surface Processes and Landforms 6(5): 459–468.

Gardner, R.A.M. and Martingell, H. 1990. Microlithic sites and their paleoenvironmental setting, Southeast India; a reevaluation. Geoarchaeology 5: 1–13.

Glennie, K.W. and Singhvi, A.K. 2002. Event stratigraphy, paleoenvironment and chronology of SE Arabian deserts. Quat. Sci. Rev. 21: 853–869. https://doi.org/10.1016/S0277-3791(01)00133-0.

Goswami, K., Krishnan, S., Kumerasan, A., Sadasivam, S.K., Kumar, P. and Jaiswal, M.K. 2019. Luminescence chronology of fluvial and marine records from subsurface core in kaveri delta, tamil nadu: implications to

sea level fluctuations. Geochronometria 46(1)(May 28): 125–137. https://content.sciendo.com/view/journals/geochr/geochr-overview.xml.
Greeley, R., Iversen, J., Leach, R., Marshall, J., White, B. and Williams, S. 1984. Windblown sand on Venus. Icarus 57: 112–124.
Grilliot, M.J., Walker, I.J. and Bauer, B.O. 2018. Airflow dynamics over a beach and foredune system with large woody debris. Geosciences 8: 147.
Hesp, P.A. 2008. Coastal Dunes in the Tropics and Temperate Regions: Location, Formation, Morphology and Vegetation Processes. pp. 29–49. Springer, Berlin, Heidelberg.
Hesp, P.A. 2012. Surfzone-Beach-Dune Interactions. *In*: University Library/University of Twente.
Jayangondaperumal, R., Murari, M.K., Sivasubramanian, P., Chandrasekar, N. and Singhvi, A.K. 2012. Luminescence dating of fluvial and coastal red sediments in the SE coast, India, and implications for paleoenvironmental changes and dune reddening. Quaternary Research 77(3)(May 1): 468–481.
Jiang, T., Liu, X., Yu, T. and Hu, Y. 2018. OSL dating of late holocene coastal sediments and its implication for sea-level Eustacy in Hainan Island, Southern China. Quaternary International 468(February 27): 24–32.
Joseph, S., Thrivikramji, K.P. and Anirudhan, S. 1999. Mud content, clay minerals and oxidation states of iron in Teris of southern Tamil Nadu. Implication to the origin of Redness. Journal of the Indian Association of Sedimentologists 18: 83–94.
Joseph, S., Thrivikramji, K.P. and Babu, D.S.S. 2002. State of alteration of ilmenite in Teris, southern Tamil Nadu. – J. Geol. Soc. India 60: 537–546.
Joseph, S. and Thrivikramji, K. 2005. Origin of calcretes in teris, southern tamil nadu. origin and paleoenvironmental implications. Journal of Geological Society of India 65: 158–168.
Joseph, Sabu, Thrivikramaji, K. and Anirudhan, S. 1997. Textural parameters, discriminant analysis and depositional environments of teri sands, Southern Tamil Nadu. Journal of Geological Society of India 50: 323–329.
Kar, R., Bhattacharya, S. and Sheraton. J.W. 2003. Hornblende-dehydration melting in mafic rocks and the link between massif-type charnockite and associated granulites, Eastern Ghats Granulite Belt, India. Contributions to Mineralogy and Petrology 145(6)(September 5): 707–729. https://link.springer.com/article/10.1007/s00410-003-0468-8.
Kale, V.S., Narayana, A.C. and Jaiswal, M.K. n.d. Wind-blown, flash flood-deposited sands of Hagari River, Anantapur district, Andhra Pradesh, India. https://doi.org/10.1016/j.dib.2020.105560.
Kocurek, G. and Lancaster, N. 1999. Aeolian system sediment states: Theory and Mojave Desert Kelso Dune field example. Sedimentology 46: 505–515.
Krishna, K.N.M., Reddy, K.S.N., Sekhar, C.R., Naidu, K.B., Rao, P.G. and Reddy, G.V.R. 2016. Heavy mineral studies on late quaternary red sediments of bhimunipatnam, Andhra Pradesh, East Coast of India. Journal of the Geological Society of India 88(5)(November 1): 637–647. https://link.springer.com/article/10.1007/s12594-016-0530-6.
Krishnan, S., Viswanathan, G. and Balachandran, K. 2001. Heavy mineral sand deposits of Kerala. Exploration and Research for Atomic Minerals 13: 111–146.
Kumaran, K.P.N., Padmalal, D., Nair, K.M., Limaye Ruta, B., Guleria, J.S., Srivastava, R. and Shukla, A. 2014. Vegetation response and landscape dynamics of Indian summer monsoon variations during Holocene: an eco-geomorphological appraisal of tropical evergreen forest subfossil logs. Plos One 9(4): 1–24.
Kunte, P.D. and Wagle, B.G. 2005. The Beach Ridges of India: A Review. Coastal Education and Research Foundation. https://agris.fao.org/agrissearch/search.do?recordID=AV20120125320.
Kunz, A., Frechen, M., Ramesh, R. and Urban, B. 2010a. Luminescence dating of late holocene dunes showing remnants of early settlement in cuddalore and evidence of monsoon activity in South East India. Quaternary International 222(1–2) (August 1): 194–208.
Kunz, A., Frechen, M., Ramesh, R. and Urban B. 2010b. Periods of recent dune sand mobilisation in cuddalore, Southeast India. Zeitschrift Der Deutschen Gesellschaft Für Geowissenschaften 161(3) (September): 353–368.
Lancaster, N. 1988. Development of linear dunes in the southwestern Kalahari, southern Africa. Journal of Arid Environments 14: 233–244.
Lancaster, N. 2002. Flux of eolian sediment in the mcmurdo dry valleys, antarctica: a preliminary assessment. Arctic, Antarctic, and Alpine Research 34(3): 318–323. https://doi.org/10.2307/1552490.
Lancaster, N. 2004. Sedimentary processes: aeolian processes. In Encyclopedia of Geology, 612–627. Elsevier Inc.
Lancaster, N. 2015. Wind-swept planets. Nature Geoscience 8(5): 330–330. Doi:http://dx.doi.org/10.1038/ngeo2428.
Mallik, T.K., Vasudevan, V., Aby Verghese, P. and Machado, Terry. 1987. The black sand placer deposits of Kerala beach, Southwest India. Marine Geology 77: 129–150.
Mauz, B., Hijma, M.P., Amorosi, A., Porat, N., Galili, E. and Bloemendal, J. 2013. Aeolian beach ridges and their significance for climate and sea level: Concept and insight from the Levant coast (East Mediterranean). Earth-Science Rev. ttps://doi.org/10.1016/j.earscirev.2013.03.003.

Maya, K., Mohan, S.V., Limaye, R.B., Padmalal, D. and Kumaran, N.K.P. 2017. Geomorphic response to sea level and climate changes during late quaternary in a humid tropical coastline: terrain evolution model from southwest India. PLoS ONE 12(5)(May 1).

Minoura, K. and Nakamura, N. 2020. Eustatic, climatic and tectonic controls on the evolution of a middle to late holocene coastal dune system in Shimokita, Northeast Japan. Geosciences 10: 410. https://doi.org/10.3390/geosciences10100410.

Mohan, P.M. 1995. Distribution of Heavy mineral in Parangipettai (Portonova) beach, Tamil Nadu. Journal of Geological Society of India 46: 401–408.

Nayak, B., Mohanty, S. and Bhattacharyya, P. 2012. Heavy minerals and the characters of ilmenite in the beach placer sands of chavakkad-ponnani, Kerala Coast, India. Journal of the Geological Society of India 79(3)(March 20): 259–266.

Nayak, G.N. 2005. Indian ocean coasts, coastal geomorphology. In Encyclopedia of Earth Sciences Series 14: 554–557. Springer Netherlands. https://link.springer.com/referenceworkentry/10.1007/1-4020-3880-1_179.

Norris, R.M. 1969. Dune reddening and time. J. Sediment. Petrol. 39: 7–11.

Novothny, Á., Frechen, M., Horváth, E., Wacha, L. and Rolf. C. 2011. Investigating the Penultimate and last glacial cycles of the sütt loess section (hungary) using luminescence dating, high-resolution grain size, and magnetic susceptibility data. Quaternary International 234(1–2) (April 1): 75–85. https://hungary.pure.elsevier.com/hu/publications/investigating-the-penultimate-and-last-glacial-cycles-of-the-sütt.

Ogura, A. and Yura, H. 2008. Effects of sandblasting and salt spray on inland plants transplanted to coastal sand dunes. Ecological Research 23(1)(January 14): 107–112. https://link.springer.com/article/10.1007/s11284-007-0347-2.

Padmalal, D., Kumaran, K., Limaye, R., Baburaj, B., Mohan, S.V. and Maya, K. 2014. Effect of holocene climate and sea level changes on landform evolution and human habitation: Central Kerala, India. Quaternary International 325.

Preusser, F., Radies, D., Driehorst, F. and Matter, A. 2005. Late quaternary history of the coastal wahiba sands, sultanate of Oman. Journal of Quaternary Science 20(4)(May 1): 395–405. http://doi.wiley.com/10.1002/jqs.922.

Pye, K. and Tsoar, H. 2009. Aeolian Sand and Sand Dunes: 476 p., Berlin (Springer).

Rajagopalan, G., Sukumar, R., Ramesh, R., Pant, R.K. and Rajagopalan, G. 1997. Late Quaternary vegetational and climatic changes from tropical peats in southern India – an extended record up to 40,000 years BP. Curr. Sci. 73: 60–63.

Rajesh, H.M. 2007. The petrogenetic characterization of intermediate and silicic charnockites in high-grade terrains: a case study from Southern India. Contributions to Mineralogy and Petrology 154(5)(November 26): 591–606. https://link.springer.com/article/10.1007/s00410-007-0211-y.

Ramakrishnan, M. and Vaidyanadhan, R. 2008. Geology of India. Bangalore: Geological Society of India.

Ramasamy, S.M., Ramesh, D., Paul, M.A., Kusumgar, S., Yadava, M.A., Nair, A.R., Sinha, U.K. and Joseph, T.B. 1998. Rapid land building activity along Vedaranniyam Coast and its possible implications. Current Science 75: 884–886.

Ramsey, C.B. 2009. Dealing with outliers and offsets in radiocarbon dating. Radiocarbon 51: 1023–1045. https://doi.org/10.1017/s0033822200034093.

Rao, S.M. 1982. Morphology and evolution of modern Cauvery delta, Tamil Nadu, India. Transactions of the Institute of Indian Geographers 4: 67–78.

Rao, N.K., Rao, U.B.C., Prakash, P.V. and Reddy, T.K. 2006. Morphostratigraphy and evolution of the Quaternary red sands near Bhimunipatnam, east coast of India. J. Geol. Soc. India 68: 857–873.

Ravindra Kumar, G.R. and Sreejith. C. 2016. Petrology and geochemistry of charnockites (felsic ortho-granulites) from the kerala khondalite belt, Southern India: Evidence for intra-crustal melting, magmatic differentiation and episodic crustal growth. Lithos 262(October 1): 334–354.

Rawat, S., Gupta, A.K., Sangode, S.J., Srivastava, P. and Nainwal, H.C. 2015. Late pleistocene-holocene vegetation and indian summer monsoon record from the lahaul, Northwest Himalaya, India. Quaternary Science Reviews 114(April 5): 167–181.

Reimann, T., Tsukamoto, S., Harff, J., Osadczuk, K. and Frechen, M. 2011. Reconstruction of holocene coastal foredune progradation using luminescence dating—an example from the Świna Barrier (Southern Baltic Sea, NW Poland). Geomorphology 132(1–2)(September 1): 1–16.

Rodrigues, R.S., Mascarenhas, A. and Jagtap, T.G. 2011. An evaluation of flora from coastal sand dunes of india: rationale for conservation and management. Ocean and Coastal Management 54(2)(February 1): 181–188.

Roonwal, G.S. 1997. Marine mineral potential in India's exclusive economic zone: some issues before exploitation. Marine Georesearch Geotechnology 15: 21–32.

Roskin, J., Tsoar, H., Porat, N. and Blumberg, D.G. 2011. Paleoclimate interpretations of Late Pleistocene vegetated linear dune mobilization episodes: Evidence from the northwestern Negev dunefield, Israel. Quat. Sci. Rev. 30: 3364–3380. https://doi.org/10.1016/j.quascirev.2011.08.014.

Roskin, J., Blumberg, D.G.N., Porat, N., Tsoar, H. and Rozenstein, O.L. 2012. Do dune sands redden with age? the case of the Northwestern Negev Dunefield, Israel. Aeolian Research 5(August): 63–75.

Sathiyaseelan, S., Panda, D.K., Banerjee, D., Ramesh, D. and Shukla, A.D. 2021. Chronology of coastal dune ridges in Vaigai prodelta region, southeastern Tamil Nadu, India. Current Science 120(2): 1812–188.

Sanil Kumar, V., Pathak, K.C., Pednekar, P., Raju, N.S.N. and Gowthaman, R. 2006. Coastal processes along the indian coastline. Indian Academy of Sciences. http://drs.nio.org/drs/handle/2264/350.

Sharma, K., Bhatt, N., Shukla, A.D., Cheong, D.K. and Singhvi, A.K. 2017. Optical dating of late Quaternary carbonate sequences of Saurashtra, western India. Quaternary Research 87: 133–150.

Sherman, D.J. and Lyons, W. 1994. Beach-state controls on aeolian sand delivery to coastal dunes. Physical Geography 15(4): 381–395.

Singhvi, A.K. and Kar, A. 2004. The aeolian sedimentation record of the Thar Desert. Journal of Earth System Science 113(3): 371–401.

Singhvi, A.K. and Porat, N. 2008. Impact of luminescence dating on geomorphological and paleoclimate research in drylands. Boreas 37: 536–558.

Singhvi, A.K., Bluszcz, A., Bateman, M.D. and Rao, M.S. 2001. Luminescence dating of loess–palaeosol sequences and coversands: Methodological aspects and palaeoclimatic implications. Earth Science Reviews 54: 193–211.

Singhvi, A.K., Deraniyagala, S.U. and Sengupta, D. 1986. Thermoluminescence dating of quaternary red-sand beds: a case study of coastal Dunes in Sri Lanka. Earth and Planetary Science Letters 80(1–2): 139–144.

Singhvi, A.K., Sauer, W., Wagner, G.A. and Kröger. K. 1986. Thermoluminescence dating of loess deposits at plaidter hummerich and its implications for the chronology of neanderthal man. Naturwissenschaften 73(4)(April): 205–207. https://link.springer.com/article/10.1007/BF00417724.

Speirs, J.C., McGowan, H.A. and Neil, D.T. 2008. Polar eolian sand transport: grain characteristics determined by an automated scanning electron microscope (QEMSCAN®). Arctic, Antarctic, and Alpine Research 40(4) (November): 731–743.

Sreejith, C. and Ravindra Kumar, G.R. 2013. Petrogenesis of high-K metagranites in the kerala khondalite belt, southern India: a possible magmatic-arc link between India, Sri Lanka, and Madagascar. Journal of Geodynamics 63(January 1): 69–82.

Sridhar, K. 2009. Bioresources of Coastal Sand Dunes - Are they Neglected? Coastal Environments: Problems and Perspectives (Ed. K.S. Jayappa & A.C. Narayana) I.K. International Publishing House Pvt. Ltd., New Delhi, India, ISBN: 978-81-80026-28-2; 2009, 53–76.

Srivastava, P., Sangode, S.J., Parmar, N., Meshram, D.C., Jadhav, P. and Singhvi, A.K. 2016. Mineral magnetic characteristics of the late quaternary coastal red sands of Bheemuni, East Coast (India). Journal of Applied Geophysics 134(November 1): 77–88.

Stoddart, D.R. and Pillai, C.S.G. 1972. Raised reefs of ramanathapuram, South India. Transactions of the Institute of British Geographersno 56(July): 111. https://www.jstor.org/stable/621544.

Sukumar, R., Ramesh, R., Pant, R.K. and Rajagopalan, G. 1993. A δ13C record of late Quaternary climate change from tropical peats in southern India. Nature 364: 703–706.

Sundararajan, M., Usha-Natesan, Babu, N. and Seralathan, P. 2009. Sedimentological and mineralogical investigation of beach sediments of a fast prograding cuspate foreland (Point Calimere), Southeast coast of India. Research Journal of Environmental Sciences 3(2): 134–148.

Szkornik, K., Gehrels, W.R. and Murray, A.S. 2008. Aeolian sand movement and relative sea-level rise in Ho Bugt, western Denmark, during the "Little Ice Age." Holocene 18: 951–965. https://doi.org/10.1177/0959683608091800.

Telfer, M.W. and Thomas, D.S.G. 2007. Late Quaternary linear dune accumulation and chronostratigraphy of the southwestern Kalahari: implications for aeolian paleoclimatic reconstructions and predictions of future dynamics. Quat. Sci. Rev. 26: 2617–2630. https://doi.org/10.1016/j.quascirev.2007.07.006.

Thomas, T.G. and Williams, J.J.R. 1985. Large eddy simulation of a symmetrical trapezoidal channel at Reynold number of 430000. Journal of Hydraulic Research 33: 825–8423.

Thomas, P.J. 2009. Luminescence dating of beachrock in the southeast coast of India-Potential for Holocene Shoreline Reconstruction. Journal of Coastal Research 25(1): 1–7.

Thrivikramji, K.P., Joseph, S. and Anirudhan, S. 2008. Teris of southern Tamil Nadu. Saga of Holocene climate change. Geol. Soc. India, Mem. 74: 351–359.

Tiwari, M., Singh, A.K. and Ramesh, R. 2011. High-resolution monsoon records since last glacial maximum: a comparison of marine and terrestrial paleoarchives from South Asia. Journal of Geological Research, 2011, Article ID 765248: 12 p.

Tsoar, H., Levin, N., Porat, N., Maia, L.P., Herrmann, H.J., Tatumi, S.H. and Claudino-Sales, V. 2009. The effect of climate change on the mobility and stability of coastal sand dunes in Ceará State (NE Brazil). Quaternary Research 71(2)(March): 217–226.

Turner, P. 1980. Continental red beds: 537 p., Amsterdam (Elsevier).

Vaidyanathan, R. and Ramakrishnan, M. 2008. Geology of India. Geological Society of India, Bangalore 2: 557–994.

Van Houten, F.B. 1964. Origin of red beds – some unsolved problems. pp. 647–661. *In*: Nair, A.E.M. (ed.). Problems in Paleoclimatology: New York (Interscience).

Vidyasakar, A., Sant'Ovaia, H., Alappat, L., Morthekai, P., Srinivasalu, S., Singhvi, A.K., Jorge, F. and Gomes, C. 2017. Paleoclimatic Registers from Semi-Arid Costal Sediments of Southeastern India: A Multi Proxy Approach. In, 3–30. Springer, Cham.

Wacha, L., Mikulčić Pavlaković, S., Frechen, M. and Crnjaković, M. 2011. The loess chronology of the island of Susak, Croatia. E&G Quaternary Science Journal 60(1)(June 6): 153–169.

Walker, T.R. 1967. Formation of Red Beds in modern and ancient deserts. – Geol. Soc. America, Bull., 78: 353–368.

Walker, T.R. 1979. Red color in dune sand. *In*: McKee, E.D. (ed.). A study of global sand seas. U.S. Geol. Surv., Prof. Pap. 1052: 62–81.

Wasson, R.J. and Hyde, R. 1983. Factors determining desert dune type. Nature 304: 337–3339.

Zhang, J., Rolf, C., Wacha, L., Tsukamoto, S., Durn, G. and Frechen, M. 2018. Luminescence dating and paleomagnetic age constraint of a last glacial loess-paleosol sequence from Istria, Croatia. Quaternary International 494(November 20): 19–33.

Zhang, J., Yuan, B. and Zhou, L. 2008. Luminescence chronology of Old Red Sand in Jinjiang and its implications for optical dating of sediments in South China. Chinese Science Bulletin 53(4): 591–601. http://doi.org/10.1007/s11434-008-0001-6.

Zheng, W. and Wei, W. 1998. Formation of "old red sands" and paleogeographic environment on South China coasts. Science in China, Series D, Earth Sciences 41: 306–313.

Chapter 21

Coastal Sand Dunes along the Western and Eastern Shores of India

Ashis Kumar Paul,[a,*] *Anurupa Paul*[b] and *Dipanjan Das Majumdar*[c]

1. Introduction

Sandy sea beaches and wide sand spits are backed by or fringed with sand dunes along the coastline of the Bay of Bengal and the Arabian Sea in India. Marine terraces as well as paleo beach ridges of the strand plains and the chenier[1] delta plains are overlain by sand dunes in many parts of the coastal region (East coast and West coast of India). Coastal sand dunes of India are studied in this work as regional units of coastal sections to identify their physical and ecological characters influenced by regional marine and coastal environments. The sand dunes of the West coast are ecologically more diverse than the East coasts of India. Physiographically, coastal sand dunes are found over marine terraces and also on raised beach platforms representing higher surfaces in the region. Older sand dunes of the backshore areas are usually oxidized or affected by reddening of sands, but also overlain by the deposition of younger sands with stronger onshore winds in the drier months (Fig. 21.1).

The prevailing onshore winds blow from the northwest, west and southwest directions on the West coast of India, however, the average wind blows from the directions of southwest, south, east and north in the alternate seasons on the East coast of India (Fig. 21.2).

The West coast of India is categorized into several regional units as the Malabar coast (Kerala and Karnataka), Konkan coast (Goa and Maharashtra), Kathiawar Peninsula-Kachchh Peninsula-Daman & Diu (state of Gujarat) to describe the nature and types of coastal sand dunes. The East coast of India, however, is separated into the Coromondal coast (Tamil Nadu, Puducherry and coastal Andhra Pradesh) and North Circar coast (Odisha and West Bengal).

Coastal sand dunes of the Indian subcontinent have been studied by several researchers particularly to explore the morphology, ecology, sediment types and composition and geochemical alterations of aeolian deposits (Paul, 1988, 1994, 1996, 2002; Paul and Paul, 2013; Gardner and

---

[a] Professor, Department of Geography, Vidyasagar University, Midnapore-721102, West Bengal, India.
[b] Senior Research Fellow (SRF), Department of Remote Sensing & GIS, Vidyasagar University, Midnapore-721102, West Bengal, India.
[c] Post-Doctoral Fellow, DACEE, Faculty of Interdisciplinary Studies, Law and Management, Jadavpur University, West Bengal India; Assistant Professor, Department of Geography, Netaji Satabarshiki Mahavidyalaya, Ashoknagar, North 24 Parganas-743223, West Bengal, India.
* Corresponding author: akpaul_geo2007@yahoo.co.in

[1] A chenier is a sandy or shelly beach ridge that is part of a strand plain. Mostly associated with shorelines characterized by generally low wave energy, low gradient, muddy shorelines and abundant sediment supply. It consists of cheniers (layers) separated by intervening mud-flat deposits with marsh and swamp vegetation. Cheniers are typically 1 to 6 m high, tens of km long, hundreds of meters wide and often wooded.

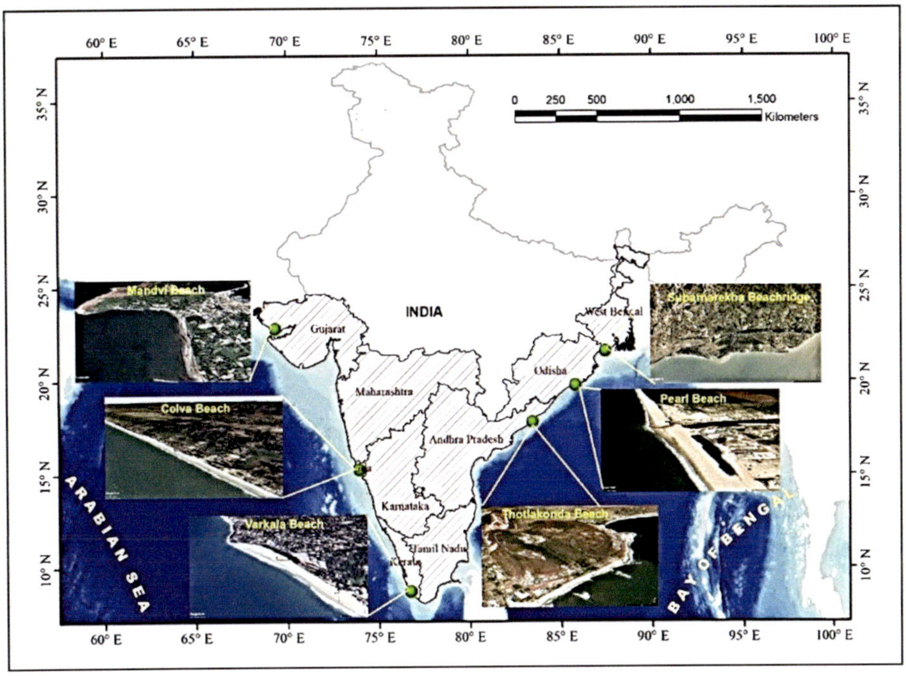

Figure 21.1. Aeolian sand deposits along the coastal belts of western and eastern shores of India.

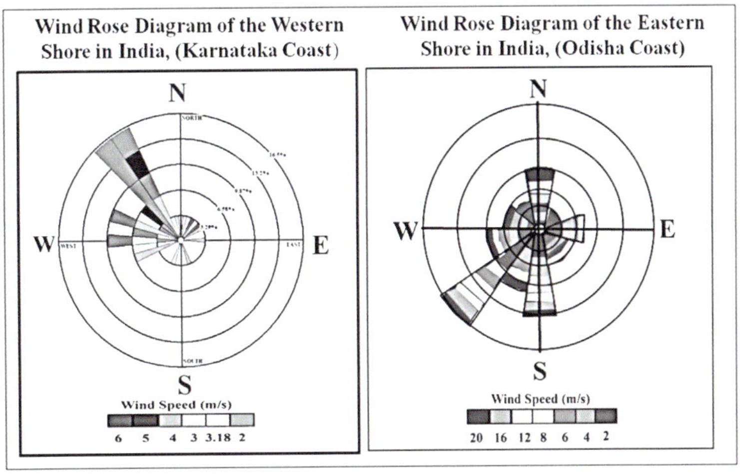

Figure 21.2. The wind rose diagrams for the West coast and East coast of India to show the dominant wind directions.

Hendry, 1995; Deswandikar and Karlekar, 1996; Fairbridge, 1995; Jayangondaperumal et al., 2012; Dalabehera et al., 2020; Alappat et al., 2013, 2017; Rodrigues et al., 2011; Kale and Rajaguru, 1983; Mascarenhas, 1990, 1996a, 1996b, 1997, 1998; Das Majumder et al., 2015; Purkait et al., 2014; Maiti, 2013). However, the present study highlights the genesis, morphology, classification, evolution, sediment characters, dynamic plant ecology and dune vulnerability index for the western and eastern shores of India.

2. Methodology of the study

Very little interest is shown in the study of coastal aeolian sand deposits in India. Coastal dunes are reddened towards the south of 15°32’ N latitude along the western and eastern shores of the subcontinent. To the north of this latitude coastal sand dunes are less old and geochemically less altered than in the tropics. Different methods are used in the present study to document the coastal dune systems of the country. The spatial distribution of coastal sand dunes and other aeolian sands, their morphology, sediments, geomorphic evolutions, geochemical alterations, dynamic plant ecology, dune vulnerability index and recent environmental changes are considered in this chapter.

Coastal dune landscapes on the west and east coasts have been inspected and monitored in the field for documenting their morphological dynamics from 1984 to 2020. The field survey is also considered for the ground truth verifications to validate the geospatial techniques used in the present study. The Survey of India's toposheets, SRTM DEM, ALOS PALSAR DEM, US Space based camera photographs of 1970s for few cases, GEBCO data, Total Station Survey are used for representing the contour plans and morphological identity of sand dunes along the coastal belt of India. The various cross-sections over the physiographic settings for coastal sand dunes are adopted on the basis of field verifications and Google Earth images. The Dune Vulnerability Index is estimated on the basis of field parameters. The climatic data used in the study are from Indian Meteorological Department and National Oceanic and Atmospheric Administration. Dune sediments from different sites were examined and analyzed in the laboratory and the existing radiometric dating and Optically Stimulated Luminescence dating records were used to estimate the ages of dunes. A few stratigraphic sections of the dune landscapes have been taken from exposure sites to identify their structures and depositional environments. A methodological flow chart was prepared to show the different inputs of primary and secondary data and resultant output of the present study (Fig. 21.3).

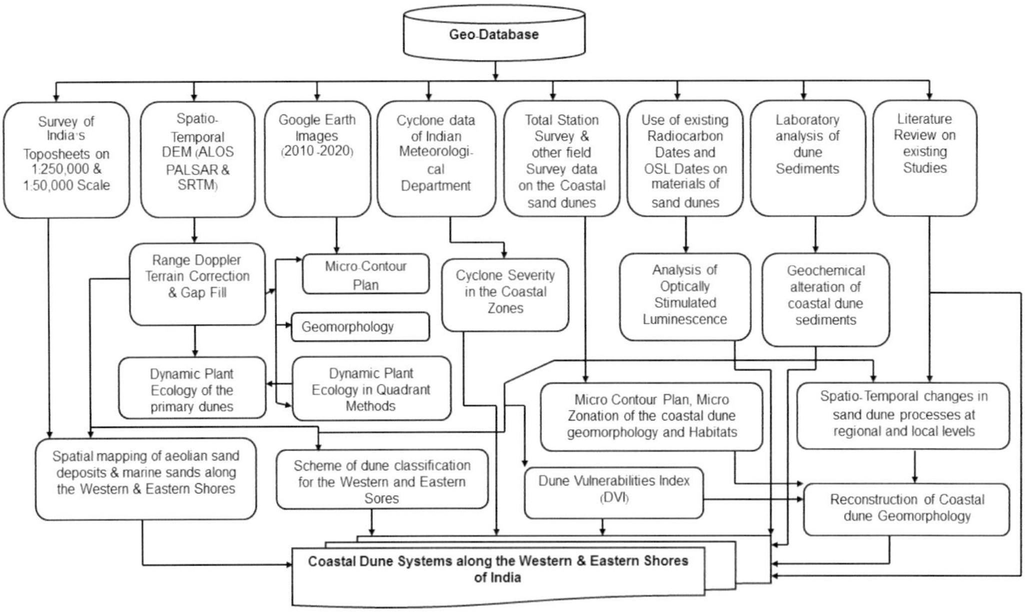

Figure 21.3. Methodological flow chart of the study area.

3. Result & Discussions

The present study highlights the distribution, characteristics and evolution of coastal sand dunes in the western and eastern shores of India. The locational characters of coastal sand dunes are described under different coastal sections of India in this part. The aeolian deposits of sands are located all along the coastal parts from Gujarat to Kerala on the western shores and from Tamil Nadu to West Bengal on the eastern shores. Coastal sand dunes of the island territories of the Bay of Bengal and Arabian Sea are also described here.

3.1 Coastal sand dunes (Gujarat or Saurashtra Coast)

Coastal areas of Gujarat are separated into the Kutch Peninsula, Kathiawar Peninsula and the eastern coast mainland unit by the extension of the Gulf of Kutch and the Gulf of Cambay (Khambhat). About 51% of the coast of Saurashtra are occupied by muddy and silty tidal flats of an estuarine environment. The wide sandy tracts, raised beach platforms, extensive sand flats and sand bars are found in the remaining 49% of areas on the coastal tracts in the states. The Mandvi and Mundra regions of the Kutch Peninsula consist of sand ridges and offshore shoals nearby the coastal fringe of Median High region and indicate the slow uplift of the surface to support the overlying sand dunes of the extensive Holocene deposits. The coastal plains of North-Western Saurashtra in the Kathiawar Peninsula also support the sand dunes over raised beach platforms along the shoreline of the Arabian Sea from Okha-Porbandar to the South-East of Junagarh (Ahmedpur, Mandvi). All the beaches of the Kathiawar Peninsula including Okha beach, Mithapur beach, Dwarka beach, Shivrajpur beach, Harshad Bandar beach, Porbandar beach, Madhavpur beach, Veraval and Somnath beach and Ahmedpur Mandvi beach (Saurashtra coast) provide sources of sands and with the prevailing strong onshore winds, sand dunes have been developed along the beach fringed shores. The younger coastal sand dunes range from 6 m to 12 m in altitude at the backshores. However, near Narayan Sarovar Kachchh the coastal sand dunes are lofty (about 18 m–27 m in altitude) and fronted by tidal shoals, present as the overlying deposits of the raised beach topography along coastal Kutch. The basement platform of the aeolian deposits has been uplifted slowly and steadily and now the system of dune environment is cut off from the windblown sand deposits due to the emerged tidal shoals in front of the earlier dune fields in the present area.

Daman and Diu on the eastern and western flanks of the Gulf of Khambhat represent the platforms of paleo beach ridge surface (Daman) and carbonate deposits (Diu) and over which coastal sand dunes are located. Small hillocks of sand dunes with maximum height of 25 meters over the central highland of Diu exhibits or indicates the ancient formation of unconsolidated sandy alluviums by windblown action in the Oligocene period (Nayak et al., 2020, Central Groundwater Board), over the marine terrace of Milliolite limestone along the sea face of the Arabian Sea (Fig. 21.4).

The moderate dune ridge topography (6.0 m–9.0 m in elevation) overlies the older beach deposits surface in the open coast tidal flat of Daman. The older beach deposits surface exhibits older positions of the active beach face around 8000 YBP and also marking the last sea level high stand of the Gujarat coast on the eastern flank of the Gulf of Khambhat (Saha et al., 2011).

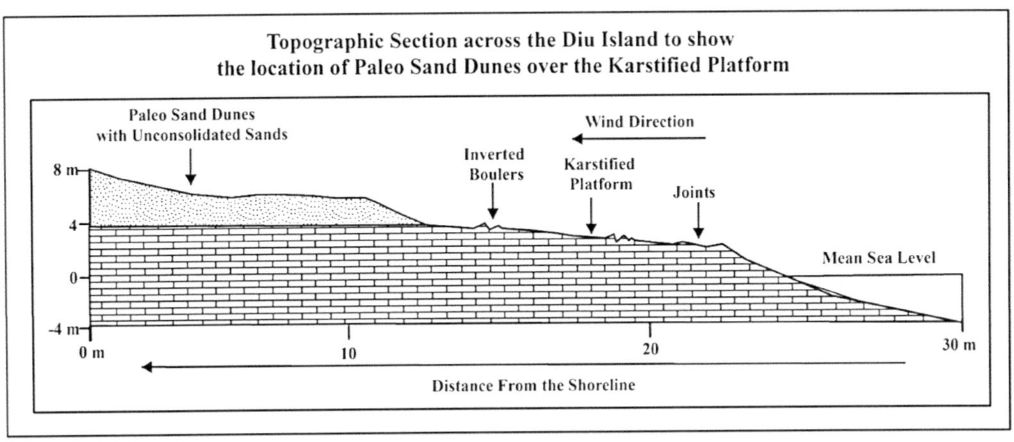

Figure 21.4. The cross sectional view of the relict sand dunes over the carbonate platform in Diu Island (Saurashtra Coast).

3.2 Coastal sand dunes (Konkan Coast of Maharashtra and Goa)

Regionally, the Konkan coast is separated into northern Maharashtra, southern Maharashtra and Goa Coasts along the Arabian Sea. They are isolated from highland Maharashtra by extensive areas of pedimented surface, marine terraces, sand dunes, tidal flats, coastal plains, numerous creeks and shoreline cliffs. Konkan coasts are broken or crenulated and characterized with alternate bays, coves, headlands, estuaries, sea cliffs and marine terraces. The sediment tight pocket beaches are located within the bays and coves, and fringed or backed by fore dunes and back dunes. They are conspicuous at Diveagar, Kelshi, Tambaldeg, Mochemad and Velaghar (Deswandikar and Karlekar, 1996). In height, they gradually increased from the fore dunes (5.0 m–6.0 m) to the extensive back dunes (> 8 m) of inland areas and separated with wide interdune sandy tracts dominated by dynamic accumulation of windblown sands from the shore fringe sandy beaches. The inland shore has parallel beach ridges away from the modern-day shoreline. These represent late Holocene deposits of unconsolidated sands (6000 YBP) in the open sea low land coasts over which the sand dunes are located with a diversity of topography. The most inland back shore dunes of Nandgaon, Diveagar, Guhagar and Jaki Mirya are exhibited as lithified fossil dunes of aeolinites and locally known as 'karal' rocks (Dikshit, 1976; Deswandikar and Karlekar, 1996; Karlekar and Gadkari, 1998) in the coastal fringe of the Arabian Sea.

The Konkan coasts from south to north represent a significant process-related environmental gradient with low tidal range (1.5 m) to high tidal range (> 3.5 m) and significant wave height (> 2.5 m) to moderate wave height (1.20 m) towards the northern shore of Maharashtra. The wide and extensive sandy sea beaches protected under sea ward projected headlands with open inland coastal plain topography of southern Maharashtra provided a favorable environment for development of diversified dune landscapes in the broken coast line of western margin of India (Ria Coast). The fossil beaches and dune ridges of the coast are found within the 700 m horizontal extent on the backshores from the High Tidal Limit of the modern day coast. Various studies depicted that the last highest sea level was at and around 3000 YBP and the similar height of present-day sea level was achieved during 6000 YBP in the region (Fig. 21.5).

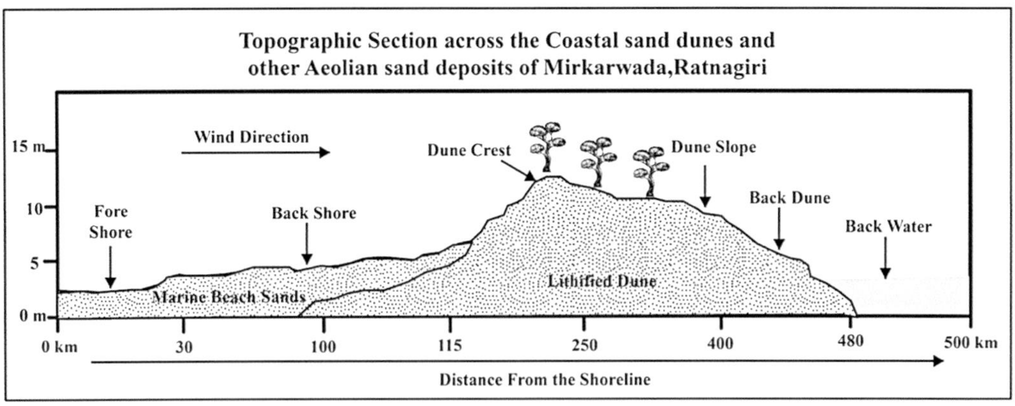

Figure 21.5. The cross-sectional view of the vegetated sand dunes at the inland back shore areas of Ratnagiri coast.

3.3 Coastal sand dunes of Goa

The southern part of the Konkan coast including Goa represents a number of sandy sea beaches within the bays, coves and estuary fringes which are backed by coastal sand dunes of topographic diversity. Windblown sands are trapped by the turf of vegetations on the backshore of the state, and the developed sand dunes include 91 genera and 106 species of coastal plants (National Centre for Sustainable Coastal Management, 2016). There are seven estuaries passing the West coast of Goa coming down from the Western Ghat mountain and flowing into the Arabian sea with huge loads of sand size sediments into the coastal zone and later are sieved by wave energy and tidal pumping processes to create sand reserves in the sediment tight beaches and estuarine shores. However, the presence of strong onshore winds across the wide sandy sea shore has provided favorable conditions for transportation and accumulation of fine to medium size sands on the roughened surface of the backshores dominated by coastal vegetation in the region.

Shore parallel linear dune ridges are extended over the coastal belt of South Goa (75 patches) and North Goa (24 patches) by windblown sand trapping activities with vegetation covers of the backshore areas. The entire coast line with bays and coves are interrupted with two major estuaries (Mandovi and Juari) and seaward projected headland of lateritic pedimented surface. The studies by National Centre for Sustainable Coastal Management (2016) revealed that, about 22.62 km length of sand dunes occupy 2.93 sq.km areas of the coastal belt in Goa. The width of these coastal sand dunes ranges from 23 to 400 m and in height ranging from 3 to 8 m respectively.

Coastal sand dunes of Goa are partially degraded due to modifications by tourism infrastructural development along the backshores. Dunes can feed the beaches by transporting sands during starvation and can improve the climate resilience capacity of beaches during the rapid rise of sea-level in the region. Artificial sand dunes are built in a few sections of the coast of Maharashtra and the coasts of Karnataka to improve the climate resilience capacity of degraded beaches and beach fringed degraded sand dunes since 2016 against the storm attacks and sea-level rise processes.

3.4 Coastal sand dunes of Karnataka

Beaches on the coast of Karnataka are backed by attractive sand dunes. There are seven major beaches (Bhatkal beach, Panambur beach, Tannirbhavi beach, Malpe beach, Om beach, Murdeshwar beach and Karwar beach) in Karnataka in between the headlands of the coastal belt. Many other unexplored (around 30) beaches are also found along the rocky coasts and sand dunes over marine terraces indicating significant topographic highs with oxygenated sands and vegetation covers. Most of the patches of sand dunes of Uttar Kannada district are largely used for sand mining activities by the local people for building purposes. Currently, to protect sea beaches from rising sea-levels, flooding, erosion and encroachment of saltwater, the local government has decided to build sand dunes and to promote favorable conditions for dune formation on the 20 km shoreline of northern Karnataka. The climate resilience capacity of beaches and sea front areas can be improved, if artificial sand dunes are built and stabilized with selective plantations along the coastal belt. About 12 patches of coastal sand dunes are considered for their bio-resources on the 300 km long coastline of the coastal state.

Rivers coming down from the Western Ghat hilly terrains of Belagavi, Dharwar, Shivamogga and Chikmagalur (e.g., Nethravati River, Gurpur River, Suvarna River, Varahi River, Kollur River, Sharavati River, Aghanashini River, Gangavati River and Kali River), have deposited huge amounts of sandy alluviums in the coastal plains along their sea mouth positions to build up shore parallel barrier bars. Beach ridges and coastal sand dunes are well developed by wave deposition and windblown deposition of sediments from the local sources of beaches and barrier surfaces with prevailing strong onshore winds in the coastal region. There are different types of embryonic sand dunes, narrow fore dunes and backshore stabilized back dunes with elevation ranging from 3 to 12 m depositional surface along the coast line of Arabian Sea.

3.5 Coastal sand dunes of Kerala

The long stretch of Kerala plains (about 530 km from north to south) of the Malabar coast is fronted by Arabian Sea (towards west and south west) and backed by the Western Ghat hills and Nilgiri hills (towards the east). The plains are narrow toward northern Kerala (about 19 km wide) and wider towards southern Kerala (about 96 km wide) and ranging in elevation from 6 to 10 m above sea-level. Many rivers have extended their courses across the coastal plains from east to west with sediment loads and fresh water supply to run the system of sedimentary depositional landforms at

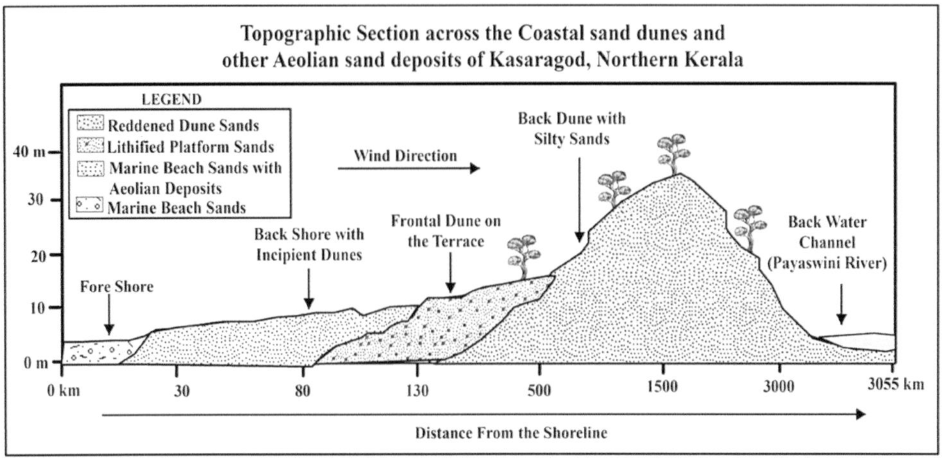

Figure 21.6. The cross-sectional view of the ancient aeolian deposits on the land ward part of Kasaragod (northern Kerala), Malabar Coast.

the sea front position of the coastal state. The coast line is characterized by landforms dominated with beaches, bars, spits, cliffs, back waters, coastal Teri sands and sand dunes. Sand dunes are found all along the coast of Kerala except for areas around Kovalam. Sand dunes of the Kollam district are very rich in placer minerals, but in other areas they are treated as building materials, thus illegally mined without realizing their role in protecting the coastal environment. Fortunately, they are covered by vegetation in many areas of dune landscapes along the coastal parts of Kerala (Bhalla et al., 2008).

Teri dune complex of coastal landscapes dominated by marine deposits of eolianite and characteristic red sand and silt dunes are visible in South Kerala as evidence of paleo environmental significance in the coastal state (Fig. 21.6). Aggradations of red sand dunes took place between 16000 YBP and 9000 YBP at the foothills of the Western Ghats, and between 9000 YBP and 3000 YBP at the west coast of Kerala due to the strengthening of South West Monsoon, sea-level changes and local recycling of earlier dunes (Alappat et al., 2013).

3.6 Coastal sand dunes of Coromondal region (Tamilnadu, Puducherry and Coastal Andhra Pradesh)

The aeolian deposits of northern Tamil Nadu coasts are distributed along the shore fringes, barrier spits of Lake Pulicat at the border states of Tamil Nadu and coastal Andhra Pradesh, beach ridges of Cauvery and the Pennar Delta and along the deltaic shorelines fringed with the Bay of Bengal. The extensive shorelines from Chennai to Mamallipuram represent marine beach sands, beach berm terraces with aeolian deposits with fringing dunes at the back shores. Archaeological sites and temple sites are excavated from the backshore with aeolian sand capping in areas of Mamallipuram. Coastal sand dunes were degraded in the deltaic areas during the tsunami of 2004, and for placer mining and tourism infrastructure development along the sandy tracts of the coastal belt.

Diversity in coastal sand dunes is visible in the southern coast of Tamil Nadu. They are characterized by Teri sands and reddening of sand dunes, aeolian sand capping on the marine terraces, aeolian deposits on the carbonate platforms, offshore islands and along the beach fringed shore lines of the southern state of India. The Teri dune complex of Southeast coastal Tamil Nadu consists of quaternary marine deposits with aeolian sands and reddening of accumulated sand and silt dunes. There is a weathering history of such red dunes with ilmenite rich aeolian sands and silts in the region since the last glacial low stand to the beginning of the Holocene marine transgression under tropical hot and humid environment and monsoon wind patterns. 'Muttom' represents the most ideal depositional sites of silt dunes in and around the south western shores and they are

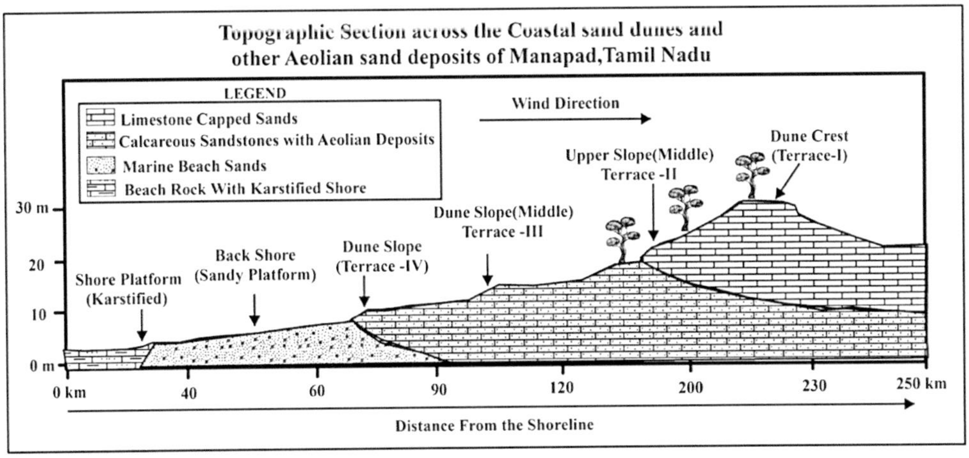

Figure 21.7. The cross-sectional view of Manapad sand dunes over the marine terraces of Limestones and Calcareous Sandstones.

now lateritized and modified into badlands by fluvial erosion. Teri sands are well distributed in Kanyakumari, Tuticorin, Tirunelveli, Ramanathapuram and Thanjavur districts of Tamil Nadu.

The Manapad and Tiruchendur coasts consist of elevated sand dunes over marine terraces built by calcareous sandstones, limestones and beach rocks along the south eastern shores of the coastal state fringed with the Bay of Bengal. Manapad is the highest elevated dune site (around 34 m in altitude) over the four successive marine terraces of the Coromondal coast in India. Beach rocks, marine beach sands and aeolian sands are found exposed over the shore platform at the foot slope of the sandy hills. A few vegetation patches are located along terrace slopes filled with aeolian sands and provide a significant topographic barrier for further accumulation of windblown sands from the south west and north east direction in respective seasons. The cross stratification of ancient aeolian deposits with foraminifers in a cemented condition (eolianite) is visible on the erosional exposure section of the sea face at present (Fig. 21.7).

Coastal sand dunes along the 22 km coastal stretch of Puducherry consist of stabilized dunes with floral diversity and degraded dunes with mobility and reactivation of aeolian sands in places. The backshore dunes are mostly exposed to the anthropogenic disturbances in the forms of stock grazing practices and bio-resource collection of medicinal plants. Large amounts of dune sands are moving inland across the wetlands where the flora has already eliminated along the coastal belt.

The tsunami inundations of 2004 exposed the buffer function or the inability of sand dunes to act as physical barriers against the incoming sea waves and tidal surges in the southern part of the Puducherry coast. The smaller dunes at the beach fringe and estuary fringe areas however, support the habitat for indigenous flora—creepers, herbs, scrubs, grasses and small trees.

The aeolian deposits of coastal Andhra Pradesh extending from Masulipatnam bay to Srikakulam district are diversified in their formations and categorized into fringing dunes, frontal dunes, aeolian sands capping over the beach ridges, barrier island and barrier spit dunes, deltaic platforms dunes, sandy hills on the marine terraces, ancient dunes of silica sand rich materials (sillimanite) with ilmenite rich mineral deposits (e.g., red hills of Vishakhapatnam), embayment fringe dunes and transgressive dunes.

The red hills of Vishakhapatnam adjacent to Thotlakonda were deposited under marine and aeolian conditions of the environment during the last glacial low stand (18000 YBP) and gradually emerged before the Holocene optimum period (12000 YBP) and modified through the weathering history under the tropical monsoon climate of the region. The silica sands with ilmenite rich minerals were hardened and lateritized by a weathering process and converted into badland topography with development of rills and gully floors and sharp divides by active down cutting process with fluvial erosion in the late Holocene period. Large scale transgressive dunes on the marine terraces and cliff

Figure 21.8. The ancient aeolian deposits of silliminite sands of Visakhapatnam (Red Hills), and Iliminite silty sands of Muttom (Kanyakumari) over the elevated surfaces are reworked by fluvial erosion and altered into indurated sand and badland topography.

top areas are visible along the backshores of Bheemunipatnam, Srikakulam and Vishakhapatnam areas in coastal Andhra Pradesh. The activities of beach and dune sands mining degraded and depleted the areas of marine and aeolian sand deposits in the coastal state. The coastal stretch between Nagavalli and Vamsadhara Rivers along the Srikakulam coast has great potential for mining of ilmenite rich sands, but the marine beach sands and frontal dunes are degraded due to unregulated large-scale mining process at present. The barrier spit sand dunes at the Godavari and Krishna Deltas are rapidly depleting due to coastal erosion along the sea face. Attempts had been made to conserve the coastal sand dunes in the Kakinada district by large scale plantation of *Casuarina equisetifolia* in the earlier decades. The primary sand dunes of deltaic shores, embayments and barrier spits, however provided good habitat for indigenous vegetation along the shorelines of the southern districts of coastal Andhra Pradesh (Fig. 21.8).

3.7 Aeolian deposits of the North Circar Coast (Odisha and West Bengal Coasts)

The north eastern shores of coastal Odisha and coastal West Bengal are fronted with the northern Bay of Bengal. The coastal stretch between the Ganjam district of Odisha and south 24 Parganas district of West Bengal represent diversified coastal sand dunes in morphology, genesis and vulnerabilities. The wide backshores, lofty sand dunes, frontal sand dunes, barrier spit sand dunes, aeolian sand capping on beach ridges, fore dune ridges, embryo dunes, aeolian sand platform terraces, transgressive dunes, over wash sand fan lobes, blowout dunes, parabolic dunes and estuary fringe dunes are distributed in the non-deltaic, deltaic and lagoonal coasts of the region. The entire coastal stretch is assessed as the most cyclone vulnerable (Mohapatra, 2015), sea-level rise vulnerable (National Disaster Management Plan, 2016) and erosion vulnerable tracts (Ramesh, 2017) in India. The significant functions of coastal sand dunes as physical barriers, sediment sinks, coastal habitats and ground water aquifers are not understood by the local people and coastal authorities in these tracts. Dunes acted as a strong physical barrier against tidal surges, storm surges and coastal currents in the near past (even before 1970s), but are now widely depleted in areas and altitudes at the coastal belts. Sand mining, coastal tourism-recreation infrastructures, coastal urbanization and over uses

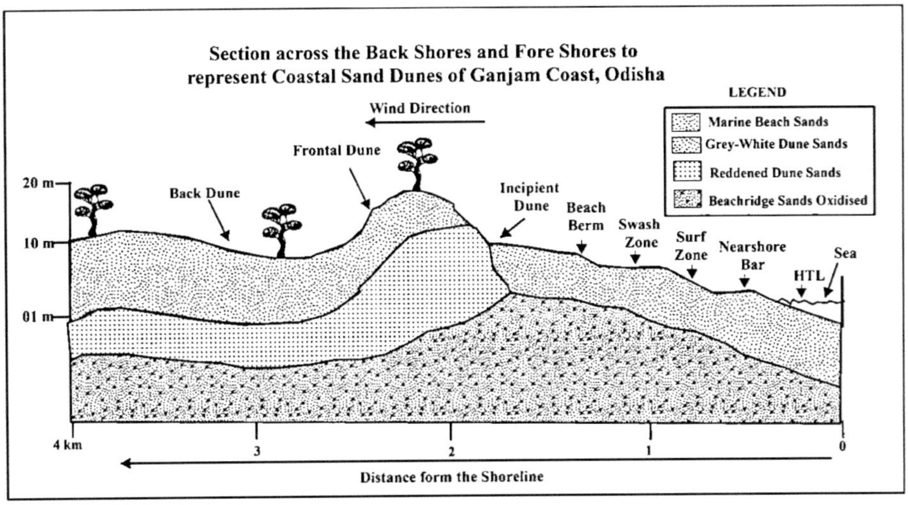

Figure 21.9. The cross-sectional views of the foreshores, immediate back shores, lofty sand dunes and inland backshores of the Ganjam Coast in Odisha.

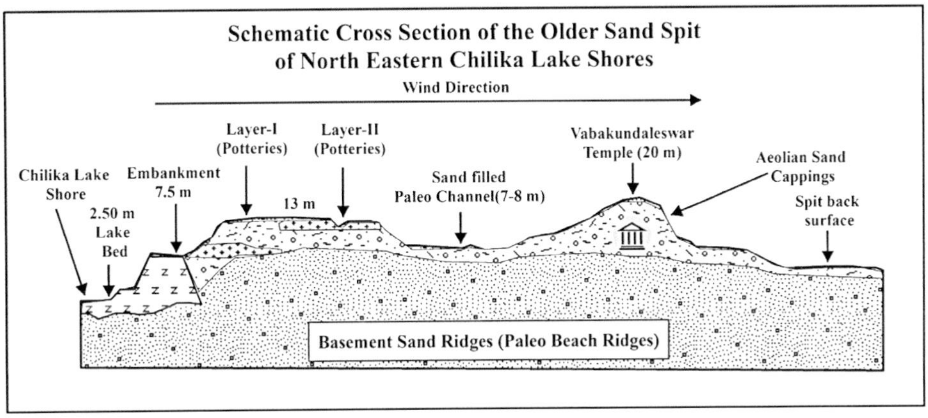

Figure 21.10. The cross-sectional view of the older spit backed dunes from the lake shore area to the inland backshores of Chilika lagoon.

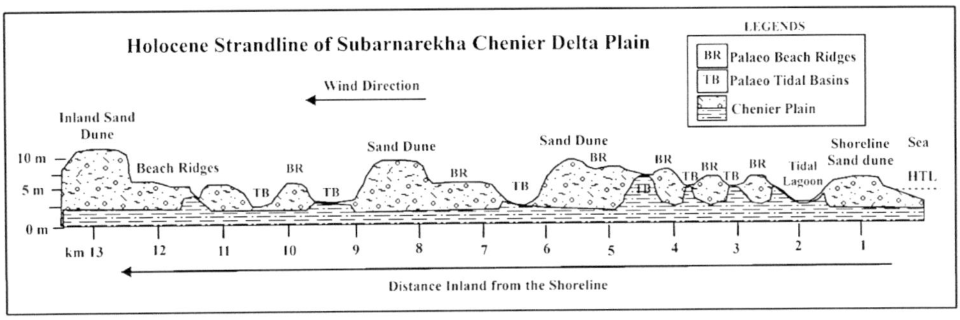

Figure 21.11. The topographic cross section of the Chenier coast of the Subarnarekha Delta across the paleo beach ridges capped by sand dunes.

of dune platforms for dry fish processing have depleted the sand dunes with large scale activities after 1990s along the coastal tracts. Climate resilient dune systems need to be rehabilitated urgently along these vulnerable coasts to protect the coastal communities from rising sea-level and impacts of storm surges (Figs. 21.9 and 21.10).

The older sand dunes are more oxidized, have more vegetation cover, and are higher in elevations. They are mostly found in the inland backshores, but also in shore parallel inner beach ridges with aeolian sand cappings in Ganjam, Puri and northern Balasore district of Odisha and the Medinipur littoral tracts of West Bengal. However, the younger dunes are located along the shore fringes of the deltaic, non-deltaic and lagoonal coasts of the region, and bear the signatures of vegetation transformations with morphological diversities in the tropical hot and humid environment. The sites of archaeological tools, temples, ancient sea ports and other evidences of ancient human settlements are explored from the depths of older sand dunes in the region.

3.8 Coastal sand dunes of the Island Territories

Aeolian sand deposits are found along the coastal fringes of Andaman and Nicobar group of islands in the oceanic environment of the Bay of Bengal and Andaman Sea. There are 25 major sea beaches in the group of islands from North Andaman to Great Nicobar district. Field documents of such aeolian sand deposits along the coastal fringes exhibit 'raised beach' terraces and the 'shore platforms' fronted by wide fore shores which hold the narrow sections of primary dunes. The Karmatang beach of North Andaman, Ramnagar and Bharatpur beach of Neil Island, Radhanagar beach of Havelock Island, Hut Bay and Buttler Bay of Little Andaman, Wandoor and Jolly Buoy Island of South Andaman represent areas of aeolian sand fields with vegetation cover by creepers, grasses and woodlands. The aeolian sand fields of the above island sections were over washed by the 2004 tsunami waves and modified into transgressive sand sheets. The silica sands and coralline sands have contributed the sediment materials with other bio-clasts for aeolian deposits under favorable conditions of strong onshore winds and stronger north-east monsoon winds in the island territory.

The islands of the Lakshadweep group of the Arabian Sea, however, occupy the aeolian sand ridges parallel to the shore line ranging from 3 to 6 m in altitude above the mean sea-level.

4. Geomorphology of coastal sand dunes: (Variable Scales of dune development)

The morphology of coastal sand dunes reflects the balance between constructive aeolian processes, complex marine processes, coastal configurations with emergent platforms, history of weathering processes and floral habitats in the subcontinent. A genetical scheme of dune classification is prepared for the western and eastern shores of India in this chapter to highlight the diversity of coastal sand dunes under the tropical coastal environments. The younger dunes are very dynamic due to constructive aeolian processes and destructive marine processes along the coastal belt (Tables 21.1 and 21.2).

The study by (Paul, 2002) reveals that under a favorable environmental gradient, the wide beach fringed platforms with regular supply of marine sands initiate dune building processes with transformation of vegetation colonies and accumulation of sands by strong on shore winds. The stages of dune development are marked by the deposition of tidal litters with seeds and rootlets of vegetations along the spring high tide shoreline; incipient fore dunes on the fore shore platform with primary vegetation colony development and accumulation of sands; hummocky shaped fore dune ridges on the back shores with next stages of vegetation colonization and a high rate of sand accumulation; frontal dunes or fringing dunes with elevated topography and vegetation covers on the back shores; and matured dunes on the inland back shores with climax vegetations controlled by small trees, bushes, long rooted grasses and heaths which remain as large sandy hills by a constant supply of aeolian sands in the coastal region. Topographically, such widespread dune landscapes can be classified as dune ridges (14–22 m), hummocky dunes (8–12 m), dune valleys (6–8 m), along the ridge slopes, dune platforms or dune flats (5–6 m) and dune furrows (2.5 m–4.5 m) in the region of primary dunes of the coasts influenced by vegetation covers (Fig. 21.12).

During the 1970s large dune belts of elevated topography (10–22 m in altitude) existed along the shorelines of Odisha and West Bengal coasts from Gopalpur to Digha. The higher dunes with

Table 21.1. A scheme of genetical classification of aeolian sand deposits on the Western shores of India (Arabian Sea fringes).

Sl. No.	Physiographic Setting	Types of Dune Morphologic Systems	Major Examples
1	Teri Dune Sands	Coastal teri; Inland teri; Marine beach sands; Ancient dune ridges, Younger dune ridges	Muttom (Tamil Nadu), Vatkal (Kerala), Kundapura (Karnataka), Agonda, Kola (Goa)
2	Protected Main land Coasts	Barrier bar sand platform,; Spit back Dunes; Transgressive dunes; Marine overwash	Neeleswar, Kannarnore,Vypeen Island barrier (Kerala), Gokarna, Kumta, Karwar, Bhatkal, Honnavar, Kodi Bengre, Kasba Bengre (Karnataka)
3	Open Coast Shore Fringe Areas	Marine beach sands; Embryo dunes; Fore dune ridges; Frontal dunes; Blow out dunes	Diveagar, Kelshi, Tambaldeg, Velaghar (Maharashtra)
4	Bays and Coves	Marine beach sands, Fore dunes; Aeolian sand platforms; Fringing dunes; Bay fill sand ridges	Kolva, Calangute, Miramar, Cola (Goa), Velneshwar (Maharashtra)
5	Beach ridge Cheniers	Beach berm platforms; Multiple dune ridges; Aeolian cappings on beach ridges	Kanhagod, Kasaragod, Kozhikode, Poovar (Kerala), Shrivardhan, Bhagwati, Purnagad fort (Maharashtra)
6	Beach Rock Platform	Cliff Top dunes; Climbing dunes; Aeolian capping on the beach rocks	Kelshi, Velas, Ratnagiri (Maharashtra), Daman (Gujarat)
7	Carbonate Plat Forms	Raised beach sands; Fore dune ridges; Relict dunes; Aeolian sand platform	Diu Island, Okha Mondal, Porbandar, Dwarka, Mandvi (Gujarat Coast)
8	Estuary Fringe	Barrier Spit dunes; Fore dune Ridges; Transgressive dunes; Aeolian sand platforms	Mandovi mouth, Zuari mouth, Talpona (Goa), Gulf of Khambhat, Gulf of Kachchh (Gujarat)
9	Emergent Coasts	Marine Beach sands; Cliff top dunes,; Climbing dunes; Aeolian sand cappings on the marine terraces; Relict dunes	Ankola, Gokarna, Kumta, Honnavar, Murdeshwar, Apsarakonda (Karnataka)
10	Mouth of the Tidal creeks or Tidal Inlets	Beach sands; Frontal dunes; Fore dune ridges; Hummocky Dunes; Blowouts; Parabolic Dunes; Marine Wash over	Mhashala Creek, Kundalika-Alibagh Creek (Maharashtra), Kali River, Gangavali, Bhatye Creek, Kalba Devi Creek (Karnataka)
11	Offshore Islands	Marine beach dunes; Fringing dunes; Aeolian sand cappings or marine terraces; Barrier spit dunes; Transgressive dunes	St. Mary's, Coconut Island (Karnataka), Ilha Grande Island (Goa), Pirotan Island, Dwarka Beyt (Gulf of Kachchh), Laksha Dweep Island

areas of cover are fronted by wide beach berm platforms and gently to moderately sloping beach faces in the region of the Ganjam district, Puri district, Balasore district and Midnapore district. Various secondary dunes are now evolved along the same coast after erosion in the form of blowout dunes, parabolic dunes, transgressive dunes, barchanoid dunes, sea facing dune cliffs, extensive aeolian sand fields, overwash sands and climbing dunes. Complete dune elimination occurred in the coastal region particularly along the Chilika New Mouth Spit, Gahirmata spit, Shankarpur-Mandarmoni coast and Sagar Island-Freserganj coast of Odisha and West Bengal after the events of repeated cyclone landfalls since 1988 to 2020 (Fig. 21.13) and (Table 21.3 and Fig. 21.14).

Aeolian sand capping over the successive beach ridges of the Chenier coast have produced inland coastal sand dunes along the paleo shoreline with elevated topography, wide-spaced inter dune paleo tidal basins with intersecting tidal rivers and subdued ridge flats particularly along the deltaic coasts of eastern shores (Fig. 21.15). Radio carbon dating records of the beach ridge capped marine shells and Optically Stimulated Luminescence date records of the aeolian sands reveal that the furthest paleo beach ridge from the present-day shoreline to the modern beach ridge of the shore front position were formed during 6900 YBP, 5700 YBP, 2900 YBP and 1000 YBP respectively in the deltaic shores with transgressive and regressive seas (Ramkumar et al., 2016; Sabyasachi

Table 21.2. A scheme of genetical classification of aeolian sand deposits on the Eastern shores of India (Bay of Bengal fringes).

Sl. No	Physiographic Setting	Types of Dune Morphologic Systems	Major Examples
1	Deltaic platforms	Marine Beach Sands, Embryo dunes; Fore dune ridges with vegetations; Frontal dunes; Transgressive dunes; Aeolian sand platforms; Parabolic dunes; Barchanoid	Bakkhali, Fresarganj, Sagar Island (West Bengal), Gahirmata, Paradweep (Odisha), Kodiya Karai, Sorla Gondi (Tamil Nadu), Guntur, Kothapalem (Andhra Pradesh)
2	Estuary fringe areas	Beach sand platforms; Fore dune ridge; Spit back dunes; Frontal dunes; Transgressive dunes; Overwash sand fan lobes	Talsari, Dagara, Chaumukh, Rushikulya Mouth (Odisha)
3	Open coast shore fringe areas	Frontal dunes; Multiple dune ridges; Transgressive dunes; Blow out dunes; Fore dune ridges; Aeolian sand platforms	Digha-Shankarpur Coast, Mandarmoni Coast (West Bengal), Puri-Konark & Gopalpur Coast (Odisha), Vishakhapatnam, Bheemunipatnam (Andhra Pradesh), Mamallipuram, Rameshwaram (Tamil Nadu)
4	Beach ridge chenier coast	Shore parallel ancient dune rows; Aeolian cappings on beach ridges; Multiple dune ridges, Aeolian sand platforms; Elevated Beach ridge platforms	Subarnarekha Delta, Penner Delta
5	Protected mainland coasts	Barrier island dunes; Spit back dunes; Transgressive sand dunes; Aeolian sand platforms; Overwash sand fan lobes; Marine Beach sands	Pulicat Lake Spit, Sriharikota, Pichavaram Spit (Tamil Nadu), Godavari Spit (Andhra Pradesh, Gahirmata Spit (Odisha)
6	Offshore Island dunes	Island fringe dunes; Transgressive dunes; Barrier spit dunes; Tombolo dunes; Embryonic dunes; marine beach sands; Blow out dunes	Rameswaram Island, Krusadai Island, Pullivasal Island, Manauali Island (Tamil Nadu), Bhitarkanika Island (Odisha)
7	Lacustrine shoreline	Barrier spit dunes; Fringing dunes; Embryonic dunes; Hummocky dunes; Beach berm platforms; Transgressive dunes, Overwash fan lobe terraces; Island Interior ancient sand dunes; Aeolian sand platforms	Chilika New Mouth, Vhabakundaleswar, Manikpatna, Brahmagiri (Odisha)
8	Embayment coasts	Fringing dunes; bay mouth barrier dunes; Marine beach sands; Aeolian sand platforms	Corringa Bay, Uppada (Andhra Pradesh), Sriharikota, Tuticorin Bay (Tamil Nadu)
9	Emergent coasts	Marine beach sands; Cliff top dunes; Beach fringe dunes; Ancient inland dune ridge platforms; Aeolian cappings on the beach rocks and marine terraces	Manapad, Tiruchendur, Kanyakumari, (Tamil Nadu), Veerampattinam, Auro Beach (Puducherry), Vishakapatnam (Andhra Pradesh)
10	Carbonate coasts	Carbonate beach sands; Fringing dunes; Raised beach platforms dunes; Transgressive dunes; Aeolian cappings on marine terraces (Reef Terraces); Tombolo dunes; Barrier Spit Dunes	Rameshwaram Island, Kanyakumari (Tamil Nadu)
11	Coastal areas of Teri sands	Inland Teri dunes; Coastal Teri dunes; Ancient dune ridges; Younger dune ridges	Tirunelveli, Tuticorin, Pamban, Kanyakumari (Tamil Nadu), Red Hills, Bheemunipatnam (Andhra Pradesh)

et al., 2009, 2011, 2013; Chakrabarti, 1991; Paul, 2002). However, the sub-surface dune sands of the paleo-beach ridges through Optically Stimulated Luminescence dating reflected the ages from top to bottom sequentially from 500 YBP-200 YBP; 1500 YBP-1000 YBP; 2900 YBP-2200 YBP (Linashree et al., 2020; Kaushik, 2013; Sabyasachi et al., 2009).

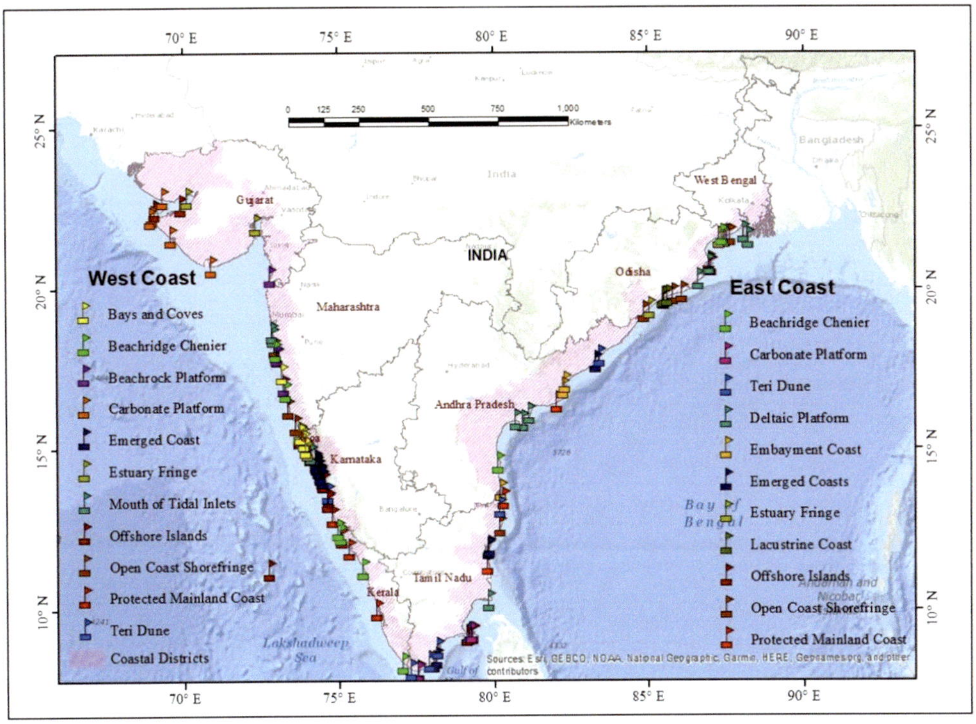

Figure 21.12. The location of coastal aeolian deposits under diversified physiographic settings along the western shores & eastern shores of India.

Figure 21.13. The aeolian sand deposits and overwash sands encroached into the mangrove wetland from the foreshores of Henry's Island, Sundarban.

Table 21.3. Increasing frequency of High magnitude cyclone events in the northern Bay of Bengal Since 1885 to 2020.

Sl No	Year	Month	Place of Landfall	Intensity	Lowest Pressure (hpa)	Maximum wind Speed (km/h)	Storm Surge (m)	Loss of Human life
1	1985	September	Paradweep	Super	918.9	252	7	5000
2	1997	October	Bangladesh	Super	922	250	< 7	175000
3	1927	November	Nellore	Super	919.3	250	6	6000
4	1942	October	Midnapore	Super	935	225	5	15000
5	1970	November	Sundarban	T7.0	920	260	< 7	500000
6	1977	November	Chirala	T7.0	911	259	5	10000
7	1989	November	Kavali	T6.5	930	235	7	56
8	1990	May	Muslipattanum	T6.5	930	235	7	1000
9	1991	April	Bangladesh	Super	932	220	< 6	138000
10	1999	October	Paradweep	T7.0	923	260	7.8	> 9500
11	2007	November	Bangladesh	Super	930	220	7	10000
12	2008	May	Yangon	Super	930	240	6	22000
13	2009	May	Sundarban	T6.0	960	170	5	191
14	2013	October	Gopalpur	T5.5	940	250	7	23
15	2019	November	Sundarban	T5.0	960	205	3	11
16	2020	May	Sundarban	T7.0	940	200	2	18

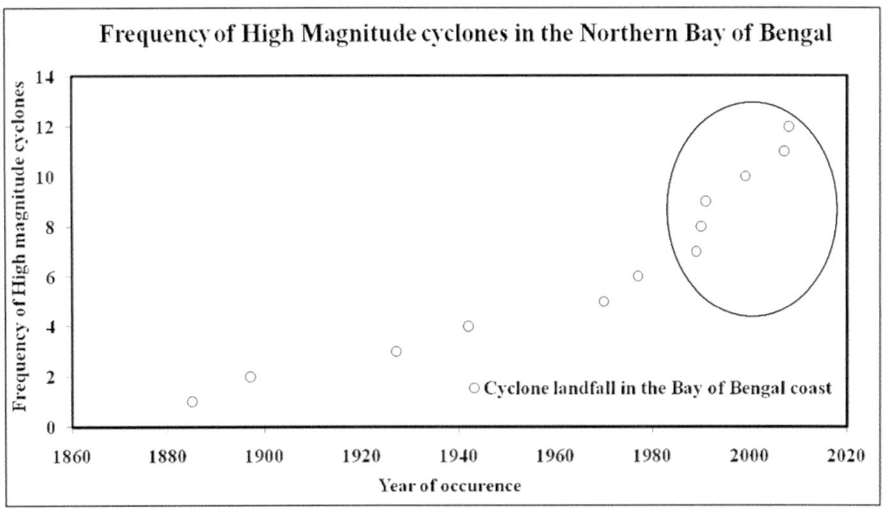

Figure 21.14. The graphical representation of the cyclone land fall occurrences in the northern Bay Bengal coastal fringes adjacent to Odisha, West Bengal and Bangladesh coasts (1885 to 2020).

The Teri dunes of Coromondal and Malabar coasts lie between 20 to 60 m above the sea level. The sands of Teri dunes were deposited in two episodes at the lower and middle parts of extensive paleo dune fields during the late Pleistocene (i.e., 16000 YBP-9000 YBP) sand cappings over upper terraces and late Holocene (i.e., 4500 YBP) sand accumulations over the lower-level back shore region (Alappat et al., 2016) on the South western coast of Tamil Nadu. The abundant sand supply was provided by lower sea level and sand-laden strong onshore winds across the wide foreshores

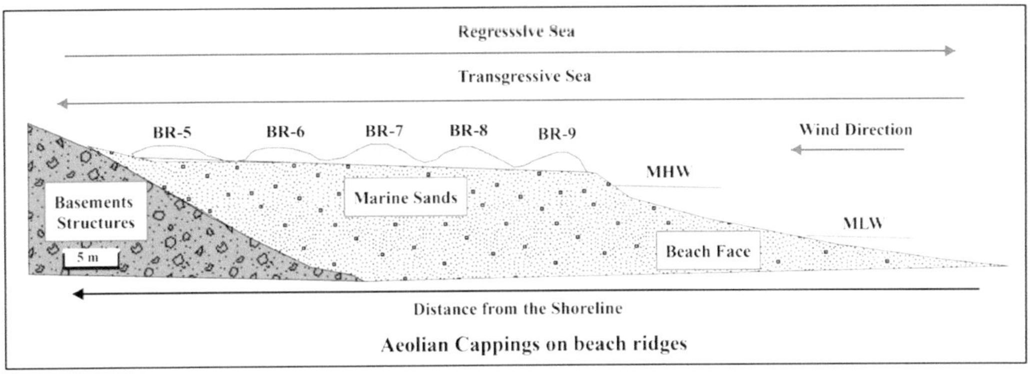

Figure 21.15. The cross-sectional views of aeolian sand cappings over paleo-beach ridges of deltaic coasts in India.

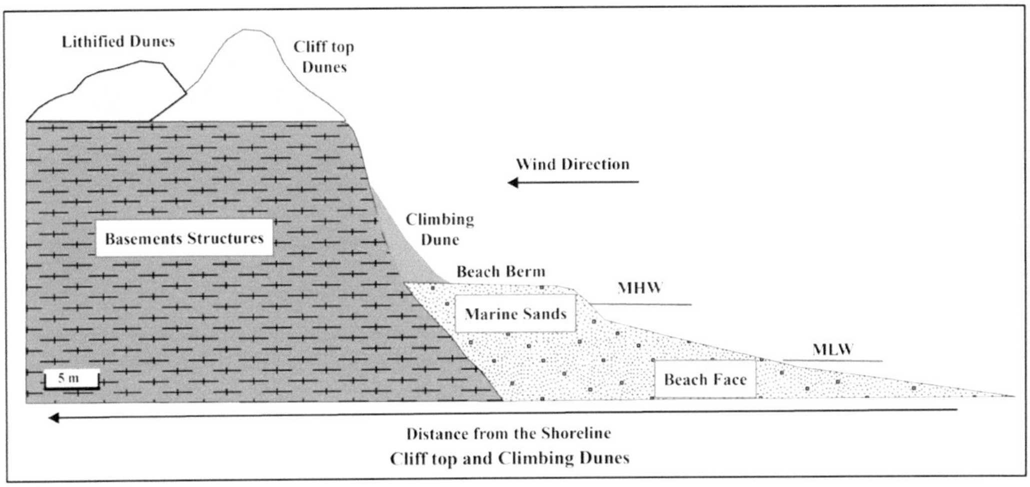

Figure 21.16. Various cliff top dunes and climbing dunes across the foreshores and backshores terraces in the western coast of India (schematic view).

and backshores in the late Pleistocene time. However, aeolian sand supply was slightly reduced in the late Holocene period due to high sea level and reflected as relatively less thick deposits over the indurated aeolian sand across the lower-level backshores. Finally, the geochemical alteration of hematite and ilmenite minerals of aeolian sands produced red colored dunes in the region. The emergent and elevated topography of such paleo dunes indicate the effects of slow and steady rise of the coastal fringe depositional surface owing to the activity of neotectonics.

Paleo dune fields of areas from Vishakhapatnam and Bheemunipatnam of coastal Andhra Pradesh represent the elevated topography of marine terraces over which the aeolian sands are composed of silliminite, ilmenite, hematite, zircon, garnet, monazite, etc. minerals. The red color, brownish red and yellow color sands of such aeolian deposits are the result of geochemical alteration of heavy minerals under oxidizing environment of sediments in the coastal belt. Indurated aeolian sands of the 'Red hills' of Vishakhapatnam are fluvially altered by rilling and gullying processes at present. Reddening of sand is also found in the underlying layers of aeolian deposits in the Ganjam coast of Odisha over which the aeolian sand cappings of younger deposits are located in the form of elevated surfaces (Fig. 21.16).

The relict dunes of Diu Island of the coast of Saurashtra and Rameswaram Island of the Tamil Nadu coast are lying over the late Pleistocene high stands fringed by raised beach platforms with carbonate rocks. Pleistocene dunes are altered into red colored sands and late Holocene dune deposits of the lower terrace represent gray colored sands in Pamban and Ramarpatham areas of

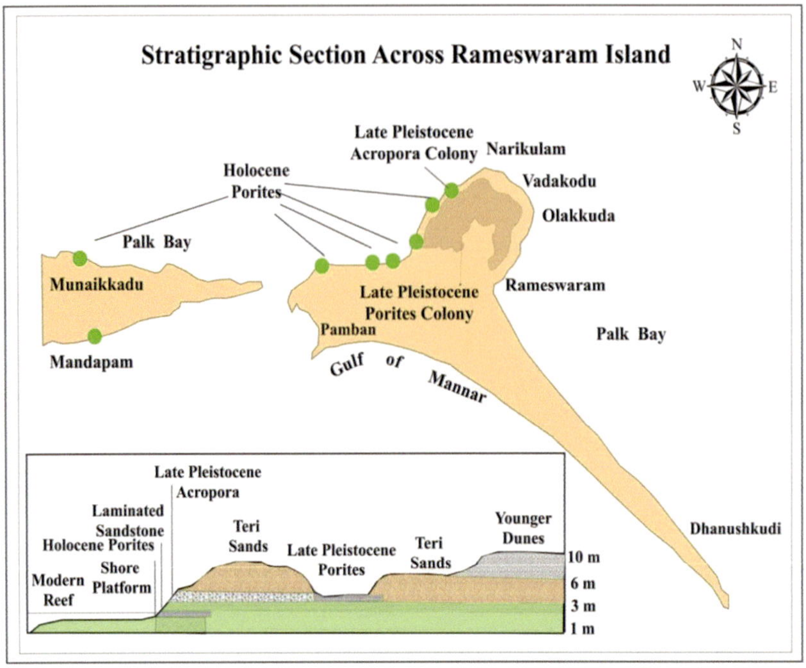

Figure 21.17. The location of older sand dunes and reddened dunes over the paleo high stands (carbonate platforms) of Rameswaram Islands.

the Rameshwaram Island. Relatively younger dunes are located on the cliff top areas of raised coral banks and beach rock platforms along the 'Palk bay' coast of the island (Fig. 21.17).

The largest and highest dune fields are located in the form of ridge crest aeolian sand cappings, upper-middle and lower slope aeolian deposits of the marine terraces in Manapad and Tiruchendur of the South eastern coast of Tamil Nadu. The maximum volume of windblown sands is trapped along the middle slope of elevated topography underlain by limestone terrain. The dune fields are affected by erosion at its sea ward side slope at present due to narrow foreshores and encroaching sea waves up to the dune base areas in the coast.

Relatively younger dunes of inland areas are found off the present-day coast line in the form of aeolian sand cappings over the indurated aeolian sands of marine terraces in northern Kerala and North Kannada coasts. Owing to the lack of abundant sand supply on the present-day shoreface, immature sand dunes are found to grow only along the back shores of sediment tight pocket beaches under the protective headlands in the south western coast of the subcontinent. Aeolian sand deposits of Goa and Ratnagiri are depleted due to erosion and anthropogenic pressures. Some dunes of the Ratnagiri coast are seen as topographic highs nearby the Mandavi beach at the south western end of the Jaki Mirya Bay. Such fringing dunes exist only in the beach fronted bays and coves in the Konkan coast where abundant sand supply is available. The presence of rocky shore platforms and sea ward projected rocky headland prevented favorable environmental regimes for growth and development of matured sand dunes along the western shores of the subcontinent. However, the beach ridges and barrier spit sand dunes of younger ages exist along the western shores of Malabar and Konkan coasts with favorable wind regimes and sand supply regimes at present.

Modern dunes and paleo dunes are found in the Saurashtra Peninsula and Kachchh (Kutch) Peninsula of Gujarat coast under favorable sand supply environment. The steep dip orientation of beddings of the lime cemented sandstones at the coastal setting suggests wind-born aeolian sand supply from the sea ward side in the geological past. Formation of such aeolianites under paleo marine environment conditions associated with gradual cemented processes of aeolian sands by available micrites along the shoreface also reflects the impact of local eustasy in the region. The

Quaternary deposits of milliolite on the Kathiawar Peninsula represent marine and aeolian genesis in the coastal belt with +4 m terrace above the mean sea level and inland occurrences of organogenic grain stones respectively (Bruckner, 1988).

Modern sand dunes are extended over the milliolitic limestone terraces along the shorelines of the Saurashtra coast. Extensive marine beach sands and aeolian sands are fringed along the shorelines of the Kathiawar Peninsula at present.

5. Long term changes in dune morphology

Natural dune barriers and aeolian sand accumulations fringe the low-level coasts and high-level coasts behind extensive sandy sea beaches in India (Fig. 21.18).

Different types of hummocky ridges, shore parallel dune ridges, dune valleys, extensive aeolian sand fields and dune platforms of the stable system have gradually changed with time. Major morphological changes of stable dune systems have been recorded through field survey methods and spatio-temporal image analysis of satellite data of the coastal sections of the Indian subcontinent. Long-term changes in coastal dunes include as:

i. Reduction in dune height and width by storm surges and washing process (Fig. 21.19);
ii. Lateral recession on the landward side with modified form and shape by blow out erosion. The shapes have been changed from hummocky ridge to turtle back surface and normal shore parallel fronted dune ridges to transgressive dune ridges (Fig. 21.20);
iii. Strong onshore wind velocity (40–50 km/hr) during dry periods (March–May) has generated hairpin dunes, parabola shaped dunes, barchanoid dunes and aeolian sand flats across the frontal dune barriers along the coastal belt (Fig. 21.21);
iv. Development of significant cliff face on the sea ward side of frontal dunes behind the beach due to wave encroachments in the dune base during high seas of the summer monsoon months (July–September) in the eastern and western shores of India (Figs. 21.22 and 21.23);
v. Lateral progradation of unconsolidated dune sands along the emerged sand spits and barriers in the low-level coasts;

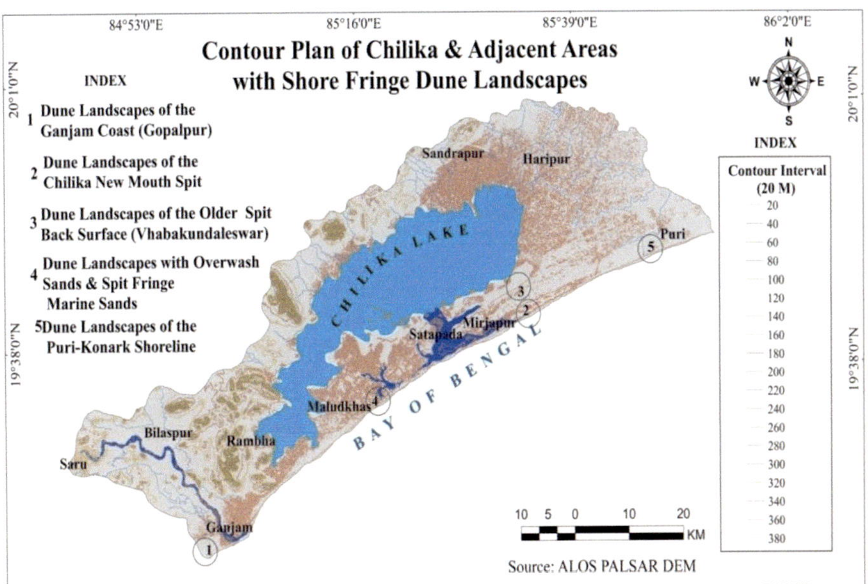

Figure 21.18. Elevated sand dunes over the younger barrier spits and older sands spits in front of the Chilka Lake has been changed over time due to erosion. The major dunes are shown by 20 m contour lines by using ALOS PALSAR DEM.

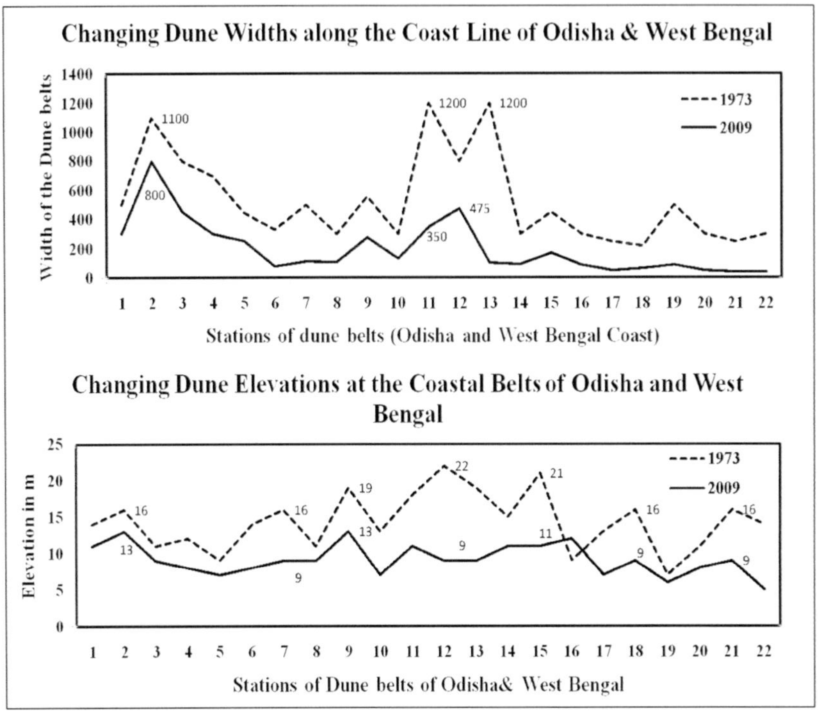

Figure 21.19. Reduction of coastal sand dunes in elevation and width assessed from coastal sites of Odisha and West Bengal (1973 to 2009).

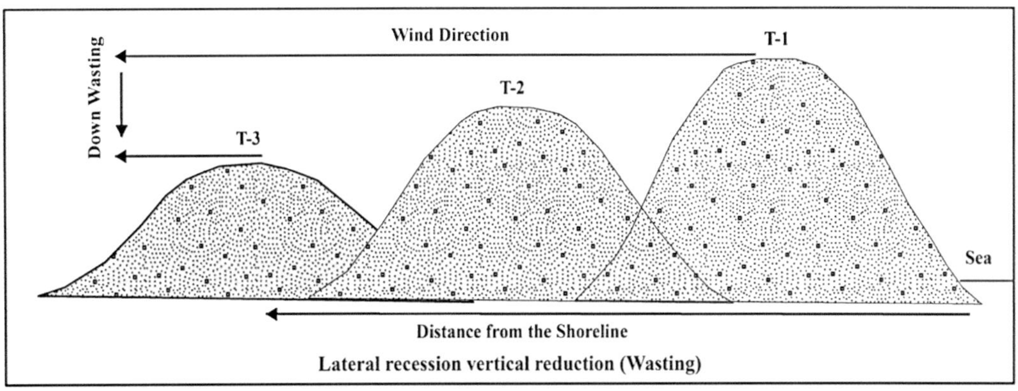

Figure 21.20. The altered forms and shapes of the coastal fore dune ridges through the lateral recession on the land ward side of the coastal belt along the states of West Bengal and Odisha.

vi. Formation of extensive aeolian sand fields with gently undulated surfaces in place of eroded dune ridges by repeated cyclone breaches and over wash activities (Fig. 21.24);

vii. Total removal of foredune ridges and elevated frontal dune ridges and formation of wave-planed sand sheets in the deltaic shores by the impact of increasing high magnitudes cyclone landfalls;

viii. The diversity along shore elevation view of the frontal dunes with the presence of high and low crest heights indicated that where low points in the dune crest coincide with sections of narrow beaches, the highest risk of dune over washing may occur in the region. Examples of

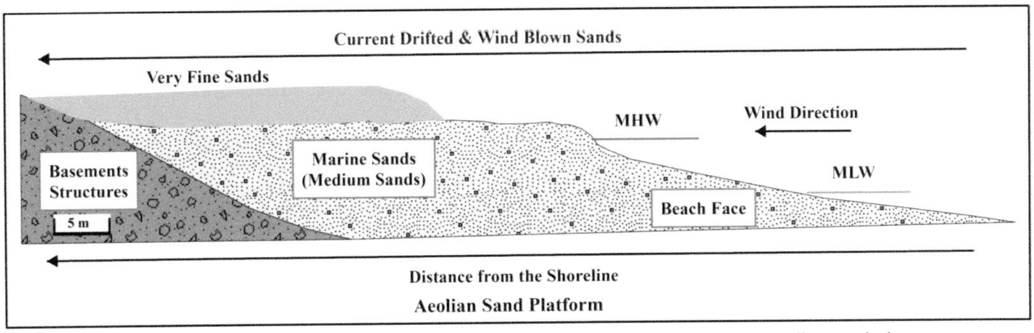

Figure 21.21. The altered aeolian sand platforms in coastal barrier spits in Andhra Pradesh.

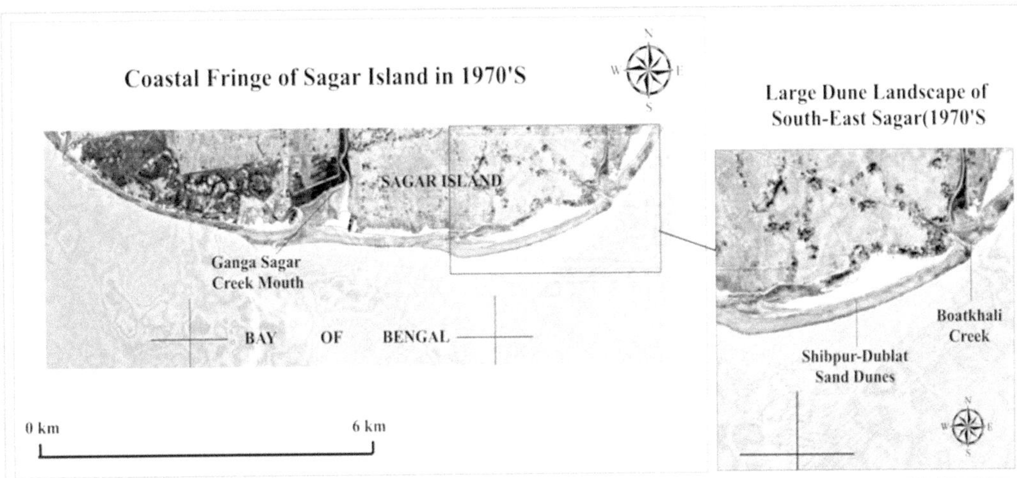

Figure 21.22. Alteration of frontal sand dunes in the deltaic Sagar Island by marine processes (West Bengal Coast). Presently the sand dunes of 1970's have been altered and eliminated (US space-based camera photograph).

Figure 21.23. The frontal dunes of 1970's (US space-based camera photograph) of the Digha-Shankarpur coast have been degraded significantly at present.

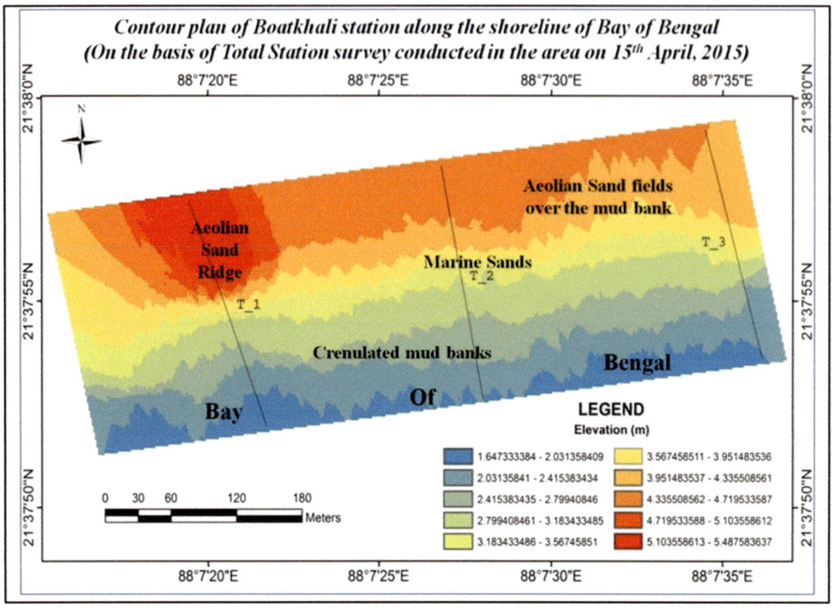

Figure 21.24. The erosive dune platforms altered from fore dunes and frontal dunes documented by Total Station Survey in the Sagar Island coast through the contour plans along the shoreline (2015).

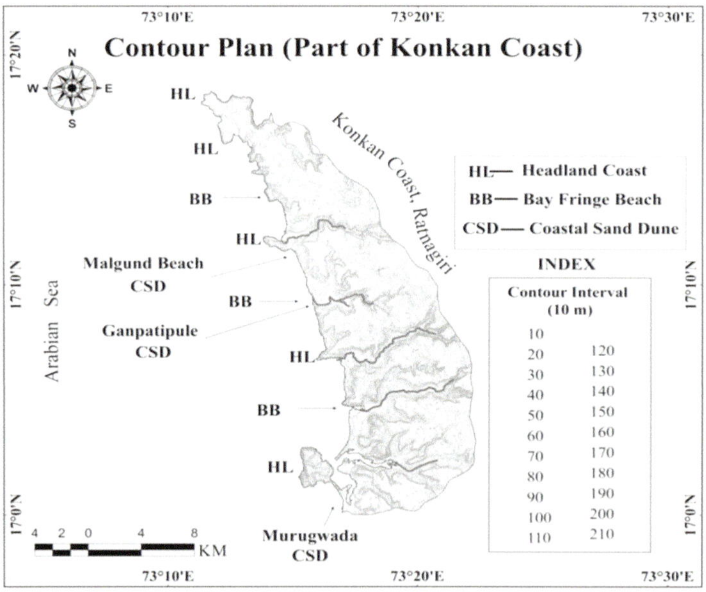

Figure 21.25. The elevated sand dunes of Ratnagiri coast fringed with the Arabian Sea on the western shores of India. The contour plan with 10 m interval shows the location of frontal dunes and backshores dunes at to different levels (10 & 30 m), the contour plan is generated with SRTM DEM.

break through points have been identified across the barrier spits of Chilika Lake, Talsari point, Gahirmata section, Godavari Delta and Krishna Delta in the eastern shores of India;

ix. On the western shores of India along the coastal stretch from Kerala to Gujarat sections, the long-term change in dune morphology represents aeolian cappings on beach ridges, aeolian sand platforms at elevated topography, cliff top dunes and climbing dunes along the marine terraces (Fig. 21.25).

6. Ecology of coastal sand dunes in India

Coastal sand dunes harbor a unique floral diversity along the sea shores of a tropical hot and humid environment. A stable dune belt with complete vegetation covers micro habitats dominated by regimes of geomorphology and climate comprises long rooted grasses, creepers, herbs, heaths, scrubs, small and large trees in the coastal zones of India. The dune crests, dune slopes, dune flats, dune valleys, dune furrows, dune slacks and dune platforms occupy different micro habitats under favorable ecological factors, inundation regimes, wind regimes and moisture regimes (Paul, 2002), in the coastal belt. Ecologically, productive dune habitats also support rich faunal diversity dominated by insects, other invertebrates, reptiles, avifauna, rodents and scavengers along the shorelines.

The coastal sand dunes of the western and eastern shores accounting for 267 and 163 plant species respectively and depicted a greater diversity in dune flora in the west coast than in the east coast (Rodrigues et al., 2011) probably due to the occurrences of higher amounts of rainfall on the western fringe on the Indian subcontinent. Climate resilient dune systems should be well vegetated to act as a natural defense against sea level rise vulnerability and storm surges vulnerability at present. Indigenous vegetation has been removed from matured coastal sand dunes in many places due to anthropogenic pressures and a number of exotic species introduced in place of them to restore the aeolian sand accumulations in the coastal belt for managing the dune habitats. The transformation of vegetation changes with elevating topography by continuous supply of sands and their accumulation rates determine the plant succession from the seaward to landward gradient and evolution of coastal sand dunes under favorable environment. Once the adjusted ecosystems of the coastal sand dunes to the local environment are affected by human modifications for construction of development projects, the entire dune system will be depleted under increased severity of storms and blow out erosion concentrations.

Historically, the coastal lowlands and wetlands behind coastal sand dunes were affected by encroachment of salt waters after incidences of tidal waves and cyclone surges and contributed ideal breeding ground of malarial diseases which killed many people in the villages after cyclone landfalls in Bengal. However, the Dutch communities in Kedogree (Khejuri) port during the 16th century Bengal realized the significance of coastal sand dunes in protection of cyclone surges and tsunami surges from the seaward side in the region. They introduced the plant *Anacardium occidentale* (cashew nut trees) to stabilize the coastal sand dunes in Midnapore littoral tract. This exotic tree species has adapted very well to the local ecological factors over the coastal sand dunes in eastern and western shores of India. Such experiments by the Dutch to make the sand dunes of coastal Bengal more climate-resilient against the cyclone surges and tidal waves protected coastal communities from the managed defense systems in the 16th century. However, the other exotic species, i.e., *Casuarina equisetifolia* (jhao trees) introduced by Forest Departments and planted over the coastal sand dunes since the 1960s, showed various limitations and were later replaced by indigenous vegetation of the coastal zones in the Indian subcontinent. Three major limitations of such dune forestry are observed along shorelines at present, i.e., (i) the undergrowth of the forest floor have been inhibited as the casuarina trees mature; (ii) the aeolian sand accumulations are restricted only along the plantation area that has created weaker and narrower dune ridges vulnerable to storm breaches and (iii) the trees did not act as a good wind break in the cyclones in the coastal states of India during the earlier decades.

The unique plant ecology of the aeolian sand deposits on the coastal zones will be eliminated if the dune systems are depleted by increasing storm severity and rapid rate of sea level rise along the western and eastern shores of India. Restoration of coastal dune habitats needs immediate conservation of indigenous plants over the aeolian sand deposits.

7. Coastal dune sediments

The aeolian sand cappings of paleo beach ridges depict the seaward dipping cross beds and the lower beach ridge stratification reflects parallel laminations with alternate bands of dark and light

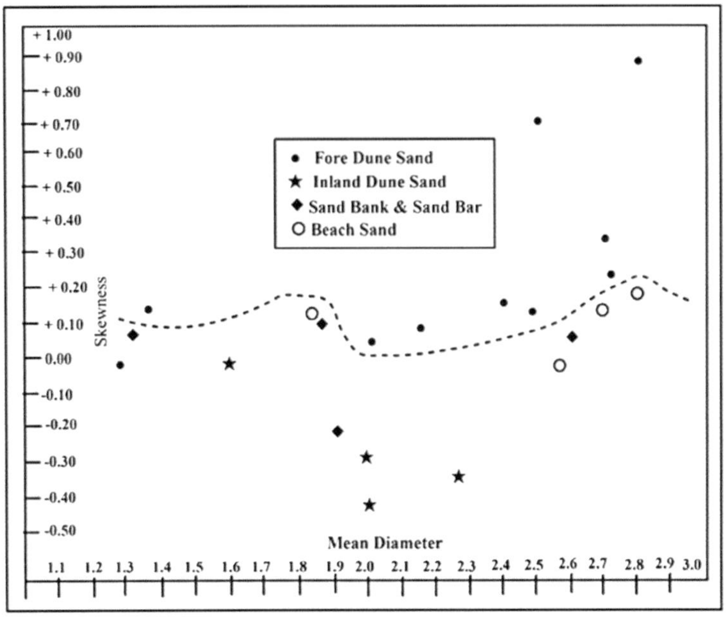

Figure 21.26. Relationships between mean sand size and skewness of four different Sites of the coastal belt of the Sagar Island.

minerals along the landward ridge of Paniparul, Bhograi and Baliapal sections in the Subarnarekha Chenier Delta. Multiple sets of cross beddings are visible on the sea facing dune cliffs along the present-day shoreline of Digha, Talsari, Shankarpur and Mandarmoni sectors of West Bengal and Odisha coasts. Massive dunes over the paleo beach ridge platform of the Talsari and Digha coast represent accumulations of sorted sands with plant roots and other organic matters. However, sorting decreases in dune sediments composed of silty sands in Bahiri near Kanthi. Beach ridge sands below the massive dune bodies contain iron coated quartz sands, shell fragments and other minerals with oxidized surface of brownish to yellowish colors (Fig. 21.26).

The coarser and medium sands of beach ridges are less well sorted. The positive skewness of the frontal dune ridges of Chuksar Island, Sagar Island and Junput dune sands are indicative of strong wind energy of the drier months when the prevailing onshore winds play a significant role in accumulation of finer sands in the West Bengal coast. The poor sorting and negative skewness of the beach ridge capped dunes of the inland areas are indicative of a quiescent environment characterized by weak energy and gravitational settling.

The Teri Red sands of South east coastal Tamil Nadu were examined by many researchers (Jayangondaperumal et al., 2012; Chandrasekharan and Murugan, 2001). These sedimentary depositional materials are classified as (i) inland fluvial Teri sediments of higher clay and silty sand component, (ii) coastal Teri sands and (iii) near coastal Teri sand dunes. The inland fluvial Teri sands were mostly generated by fluvial process under strong winter monsoon around 15000 YBP, the coastal Teri dune sediments were deposited at around 11500 YBP and the near coastal dune sands accumulated around 5600 YBP under strong summer and winter monsoons. The coastal and near coastal Teri sands are of aeolian origin and reddened by the weathering process during the post depositional environment after stabilization of dune sediments in the coastal tract.

The reddening processes of sands in the aeolian deposits are restricted to the southern parts of Indian coasts due to following reasons;

i. The charnockitic rocks from the Eastern and Western Ghats contributed heavy minerals and placer minerals deposits into the coastal zone sediments after weathering and fluvial transport in the region towards south of 15°32 N Latitude in India.

ii. The south-west monsoon drift currents are relatively stronger than the north-east monsoon drift currents and have transported the minerals sands from Coromondal coasts to Lake Chilka on east coast and from Malabar coasts to the south of Goa coasts on the western shores (influenced by the angle of shoreline configurations of India from South to North).

iii. Over the geological time since the last glacial maximum (22000 to 18000 YBP) the mineral sands were transported inland by aeolian deposits and marine deposits with prevailing strong on shore winds and Late Quaternary fluctuations of the sea level in the region.

iv. The long history of weathering processes have altered the aeolian deposits geochemically, to develop them as indurated sands and reddened dunes with higher rainfalls, temperatures and moisture regimes along the coastal belt.

v. The shoreline orientations of the east coast and west coast (4 orientations and orientations on the east coast and on the west coast respectively) have guided the coastal current velocity and directions of their movement in the period of active north east monsoon drift and active south west monsoon drift in India for achieving such diversity of mineral sands distribution along the coastal belt.

The reworking of the near coastal Teri sands, that were laid down between 4100 YBP to 2300 YBP in the region owing to receding of sea level, is now occurring under present day semiarid conditions. Other modern coastal sediments of sand dunes and beaches deposited after 2300 YBP up to the present stage. Thus, coastal dune sediments and beach ridge sediments are very good records of ancient environmental changes, depositional patterns, stabilization of sediments, weathering processes and sea level fluctuations. The following stratigraphic section of the beach ridge capped sand dunes of Subarnarekha Chenier Delta represents the records of paleo beach ridge formations and aeolian deposits in the eastern coasts (Fig. 21.27).

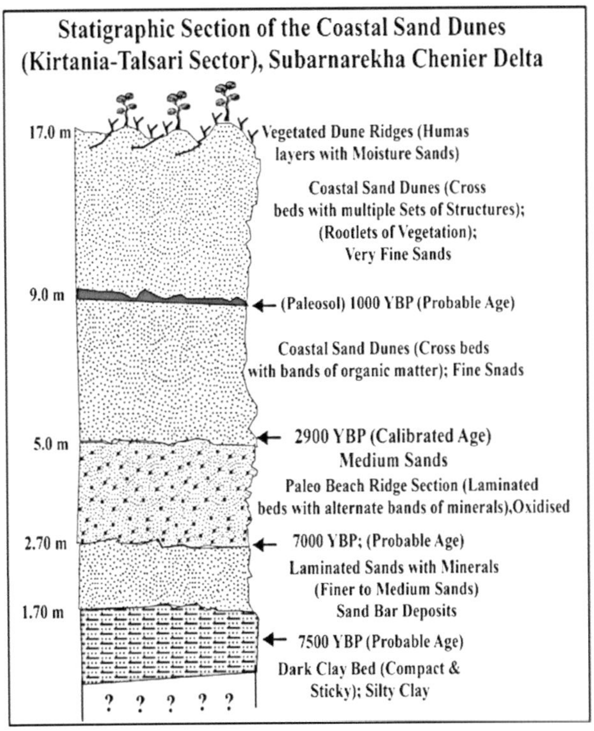

Figure 21.27. The beach ridge capped sand dunes and their sedimentary facies in the stratigraphic sections (Subarnarekha Chenier Delta, Odisha coast). The relative age and calibrated age are shown in the vertical section of stratigraphic layers.

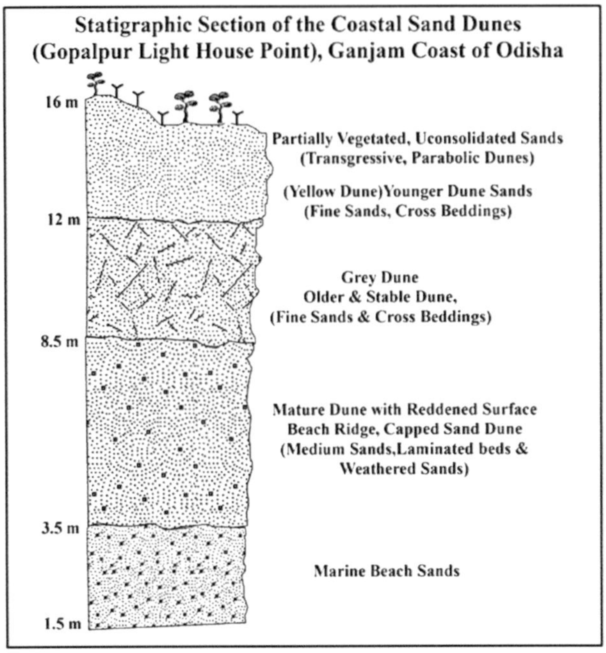

Figure 21.28. The stratigraphic section of the coastal sand dunes (Gopalpur) showing the sedimentary facies. The reddened sand dune is underlain by the overlying gray dunes and yellow dunes.

The detailed description presented by various photographs explains that the beach ridge capped sand dunes of Chaumukh-Dagara section overlies the sticky clay banks of the delta plain surface enriched with organic muds. Dune sediments of the Gopalpur coast are well represented in the stratigraphic section of the erosive dune cliff near by the Light House point (Odisha coast). The older dunes are oxidized, reddened and underlain by the younger sand dunes in the Ganjam coast. However, the younger dunes are converted into secondaries by erosion and blow out sand drifts (Fig. 21.28).

8. Evolution of sand dunes along the Western and Eastern Shores of India

The aeolian deposits of sand dunes are evolved through different stages of development on the basis of marine processes, sea level fluctuations, sediment supply, biological processes, long term weathering processes of the sediments and coastal neotectonics along the western and eastern shores of India. The cross sectional forms of the primary dunes, secondary dunes and older dunes resulted from the development of vegetation succession, accumulation rate of sands, stabilization of sediments, erosion rate, and elevation changes across the coastal inlands of the region. Topographically, the sequences of dune development have been initiated from the foreshore areas in the forms of embryo dunes, fore dune ridges and extended towards inland areas in the forms of yellow dunes, gray dunes, coastal Teri dunes and inland Teri dunes with increasing time of development and environmental changes in the coastal belt. Such ideal cross-sectional forms of the coastal sand dunes are diagrammatically represented (Fig. 21.29).

The evolutionary processes of the coastal aeolian deposits and their development dynamics are shown in a tested hypothesis for the western and eastern shores of India (Fig. 21.30). The major factors, sub-factors, consequences and time of dune development are estimated in the following ways on the basis of coastal physiographic settings, dune morphology, genesis, sediment characters, radio carbon dates, climatic role and coastal configurations of the Indian subcontinent. The dunes of the coastal belts have been evolved from the last glacial maximum (18000–22000 YBP) to the

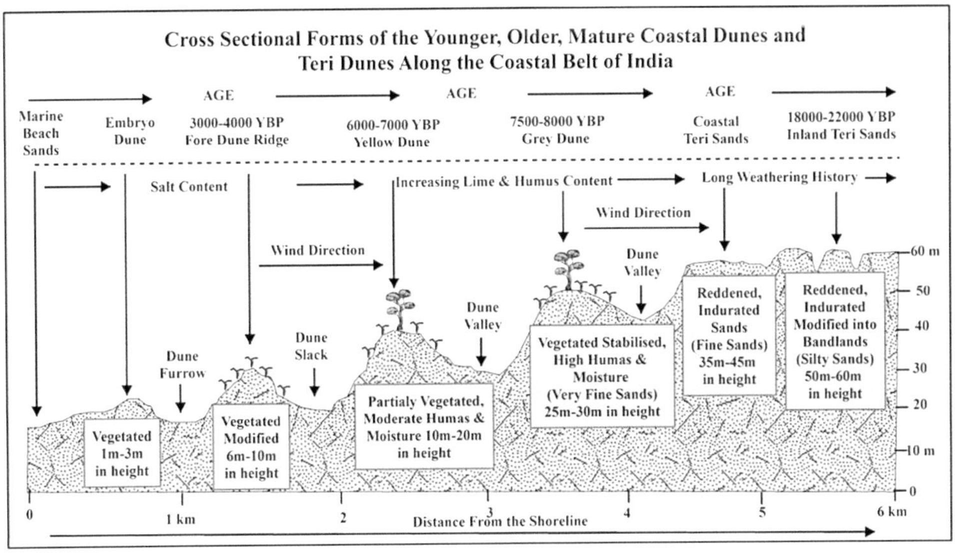

Figure 21.29. The evolutionary stages of dune development across the coastal belts of western shores and eastern shores of India.

late Holocene stage and up to present-day modifications by marine processes and anthropogenic stressors.

9. Dune vulnerabilities index (DVI)

The concern of coastal management is based on the fact that coastal systems still experience intensive and sustained pressures from a range of driving forces and that these 'drivers' are likely to be operative for many decades to come (Mc Fadden and Green, 2007). The vulnerability of coastal dunes is the resultant outcome of interaction between aeolian and marine processes, vegetation, human pressure and the geomorphology of the system. Coastal dune vulnerability assessment is necessary to evaluate the conservation status and determine the most relevant disturbance events. It is multi-dimensional, covering natural bio-geophysical response of the coast, but also involves economic and socio-cultural aspects.

Dune vulnerability has been defined as the loss of ability of the dune system to withstand the threshold events (e.g., storm surge induced flood, cyclonic wind, tsunami, etc.) after displacement from the original system steady state. Two definite dune vulnerability states as described by (Davies et al., 1995) are the *potential vulnerability* and the *conditional vulnerability*. The potential vulnerability is determined by the parameters which constitute the physical environment, e.g., geomorphological, ecological, etc., whereas, the conditional vulnerability is controlled by the levels of the management activities being operated in a particular region. In areas, where the management options are less active the conditional vulnerability is high leading to the degradation of the system. On the other hand, places with high protection reduce the chances of conditional vulnerability (Williams et al., 1993). In environmental monitoring and knowledge gathering contexts Davies et al. (1995) proposed a Dune Vulnerability Checklist (DVC) for monitoring the vulnerability and protection measures for coastal sand dunes in varying spatial and temporal scales. In the present study, the usefulness of the DVC has been critically verified over a continuous shoreline of Odisha and West Bengal with an aim to classify the dune vulnerability, to analyze the system responses to environmental perturbations and to prescribe probable management strategies for dune conservation.

The DVC used in this chapter as a problem delineation procedure (Taylor J.W., 1961) in which parameters solely responsible for dune degradation have been tabulated, rated and evaluated with respect to dune vulnerability and protection (Majumdar et al., 2014). The checklist parameters used

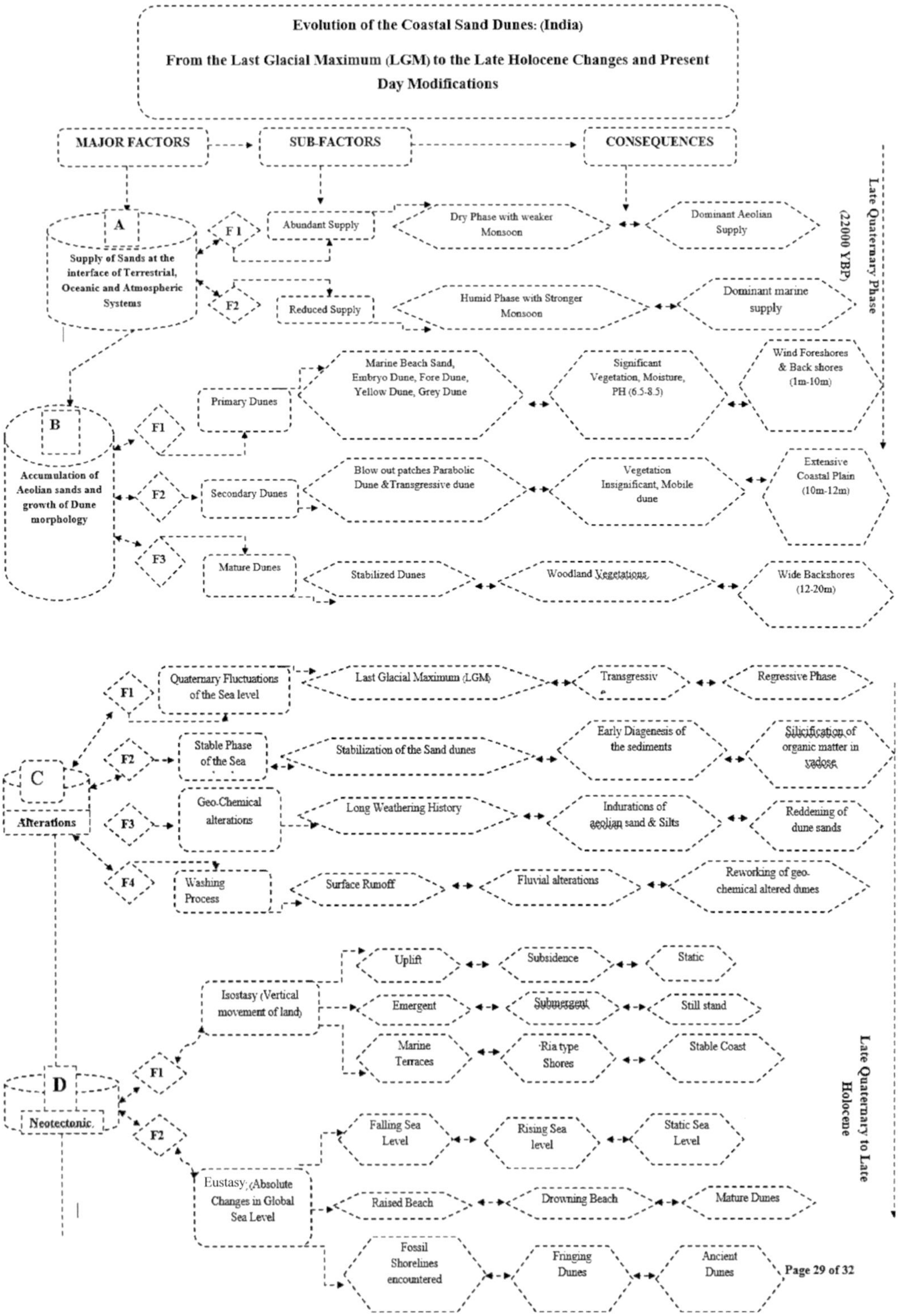

Figure 21.30 contd. ...

...Figure 21.30 contd.

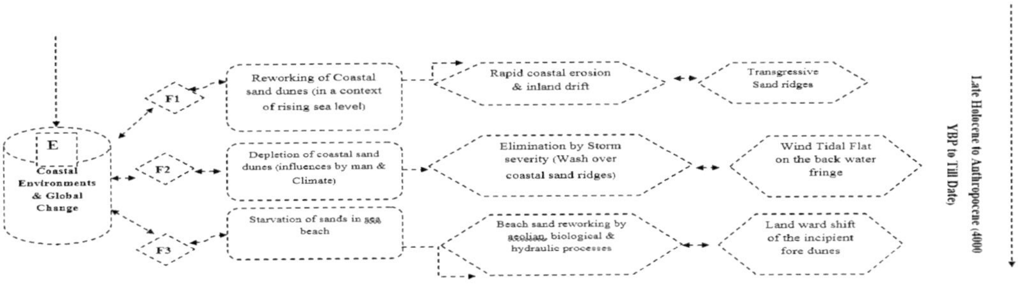

Figure 21.30. The evolutionary history of coastal aeolian sand deposits influenced by major factors, sub-factor, consequences and time (Estimated based on the field studies in western shores and eastern shores of India).

here were modified from the standard checklist of (García Mora et al., 2001) and (Williams et al., 2001) as the parameters included in the previous work may not fit for the present study area. The present checklist comprises 64 variables grouped into five Partial Vulnerability (PV) indices, e.g., Geomorphological Condition (GC)–10 variables, Aeolian Influence (AE)-11 variables, Vegetation Condition (VC)-9 variables, Anthropogenic Effect (AE)-22 variables and Marine Influence (MI)-12 variables. Detailed breakdown of the Dune Vulnerability Checklist (DVC) has been given elsewhere in (Majumdar et al., 2014). Each parameter was measured separately and the rating scale was recorded by ticking the appropriate box in the checklist. Each selected variable was associated with a five-point sliding scale graded from 0 to 4 with the last-named being the most vulnerable. Partial vulnerability indices (GC, MI, AI, VC and AE) were calculated as the ratio between the addition of given variable ranks within each variable class (PV_i) and the total maximum possible rank within the class (PV_{max}) (Majumdar et al., 2014; Williams et al., 2001). A total DVI (i.e., total Dune Vulnerability Index) was calculated as the average of the five partial vulnerability indices (PV) as per (Garcia Mora et al., 2001) and (Williams et al., 2001). The Dune Vulnerability Index ranged between 0 and 1, and as the index increased, the ability of the dune system to withstand further intervention decreases (Table 21.4).

A total of 46 coastal dune sites have been sampled for vulnerability assessment from six coastal sectors of which three from West Bengal (Digha, Mandarmani and Shankarpur-Chandipur) and three from Odisha (Chilika, Puri and Konark). The present investigation indicates that average dune vulnerability (0.675) is highest at Puri whereas the average vulnerability (0.386) of Chilka is the lowest. The results have been presented graphically on a five-axis figure, where each axis corresponds to a vulnerability class. Linking of all five partial vulnerability indices result in a polygon area as a function of DVI. The greater the polygon area represents higher the degree of vulnerability.

Table 21.4. Results of the Total DVI and partial vulnerability indices for each coastal sector. The output are averaged for all the stations under each sector.

Sl.No.	Sectors	GC	MI	AI	VC	AE	DVI
1	Digha	.504	.558	.525	.614	.681	.576
2	Shankarpur-Chandpur	.591	.549	.465	.623	.528	.551
3	Mandarmani	.631	.497	.626	.585	.558	.579
4	Chilika	.408	.425	.442	.401	.283	.386
5	Konark	.597	.559	.642	.614	.645	.611
6	Puri	.711	.589	.681	.682	.711	.675

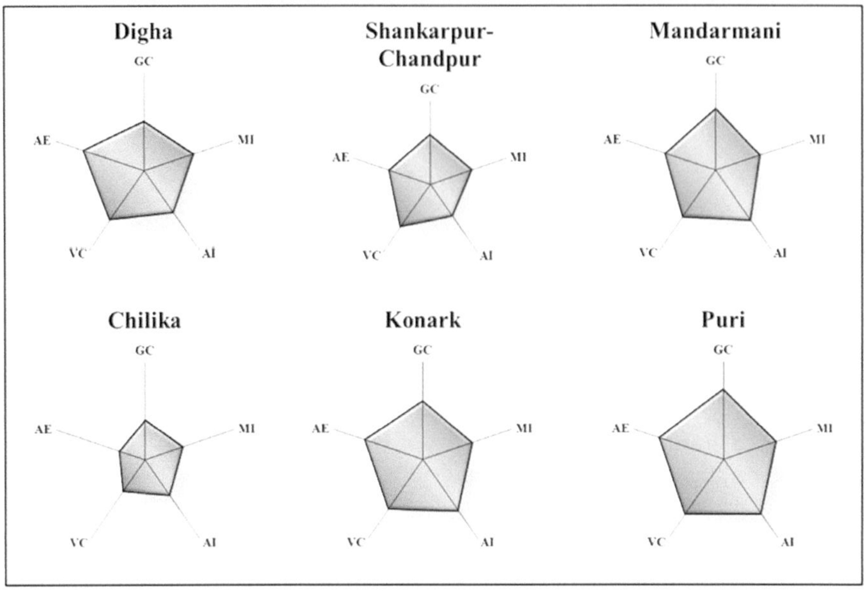

Figure 21.31. Five axis plots of the partial vulnerabilities represent a polygon, where larger the area of the polygon signifies greater vulnerability for the dune system.

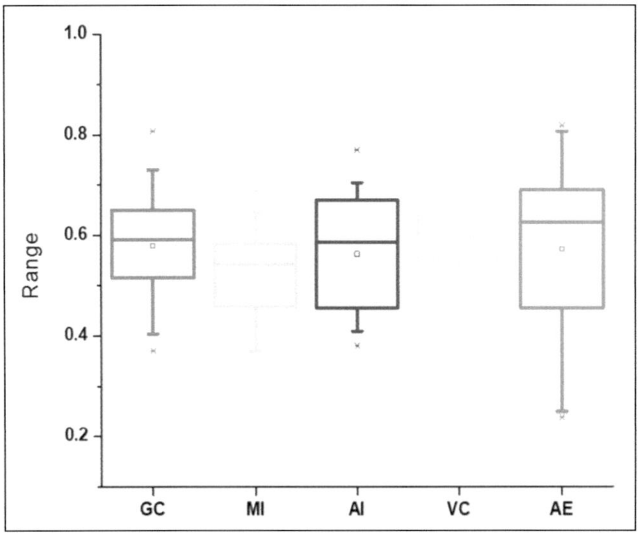

Figure 21.32. Box plot showing the inter-quartile range (Q3–Q1) for the calculated partial vulnerability (PV) indices. Variables with high inter-quartile range indicating higher variation of impact magnitude in the study area and vice-versa.

The high range for AE (Figs. 21.31 and 21.32) shows high spatial variability among the dune sites. Low range of VC and GC indicates that vulnerability of the entire study area is consistently powered by the variables associated with these two PV classes. Ranking of high vulnerability reveals that the AE and GC contribute the maximum share to the total DVI. Throughout the year the pressure from visitors is high along the study area due to its popularity as an important tourist places of eastern India. The AE acts as an important destabilizing factor of the dune system, directly related to the dune instability and degradation (Majumdar et al., 2014; Paul, 2002). The annual tourist flow of this region is approximately 10 million. Maximum replacements of active dune fields, agricultural lands, over dune forest land have been found along these coastal stations of Puri,

Konark and Digha in the last few decades. Routine sand mining from the beach disturbs the beach-dune sand sharing system and often causes erosion in and around the mining places that had been in stable equilibrium for a long time in the past. Several localized and intensive human activities like creation of settlements, removal of vegetation by intense stock grazing practices and making of tracks and trails over the dunes enhance the fragility of the dune system.

Cliff faces are found along the seaward side of the fore-dune ridge resulting by cutback processes of storm waves during high spring tide phase. Later with successive crestal slumps and active landward transport of basal sands by high water waves and currents, the retreat of dune cliffs is triggered. The presence of wash-over fan lobes along with retreating dune cliffs clearly indicates that the area is under an active erosive phase. Uncontrolled movements of tourists through the dunes to the beach led to the creation of deflation breaches that cut across and fragmented the fore-dunes. Such morphological alterations of the dune system greatly contribute to increment negative effects of natural disturbances such as wind erosion and storm tides. Extraction of groundwater along the study area is so high that the upper portion of the dunes has become drier and loosened, which in turn makes the system unstable against erosive agents. High level of trampling across dunes causing soil compaction and loss of organic matter that directly affects vegetal cover and alters species composition of the dune system and as a result the sand binding ability of the vegetation reduces.

In the present chapter, the Dune Vulnerability Checklist has been applied in the context of dune management. Decision making should be rational and as far as possible utilize both the qualitative and quantitative data. The application of DVC at a regional scale proves helpful in identifying the major causes of dune degradation as well as facilitating comparison between dune locations (sectors) and addressing the importance of management needs. For example, after determining the vulnerability status of each dune sector, it would be easier to balance the competing management demands of the coast at a regional scale, i.e., which areas should be maintained and which should be restored? As shown in the present study, it is obvious that high vulnerability sites need to be restored in order to avoid or ameliorate the negative factors responsible for the total disruption of coastal dunes.

10. Conclusions

The following conclusions may be drawn on the basis of these studies.

Coastal sand dunes along the western and eastern shores of India provide multiple areas of younger dunes, older dunes and relict dunes (ancient dunes) for diversified geological, geomorphological and climatological settings of the subcontinent as well as for the marine processes and sea level fluctuations of the northern Indian Ocean fringe environment. Such sedimentary depositional landforms of unconsolidated and indurated materials along the shorelines represent a very good indicator of the Pleistocene–Holocene climate changes and sea level fluctuations around the regions. The physiographic functions, habitat functions, coastal defense functions, heritage functions and coastal aquifer functions of the aeolian deposits are not understood by the coastal communities and coastal authorities in India. Depletion, degradation and elimination of coastal sand dunes along the low-level and high-level coasts have exposed the coastal communities to the effects of marine processes under cyclone vulnerabilities and sea level rise vulnerabilities.

The aeolian sand deposits of the Indian coastal landscape along the shore fringes of the Arabian Sea and Bay of Bengal are classified on the basis of their physiographic settings and morphologic types. The scheme of classification has provided information about coastal sand dunes for their diversities in origin, morphology and processes of marine environment in the Indian Subcontinent. Ancient and modern aeolian deposits of the coastal zones have been assessed for the western and eastern shores to identify their level through geospatial techniques and field survey records.

The aeolian sand deposits along the back shores have produced larger coastal dune landscapes in many sections of the eastern shores of India from Tamil Nadu to West Bengal. The annual average rainfall varies from 998 mm to 1050 mm in the coasts of Tamil Nadu and in coastal Andhra

Pradesh, ranges from 1200 mm to 1500 mm in Coastal Odisha and in Coastal West Bengal on the eastern shores. Even moderate amounts of rainfall provide ideal conditions for windblown sand accumulations along the coastal zones. However, the annual average rainfall varies from 2500 to 2600 mm in the Malabar and Konkan Coasts of the western shores in India. Though the ancient lithified or fossil dunes of the landward parts are higher and elevated (30 to 50 m in altitude) the beach fringe fore dunes and frontal dunes are relatively lower (6 to 12 m in altitude) on the western shores of the sub-continent due to availability of high moisture and species diversity of plants.

Finally, a report is built regarding the aeolian sand deposits and their evolutionary history along the western and eastern shores of India with influencing factors and related consequences of the sedimentary depositional landforms. The hypothesis is tested in the field sites of eastern coasts and western coasts of India. Geospatial techniques are also used with spatio-temporal satellite images of the coastal dune landscapes for further validation of the hypothesis.

Geomorphologically, the coastal sand dunes are reduced in width and elevation, affected with lateral recession, elimination, significant cliffing on the seaward slopes and erosion; transformed with mobility on the landward sides, transgressive sand ridges and many other secondary origins at present. Extensive aeolian sand fields, overwash sands, transgressive sand sheets and primary dunes with vegetation colonies exist only on the barrier bars and sand spits along the river mouth sections. Older sand dunes with lithified and indurated sands, red color Teri sand dunes, stabilized sand dunes and lofty dune systems are found on the marine terraces and under the protective favorable coastal configurations. Mangrove wetlands behind the dune barriers are trapped by over wash sand fans by cyclone surges and tidal waves in the deltaic and barrier coasts.

Alternate dry phases, humid phases, stabilized sea level phases, transgressive and regressive seas, weaker and stronger monsoon during Pleistocene-Holocene climate changes and weathering history are responsible for the evolution of coastal sand dunes. However, cyclone severity, sea level rise vulnerability, sand starvation on the sea beaches and diversity of marine processes along the western and eastern shores have depleted, eliminated and reactivated the dune fields in the subcontinent at present. Coastal sand dunes are largely degraded due to the construction of tourism infrastructures, human modifications for the spread of coastal urbanization, placer minerals mining, sand mining, livestock grazing over the vegetative dunes and also for the use of aeolian sand fields for dry fish processing by the fishermen community in the coastal belts of India.

An endeavor is needed for the conservation of a few plant species, i.e., long rooted tussock forming grasses/sedges like *Phragmites* spp. *Cyperaceae* spp.*;* creepers like *Ipomoea* spp.*, Spinifex* spp.*;* Scrubs like *Pandanacia* spp.*, Opuntia* spp.*, Calotropis* spp.*;* trees like *Anacardium* spp.*, Tamarix* spp. for stabilization of dune surfaces along the coastal belts. The climate resilient dune systems may be created or may be promoted along the western and eastern shores of India to withstand storm surge inundations, erosion and sea level rise effects.

Understanding the significance of coastal sand dune processes and management for coastal floods and defense measures by coastal authorities will resolve the issues on conservation, restoration and sustainability of coastal heritage landscapes and associated habitats in the western and eastern shores of India. The information presented here was compiled by researchers over many years on coastal sand dunes will provide new opportunities for research at the regional level to understand the geomorphology, habitats and paleo-environment of the tropical coast.

Acknowledgements

Our gratitude to the students and researchers of coastal management of Geography Department and Remote Sensing & GIS Department of Vidyasagar University for immense assistance in the field works and survey works along the western and eastern shores of India. Thanks to Joydeb Sardar of the Centre for Environmental Studies (CES) of Vidyasagar University for the geospatial application.

References

Alappat, L., Seralathan, P., Shukla, A.D., Thrivikramji, K.P. and Singhvi, A.K. 2013. Chronology of red dune aggradations of South India and its Palaeo-environmental significance. Geochronometria 40(4): 274–282.

Alappat, L., Joseph, S., Tsukamoto, S., Kaufhold, S. and Frechen, M. 2016. Chronology and weathering history of red dunes (Teri Sands) in the southwest coast of Tamil Nadu, India. Zeitschrift der Deutschen Gesellschaft für Geowissenschaften 168(1): 1–16.

Alappat, L., Joseph, S., Tsukamoto, S., Kaufhold, S. and Frechen, M. 2017. Chronology and weathering history of red dunes (Teri Sands) in the southwest coast of Tamil Nadu, India. Zeitschrift der Deutschen Gesellschaftfü Geowissenschaften 168(1): 183–198.

Bhalla, R.S., Ram, S. and Srinivas, V. (eds.). 2008. Studies on Vulnerability and Habitat Restoration along the Coromandel Coast. 1st ed. Pondicherry, India: FERAL, UNDP-UNTRS.

Bruckner, Helimut. 1988. Late quaternary shorelines in India. pp. 169–194. *In*: Late Quaternary sea level correlation and Application. Springer link.

Chakrabarti, P. 1991. Morphostratigraphy of coastal quaternaries of West Bengal and Subarnarekha delta, Orissa. Indian Journal of Earth Sciences 18(3-4): 219–225.

Chandrasekharan, S. and Murugan, C. 2001. Heavy minerals in the beach and the coastal red sands (Teris) of Tamilnadu Spl. issues on 'Beach and Inland Heavy Mineral sand Deposits of India'. Exploration and Research for Atomic Minerals 13: 87–109.

Deswandikar, Abhay and Karlekar, Shrikant. 1996. A geomorphic significance of coastal dune complex at Diveagar. Indian Journal of Geomorphology 1&2: 125–135.

Dalabehera, L., Hazra, A., Pattanayak, S. and Pal, T. 2020. The Paleo beach Ridges of Digha Coastal Tract, West Bengal, India: Observation and Implication for Sea Regression from 500 YBP to 200 YBP. Journal of the Geological Society of India 95(2): 131–144.

Dikshit, K.R. 1976. Geomorphic features of the west coast of India between Bombay and Goa. Geographic Review of India 38: 260–281.

Davies, P., Williams, A.T. and Curr, R.H.F. 1995. Decision making in dune management: theory and practice. Journal of Coastal Conservation 1(1): 87–96.

Fairbridge, R.W. 1995. Eolianites and eustasy: early concepts on Darwin's voyage of HMS Beagle. Carbonates and Evaporites 10(1): 92–101.

García Mora, M.R., Gallego Fernández, J.B., Williams, A.T. and García Novo, F. 2001. A coastal dune vulnerability classification. A case of study of the SW Iberian Peninsula. Journal of Coastal Research 17(4): 802–811.

Garzanti, E. 2018. Diagenetic Control on mineralogical suites in sand, silt and mud. https://wwww.sciencedirect.com>pii.

Gangopadyay, K. 2013. Report of the excavation at Bahiri (east Medinipur district) CAST, Unpublished paper pp. 1–4.

Jayangondaperumal, R., Murari, M.K., Sivasubramanian, P., Chandrasekar, N. and Singhvi, A.K. 2012. Luminescence dating of fluvial and coastal red sediments in the SE coast, India, and implications for paleo environmental changes and dune reddening. Quaternary Research 77(3): 468–481.

Karlekar, S.N. and Gadkari, D. 1998. Fossil littoral deposits of beach and dune origin at Nandgaon (coastal Maharashtra). Indian Journal of Geomorphology 3(2): 173–190.

Kenneth Pye. 1983. Coastal dunes. https//doi.org/10.1177/030913383007004 03.

Kale, V.S. and Rajaguru, S.N. 1983. Mid-Holocene fossil wood from Colva, Goa, Current Science Association, Bangalore 52(16): 778–779.

Maiti, S. and Bhattacharya, A.K. 2009. Shoreline change analysis and its application to prediction: A remote sensing and statistics based approach. Marine Geology 257(1-4): 11–23.

Maiti, S. and Bhattacharya, A.K. 2011. A three-unit-based approach in coastal-change studies using Landsat images. International Journal of Remote Sensing, 32(1): 209–229.

Maiti, Sabyasachi. 2013. Interpretation of coastal morpho dynamics of Subarnarekha estuary using integrated cartographic and field techniques. Current Science 1709–1714.

Mascarenhas, A. 1990. Why sand dunes are needed. Herald, 21 December, p.4.

Mascarenhas, A. 1996a. The fate of sand dunes of Goa. *In*: Rajagopalan, G. (ed.). Voices for the Oceans. International Ocean Institute (India), p. 111.

Mascarenhas, A. 1996b. Some observations on the coastal zone management plans of Goa (unpublished report), 25 pp.

Mascarenhas, A., Sawkar, K. and Chauhan, O.S. 1997. The coastal zone of Goa: then and now (Abstract 49). Seminar on Coastal Zone Environment and Management Mangalore University, February 12–14, pp. 50–52.

Mascarenhas, A. 1998. Coastal sand dune ecosystems of Goa: significance, uses and anthropogenic impacts. National Institute of Oceanography, Dona Paula, Goa - 403 004, India, pp. 1–58.

Mohapatra, M. 2015. Cyclone hazard proneness of districts of India; Indian Metrological Department (IMD); Journal of Earth System Science 124: 515–526.

Majumdar, D.D., Bera, S., Purkait, B., Paul, A.K. and Bhandari, U. 2014. Insights into the dichotomy of coastal dune vulnerability and protection measures from multi-criteria decision analysis: a case study of West Bengal Coast, Bay of Bengal, India Journal of Coastal Sciences, CERF.inc.

Mc Fadden, L. and Green, C. 2007. Defining 'vulnerability' conflicts, complexities and implications for Coastal Zone Management. Journal of Coastal Research, 120–124.

Nayak, K.M. and Dhokia Sh. Nilesh. 2020. Ground Water Year Book-2018-19 UT of Daman & Diu, Regional Office Data Centre; Central Ground Water Board (CGWB), Ahmedabad; pp. 1–44.

National Disaster Management Plan. 2016. Published by National Disaster Management Authority, Govt. of India, New Delhi pp. 1—192.

National Centre for Sustainable Coastal Management (NCSCM-2016). COASTAL SAND DUNES OF GOA. Pp-1-31. http://dstegoa.gov.in/Sand%20dune%20report.pdf.

Paul, Ashis Kr. 1988. Coastal dunes systems of the Sundarban. Geographical Review of India. Calcutta 50(1): 14–23.

Paul, Ashis Kr. 1994. Dune environment of West Bengal coastal plain. pp. 314–351. *In*: Dikshit, K.R., Kale, V.S. and Kaul, M.N. (eds.). India Geomorphological Diversity. Rawat Publications, Jaipur.

Paul, Ashis Kr. 1996. Chenier beach ridge and chenier sand ridge formations around Subarnarekha estuary. National Geographer. XXXI(1 & 2): 143–153. Allahabad.

Paul, Ashis Kr. 2002. Coastal sand dunes. pp. 165–187. *In*: Coastal Geomorphology & Environment, Ashis Kr. Paul. acb publications Calcutta.

Paul Ashis Kumar. 2005. Tsunami; the disasters in Indian Ocean Nations, ACB Publication Kolkata, pp. 125.

Paul, Ashis Kr. and Paul Anurupa. 2013. Coastal Environment & Dune Ecology. pp. 22–69. *In*: Devmita Dinda (ed.). Phytography of Saltmarshes, Mangroves and Sand Dunes along the Coastal Belt of West Bengal. Mugberia Gangadhar Mahavidyalaya.

Purkait, B. and Majumdar, D.D. 2014. Distinguishing different sedimentary facies in a deltaic system. Sedimentary Geology 308: 53–62.

Purkait, B., Majumdar, D.D. and Paul, A.K. 2015. Grain Sorting on a Coastal dune of Orissa, Bay of Bengal and a comparison with the Desert Dune of Rajasthan, India. International Research Journal of Earth Sciences 3(8): 14–31.

Ramesh, R. 2017. Coastal Sand Dunes of Goa; National Centre for Sustainable Coastal Management (NCSCM), Ministry of Environment, Forest and Climate Change, Chennai pp. 1–31.

Rouchelle, S., Rodrigues, Antonio, Mascarenhas, Tanaji G. Jagtap. 2011. An evaluation flora from coastal sand dunes of India: Rationale for conservation and management, Ocean & Coastal Management, Elsevier pp. 181–188.

Ram Gardner and Hendry, D.A. 1995. Early Silica diagenetic fabrics in Late Quaternary Sediments, South India, Journal of the Geological Society 152: 183–192.

Ramkumar, M., Menier, D., Mathew, M. and Santosh, M. 2016. Geological, geophysical, and inherited tectonic imprints on the climate and contrasting coastal geomorphology of the Indian peninsula. Gondwana Research 36: 65–93.

Saha, S., Ghosh, A., Banerjee, S., Saraswati, P. K. and Burley, S.D. 2011. Characteristics of an open coast tidal flat: Example from Daman, west coast of India. Journal of the Geological Society of India 77(5): 409–418.

Taylor, J.W. 1961. How to Create Ideas. Englewood Cliffs, Prentice Hall, NJ.

Williams, A.T., Davies, P., Alveirinho-Dias, J.M., Pereira, A.R., García-Mora, M.R. and Tejada, M. 1994. A re-evaluation of dune vulnerability checklist parameters. Gaia 8: 179–182.

Williams, A.T., Alveirinho-Dias, J., Novo, F.G., Garcıa-Mora, M.R., Curr, R. and Pereira, A. 2001. Integrated coastal dune management: checklists. Continental Shelf Research 21(18-19): 1937–1960.

Chapter 22

The Coastal Zone of Sri Lanka

Characteristics, Problems and Prospects

U. de S. Jayawardena

1. Introduction

1.1 The Island of Sri Lanka

The island of Sri Lanka lies between latitudes 5°55' and 9°51' north and longitudes 79°41' and 81°53' east, situated 32 kilometers east of the southern tip of India, separated by the Palk Strait and the Gulf of Mannar (Fig. 22.1). The total land area measures 65,580 square kilometers in extent 430 km in length from north to south and 225 km in width at the broadest part.

The island's physiography can be described as consisting of a central mountainous mass rising from a low flat plain surrounding it in all sides and extending to the sea. Based on the height and slope characteristics, the island of Sri Lanka can be divided into three main morphological regions which show a series of irregular scarps or breaks in slopes at different elevations. These are namely (i) The Coastal Lowlands (from mean sea level to 270 m, MSL), (ii) The Uplands (from 270 m to 1060 m, MSL) and (iii) The Highlands (from 910 m to 2420 m, MSL) (Vitanage, 1970). The boundaries are very clear escarpments especially in the southern and southeastern parts of the island. Both uplands and highlands are considered as the central highlands with undeveloped drainage patterns and marked relief appearing in numerous strike ridges, hills and mountains with waterfalls, deep valleys, escarpments, high plains and plateaus. It is believed that these three geomorphological units are the formation of three peneplains in different periods as the results of earth tectonic activities. The streams starting from the central mountains flow to the sea in all the directions crossing the country.

Sri Lanka has a humid tropical climate. The average annual rainfall varies from below 1,270 mm in the north-west and south-east parts of the lowland zone to 5,076 mm in the south west slopes of the central hill country. The mean annual rainfall for the island is 2,030 mm (NARESA, 1991). On the basis of rainfall, the dryness and topography of Sri Lanka can be divided into three climatic zones namely the Dry Zone (rainfall less than 2,000 mm), the Intermediate Zone (rainfall between 2,000 mm to 3,000 mm) and the Wet Zone (rainfall above 3,000 m) (Walker, 1962) (Fig. 22.2). The temperature ranges from 14°C to 32°C within the country. Weather patterns are influenced by the South-West monsoon (May-early October) and North-East monsoon (November to February). Strong winds usually occur during the South-West monsoon. The relative humidity is high throughout the year.

Department of Civil Engineering, Faculty of Engineering, University of Peradeniya, Peradeniya, Sri Lanka.
Emails: udesja@gmail.com, udsj@pdn.ac.lk

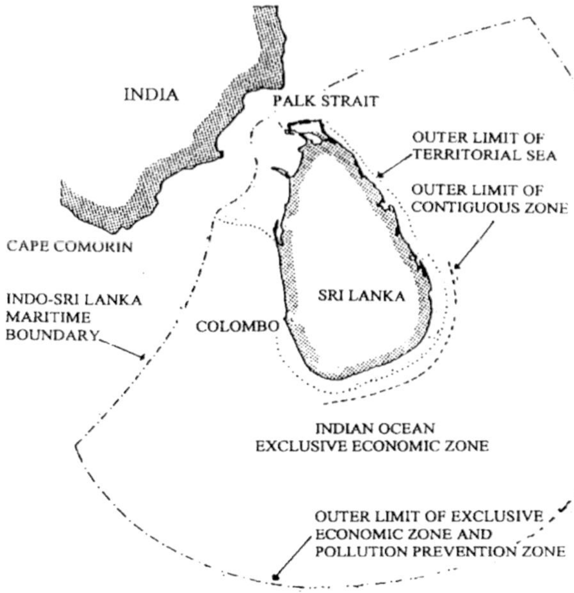

Figure 22.1. Coastal and maritime jurisdiction of Sri Lanka (NARESA, 1991).

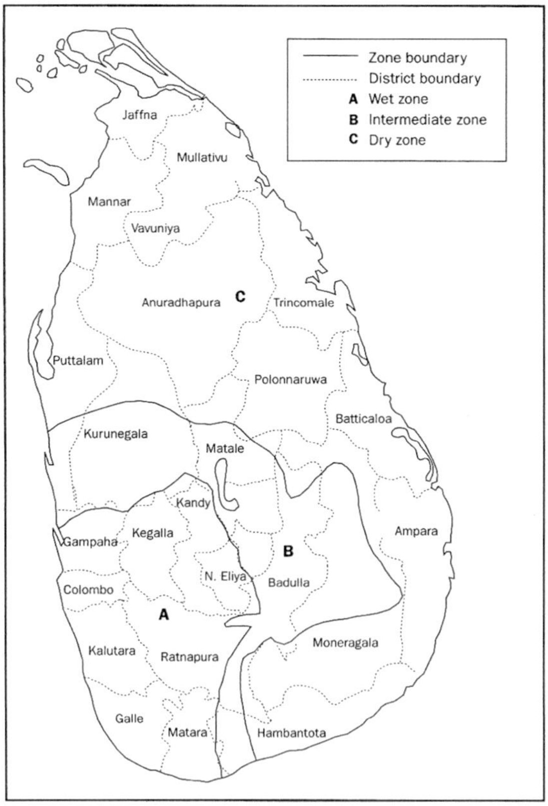

Figure 22.2. Climatic zones and administrative district boundaries of Sri Lanka.

Sri Lanka's location close to the Indian subcontinent, gives it a predominantly monsoonal and tropical climate. The central hills of the island that rise above 2,000 meters intercept the moisture-laden monsoonal winds creating rainfall patterns in the country. The radial drainage carries surface water down towards the sea, crossing the country from high watersheds and central mountains it includes 103 distinct natural river basins that cover over 90% of the island (Arumugam, 1969). The eroded materials are transported by rivers and settled at the base of reservoirs or moved to the sea with river water. Suspension materials, bed loads and the dissolved ions are materials that move with river water to the sea.

2. The coastal zone of Sri Lanka

2.1 The geographical details

The length of the coastline, open to the sea, including bays is 1639 km and the coastline of the other small islands belonging to Sri Lanka is about 281 km. Sri Lanka is separated from India by a shallow sea. There are beach sands, beach rocks and large dune sand hills along some coastal stretches. Most of dune sand ridges are a few meters high above the normal ground level parallel to the coast. Beach sands and dune sands are the most common unconsolidated sediments. But there is a large deltaic sedimentary deposit at the mouth of River Mahaweli (Table 22.1).

Table 22.1. Major geographical details of the coastal areas in Sri Lanka.

Item	Description
Coastline	Length 1,639 km, 45 lagoons and estuaries are 45 (420 sq. km)
Climate	Rainfall from below 1,000 mm to more than 2,000 mm, 70–90% humidity, average temperature 26–33°C, Two monsoons N-E and S-W and two inter monsoons, average wind speed 12–40 km/hour and gust 50 km/hour
Coastal area	Continental shelf: 2.5–23 km, 30–90 meters depth variation and 26,000 sq. km area. This south west coastal line is the closest shoreline to the deep sea.
Tidal range	Maximum 0.85 m, more than 50% water level goes up 0.53. The water level exceeds the mean sea level more than 65% times, current speed 3–15 cm/s
Economy zone	Over 230,000 sq. km of the ocean
Vegetation	60–100 sq.km covers by mangroves (23 species) and 238.2 sq.km covers by salt marshes and 12 species of sea grasses.
Coastal land	About 45% are urban areas and the rest is crop lands
Agriculture	17% of the country agriculture production
Live corals	Coral reefs occur along the south-west, south and east coastal areas. At present 171 species of stony coral have been recorded in the major reef types.
Sedimentation	From the erosion of country rock and discharged by rivers (total 103)
Fish	Annual production for food exceeds 200,000 metric tons/year. One hundred and thirty nine species are ornamental fish for export.
Dune sands	No accurate estimate of the rate of annual deposition.
Resources	Mineral sands, shell and coral limestone, silica sand, peat, salt and alluvium
Tourism	The average annual tourist arrival is over 1 million. About 85% of tourist revenue comes from coastal areas. Many graded hotels are located there.

Source: Coast Conservation Dept. (1991), Dayanande (1992), MaddumaBandara (1989), NARESA (1991).

2.2 Coastal habitats

For management and conservation purposes of the coastal environment, the coastline is defined as the "Coastal Zone", which is described in the coast conservation act of 1981 as "the area lying within a limit of 300 m landward of the mean high-water level and a limit of 2 km seaward of the mean low water level". In the case of rivers, streams, lagoons or any other body of water connected to the sea either permanently or periodically the landward boundary extends to a limit of 2 km measured perpendicular to the straight base line drawn between the natural entrance points. These include the waters of such rivers, streams and lagoons or any other body of water so connected to the sea (CCD, 1997).

The main feature of the coastal zone is the presence of a number of ecologically and economically important ecosystems characterized by high rates of production, consumption and exchange. These ecosystems serve as habitats for a wide diversity of plant and animal life. They also act as buffers against ocean waves that cause coastal erosion. These natural coastal habitats include coral reefs, estuaries and lagoons, mangroves, sea grass beds, tidal flats or salt marshes and barrier beaches, spits and dunes. There are five major large lagoons in the island but these are really basin estuaries (MaddumaBandara, 1989). Table 22.2 shows the extent of coastal habitats in the different districts along the coastline in the island (Fig. 22.2 for district locations).

Table 22.2. Extent of Coastal habitats (by district in sq.km) in Sri Lanka (Samaranayake, 2006).

District	Mangroves	Salt marshes	Dunes	Beaches barriers and spits	Lagoons basins and estuaries	Other Water bodies	Marshes
Colombo	-	-	-	1.12	-	4.12	0.15
Gampaha	1.22	4.97	-	2.07	34.42	2.05	16.04
Puttalam	22.64	34.61	26.89	27.72	391.19	34.28	25.15
Mannar	12.61	51.79	14.58	9.12	38.28	23.71	3.08
Kilinochchi	3.12	49.75	5.09	4.20	119.17	12.56	10.46
Jaffna	2.60	49.63	21.45	11.03	455.25	18.62	1.49
Mullativu	4.63	5.17	-	8.64	92.33	5.70	1.94
Trincomalee	14.91	14.01	-	6.71	183.17	21.80	11.26
Batticaloa	14.21	21.96	-	14.89	1368.22	23.65	9.68
Ampara	2.92	1.27	3.57	13.98	72.35	11.71	8.94
Hambantota	5.39	3.18	4.44	10.99	44.88	15.26	2.00
Matara	0.06	-	-	1.91	-	2.34	0.80
Galle	1.87	1.85	-	4.85	11.44	7.83	5.61
Kalutara	0.70	-	0.04	0.77	0.87	4.76	0.91
TOTAL EXTENT	86.88	238.19	76.06	118.00	1580.17	188.39	97.51

2.3 Mineral deposits in coastal land

The dominant mineral resources of the coastal region in Sri Lanka are heavy mineral sands (ilmenite, rutile, monazite, garnet and zircon), silica sands, sand from shallow seas, salt, Miocene limestone, coralline limestone, shell beds, brick and tile clays, laterite, iron ore and peat (Cooray, 1984). The most important material along the beach is the mineral sand deposit at the Pulmudai coast. Large quantities of ilmenite, rutile, zircon and monazite can be obtained from this source (Fig. 22.4).

2.4 Geomorphology of the coastal zone of Sri Lanka

Geomorphologically the nature of the coastline can be grouped according to the existing features. These are straight coastlines, sand spits, lagoons, bays, estuaries, deltas, islands, harbors, headlands, cliffs, reefs and rock outcrops. Reefs are beach rocks, corals and calcareous algae. The coastline can also be grouped according to the nature of materials (Fig. 22.3).

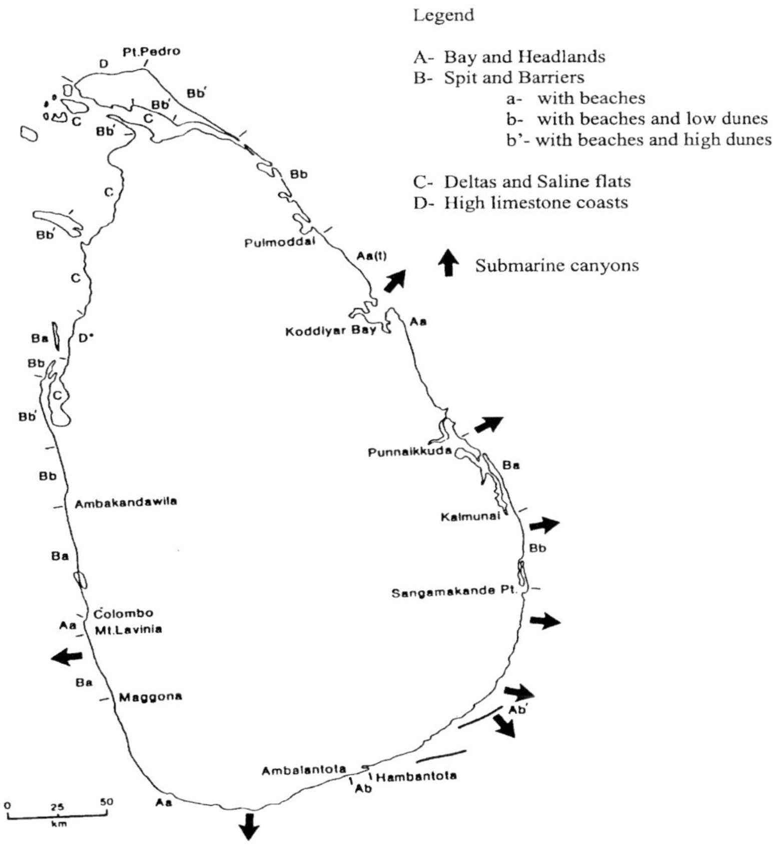

Figure 22.3. Principal coast and shoreline classes (Swan, 1983).

2.4.1 Sedimentation

The major rivers which start from the central hills supply sediments to the coast. Sediments are supplied in a cyclic manner with maximum supply coinciding with heavy rains, and stronger wave climates which occur during the South-West monsoon. Sediments supplied in this way are coarse and have typical grain diameters in the range of around one millimeter. Smaller streams supply finer sediments. The other sources of sediments have marine origins. These consist of calcareous substances derived from live corals reefs and shells of marine organisms. In Sri Lanka these types of live reefs are found in many areas. The reef that extends from Colombo to northwards is dead.

Spit formation depends on the average littoral drift on the coast and the quantity of sediment discharged by the river to the coast. Owing to the varying sediment supply by rivers the coastal dynamics at the outlets are unpredictable. Waves are the main source of energy for sediment movement along the coasts of Sri Lanka. There is no accurate estimate of the rate of annual deposition.

During the northeast monsoon, wind waves are dominant. This affects only the northeast and the east coasts of Sri Lanka. During this time waves on the east coast are produced by local winds which approach from east to northeast. Along the west coast, during the southwest monsoon the

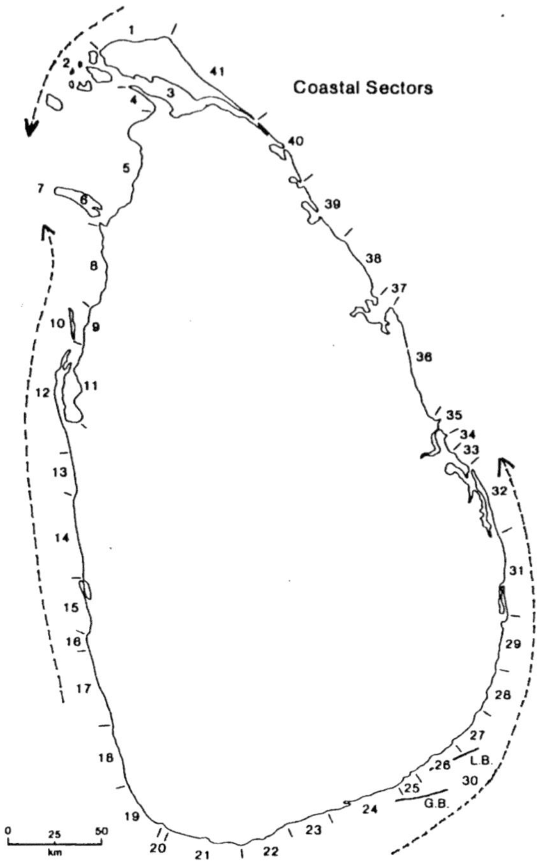

Figure 22.4. Geomorphological Sectors or subdivisions of Sri Lanka. The arrows indicate the directions of sediment transport along the coast (Swan, 1983).

waves move sediments northwards with a very high capacity. During this time monsoon fed rivers on the west coast also have large runoff and sediment discharges. The coastal dynamics are therefore complicated and very intense. Tidal currents help a portion of the sediments to cross the inlets of the lagoons and estuaries. Similarly, sediments along the east coast move during the northeast monsoon period. The wave forces are generally dissipated by partially or fully submerged reefs. Due to various reasons coastal areas are extremely variable and complicated (Fig. 22.4).

3. Dune sand deposits of Sri Lanka (Aeolian deposits)

3.1 General

A dune is a hill of loose sand or accumulation of sand built by aeolian processes (wind) and lies behind the part of the beach affected by tides. They are formed over many years when windblown sand is trapped by beach vegetation or other stationary objects such as rock outcrops, boulders or cliffs. These act as obstructions to the wind flow. Some are self-accumulation of sand into piles due to aerodynamics of wind. In addition, other factors, such as variation in wind direction, climatic changes manifested by dramatic shift in wind direction and magnitude, velocity and frequency of storms, sand availability and thickness of sand cover and the sudden removal of vegetation cover can have an effect on dune morphology. Dunes occur in different shapes and sizes, formed by interaction with the flow of air. Light and rough sand grains are transported inlands by wind thus causing dune formation. Sand is constantly moving back and forth in a dune and this everlasting replenishment process works naturally. Dune sands are essentially fine sands deposited in certain coastal areas and arid zones where conditions are favorable for dune formation. The origin of these sands may be traced to the eroding land surface. About 63% beach materials are sandy and 14% are silty. The silty type sediments can be seen along the north-north western coastline which shows some tidal flat environments. Sandy beaches and spits, raised beaches, sand dune hills and barrier beaches are common in N-W and S-E coastline areas of the island.

3.2 Locations of major dune sand deposits

Development of sand dunes occurs along about one fifth of the coastline (nearly 350 km), in the dry and wet zones, and both seasonally high and perennially low wave energy sectors of Sri Lanka. Most of the thick dune hills occur on northwest, north and southeast coasts. The formation of dune deposits in Sri Lanka depends on the climate, particularly rainfall distribution, wind strength, wind direction and sand supply. Local control of wind intensity such as the presence of a dense growth of tall trees behind the berm, and bedrock relief affecting beach stability play a significant role. The best development of sand dunes occurs between Mullativu and Point Pedro (north-northeast coast), between Elephant Pass and Chavakachcheri (north coast), across Mannar Island and the Poonaryn Peninsula (northwest coast), intermittently between Ambakandawila and Dutch Bay settlement on the Kalpitiya Peninsula (northwest coast), and along the southeast coast from Kirinda to Sangamankande Point. Less conspicuous dunes occur elsewhere in northwest, north and southeast. Highly weathered reddened old dunes occur between Aruwakkalu and Kudiramalei point in northwest.

3.3 Factors affecting to the formation of dunes

Climate Zones: Better developed dune systems are confined to parts of the country which have been marked as dry zones. They are absent and localized on wet zone coasts and there is no marked dry season. Generally, the dry season occurs between March to August. The sand grains in NW, N, NE and SE coastal zones are free to transport them if the wind velocity is enough during this period. According to Swan (1983) the number of months for sand mobilization in these regions lies between 5 and 8 per year.

Rainfall: More than four months are consecutively dry in north, northwest and southeast. This climatic condition is a significant reason to mobilize sand transfer from beach to landward and best dune sand development. The dune system is poorest in the southwest where the number of dry months is one or less. But some dry climatic conditions in the east do not show dune formations due to other reasons.

Wind speed, strength and directions: Wind speed is a decisive factor because threshold velocities have to be exceeded before sand grains are mobilized. These are lowest in very fine sands and vary with particle size. Coastal observations in Sri Lanka are made at varying elevations above sea level and the instruments recording wind speed are usually placed several meters above the ground level. Therefore, wind behavior just above the beach is not known and cannot be used for the study of sand mobilization. The annual normal wind velocity varies between 8.6 km/h and 19.8 km/h in the country and the maximum is 26.6 km/h under normal conditions. The wind velocity above 14.4 km/h exists for 11 months in southeast, 7 months in north and 4 to 7 months in northwest (Swan, 1983). The average wind speed during the NE and SW monsoon periods occurs as 12–40 km/h and gust 50 km/h. It allows the higher wind velocities to exist for more months in the locations where the dune sands are developing. These wind velocities are capable of mobilizing fine and fine to medium dry loose sand grains. The trend of dunes depends on the locally dominant winds. In generally it follows the NE and SW monsoonal winds of the coastal zone. Southern and western coast dunes respond to the SW monsoon and north and north eastern dunes respond to the NE monsoon. Almost all dunes are parallel to the NE-SW direction with slight changes. Variations are induced by bed rock relief and local breezes. Dunes are absent in the east because winds are not strong enough to mobilize dry sand. In southwest sand dunes are less because the climate is wet even in the higher wind speed period.

***Humidity*:** High humidity helps to moisten hygroscopic salt coatings on sand particles, increasing the adhesiveness of grains. It reduces the wind ability to move sand grains.

***Geomorphology*:** On Sri Lankan beaches it has been noted that where berms are high, they tend to dry out more quickly than when low and closer to the water table. Where berms are generally low lying or lacking, dunes are absent. The height of the berm rather than its width is decisive in dune building, where other factors seem to lead to dune development. The bedrock relief such as rocky headlands, low inselbergs and rock knobs adjacent to beach are the other ground features to form dunes.

***Vegetation*:** Active dunes are not in evidence in areas where tall trees and coconut trees are growing near the shore. Wind-blown sand is noticeable where trees are scattered and their crowns are not contiguous. There is a dune belt with less vegetation in a section of eastern part of north beach (Manalkadu dune). Sand migration is very high here during dry southwest monsoon period.

***Other factors*:** Due to low wave energy conditions the delivering and sorting of sand from clays and silts are poor in some sections in the northwest beach and hence dunes are absent. However during the southwest monsoon, the affected tract of the small wave regions deposits only sand on the beach and later during drought strong winds carry this towards the inland.

Poor sand supply, higher coastal erosion and absence of berms and wet climatic conditions create a negative environment to development of dunes along the southwest coast. The coarseness favors sand mobility in Sri Lanka. Large grain sizes with the greatest dune activity can be seen at the Manalkadu area in the northeastern zone, Transverse primary dunes leaving behind transgressive secondary dunes in the northwest coast indicates coastal changes. Sand from old weathered Pleistocene sand dunes, raised beach, red earth—along with the erosion of the foundation of the calcareous sandstone and limestone hillocks are a source of dune-building sediments.

Dunes are not developing at the old raised beach in some places along the northwest coast. These are partially weathered old transverse dunes and the beach is at the base of the cliff. Some dunes can be seen in the crevices of rock masses. Some old dunes occur as flat lands about 1 km away from the present beach. Some are hardened and cemented. This feature occurs in exposed rocky masses along the southeastern coast. The presence of sand dunes along sectors of Sri Lanka's coasts indicates that such features occur in sub–equatorial latitudes in a variety of environments. Local control of wind intensity such as the presence of a dense growth of tall trees behind the berms, and bedrock relief affecting beach stability also play a significant role.

3.4 Classification of Sri Lankan dunes

3.4.1 Classification on the geological time

Coastline deposits in Sri Lanka belong to Quaternary period. There are two groups, known as younger and older.

***Younger group*:** The younger group occupying barrier beaches, bars, sand dunes, lagoons, younger raised beaches and estuaries adjacent to the coastline. These are Holocene deposits. Sand dunes can be seen in northwestern, northern and southeastern coasts with different sizes and shapes. These are largely active.

Low dunes zone up to 800 m wide and less than 1m height. These form a low, hummocky surface on which only grass and ground creepers can grow and trees are absent. These are recent dunes. High dune zone up to several km wide and the height goes up to 12 m. They are formed on the younger raised beach and taller shrub vegetation can often be seen.

Older group: Old sand dune deposits and beach deposits (beach rocks) which occur on slightly higher land along the northwest coastline at Kudiramalei are called red earth. It is a red or brick red color clayey sand or loam due to highly weathering and oxidation. These old sand dunes occur as low, narrow, elongate ridges or domes, covered by shrub jungles with about 2–3 m high trees. The ridges are generally 2 to 5 km long, 2 to 3 km width and the average height about 25 m. The height of ridges goes up to 70 MSL in some places. They are aligned in a north south direction. It was formed above the limestone bedrock and mixed with weathered latosols. These are Pleistocene deposits which are fixed by dense forests. This red earth zone goes up to 5 kms in width, rising probably up to 20 m usually and the thickness of red earth deposit and latosols may go up to 35 m in the Aruwakkalu area. These are older dunes on the former raised beaches (Cooray, 1984)

Another red earth deposit was found near Hambantota and Bundala (Dorawa) area, in the Southeast of Sri Lanka. It lies above 10 m MSL and may be an old sand dune deposit. But it is not confirmed as a Pleistocene deposit (Weerakkodi, 1990).

3.4.2 Classification on the morphology (shape and size)

Low flat dunes: These are slightly undulating, isolated platforms of sand less than 1 m high, occurring mainly in the wet zone coast from Matara to north of Colombo. They spread behind narrow sheltered beaches, where no coconut or large trees are found. The maximum width is about 30 m. In the other locations this may be called an incipient foredune.

Transverse primary dunes: These consist of single foredune ridges or undulating sand masses or of parallel sets of such dunes where repeated progradation has occurred. These are active dunes and are associated with stable beaches whose plan curvature ranges between 16° and 30°. This can be seen in south side of the Mannar Island, southern shores of Poonarinyn Peninsula (in a low wave energy environment) and southern portion of Kalpitiya Peninsula. Different primary dunes are found as single or multiple sand ridges parallel to the shore principally in the southeast coast. The height exceeds 5 m.

Secondary transgressive dunes: Secondary dunes are the result of the subsequent modification of the "primary dune" by continued aeolian processes, and are generally located further inland, separated from near shore processes. These are mostly longitudinal but some are parabolic, blowouts depressions and hollows, and a few more are complex. The height exceeds 3 and it may go up to 12 m, and are widespread in the Mannar Island, Kalpitiya and Poonaryn Peninsulas. Due to the effect of wind directions as the northeast monsoon, southwest monsoon and north-northeast wind during the northeast monsoon, the transgressive dunes in the eastern coast of Jaffna peninsula are very clear.

Different transgressive type dunes are found in the southeast coast. Transgression has taken place along the axis of dunes on bedrock relief such as rocky headlands, low inselbergs and rock knobs. Hilltop dunes have resulted from such longitudinal transgression. The actual height may reach above 30 m MSL (Swan, 1983).

Back dune: Back dunes are those dunes located beyond the secondary transgressive dunes and are typically more stable and covered with intermediate woody and herbaceous species. Taller shrubs and trees grow here forming coastal forests.

Other features: Blowout depressions between incipient dunes and back dune locations are formed as a part of the dune process.

3.5 Volume of dune sand deposits

Development of sand dunes occurs along approximately 350 km length of coastline, in the dry and wet zones in Sri Lanka. Most of thick dune hills occur on northwest, north and north-northeast east covers about 90% and southeast covers less than 10%. All other coastlines, such as southwest coast may be less than 1%. No accurate estimate of the rate of annual deposition of dune sands has been calculated. However, it has been found that the dune sand volume in the Kalpitiya Peninsula is approximately 14×10^6 m^3 in a 2.8×10^3 m^2 area. Here the average height is 5 m and the maximum is 10 m.

3.6 Properties of dune sand

Particle size distribution in dune sand shows about 60% fine to medium sand (62.5–500 microns), about 30% silt and about 5–10% clay parts and dust particles. The Uniformity Coefficient is 1.8. It belongs to the fine zone as defined by the British Standards classification system. It shows that dune sand is poorly graded or is a more uniform type of material. There are no coarse sands (or very small amounts). This implies that with respect to the particle size distribution, dune sand in its natural condition is suitable for wall plastering and some other masonry works (Jayawardena and Indratilaka, 2014).

Microscopic analysis shows that dune sands consist of 95% quartz and 5% garnet, feldspar, ilmenite and other heavy minerals with clay, fine dust, fine shell fragments and about 0.1% organic matters. Grains are angular to sub-rounded and tabular shapes. The color of younger/Holocene dune sands vary between light to dark yellowish brown and the old Pleistocene dune grains are reddish brown or red due to weathering. Specific gravity lies between 2.74 and 2.76 and bulk density of loose sample is 1633 kg/m^3. Water absorption is 0.26% (Jayawardena and Indratilaka, 2014)

The average 28-day compressive strength of concrete with river sand is 40 N/mm^2 and is 36 N/mm^2 with dune sand (28 days, Grade 25). This may be due to the low percentage of coarse sand in dune deposits. When dune sand is mixed with one third of crusher dust the compressive strength increases up to 40 N/mm^2. It indicates that the compressive strength of dune sand can be increased by mixing of coarser grain sands. Tests with additional 15% cement, dune sand and rock aggregates show a much higher compressive strength of 42 N/mm^2. Compressive strength of mortar with river sand and dune sand were 29 N/mm^2 and 27 N/mm^2 respectively. The compressive strength of mortar with dune sand and cement can be increased by adding more river sand. It can also be used for concrete mixes in various constructions by adding the required amount of coarse sand either from the river or from crusher dust (Jayawardena and Indratilaka, 2014).

3.7 Importance of sand dunes in Sri Lanka

In general, sand dunes provide natural coastal protection against storm surges and high waves, preventing or reducing coastal flooding, coastal erosion and structural damage, as well as providing important ecological habitat. It protects people and their homes from uncontrollable destructive forces of sea waves. This means sand dunes are natural barriers to the destructive forces of wind and waves. Sand dunes are the first line of defense against coastal storms and beach erosion. They also act as sand storage areas, supplying sand to eroded beaches.

The effect of tsunami created severe damage to the people and their homes in the coastal zones of Sri Lanka, mainly south and southwest, east and north east, and some parts of north in the country on December 26, 2004. The damage in the dune sand areas was much lower than the other coastal zones where the sea waves easily moved into the land. Most of the damaged areas were almost flat lands with coconut and paddy cultivation and some urban areas including major cities. There are no developed dune sand ridges or hills in the affected zones. The sea waves moved freely about 2 km into the land.

Furthermore, sand dunes also provide an important habitat for a number of plants and animals and have environmental, social and economic significance. Sand dunes are also a rich habitat for specific vegetation and wildlife. The dune vegetation shows creepers to trees from primary dunes to back dunes or hind dunes. The first wall of defense is the primary vegetation found on the incipient foredune. The grasses try and hold much of the sand in place and protect the beach. It provides a breeding ground for all dune animal species.

In Sri Lanka, plants in coastal zones are mainly used for the collection of firewood by nearby inhabitants, the supply of timber, sticks, bushes, etc., foods for goats and cattle feeding and some plants species are used as medicinal plants.

Sand dunes serve to protect inland areas from coastal water intrusion. The fishermen and coastal dwellers obtain groundwater from the shallow wells in the incipient or primary foredune areas in Sri Lanka. The water volume and the quality depend on micro-geomorphology and the rainfall pattern.

3.8 Degradation/destruction of sand dunes

The biggest and most common threats to sand dunes have a human origin. The dune environment is sensitive and vulnerable even to a slight environmental stress. When humans decide to alter this sensitive grainy environment, the beach and the entire coastline change almost overnight. Coastal sand dune systems are highly vulnerable to disturbances by trampling. Walking, running, and playing can irreversibly damage a beach dune.

Field investigations revealed that several segments of the existing dune system in some coastal areas have been irreversibly damaged due to human activities both by incorrect design and uses. Two reasons can be identified for the high demand for sand dune lands closer to the main road and urban city areas. First, there is a tendency for people to settle closer to coastal urban areas in order to use the existing infrastructure facilities. Second, almost all the lands where most of the dunes are located, belong to the Government. These lands can be obtained for long-term lease or temporary occupation licenses by paying a minimal fee to the Government. This increases the chances of settlements on sand dunes. There are also many hotels being built along the coastlines. With the construction of houses and roads, car parks, play grounds as well as small shops being constructed, including other facilities for the residents. The removal of sand (to prepare the land) and landscaping the dunes for housing construction are two major factors for dune degradation in Sri Lanka. The tsunami waves moved into the land through a man-made passage across a sand dune ridge (a path made to walk to the beach) and created severe damage close to salt lakes/salterns at Hambantota.

Illegal sand mining for construction purposes and converting sand dune areas for farming are the other adversaries of sand dunes. These are the features in rural coastal areas. Some of the present prawn farms in Kalpitiya, Northwest of Sri Lanka were lands of dune sand a few decades ago. In addition, the linear sand dunes were cut-down a few years ago to control floods, which has resulted in the discontinuation of sand dunes. In some cases, the growth of backdunes has been disturbed due to the existence of salterns.

Other main causes of sand dune degradation are: unauthorized housing construction or encroachment, making paths and tracks across the dunes, beaching of fishing crafts and drying of nets, clearing of dunes for fishing activities, dumping of solid wastes, livestock on dunes, extraction of vegetation for cottage industries, lighting fires on dune vegetation, driving motor cycles and 4WD vehicles on dunes for tourisms or personal. Removal of sand from the dunes for rural road construction and funeral services have also been responsible for dune destruction.

The sea level rise by natural earthquakes or as a direct result of global warming is also destroying these extraordinary natural fences, however they do act as natural barriers.

3.9 Flora and fauna in dune sand areas

Coastal sand dunes provide habitat for plants and animals, including rare and endangered species. The structure of sand dunes depends largely on stabilizing vegetation such as native dune grass. The water table is deeper beneath dunes and raised beaches. This is due to the freely and rapidly draining water through sand grains and less rainfall in local dry climate. Therefore, dune morphology has an important effect on vegetation above the higher water table. There are three overlapping vegetation zones. The vegetation cover varies with the stability of the dunes. Less vegetated portions are associated with mobile sands. These are partially covered by plants which are less wind resistant. Further landward there is an almost complete vegetative cover. Some grasses play a major role in fixing the sands of dunes. More inland where sands are stable and there are no effects from waves and tides, low herbaceous plants, bushes and small trees appear. In the innermost zones the sands are more weathered and form different soils where various types of littoral trees are growing. This vegetation has been replaced by coconut, palmyra and cashew nut and home gardens. Some of these trees also partially bury if dunes are unfixed by vegetation. Nearly 70 species have been identified as dune vegetation in Sri Lanka (Katupotha, 2018). A total of 71 faunal species were recorded representing butterflies, reptiles, birds and mammals. There are seven species of migrant birds. Two species of birds observed at the site are listed as nationally near threatened (Katupotha, 2018). The mammals observed at the site included domestic and feral species.

There are Pleistocene old dunes in the northeast and northwest ridges which shows reddish color due to iron rich minerals as a result of weathering effect. These are undisturbed areas and support for forest covers. There was enough time for plant succession in different climatic stages in the past. Similar red earth deposit in a higher elevation area occurs with different vegetation in southeast coast also.

3.10 Future of sand dunes in Sri Lanka

Sand dunes must be protected and defended like national treasures because sand dunes protect inland areas from swells, tides and winds. A strong and healthy beach dune is a powerful buffer against coastal erosion. Furthermore, sand dunes also provide an important habitat for a number of plants and animals and have environmental, social and economic significance INSERT (see above).

According to the national environment policy and the national policy on wild life conservation in Sri Lanka, the national biodiversity of sand dunes should be conserved and the resources of dune sand should be protected, while conserving the coastal environment. Although there is a coast conservation department their attention is mainly to protect the coast from erosion and other activities and there is no provision for sand dune conservation and protection (CCD, 1991). Human activities are the main reasons for sand dune degradation. This is not only by the dwellers and coastal communities but also by other outsiders, such as farmland owners, business people, tourists and hotel owners, contractors and politicians, etc., Natural incidences of tsunamis are very rare in Sri Lanka. The effect of swells and tides are also not strong enough to damage coastal sand dunes, but there are some locations where the coastal erosion is very high. Therefore, degradation from humans should be controlled and minimized.

A few years ago, there was a project proposal for rehabilitation and restoration of sand dunes in north western, southern and eastern provinces of Sri Lanka after the tsunami disaster in 2004, but it was not implemented correctly. Therefore, a proper plane for rehabilitation and management of sand dunes is very important to protect the resources and conserve the natural biodiversity environment in dune sands in Sri Lanka. It should include a simple effective dune restoration mechanism, all conservation activities, stop all illegal activities, do postgraduate level research on dune sand formation and annual expansion, plant sand dune vegetation while providing employment opportunities, selecting areas for tourism and other environment protecting industries, welfare for coastal communities, alternative income sources and a method to assess the environmental impact, etc.

4. Environmental problems in the coastal zone

4.1 Environmental problems

The activities in the coastal strip of land and in the near shore area are numerous. The activities at any coastal location may cause instabilities at the locality or sometimes in distant coastal areas. Natural processes cause most shoreline erosion, but various human activities contribute a lot.

Table 22.3. Categories creating problems to the environment.

Event	Description	Quantity
Coastal Erosion	Annual rate, very high in west, south west and south coasts than other areas.	200,000–300,000 square meter/year or nearly 1 meter for three years.
Coral mining	Annual use of coral limestone (west coast)	About 28,000 metric tonnes (58% land, 30% beach and live reefs 12%)
Sand mining	Volume of sand extracted annually	48704 cubes by the beach and total 500,000 cubes from the rivers near coastal zones and annual increase about 14%.
Coastal vegetation	Mangrove cover 60–100 sq.km and 23.82 sq.km of salt marshes	Gradually decreasing due to land reclamation and pollution but it has not yet been estimated .
Limestone mining	Inland limestone deposit along the north-west and north coastal zone for cement and lime industries.	Annual cement production is about 600,000 metric tons
Mineral sands	Ilmenite, rutile, zircon, monazite mixed beach sands from north east coast.	Annual production is about 110,000 metric tons.
Dune sands	Volume of sand cut annually	Use for building construction but no accurate estimation.

Coast Conservation Dept (1991), Dayanande (1992), NARESA (1991).

Table 22.3 gives a brief description about the environmental problems around the coastline in Sri Lanka.

Coast erosion in Sri Lanka is not unusual in terms of world trends. The erosion hazard is the greatest when monsoons are onshore. Erosion along Sri Lanka's land beaches is highly localized. The southwest monsoon, extends over the breadth of the Indian Ocean, generates waves with high energy. This monsoon distributes its greatest energy to the areas in west, south and southeast. The northeast monsoon blows over a more limited fetch of the Bay of Bengal and accordingly delivers somewhat less energy to the shore. Its effects are mainly felt along the north coast, the east coast as far south east. Information based on qualitative estimates of coastal residents suggests that on the average the coastline of Sri Lanka receded by about one meter within a three-year period. The wind generated local waves cause severe erosion problems at the coastline.

The most significant of these include sand mining from river beds and beaches, coral mining and destruction of reefs, removal of mangroves and other coastal vegetation and improper location or construction of maritime structures such as shoreline protection works and fishery harbors. There are also some environmental problems due to accretion. These include obstructing of outlets affecting: escape of flood waters, navigation, migration of fish and shellfish species between an estuary and lagoon and the sea, silting of harbors, outlets for urban and industrial pollutants, intake sites for seawater required salterns and some fishery activities. Accretion is also caused by wind and aggravated by removal of dune vegetation. In certain areas sand blown away during the dry season is deposited on paddy lands and coconut plantations.

Fishing is the major activity all around the coast except in the coastlines bordering natural reserves. However, the construction of breakwaters for fishing or commercial harbors could cause instabilities in a large area adjacent to such constructions.

Urbanization and tourism are directly related to population increase and various environmental problems arising as a result within the zone. Pollution and the destruction of most wanted natural environment are the worst effects due to these activities

4.2 Sea level rise

Global warming and the climatic change are the main reasons for the rise in sea level. The impact of sea level rise to the habitats in coastal areas is well understood to a large extent. The UNDP project report mentions about 0.958 m of sea level may rise within next 100 years around Sri Lanka. This is a change of the sea level since the establishment of the datum, due to climate change and high tide. As a result, the total inundated area will be more than 780 sq.km and the coastal erosion extents about 95 m (UNDP Report, 2018). This future prediction may create adverse impacts to the coastal habitats. Most of coastal community people are living in the lands around 1m sea level land areas. This effect is high in NW, N and SE coastal areas of Sri Lanka. Recent analysis indicates that the Jaffna Peninsula in north is projected to lose approximately 106.3–367.8 sq.km of paddy fields by 2100 due to inundation as a result of gradual sea level rise (Gopalakrishnan and Kumar, 2020). In addition, lagoons and the adjacent islands are identified as being the most vulnerable to sea level rise.

4.3 Development of coastal zone

The coastal zone represents about 24% of the country land area. More than 80% of the country's industrial units are located within the coastal zone. About 85% of tourists' revenue comes from facilities in coastal areas. As a result, this zone has been highly polluted and damaged by different activities during last the 30 years. But new constructions could be considered as development. The number of harbors has been increased for fish landing, boat repair works and commercial purposes and more bare lands have been converted into residential and industrial areas. Some breakwaters have been constructed for fishing. New roads have been opened in and around the coastal zone. The railway lines parallel to the coast were constructed by the British without thinking of the effect of it to the coastal environment. Drainage facilities, sewage disposal facilities, communication facilities, schools, religious places, etc., are being built within the coastal zone. All these new constructions may be part of new development but definitely not as the coastal zone development.

The coastal zone has been clearly defined by the Sri Lanka Coastal Conservation Department by a Parliament Act. And there is also a development plan for the coastal zone (CCD, 1997). The development plan of the coastal zone of Sri Lanka was unable to be implemented due to the various objections of the higher-level administration. At present the definition of the coastal zone may not be valid for future development.

4.4 Coastal zone management in future

Sands are moving from south and south west coasts toward the north west coast and east coast. Sands from the north also moving to the north west coast. Therefore, the north-northwest and north-east and east coasts are stable and accreting coasts.

South-west and south coastlines are subjected to severe erosion by sea waves. This is a major natural hazard along the coastline. All others are seasonal and may or may not occur as climatic conditions worsen. However, man-made activities either legal or illegal, are increasing with some commercial and industrial plans operating now. As a result, at present the environmental problems along the coastal zones in Sri Lanka are growing beyond the controlling limit.

Control of erosion due to the action of waves is the most important activity under the coastal zone development in Sri Lanka. The coastal zone must be redefined by considering the upper limits of the sea level during the 2004 tsunami. Highly elevated zones such as old beaches and headland may be suitable to define the boundary levels. This zone can be used for some agricultural development and fisheries activities, but should not be allowed for any other development activities. The administration in this coastal zone should be under one ministry. A separate security unit should be operated only for coastal zones to protect the law and order and avoid the illegal constructions and other activities which may be helpful to protect the existing eco-environment.

The coastal zone should be defined again with new boundaries. The main purpose to redefine is to keep the coastal equilibrium, protect the natural environment and protect the lives and homes of the people living in adjacent areas. Preparation of boundaries of the areas affected by the tsunami should be considered. Geomorphologically highly elevated zones such as old beaches and headland may be suitable to define the boundary levels. To implement a new coastal zone development plan, the leaders of the ruling government and the opposition parties, local political leaders and the administrative officers of the various departments and institutions should be of one opinion.

At the moment each activity section belongs to different government ministries. Therefore, their activities are different. Although there is enough legal power to protect the coastal environment, the activities of different ministries are unable to reduce the destruction of the coastal zone environment. Every ministry is trying to provide more employment to the people. Therefore, to protect the eco-environment around the estuaries and coastal line, all the administrative activities along the coastal zone should be within one ministry. A separate police unit must be within the same ministry to protect law and order.

5. Concluding remarks

While describing the various issues in the coastal zone of Sri Lanka special attention was made for dune sand deposits. Sand dunes, considered as one of the natural resources in Sri Lanka are developing mainly in northwest, north and southeast coastal zones where the hot dry climate and comparatively higher wind velocities are recording of the island. It provides a natural biodiversity environment for flora and fauna. Due to lack of research on dune sand development there are no recorded data or useful information.

Use of natural resources by the coastal community and the various outsiders for different purposes damage this dune ecosystem. To stop all these illegal activities and protect the resources and ecosystem for future a proper plan for rehabilitation and management of sand dunes should be implemented. There should be some ways to improve the income and welfare facilities of the coastal community providing some avenues for employment. Dune sand is good for the construction industry as an aggregate. It can be extracted for small scale constructions or some other purposes only from the selected or recommended locations where there are no impacts to the ecosystem or environment. This is at present a requirement.

Many countries have already developed policies on sea front management to reduce the damage due to inundation from sea level rise. This island has become smaller day by day. Despite this, there is no national plan to reduce the adverse effects due to sea level rise. For better management of the coastal zone of Sri Lanka in the future, new policies are urgently required. In addition, the law and order should be implemented strictly to protect and better manage the coastal zone.

References

Arumugam, S. 1969. Water Resources of Ceylon, Water Resources Board publication, Colombo, Ceylon.

Coast Conservation Department Publication (CCD). 1991. Proceedings of the Seminar on Causes of Coastal Erosion in Sri Lanka, 9–11 February 1991, Colombo, Sri Lanka.

Coast Conservation Department Publication (CCD). 1997. Revised Coastal Zone Management Plan, Sri Lanka, ISBN 955 9108 13 1, Colombo, Sri Lanka.

Cooray, P.G. 1984. An Introduction to the Geology of Sri Lanka, Dept. of National Museums, Govt. Press, Colombo, Sri Lanka.

Dayananda, H.V. 1992, Shoreline Erosion in Sri Lanka's Coastal Areas, Coast Conservation Department Publication, State Printing Corp., Colombo, Sri Lanka.

Gopalakrishnan, T. and Kumar, L. 2020. Potential impacts of sea-level rise upon the Jaffna Peninsula, Sri Lanka: How climate change can adversely affect the coastal zone. Journal of Coastal Research 36(5): 951–960. Coconut Creek (Florida), ISSN 0749-0208.

Jayawardena U.de S. and Indratilaka, H.M.L. 2014. Use of dune sand as an alternative for river sand for construction industry in Sri Lanka, Proc. of the XII International IAEG Congress, 15–19 September, 2014, Torino, Italy, Vol.5,part XXI: 1277–1280.

Katupotha, J. 2018, Sand dune formation, Biodiversity and ecological significance, Proceedings on Workshop on Coastal Resource Management, Anuradhapura, Sri Lanka.

Madduma Bandara, C.M. 1989. A Survey of the Coastal Zone of Sri Lanka, Coast Conservation Department Publication, State Printing Corp., Colombo, Sri Lanka.

Samaranayake, R.A.D.B. 2006. Pre- and post-tsunami coastal planning and land-use policies and issues in Sri Lanka, Proceedings of the workshop on coastal area planning and management in Asian tsunami-affected countries, September 27–29, 2006, Bangkok. Thailand.

Swan, B. 1983. An Introduction to the Coastal Geomorphology of Sri Lanka, Dept. of National Museums, Govt.Press, Colombo, Sri Lanka.

The Natural Resources, Energy and Sciences Authority of Sri Lanka (NARESA), 1991, Natural Resources of Sri Lanka, Published by the Institute, Colombo, Sri Lanka, p280.

UNDP support project report. 2018. Strategic support to operationalize the Sri Lanka Comprehensive Disaster Management Programme (SS2CDMP).

Vitanage, P.W. 1970. A study of the geomorphology and morphotectonics of Ceylon, Proceedings of the Second Seminar on Geological Prospecting Methods and Techniques, New York, USA, pp. 391–405.

Walker, R.C. 1962. The Hydrometeorology of Ceylon Part 1, Canada-Colombo plan project, Government press, Ceylon.

Weerakkodi, Upali. 1990. Coastal sand dunes of southeast of Sri Lanka, Jour. Natn. Sci. Coun. Sri Lanka, Colombo, Sri Lanka 18(2): 133–152.

Chapter 23

Integrated Coastal Zone Management

An Approach to Protect and Conserve the Morphology and Ecology of Coastal Dune Systems of Bangladesh

M. Shahidul Islam

1. Introduction

Coastal dunes are widely distributed in many countries of tropical regions (Hesp, 2004), among which are regions closer to Bangladesh, such as India (Sanjeevi, 1996), Sri Lanka (Swan, 1979), Thailand (Pitman, 1985) and Vietnam (Dinh, 1998), where dunes of varying sizes and shapes are found. In two coastal districts of West Bengal (India), which is much closer to the south-western border of Bangladesh, sand dune landscape has been reported by Chakraborty (2010) and Paul (2002) and this volume. The sixty kilometer long Midnapur coast is characterized by beach, dunes and mudflats. Beach and dune erosion are the major threats to this coast.

In Bangladesh records on dune morphology and dune ecology are very rare and patchy. Only a few studies on coastal dune systems are available (e.g., Islam and Islam, 2010; Bushra et al., 2014 and 2016). Descriptions on dune morphology and ecology along Bangladesh coast are also sporadically found in Alam et al. (1996), Alam et al. (1999a), Alam et al. (1999b), Islam et al. (2001), Rajib et al. (2007), Haque et al. (2011), Islam and Akter (2012), Hossain et al. (2016), Islam and Saha (2017), and Kabir et al. (2018).

The present chapter is an overview of coastal sand dune system of Bangladesh based on available documents and field observations. The major objectives of this chapter is to bring together the existing knowledge on morphological and ecological characteristics of the coastal dune system of Bangladesh, and to explore how this unique habitat is being degraded due to unwise human interventions. The protection and conservation approach has largely been ignored both at policy and implementation stages. This chapter aims to highlight the importance of an integrated approach for protection, conservation and sustainable management of dune habitats.

Coastal Research Unit (CRU), Department of Geography and Environment, University of Dhaka, Bangladesh.
Email: shahidul.geoenv@du.ac.bd

2. Characteristics of the Bangladesh coast

The coastal zone of Bangladesh is characterized by a vast and complex network of rivers, enormous discharge heavily laden with suspended sediments, numerous temporary, semi-temporary and permanent off-shore islands, strong tidal and wind action and cyclone-induced storm surges. The average discharge from the coastal rivers into the Bay of Bengal is 29.07 m^3/s in winter and 65396.12 m^3/s during the monsoon. The wind direction is from north-east during the winter, with an average speed of 1.3 km/hour and from south-west direction during the monsoon, with an average wind speed of 13.8 km/hour (BMD, 2016). However, about 70% annual wind blows during the monsoon period and maximum wind speed of up to 260 km/hour (72 m/sec) at the time of a cyclone.

The coastal belt of Bangladesh is diverse in its physical settings and hydro-meteorological characteristics and includes a wide range of ecosystem, such as mangrove forests, salt marshes, coral bearing islands, estuaries, mudflats, sandy beaches and coastal dunes, each has its own distinct biotic and abiotic properties. The belt is one of the severe disaster-prone areas of the world, which is being affected by a series of hydro-morphological hazards almost every year. Both erosion and accretion rate in the coastal belt are very high and at an accelerating rate, but of different scales at various locations. Ahmed et al. (2018) showed that the sandy beach and dunes landscapes along the coastal belt are highly to moderately susceptible to coastal erosion.

The coastal belt of Bangladesh is broadly divided in to three distinct hydro-geomorphic regions, which are the eastern coast, central coast and western coast (Pramanik, 1983). The eastern coastal belt extending from Feni River to the Teknaf tip is a regular, relatively stable, unbroken strip of land, backed by the Tertiary hills in the landward side, and mud flat and submerged sands in the sea-ward side. The important geomorphic landforms are sandy beaches, mudflats and coastal sand dunes. Along this coast, there is a 145 km long continuous sand beach, extending from Cox's Bazar to the Teknaf Peninsula, and is one of the tourist attractions of the country.

The central coastal zone is one of the largest hydrological regimes of the world, receiving enormous amounts of sediment from up-steam region through the Ganges-Brahmaputra-Meghna (GBM) river system. A part of these sediments is deposited both on-shore and off-shore to continue the delta building process and any remaining sediments are drifted out to the deep Bay of Bengal through the Swatch of No Ground. The shoreline is broken and consists of series of submerged and newly formed *chars* (vegetated islands) due to off-shore sediment deposition. These newly formed islands are very unstable, comprise of complex network of tidal creeks, mangroves and saltmarsh vegetation, mudflats, sand flats, sand beaches, sand dunes and deeper channels.

The beachfront of the western coast, in general, is transverse to the structure of the continental margin. This coast is relatively stable, and the flat topography is criss-crossed by a complex network of tidal rivers. The major geomorphological features are tidal-flats, sandy beaches, dunes and beach ridges (Khan, 1990). This coast includes the world's largest single mangrove forest-the Sundarbans, which acts as a natural defense against cyclonic events.

3. Dunes along Bangladesh coast

Dunes are unique coastal habitats along the Bangladesh coast, which change their shape frequently due to storm generated winds and wave actions. Some important dune sites along the coast are illustrated below:

3.1 Dunes in the Eastern coastal belt

In the eastern coast, dunes are more distinctive, well developed and rich in bio-diversity. Some important locations are, St. Martins Island, Teknaf Peninsula, Cox's Bazar beach, Sonadia Island and Chittagong City beach (Fig. 23.1).

St. Martins is the only coral bearing off-shore island of Bangladesh with an area of 8 km^2. The main shoreline habitats of the island are sandy beaches and sand dunes, which are primarily

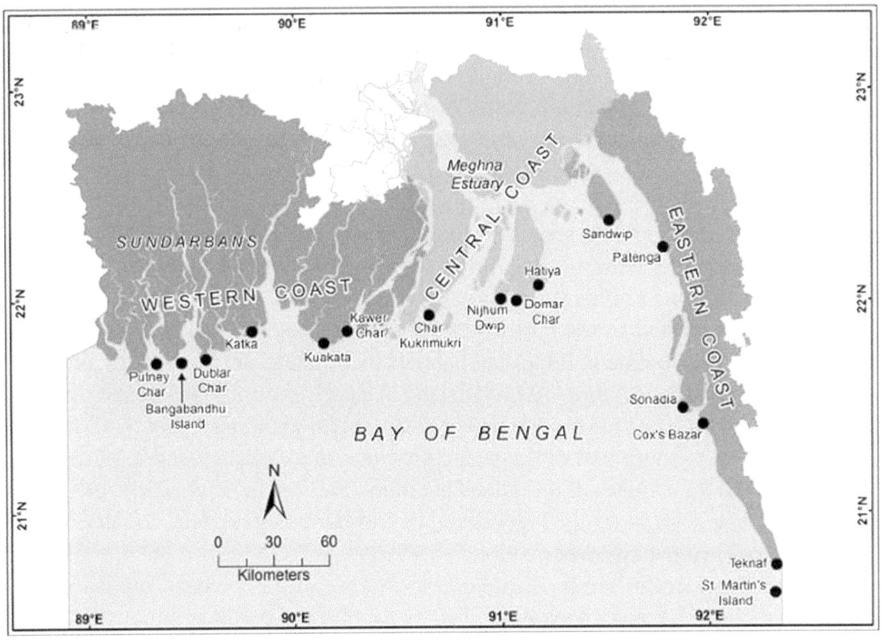

Figure 23.1. Map showing dune sites along Bangladesh coast.

composed of alluvial sands. The dune of the island is classified into high and low types (Islam et al., 2001). The high dunes of more than 5 m high are ridge-like topographies with very steep windward and gentle leeward slopes, and are parallel to the shore. They are mostly found in the north-western side of the island. The low dunes are less than 2 m in height, have an undulating topography, gentle sloping both landward and seaward, and are widely distributed along the shoreline of the island (Islam et al., 2001). The sandy beach in the north and north-east stretches up to 400 m to the sea. The sandy beach and sand dunes of the island are the best nesting ground in Bangladesh for globally threatened marine turtles (Islam, 2002).

A survey by UNDP (2010) shows that the common dune vegetation in the island are *Cocos nucifera* (coconut palm), *Streblus asper* (locally called shaora), *Pandanus odoratissimus* (keya), *Vitex trifolea* (nil-nishinda) and *Vitex negundo*, *Ipomoea pescaprae* (shagor lota), *Allium* sp., *Panicum repens, Paspalum vaginatum, Cyperus* spp. and *Fimbristylis* spp. Deforestation in St. Martins Island is at an alarming stage in the recent years due to expansion of agricultural land, and collection of sand dune vegetation for fodder and fuel wood. The main drivers are an increase of population density, increasing number of tourists and more frequent natural disasters.

The Teknaf Peninsula is the south-easternmost tip of Bangladesh. Sandy beaches and coastal dunes are two unique geomorphological and ecological features at Teknaf, which provide breeding grounds for two globally threatened species of turtles. Sand dunes occur along the beach, with sparsely distributed dune vegetation. Vegetation is adapted to different zones and is denser in the herbaceous zone with some mat forming herbs. Mixture of herbaceous plants and shrubs, including climber species are found at the top and back dune zones.

The Cox's Bazar beach is the most popular tourist spot in Bangladesh. Sandy beaches and sand dunes are two important geomorphological features in the coastal belt of Cox's Bazar. These two features indicated the infilling of the sheltered basin in front of the Tertiary hill-ranges by high wave, micro- to meso-tidal conditions, and also triggered by cyclonic events and storm surges (Alam et al., 1999a). The beach faces the open sea, and the impact of cyclone-induced storm surges, wave and wind actions play a greater role shaping the beach morphology than that in dune formations located further landward. This is due to a considerable reduction of strength of these events as they move inland. The beach plain of this coastal belt is 2–3 m above the MSL, and the dunes are located on the beach side by side, separated by tidal creeks. Dunes in this beach are located in north-south direction almost parallel to the beach plain and shoreline, and having a maximum width of 200 m (Alam et al., 1999a). The boundary between dunes and the upper beach is very dynamic and shifting in nature due to continuous interaction with the prevailing wind-blown sands and the growth of creeper vegetation ganga-lata (*Ipomea*). Dunes are dissected by micro-channels (dune valleys) and unlike dune features, such valleys are wind-cut erosional features of the old dunes, created at the time of storm wind blows. Dune features are immature and unstable, and are highly susceptible to erosion and migration due to storm surges, spring tides and cyclonic winds. However, some dunes have recovered and rebuilt within the last 3–6 months and the ganga-lata (*Ipomea*) vegetation acts as an anchor to trap airborne sediment and promote the growth of the dune landform.

Sonadia Island is located in the north-west of Cox's Bazar and has an area of 4.9 km^2. The island is owned by the Government. Important habitats in the island are mangroves, mudflats, canals and lagoons, beaches and sand dunes. Sonadia Island is largely the combined result of erosional and deposition process under a transgressive episode of the recent geological past. A 13 km long sandy beach from north-west to south-east of the island provides an excellent location of nesting grounds for marine turtles. The dune ridges of this island are more than 7 m in height (pers. obs.). The coastal sand dunes of this island are so far the highest and largest dune ridges of the country, which act as a strong natural barrier against cyclonic events.

In the western segment of the beach (about 4 km long), no permanent human settlements or any other infrastructures have yet been built. This portion of the beach is famous for turtle nesting and can be dedicated as *in-situ* conservation site for nursing juvenile turtles and nesting ground for marine turtles. This part of the beach represents the core value of Ecologically Critical Area (ECA). The central part of the beach is characterized by about 500 m wide sandy beaches and well-defined dune fields. In this portion of the island, there are two small villages and some important infrastructures, such as a cyclone shelter, mosques and permanent houses. Some patches of jhau Tamarix plantation in the dune fields with the active participation of village conservation groups are also found. The eastern part of the beach is very wide (up to 500 m) and includes sand dunes and mangrove shrubs, and provides a good ground for local and migratory birds. Some temporary settlement, particularly fish drying activities are found in this zone. In the Sonadia Island, sand dunes covers 4% of the total vegetation species and the important species are *Ipomea pes-caprae, Vitex trifolia, Pandanus foetidus* and *Casuarina equisetifolia.*

Islam and Akter (2012) identified 12 distinctive landform types, along the west and south coast of Chittagong city, which are mud plains, mud flats, mangrove plains, back swamps, tidal plains, sandy tidal plains, salt marshes, saltpans, sandy plains, sandy beaches, sand ridges and sand dunes. Sand dune landscapes are prominent at Uttar Patenga area (south of the city), in a north-south alignment and running parallel to the shoreline. The dune ridges are 25–70 m inland from the shoreline. The common vegetation types are *Cocos nucifera* and *Bumbosa arundinacea*, although many dunes are barren. Dunes in the northern part are larger with elevations that are 2–3 m higher than those of the south. Dune ridges in the north are steeper (up to 25°) in the windward side than the lee side (less than 15°).

3.2 Dunes in the central coastal belt

It was mentioned earlier that the central coastal zone is largely dynamic due to strong tidal and wave actions, and is vulnerable to a set of natural disasters, such as cyclones and storm-surges (Brammer, 2014; Hossain et al., 2016). The tidal range varies from less than one meter during neap tide to more than 4 m during spring, with a maximum of 7 m at Sandwip (Rashid, 2019). The tidal current in the Sandwip and Hatia channels is 10.8 km/hour (Barua, 1997). Despite adequate supplies of sediments as materials for dune formation, in the central zone the beach profile is not supportive enough to form new dune landscape and stabilization of existing dune topography. Strong winds, tidal and wave actions are not conducive However, dunes are found at Kuakata, Nijhum dwip, char Kukri

Mukri, Domar char, Kawar char and many other off-shore islands. Except Kuakata, dunes in this coastal zone are small in size and the height is less than 3 m.

Detailed descriptions of the morphology and ecology of coastal sand dunes at Kuakata are available in Bushra et al. (2014 and 2016). Kuakata is located on the west of the central coast of Bangladesh, and has a 30 km long and 120–350 m wide extensive sandy beach with a panoramic view of dune landscape. The beach has a gentle slope into the Bay of Bengal and its maximum altitude is 3 m (Rasheed, 2008). There are a series of dune ridges, which are degraded due to massive erosion and human interventions. Most of the dunes have longer windward slopes, where sands are pushed upward, and a shorter 'slip face' in the leeward sides. Dune height varies from less than 1m up to 6.5 m. Some of the dunes are embryonic in nature and others are established foredunes. Slacks and blows are very common in these dune fields.

3.3 Dunes in the Western coastal belt

The greater part of western coastal belt is occupied by the Sundarbans. In the Sundarbans area, the eroded materials from river banks and islands are blown above high-water marks by strong southwest monsoon winds to form the dune landscape. Muds are accumulated in the leeward side of the dune to form mud flats, which remain protected from wave actions. Later these were overlain by drifted sand from the dune and finally developed into grassy meadows. This process continues until the windward side of the dune is exposed to wave actions. In the newly emerged islands in Sundarbans, sand dunes fringing the sea are primarily colonized by grasses and important species of grasses are *Paspalum vaginatum, Panicum repens, Aeluropus lagopoides* and *Phragmites karka.* These grasses are followed by *Sesuvium portulacastrum* and *Ipomoea pescaprae* as the climax dune vegetation (Gopal and Chauhan, 2006). Two important locations for dune landscape are at Kotka-Kochikhali beach and Bangabandhu Island.

Rashid et al. (2009) had shown that the at Kotka-Kochikhali area of the Sundarbans, sand dunes are well developed and are exposed. These dunes experience the most severe tidal action and are very unstable. The dune soil at this area is characterized by salinity range 5–10.4 mS/cm, daily inundation varying from 0.5–1 m in the monsoon but no inundation during winter, soil texture is sand-54.4%, silt-30.90%, and clay-14.70%. Soil OM is 1.38%, pH is 7.6, and EC is 39.30 mS/cm. Sand dunes are dominated by grasses and sedges with sporadic occurrence of mangrove species (*Phragmites karka, Imperata cylindrica, Sporobolus tremulusa, Acrostichum aureum, Acanthus ilicifolius*). Some common dune species are *Borassus flabellifer, Syzygium fruiticosum, Porteresia coarctata, Pandanus foetidus, Saccharum spontaneum, Clerodendrum inerme, Tamarix indica, Erythrina variegate, Caesalpinia bonduc, Eugenia fruticosa, Leea crispa, Streblus asper,* and *Ipomoea pes-caprae*

Bangabandhu Island is a newly emerged off-shore island in the western shelf zone of Bangladesh. It was first visible above the tide level in 1976 and at present it covers an area of about 8 km^2 and is about 2 m above the spring tide level. Islam and Saha (2017) gave an elaborate description about the origin and evolution of the island. They identified six major geomorphic units, which are tidal creeks/water bodies, tidal flats, supratidal flats, mudflats, sandy beaches and sand dunes. About 9 km long and 60–110 m wide extensive beach land has been formed in the east, south and sides of the island. Well-developed dune landscapes are found along the landward edge of the beach fields. Vegetation is not well developed in the dune fields and most of the dunes are in an embryotic to juvenile stage.

4. Dune forming processes along Bangladesh coast

The dune forming process and ecology of coastal dune system of Bangladesh is very poorly understood. The evolution of coastal sand dunes, in longer timespans have largely been controlled by relative sea-level changes. In a regressive sea-level phase, the exposure of subaerial beach land increases, sand budget significantly increases and the beach progresses towards the sea. As a consequence, new dune chains are formed, leaving behind the inner limit of the new dune strip as a paleo-shoreline. In a transgressive phase of sea-level, beaches are flooded and eroded, which narrow down the beach limit, sediment supply is negative, hardly any new dune strip is formed, the existing dune, particularly the dune berms are washed away by waves and high waves often wash over the main dune field. In case of a stable phase of relative sea-level, the sand budget in the beach maintains an equilibrium, the beach is relatively stable and as also the adjacent dune belt.

The prerequisite for the formation of coastal sand dunes is wind, sand and vegetation, and their complex interactions. However, some important dune forming processes along Bangladesh coast are described below:

4.1 Oceanographic process

The topography of the sea floor in the northern Bay of Bengal plays an important role in sediment transportation and redistribution, and frequent changes of coastal geomorphology, including dune morphology. Large amounts of sediments are carried to deeper waters in the Bay of Bengal, which later on drift back to the near shore zone by underwater current (Hossain, 2015). This dynamic process in the shallow shelf zone of northern Bay of Bengal demonstrates the unstable characteristics of the coastal landscape, including the dune features of Bangladesh, particularly in the central coast.

Krantz (1999) showed that the long-shore current in the off-shore coastal belt of Bangladesh travels anticlockwise—being influenced by waves and tidal bores. Sediment transport to the near-shore zone of the coastal belt and the formation of dune morphology is a complex function of long-shore and cross-shore currents and wave actions. Sediments are transported to the near-shore zone by long-shore currents. Wave actions along with cross-shore current are used to carry sediments to the onshore and up to high water mark depending on wave energy and sea floor condition of the near-shore zone. Moreover, the eroded materials from the beach and dune front are carried away, and bottom sediments are lifted by wave action and add sediment in the long-shore current.

4.2 Climatic conditions

In humid tropical countries like India and Bangladesh, climate plays a significant role in coastal dune formation as an agent of sediment transfer by prevailing winds, dune soil formation and vegetation stabilization (Bandyopadhyay, 1997; Bushra et al., 2014). The direction, frequency, duration and speed of wind are the most important contenders in dune formation along the Bangladesh coast. It requires wind speed greater than 16 km/hour (4.5 m^{sec}) to lift, displace and transport dry sand particles (Mascarenhas, 2002). Such winds are very common in the northern Bay of Bengal during most parts of the year, except in the winter season. Wind is more active along low gradient beaches, particularly along the east coast, when they are sub-aerially exposed during the low tide, and strong wind blows perpendicular to the beach.

4.3 Vegetation cover

Sand particles, either carried by waves or winds to the beach are the main elements for the formation of dune morphology. Usually, beach sands are drifted and blown landward by onshore wind, which are finally trapped by vegetation, that act as a nucleus to sediment accumulation and dune stabilization. Vegetation that grows close to the beach plays the key role in dune formation by acting as a wind breaker, reducing the wind speed and forcing the wind to drop some sand particles along its path, as has been found in all major dunes sites of Bangladesh. Vegetation thus traps the moving sands. This first step of sand accumulation by a salt tolerant plant is called a berm. Dune vegetation is highly tolerant to salt-laden winds and helps to trap windblown salty sands in the foredune, which continue upward to the dune crest. The combination of dune height, dune shape and vegetation coverage create a protective system to repair the damage, either created by natural processes or human impacts.

Untawale (1994) showed that herbaceous species are the pioneer vegetation at the base of a dune. Along the Bangladesh coast, the main pioneer species are *Ipomea pes-caprae* (sagor lata), *Ipomea* sp. (ganga-lata), *Lathyrus japonicas* (dune pea), *Ammophila host* (marram grass); in the mid-shore zone naturally grown herbs and shrubs with deeper root system are *Vitex trifolia* (nil nishinda), *Lumnitzera racemosa* (kipra); and in the backshore, dune trees are found, which are *Pandanus fascicularis* (keyakata) *Pandanus odoratissimus* (keya), *Tamarix dioica* (nona-jhau) *Tamarix dioica, Allium* sp., *Panicum repens, Paspalum vaginatum, Cyperus* spp. and *Fimbristylis* spp. Bandyopadhyay (1997) showed that at Sagar Island in West Bengal (India) major dune species are *Ipomoea pes-carpae, Launaea, sarmentosa, Cyperus exaltatus* and *Opuntia dillenii* which play very important roles in dune stabilization. In the Medinipur coast it was found that dunes work as an effective barrier against storm surges (Chakraborty, 2010).

5. Coastal dune morphology

The morphology of coastal sand dunes, particularly the size, shape and height of foredunes depends on plant diversity, wind speed and rate of sediment transport. Moreover, the rate of occurrence of swash inundation, storm wave erosion, over-wash incidence and wind direction are important for further growth and evolution of coastal dunes (Hesp, 2002). van Dijk et al. (1999) demonstrated that a plant canopy has a great impact on sand flow and controls dune morphology (shape and height). They showed that with the increase of plant height, the height of the dune increases, but the dune length decreases.

Foredunes along the Bangladesh coast can be divided into incipient foredunes and established foredunes. Incipient or embryotic sand dunes are the initial stage of dune formation, which is formed due to accumulation of sand particles anchoring a single plan, a discrete bunch of vegetation or any physical obstacle (Islam and Islam, 2010; Bushra et al., 2016). Many of such coastal dunes are temporal and seasonal, particularly along the off-shore islands, which largely depend on seasonal wind variations and limit of spring high tide.

Dune morphology along the Bangladesh coast is site specific, and depends on the predominant direction of onshore wind and onshore current with the shoreline, and with the supply of sediment flux (Table 23.1). In the eastern coast, dunes are unique in their morphology and ecology. Alam et al. (1999a) showed that along the Cox's Bazar beach, dunes are variable in shape, from 2–4 m high, with a slope of about 10° and are parallel to the plain land (Fig. 23.2). Waves and wind are the two major controlling factors for dune growth. The textural composition of coastal dune sediment in the eastern belt shows that dune soils are well sorted.

Dune sandy soils are more porous and leached. The area between the dune ridges in the seaward side and Tertiary hills in the landward site acts a buffer zone, where remnants of paleo-dune ridges are found; either buried or exposed (Monsur and Kamal, 1994). The dune morphology along this coast is threatened by cyclonic storm-surges and anthropogenic interventions, particularly by tourists.

In the central zone, dune sandy soil is mixed with clayey particles and is more retentive. However, in a typical coastal dune system in this belt, the bulk density and salinity are higher in the dune front and dune crest than that of the dune back, although the hardness is higher. Due to such soil properties, the front and crest areas of a dune system are less stable than that of the dune back. In the off-shore islands, the alluvial deposits show alternating layers of clay, silt and sand. In a newly formed off-shore island, the shoreline, including the dune features are constantly being eroded. The eroded sand particles are blown to the beach above the high-water mark to form dune topography. Mudflat in the lee side of the dunes is eventually overlain with sand deposits and covered in grasses

In the western coast, dunes are composed of coarse sands of light-brown to white in color, and can be characterized by small undulation dune ripples, parallel/sub-parallel to wind direction. Vegetation has not been well developed in the dune fields and most of the dunes are in an embryotic to juvenile stage. The upper layer of the beaches and sand dune fields is dominated by loose sand

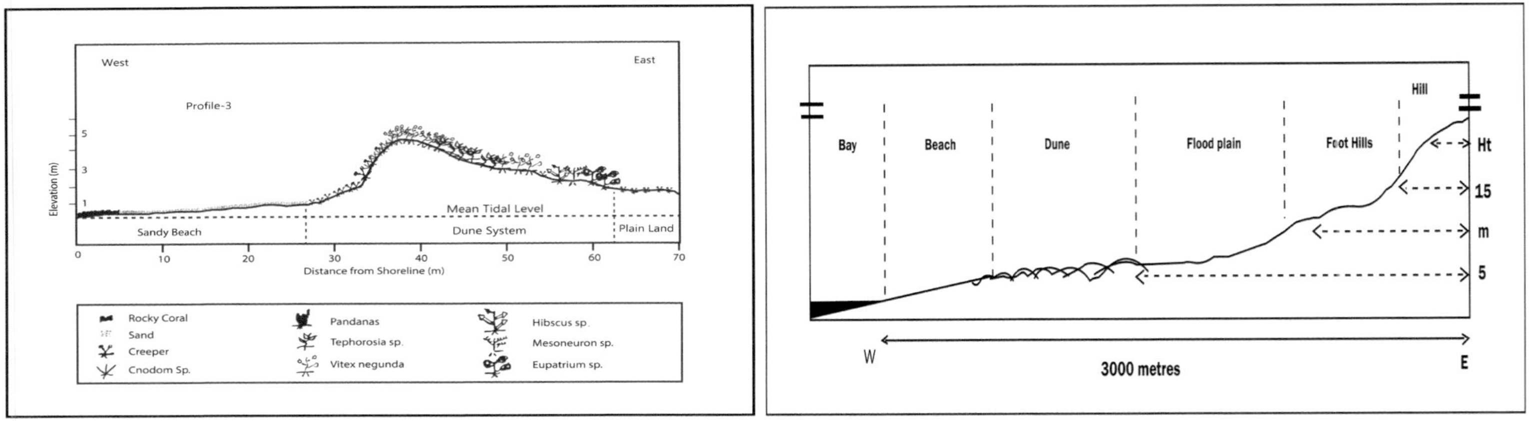

Figure 23.2. Left: Dune Profile at St. Martins Island (source: Islam et al., 2001); Right: Beach Profile at Cox's Bazar (Source: Alam et al., 1999a).

Table 23.1. Site specific morpho-ecological characteristics of coastal sand dunes in Bangladesh.

Coastal Belt	**Dune Sites**	**Dune Morphology**	**Dune Ecology**	**Human Interventions**
East Coast	St. Martins	Height-up to 5 m; windward slope 45°; narrow beach, soil poorly formed; poor organic content	Well vegetated; Major species: *Cocos mucifera, Pandanus odoratissimus Bambusa catechu, bambusa arundinacea;* nesting ground of sea turtles.	Human settlements; hotels/motels; tourist activities; fish drying; govt structures
	Cox's Bazar	Height: up to 4 m; slope 10°; undulating; wide beaches; high concentration of heavy minerals; soil well sorted	Sparsely vegetated; Major species ganga-lata (*ipomea*);	Hotels/motels; tourist activities; marine drive; shrimp hatcheries; urbanization;
	Sonadia Island	Height: up to 7 m; steep windward slope; dune ridge well formed; dune slacks are well developed; wide beaches; well sorted fine to medium sand; presence of heavy minerals;	Well vegetated; Major species- *Ipomea pes-caprae Ammophila host. Vitex trifolia, Casuarina equisetifolia;* nesting ground of sea turtles.	Two small villages; fishing activities; fish dry; melon cultivation
	Ctg City coast	Height-2–3 m; windward slope 25°; narrow beaches; undulating;	Sparsely vegetated; Major species *Cocos nucifera* and *Bumbosa arundinacea*	Tourism; Fishing;
Central Coast	Kuakata	Height-up to 6.5 m; dune ridges are well formed; slacks and blows are common; highly eroded	Sparsely vegetated; Major species- *Ipomea pescaprae*	Tourism; housing plots; road construction;
	Nijhum Dwip	Height up to 2.5 m; undulating; juvenile dunes; steep to gentle slope;	Sparsely vegetated; Major species- *Ipomea pescaprae*	Not much disturbed; limited fishing;
West Coast	Kawar Char	Height-up to 1.5 m; embryotic dunes; gentle slope	Sparsely vegetated; Major species- *Ipomea pescaprae*	Remains undisturbed; few fishing activities
	Kotka-Kochikhali	Height-up to 1.5 m; well developed and exposed; soil highly saline; sand mixed with silt and clay; high organic content; highly eroded; palaeo-dunes are found	Densely vegetated; Mangrove associate species are common; Major species are *Phragmites karka, Acrostichum aureum, Pandanus foetidus, Saccharum Tamarix indica, Caesalpinia bonduc, Eugenia fruticosa, Ipomoea pescaprae*	Limited fishing; Occasional tourist visits
	Bangabandhu Island	Height up to I.5 m; juvenile dunes; undulating, gentle slopes; extensive beaches; dominated by loose sands	Sparsely vegetated; Few mangrove associates; common species- *Acrostichum aureum, Ipomoea pescaprae;* large number of red crabs.	Remains unaltered by human n activities;

Sources: Alam et al., 1999a; Islam et al., 2001; Islam and Islam, 2010; Bushra et al., 2014; Islam and Saha, 2017.

and there is not enough resistance to shoreline erosion. Erosion is a major threat to dune landscape in this zone.

6. Paleo dunes along Bangladesh coast

Along the eastern coast, in the area between the foredunes and eastern hills, the elongated paleo-dune ridges are sparsely distributed. Monsur and Kamal (1994) suggested that during the high stand sea-level about 6000 years BP, the coastline along the Cox's Bazar-Teknaf belt was closed to the foot of the hillocks. The remnants of mid-Holocene paleo-dune ridges along the east coast indicate the seaward development of the beach area and the formation of dune fronts. At present the paleo-dune ridges are not as dynamic as the foredune belt, because of their disconnection with the sea. They are now sheltered from the wind by a front dune. These paleo-dune areas along the east coast are now partly populated, and being used for agriculture and human settlements.

Paleo-dunes in the east coast are composed of medium grained sand and contain lower concentration of heavy minerals than front dunes. Zaman et al. (2012) found the concentration of heavy minerals, such as zircon, garnet, rutile, ilmenite and magnetite in the paleo-dune areas. The radioactive doses are quite elevated both in fore-dunes and paleo-dunes (Alam et al., 1999b). The fore-dunes are mostly free from human settlement and occupied by fishermen. However, the paleo-dune areas are mostly occupied by human settlement and because of such occupancy on top of the heavy mineral deposits, people are more exposed to radioactive doses.

Islam (2001) argued that during the Holocene there were five phases of marine transgressions and associate shifting of the shorelines. At about 7000 yrs BP, the coastline of Bangladesh was close to Satkhira and Khulna and close to the shoreline of Kolkata and Dum Dum as suggested by Sen and Banerjee (1988). Both reconstructions were completed based on biological and ^{14}C records. It is assumed that with the shifting of shorelines, the coastal dune fields also moved parallel to the beaches, but until now, no field evidence of such paleo-dunes along this part in the northern Bay of Bengal, either in West Bengal or in south-western Bangladesh, has been recorded. However, at the Cauvery delta coast (India), paleo-dunes have been recorded by Ramasamy et al. (1989) at about 35 km inland from the present shore. Paleo sand dunes at Goa (India) of about 5000–6400 years old have also been found and in many places such dunes are as high as 10 m even far from from the shoreline (Mascarenhas, 2002).

During recent field visits to the Sundarbans areas, the evidence of paleo-dune landscape have been recorded (pers. obs.). At Kotka-Kochikhali area of the Sundarbans, at least four distinctive paleo-dune barriers, each length varies from 1.5 to 3.5 km long and about 1.5 km inland from the shore, have been observed. The height of each of the paleo dune ridge is well above the spring tide level and is not inundated by tidal water. This undulating dune topography is mostly barren, or sparsely occupied by some grasses, shrubs and small woody trees. The famous tower point of Kotka is built on a paleo dune surface. In the dune valley areas (dune slacks) rain water remains stored and these are the source of fresh water for deer and other wild animals of the forest to drink. At the south-east corner of the Jamtala beach, there are at least five parallel paleo-dune ridges of the recent past, which are exposed to the storm-surge events. At Kotka point, west of Kotka khal, many of such paleo-dune ridges are washed away by earlier cyclones, leaving behind some relic features.

7. Coastal dune ecology in Bangladesh

Vegetation plays an important role to stabilize dune sands by impeding sand movement through their newly produced stems and roots. Vegetation also protects coastal sand dunes from erosion by winds and waves, and plays a vital role for dune conservation (Lee et al., 2012). Once the dune is stabilized, it acts as a barrier and helps in the succession of more complex plant species. Griffiths (2006), Perumal and Maun (2006) and Landi and Angiolini (2015) showed that vegetation in the coastal dune system follows a distinct zonal pattern, which is related to the distance from the sea and height of the tide level.

Along the Bangladesh coast, a total of 128 species of dune vegetation, such as grasses, creeping plants, herbs, shrubs and tree species have been recorded. *Canavalia cathartica* is a common dune species found not only in Bangladesh, but also widely distributed in many other countries of tropical Asia, such as India, Kenya, Tanzania, Japan, Taiwan, Cambodia, Indonesia, Malaysia, Myanmar, Papua New Guinea, the Philippines, Sri Lanka, Thailand, Vietnam and Australia (Silc et al., 2016).

Figure 23.3. a. Sagar lata (*Ipomea pescaprae*) in Cox's bazar; b. Keyakata (*Pandanus fascicularis*) in St. Martins Island; c. Marram grass (*Ammophila host*) and creepers in St. Martins Island; d. Jhao (*casuarina*) plantation in Kutubdia Island (Photograph: By the author).

Canavalia seeds are an excellent source of dietary fibers; it helps in preventing certain intestinal diseases.

Along the eastern coast of Bangladesh, major dune species are *Cocos nucifera* (coconut palm), *Streblus asper* (shaora), *Pandanus odoratissimus* (keya), *Pandanus foetidus, Vitex trifolea* (nil-nishinda), *Vitex negundo, Ipomoea pes-caprae* (shagor lota), ipomea (ganga-lata), *Panicum repens, Paspalum vaginatum, Casuarina equisetifolia, Bumbosa arundinacea, Cyperus* spp. and *Fimbristylis* spp. *Allium* sp. (Alam et al., 1999a). In the central coast important species are *Paspalum vaginatum, Panicum repens, Aeluropus lagopoides, Sesuvium portulacastrum, Ipomoea pescaprae* and *Phragmites karka.* In the western zone, including the Sundarbans area the major species are *Phragmites karka, Imperata cylindrica, Sporobolus tremulusa, Acrostichum aureum, Acanthus ilicifolius.* Some common dune species are *Borassus flabellifer, Syzygium fruiticosum, Porteresia coarctata, Pandanus foetidus, Saccharum spontaneum, Clerodendrum inerme, Tamarix indica, Erythrina variegate, Caesalpinia bonduc, Eugenia fruticosa, Leea crispa, Streblus asper,* and *Ipomoea pes-caprae.*

Beach sands and sand dunes along the Bangladesh coast from the Sundarbans to St. Martin's Island are a suitable nesting grounds for sea turtles. For nest sites sea turtles prefer an elevated location bordering the beach, closer to vegetation and away from the anthropogenic structures. The preference of an elevated beach and dunes fields rather than lower elevation areas are to avoid the risk of tidal flooding during the hatching period.

8. Integrated coastal zone management in Bangladesh

Integrated Coastal Zone Management (ICZM) in Bangladesh is primarily imposed by the doctrine of pro-development approach and the uses of coastal resources, with little attention to protection and conservation of landscape ecology. A number of institutes, particularly related to agriculture, fisheries and forestry are engaged to promote and utilize coastal resources, and there is always a missing gap and lack of inter-institutional coordination. The Coastal Policy-2005 of Bangladesh is an engagement of various agencies and stakeholders to take part in the ICZM initiatives in Bangladesh. However, there remain many limitations and criticism of this ICZM effort, particularly the delineation of coastal zone of Bangladesh. The coastal belt is a hydro-ecological regime and its areal extension largely depends on topography, slope of landscape, fluvial dynamic, inter-actions between land-ocean interfaces, tidal range, the extents of tidal inundation, soil and water salinity and vegetation types. Defining 19 administrative districts as coastal belts in Bangladesh is a politically driven unit, which might have the advantage to managing coastal resources and people, but is not a scientific attempt of hydro-ecological consideration. For any operative ICZM effort, the redefinition of the coastal belt of Bangladesh in the context of hydro-ecological dynamics is a prerequisite. However, the classification suggested by Pramanik (1983)-the eastern, central and western coast-till now is well cited and accepted, although it is very generalized, and requires further subdivision and illustration of each of the coastal regimes. To extract coastal resources, to foster economic growth and to improve the livelihood of coastal people, the doctrine of pro-development initiative, as suggested in the Coastal Policy-2005, requires a rethink. The P^3C doctrine (Protection, Preservation, Prevention and Control) suggested by Paleri (2009) might have more durability and sustainability for ICZM of Bangladesh in general and dune landscape in particular.

9. Management of dune morphology and ecology

Coastal dunes are ecologically sensitive landscapes and requires P^3C doctrine of management (Paleri, 2009). Aided by ample sediment supply, prevailing wind system, tidal range and wave activities, the coastal belt of Bangladesh, particularly the eastern belt, is ideal for the growth and development of a sand dune landscape. However, major threats to dune habitats are both natural and anthropogenic in origin. Frequent cyclone-induced tidal surges, wave cut erosion and potential rise of tidal inundation, together with an increase of population densities, expansion of tourist spots and facilities near the dune fields and alteration of dune habitats by human activities are some major challenges to manage dune morphology and ecology.

9.1 Importance of coastal dunes

Coastal dunes are considered as the first line of natural defense against tidal waves, storm surges, tsunami surges and coastal erosion (Mascarenhas and Jayakumar, 2008; Chakraborty, 2010; Islam et al., 2011). Elevated dunes ridges along the Bangladesh coast have already proven that dunes provide strong natural protection against cyclonic events. For example, during 1991 super cyclone, the Sonadia Island was protected by such dune ridges, and many people were able save their lives and homes (Islam et al., 2011). Spalding et al. (2014) suggested that the combined effect of coastal ecology, such as mangroves, saltmarshes and sand dunes can reduce disaster risks manifold, by keeping out of harm's reach, slowing transport, reducing wave heights, improving quality of life and conserving biodiversity. Under strong wave actions, the beaches and foredunes along the Bangladesh coast may retreat considerably by losing their sands to form off-shore bars, which under calm weather condition is recovered by normal waves through resupplying sands from off-shore islands to the beach. The remnants of vegetation help to trap and recapture the sediment flux.

The importance of dune flora is immense. Dune plant species have value added nutritional, medicinal and agricultural uses. Many dune plants are used a food, fodder and green manure. Coastal dunes are also important as a source of bioactive compound of pharmaceutical and veterinary values (Sridhar and Bhagya, 2007). With a high supply of nutrients, dune legumes play a vital role in nitrogen and phosphorous fixation in dune sand with the help of mycorrhizal fungi. Mridha and Mohammed (1989) studied the mycorrhizal associations of a large number of dune plant from the Bangladesh coast and showed that such plants are highly mycotrophic. Mycorrhizal fungi in dune sands of Bangladesh have the beneficial effects on soil aggregation and have the capacity to bind dune sands (Mirdha and Xu, 2008; Yoon et al., 2017).

The physico-chemical properties of dune soil are very important for the growth and distribution of flora and fauna. Parameters, such as nutrient level, soil salinity, salt spray, organic matter and pH are the determinants for plant succession and dune stabilization (Hesp, 1991) Pioneer plants in the dune are adapted to harsh conditions and their deep root system reach up to the water table, which help to anchor and hold dune sands in place. During the rainy season in Bangladesh, the dunes surfaces are eroded and leaching occur in dunes, washing nutrient matters in to the slacks and sometime the dune slacks are waterlogged, as has been found in Sonadia Island. It is common that in dune slacks vegetation is sparse, but in some cases, where the humus influx is high, the slacks area supports a wider collection of vegetation than the exposed dune crest (Ranwell et al., 1959).

Sand dunes along the Bangladesh coast are closely related with the lives and livelihood of the coastal people. The open beach space and nearby dune areas are widely used by fishermen for landing of fish boats, sewing nets, drying nets, sorting their catch, auctioning fish, fish drying, taking rest and also for temporary settlements. Fishermen prefer elevated sites, preferably dune landmass for building their temporary houses for better protection from storms and waves, as has been seen in Nijhum dwip and Doblar char areas.

9.2 Human interventions on coastal dunes

The beach profile survey shows that the morphology and ecology of half of the beaches in the Bangladesh coast is disturbed by anthropogenic causes, such as road construction, human settlements and coastal tourism (Mahbub et al., 2020). For example, the construction of 80 km long marine drive road along the Cox's Bazar to the Teknaf beach is a serious intervention to beach morphology, ecology and sea turtle habitats.

Shrimp hatcheries, particularly along the Cox's Bazar-Teknaf coast is a main intervention that impacts on dune morphology and ecology. There are about 60 hatcheries along the beach, which have been established either in or adjacent to sand dunes. These hatcheries adversely affect the nesting habitat of sea turtles. Moreover, the disposal of untreated hatchery water is the main source of pollution to dune habitats and adjacent areas.

In Bangladesh beach tourism is the major sector of tourism, particularly during the winter season. St. Martin's Island, Teknaf, Inani, Cox's Bazar, Sonadia Island, Parki, Kuakata, and Kotka beaches are the prime choices for tourists to visit. From November to February, nearly 5000 tourists visit the St. Martins Island every day to enjoy its natural beauty and many of them stay overnight. Construction of hotels and beach resorts for tourists lead to the destruction of natural dune landscape, which has been increasing worldwide (Namboothri et al., 2008). Construction of new roads to allow the tourists to travel to the beach could lead to serious damage of dune flora and fauna (Stephenson, 1999). In Bangladesh, more than hundred hotels/motels and resorts have been constructed on the dune landscape at Cox's Bazar, St. Martins Island and Kuakata beach leading to destruction of dunes, their morphology and ecology.

Urbanization on dune lands is also a threat to dune morphology and habitat, because it requires massive land leveling and large-scale removal of dune sands. Flattening of dunes accelerates coastal erosion, shoreline recession, submergence of beaches and reduction of the water absorption capacity of coastal dunes. At Kuakata beach many plots on dune land are now being altered and leveled for housing purposes. Because of the expansion of the city of Cox's Bazar towards the sea, there has been a significant change of beach topography and dune morphology, particularly in the south-west side to the city.

Sand mining from dune fields either for heavy mineral extraction or for construction purposes is another threat to dune morphology and ecology. Beaches along the Bangladesh coast are rich with heavy mineral deposits (Bari et al., 2011). Beach Mineral Exploration Centre (BMEC) has identified 17 potential reservoirs of heavy mineral deposits along the coastal belt of Bangladesh (Rajib et al., 2007). The extraction of beach sand from those reservoirs, particularly from the beaches and dune fields at Cox's Bazar to separate heavy minerals by BMEC at Cox's Bazar will alter beach morphology and ecology.

Dunes fields of remote off-shore islands, where there is limited destruction of dune morphology by permanent constructions and interventions, the pedestrian connection between fishing villages and beaches has caused considerable erosion of dune morphology, as has been seen in many dune sites in Bangladesh (e.g., Inani beach, Sonadia Island, Nijhum dwip, Char Kukri Mukri, Domar Char, Kawar char, Mustakpara Shutkighat of Cox's Bazar and Dublar char).

Based on the degree of stabilization Namboothri et al. (2008) classified coastal dunes as

i) *highly degraded dune*: when many parts of dune topography have been altered and flattened, and large portions of vegetation is degraded due to natural and human-induced causes. Such dune sites need immediate restoration programs to bring then back to their original stage.
ii) *Partially degraded dune*: when dune morphology and ecology have undergone considerable degradation by natural or human processes, but are able to come back to their original state when the disturbance is stopped. Such dune sites require prevention and protective measures. and
iii) *Undisturbed dune*: when dunes in remote coastal areas, particularly in off-shore islands are still free from any disturbances. These dunes can be used as a reference for dune restoration programs.

According to this classification, most of the dune fields in the eastern coast of Bangladesh are in category one. At Sonadia Island and onshore dunes at the Kuakata site are mostly in category two. Dunes in the off-shore island of Bangladesh, particularly in the central coast and also in the Sundarbans areas are not considerably altered by human activities. For example, the dune fields of Nijhum dwip, Char Kukri Mukri, Domar Char, Kawar char, Kotka beach and Bangabandhu island are not notably destroyed by human activities.

9.3 Protection of dunes of Sonadia Island

Sonadia Island is one of the important biodiversity hotspots in Bangladesh. The island was leased to Bangladesh Economic Zones Authority (BEZA) in 2017 by the Ministry of Land with the aim to developed ecotourism on the island, with a guideline to use up to 30% of its land under the development plan (https://www.beza.gov.bd/special-tourism-park/), which could cause major physical damage to landscape, including morphology, An Indian business firm-Mahindra Group-will implement the project at a cost of Tk 1.5 billion,[1] and have submitted a draft plan for tourism development to the Government (Financial Express). The island will be divided into five thematic zones, and will involve a series of development works (Prothomalo; Financial Express). All this physical development would obviously alter the topography of the island, particularly the beach and dune morphology, increase the noise level and pollute the surroundings. There is a high risk of destruction and the disappearance of unique dune topography and habitat of endangered species of turtles, once the BEZA project on tourism deployment in the island is implemented.

It was mentioned earlier that the coastal sand dunes of Sonadia Island are so far the highest and largest dune ridge system of the country, which not only act as a strong natural barrier against cyclonic events, but also is the unique habitat for endangered sea turtles to hatch in the dune sands. With the implementation of Sonadia tourism development plan, the dune ecology will be seriously altered. This ecologically virgin island will lose its credential as an eco-sensitive habit and rich biodiversity, and will turn into a noisy and polluted commercial hub at the edge of the Bay of Bengal. The sensitive and precious ecology of the St. Martin's Island has already been lost and has now turned into one of the pollution hotspots in the coastal belt of Bangladesh. With the implementation of BEZA project on tourism development, the fate of the Sonadia Island would be even worse. Due to alteration of landscapes, increase of noise level and pollution of beach and mangrove habitats, the eco-sensitive characteristics of the island will not be sustained. It is not wise to destroy the unique dune landscape of the island, rather to protect and conserve the dune ecology and habitat.

[1] About 17.6 million $US at time of writing.

9.4 Conservation of dune biodiversity

Biodiversity conservation in Bangladesh primarily focuses on forest areas, giving little attention to conserve and protect dune vegetation, although the dune areas offer immense and irreplaceable value in biological diversity, uniqueness, beauty and contribution to livelihood.

Considering the ecological importance and sensitivity of the coastal zone, the Government of Bangladesh has identified a few sites, such as mangroves, coral reefs, sandy beaches and sand dunes, mudflats, marine wildlife protected areas, coastal freshwater bodies, salt marshes, turtle nesting grounds, horseshoe crab habitats, sea grass beds, seaweed beds and nesting grounds of birds as Ecologically Critical Area (ECA) and Protected Areas (PA), although there is a scarcity of scientific information on such ecosystem and habitats (CWBMP, 2006; BOBLME, 2011).

Dune vegetation, either naturally grown or planted, not only contributes to recreational and aesthetic values, but is also important for beach and dune stabilization. For conservation and management of dune landscape and ecology, zoning is an integral part, which means dividing the dune landscape into some logical units or belts based on its existing ecological settings, mainly for management purposes. Zoning is also a tool to guide and control human activities. The landward side of the dune landscape in Bangladesh is ideal for plantation of Jhao tree (*Casuarina equisetifolia*). At Parki beach (near Chittagong city) about 35 ha of *casuarina* shelterbelt was created in 1993 to protect the coastal communities from cyclone and storm surges (Miah et al., 2013). It has been found that the *casuarina* plantation is a successful program to reduce wind speed significantly and to increase the size of sand dunes. The plantation helps not only to increase the existing dune up to 1 m high, but also to create new dunes ridges between the sea and the forest coverage. Along the coastal belts of Sri Lanka, it has been found that the *Casuarina* shelterbelt could reduce wind speed considerably, increase dune sizes and retard dune movements (Takle et al., 2007), as also found in China (Bai, 2010). Carter (1991), and Junaidi and Aoki (2009) had shown that dune forest and vegetation play a significant role to stabilize dune sediment.

Under the CBA-ECA project a total of 62 ha of dune land was brought under dune plantation at seven different locations at Sonadia and beach strip from Cox's Bazar to Teknaf. Common species planted in the sand dunes are *Pandnaus furcatus* (keya), *Vitex negundo* (nishinda), *Saccharum sp.*(haora/khair), *Ipomoea pes-caprae* (sagar lata), *Calotropis gigantea* (akonda) and *Ipomoea fistulosa* (kolmi). Once-degraded sand dune habitats of these locations have now been restored to a great extent. Because of human interventions and alteration of dune topography, the natural protective role of coastal dunes against extreme events has declined considerably, as has been observed globally (Nordstrom, 2000), and regionally along the coast of the northern Bay of Bengal (Paul, 2001). The protection and conservation of dune sites and their ecology requires appropriate regulatory mechanisms. Unfortunately, none of the regulatory instruments has given adequate attention to protect and conserve dune morphology and ecology. Sand mining and alteration of dune topography and other destructive activities are completely prohibited. It means that several open sandy beaches and all coastal sand dunes in the east and west coast of India are protected by law. Bangladesh should follow suit.

10. Conclusions

Dune conservation and the restoration program has been carried out in many parts of the world. Both should include; the protection of the dune fields from human and livestock encroachment, creation of micro-wind breaks on the dune slope by shrubs and materials, direct seeding or transplantation of indigenous and exotic species, plantation grass slip on the leeward side and management of re-vegetated sites.

To conserve the dynamic habitat of coastal dunes, the complex relationship between the natural process and human interventions should be correctly assessed and studied. The adverse impacts of anthropogenic interventions on coastal morphology and ecology have already been seen at different locations along the Bangladesh coast. The construction of hotels, resorts and buildings for tourism, have also led to irreversible impacts on dune morphology and habitat, and need to be controlled by concerned authorities urgently.

Since dunes are aeolian landforms and act as a natural barometer to sensitize coastal ecosystems, they should be allowed to evolve naturally without any human interventions. The natural equilibrium of sand budgeting in the beach area is a matrix of vegetation cover and dune topography, which requires adequate attention to monitor sustainable management of the coastal ecosystem.

As there is a threat that the majority of dune fields along the Bangladesh coast might retreat landward under the projected sea-levels scenarios, immediate measures should be taken to assess the future fate of these coastal features. The experience of local people needs to be considered for dune field management. Effective and realistic measures should be taken to ensure the participation of local people during different stages of program implantation and monitoring the dune ecology. The local community needs to get the ownership of the restoration program and it should become a source of economic return for them under an umbrella of social and community afforestation.

The scientific understanding of coastal dune habitat in Bangladesh is poor. Dune landscape reflects the complex interactions between the off-shore and near-shore environment, and fluvio-dynamic nature of the mainland river system. The barometric interactions between oceanic and terrestrial processes can be easily documented by a program seeking to gain a better scientific understanding of dune dynamics along the coastal belt of Bangladesh.

Acknowledgement

The chapter is an outcome of the ongoing project on dune dynamics along Bangladesh coast, with technical and financial supports from BACCOS (Bangladesh Center for Coastal and Ocean Studies). The support from Premanondo Debnath of CEGIS at different stages of the work, including joining with me during fieldworks at Sonadia Island and Nijhum dwip are sincerely acknowledged.

References and further readings

Ahmed, A., Nawaz, R., Drake, F. and Woulds, C. 2018. Modelling land susceptibility to erosion in the coastal area of Bangladesh: a geospatial approach. Geomorphology. DOI:10.1016/j.geomorph.2018.08.004.

Alam, M., Salam, A. and Haque, M. 1996. Geomorphological features of the Hatia and Sandwip islands of the Bay of Bengal. Journal of NOAMI 13(1&2): 61–72.

Alam, M.S., Huq, N.E. and Rashid, M.S. 1999a. Morphology and sediments of the Cox's Bazar coastal plain, south-east Bangladesh. Journal of Coastal Research 15(4): 902–908.

Alam, M.N., Chowdhury, M.I., Kamal, M., Ghose, S., Islam, M.N., Mustafa, M.N., Miah M.M.H. and Ansary. M.M. 1999b. The ^{226}Ra, ^{232}Th and ^{40}K activities in beach sand minerals and beach soils of Cox's Bazar, Bangladesh. Journal of Environmental Radioactivity 46: 243–250.

Bai, J. 2010. The role of casuarina in environmental amelioration in China. *In*: Zhong, C., Pinyopuarerk, K., Kalinganire, A. and Franche, C. (eds.). Improving Smallholder Livelihoods through Casuarina Productivity. Haikou: China Forestry Publishing House.

Bandyopadhyay, S. 1997. Natural environmental hazard and their management: a case study of Sagar island, India". Singapore Journal of Tropical Geography 18(1): 20–45.

Bari, Z., Rajib, M. and Ameen, S.M.M. 2011. Heavy mineral assemblages of the beach sands of Kuakata, southern Bangladesh. Jahangirnagar University Journal of Science 34(2): 143–158.

Barua, D.K. 1997. The active delta of the Ganges-Brahmaputra rivers: dynamics of its present formations. Marine Geodesy 20(1): 1–12. http://dx.doi.org/10.1080/01490419709388091.

Brammer, H. 2014 Bangladesh's dynamic coastal regions and sea-level rise. Climate Risk Management. 1: 51–62.

BOBLME. 2011. Country report on pollution in the BOBLME – Bangladesh. BOBLME-2011-Ecology-01.

Bushra, N., Khan, A.U. and Islam, M.S. 2014. Natural process on coastal sand dune dynamics of Kuakata: A geographical analysis. Dhaka University Journal of Earth and Environmental Sciences: 3(1): 9–14.

Bushra, N., Islam, M.S. and Khan, A.U. 2016. Dune morphology of Kuakata coast in Bangladesh. Journal of NOAMI: 33(1&2): 91–102.

Carter, R.W.G. 1991. Near-future sea level impacts on coastal dune landscapes. Landscape Ecology 5: 29–39.

Chakraborty, S.K. 2010. Coastal environment of Midnapore, West Bengal: potential threats and management". Journal of Coastal Environment 1(1): 27–41.

CWBMP. 2006. Teknaf Peninsula ECA Conservation Management Plan. Coastal and Wetland data based approach for coastal landuse planning. Department of Environment (DoE), Ministry of Environment and Forests, Dhaka, Bangladesh.

Dinh, L.C. 1998. Fixation des dunes vives par Casuarina equisetifolia au Vietnam. Landform 24: 319–333.

DoE. 2015a. Community Based Ecosystem Conservation and Adaptation in Ecologically Critical Areas of Bangladesh. Responding to Nature and Changing Climate. Department of Environment (DoE), Ministry of Environment and Forests, Dhaka, Bangladesh.

DoE. 2015b. Fifth National Report to the Convention on Biological Diversity. Department of Environment (DoE), Ministry of Environment and Forests, Dhaka, Bangladesh.

Financial Express: https://thefinancialexpress.com.bd/economy/bangladesh/islands-virginity-at-risk-1574483880, Nov 232019.

Gopal, B. and Chauhan. M. 2006. Biodiversity and its conservation in the Sundarban mangrove ecosystem. Aquatic Science. 68: 338–354. DOI 10.1007/s00027-006-0868-8.

Griffiths, M.E. 2006. Salt spray and edaphic factors maintain dwarf stature and community composition in coastal sand plain heathlands. Plant Ecology 186: 69–86.

Haque, A., Islam, M.S. and Paul, A. 2011. Geomorphological evolution of Kuakata coastal plain in Bangladesh", Oriental Geographer: 52(1 & 2): 11–24.

Hesp, P.A. 1991. Ecological processes and plant adaptations on coastal dunes. Journal of Arid Environment 21: 165–191.

Hesp, P.A. 2002. Foredunes and blowouts: initiation, geomorphology and dynamics. Geomorphology. 48: 245–268.

Hesp, P.A. 2004. Coastal dunes in the tropics and temperate regions: location, formation, morphology and vegetation processes. pp. 29–49. *In*: Martínez, M.L. and Psuty, N.P. (eds.). Coastal Dunes, Ecology and Conservation. Springer-Verlag Berlin Heidelberg.

Hossain, K.T., Salauddin, M. and Tanim, I.A. 2016. Assessment of the dynamics of coastal island in Bangladesh using geospatial techniques. Domar char. Journal of Asiatic Society of Bangladesh, Science 42(2): 219–228.

Hossain, S. 2015. Assessment of sediment movement pattern along nearshore coastal water of Cox's Bazar. M.S. Thesis. Department of Water Resources Engineering Bangladesh University of Engineering and Technology, Dhaka.

Islam, M.S. 2001. Sea-Level Changes in Bangladesh: The Last Ten Thousand Years. Asiatic Society of Bangladesh.

Islam, M.S., Haque, A. and Uzzaman, M.R. 2001. Quaternary geomorphic evolution of the St.Martin's island in Bangladesh. Indian Journal of Geography and Environment 6(1): 1–24.

Islam, M.S. and Islam. M.T. 2010. Morpho-ecological characteristics of sand dunes along the coastal belt of Bangladesh. Oriental Geographer: 54(1 & 2): 23–40.

Islam, M.S. and Akter, N. 2012. Geomorphological classification of Chittagong city Coast. Journal of NOAMI 29(2): 29–46.

Islam, M.S. and Saha, C.R. 2017. The emergence of Bangabandhu island in the western shelf zone of Bangladesh and its morphological changes. Journal of NOAMI 34(2): 25–36.

Islam, M.Z. 2002. Threats to sea turtles in St. Martin's island, Bangladesh. Kachhapa 6: 8–12.

Islam, S.A. and Rahman. M. 2015. Coastal afforestation in Bangladesh to combat climate change induced hazards. Journal of Science, Technology and Environment Informatics. 2(1): 13–25. http://www.journalbinet.com/jstei-volume-02.html.

Islam, S.M. 2015. A critical analysis of integrated coastal and ocean management in Bangladesh with lessons from global practices. M.S. Dissertation, World Maritime University. http://commons.wmu.se/all_dissertations/489.

Islam. M.A., Maitra, M.K., Majlis, A.B.K., Murshed, S. and Rahman, S. 2011. Spatial changes of land use/land cover of Moheshkhali island, Bangladesh: a fact-finding approach by remote sensing analysis. Dhaka University Journal of Earth and Environmental Science. 2: 43–54.

Junaidi and Aoki, S. 2009. Wind-blown sand and topographic changes of the coastal dune at the eroded beach". Journal of Coastal Research (Spec. Issue). 56: 322–326.

Kabir, M.Z., Deeba, F., Rasul, M.G., Majumder, R.K., Khalil, M.I. and Islam. M.S. 2018. Heavy mineral distribution and geochemical studies of coastal sediments at Sonadia Island, Bangladesh. Nuclear Science and Application. 27(1&2): 1–5.

Khan, S.R. 1990. Geomorphology and holocene geology of the patuakhali and barguna districts-a part of the lower deltaic plain in the recent ganges Delta, Bangladesh. Journal of Geology, 9: 9–18.

Krantz, M. 1999. Coastal erosion on the island of Bhola, Bangladesh. Earth Science Center, Goteborg University, Sweden.

Landi, M. and Angiolini, C. 2015. Soil-plant relationships in Mediterranean salt marshes across dune-cultivated land gradient. Journal of Coastal Research 31: 588–594.

Lee, J.S., Jeon, J.Y,, Ihm, B.S. and Myeong, H.H. 2012. The classification of biotope type and characteristics of naturalized plant habitat on the coastal dune ecosystem. Journal of Ecology and Field Biology 35: 165–172.

Mahbub, R.B., Ahmed, N. and Yeasmin, F. 2020. Towards reducing the data gap in the conservation efforts for sea turtles in Bangladesh. Regional Studies in Marine Science. DOI: https://doi.org/10.1016/j.rsma.2020.101151.

Majlis, A.B.K., Islam, M.A., Khasru, M.H. and Ahsan, M.K. 2011. Protected to open basin depositional system: an appraisal for Moheshkhali-Kutubdia coastal plain (Abstract). Bangladesh Coast. Geology, Hazards and Res., Dhaka Bangladesh, 3.

Mascarinhas, A. 2002. Coastal sand dunes of Goa: comments on evolution and anthropogenic impacts. pp. 52–64. *In*: Biodiversity of Benthic Marine Algae along the Indian Coast.

Mascarenhas, A. and Jayakumar, S. 2008. An environmental perspective of the post-tsunami scenario along the coast of Tamil Nadu, India: Role of sand dunes and forests. Journal of Environmental Management. 89: 24–34.

McLachlan, A. and. Brown. A.C. 2006. The Ecology of Sandy Shores. Academic Press. 373.

Miah, M.D., Siddik, M.A. and Shin, M.Y. 2013. Socio-economic and environmental impacts of casuarina shelterbelt in the Chittagong coast of Bangladesh. Forest Science and Technology 9(3): 156–163, DOI: 10.1080/21580103.2013.814592.

Monsur, M.H. and Kamal, A.S.M.M. 1994. Holocene sea-level changes along the Maiskhali and Cox's Bazar-Teknaf coast of the Bay of Bengal. Journal of NOAMI 11: 15–22.

Mridha, M.A.U. and Mohammed, A. 1989. Mycorrhizal association of some crop plants of Bangladesh. Paper presented in the Indian National Conference of Mycorrhizae, Harayana, India.

Mridha, M.A.U. and Xu, H.L. 2008. Nature farming with vesicular-arbuscular mycorrhizae in Bangladesh. Journal of Crop Producion. 3(1): 303–312. DOI: 10.1300/J144v03n01_25.

Namboothri, N., Subramanian, D., Muthuraman, B., Sridhar, A., Rodriguez, S. and Shanker, K. 2008. Beyond the tsunami: coastal sand dunes of Tamil Nadu, India—an overview. UNDP/UNTRS, Chennai and ATREE, Bangalore, India. 56.

Nordstrom, K., Lampe, R. and Vandemark, L. 2000. Reestablishing naturally functioning dunes on developed coasts. Environmental Management 25: 37–51.

Paleri, P. 2009. Marine Environment: Management and People's Participation. KW Pub. New Delhi, India. pp. 336.

Paul, A.K. 2001. Cyclonic storms and their impacts on West Bengal coast. pp. 8–31. *In*: Rajamanickam, G.V. and Tooley, M.J. (eds.). Proceedings of the International Seminar on Quaternary Sea Level Variation, Shoreline Displacement and Coastal Environment, New Academic Publishers, Delhi.

Paul, A.K. 2002. Coastal Geomorphology and Environment. ACB Publications. 575 pp.

Perumal, V.J. and Maun, M.A. 2006. Ecophysiological response of dune species to experimental burial under field and control conditions. Plant Ecology 184: 89–104.

Pitman, J.I. 1985. Thailand. *In*: Bird, E.C.F. and Schwartz, M.L. (eds.). The World's Coastline.

Pramanik, A.H.M. 1983. Remote Sensing Application to Coastal Morphological Investigation in Bangladesh", Un Pub. Ph.D. Thesis, Jahanr Nagar University.

Prothomalo- https://en.prothomalo.com/bangladesh/Sonadia-thrust-into-tourism-tangle, July 20 2019.

Rajib, M., Kabir, M.Z.M.. Deeba, F., Zaman, M.M. and Rana, S.M. 2007. Distribution of five major heavy minerals along the recent beach areas of Bangladesh. Journal of NOAMI. 24(1): 1–9.

Ramasamy, S.M., Pavanivelu, R. and Sanjeevi. S. 1989. Remote sensing and ecosystem modelling of Coromandel coast, South India. *In*: Coastal Zone '89. Proceedings of the 6th Symposium on Coastal Zone Management, Charleston, South Carolina, American Society of Civil Engineers, New York, U.S.

Ranwell, D., Warren, N. and Anglesley, I. 1959. The dune system and dune slack habitats. Journal of Ecology 47: 571–601.

Rasheed, K.B.S. 2008. Bangladesh; Resource and Environmental Profile. AHDPH, Dhaka

Rashid, S.H., Biswas, S.R., Bocker, R. and Kruse, M. 2009. Mangrove community recovery potential after catastrophic disturbances in Bangladesh. Forest Ecology and Management 257: 923–930.

Rashid, S.M.A. 1986. Exploitation of marine turtles in Bangladesh. *In*: Groombridge, B. and Luxmoore, R. (eds.). The Green Turtle and Hawksbill Turtle (Reptilia: Chelonia), World Status, Exploitation and Trades. CITES Lausanne. 601.

Rashid, S.M.A. 1997. Bangladesh national report for the northern Indian Ocean sea turtle workshop and strategic planning. 13–18. Bhubaneswar, India, 16pp.

Rashid, S.M.A. and Islam, M.Z. 1999. Establishing marine turtle hatchery in Saint Martins Island, Bangladesh. Proceedings of the 4th Asia-Pacific NGOs Environmental Conference, Singapore, Published by the Department of Biological Sciences, National University of Singapore, January, 255–264.

Rashid, S.M.A. 2019. Coastal biodiversity—a review. Report prepared for Long Term Monitoring Research and Analysis of Bangladesh Coastal Zone. 30.

Sanjeevi, S. 1996. Morphology of dunes of the Coromandel coast of Tamil Nadu: A satellite data based approach for coastal landuse planning. Landscape Urban Planning, 34: 189–195.

Sen. P.K. and Banerjee, M. 1988. Palaeoenvironment of Bengal Basin During the Holocene Period. Geographical Review of India 50(4): 21–38.

Silc, U., Caković, D., Küzmič, F. and Stešević, D. 2017. Trampling impact on vegetation of embryonic and stabilised sand dunes in Montenegro. Journal of Coastal Conservation. 21: 15–21. DOI 10.1007/s11852-016-0468-2.

Silc, U., Küzmič, F., Caković, D. and Stešević, D. 2018. Beach litter along various sand dune habitats in the southern Adriatic (E Mediterranean). Mar. Pollut. Bull. 128: 353–360.

Spalding, M.D., McIvor, A.L., Beck, M.W., Koch, E.W,. Möller, I., Reed, D.J., Rubinoff, P., Spencer, T.J., Tolhurst, T.V., Wamsley, B.K., van Wesenbeeck, E., Wolanski, E. and Woodroffe, C.D. 2014. Coastal ecosystems: A critical element of risk reduction. Conservation Letter 7(3): 293–301.

Sridhar, K.R. and Bhagya, B. 2007. Coastal sand dune vegetation: a potential source of food, fodder and pharmaceuticals. Livest. Research in Rural Development 19(84). http://www.cipav.org.co/lrrd/lrrd19/6/srid19084.htm.

Stephenson, G. 1999. Vehicle impacts on the biota of sandy beaches and coastal dunes: a review from a New Zealand perspective. Science for Conservation, Department of Conservation, Wellington, New Zealand.

Swan, B. 1979. The presence of sand dunes in a tropical low energy zone. Friday Island.

Takle, E.S., Chen, T.C. and Wu, X. 2007. Protecting functions of coastal forests and trees against wind and salt spray. *In*: Braats, S., Fortuna, S., Broadhead, J. and Leslie, R. (eds.). Regional Technical Workshop. Coastal Protection in the aftermath of the Indian Ocean Tsunamis: what role for Forests and Trees? RAP Publication 2007/07. Bangkok (Thailand): Food and Agriculture Organization (FAO) of the United Nations. 65102.

UNDP. 2010. Environmental profile of St. Martin's Island. Coastal and Wetlands Biodiversity Management Project. United Nations Development Programme, Dhaka.

Untawale, A.G. 1994. Coastal Vegetation of India and its adaptation during the climate change. pp. 19–42. *In*: Biodiversity of Benthic Marine Algae along the Indian coast.

van Dijk, P.M., Arens, S.M. and van Boxel, J.H. 1999. Aeolian processes across transverse dunes: modelling the sediment transport and profile development. Earth Surface Process Landform 24: 319–333.

Yoon, H.S., Yi, Y.M., Sung, K. and Park. S. 2017. Vegetation roles on changes of soil properties and microbial activities in a coastal sand dune: A case study. Journal of Coastal Conservation 21: 145–152. DOI: 10.1007/s11852-016-0483-3.

Zaman, M., Schubert, M. and Antao, S. 2012. Elevated radionuclide concentrations in heavy mineral-rich beach sands in the Cox's Bazar region, Bangladesh and related possible radiological effects. Isotopes in Environmental and Health Studies 48(4): 512–525.

Index

Dune Types

Remote sensing

Sand

Author Index

V

W

X

Y

Z

About the Editors

Professor Lu Qi is Chief Scientist and Director of the Institute of Desertification Studies (IDS) in Beijing. The IDS is one of two such Institutes in the world. Dr. Lu has a PhD in Ecology from the Chinese Academy of Sciences, Beijing and is a specialist in Dryland Sustainable Development; Integrated ecosystem management/Desert Ecology and Combating desertification and Land degradation. He is the recipient of several prestigious awards: Three National Awards on Scientific & Technological Progress (Second Class, 2006/2009/2010) PRC National Award on International Science & Technology Cooperation (as Chinese Partner), 2008; Ningxia Hui Autonomous Region Award on Scientific & Technological Progress (First Class), 2008; and UNDP Best Practices Award on indigenous technology in combating desertification and mitigating the effect of drought. Dr. Lu is author/editor of 20 books notably *A Study of Kumtagh Desert* (in Chinese). Beijing: Science Press, 2012; *Desert Plants in China* (Illustrated Handbook, in Chinese). Beijing: China Forestry Publishing House, 2012; *Rangeland Degradation and Recovery in China's Pastoral Lands*, CABI, UK, 2009; *Studies on Populus euphratica Forest in Ejina Oasis*. Beijing: Science Press, 2009. *Regional Review of UNCCD Implementation and Best Practices in Asia and the Pacific*—Case Studies from Australia, China, Mongolia, Thailand and Viet Nam. Beijing: Chinese Environmental Science Press, 2005 and 180 papers (mostly in Chinese).

Dr. Mahesh K. Gaur is an Indian, who is working at the ICAR-Central Arid Zone Research Institute, Jodhpur, India as a Principal Scientist, since September 2020. He worked as Senior Scientist from 2012 to 2020. He specializes in *aridlands geography* and the application of satellite remote sensing, GIS and digital image processing for natural resources mapping, management and assessment. He also researches *drought, desertification, land degradation, indigenous knowledge systems, and the socio-economic milieu of the Thar Desert of India*. Prior to this, he has worked as an Associate Professor for more than 13 years with the Department of Higher Education, Government of Rajasthan State, India and has also worked at Sandhan, Jaipur, IIM Ahmedabad, IHS Hyderabad and IN-RIMT in various capacities. He has also completed sponsored studies for ICAR, ISRO, SAC, UGC, and ICARDA. He has published 50 research papers in national and international research journals, 50 conference papers, 45 book chapters, five popular articles, and eight books. He is a member of a number of national and international scientific organizations, such as the Association of American Geographers and the Society for Conservation Biology, and several editorial boards of journals. He has been to South Africa, Namibia and Costa Rica on academic assignments. He is awarded the *Citizen Karamveer Award 2011* by iCONGO for working in the field of Higher Education, Environment and Technology Applications for Community Upliftment; the *Millennium Award* for working in the field of Environment, and prestigious recognitions by the UGC of India and Scientific Assembly of the International Committee on Space Research (COSPAR).

Dr. Victor R. Squires is a Distinguished Guest Professor in the Institute of Desertification Studies, Beijing. He is an Australian who as a young man studied animal husbandry and rangeland ecology. He has a PhD in Rangeland Science from Utah State University, USA. He is a former Foundation Dean of the Faculty of Natural Resource Management at the University of Adelaide, Australia where he worked for 15 years after a 22 year career in Australia's CSIRO. He is author/editor of 22 books on Drylands, Desertification, Ecological Restoration from such publishers as CRC Press, Springer, Routledge and Kluwer Academic and numerous research papers and invited book chapters. Since retirement from the University of Adelaide Dr. Squires was a Visiting Fellow in the East West Center, Hawaii, and an Adjunct Professor in the University of Arizona, Tucson and at the Gansu Agricultural University, Lanzhou, China. He has been a consultant to World Bank, Asian Development and various UN agencies in Africa, China, Central Asia and the Middle East. He was awarded the *2008 International Award and Gold Medal for International Science and Technology Cooperation* by the Government of China and in 2011 was awarded the *Friendship Award* by the government of China. The Gold Medal is the highest award for foreigners. In 2015 Dr. Squires was honoured by the Society for Range Management (USA) with an *Outstanding Achievement Award.* Dr. Victor Squires was recognised in the 2021 Queen's birthday Honours with a member of the Order of Australia (AM), for services to ecology, especially rangelands.